D0983377

Scientific Visualization

Overviews, Methodologies, and Techniques

Gregory M. Nielson
Hans Hagen
Heinrich Müller

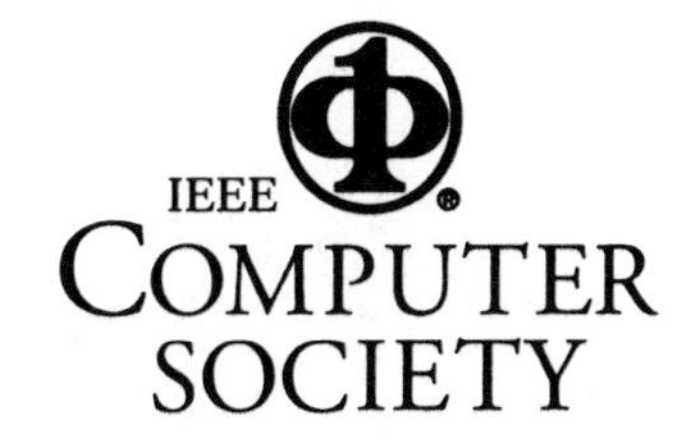

Los Alamitos, California

Washington • Brussels • Tokyo

Library of Congress Cataloging-in-Publication Data

Nielson, Gregory M.
Scientific visualization: overviews, methodologies, and techniques / Gregory M. Nielson, Hans Hagen, and Heinrich Müller.
p. cm.
Includes bibliographical references and index.
ISBN 0-8186-7777-5
1. Science—Methodology. 2. Visualization—Technique. 3. Algorithms. I. Hagen, H. (Hans), 1953– . II. Müller, Heinrich. III. Title.
Q175.N438 1997
501 '. 1' 366—dc21
97-5922
CIP

IEEE Computer Society Order Number BP07777
Library of Congress Number 97-5922
ISBN 0-8186-7777-5

Additional copies may be ordered from:

IEEE Computer Society Press
Customer Service Center
10662 Los Vaqueros Circle
P.O. Box 3014
Los Alamitos, CA 90720-1314
Tel: +1-714-821-8380
Fax: +1-714-821-4641
Email: cs.books@computer.org

IEEE Service Center
445 Hoes Lane
P.O. Box 1331
Piscataway, NJ 08855-1331
Tel: +1-908-981-1393
Fax: +1-908-981-9667
mis.custserv@computer.org

IEEE Computer Society
13, Avenue de l'Aquilon
B-1200 Brussels
BELGIUM
Tel: +32-2-770-2198
Fax: +32-2-770-8505
euro.ofc@computer.org

IEEE Computer Society
Ooshima Building
2-19-1 Minami-Aoyama
Minato-ku, Tokyo 107
JAPAN
Tel: +81-3-3408-3118
Fax: +81-3-3408-3553
tokyo.ofc@computer.org

Publisher: Matt Loeb
Acquisitions Editor: Bill Sanders
Developmental Editor: Cheryl Smith
Advertising/Promotions: Tom Fink
Production Editor: Lisa O'Conner
Printed in the United States of America by BookCrafters

Contents

Preface

The theme of this book is scientific visualization. This is a very active and vital area of research, teaching, and development these days. The success of scientific visualization is mainly due to the soundness of the basic premise behind it; that is, the basic idea of using computer-generated pictures to gain information and understanding from data (geometry) and relationships (topology). This is an extremely simple, but very important concept which is having profound and widespread impact on the methodology of science and engineering.

The intent of this book is to present the state of the art in visualization techniques, both as an overview for the inquiring scientist, and as a basic foundation for developers. The book contains some chapters dedicated to surveys and tutorials of specific topics as well as a great deal of original work not previously published, but in all cases the emphasis has been on presenting the extensive detail necessary for others to reconstruct the techniques and algorithms.

Another goal for this book is to provide the basic material for the teaching of state of the art techniques in scientific visualization. The extensive bibliographies included with many of the chapters point out where to obtain further information to complete the course material. The index also aids the learning and discovery process.

The first five papers of the book are overview/survey papers on some of the most important topics of scientific visualization. In an area as active as visualization, where most researchers are spending their efforts expanding the field, it is rare, but fortunate to have leading experts take the time and effort to write such papers.

The next section on framework/methodologies contains six papers which establish foundation material for scientific visualization. Papers of this variety have rarely appeared in visualization previous to this volume. The field has now matured to the level where these types of papers are appropriate.

The final section of the book is devoted to techniques and algorithms for scientific visualization. For the most part, these are cutting edge research papers containing the latest and greatest research in scientific visualization but there are also some papers which have very strong tutorial components.

The initial planning for this book took place at the Dagstuhl seminar on scientific visualization during May of 1994. The IBFI Schloss Dagstuhl was founded in 1990 and is located in southwest Germany between Saarbruecken and Trier. It offers the opportunity of one-week meetings which bring together the most significant worldwide researchers on topics of importance in computer science. The Dagstuhl seminar on scientific visualization was attended by some forty participants from the United States, Germany, the Netherlands, France, Russia, and Austria. The participants were invited to contribute to this volume and a list of these authors and their addresses can be found immediately after the index. The editors wish to express special thanks to them for their impressive efforts.

Gregory M. Nielson, Hans Hagen, and Heinrich Müller

Part I

Overviews and Surveys

Chapter 1

30 Years of Multidimensional Multivariate Visualization

Pak Chung Wong and R. Daniel Bergeron

Abstract. *Multidimensional multivariate (mdmv) visualization has been an active research field for more than three decades. This subfield of scientific visualization deals with the analysis of data with multiple parameters or factors, and the key relationships among them. The course of development is roughly organized into four stages: the searching stage, the awakening stage, the discovery stage, and the elaboration and assessment stage. The searching stage was mostly concerned with the graphical presentation of either one or two variate data. The awakening stage was dominated by Tukey's exploratory data analysis. The discovery stage was characterized by a remarkable number of new techniques being proposed. The current stage is concerned with the elaboration and assessment of visualization techniques. Recently developed techniques including static representations as well as dynamic animations are explored with appropriate examples.*

1.1 Introduction

Multidimensional multivariate (mdmv) visualization is an important subfield of scientific visualization. It was studied separately by statisticians and psychologists long before computer science was deemed a discipline. The appearance of low-priced personal computers and workstations during the 1980s breathed new life into graphical analysis of mdmv data. This research topic was among one of the short-term goals included in the 1987 National Science Foundation (NSF) sponsored workshop on *Visualization in Scientific Computing* [39]. The quest for effective and efficient mdmv visualization techniques has expanded since then.

This paper attempts to trace three decades of intensive development in this visualization field. It is by no means a comprehensive survey. We provide a brief history along with a description of the principal concepts of many mdmv visualization techniques. Recently developed mdmv visualization techniques are discussed in detail with examples.

1.2 Four Stages of Multidimensional Multivariate Visualization Development

The last three decades of mdmv visualization development can be characterized roughly into four stages. The classic exploratory data analysis (EDA) book by Tukey [58], the 1987 NSF workshop on Visualization in Scientific Computing [39], and the IEEE Visualization '91 conference [44] are the watersheds defining these stages. The first stage was primarily concerned with the graphical presentation of either one or two variate data. The second stage was dominated by Tukey's exploratory data analysis. Scientists started looking at graphical data with a different perspective. Although most of the graphics was still two-dimensional, scientists were able to encode data with multiple parameters, that is, multivariate, into meaningful two-dimensional plots. The momentum of this work carried on through the next stage when NSF recognized the importance of mdmv data visualization. The involvement of computer scientists accelerated the growth of the research by computerizing many of the old ideas and developing many new ones. The mission was formally defined and many promising concepts were developed during the subsequent few years. The final (current) stage is concerned with the elaboration and assessment of mdmv visualization techniques. It remains to be seen whether the existing mdmv visualization concepts can lead to better visualization of a problem and better understanding of the underlying science. This discussion of mdmv visualization is far from complete. There are other important topics including volume visualization and vector/tensor field visualization that are not covered. The principal concepts and research issues related to these subjects can be found in [43, 48, 30, 26].

1.2.1 Pre-1976: The Searching Stage

Scientists have studied multivariate visualization since 1782 when Crome used point symbols to show the geographical distribution in Europe of 56 commodities [18]. In 1950, Gibson [24] started the research on visual texture perception. Later, Pickett and White [45] proposed mapping data sets onto artificial graphical objects composed of lines. This texture mapping work was further investigated by Pickett [46], and was eventually computerized [47]. Chernoff [15] presented his arrays of cartoon faces for multivariate data in 1973. In this well-known technique, variates are mapped to the shape of cartoon faces and their facial features including nose, mouth, and eyes. These faces are then displayed in a two-dimensional graph.

The searching stage can be characterized by relatively small-sized data, and tools for data visualization that usually consisted of color pencils and graph paper. The graphical output was mostly two-dimensional *xy* displays. Statisticians were the dominant research force during this period. Graphics was used to bring out the key features of the data, suggest statistical analysis methods that are applied to the data, and present the conclusions [22].

1.2.2 1977–1985: The Awakening Stage

Tukey's exploratory data analysis signified a new era of scientific data visualization. Exploratory data analysis is more than a tool; it is a way of thinking. It teaches people how to visually decode information from the data. When the personal computer arrived, it became the scientist's most powerful tool ever. Now scientists could visualize data beyond two dimensions interactively. The painfully long calculations suddenly became available in real time. Statisticians could visualize data during each stage of the analysis instead of waiting until the final results were available. The availability of other computer hardware such as high-resolution color displays also gave the study of mdmv visualization new opportunities.

During this stage, two- and three-dimensional spatial data were the most common data types being studied, although multivariate data started gaining more attention. Asimov [2] presented the grand tour technique for viewing projections of multivariate data on two-dimensional planes. Earth resource satellites sent out decades ago are still transmitting data continuously. Gigabyte-sized multivariate data had arrived.

1.2.3 1986–1991: The Discovery Stage

The 1987 NSF workshop formally declared the need for two- and three-dimensional spatial object visualization. The two-dimensional projections of multivariate data sets is also included as one of the short-term potential targets for scientific visualization research. Once the mission was defined, scientists started pushing hard on the representation and visualization of mdmv data. The limited availability of high-speed graphics hardware during the previous stage was gradually conquered. A majority of research was directed away from the development of exploratory data analysis tools, which lay heavily on statistical measures, towards colorful high-dimensional graphics that required high-speed computations. Some of the mdmv visualization concepts developed during this stage include: grand tour methods [12], parallel coordinates [27, 28, 29], iconography [47, 8, 4, 37], worlds within worlds [20, 21], dimension stacking [36], hierarchical axis [40, 41, 42], hyperbox [1], and various ideas collected in [16, 17]. Some of these techniques attempt to show all dimensions and all variates visually as one display, whereas others aim at *direct manipulation graphics*, in which the user interactively selects subsets for display by using an input device such as a mouse. Virtual reality [20, 21] began to appear in the mdmv visualization literature.

1.2.4 1992–Present: The Elaboration and Assessment Stage

In 1990 and 1991, there were at least fourteen mdmv-related papers published in the IEEE visualization conferences. A total of seven have been published in the four visualization conferences since then. This stage so far has been a period of retrenchment in the development of new mdmv visualization techniques. Some of the most recently developed tools are elaborations, each in a different way, of work done in previous stages. For example, HyperSlice [60] is an attempt to combine the panel matrix of scatterplot matrix with direct manipulation of brushing [3]. AutoVisual [9] is an extended version of worlds within worlds with new rule-based interfaces. XmdvTool [61, 38] integrates four existing mdmv visualization tools: dimension stacking, scatterplot matrix, glyphs, and parallel coordinates into one system with enhanced n-dimensional brushing.

The research in mdmv visualization has also been diversified into multidisciplinary collaborations. Attempts to combine sound with graphics [53, 51] are currently being made.

The concept of a rule-based queue [9, 10] was also introduced. One of the latest research issues of mdmv visualization is the need to evaluate the correctness, effectiveness, and usefulness of mdmv visualization techniques. Similar concerns also appear in the other fields of visualization research [49, 26].

1.3 Terminology

Unfortunately, the mdmv literature suffers from ill-defined and inconsistent terminology. The term *dimensionality* is especially overloaded. Mathematicians consider dimension as the number of independent variables in an algebraic equation. Engineers take dimension as measurements of any sort (breadth, length, height, and thickness). Even the prefix *multi* is frequently interchanged with another prefix *hyper*. In statistics literature, the prefix *multi* means two or more, indicating a natural breakpoint between one and two dimensions in probabilistic methods. For the breakpoint between three and four (or beyond), the prefix *hyper* is used [16]. We use the prefix multi to refer to dimensionality of two or more.

Beddow [5] points out the difference between multidimensional *objects* and multidimensional *data*. Multidimensional *objects* are *spatial* objects, and our goal is to understand their geometry. The most common form are two-dimensional images and three-dimensional volumes. They can best be described as n-dimensional Euclidean spaces $\mathbb{R}^n$. Multidimensional *data*, on the other hand, refers to the study of relationships among multiple parameters. Mathematically these parameters can be classified into two categories: *dependent* and *independent* [35]. Some statisticians prefer the terms *factor* and *response* [16]. A variable is said to be dependent if it is a function of another variable, the independent variable. The relationship of an independent variable x and a dependent variable y can best be described by the mathematical equation $y = f(x)$. We adopt the convention that the term multidimensional refers to the dimensionality of the independent variables, while the term multivariate refers to the dimensionality of the dependent variables [6]. This is by far the most popular way to describe the dimensionality of mdmv data sets in scientific visualization literature. For example, a three-dimensional volume space in which both temperature and pressure are observed and recorded in various locations produces 3d2v data.

Beddow [5] argues that analytic methods used to explore n-dimensional Euclidean spaces $\mathbb{R}^n$ are not appropriate for general multivariate analysis. In mdmv visualization research, the emphasis shifts away from the strong mathematical definition of dependent and independent variates towards the broader definition of multiple variables or factors. This happens not only in mdmv scientific visualization research but also in statistical studies. The tools are different, but the goal is the same: to find the hidden relationships between the variables (also known as *fitting* in statistics).

In general, raw *scientific data* can be categorized into a hierarchy of data types. The most general and the lowest of the hierarchy is the *nominal* data, whose values have no inherent ordering. For example, the names of the fifty states are nominal data. The next-higher type of the hierarchy is *ordinal* data, whose values are ordered, but for which no meaningful distance metric exists. The seven rainbow colors (red, orange, and so on) belong to this category. The highest of the hierarchy is *metric* data, which has a meaningful distance metric between any two values. Times, distances, and temperatures are examples.

If we bin metric data into ranges, it becomes ordinal data. If we further remove the ordering constraints, the data is nominal. Some of the visualization techniques included in this survey are specially designed to handle metric data (see the sections on HyperSlice and worlds within worlds).

The above 3d2v temperature/pressure example more or less implies that each three-dimensional coordinate contains *simple* (that is, neither a set nor an interval) and *atomic* (that is, not composite) values of pressure and temperature. This is different from the case when we measure, for example, the chemical contents of a volume. Each coordinate now has a set (instead of a simple value) of composite data (the chemical elements). The varying numbers of values of a variate plotted in any single-dimensional point is known as the *density* of that coordinate.

1.4 Fundamental Objectives and Approach

The main objectives of mdmv visualization are to visually summarize an mdmv data set, and find key trends and relationships among the variates. Different properties and characteristics of the data may change the way we carry out visualization, but not its goals.

The traditional two-dimensional point and line plots are among the most commonly used visualization techniques for data with lower numbers of variates. This technique can be enhanced by putting an array of plots into one display, so as to add another variate to the visual presentation. This approach is discussed in the sections on the scatterplot matrix and HyperSlice.

We can also map the variates of the data into graphical primitives of different colors, sizes, shapes, and locations (see the sections on stick figure icons, autoglyphs, and color icons). The display of all dimensions and all variates creates some kinds of texture patterns, and provides critical insights needed for scientific discovery.

For large (larger than the number of pixels of a display) scientific data, we can display a certain portion of data and allow the user to navigate the rest interactively. This is described in the sections on HyperSlice, hierarchical axis, dimension stacking, and worlds within worlds.

Most of the visualization techniques assume a Euclidean space environment. Orthogonal axes, however, are not always the best choice to plot data. The sections on hierarchical axis, parallel coordinates, the VisDB system, and recursive pattern give some alternatives.

A powerful visualization technique is to display the data frame by frame according to a time variate. This animation approach is discussed in the sections on grand tour methods, Exvis on a supercomputer, and a scalar visualization model.

1.5 Multidimensional Multivariate Visualization and Concepts

The body of this paper covers the principal concepts and brief history of some of the popular mdmv visualization techniques. During the last decade, hundreds of so-called *new* mdmv visualization techniques have been invented. (Refer to [35] for more details in this regard.) A majority of them are designed for special purposes such as volume visualization and vector/tensor field visualization, which are not covered in our discussion. Some of the rest are merely ad hoc tools that produce pretty pictures. They are difficult to create and their results are hard to interpret. We are interested in techniques that are founded on a solid basis and that have potential for practical value.

Categorizing mdmv visualization techniques is a difficult task. Possible criteria for such a categorization include the goal of the visualization, the type and/or dimensionality of the data, the dimensionality of the visualization technique, and so forth. We have not found a convincing set of criteria that cleanly differentiate the visualization techniques we wish to describe. We have chosen to group the techniques into those based on 2-variate displays, those based on multivariate displays, and those using time as an animation parameter:

- *Techniques based on 2-variate displays* include the fundamental 2-variate displays and simultaneous views of 2-variate displays. Most of these were developed in the statistics world. Both visual perception and statistical fitting of the data are of major concern. The data size is relatively small, usually on the order of hundreds of items. The graphics are mostly variations on two-dimensional point and line plots.

- *Multivariate displays* are the basis for many recently developed mdmv techniques, most of which use colorful graphics created by high-speed graphics computation. A majority of the techniques were developed within the period of 1987–1991.

- *Animation* is a powerful tool for visualizing mdmv scientific data. Various movie animation techniques on mdmv data, and a scalar visualization animation model are presented. In principle, any single-frame visualization technique can be extended to animation if the data can be represented as a time series showing two-way correlations.

1.5.1 Techniques Based on 2-Variate Displays

This section highlights some of the tools and summarizes the general approach developed for 2-variate displays. The discussion is based upon the book by Cleveland [16], which has a good collection of elegant visualization techniques developed by Cleveland, Tukey, and others throughout the 1980s. Tukey's exploratory data analysis [58] is an important milestone of data visualization; most of the techniques were developed with pencil and paper during the early 1970s. Cleveland's work emphasizes the structure of data and the validity of statistical models fitted to data. A majority of the visualization techniques are two-dimensional, with the exception of isosurface plotting. Color is rarely used. Most of the tools show correlations between two variates. Our discussion skips the formulas, algorithms, and theories; only the concepts and techniques are presented.

Data Types

The basic data types for statistical data analysis are *univariate*, *bivariate*, *trivariate*, and *hypervariate* which represent data with one dimension and one, two, three, and four or more variates. Cleveland also describes the *multiway* data type for data with higher dimensionality.

Reference Grids

The most common display unit in statistics visualization is a two-dimensional scatterplot, as depicted in the left panel of Figure 1.1. In the middle panel, simple grid lines are drawn

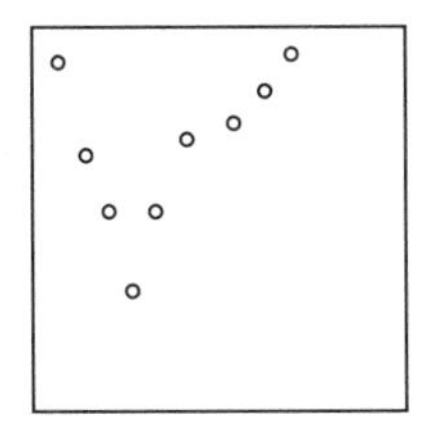
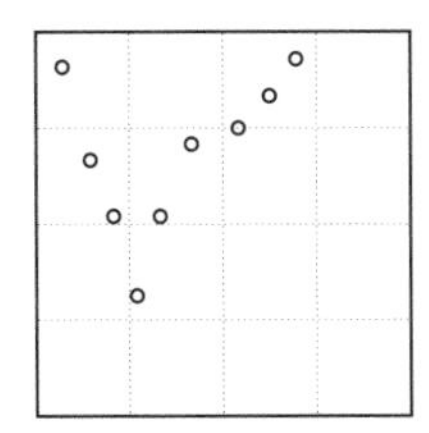
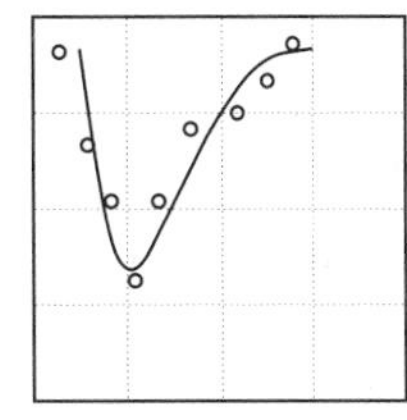

Figure 1.1: *Left:* A simple 2D scatterplot. *Middle:* A scatterplot with visual reference grids. *Right:* A fitted curve is included in the plot.

for enhancement of pattern perception, not for plotting accuracy. Grid lines are drawn in equal intervals and are particularly powerful when we need to do scanning and matching of a matrix of scatterplots.

Fitted Curve

In statistics, fitting means finding a description of a data set. For example, if a data set fits into a normal distribution, the whole data set can then be described by two numbers: its mean and standard deviation. In statistics visualization, fitting means finding a smooth curve that describes the underlying pattern. In the right panel of Figure 1.1, a curve fit to the data is plotted; a pattern not apparent from the scatterplot before may suddenly emerge. Fitting formulas are not discussed in this paper; [57, 16] are good references for this topic.

Banking

The perception of the orientations of line segments can be enhanced by adjusting the *aspect* ratio of the graph. The aspect ratio of a graph is defined as the height of the data rectangle divided by the width. A line segment with an orientation of 45° or −45° is the best to convey linear properties of the curve. This technique is known as the *banking to* 45° *principle* [17]. In Figure 1.2, the same curve is plotted in three different aspect ratios. Only the

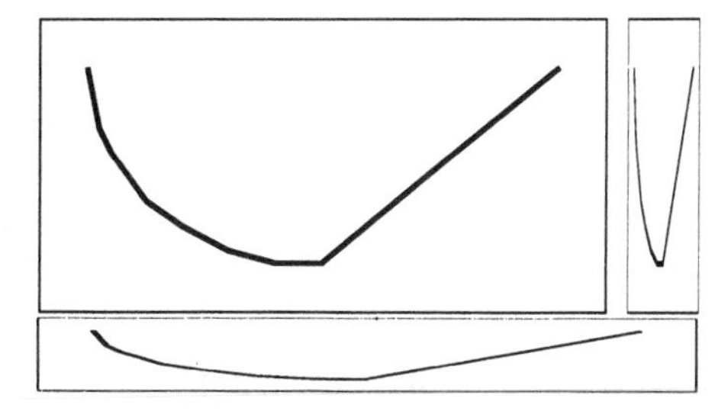

Figure 1.2: The same curve is plotted in three different aspect ratios. The upper-left one conveys more information than the other two.

upper-left panel shows both a curve on the left and a straight line on the right. The banking method is covered in [16].

Scatterplot Matrix

One of the more popular statistics mdmv visualization techniques is the scatterplot matrix which presents multiple adjacent scatterplots. Each display panel in a scatterplot matrix is identified by its row and column numbers in the matrix. For example, the identity of the upper-left panel of the matrix in Figure 1.3 is $(3, 1)$, and the lower-right panel is $(1, 3)$. The

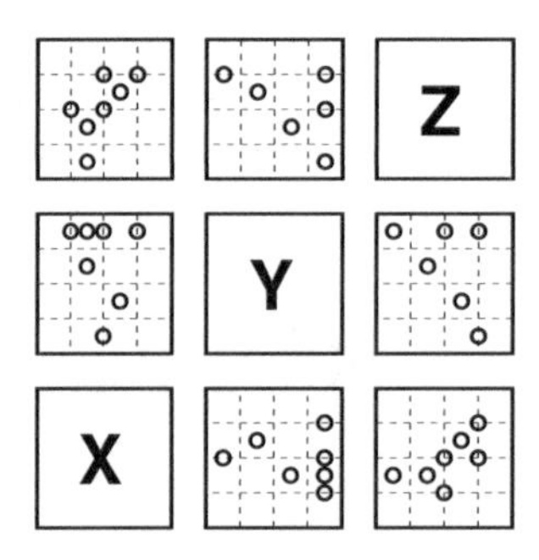

Figure 1.3: Scatterplot matrix displays of data with three variates X, Y, and Z.

empty diagonal panels denote the variable names. Panel $(1, 2)$ is a scatterplot of parameter X against Y while panel $(2, 1)$ is the reverse, that is, Y versus X. In a scatterplot matrix, every variate is treated identically. The basic idea is to visually *link* features in one panel with features in others. The redundancy is designed to improve the effect of visual linking. The technique is further enhanced with the help of reference grids. The pattern can be detected in both horizontal and vertical directions. The concept of linking is also discussed in [13].

The idea of pairwise adjacencies of variables is also a basis for the hyperbox [1], hierarchical axis [40, 41, 42], and HyperSlice [60]. Despite its popularity in mdmv visualization applications, nobody knows the identity of the original inventor [16]. The technique was first presented in [14]. A variety of powerful tools using this kind of multipanel display are presented in [16]. The scatterplot matrix is also implemented in XmdvTool [61, 38].

Other Two-Dimensional Analytical Techniques

Cleveland's book [16] also includes other powerful graphical techniques such as *medium-difference plot*, *quantile-quantile plot*, *spread-location plot*, *given plot*, and *conditional plot*; fitting tools such as *loess* and *bisquare*; and visual perception techniques such as *jittering* and *outlier deletion*.

1.5.2 Multivariate Visualization Techniques

The scatterplot matrix uses multiple two-way displays in an effort to provide correlation information among many variates simultaneously. The techniques described in this section are aimed, however, at extending the possibilities of multivariate correlation. All the techniques, with the exception of brushing and parallel coordinates, were developed after the 1987 NSF workshop. All of them claim positive results with real-life mdmv scientific data. These techniques are also aimed at presenting much larger data sets than those appropriate for the statistical visualization techniques. Today's scientific data is huge; terabyte-sized data will soon be common. A static scatterplot is just not big enough to display more than a few hundred data items. These techniques are broadly categorized into five subgroups:

- *Brushing* allows direct manipulation of an mdmv visualization display. Both conventional brushing and high-dimensional brushing are described.
- *Panel matrix* involves pairwise two-dimensional plots of adjacent variates. Techniques included are HyperSlice and hyperbox. Both of them are elaborations of the scatterplot matrix.
- *Iconography* uses variates to determine values of parameters of small graphical objects, called *icons* or *glyphs*. Thousands of data points are represented by thousands of these icons which create a visual display characterized by varying texture patterns determined by the data. The mappings of data values to graphical parameters are usually chosen to generate texture patterns that hopefully bring insight into the data. Three iconographic techniques are described: stick figure icon, autoglyph, and color icons.
- *Hierarchical displays* map a subset of variates into different hierarchical levels of the display. Hierarchical axis, dimension stacking, and worlds within worlds belong to this group. These techniques support, or at least enable, dynamic interactive analysis.
- *Non-Cartesian displays* map data into non-Cartesian axes. They include parallel coordinates, VisDB, and recursive pattern. Parallel coordinates is the only technique that is capable of studying both multidimensional objects and multidimensional data.

Brushing

Brushing was first presented in [3] and is included as one of the many direct manipulation techniques in [16]. Buja et al. [13] used the terms *focusing* and *linking* to describe various brushing techniques. Focusing involves data selection, dimension reduction, and data layout manipulation such as zooming. A sequence of focusing views are linked together so that the information of individual views can be integrated into a coherent image.

Cleveland [16] describes two kinds of brushing for a scatterplot matrix: labeling and enhanced linking. Labeling involves an interactive brush (such as a mouse pointer) that causes information label(s) to pop-up for particular display item(s). In enhanced linking, the brush is an adjustable rectangle. It is used to cover a set of points in one of the panels. Figure 1.4 shows a rectangle brush in panel $(2, 3)$. Data inside the rectangle is displayed with a "+" instead of a "$\circ$." The same changes are applied to the corresponding data points in the other panels. By looking at different panels and comparing the vertical and horizontal extent of the brush, this enhanced linking technique provides a powerful direct manipulation tool for visual conditioning analysis. The effect of brushing is more intense in a dynamic interactive display. More applications can be found in [16].

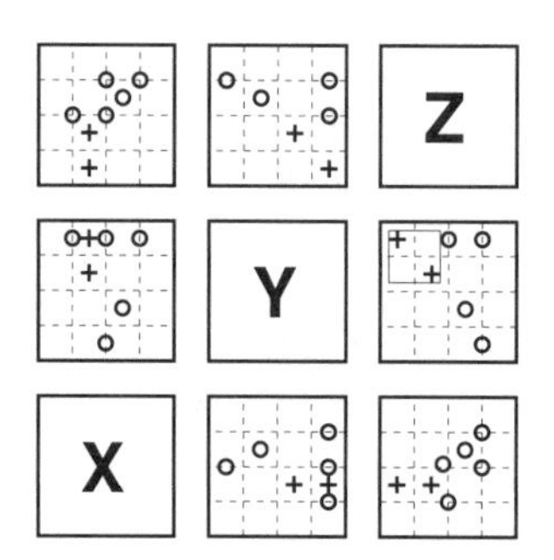

Figure 1.4: Enhanced brushing with the square brush located on panel $(2, 3)$.

Ward [61, 38] expanded the direct manipulation concept and created *high-dimensional* brushing in XmdvTool. In additional to the scatterplot matrix technique, the concept of brushing is applied to star glyphs [50], parallel coordinates [27, 28, 29], and dimension stacking [36]. The multidimensional brushing supports the notion of *demand-driven brushing* in which the data brush is used as a tool to visually *query* the multivariate data. The system is design to visualize both scientific and nonscientific data. A detailed discussion on XmdvTool can be found below.

HyperSlice

HyperSlice [60] is one of the techniques invented during the elaboration and assessment stage. Like the scatterplot matrix, it has a matrix of panels, although each individual scatterplot is replaced with color or gray shaded graphics representing a scalar function of the variates. Furthermore, panels along the diagonal show the scalar function in terms of a single variate.

HyperSlice defines a focal point of interest $c = (c_1, c_2, \cdots, c_n)$ and a set of scalar widths w_i, where $i = 1, \cdots, n$. Only data within the range $R = [c_i - w_i/2, c_i + w_i/2]$ are displayed in the panel matrix. The rest of the data appears only if the user steers the focal point near it. Figure 1.5 shows the display of a HyperSlice of four variates. Like

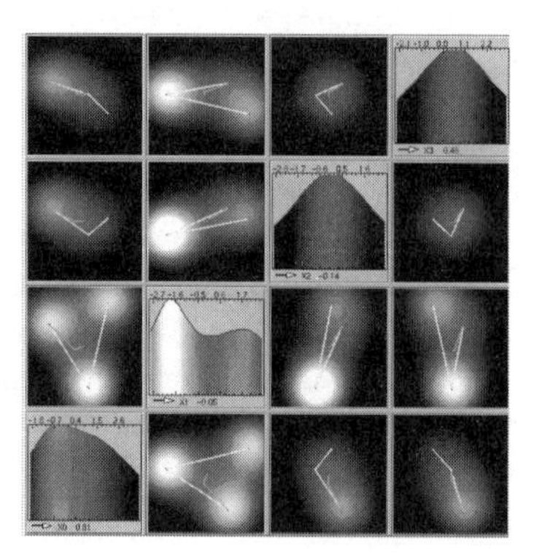

Figure 1.5: See Color Plate 1.

the coordinate system used in the scatterplot matrix, a HyperSlice panel is identified by a horizontal and a vertical coordinate. For an off-diagonal panel (i, j) such that $i \neq j$, the color shows the value of the scalar function that results from fixing the values of all variates except i and j to the values of the focal point, while varying i and j over their ranges in R. The diagonal panels show a graph of the scalar function versus one variate which changes over its range in R.

The most important improvement of HyperSlice over the traditional scatterplot matrix is the idea of interactively *navigating* in the data around a user-defined focal point. The user changes the focal point by interacting with any of the panels, as shown in Figure 1.6. The user moves the mouse into any panel and defines a direction by button down, move,

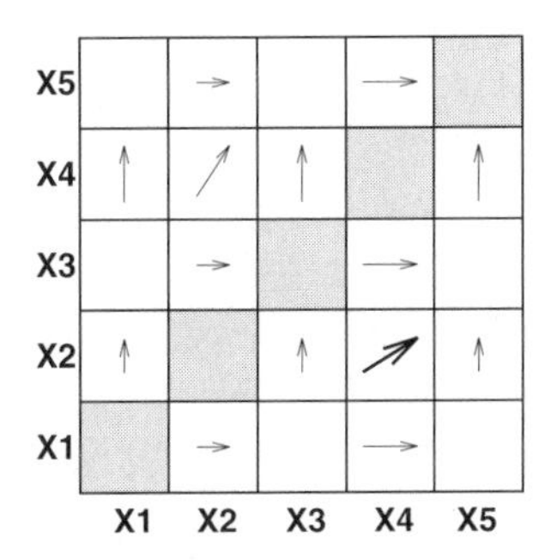

Figure 1.6: Navigate a five-variate HyperSlice by dragging panel $(2, 4)$.

and up. For example, the boldface arrow in panel $(2, 4)$ represents such an interaction. The direction of each arrow shows the motion of the focal point when the focal point is dragged in panel $(2, 4)$. Notice that the length (magnitude) of the vertical arrows across the X_2 row is the same as the vertical component of the arrow in $(2, 4)$. Similarly, each horizontal arrow in column X_4 has the same length as the horizontal component of the arrow in panel $(2, 4)$. Panels solely related to X_1, X_3, and X_5 move perpendicular to the image plan. Since the matrix is somewhat similar to an orthogonal matrix (along the gray diagonal panel), the motion on the upper-left half is the mirror projection of the lower right.

Interactive data navigation is a welcome addition to direct manipulation graphics. The use of the width scalar supports the notion of multiresolution analysis, and begins to address more than two-way correlations. Changing the focal point in one panel affects two variates which in turn results in simultaneous visual changes in displays of these variates with others. HyperSlice is an example of a successful elaboration which builds on another successful tool.

The basic ideas of HyperSlice can be extended to discrete data sets using data projection in a technique called *prosection* [54, 59]. Wong et al. [62] present a dual multiresolution HyperSlice which allows a user to control the *physical data* resolution using orthogonal wavelets [55, 56], as well as the *logical display* resolution using *norm* projections. The system provides a progressive refinement environment to support very large data visualization.

Hyperbox

We place the discussion of hyperbox [1] here because it works very much like the scatterplot matrix and HyperSlice. It too involves pairwise two-dimensional plots of adjacent variates, yet hyperbox is very different from the other two.

A hyperbox is a two-dimensional depiction of an n-dimensional box. Figure 1.7 shows

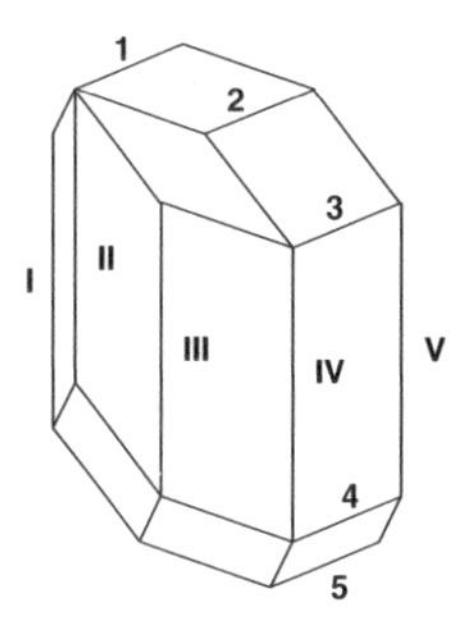

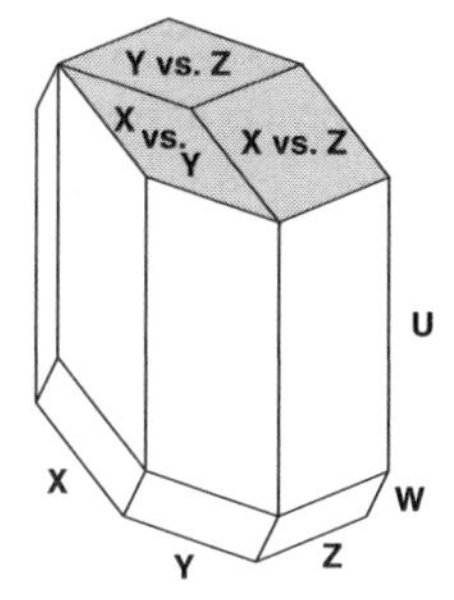

Figure 1.7: *Left:* Hyperbox of dimension five. *Right:* Variates x, y, z, w, and u are mapped to different dimensions. The two-dimensional plots of x versus y, y versus z, and x versus z are shown in gray.

a simple hyperbox of dimension five. An n-dimensional hyperbox is made up of n^2 *lines* and $n(n-1)/2$ *faces*. There are $5^2 = 25$ lines and $5(5-1)/2 = 10$ faces in a hyperbox of dimension five. For each line in a hyperbox, there are $n - 1$ other lines with the same length and orientation. The length and slope of the lines are arbitrary. Both of them can be mapped to the data variates for visualization. Lines with the same length and orientation form a *direction set*. In Figure 1.7, lines 1, 2, 3, 4, and 5 form one direction set while lines I, II, III, IV, and V form another. Given a five-variate data set, x, y, z, w, and u, each variate is mapped to one direction, as shown in the hyperbox on the right-hand side of Figure 1.7. Each face of the hyperbox can now be used to plot data of two variates such as a scatterplot or a line plot.

To support data analysis, variates can be selected by *cutting* the hyperbox along each direction set. A hyperbox cut is similar to a *discrete table selection* in relational databases, except that the variates are discarded instead of selected. Suppose the data we have is time-series related and u is the time variate. The time variate can then be cut and grouped into periods analogous to a histogram. Figure 1.8 depicts an array of hyperboxes with variate u and w being cut. Each hyperbox depicts all values of variates x, y, and z that occur for all u in the range defined for that hyperbox. With a hyperbox, and any other panel-matrix-type mdmv technique, occlusion occurs when the most recent value replaces the earlier displayed data value at the same spot. This is one of the reasons why an array of hyperboxes is shown instead. An example hyperbox visualization is shown in Figure 1.9.

The design of the hyperbox requires a little practice to understand. It is a more powerful tool compared to the scatterplot matrix in the sense that it is possible to map variates to both the size and shape of each facet. It also gives scientists the option to emphasize some of the more important variates and de-emphasize others. On the other hand, it violates Cleveland's banking to 45° principle. The arbitrary setting of length and orientation makes

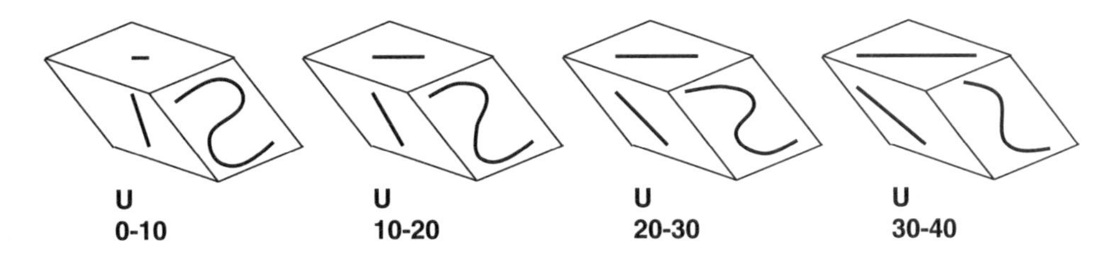

Figure 1.8: An array of hyperboxes excluding dimension u and w is shown in different time periods.

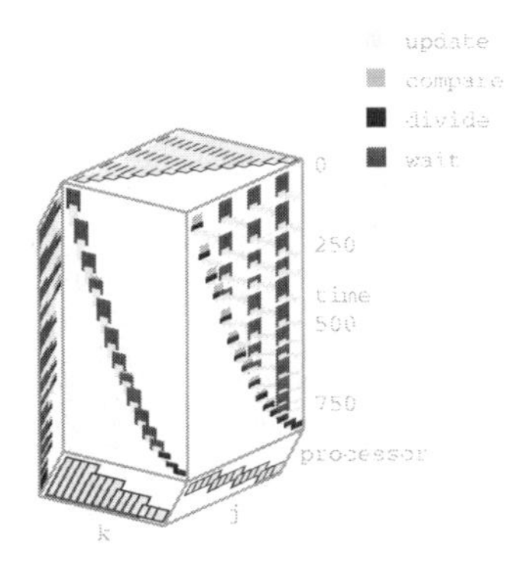

Figure 1.9: See Color Plate 2.

it impossible to do the right banking on all facets. This means that some of the plots may not be able to convey the *right* information.

The idea behind Cleveland's reference grid in the previous discussion is to maximize the visual perception of spatial locality during data scanning and matching. Once again, the arbitrary assignment of length and orientation of each facet makes it impossible for the hyperbox display to take full advantage of this technique.

Stick Figure Icon

Pickett and Grinstein [47] developed the stick figure icon visualization technique. It is rich in concepts, and practical development [25, 52]. The basic stick icon is a five-stick figure with controllable limb angles. Figure 1.10 depicts a family of twelve of them. This icon family is designed to display data with up to five variates. Four of them can be mapped to the orientation of each limb and the fifth can be mapped to the inclination of the body. Other variates can map to the length, thickness, or color of the limbs. An example stick figure icon visualization with seven variates is shown in Figure 1.11. A two-stick icon is also presented in [52] to display bivariate MRI data. The notion of icon construction is to allow humans to exploit the capacity to sense and discriminate the texture in a complex image. In the bivariate data example, the stick icon successfully locates a hot spot not seen in the original MR images. The authors report that these multiline icons have been used to

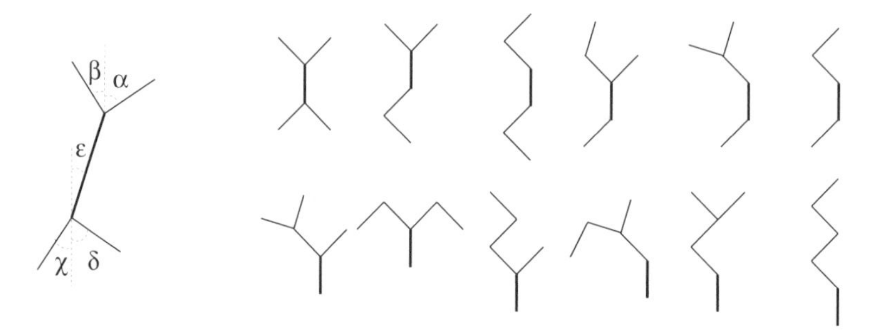

Figure 1.10: *Left:* A five-stick figure icon with orientation plotted according to some values. *Right:* A stick figure icon family. Each one has a body and four limbs.

Figure 1.11: See Color Plate 3.

display as many as thirty variates. However, the flexibility of this technique can also be a weakness. The visual discernment of an important pattern can be highly dependent upon the selection of an appropriate mapping of the data parameters to the visual parameters. Since the number of potential mappings grows as the factorial of the number of variates, the selection process can become a bottleneck.

Autoglyph

Beddow [4] describes what he calls an *autoglyph*. The first generation of his glyph, known as Datapix, was comprised of two rectangles, a filled arc, and a circle. A few examples are discussed in [5]. Based on the results of Datapix, a second generation of autoglyph, which contains only a circle and a box, was developed. Suppose we have twelve-variate data. Each variate is first normalized and then divided into three groups based on the standard deviation function, and a group representing missing values. The high and low groups are assigned to black and gray, and the rest are white. Each variate can then be mapped into one square tile inside the rectangle glyph, as shown in Figure 1.12. The paper also presents examples using primary colors instead of black/gray/white. Beddow [4, 5] provides a brief

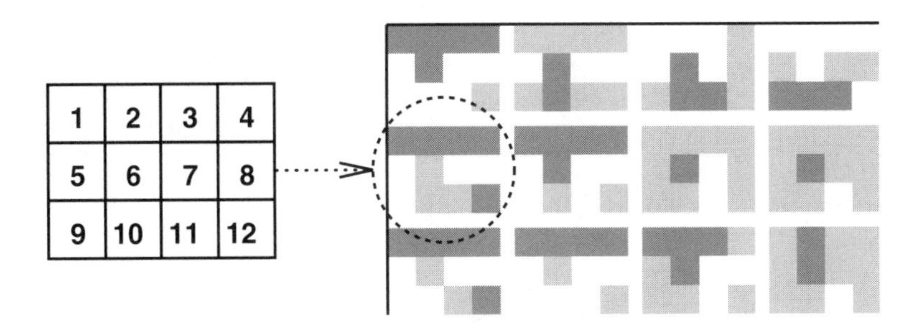

Figure 1.12: Autoglyph designed for twelve-variate data.

analysis of the use of color in such glyphs. Autoglyph was originally designed to study the *correlation* among large numbers of variates. It has a very compact display and can be extended to using threshold functions other than standard deviation. Keller and Keller [35] do not recommend this technique for group presentation because it requires time to study.

Color Icon

Levkowitz [37] describes a *color icon* that merges color, shape, and texture perception for iconographic integration. Color and shape are perceptually orthogonal, and are particularly good at *unmixing* parameters of an mdmv display. A color icon is an area on the display to which color, shape, size, orientation, boundaries, and area subdividers can be mapped by multivariate data. Figure 1.13 shows a square color icon which maps up to six variates.

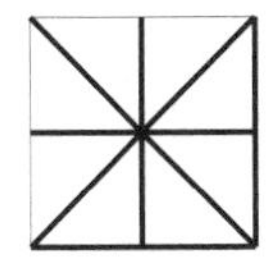

Figure 1.13: A square color icon.

Each variate is mapped to one of the six thick lines. The thin lines serve as boundaries to separate neighboring icons.

There are two different ways to paint a color icon. The first approach requires color shading. A color is assigned to each thick line according to the value of the mapping variate. The color of the color icon is then computed by interpolating the colors assigned to all thick lines. A second way to paint a color icon is to assign color to each pie-shaped subarea according to the values of the mapping variates. The first approach provides better parameter blending, while the second one gives better parameter separation. Figure 1.14 depicts the mapping of color icons from data with one to six variates. The number of variate mappings can be tripled by having each variate control one of the HSV values. Icons with different shapes (such as a hexagon) can be used in place of the square shaped icon. An example of color icon visualization is shown in Figure 1.15.

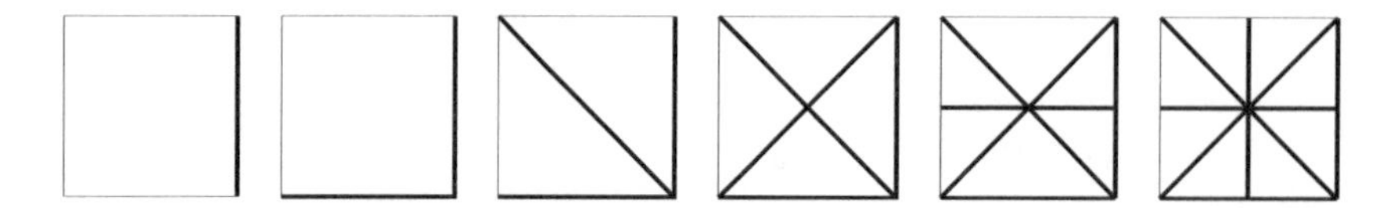

Figure 1.14: Mapping mdmv data with one to six variates to color icons. The value is mapped to the thick line only.

Figure 1.15: See Color Plate 4.

Hierarchical Axis

The conventional way to describe a three-dimensional Euclidean space is by using three orthogonal axes. In hierarchical axis [40, 41, 42], axes are laid out horizontally in a hierarchical fashion. For example, assume we have three variables x, y, and z, and the domain of each variable is $\{0,1,2\}$. The Euclidean 3D space and the hierarchical axes are depicted in Figure 1.16. This technique depends upon a relatively coarse discretization of the data.

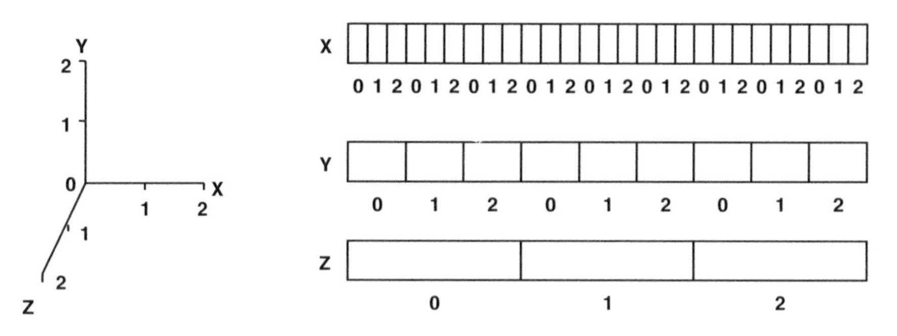

Figure 1.16: *Left:* Orthogonal axes. *Right:* Hierarchical axes.

The term *speed* is used to describe the hierarchical axes, with x being the *fastest* axis and z being the slowest [40, 36, 41]. Once these axes are defined, a variety of potential display options can be used.

One simple example is the histogram (or histograms within histograms) plot. Given a set of data $\{(x, y, z)\} = \{(0, 0, 1), (1, 0, 0), (2, 0, 1)\}$, the left-hand side of Figure 1.17 shows a simple histogram. The height of the dark gray rectangle shows the value of the

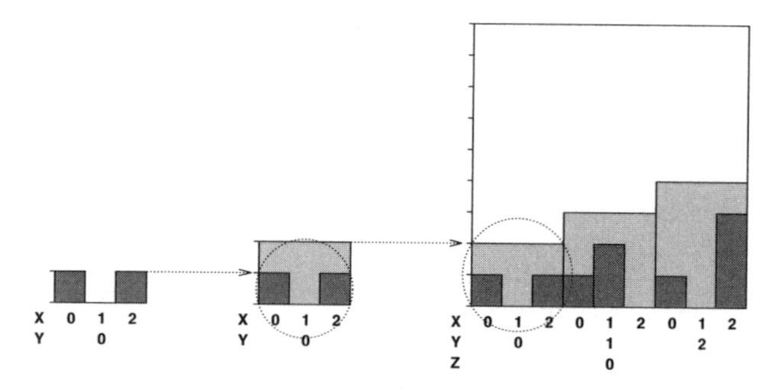

Figure 1.17: *Left:* A simple histogram plot. *Middle and Right:* Histograms within histograms plots.

dependent variate z. The independent variates x and y are described by the horizontal hierarchical axes located at the bottom. In the middle plot, the height of the light gray rectangle is defined as the *sum* of all z variables, that is, the dark gray rectangles, enclosed inside. In this example, the sum equals two. The use of this histogram can be clarified in the last plot on the right, which includes additional data. The histogram in the middle becomes embedded inside another histogram, which shows the sum of the values of each inner histogram. The sum function is only one of the many possible options used in this system. Other possible functions are *min/max* and *mean/standard deviation.*

The hierarchical axis technique can plot as many as twenty variates in one screen. For data with a larger number of records, that is, larger than the number of columns of pixels on the display screen, a technique called *subspace zooming* is introduced. A display of multivariate data involves a series of panels (the number of panels is equal to the number of variates). Each panel displays data from two hierarchical axes, ordered from the slowest to the fastest. This can be considered as a *tree* structure with the panel showing the slowest axis as the root. The other panels are nodes. A subspace is a subtree of the root. A series of panels is a *path* of the tree. Only the root panel is static, and only one path is shown at a time. The panels of the other paths are hidden until the user interactively clicks the specific data of any nonterminal panel to select another subspace (subtree).

A panel matrix of histograms is also presented in [40, 41, 42] to gain visual perception similar to the scatterplot matrix. It is suggested that brushing (that is, direct manipulation graphics) can also be implemented by mapping all or some of the variates to axes with different speeds according to selection conditions. An example visualization is shown in Figure 1.18.

Dimension Stacking

Dimension stacking [36] is a variant of hierarchical axis. In the hierarchical axis technique, each element of the fastest axis is a one-dimensional histogram. In dimension stacking, each element is a two-dimensional *xy* plot. If the data has an odd number of variates, a dummy variate is added. Dimension stacking also has the flavor of iconographics [47].

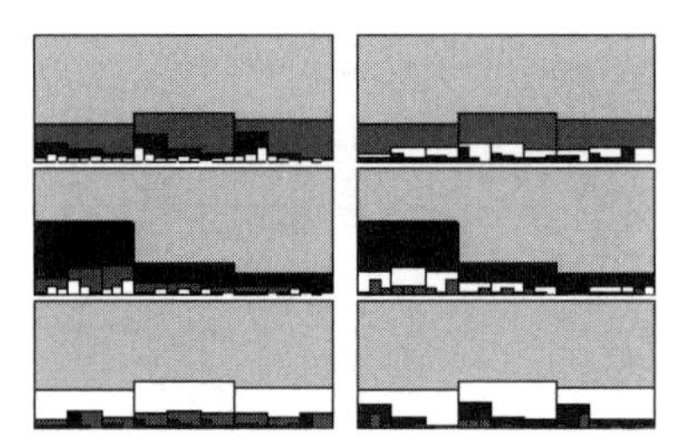

Figure 1.18: See Color Plate 5.

Each two-dimensional xy plot element forms a texture pattern for data visualization and analysis. A major advantage of dimension stacking over hierarchical axis is that no extra function or rule, such as the use of sum in our previous example, is needed to plot the data.

Assume we have a data set with four independent variates x, y, z, and w, such that the corresponding sizes of the variates are 2, 3, 5, and 6. Figure 1.19 shows two different variate

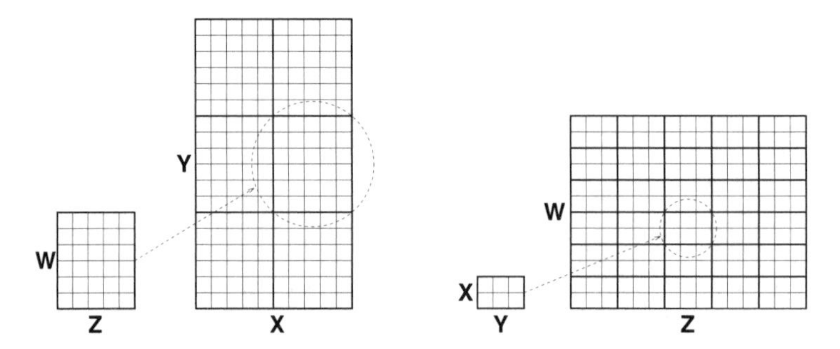

Figure 1.19: Two different variate mappings of the same data in dimension stacking.

mapping schemes. In the first example (left), w and z are the fastest axes while x and y are slower. The second rectangle (x versus y) contains six (two by three) smaller rectangles, each of which is a two-dimensional plot of z versus w, like the leftmost rectangle. If there is a dependent variate available, its values can be plotted as color/gray intensity in each of the squares. Otherwise, each of the two-dimensional plots can be a simple scatterplot. The second example shows a different mapping with the x and y axes being the fastest ones. This hierarchical display technique is also implemented in XmdvTool [61, 38]. An example of dimension stacking visualization is shown in Figure 1.20.

Worlds Within Worlds

All mdmv visualization techniques we have discussed so far, with the exception of the subspace zooming option of the hierarchical axis techniques, involve the generation of *static* objects such as texture maps. The idea behind worlds within worlds [20, 21] is somewhat different. Dimensions are nested together with at most three variates being shown at each level, so as to generate an interactive hierarchy of displays. The three slowest axes are represented only by a display of three orthogonal axes, as shown in Figure 1.21. A

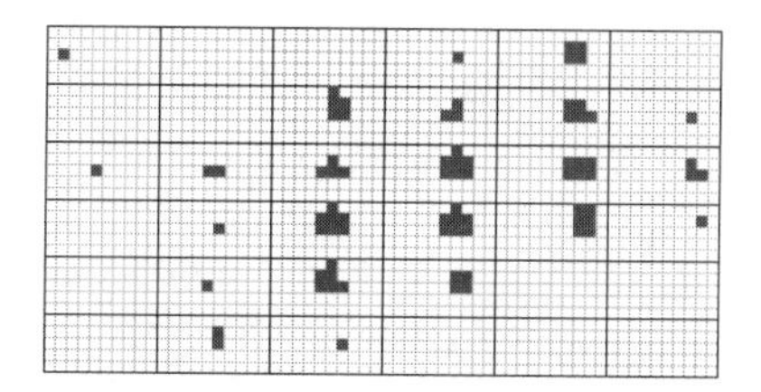

Figure 1.20: See Color Plate 6.

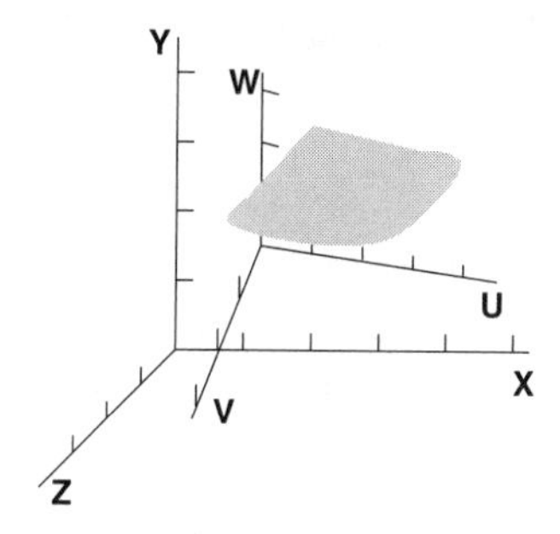

Figure 1.21: Worlds within worlds. Variates x, y, and z are plotted initially. Variates u, v, and w are plotted after all previous variates are defined.

three-dimensional power glove is used interactively to define a position in the space defined by these three axes. A new set of these axes appears at this point. The glove can then be used to pick a point in this space. This continues until all variates are defined. At the lowest level the final variate can be displayed as a surface in the innermost *world* as shown in Figure 1.21. The output is a three-dimensional stereo display of virtual worlds within which users can explore data. This dynamic data visualization technique gives a new meaning to *direct manipulation* graphics. Different variate mappings give different views of the data. The direct manipulation process is in fact a data retrieval query. Users have to know what they are looking for as most of the information is not visible in the initial display. This interactive process tends to be difficult and tedious because there are too many possible combinations of variate mappings. A new generation of worlds within worlds, AutoVisual [9, 10], adds a rule-based user interface. Now users can specify the task through the interface, and the system generates virtual worlds accordingly. An example of worlds within worlds visualization is shown in Figure 1.22.

Parallel Coordinates

All techniques we have discussed so far are designed to do data analysis on multidimensional *data*. They are not really aimed at studying the *geometry* of multidimensional *objects*. Parallel coordinates [27, 28, 29], on the other hand, can do both.

In a parallel coordinate system, the axes of a multidimensional space are defined as parallel vertical lines separated by a distance d. A *point* in Cartesian coordinates corresponds to a *polyline* in parallel coordinates. To avoid confusion, we use lowercase letters for lines,

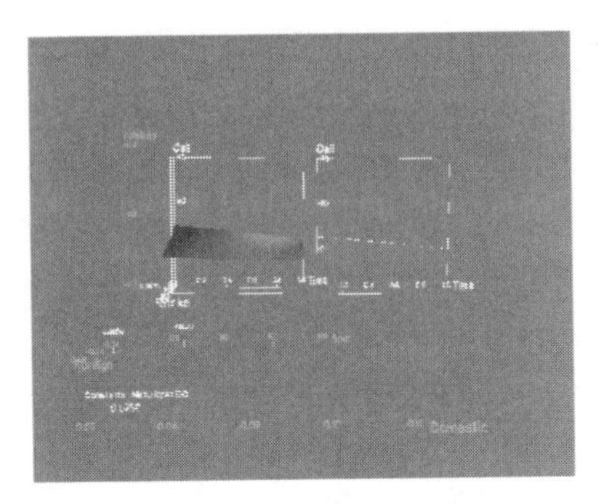

Figure 1.22: See Color Plate 7.

and uppercase letters for points in Cartesian spaces. In parallel coordinates, we use similar conventions except we put a superscript bar on all letters.

To see how a multidimensional object is represented in parallel coordinates, consider a simple two-dimensional straight line

$$l : x_2 = mx_1 + b$$

where $m < \infty$. In Figure 1.23, the continuous straight line in Cartesian coordinates is sam-

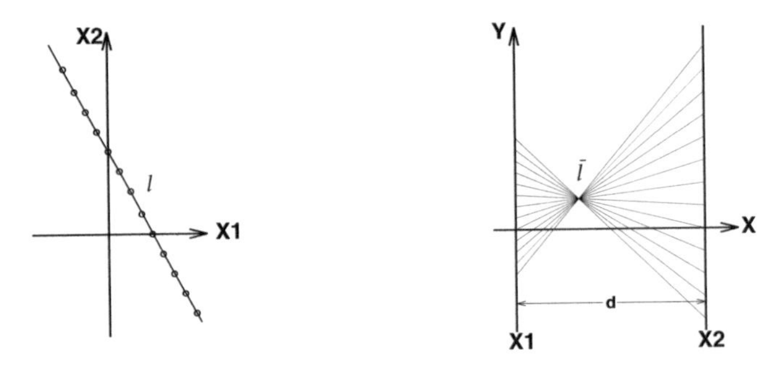

Figure 1.23: *Left:* Two-dimensional Cartesian coordinates. *Right:* Parallel coordinates.

pled, and the values are plotted on corresponding axes in parallel coordinates. A collection of points, A, sampled from a straight line in Cartesian coordinates, corresponds to a set of lines $\overline{A}$ in parallel coordinates that intersect at the point $\bar{l}$,

$$\bar{l} : \left(\frac{d}{1-m}, \frac{b}{1-m} \right)$$

for $m \neq 1$. For example, given a straight line $x_1/2 - x_2 + 1 = 0$, and $d = 5$, we get $m = 1/2$ and $b = 1$. When the sampled points are plotted in parallel coordinates, all lines intersect at $(\frac{5}{1-1/2}, \frac{1}{1-1/2}) = (10, 2)$, as shown in Figure 1.24. Notice that the location of the intersection point shows an important property of the data. In Figure 1.23, where two variates x_1 and x_2 are inversely proportional to each other, that is, $x_1 \propto 1/x_2$, the

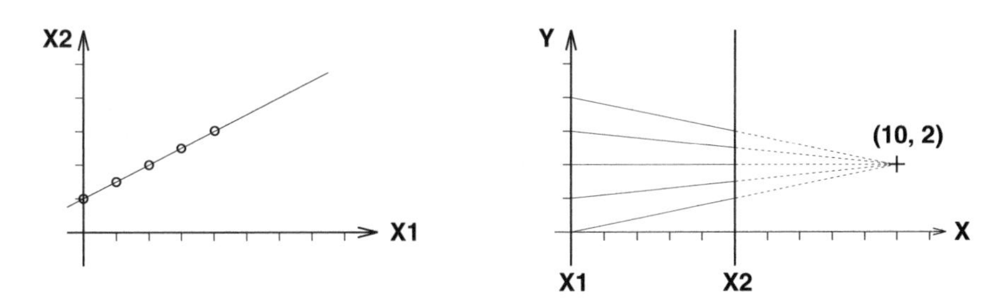

Figure 1.24: *Left:* The straight line, $x_1/2 - x_2 + 1 = 0$, is plotted in Cartesian coordinates. *Right:* The same line is plotted in parallel coordinates. The intersection point, $(10, 2)$, is located outside the two axes.

intersection point is in between the two parallel axes. In Figure 1.24, where variate x_1 is directly proportional to x_2, that is, $x_1 \propto x_2$, the intersection point is located outside the two parallel axes.

Parallel coordinates allow humans to visualize three-dimensional time-series data better than Cartesian coordinates. A simple application is aircraft collision checking [29]. In Figure 1.25, the locations of two aircraft are displayed in Cartesian coordinates and parallel

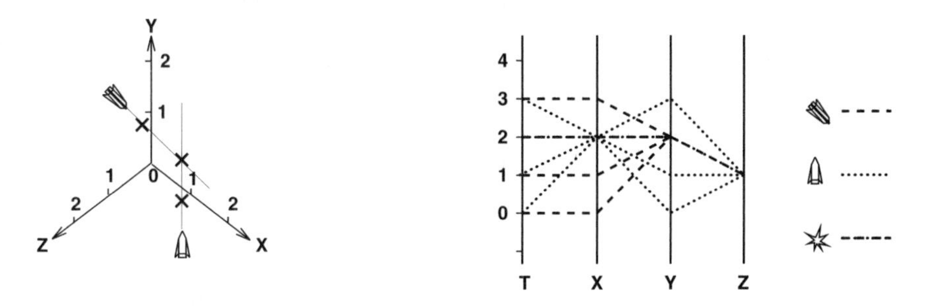

Figure 1.25: *Left:* Locations of two aircraft in three-dimensional Cartesian coordinates at time 0. *Right:* The trajectory is plotted with time axis, T, in parallel coordinates.

coordinates. It is almost impossible to confirm a collision solely by judging the locations of two aircraft in any one single view of a three-dimensional Cartesian plot. For example, suppose we have an isometric projection [23] with three aircraft located at coordinates $(0, 0, 0)$, $(1, 1, 1)$, and $(2, 2, 2)$ at time t. They are all displayed at the same spot, yet no collision occurs. On the other hand, a parallel coordinate plot including time t and the coordinates x, y, z is shown on the right-hand side of Figure 1.25. Two aircraft collide if and only if they are in the same location at the same time. That means there will be a collision in location $(2, 2, 1)$ at $T = 2$. A four-dimensional intersection can be detected by searching for any overlapping dashed lines. In our example, an overlap is detected at $(2, 2, 2, 1)$ of the parallel coordinate plot.

To help avoid collision, *parallelograms* can be defined along with the trajectory of the aircraft. The number, size, and shape of the parallelogram are computed according to its relative velocity and locations with respect to others, as shown in Figure 1.26. If the two aircraft are flying at the same velocity, the lower-right-hand aircraft must avoid any contact

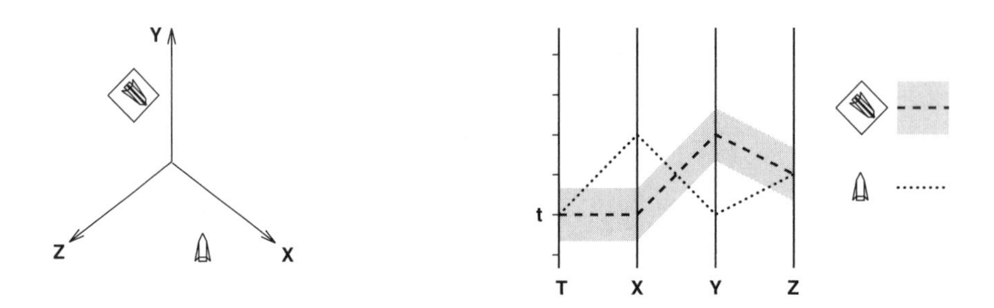

Figure 1.26: *Left:* Conflict parallelogram. *Right:* At time t, the graph shows that the location of one aircraft is not entirely inside the gray area of the others, so no collision occurs.

with the parallelogram of the upper-left-hand aircraft. Like our previous example, it is almost impossible to spot the conflict in the three-dimensional Cartesian plot. In Figure 1.26, the safety zone is indicated in gray in the parallel coordinate plot. There is a conflict if at any time the plotting of location/time of one aircraft is entirely inside the gray area of another.

Parallel coordinates can also be used to study correlations among variates in mdmv data analysis. By spotting the locations of the intersection points (see Figures 1.23 and 1.24), we can have a rough idea about the relationships between each pair of variates. This is one of the more promising applications of parallel coordinates in mdmv visualization. The problem with this technique is the limited space available for each parallel axis. The display can rapidly darken with even a modest amount of data. An implementation of parallel coordinates is also available in XmdvTool [61, 38] with brushing.

The VisDB System

Keim et al. [33, 34, 32] describe a multivariate visualization technique that is motivated by the desire to visualize information from a database. They use a combination of *distance functions* and *weighting factors* to visualize the *relevance factor* of very large multivariate data. The user issues a database query which identifies a *focal point* in multidimensional space. The data is arranged using a function that represents the relevance factor of that item with respect to the focal point defined by the query. Each display item is represented by a single pixel whose color is also determined by the relevance factor.

Once the relevance factors of the extracted data are computed, several display methods are possible. In a normal arrangement, data are sorted and arranged such that data with the highest relevance factors (smallest distance) are centered in the middle and the rest are placed in a rectangular spiral around this region, as shown in Figure 1.27.

In this basic form the information is somewhat limited since the position and color of each data item is determined by the same value—the distance of the item from the focal point. The overall pattern of color variation gives a general sense of how all the items in the database (or at least all that fit in the display) relate to the focal point, but it does not provide any visual insight regarding the relationships among the multiple variates. Such a comparison can be accomplished by generating additional displays that are viewed simultaneously. There is one display for each variate that contributes to the query; the

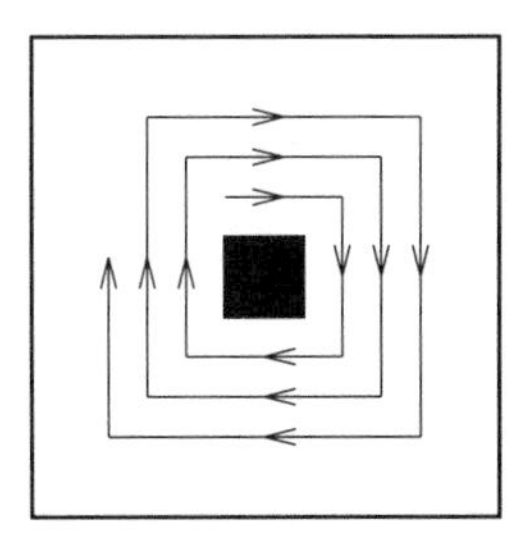

Figure 1.27: Normal arrangement.

position of each data item is the same as in the first display, but the color is based on only the distance of that variate from its value at the focal point. By comparing the color patterns of two or more variate displays, it may be possible to infer correlations of these variates with respect to the query that defines the focal point. See Figure 1.28.

Figure 1.28: See Color Plate 8.

The system has a heavy database orientation. The definition of the distance function becomes somewhat arbitrary for nonmetric variates and more complex for nested queries with Boolean operators. This approach is different from all the other techniques described in that it does not display the data directly, but instead presents the data based on the computed relevance factors for a particular query.

Recursive Pattern

Keim et al. [31] describe the recursive pattern visualization technique which extends the VisDB layout design. This technique, which is designed for time-series data visualization, requires only the *width* and the *height* of the display layout as input. Like the spiral layout in VisDB, each data item is represented by one display pixel with its pixel color indicating the data value. The data is displayed in a back-and-forth arrangement. The basic concept of this technique is also flexible enough to visualize data without inherent structure. Some of the suggested arrangements are depicted in Figure 1.29.

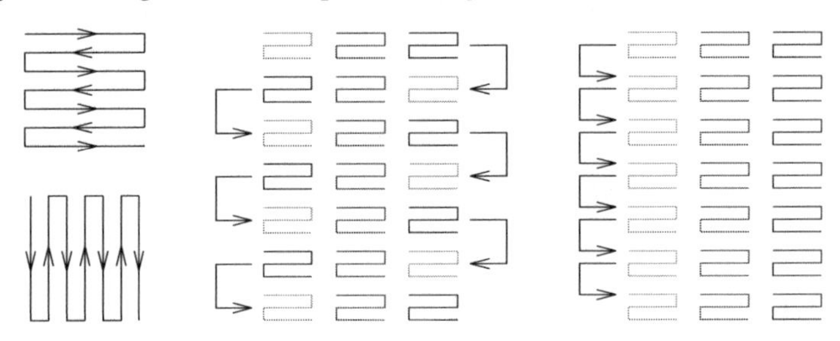

Figure 1.29: Examples of recursive patterns.

XmdvTool

Cleveland begins and ends his book [16] with the same quote, "Tools matter." The idea is that you have to pick the right technique to visualize mdmv data. The implication is that no technique alone is powerful and flexible enough to handle all mdmv scientific data. Ward [61, 38] integrates four popular static mdmv visualization techniques, one from each of the four static subgroups defined earlier, into a single analysis system, XmdvTool. The four tools included are: scatterplot (panel matrix), dimension stacking (hierarchical display), star glyph (iconography), and parallel coordinates. The original brushing [3], which relies heavily on mouse clicking interaction, is modified into a more complicated tool with user-controlled parameters. A user can control the shape, size, boundary, position, motion, and orientation of the n-dimensional brush. Some of these parameters are customized while the others can be controlled with slider widgets. The original brushing was implemented on the scatterplot matrix only. Brushing in XmdvTool is implemented on all four options. One of the major differences compared to the original version is that the brush itself is also displayed with the brushed data. In the case of parallel coordinates, the brush is just like the gray region of Figure 1.26, and the brushed data are the polylines inside the brush.

1.5.3 Animation of Multidimensional Multivariate Data

Data animation is no longer considered merely a means to present results. It can also be used as an exploratory tool to look for known phenomena of the data, as well as to investigate unpredicted and significant effects hidden behind the data [7, 19].

The development of scientific data animation has not been as intensive as static techniques. However, researchers have tried to use animation throughout the course of mdmv visualization development. Any of the techniques we have described can be applied to each step of time-series data and then displayed as an animated sequence. Below we discuss three approaches that in some way go beyond just presenting time series of static displays—grand tour methods [12], Exvis on a supercomputer [52], and a scalar visualization animation model [11].

Grand Tour Methods

Asimov [2] created the grand tour technique to project multidimensional data onto two-dimensional planes. Buja and Asimov [12] later developed grand tour animation methods. To quote from their paper, "Our methods are based on the simple idea of moving projection planes in high (4–10) dimensional data spaces. That is, we design 1-parameter families of 2-planes in p-space, with the parameter being thought of as time." That is to say, suppose we have data with $p + 1$ variates, we take one variate out as a time parameter for animation, and project the rest of the p variates onto two-dimensional planes in rapid succession according to the changes of the time parameter. This technique encodes data in the smooth motion of scatterplots, and provides a multitude of aspects among variates simultaneously.

Grand tour methods were tested with modeled data, as well as real-life data from a government housing project and particle physics research. A few observations presented in the paper include:

- Smoothness of motion is a basic requirement for scientific data animation. Jittering causes fatigue for human eyes.
- Excessive within-screen-spin (that is, rotation that takes place within the projection plane surface instead of in the projection spaces) is uninformative and may disturb human eyes. Therefore it should be avoided.
- The two-dimensional plane which encodes the velocity, that is, the speed vector of the data point, needs to be orthogonal to the two-dimensional projection plane to avoid confusing location (projection plane) and speed (velocity plane) of the dynamic scatterplot point.

Exvis on a Supercomputer

The implementation of the stick figure icon was extended from workstations to a data parallel supercomputer [52], so as to gain sufficient computational power for real-time icon generation. For data with n variates, there are $n!$ possible variate mappings to display parameters in iconography. In this version of Exvis, a set of display controls are included to support real-time interactive analysis. These include the size of the bounding box of each icon, the orientation of each icon with respect to a reference axis, and the maximum amount of random displacement of each icon in the xy directions. Scientists can adjust these control parameters until interesting spots are detected visually from the texture patterns. This kind of continuous parametric interaction can be considered as fine-tuning the physical representation of the data.

A Scalar Visualization Animation Model

Bragatto et al. [11] describe a scientific animation model designed and implemented to animate scalar fields of mdmv data. The data supported are multidimensional volumetric, that is, three-dimensional multivariate data defined on a uniform grid. For example, given a three-dimensional space, internal pressure is measured regularly over a period of time to create time-dependent scalar data. Three of the possible animation techniques are: a sequence of three-dimensional volumes animated according to the time variate, a single static volume if time is not included, or a series of two-dimensional planes if one of the three spatial dimensions is dropped. Data can be selected spatially (that is, a subset of three spatial dimensions) or temporally (that is, a particular time interval). Data is mapped to either three-dimensional shaded blocks with pseudocolor, three-dimensional isosurface blocks, or individual blocks characterized by color and textures for each grid point. An animation is composed of one or more *scenes*. Each scene contains *actors* (including objects from the data, light, and camera classes), and *actions* (group of events or actions). There is also a system *clock* to control the speed of the animation. Animation is described by a linear-list-type language. The system can be extended to multiple scalar data sets and vector fields.

1.6 Conclusions

Most of the mdmv visualization techniques described in this paper were developed and implemented before 1992. Although there are newer tools such as AutoVisual [9, 10], HyperSlice [60], and XmdvTool [61, 38], the intensity and variety of the recent activities are not comparable to the period from 1987 to 1991. An interesting fact that emerges from this survey is the remarkable stability of some of the two-way correlation techniques such as the scatterplot matrix, as compared to the generally unverified success of the multiway correlation methods.

We believe that the emphasis of mdmv visualization research must shift away from the design of yet more visual displays towards the rigorous evaluation of experimental visualization techniques. We must learn what approaches actually lead to more accurate results, enhanced productivity, and better understanding of the underlying science.

Finally, we believe that future mdmv visualization research must be integrated with developments in many other areas. Terabyte mdmv data is here. Scientists have to deal with data that is many thousands of times bigger than the number of pixels on a display. Visualization techniques initially motivated solely by visual perceptions are diversified. There is a great deal of current work on the design and implementation of scientific databases, multimedia systems, virtual worlds, and multiresolution analysis supporting mdmv visualization.

Acknowledgment

This work has been supported in part by the National Science Foundation under grant IRI-9117153. We wish to thank our colleagues who provided the sample images shown in Color Plates 1–8: Robert van Liere, Bowen Alpern, Georges G. Grinstein, Haim Levkowitz, Ted Mihalisin, Matthew O. Ward, Clifford Beshers, and Daniel Keim.

Bibliography

[1] Bowen Alpern and Larry Carter, "Hyperbox," *Proceedings of IEEE Visualization '91*, Gregory M. Nielson and Larry Rosenblum, editors, IEEE Computer Society Press, Los Alamitos, Calif., Oct. 1991, pp. 133–139.

[2] Daniel Asimov, "The Grand Tour: A Tool for Viewing Multidimensional Data," *SIAM Journal on Scientific and Statistical Computing*, Vol. 6, No. 1, Jan. 1985, pp. 128–143.

[3] Richard A. Becker and William S. Cleveland, "Brushing Scatterplots," *Technometrics*, Vol. 29, 1987, pp. 127–142.

[4] Jeff Beddow, "Shape Coding of Multidimensional Data on a Microcomputer Display," *Proceedings of IEEE Visualization '90*, Arie Kaufman, editor, IEEE Computer Society Press, Los Alamitos, Calif., Oct. 1990, pp. 238–246.

[5] Jeff Beddow, "An Overview of Multidimensional Visualization: Elements and Methods," *Designing a Visualization Interface for Multidimensional Multivariate Data*, Jeff Beddow and Cliff Beshers, editors, IEEE Visualization '92 Tutorial 8, Oct. 1992.

[6] R. Daniel Bergeron, William Cody, William Hibbard, David T. Kao, Kristina Miceli, Lloyd Treinish, and Sandra Walther, "Database Issues for Data Visualization: Developing a Data Model," *IEEE Visualization '93 Workshop on Database Issues for Data Visualization*, John P. Lee and Georges G. Grinstein, editors, Springer-Verlag, Oct. 1995, pp. 3–15.

[7] R. Daniel Bergeron and Georges G. Grinstein, "The Impact of Scientific Visualization on Workstation Development," *Workstations for Experiments*, J. L. Encarnação and Georges G. Grinstein, editors, Springer-Verlag, 1989, pp. 3–11.

[8] R. Daniel Bergeron and Georges G. Grinstein, "A Reference Model for Scientific Visualization," *Proceedings Eurographics '89*, 1989, pp. 393–399.

[9] Clifford Beshers and Steven Feiner, "Automated Design of Virtual Worlds for Visualizing Multivariate Relations," *Proceedings of IEEE Visualization '92*, Arie E. Kaufman and Gregory M. Nielson, editors, IEEE Computer Society Press, Los Alamitos, Calif., Oct. 1992, pp. 283–290.

[10] Clifford Beshers and Steven Feiner, "Autovisual: Rule-Based Design of Interactive Multivariate Visualizations," *IEEE Computer Graphics and Applications*, Vol. 13, No. 4, 1993, pp. 41–49.

[11] Paolo Bragatto, Naida Mazzino, and Patrizia Palamidese, "Animated Visualization of Scalar Fields," *Conference Proceedings of Second Eurographics Workshop on Animation and Simulation*, Vienna, Austria, Sep. 1991, pp. 115–127.

[12] Andreas Buja and Daniel Asimov, "Grand Tour Methods: An Outline," *Proceedings of the 18th Symposium on the Interface*, American Statistical Association, 1986, pp. 63–67.

[13] Andreas Buja, John A. McDonald, John Michalak, and Werner Stuetzle, "Interactive Data Visualization Using Focusing and Linking," *Proceedings of IEEE Visualization '91*, Gregory M. Nielson and Larry Rosenblum, editors, IEEE Computer Society Press, Los Alamitos, Calif., 1991, pp. 156–163.

[14] J. M. Chambers, William S. Cleveland, B. Kleiner, and P. A. Tukey, *Graphical Methods for Data Analysis*, Chapman and Hall, New York, 1983.

[15] H. Chernoff, "The Use of Faces to Represent Points in k-Dimensional Space Graphically," *Journal of American Statistical Association*, Vol. 68, 1973, pp. 361–368.

[16] William S. Cleveland, *Visualizing Data*, Hobart Press, Summit, N.J., 1993.

[17] William S. Cleveland, M. E. McGill, and R. McGill, "The Shape Parameter of a Two-Variable Graph," *Journal of American Statistical Association*, Vol. 38, 1993, pp. 289–300.

[18] Brian M. Collins, "Data Visualization—Has It All Been Seen Before?" *Animation and Scientific Visualization—Tools and Applications*, Rae A. Earnshaw and David Watson, editors, chapter 1, Academic Press, 1993, pp. 3–28.

[19] Jose L. Encarnação, Detlef Kromker, Jose Mario de Martino, Gabriele Englert, Stefan Haas, Edwin Klement, Fritz Loseries, Wolfgang Muller, Georgios Sakas, and Ralf Rainer Vohsbeck Petermann, "Advanced Research and Development Topics in Animation and Scientific Visualization," *Animation and Scientific Visualization—Tools and Applications*, Rae A. Earnshaw and David Watson, editors, chapter 3, Academic Press, 1993, pp. 37–73.

[20] Steven Feiner and Clifford Beschers, "Worlds Within Worlds: Metaphors for Exploring n-Dimensional Virtual Worlds," *Proc. UIST '90, ACM Symposium on User Interface Software and Technology*, Snowbird, Utah, Oct. 1990, pp. 76–83.

[21] Steven Feiner and Clifford Beshers, "Visualizing n-Dimensional Virtual Worlds with n-Vision," *Computer Graphics*, Vol. 24, No. 2, Mar. 1990, pp. 37–38.

[22] Sir Ronald A. Fisher, *Statistical Methods for Research Workers*, Oliver and Boyd, 14th edition, 1970.

[23] James Foley, Andries van Dam, Steven K. Feiner, and John F. Hughes, *Computer Graphics—Principles and Practice*, The Systems Programming Series. Addision-Wesley, second edition, 1990.

[24] J. J. Gibson, *The Perception of the Visual World*, Houghton Mifflin Co., Boston, Mass., 1950.

[25] Georges G. Grinstein, Ronald M. Pickett, and Marian G. William, "EXVIS: An Exploratory Visualization Environment," *Proceedings of Graphics Interface '89*, 1989, pp. 254–261.

[26] Lambertus Hesselink, Frits H. Post, and Jarke J. van Wijk, "Research Issues in Vector and Tensor Field Visualization," *IEEE Computer Graphics and Applications*, Vol. 14, No. 2, Mar. 1994, pp. 76–79.

[27] A. Inselberg, M. Reif, and T. Chomut, "Convexity Algorithms in Parallel Coordinates," *Journal of ACM*, Vol. 34, No. 4, Oct. 1987, pp. 765–801.

[28] Alfred Inselberg and Bernard Dimsdale, "Parallel Coordinates for Visualizing Multi-Dimensional Geometry," *Proceedings of Computer Graphics International '87*, T. L. Kunii, editor, Springer-Verlag, Tokyo, 1987.

[29] Alfred Inselberg and Bernard Dimsdale, "Parallel Coordinates: A Tool for Visualizing Multi-Dimensional Geometry," *Proceedings of IEEE Visualization '90*, Arie Kaufman, editor, IEEE Computer Society Press, Los Alamitos, Calif., Oct. 1990, pp. 361–375.

[30] Arie Kaufman, Karl Heinz Hohne, Wolfgang Kruger, Lawrence Rosenblum, and Peter Schröder, "Research Issues in Volume Visualization," *IEEE Computer Graphics and Applications*, Vol. 14, No. 2, Mar. 1994, pp. 63–67.

[31] Daniel A. Keim, Mihael Ankerst, and Hans-Peter Kriegel, "Recursive Pattern: A Technique for Visualizing Very Large Amounts of Data," *Proceedings of IEEE Visualization '95*, Gregory M. Nielson and Deborah Silver, editors, IEEE Computer Society Press, Los Alamitos, Calif., Oct. 1995, pp. 279–286.

[32] Daniel A. Keim and Hans-Peter Kriegel, "VisDB: Database Exploration Using Multidimensional Visualization," *IEEE Computer Graphics and Applications*, Vol. 14, No. 5, Sep. 1994, pp. 40–49.

[33] Daniel A. Keim, Hans-Peter Kriegel, and Thomas Seidl, "Visual Feedback in Querying Large Databases," *Proceedings IEEE Visualization '93*, Gregory M. Nielson and R. Daniel Bergeron, editors, IEEE Computer Society Press, Los Alamitos, Calif., Oct. 1993, pp. 158–165

[34] Daniel A. Keim, Hans-Peter Kriegel, and Thomas Seidl, "Supporting Data Mining of Large Databases by Visual Feedback Queries," *Proceedings Tenth International Conference on Data Engineering*, Feb. 1994, pp. 302–313.

[35] Peter R. Keller and Mary M. Keller, *Visual Cues, Practical Data Visualization*, IEEE Computer Society Press, Los Alamitos, Calif., 1993.

[36] Jeffrey LeBlanc, Matthew O. Ward, and Norman Wittels, "Exploring n-Dimensional Databases," *Proceedings of IEEE Visualization '90*, Arie Kaufman, editor, IEEE Computer Society Press, Los Alamitos, Calif., Oct. 1990, pp. 230–237.

[37] Haim Levkowitz, "Color Icons: Merging Color and Texture Perception for Integrated Visualization of Multiple Parameters," *Proceedings of IEEE Visualization '91*, Gregory M. Nielson and Larry Rosenblum, editors, IEEE Computer Society Press, Los Alamitos, Calif., Oct. 1991, pp. 164–170.

[38] Allen R. Martin and Matthew O. Ward, "High Dimensional Brushing for Interactive Exploration of Multivariate Data," *Proceedings IEEE Visualization '95*, Gregory M. Nielson and Deborah Silver, editors, IEEE Computer Society Press, Los Alamitos, Calif., Oct. 1995, pp. 271–278.

[39] Bruce H. McCormick, Thomas A. DeFanti, and Maxine D. Brown, "Visualization in Scientific Computing," *Computer Graphics*, Vol. 21, No. 6, Nov. 1987, pp. 1–14.

[40] Ted Mihalisin, E. Gawlinski, John Timlin, and John Schwegler, "Visualizing a Scalar Field on an n-Dimensional Lattice," *Proceedings of IEEE Visualization '90*, Arie Kaufman, editor, IEEE Computer Society Press, Los Alamitos, Calif., Oct. 1990, pp. 255–262.

[41] Ted Mihalisin, John Timlin, and John Schwegler, "Visualization and Analysis of Multi-Variate data: A Technique for All Fields," *Proceedings of IEEE Visualization '91*, Gregory M. Nielson and Larry Rosenblum, editors, IEEE Computer Society Press, Los Alamitos, Calif., Oct. 1991, pp. 171–178.

[42] Ted Mihalisin, John Timlin, and John Schwegler, "Visualizing Multivariate Functions, Data, and Distributions," *IEEE Computer Graphics and Applications*, Vol. 11, No. 3, May 1991, pp. 28–35.

[43] Gregory M. Nielson, "Modeling and Visualizing Volumetric and Surface-On-Surface Data," *Focus on Scientific Visualization*, Hans Hagen, Heinrich Muller, and Gregory M. Nielson, editors, Springer-Verlag, 1992, pp. 191–242.

[44] Gregory M. Nielson and Larry Rosenblum, editors, *Proceedings IEEE Visualization '91*, IEEE Computer Society Press, Los Alamitos, Calif., Oct. 1991.

[45] R. M. Pickett and B. W. White, "Constructing Data Pictures," *Proceedings of Society for Information Display Seventh National Symposium*, 1966, pp. 75–81.

[46] Ronald M. Pickett, "Visual Analyses of Texture in the Detection and Recognition of Objects," *Picture Processing and Psycho-Pictorics*, B. S. Lipkin and A. Rosenfeld, editors, Academic Press, New York, 1970.

[47] Ronald M. Pickett and Georges G. Grinstein, "Iconographics Displays for Visualizing Multidimensional Data," *Proceedings IEEE Conference on Systems, Man, and Cybernetics*, PRC, Beijing and Shenyang, May 1988, pp. 514–519.

[48] Frits H. Post and Theo van Walsum, "Fluid Flow Visualization," *Focus on Scientific Visualization*, Hans Hagen, Heinrich Muller, and Gregory M. Nielson, editors, Springer-Verlag, 1992, pp. 1–40.

[49] Philip K. Roberston, Rae A. Earnshaw, Daniel Thalmann, Michel Grave, Julian Gallop, and Eric M. De Jong, "Research Issues in the Foundations of Visualization," *IEEE Computer Graphics and Applications*, Vol. 14, No. 2, Mar. 1994, pp. 73–76.

[50] J. H. Siegel, E. J. Farrell, R. M. Goldwyn, and H. P. Friedman, "The Surgical Implication of Physiologic Patterns in Myocardial Infarction Shock," *Surgery*, Vol. 72, 1972, pp. 126–141.

[51] Stuart Smith, R. Daniel Bergeron, and Georges G. Grinstein, "Stereophonic and Surface Sound Generation for Exploratory Data Analysis," *Multimedia and Multimodal Interface Design*, M. Blattner and R. Dannenberg, editors, ACM Press, New York, 1992.

[52] Stuart Smith, Georges G. Grinstein, and R. Daniel Bergeron, "Interactive Data Exploration with a Supercomputer," *Proceedings IEEE Visualization '91*, Gregory M. Nielson and Larry Rosenblum, editors, IEEE Computer Society Press, Los Alamitos, Calif., Oct. 1991.

[53] Stuart Smith, Ronald M. Pickett, and Marian G. Williams, "Environments for Exploring Auditory Representations of Multidimensional Data," *Proceedings of the International Conference on Auditory Display*, Santa Fe Institute, Santa Fe, N.M., 1992.

[54] Bob Spence, Lisa Tweedie, Huw Dawkes, and Hua Su, "Visualization for Functional Design," *Proceedings of IEEE Information Visualization '95*, Nahum Gershon and Steve Eick, editors, IEEE Computer Society Press, Los Alamitos, Calif., Oct. 1995, pp. 4–9.

[55] Eric J. Stollnitz, Anthony D. DeRose, and David H. Salesin, "Wavelets for Computer Graphics: A Primer, Part 1," *IEEE Computer Graphics and Applications*, Vol. 15, No. 3, May 1995, pp. 77–84.

[56] Eric J. Stollnitz, Anthony D. DeRose, and David H. Salesin, "Wavelets for Computer Graphics: A Primer, Part 2," *IEEE Computer Graphics and Applications*, Vol. 15, No. 4, July 1995, pp. 75–85.

[57] John Keenan Taylor, *Statistical Techniques for Data Analysis*, Lewis Publishers, Boca Raton, Fla., 1990.

[58] John W. Tukey, *Exploratory Data Analysis*, Addison-Wesley, 1977.

[59] Lisa Tweedie, Robert Spence, Huw Dawkes, and Hua Su, "Externalising Abstract Mathematical Models," *Proceedings of CHI 96*, Michael J. Tauber, editor, IEEE Computer Society Press, New York, N.Y., Apr. 1996, pp. 406–412.

[60] Jarke J. van Wijk and Robert van Liere, "HyperSlice," *Proceedings IEEE Visualization '93*, Gregory M. Nielson and R. Daniel Bergeron, editors, IEEE Computer Society Press, Los Alamitos, Calif., Oct. 1993, pp. 119–125.

[61] Matthew O. Ward, "XmdvTool: Integrating Multiple Methods for Visualizing Multivariate Data," *Proceedings IEEE Visualization '94*, R. Daniel Bergeron and Arie E. Kaufman, editors, IEEE Computer Society Press, Los Alamitos, Calif., Oct. 1994, pp. 326–336.

[62] Pak Chung Wong, Andrew H. Crabb, and R. Daniel Bergeron, "Dual Multiresolution Hyperslice for Multivariate Data Visualization," *Proceedings of IEEE Information Visualization '96*, Nahum Gershon and Steven Eick, editors, IEEE Computer Society Press, Los Alamitos, Calif., Oct. 1996.

Chapter 2

Immersive Investigation of Scientific Data

Helmut Haase, Fan Dai, Johannes Strassner, and Martin Göbel

Abstract. *Scientific visualization systems normally use keyboard or mouse as input devices and desktop screens or printers/plotters for output. Virtual environment systems, on the other hand, use devices like gloves or head-mounted displays and a number of software techniques in order to achieve realistic, immersive real-time graphics. This chapter discusses the techniques used for virtual environments and investigates possibilities to construct systems that use virtual environment techniques for scientific visualization. Several examples are given of immersive scientific visualization.*

Keywords: scientific visualization, virtual reality, immersive exploration, real-time interaction

2.1 Introduction

Scientific visualization is the technique to explore numerical data by means of visual, graphical objects. Through visualization, results of complex numerical computations and measurements become intuitively understandable. Interaction methods like 3D viewing, data probing, and such, allow us to investigate all relevant aspects of a data set.

Nevertheless, 3D presentation techniques and interaction methods using a "conventional" (mouse-menu-driven) user interface still have some disadvantages: spatial relationships are often difficult to recognize and navigation is only indirectly possible.

Now, the so-called virtual environment (VE), or virtual reality techniques offer new possibilities of presentation and investigation of multidimensional data. Virtual environments are characterized by the simulation of an environment which behaves as if it were real. This realism is not only the photo-realistic presentation of the objects, but also the human user's impression of being immersed in virtual objects. From its very beginning, handling of 3D graphics and interaction with 3D objects strived for being as intuitive as possible. Immersion offers the highest degree of intuitive interaction. Scientific visualization, specifically, can benefit from the possibility of immersion [9, 25, 23]. A discussion of human factors in immersive scientific visualization may be found in [5].

In an immersive presentation, interactions are performed in real time, intuitively, simply, and directly. Of course, some engineering tasks require precise positioning and parameterizing of data objects. For these tasks, keyboards and mouse-menu-driven user interfaces seem to be useful. But often, intuitive, easy-to-use interaction tools are needed. This is especially true if several people with different technical backgrounds are discussing results of complex data analyses. In such cases, realistic presentation and intuitive interactions like "point and fly," "grab and move," and so on, are required.

Immersive exploration of scientific data is the natural next step after interactive real-time visualization. Visualization systems as we know them will probably not disappear, but they will be augmented and complemented by VE technologies. Furthermore, immersive scientific visualization will lead to a number of new application areas due to its capability of handling scientific data in a formerly unknown manner.

This analyzes existing techniques for virtual environments and presents some possibilities for integrating scientific visualization and virtual environments. First experiences are described with discussions on realized examples. Finally, remarks on future directions of research and development are made.

2.2 Techniques for Virtual Environments

This section gives an overview of the state of the art in hardware devices and software techniques for creating virtual environments. It will also point out their strengths and weaknesses.

2.2.1 Devices for Virtual Environments

In the following, we will give an overview of device technology for virtual environments which is currently available commercially. We group these devices into three categories: input devices, acoustic output devices, and visual output devices. Other output devices such as force feedback, or even olfactory devices, are still subject to research and not generally available.

Input Devices

Typical VE input devices allow multidimensional interaction of a user with an application. By means of tracking devices, a system is provided with 6D information, that is, position and orientation within 3D space. This allows direct processing of information about the user's body, for example, the position of head or hand. Tracking units mostly use one of three possible technologies: electromagnetics, ultrasound, or mechanics.

Furthermore, gloves are being used to detect the angle of certain finger joints. This may be achieved by means of special sensors within the glove (such as optical fibers), or by means of a mechanical skeleton surrounding the glove. So-called "flying joysticks" (a joystick equipped with buttons and a tracking sensor) are increasingly used to control applications.

In nonimmersive desktop applications, other devices like SpaceBall or SpaceMouse may be used alternatively. These devices measure pressure and torsional stress along the three coordinate axes.

Last but not least, speech input can be used. Speech input systems are very well suited for immersive environments because hands or eyes which may be engaged by data gloves or head-mounted displays are not needed in order to utilize them. Today, if speech recognition is to be independent of the user, it is only suitable for simple tasks, for example, pressing a button or making a selection from a list. Research is underway to develop more general speech input systems. Another topic, namely, direct analysis of human biosignals (such as ECG and EEG, for example) is also still being researched.

Acoustic Output Devices

Output of acoustic signals can be realized either by means of acoustic hardware included in multimedia workstations or by means of external devices (for example, synthesizers). After generating acoustic signals, additional hardware may be used to subsequently alter these signals. For example, effect generators and signal processors can filter the signal in order to allow a listener to localize the virtual sound source in 3D.

Visual Output Devices

There are many different kinds of output systems for presenting virtual worlds visually. In selecting an output device, the desired image resolution and image quality must be considered, as well as the degree of immersion which is to be achieved. Immersion stands for the illusion of presence in a virtual world. In general, VE utilizes stereoscopic displays for improved depth perception and a wide-angle field of view for immersion. Typical systems are:

- traditional desktop monitor,
- large projection screen (sometimes several simultaneously),
- head bound systems (for example, helmets or glasses).

This list is sorted roughly according to increasing degree of immersion. Nowadays, many graphics workstations offer means of displaying stereoscopic images. Mostly, this is achieved by means of shutter systems in time multiplex mode. These systems typically have a frequency of 120 Hz and switch between perspective images for the left and right eye. A shutter mechanism (for example, a pair of glasses with LCDs) is used to deliver each image to the appropriate eye, resulting in 60 Hz images for each eye.

Such systems allow stereoscopic viewing of high-resolution true-color images, but due to the limited field of view, immersion is not achieved. Latest research aims at autostereoscopic viewing without glasses. Applications of shutter technology in combination with desktop monitors are found in the fields of CAD and of modeling in general; for vehicle simulators or VE systems they are less appropriate.

Large screen projection can extend the field of view, thus increasing the perception of immersion. By combining several images or by means of special techniques, convincing panoramic views can be achieved (such as in IMAX theatres or in planetariums). Vehicle simulators often employ several large screen projections.

In VE systems, large screen projections often present stereoscopic images. Head tracking allows us to "walk around" scene objects, which further increases the impression of immersion. Stereoscopic large screen projection can be achieved by means of a shuttering technique or by means of a time parallel system (that is, simultaneous display of perspective image for the left and right eye and image separation by means of polarized light). Both techniques can achieve high resolution, stereoscopic full-color images with large or very large fields of view.

The highest degree of immersion is achieved by head-bound systems. We distinguish systems which rest directly on the head (helmet and glasses) from systems which, due to their weight, are fixed to a mechanical system and which are merely held in front of the head (for example, BOOM). Both systems use a separate screen for each eye in order to achieve a stereoscopic image. In both cases, the output systems are attached to the head and follow its movements. Liquid crystal displays (LCD), cathode-ray tubes (CRT), and glass fiber optics are used. Special optics allow us to achieve a field of view of approximately 100 degrees. Some head-bound systems offer see-through capabilities, that is, the "real world" can still be seen "behind" the transparent computer-generated image.

Research is underway for "retinal scanners," which will project the computer-generated image directly onto the user's retina by means of laser light. Research is also being done on volumetric output devices which, similar to holography, present a new medium for displaying 3D images.

2.2.2 Software Techniques for Virtual Environments

Software techniques used for virtual environments often are suited for any kind of realistic, interactive real-time graphics. Therefore, they are also of interest for "traditional" scientific visualization.

Hardware Rendering/Shading

Today's graphics workstations offer many hardware features supporting fast rendering of 3D scenes in high quality. Virtual environment systems utilize these features in order to achieve real-time rendering of realistic images.

Light sources can be defined in 3D, and 3D polygons will be rendered by a hardware pipeline using flat shading or smooth shading (Gouraud). Hidden-surface removal normally is performed by means of a so-called Z-buffer.

Often, a huge number of polygons has to be rendered, most of which have neighbors at their edges. The hardware has to triangulate these polygons in order to render them. Therefore, if triangles are given to the hardware instead of complex polygons, rendering speed will increase. Furthermore, if a number of triangles are chosen to form a stipe (that is, each new triangle shares one edge with the previous one), each new triangle is precisely defined by just one point in 3D, since the other two points are already known to be the ones of the predecessor. If only this one vertex of each new triangle is given, rendering speed again is increased drastically (typically by a factor of at least two on high-end hardware).

On many machines, transparency of polygons is also realized in hardware, even though artifacts often occur if the polygons have not been sorted. Texture mapping (see "Textures" below) is also available on a number of machines.

Sometimes, even 3D spheres are offered for drawing primitives with high performance (for example, for molecular modeling).

Hierarchical Data Structures/Viewing Culling

Grouping objects of a scene into hierarchies is the best way of managing the data associated with complex scenes. Objects next to each other are grouped recursively to form larger objects. For example, a factory hall may consist of a number of machines, each machine of screws and other parts, and each screw of some polygons. Object hierarchies can be defined by the user or by the system based on spatial information. Bounding volumes are assigned to each node of the hierarchy tree surrounding all the objects of the corresponding subtree. The bounding volumes can be used, for example, for collision detection (see below).

Perhaps the main benefit of object hierarchies is for viewing culling. In viewing culling, in order to determine if any object of the subtree has to be considered for rendering, the associated bounding volume is tested against the viewing frustrum (or cone of vision). If the bounding volume lies completely outside the viewing frustrum, the whole subtree associated with it can be disregarded for rendering.

A similar technique uses portal nodes (also called switching nodes). In architectural presentations, normally only one room is visible at a time. A so-called portal node is associated with each room. Only the objects belonging to the subtree of the current portal node need to be considered for rendering. If the user moves from one room to another, the current portal node is changed.

Level of Detail

Realism is an important but also controversial issue for virtual worlds. It can be achieved by employing very complex models. But, the number of polygons which can be rendered in real time is limited. Therefore, a technique called "level of detail" is used in order to get a compromise between realistic appearance and scene complexity.

The farther away an object is from an observer, the fewer details can be distinguished. This is due to perspective shrinking of an object and limited screen resolution.

Therefore, the complexity of an object can be reduced according to its distance from the observer without loss of realism. If an object occupies the full field of view at a distance d_0, then it will occupy $\frac{1}{i^2}$ of the field of view at distance $d_i = i * d_0$. Figures 2.1 and 2.2 show how the apparent size of an object is decreasing with increasing distance from the observer. Since the apparent size of an object decreases dramatically up to a distance of $4d_0$, it is recommended to supply most levels of detail for this interval. This results in a relatively high frame rate in spite of complex objects. The marks in the figures present a suggestion for assigning eight levels of detail to distances.

In order to generate reduced models which resemble the high-resolution model as closely as possible, automatic reduction algorithms have been developed.

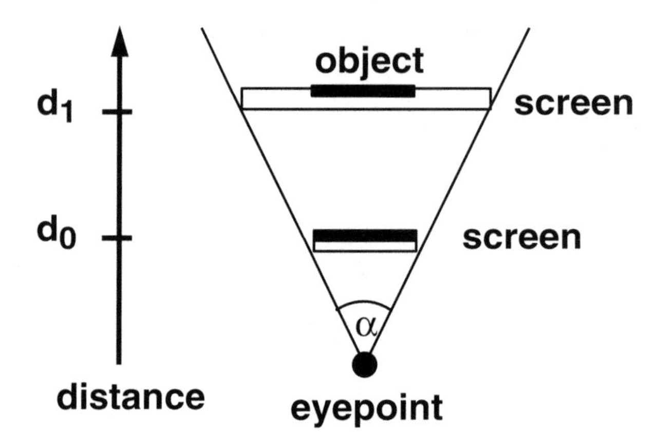

Figure 2.1: Level of detail: distance of an object and associated apparent size—schematic view.

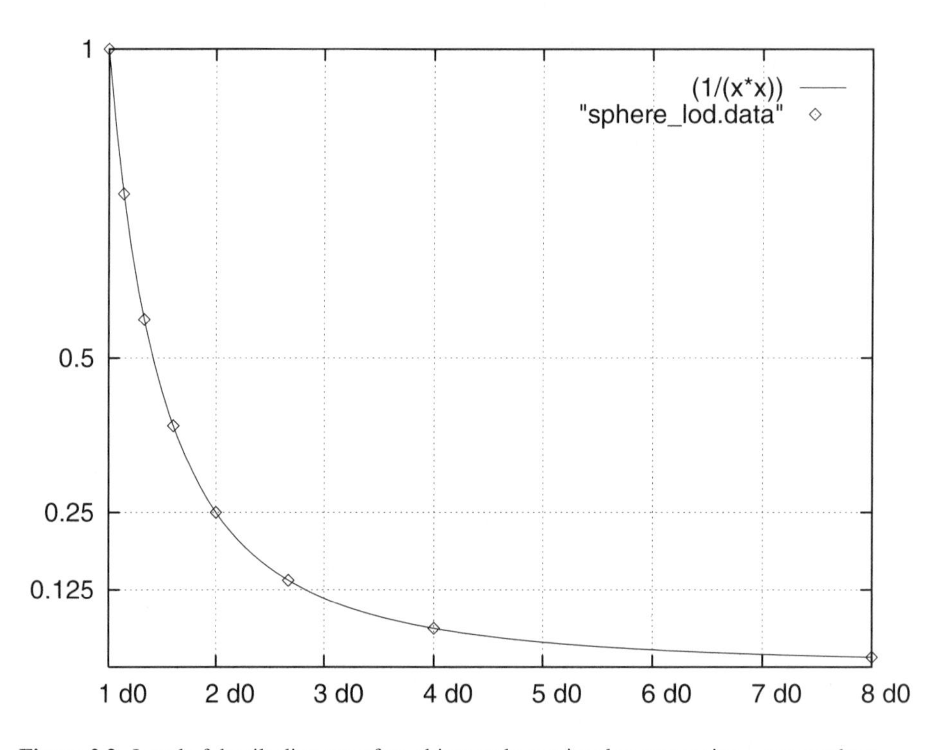

Figure 2.2: Level of detail: distance of an object and associated apparent size—exact values.

Level of detail is a technique that is used whenever smooth interaction with high frame rates is more important than precise representation of an object. For example, during movement of an object representing a complex data set, it can be useful to choose a reduced model complexity. Thus, the system will adapt to the speed of the user and not vice versa.

If object hierarchies are used, each level of detail is represented by a subtree. Several subtrees are stored for each object and, at rendering time, the subtree with appropriate complexity is chosen according to the observer's distance, movement criteria, or the like.

Successive Refinement

Successive refinement is similar to level of detail in that it offers several models of an object with varying complexity, which are rendered according to some well-defined rule. Successive refinement is used for objects which cannot be precomputed but which must be computed at runtime (for example, isosurfaces of a 3D data set). In order to present fast response to the user's actions, only a rough approximation of the desired geometry is computed first, which can be rendered very quickly. If the user decides to change a parameter used for the computation (for example, threshold of isosurface), another rough approximation is computed, now with the new parameter. If the user does not do another interaction, computation proceeds to a more detailed solution, which is rendered as soon as available.

Thus, response time is very short during user interaction, and the quality of the output is very high if the system is idle.

Rendering Caches

An object hierarchy is a complex data structure which needs some time to be traversed. The rendering hardware, on the other hand, accepts only a display list, which is a very simple, flat data structure. Thus, for each rendering, the object hierarchy must be transformed to a display list, costing valuable CPU time. On the other hand, if part of a scene is static, the display list for the appropriate subtree will always be the same. Therefore, this display list can be stored (cached) and reused without traversing the object hierarchy, as long as no changes occur. This can speed up rendering performance drastically in static scenes.

Multiprocessing

For dynamic objects, caching the display list does not work because it changes constantly. In this case, a multiprocessor system can still result in a speedup. One process traverses the hierarchy and performs viewing culling. Another process on another processor gets the subtrees which need to be rendered and produces display lists. For stereoscopic rendering, two such processes can be used. Another process can control the dynamics and modify the object hierarchy while the other processors are rendering. Inconsistencies introduced by this mechanism are hardly perceivable at real-time frame rates of 25 frames per second.

Collision Detection

Collision detection is a very complex task and much research has been done in this field. Often, it is important to know if any object intersects with any other object in the scene, for example, in an assembly simulation. Collision detection is also used for probing or if objects are to be grabbed. A very naive algorithm would check each face of each object against each face of all other objects, resulting in a computational complexity of $O(N^2)$. With typical scenes, this cannot be done in real time. Therefore, bounding boxes or bounding spheres around objects are normally used which can be checked much more efficiently than polygons due to their small number. On the other hand, such approaches can lead to very unnatural behavior, for example, if a mug with a handle is to be grabbed. Thus, the problem of collision detection is still under research.

For a more comprehensive survey of dynamic collision detection see [24].

Textures

By texture we mean a raster image which is mapped onto the surface of a 3D object. Textures are mainly used to increase realism of a scene or of an object. For virtual environments, textures have a number of merits:

- *Real-world materials:* Textures may be used to give objects the appearance of being made of real-world materials, for example, a table can be rendered using a wood texture. Textures normally are stored in a special memory area (which is of limited size) in graphics workstations. Therefore, the texture should be an appropriate size. Surfaces with regular structure can be tiled by repeatedly applying one small texture.

- *Billboards:* Objects that look similar from all sides and that are mostly seen from a distance can be rendered very quickly by using so-called billboards. A billboard is a textured polygon. Pixels of the texture which do not belong to the object are rendered completely transparent. In order to ensure that the viewer always looks perpendicular to the polygon, the billboard is rotated accordingly. This technique is useful for rotating symmetric objects like trees or clouds.

- *Rapid scene prototyping:* During model generation, textures can be used to determine if an object fits in a scene or to give an early preview of the final scene. A real-world object can be photographed in front of a neutral background, and all background pixels can be set to maximum transparency. Objects textured this way can help in making decisions about the final model. Later, they can be replaced by more complex geometries.

- *Projected light:* Textures can be projected on surfaces in order to show slides or a movie within the virtual world. If a texture contains lightness according to a virtual lamp, it can be moved over a smooth surface in order to simulate the movement of the corresponding light source. Further applications of this technique are the simulation of shadows (see "Fast Shadows" below).

- *Textures for scientific visualization:* Properties of objects can be computed and visualized on the surface, for example, by means of textures containing isolines or color maps.

A more detailed survey on uses of texture mapping may be found in [22].

Radiosity

Radiosity is a technique used for calculation of light distribution in a static scene. This can be done independently of any rendering.

Several methods are known for performing radiosity computations. All of them are computationally very expensive, therefore radiosity is normally used during model generation to calculate the illumination of the objects in a scene. This results in photorealistic appearance of the model. Therefore, this technique is mainly used in architecture to investigate lighting conditions within a building or for impressive presentations.

Up to now, radiosity is commonly used for static scenes because update of shadows of moving objects in real time is still subject to research. But, if the photorealistic appearance of a model is more important than shadows of moving objects, these objects may be omitted during the radiosity computation and rendered into the scene at runtime.

Fast Shadows

Shadows can be used to improve realism of a scene, and also to give an idea of size and position of an object floating in 3D space. Radiosity already has been mentioned as a way of precomputing realistic, smooth shadows. If radiosity computation takes too long, especially for moving objects, or if shadows mainly serve for orientation and not so much for high realism, other techniques which meet the following conditions are needed:

- The shadow should move according to the movement of the corresponding object.
- The outline of the shadow does not need to represent the object perfectly.
- The shadow may be colored with a uniform gray value, as would be the case if one light source at infinite distance was present.
- If the shadow is to be used for size estimation, it should have constant size no matter how far the distance of the object to the "ground."

Two interesting techniques with such properties are presented here: faked shadows and precomputed textures.

- *Faked shadows:* This method computes the shadow by projecting the corresponding object to a surface. If an object is lit by a light source at infinite distance, all light rays are parallel and computation of the shadow is simple. If the light is even perpendicular to the surface, an instance of the object can be compressed to height 0, colored black, and rendered onto the surface. In order to further speed up the algorithm, a reduced, less complex version of the object can be used.
- *Precomputed textures:* Computing faked shadows at runtime may still be too expensive computationally. Therefore, an exact shadow of an object can be precalculated. It is assigned as a texture to a shadow polygon which is shown approximately below the object. If the object is rotated about the vertical axis or if it is translated perpendicular to this axis, these operations are also performed on the shadow polygon. But, if the object is rotated about another axis, the shadow will no longer be realistic and

may be exchanged for another precomputed shadow. This problem can be avoided if a circle or an ellipse is used as texture. Of course, such an idealized shadow no longer represents the object exactly.

For a more comprehensive discussion of shadow algorithms see [37].

Physically Based Modeling

Physically based modeling is used to describe dynamic objects by giving physical constraints for the motion of these objects.

Objects move according to physical laws, for example, Newton's laws of motion. The task is to describe the force or force field which causes the movement of an object. The motions are then calculated using the object's mass properties. Objects within a scene influence each other's movements, for example, if they are attached to each other or if they collide with each other. All this information must be considered to simulate realistic movement of objects.

If objects possess a complex mechanical structure, this structure must be taken into account since all the parts may interact with each other.

Generally speaking, each object in a dynamic scene interacts with any other object of this scene. According to the type of these relations, different motion rules and constraints are derived: geometric contact directs the motion of objects; kinematic links require calculation of a kinematic solution (forward and inverse kinematic); a magnetic relation leads to a force which depends on the distance between the two objects, and so forth. Physical and heuristic rules are used to simulate these systems. For further information, please refer to [2].

Sonification

Sonification is another technique which is pushed by virtual environments. Sound samples which may be altered according to the acoustic properties of a simulated world and of the virtual transmitters and receiver of the sound are being used to increase realism and to provide additional information to the user. Examples are the sound of an approaching object or the acoustic feedback during an assembly task. It has been shown that sonification is also a powerful means to aid scientific visualization [1].

2.3 Classification of Immersive Scientific Visualization

Similar to scientific visualization, immersive visualization can be done at different levels of dynamics and interactions (see also [14]):

- *Static scene without interaction:* A static scene is the visualization of an individual "slice" from a dynamic data set. Using virtual environment techniques, such scenes can be viewed with improved 3D effects. Just like walkthrough applications in interior design, spatial relationships can be presented much better than by using conventional 3D techniques.

- *Static scene with interaction:* Presentation attributes like color, reflection, transparency, and so on, can be changed interactively in VE, for example, by pointing onto a 3D menu. Furthermore, objects can be made hidden or visible simply by applying direct manipulation and/or gestures. Techniques like data probing can also be realized this way.

- *Dynamic scene without interaction:* Dynamics in scientific visualization are usually presented as animations. Since animations can be described with a series of transformation matrices or vertex-color pairs, dynamics can be presented in VE by preloading such "scripts."

- *Dynamic scene with interaction:* Full functionalities of interactive scientific visualization are integrated with VE techniques. VE presentation and interaction methods replace "pseudo-3D" features of traditional mouse-menu-driven user interfaces.

The last scene, providing dynamic, highly interactive scenes in VE for scientific visualization, is the most promising of these alternatives. But for its realization, a lot of problems need to be solved—specifically, the accuracy of VE interactions and the speed of visualization.

On the other hand, less complex categories of VE may also be satisfactory depending on the specific application. Furthermore, from a software techniques point of view, there are different possibilities of realization. Analyzing existing systems for scientific visualization and those for VE, we see the following possibilities of realizing immersive visualization (see Figure 2.3):

1) Extending existing, complex visualization system with some additional VE capabilities;

2) Extending existing, complex virtual environment system with some additional scientific visualization capabilities;

3) Writing a new system from scratch, using some scientific visualization and some virtual environment capabilities;

4) Using existing, complex VE system to investigate objects precomputed by some system for scientific visualization;

5) Integration of complex VE and scientific visualization system.

Extending existing visualization systems has two major drawbacks. First, the user interface is not feasible for VE. Furthermore, VE tools must be reimplemented. Similarly, when existing VE systems are extended, techniques for scientific visualization must be reimplemented.

Possibility (2) is the simplest way, and sufficient for many applications. Both static scenes and precomputed dynamic scenes can be investigated. But techniques like data probing or particle tracing can be applied only with a number of restrictions, since all possible situations must be precomputed for VE presentation.

To use all benefits of VE and scientific visualization features, possibility 3 is necessary, that is, integration of full-size VE and scientific visualization systems. In this case, communication between the two systems is a major issue. Often, for example, in data probing

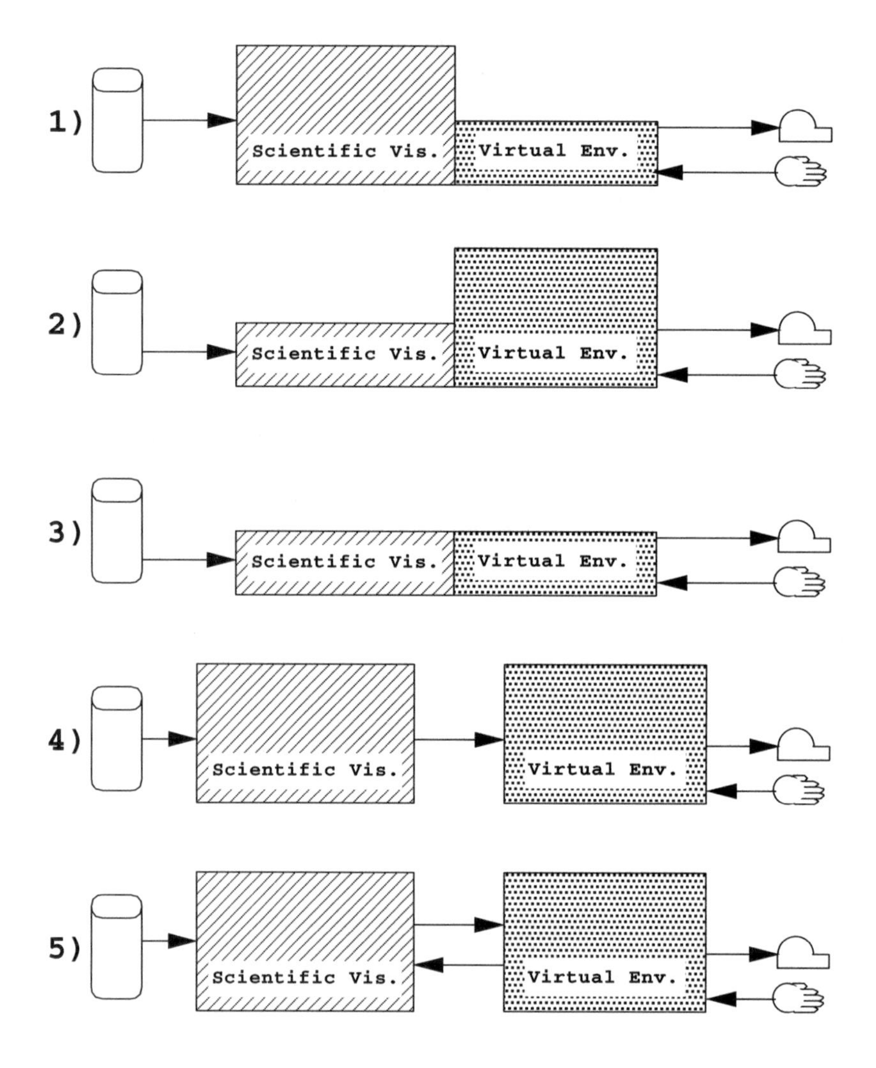

Figure 2.3: Categories of systems for immersive scientific visualization. See text for explanation.

or in particle tracing, the amount of data to be transmitted between the two systems is rather small. But in other cases, for example, a new threshold for an isosurface or a new mapping function for surface color, huge amounts of data must be transmitted. Thus, it makes sense to characterize the communication between a VE and a scientific visualization system by the available bandwidth as follows:

5a) Communication via pipe/socket/file system:

+ easy to distribute in network
- low data bandwidth

5b) Communication via shared memory:

+ high data bandwidth, good for shared memory machines
- difficult to distribute in network

2.4 Examples for Immersive Scientific Visualization

In this section, we will present several cases of immersive scientific visualization which were reported in the literature. Overviews of such work can also be found in [36] and [29]. Some additional works on this topic which cannot be discussed in detail here are, for example, [6, 33, 15, 26]. At the end of this section, two examples of our own work will be outlined. A more detailed discussion of another system will follow in Section 2.5.

2.4.1 Virtual Environment Extension to Dataflow Systems

Application builders based on the dataflow paradigm have become widely used for scientific visualization in recent years. They offer a magnitude of functions to fulfill most needs of scientific visualization and they can easily be adapted to many problems. Basically, there is a one-way flow of data from the data import module to the rendering module, with filtering and mapping modules in between. These systems mainly use mouse interaction for configuration of the processing pipeline, for control of processing parameters, and for control of the viewing parameters. Virtual environment devices generally are not supported.

In [31], possibilities for incorporating virtual environment devices in application builders were investigated. According to our characterization, such systems belong to category (1). Results showed that VE devices indeed can be useful in dataflow systems.

2.4.2 Virtual Windtunnel

In [10], a monolithic system for exploring numerically generated 3D unsteady flow fields is presented. The system employs virtual environment techniques. The system was designed for the very purpose of a "walk-around inside three-dimensional single-grid steady flow tracking a streamline from the hand at frame rates" [11]. When it was presented in 1992 it was revolutionary in the way that it allows investigation of flow fields in VEs at reasonable frame rates. The fact that it was developed by NASA Ames Research, and noting the applications it is being used for (for example, flow around Space Shuttle) made it clear that VE techniques can indeed be used for applications other than architectural walk-throughs.

Yet, by trying to gain maximum performance, a very special system was designed which lacks many techniques used for scientific visualization or for virtual environments, for example, it lacks level of detail. In our characterization of systems for immersive scientific visualization, it falls under category (3).

2.4.3 Scientific Visualization in the CAVE Environment

The CAVE (CAVE Automatic Virtual Environment) is an assembly of typically four large rear-projection screens and a computer-controlled audio system with multiple speakers [13]. The projection screens are arranged in a cube-like manner on three walls and on the floor, each providing the user with a high-resolution, full-color computer-generated image. Stereo perception is achieved by means of stereo shutter glasses. A user's head and hand are tracked with electromagnetic sensors. This setup allows for a highly immersive experience without having to wear a heavy head-mounted display.

A large number of applications from the field of scientific visualization have been tested within this environment [12]. One of them, called the Cosmic Explorer, is explained in more detail in [32]. It is a computer program for exploring the stages of the evolution of the universe. The results of numerical simulations are visualized in order to investigate the formation of the universe. Each galaxy is represented by a point in three-dimensional space, colored according to the age of the galaxy. An animation shows the evolution of the universe since the Big Bang, expanding in front of the viewer until it completely surrounds him. The user can pause the animation and travel through space to explore faraway galaxies or to get an overall view of the whole universe structure.

The Cosmic Explorer system is best characterized as belonging to category (3) of our classification of immersive scientific visualization systems.

2.4.4 Small-Scale Cellular Radio Networks

In a project for German Telekom, a VE system is being used to investigate measured and simulated propagation data of small-scale cellular radio networks. An X windows-based mapping program is employed to map propagation data to geometric objects. Data such as received power or delay spread determine the size and color of spheres, cylinders, and so on, representing locations of measured or simulated data.

These objects are stored in a format which is read by a VE system. Thus, the mapping program serves as a postprocessor to the measurement or simulation and as a preprocessor to the VE system (category (4) in our classification). The VE system allows us to investigate the data by viewing the resulting geometric objects in their proper context, that is, a detailed model of the buildings, trees, and so on, which influence propagation. The great advantage of this 3D presentation over visualization of the data in 2D is that heights of buildings, trees, and so forth and their influence on received signals are much easier for scientists to perceive. Furthermore, this technique may also be used to present results to nonexperts.

Figure 2.4 shows an image created by the system.

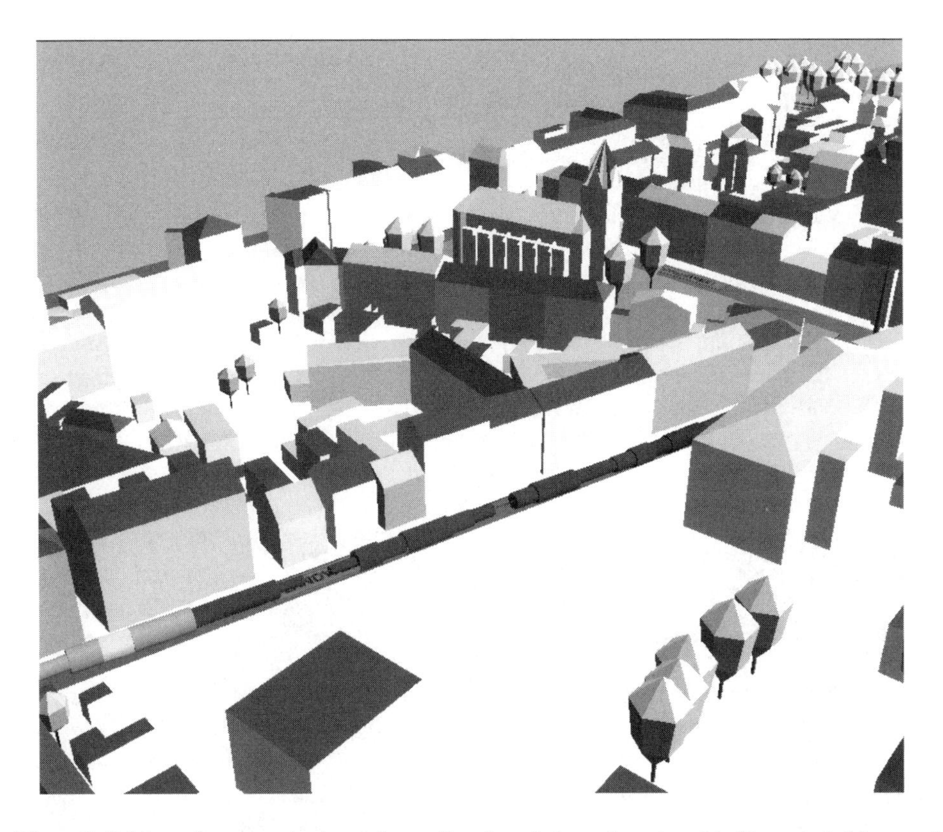

Figure 2.4: View of measured values of a small-scale cellular radio network in Darmstadt. Measured data are represented by cylinders colored according to received power. The diameter of the cylinders is proportional to delay spread. Data courtesy of German Telekom, Forschungs und Technologiezentrum, Darmstadt. See Color Plate 9.

2.4.5 Combination of Full-Size VE and Scientific Visualization Systems

Integration of full-size VE and scientific visualization (category (5)) systems has also been explored recently by our group. First results are very promising and show that there is great potential in this approach.

To be able to utilize as many VE and scientific visualization techniques as possible in one application, two systems designed at IGD, ISVAS and Virtual Design, have been linked.

ISVAS[1] [16] is a monolithic visualization system for finite element and volume data. It supports methods that allow interactive analysis of very large data sets. Among the features offered by this system are support for time-dependent data sets, user-defined mapping transfer functions, flexible means for mathematical manipulation of the data, data probing, cutting, flow field analysis, and hybrid display of FE and voxel data.

[1] ISVAS is a registered trademark of Fraunhofer IGD.

Virtual Design is a general-purpose VE system allowing highly diverse applications [4]. It consists of a toolkit of VE functions including an interaction toolkit for access to VE devices, import of many data formats, fast rendering, radiosity simulation, sonification modules, and multiple users. In several projects it has shown to be easily adaptable to new requirements and application demands.

The two systems run separately as individual processes. Communication between them is achieved by means of a UNIX pipe, allowing for two-way communication.

In a simple application which has been realized for test purposes, the user is immersed within a finite element data set in virtual design. The user can position a particle source interactively with a data glove. This position is transmitted to ISVAS, which uses the pre-computed flow field to simulate the movement of a particle within this finite element data set. The particle positions are returned to Virtual Design, which visualizes the moving particle in its proper context (see Figure 2.5). Another possible application of this configuration is data probing.

Figure 2.5: Particle tracing in virtual environment: A finite element data set containing the air flow around a vehicle (100,000 elements) is presented in VE. A particle source can be positioned interactively with the data glove. This position is transmitted to a visualization system which simulates the movement of the particle and transmits the resulting positions back to the VE system for visualization. See Color Plate 10.

The advantage of this approach is that the full power of an advanced visualization system as well as of a state-of-the-art VE system is available without having to reimplement every feature for a given application.

Of course, communication via a UNIX pipe is a bottleneck if large amounts of data are to be transmitted. This may be the case if the mapping of values to color or to geometry is to be changed, as is frequently the case for time-dependent data sets. Therefore, we have investigated data exchange via shared memory. The results of this work are reported in [17].

2.5 Extensive Example: Molecular Modeling

Finally, we want to present one example for immersive scientific visualization in more detail. This example utilizes many of the virtual environment techniques discussed previously and thus illustrates their usefulness for scientific visualization. The system was realized by extending an existing virtual environment system (Virtual Design) accordingly (category (2) of our characterization of immersive scientific visualization systems).

2.5.1 Introduction

The aim of this system, which was designed at Fraunhofer IGD in Darmstadt, is to investigate the relationship of complex molecules interactively in an immersive environment [20, 21]. Visualization of molecular data began very early [27] and has been applied since then (see, for example, [35, 7]). VR techniques have been utilized for visualization of molecular data as reported in [8] and in [34].

In chemistry, several different models for molecules are known which help in understanding 3D structure and properties of a molecule. Each of these models stresses certain properties and is more or less appropriate for certain tasks.

Ball-and-stick models represent atoms by spheres and linkages by sticks (see Figure 2.6). Types of atoms are distinguished by characteristic colors (for example, red for oxygen). The length and the angle of linkages become apparent.

Surface models allow us to project properties like the electrostatic potential onto the shell by means of a color-coding scheme.

Enzyme-substratum complexes play an important role in biochemical processes. Enzymes are catalysts of chemical reactions within cells. If they are present, the activation energy of a reaction is lowered. The enzyme reacts with its substratum, which is a certain chemical substance. It is important to note that the substratum fits to the enzyme like a key to its lock. By means of the enzyme-substratum complex, the substratum is induced to participate in a specific chemical reaction.

Some enzymes can be deactivated by inhibitors, which modify the surface of the enzyme in a way that makes it impossible for it to form an enzyme-substratum complex.

In the following we will present a VE system which allows interactive investigation of molecules and of the relationships between enzyme, substratum, and inhibitor.

2.5.2 Realization

Molecule data can be imported in the form of the common pdb file format (Broohaven Protein Data Bank). Since this file does not contain the linkages directly, they must be inferred from the data available. This makes it necessary to include a complete table of all possible linkage atoms for each amino acid.

The corresponding surface model is read from an additional file. Several molecules can be loaded into the system in order to compare them or to combine them at will.

Figure 2.6: Detail view of a Trypsin molecule in ball-and-stick representation. The "sticks" consist of only five surfaces; smooth appearance is achieved by Gouraud shading. Spheres have varying resolution depending on their distance from the viewer (level of detail). Data courtesy of Wolfgang Heiden, Department of Chemistry, Technical University Darmstadt. See Color Plate 11.

Rendering

The molecules are displayed in any combination of ball-and-stick, calott, and surface model. Atoms are represented by colored spheres, linkages by approximated "cylinders" with only five polygons. Smooth appearance of these "cylinders" is achieved by Gouraud shading.

The surface model is rendered transparently with colors depicting properties of the shell.

A technique with several levels of detail is used to achieve fast rendering while still showing as many details as possible due to resolution.

Size and position of a molecule can be perceived due to a shadow which is cast onto a "floor" (see Figure 2.7). For several standard orientations of the molecule, exact shadows have been precomputed. One of them is shown under the molecule. Smooth transition from one precomputed shadow to the next is desired if the molecule is rotated about an axis other than the vertical one. Even though some ideas exist to solve this problem (for example, interpolation between two images), no satisfactory solution has been implemented yet.

Figure 2.7: Overall view of an enzyme and a substratum (Trypsin and Arginin). Ball-and-stick representation and surface representation combined. Size and shape of the molecules are also shown by the shadows which are implemented by means of precomputed textures. Data courtesy of Wolfgang Heiden, Department of Chemistry, Technical University Darmstadt. See Color Plate 12.

In order to be able to estimate surface colors accurately (and thus surface properties), the molecule under investigation must always be properly lit. Increasing the ambient light would lead to a brighter image, but colors would look less intense and perception of differences in shell properties would be more difficult. Therefore, a point light source slightly above the eye point is used, which results in deep colors even in indentations that normally would be in the shade.

The numerical value associated with a color can be interesting to inspect in order to decide if a substratum fits to an enzyme. Currently, this information is given in a legend which can be shown at the left border of the screen if needed. Data probing (for example, with the index finger of the data glove) would also help in this task but is not yet implemented.

Input and Output Devices

Since the properties of the molecule as indicated by the surface model are of major importance for assessment of a molecule, a large-screen stereo projection is used instead of a head-mounted display. A head-mounted display would allow much more intuitive navigation, but the low resolution of the device available for this project, as well as its weight, make the projection superior to the head-mounted display.

A data glove is used for navigation, that is, to change the position of the observer. An additional space mouse is used to rotate and translate the molecule. Translation of the space mouse can be switched off in order to ensure that the molecule rotates only at a fixed position. Even an unexperienced VE user can easily view a molecule by means of the space mouse.

Due to these two input devices it is also possible to rotate the molecule while changing the point of view simultaneously.

Interaction

Control of the VE system is done mainly via certain data glove gestures. They were selected with special consideration of intuitive handling. By moving the index finger, the center of interest can be changed and by making a fist, an object can be grabbed.

For the assessment of molecules, viewing from several directions as well as fast switching between different representations for each of them is needed.

A virtual 3D menu floating in space is provided for selecting different representations which can be switched on or off. It offers a choice of all molecules currently available. Furthermore, ball-and-stick, calott, or surface model can be toggled independently, and the surface can be rendered as polygons or as dot cloud. A toggle between opaque and transparent appearance is also available. Of course, if the surface is rendered opaquely and the molecule is seen from a distance, the ball-and-stick model is not visible and can be automatically disabled for rendering to save time.

In order to increase interactivity of the system, larger molecules are not rendered while the menu is active.

In addition to this main menu, individual menus are available for each currently displayed molecule, which are used to set properties of each molecule.

Visual examination of the surface structure of molecules can be used to find out which substratum and which inhibitor fit a certain enzyme. Therefore, a molecule can be grabbed with the data glove and moved arbitrarily.

The echo of the glove is scaled in order to allow navigation within a molecule. Thus, a molecule can only be grabbed directly when the observer is close to it. Often, one wants to see the whole scene while moving a molecule. This can be achieved by means of a "ray" controlled by the hand. If this ray points to a molecule, any gesture done with the hand will be applied to the molecule. Thus, a molecule may be moved or a menu for it may be opened from a distance.

Another way of moving a molecule is by means of a space mouse. If a molecule is touched by the hand echo or by the ray, it will be activated. It then will move according to the space mouse. Selection of a molecule can also be performed by pressing a key of the space mouse. This last method has the disadvantage of being limited to up to eight molecules due to the number of keys on a space mouse.

The system presented here is very useful for analyzing individual molecules as well as for investigating pairs of molecules.

A useful extension would be data probing in order to determine properties on the shell of a molecule exactly. Collision detection is necessary in order to have a more realistic handling of scenes containing more than one molecule and in order to investigate the fitting of molecules. Force feedback devices could eventually be used for the fitting process once they become available. They could also give a feeling of the stiffness of molecules and of the degrees of freedom in bending them.

2.6 Conclusion and Future Work

In this chapter we gave an overview of technologies and techniques for virtual environments and we motivated—mainly by a number of examples—that scientific visualization will benefit from these new developments. Immersive scientific visualization, that is, scientific visualization giving the user the impression of being within a highly realistic but actually computer-generated environment, will supplement traditional, keyboard-and-mouse-based scientific visualization.

We are only at the beginning of immersive scientific visualization, which will give a new quality to traditional scientific visualization, leading to a large number of new applications in such fields as industry (for example, assembly simulation), physics (deforming FE data sets), chemistry (fitting of enzyme and inhibitor), and medicine (operation planning and training).

Future research will focus on how to best combine virtual environment and scientific visualization technologies. We will find that many applications will benefit or just become possible due to these new approaches. Still, research needs to be done in many fields, for example, human perception [18, 5], force feedback [28], fast collision check, physically based modeling, and so forth. It will be worth it.

Acknowledgments

We thank our colleague Wolfgang Felger for giving us information on VE devices, all the colleagues and students at our lab who contributed to the systems presented in this chapter, and Professor J. L. Encarnação for providing the environment which made this work possible.

Bibliography

[1] P. Astheimer, "Sonification Tools to Supplement Dataflow Visualization," *Third Eurographics Workshop on Visualization in Scientific Computing*, Viareggio, Italy, Apr. 1992, pp. 39–69.

[2] P. Astheimer and F. Dai, "Dynamic Objects in Virtual Worlds: Integrating Simulations in a Virtual Reality Toolkit," *European Simulation Symposium '93*, Delft, Oct. 1993.

[3] P. Astheimer, F. Dai, M. Göbel, R. Kruse, S. Müller, and G. Zachmann, "Realism in Virtual Reality," *Artificial Life and Virtual Reality*, N. Thalmann and D. Thalmann, editors, John Wiley & Sons, 1994, pp. 189–210.

[4] P. Astheimer, W. Felger, and S. Müller, "Virtual Design: A Generic VR System for Industrial Applications," *Computers & Graphics*, Vol. 17, No. 6, 1993, pp. 671–677.

[5] M.P. Baker and C.D. Wickens, *Human Factors in Virtual Environments for the Visual Analysis of Scientific Data*, Technical Report NCSA-TR032, NCSA, 26 pages, in WWW at URL: file://ftp.ncsa.uiuc.edu/ncsapubs/preprints/TR032.ps as well as http://monet.ncsa.uiuc.edu/~baker/PNL/paper.html, Aug. 1995.

[6] M. Bajura, H. Fuchs, and R. Ohbuchi, "Merging Virtual Objects with the Real World: Seeing Ultrasound Imagery within the Patient," *Computer Graphics, Proc. SIGGRAPH*, Vol. 26, No. 2, July 1992, pp. 203–210.

[7] L.D. Bergman, J.S. Richardson, D.C. Richardson, and F.P. Brooks, "VIEW—An Exploratory Molecular Visualization System with User-Definable Interaction Sequences," *Computer Graphics, Proc. SIGGRAPH*, Anaheim, Calif., Aug. 1993, pp. 117–126.

[8] F.P. Brooks, "Project GROPE—Haptic Displays for Scientific Visualization," *Computer Graphics*, Vol. 24, No. 4, 1990, pp. 177–185.

[9] S. Bryson, "Real-Time Exploratory Scientific Visualization and Virtual Reality," *Scientific Visualization—Advances and Challenges*, L. Rosenblum and R. Earnshaw, editors, Academic Press Ltd., London, UK, 1994, pp. 65–85.

[10] S. Bryson and C. Levit, "The Virtual Windtunnel," *IEEE Computer Graphics and Applications*, Vol. 12, No. 4, 1992, pp. 25–34.

[11] S. Bryson and C. Levit, "Lessons Learned While Implementing the Virtual Windtunnel Project," *Visualization '92, Tutorial # 2*, 1992, pp. 4.1–4.7.

[12] C. Cruz-Neira, J. Leight, M. Papka, C. Barnes, S.M. Cohen, S. Das, R. Engelmann, R. Hudson, T. Roy, L. Siegel, C. Vasilakis, T.A. DeFanti, and D.J. Sandin, "Scientists in Wonderland: A Report on Visualization Applications in the CAVE Virtual Reality Environment," *Proc. IEEE Symposium on Research Frontiers in VR*, San Jose, Calif., Oct. 1993, pp. 59–66.

[13] C. Cruz-Neira, D.J. Sandin, and T.A. DeFanti, "Surround-Screen Projection-Based Virtual Reality: The Design and Implementation of the CAVE," *Computer Graphics, Proc. SIGGRAPH*, Anaheim, Calif., Aug. 1993, pp. 135–142.

[14] J.L. Encarnação, P. Astheimer, W. Felger, Th. Frühauf, M. Göbel, and S. Müller, "Graphics & Visualization: The Essential Features for the Classification of Systems," *Proc. International Conference on Computer Graphics ICCG '93*, Bombay, India, Feb. 1993.

[15] K.M. Fairchild, L. Serra, N. Hern, L.B. Hai, and A.T. Leong, "Dynamic FishEye Information Visualizations," *Virtual Reality Systems*, R.A. Earnshaw, M.A. Gigante, and H. Jones, editors, Academic Press, 1993, pp. 161–177.

[16] Th. Frühauf, M. Göbel, H. Haase, and K. Karlsson, "Design of a Flexible Monolithic Visualization System," *Scientific Visualization—Advances and Challenges*, L. Rosenblum and R. Earnshaw, editors, Academic Press Ltd., London, UK, 1994, pp. 265–285.

[17] H. Haase, "Symbiosis of Virtual Reality and Scientific Visualization System," *Computer Graphics Forum*, Vol. 15, No. 3, 1996, *also: Proc. Eurographics '96*, Poitiers, France, Aug. 1996.

[18] H. Haase and C. Dohrmann, "Doing It Right: Psychological Tests to Ensure the Quality of Scientific Visualization," *Proc. Eurographics Workshop on Visualization in Scientific Computing (EGVISC'96)*, Prague, Apr. 1996.

[19] H. Haase, M. Göbel, P. Astheimer, K. Karlsson, F. Schröder, Th. Frühauf, and R. Ziegler, "How Scientific Visualization Can Benefit From Virtual Environments," *CWI Quarterly*, Amsterdam, Vol. 7, No. 2, 1994, pp. 159–174.

[20] H. Haase, J. Strassner, and F. Dai, "Virtual Molecules, Rendering Speed, and Image Quality," *Virtual Environments '95*, M. Göbel, editor, Springer-Verlag, Wien, 1995, pp. 70–86.

[21] H. Haase, J. Strassner, and F. Dai, "VR Techniques for the Investigation of Molecule Data," *Computers & Graphics, Special Issue on Virtual Reality*, Elsevier Science Ltd., Vol. 20, No. 2, 1996.

[22] P. Haeberli and M. Segal, "Texture Mapping as a Fundamental Drawing Primitive," *Proc. Fourth Eurographics Workshop on Rendering*, M. Cohen, C. Puech, and F. Sillion, editors, Paris, June 1993, pp. 159–266.

[23] M. Jern and R.A. Earnshaw, "Interactive Real-Time Visualization Systems Using a Virtual Reality Paradigm," *Visualization in Scientific Computing*, M. Göbel, H. Müller and B. Urban, editors, Springer-Verlag, 1995, pp. 174–189.

[24] V.V. Kamat, "A Survey of Techniques for Simulation of Dynamic Collision Detection and Response," *Computers & Graphics*, Vol. 17, No. 4, 1993, pp. 379–385.

[25] M.W. Krueger, *Artificial Reality II*, Addison-Wesley, 1991, pp. 186–187.

[26] J. Leigh, C.A. Vasilakis, T.A. DeFanti, R. Grossman, C. Assad, B. Rasnow, A. Protopappas, E.D. Schutter, and J.M. Bower, "Virtual Reality in Computational Neuroscience," *Proc. Virtual Reality Applications*, Leeds, UK, June 1994.

[27] C. Levinthal, "Molecular Model-Building by Computer," *Scientific American*, Vol. 124, No. 6, 1966. pp. 42–52.

[28] W.R. Mark, S.C. Randolph, M. Finch, J.M. Van Verth, and R.M. Taylor II, "Adding Force Feedback to Graphics Systems: Issues and Solutions," *Computer Graphics, Proc. SIGGRAPH 96*, Aug. 1996, pp. 447–452.

[29] K. Pimentel and K. Teixeira, *Virtual Reality—Through the New Looking Glass*, Intel/Windcrest/McGraw-Hill, 1993, pp. 161–208.

[30] F. Schröder and P. Roßbach, "Managing the Complexity of Digital Terrain Models," *Computers and Graphics Special Issue on Modelling and Visualization of Spatial Data in GIS*, Nov. 1994.

[31] W.R. Sherman, "Integrating Virtual Environments into the Dataflow Paradigm," *Fourth Eurographics Workshop on Visualization in Scientific Computing*, Abingdon, UK, Apr. 1993.

[32] D. Song and M.L. Norman, "Cosmic Explorer: A Virtual Reality Environment for Exploring Cosmic Data," *Proc. IEEE Symposium on Research Frontiers in VR*, San Jose, Calif., Oct. 1993, pp. 75–79.

[33] A. State, M.A. Livingston, W.F. Garrett, G. Hirota, M.C. Whitton, D.P. Etta, and H. Fuchs, "Technologies for Augmented Reality Systems: Realizing Ultrasound-Guided Needle Biopsies," *Computer Graphics, Proc. SIGGRAPH 96*, Aug. 1996, pp. 439–446.

[34] R.M. Taylor, W. Robinett, V.L. Chi, F.P. Brooks, W.V. Wright, R.S. Williams, and E.J. Snyder, "The Nanomanipulator: A Virtual-Reality Interface for a Scanning Tunneling Microscope," *Computer Graphics, Proc. SIGGRAPH*, Anaheim, Calif., Aug. 1993, pp. 127–134.

[35] M. Waldherr-Teschner, T. Goetze, W. Heiden, M. Knoblauch, H. Vollhardt, and J. Brickmann, "MOLCAD—Computer Aided Visualization and Manipulation of Models in Molecular Science," *Advances in Scientific Visualization*, F.H. Post and A.J.S. Hin, editors, Springer-Verlag, 1992, pp. 58–67.

[36] D.M. Weimer, "Brave New Virtual Worlds," *Frontiers of Scientific Visualization*, John Wiley & Sons, New York, 1994, pp. 245–278.

[37] A. Woo, P. Poulin, and A. Founier, "A Survey of Shadow Algorithms," *IEEE Computer Graphics & Applications*, C.A. Pickover and S.K. Tewksbury, editors, Nov. 1990, pp. 13–32.

Chapter 3

A Survey of Grid Generation Methodologies and Scientific Visualization Efforts

Bernd Hamann and Robert J. Moorhead II

Abstract. *This survey chapter discusses recent and ongoing projects in grid generation and scientific visualization at the National Science Foundation (NSF) Engineering Research Center for Computational Field Simulation. The chapter provides a brief overview of standard grid generation techniques and summarizes the design requirements and current functionality of the National Grid Project, a universal, interactive grid generation system. Several projects that have led to powerful visualization systems are discussed.*

Keywords: advancing front, approximation, B-spline, CAD, Chimera grid, concurrent visualization, Delaunay triangulation, edge detection, elliptic grid generation, finite element method, hierarchical grid, hybrid grid, multiblock grid, NURBS, partial differential equation, ray casting, region growing, scattered data interpolation, scalar field visualization, structured grid generation, unstructured grid generation, vector field visualization, volume visualization, wavelets.

3.1 Introduction

The NSF Engineering Research Center for Computational Field Simulation (ERC-CFS) is an interdisciplinary research institution devoted to the simulation of field phenomena. Examples of field phenomena being studied are fluid flow, heat transfer, acoustic propagation, and electromagnetic radiation. Faculty members as well as full-time researchers from a variety of disciplines contribute to the mission of the ERC-CFS. Areas of expertise include aerospace engineering, electrical engineering, mechanical engineering, computer engineering, mathematics, and computer science. The ERC-CFS is divided into several research thrusts, among them the Grid Generation and the Scientific Visualization thrusts.

This survey chapter provides an overview of some of the efforts in these two thrusts and briefly outlines common grid generation methodologies.

Grid generation is the step required for solving the partial differential equations (PDEs) governing a model of some physical field phenomenon. Usually, CAD data describing some geometry is the input, and a finite set of points discretizing the given curves, surfaces, and possibly a surrounding volume, are the output. Scientific visualization techniques are required to analyze and display the—often time-varying—solution of the phenomenon being investigated. It is useful for visualization experts to have a thorough understanding of the grid types that are necessary to solve the PDEs and the ways different grid types are generated. This chapter describes the most common grid generation methods and grid types, discusses a universal grid generation system being developed at the ERC-CFS, and reports on various scientific visualization efforts. The chapter is divided into these sections:

- Grid generation methods and grid types used in CFS
- The National Grid Project (NGP)—A universal system for the generation of structured and unstructured grids (including geometry correction for CAD data with "gaps," intersecting surfaces, or overlapping surfaces)
- Visualization efforts at the ERC-CFS (scalar and vector field visualization with applications in aerodynamics, oceanography, acoustics, meteorology, and site characterization; feature extraction, wavelet-based data compression, and concurrent visualization)

3.2 Survey of Numerical Grid Generation Techniques

3.2.1 List of Grid Generation Methods and Grid Types

In this section, we discuss the main techniques used in numerical grid generation and provide a classification of the resulting grid types (see Figure 3.1). Essentially, one can distinguish between two grid types:

- *Structured grids*, classically generated by a combination of transfinite interpolation (TFI) and solving elliptic (or hyperbolic) systems of PDEs
- *Unstructured grids*, typically generated by a triangulation/tesselation algorithm (for example, Delaunay triangulation and Vornoï diagram) or an advancing front approach

In the grid generation community, the term *structured grid* is associated with a grid whose constituting elements (or cells) are topologically equivalent to a square (2D) or a cube (3D). In grid generation, the terms quadrilateral and hexahedron usually refer to a continuously deformed square or cube. When discretizing a single deformed square or cube by a structured grid, one usually chooses a curvilinear grid, that is, a grid whose edges are all straight line segments.

The connectivity among nodes in a structured grid is completely defined by the nodes' indices. For example, a node $\mathbf{x}_{i,j}$ in a 2D structured grid is connected with the nodes $\mathbf{x}_{i-1,j}$, $\mathbf{x}_{i+1,j}$, $\mathbf{x}_{i,j-1}$, and $\mathbf{x}_{i,j+1}$. This implied connectivity helps in designing highly efficient solution algorithms for the PDEs describing some phenomenon.

The term *unstructured grid* refers to any kind of grid that is not a structured grid, that is, is not of the finite difference type. The only unstructured grids of practical importance consist of triangles and/or quadrilaterals (2D) and tetrahedra, pentahedra (prisms), and/or hexahedra (3D). Thus, quadrilaterals and hexahedra can constitute an unstructured grid but the number of edges sharing a node is not restricted.

Combinations of structured and unstructured grid types are used as well. The most common grid types and combinations of various grid types are:

- *Structured grids*, consisting of quadrilateral and hexahedral elements whose node connectivity is implicitly defined by the nodes' indices
- *Multiblock structured grids*, consisting of multiple structured grids, each one associated with one of many *blocks* (connectivity among blocks not necessarily structured)
- *Unstructured grids*, consisting of quadrilateral, triangular, hexahedral, pentahedral, tetrahedral, and other types of polygonal/polyhedral elements; node connectivity explicitly defined for each node
- *Hybrid grids*, consisting of structured and unstructured grid regions
- *Chimera grids*, consisting of multiple structured grids with partially overlapping grid elements; overlap regions typically "resolved" using an appropriate interpolation schemes
- *Hierarchical grids*, generated by quadtree- and octree-like subdivision schemes (also referred to as *embedded grid* or *semi-structured grids*)

These grid types are illustrated for 2D in Figure 3.1.

Detailed information about these and other existing grid types and generation techniques/systems is provided in [5, 12, 19, 26, 53, 54].

3.2.2 Structured Grid Generation using TFI and Elliptic PDEs

One of the classical approaches for generating multiblock structured 2D (3D) grids for a finite space surrounding a geometry is based on TFI (see [10] and [15]) and solving elliptic systems of PDEs (see [53]). Once the block configuration around a geometry is established, an initial grid is generated inside each block by performing bilinear (trilinear) TFI of each block's boundary edges (faces). The TFI algorithm is performed in a discrete manner for a finite set of points on the boundary edges (faces) of each block. The points on the boundary edges (faces) are spaced according to a specified distribution function (for example, uniform in parameter space, uniform with respect to arc length, uniform with respect to integrated absolute curvature, and so on).

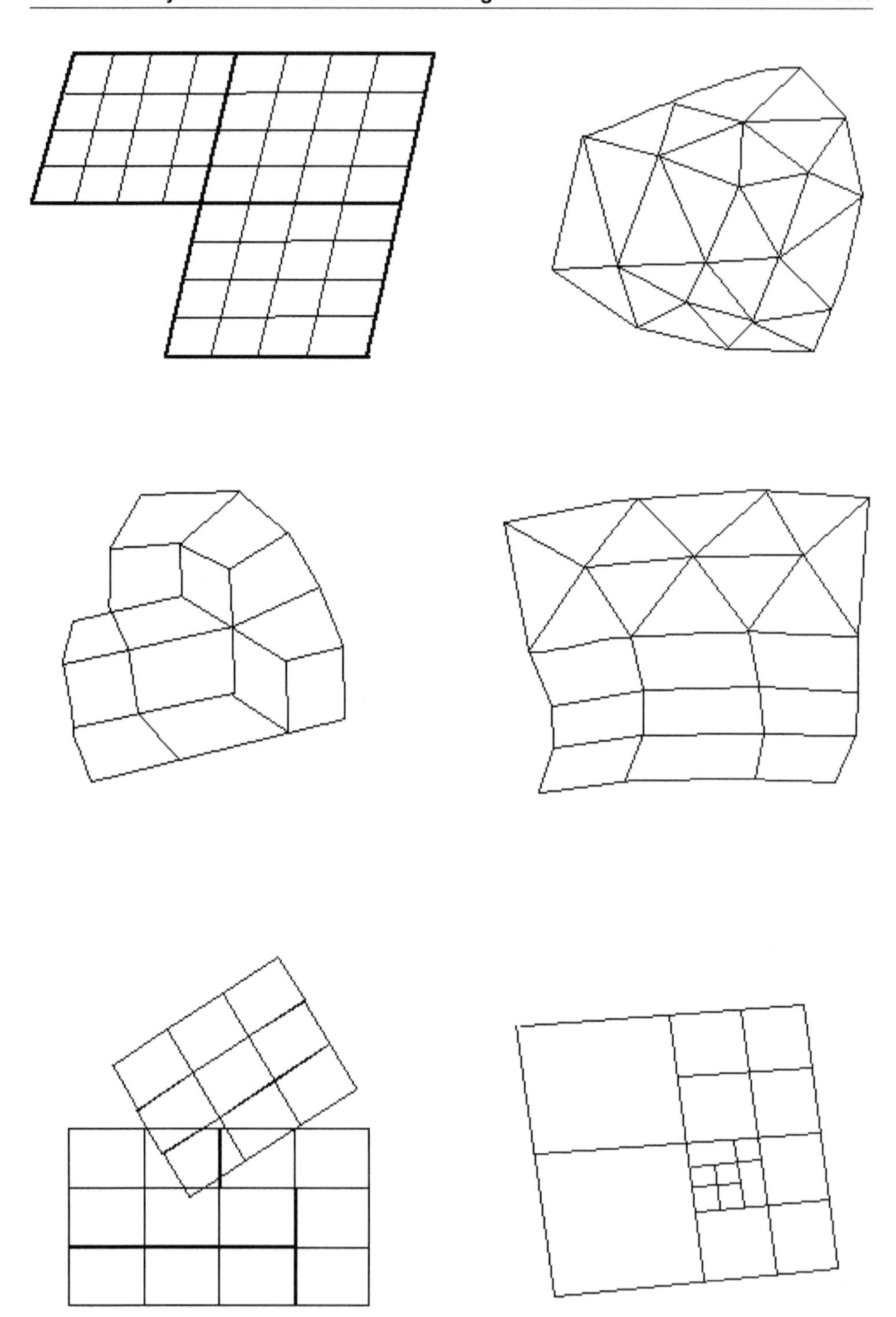

Figure 3.1: Multiblock structured *(upper-left)*, unstructured-triangular *(upper-right)*, unstructured-quadrilateral *(middle-left)*, hybrid *(middle-right)*, Chimera *(lower-left)*, and hierarchical grid *(lower-right)*.

In the 2D case, for example, the boundary curve points $\mathbf{x}_{I,0}$, $\mathbf{x}_{I,N}$, $\mathbf{x}_{0,J}$, and $\mathbf{x}_{M,J}$ associated with parameter values $(u_{I,0}, v_{I,0})$, $(u_{I,N}, v_{I,N})$, $(u_{0,J}, v_{0,J})$, and $(u_{M,J}, v_{M,J})$, $I = 0, \ldots, M$, $J = 0, \ldots, N$, define the initial interior grid points by applying TFI:

$$\begin{aligned} \mathbf{x}_{I,J} &= [(1-u_{I,J}) \quad u_{I,J}] \begin{bmatrix} \mathbf{x}_{0,J} \\ \mathbf{x}_{M,J} \end{bmatrix} + [\mathbf{x}_{I,0}\ \mathbf{x}_{I,N}] \begin{bmatrix} (1-v_{I,J}) \\ v_{I,J} \end{bmatrix} \\ &- [(1-u_{I,J}) \quad u_{I,J}] \begin{bmatrix} \mathbf{x}_{0,0} & \mathbf{x}_{0,N} \\ \mathbf{x}_{M,0} & \mathbf{x}_{M,N} \end{bmatrix} \begin{bmatrix} (1-v_{I,J}) \\ v_{I,J} \end{bmatrix}, \end{aligned} \tag{3.1}$$

where $u_{I,J}$ varies linearly between $u_{I,0}$ and $u_{I,N}$ and $v_{I,J}$ varies linearly between $v_{0,J}$ and $v_{M,J}$. The computation of surface grid points from a finite set of boundary curve points using TFI is illustrated in Figure 3.2.

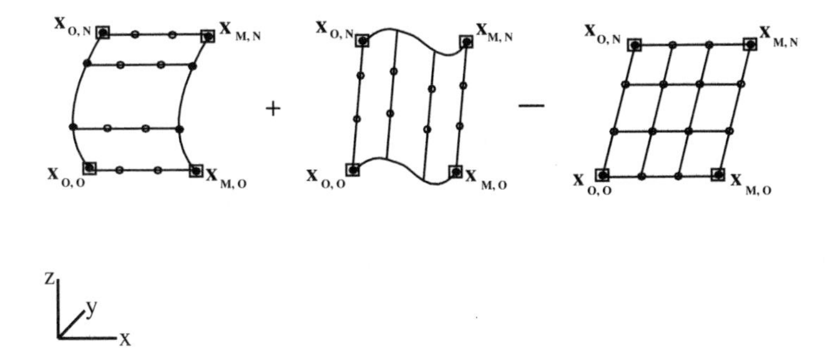

Figure 3.2: Principle of TFI: Two linear and one bilinear interpolation step.

The elliptic equations used for surface grid generation are a generalization of Laplace equations for harmonic mappings of planar regions. Let $\mathbf{x}$ be a simply connected parametric surface, written as

$$\begin{aligned} \mathbf{x}(u,v) &= (x(u,v), y(u,v), z(u,v)) \\ &= (x(u(\xi,\eta), v(\xi,\eta)), y(u(\xi,\eta), v(\xi,\eta)), z(u(\xi,\eta), v(\xi,\eta))), \end{aligned} \tag{3.2}$$

where $0 \leq u, v \leq 1$. This notation implies that the *parametric variables* u and v are functions of the *computational variables* ξ and η. A rectilinear grid in the computational square ($\xi\eta$ space) generates a curvilinear grid in the parametric square (uv space), which maps to a curvilinear grid on the surface (xyz space or *physical variables*). Thus, a uniform grid in the computational space generates a curvilinear grid on the surface. Figure 3.3 shows the physical, parametric, and computational spaces associated with a single surface.

The elliptic system of partial differential equations defining the transformation between computational variables and parametric variables is related to conformal mappings on surfaces (see [32]). The derivation of the elliptic system of PDEs is based on conformal mappings. The approach utilizes the intrinsic orthogonality and uniformity properties that are inherent to a grid generated by a conformal mapping. Another advantage of using conformal coordinates is the fact that they allow solutions of differential equations on surfaces with the same ease as on a rectangle in the Cartesian plane.

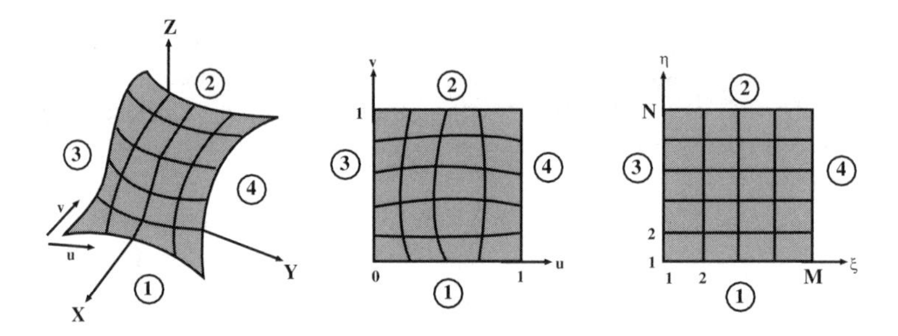

Figure 3.3: Physical, parametric, and computational spaces.

A surface grid generated by the conformal mapping of a rectangle onto the surface $\mathbf{x}$ is orthogonal and has a constant aspect ratio. These two conditions can be expressed as

$$\begin{aligned} \mathbf{x}_\xi \cdot \mathbf{x}_\eta &= 0 \quad \text{and} \\ M\,|\mathbf{x}_\xi| &= |\mathbf{x}_\eta|, \end{aligned}$$

where M is the *grid aspect ratio*. These two equations can be rewritten as

$$\begin{aligned} x_\xi x_\eta + y_\xi y_\eta + z_\xi z_\eta &= 0 \quad \text{and} \\ M^2(x_\xi^2 + y_\xi^2 + z_\xi^2) &= x_\eta^2 + y_\eta^2 + z_\eta^2. \end{aligned}$$

Using the chain rule for differentiation, this system of equations can be expressed in the form

$$\begin{aligned} Mu_\xi &= av_\eta - bu_\eta \quad \text{and} \\ Mv_\xi &= bv_\eta - cu_\eta, \end{aligned}$$

where

$$a = \frac{\overline{g}_{22}}{\overline{J}}, \ b = -\frac{\overline{g}_{12}}{\overline{J}}, \ c = \frac{\overline{g}_{11}}{\overline{J}}, \ \overline{J} = \sqrt{\overline{g}_{11}\overline{g}_{22} - \overline{g}_{12}^2},$$
$$\overline{g}_{11} = \mathbf{x}_u \cdot \mathbf{x}_u, \ \overline{g}_{12} = \mathbf{x}_u \cdot \mathbf{x}_v, \quad \text{and} \quad \overline{g}_{22} = \mathbf{x}_v \cdot \mathbf{x}_v.$$

If the parametric and computational variables are exchanged such that the parametric variables become the independent variables, the system becomes

$$\begin{aligned} -M\eta_u &= b\xi_u + c\xi_v \quad \text{and} \\ M\eta_v &= a\xi_u + b\xi_v. \end{aligned}$$

This first-order system is exactly *Beltrami's system of equations for the quasi-conformal mapping of planar regions*. It follows that the computational variables ξ and η are solutions of the following second-order linear homogeneous elliptic system ($\Phi = \Psi = 0$):

$$\begin{aligned} \overline{g}_{22}\xi_{uu} - 2\overline{g}_{12}\xi_{uv} + \overline{g}_{11}\xi_{vv} + (\Delta_2 u)\xi_u + (\Delta_2 v)\xi_v &= \Phi \quad \text{and} \\ \overline{g}_{22}\eta_{uu} - 2\overline{g}_{12}\eta_{uv} + \overline{g}_{11}\eta_{vv} + (\Delta_2 u)\eta_u + (\Delta_2 v)\eta_v &= \Psi. \end{aligned}$$

The Beltramians $\Delta_2 u$ and $\Delta_2 v$ are given by

$$\Delta_2 u = \overline{J}(a_u + b_v) = \overline{J}\left[\frac{\partial}{\partial u}\left(\frac{\overline{g}_{22}}{\overline{J}}\right) - \frac{\partial}{\partial v}\left(\frac{\overline{g}_{12}}{\overline{J}}\right)\right] \quad \text{and}$$

$$\Delta_2 v = \overline{J}(b_u + c_v) = \overline{J}\left[\frac{\partial}{\partial v}\left(\frac{\overline{g}_{11}}{\overline{J}}\right) - \frac{\partial}{\partial u}\left(\frac{\overline{g}_{12}}{\overline{J}}\right)\right].$$

This system is the basis for most elliptic grid generation methods. The source terms (or control functions) Φ and Ψ are added to allow control over the distribution of grid points on the surface. Typically, points are given in computational space and points in parametric space must be computed. Therefore, it is convenient to exchange variables again such that the computational variables ξ and η are the independent variables. This leads to the following quasi-linear elliptic system:

$$\begin{aligned} g_{22}(u_{\xi\xi} + Pu_\xi) - 2g_{12}u_{\xi\eta} + g_{11}(u_{\eta\eta} + Qu_\eta) &= J^2\Delta_2 u \quad \text{and} \\ g_{22}(v_{\xi\xi} + Pv_\xi) - 2g_{12}v_{\xi\eta} + g_{11}(v_{\eta\eta} + Qv_\eta) &= J^2\Delta_2 v, \end{aligned} \tag{3.3}$$

where

$$\begin{aligned} g_{11} &= \overline{g}_{11}u_\xi^2 + 2\overline{g}_{12}u_\xi v_\xi + \overline{g}_{22}v_\xi^2, \\ g_{12} &= \overline{g}_{11}u_\xi u_\eta + \overline{g}_{12}(u_\xi v_\eta + u_\eta v_\xi) + \overline{g}_{22}v_\xi v_\eta, \\ g_{22} &= \overline{g}_{11}u_\eta^2 + 2\overline{g}_{12}u_\eta v_\eta + \overline{g}_{22}v_\eta^2, \quad \text{and} \\ J &= u_\xi v_\eta - u_\eta v_\xi. \end{aligned}$$

The elliptic system is solved using a finite difference approach using either Dirichlet (or Neumann) boundary conditions. Since P and Q control the grid point distribution, $P = Q = 0$ leads to a uniformly spaced grid in the absence of boundary curvature. These two control functions are estimated from an initial grid having a point distribution that is close to the desired one. It is possible to compute the functions P and Q by performing TFI from user-specified boundary curve distributions. They are commonly defined in terms of relative spacing between grid points. When grid line orthogonality is desired, the grid points are allowed to move on the boundary curves.

A detailed discussion of elliptic structured grid generation using NURBS surfaces is given in [25]. Parametric surfaces/volumes and their relation to grid generation applications are dealt with in [2, 10, 22].

Structured grids can be generated by using either an elliptic or a hyperbolic approach. Earlier approaches use elliptic systems of PDEs (see [53]). In this case, the outer (or far-field) boundary must be known. The alternative approach uses hyperbolic systems (see [6]), where the outer boundary is not prescribed but is a result of the grid generation process.

Sometimes it is advantageous to decompose a field to be discretized into multiple structured grids (multiple blocks). The entire field is first divided into blocks (that is, boundaries of each block are defined), and a structured grid is computed for each block. Certain continuity conditions (for example, C^0, C^1, or C^2 continuity) are often imposed at block interfaces. The curves and surfaces used for the block boundaries can consist of multiple curve segments and surface patches. Figure 3.4 shows a multiblock configuration around a 2D geometry.

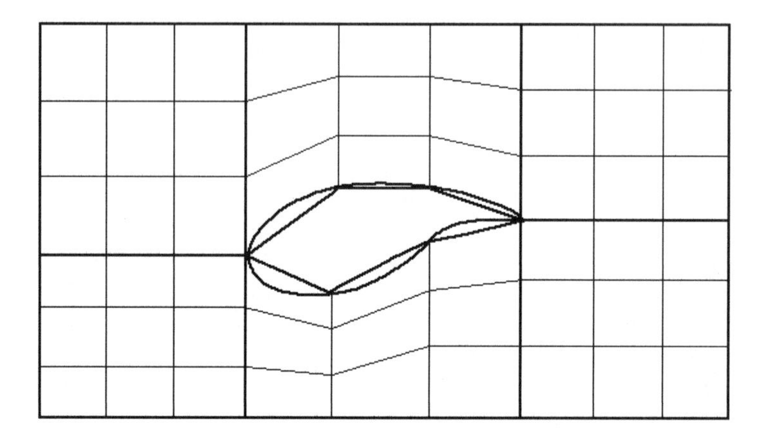

Figure 3.4: Multiblock structured grid around 2D geometry.

3.2.3 Unstructured Grid Generation Based on the Delaunay Triangulation

Most algorithms for generating unstructured-triangular and unstructured-tetrahedral grids are based on the *Delaunay triangulation* or the so-called *advancing front* method. Unstructured grids consisting of types of elements other than triangles and tetrahedra are not considered here.

The Delaunay triangulation has been used for various applications, including scattered data interpolation. In that context, function values f_i are given at scattered locations $\mathbf{x}_i$ for which no connectivity is given. Typically, the Delaunay triangulation is computed for the points $\mathbf{x}_i$ and triangular (tetrahedral) interpolants are constructed. The Delaunay triangulation is characterized by the fact that the circle (sphere) passing through the vertices of any triangle (tetrahedron) does not contain any point of the original point set in its interior. In the 2D case, the Delaunay triangulation is the max-min angle triangulation of the given point set, that is, it maximizes the minimum angle in the triangulation (see [42]). This property is very desirable for many applications, particularly for grid generation.

The Delaunay triangulation has found wide popularity in the finite element method (FEM) and unstructured grid generation communities. In general, triangles (tetrahedra) must be generated for a simply connected region in 2D (3D). They are generated in two steps. The first step is the generation of a boundary grid for the boundary curves (boundary surfaces) of the space to be discretized, that is, a set of boundary conforming line segments (triangles) is computed. The second step is the generation of triangles (tetrahedra) inside this boundary grid (see [57, 58, 60]).

The unstructured grid can be generated by inserting points and performing local retriangulation iteratively or by first generating the entire point set and then triangulating it, possibly using parallel programming paradigms. Grid points are placed such that certain quality measures (edge lengths, areas, volumes, angles, ratios thereof, and such) and specified distributions are satisfied.

The grid points are typically chosen according to geometrical properties of the boundary (for example, arc length or absolute curvature) and distribution functions. Spacing parameters are used for the boundary point distribution, and so-called *sources* (that is, point, line, and plane sources) are used to further control grid point distributions in the interior. Grid point densities decrease in a predefined fashion with increasing distance from the sources. Grids must be *graded* in this fashion due to the sometimes sudden occurrence of discontinuities in certain field parameters, for example, shocks. Figure 3.5 shows 2D triangular grids with point and line sources.

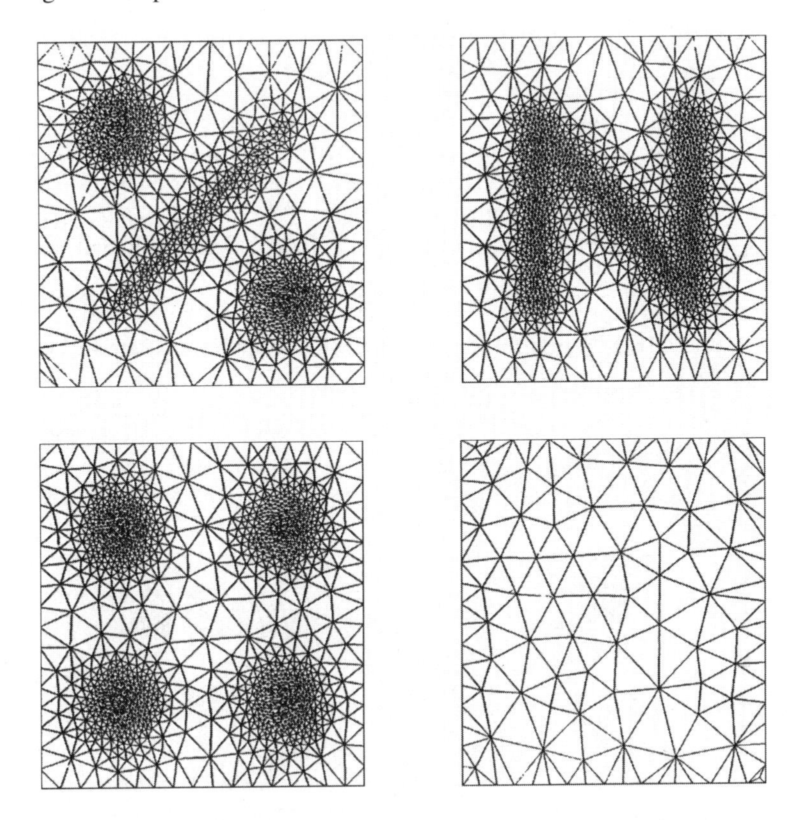

Figure 3.5: Unstructured-triangular grids with point and line sources (courtesy of N.P. Weatherill, University College of Swansea, U.K.).

3.2.4 Unstructured Grid Generation Based on Element Size Optimization

An automatic algorithm for the generation of 3D unstructured-tetrahedral grids is described in [18]. This algorithm can be applied to any closed geometry, that is, a geometry consisting of surfaces whose boundary curves are each shared by exactly one other surface boundary curve. The method is based on intersecting the edges of an initial (coarse) tetrahedral grid with the given geometry, clipping this initial grid against the geometry by extracting the (parts of) tetrahedra on the outside (or inside) of the geometry, and iteratively inserting grid points.

The initial tetrahedral grid consists of a uniform density of points and tetrahedra, and the triangulation is iteratively improved by inserting grid points until the tetrahedral volumes satisfy a specified condition. This condition considers the distance to and (if desired) the absolute curvature of the geometry. The desired volume V of a tetrahedron at surface distance d is

$$V(d) = \left(1 - \left(\kappa \frac{d}{d_1}\right)^p\right) V_0 + \left(\kappa \frac{d}{d_1}\right)^p V_1, \quad 0 \le p < \infty, \tag{3.4}$$

where V_0 and V_1 are specified tetrahedral volumes at distance 0 and distance $d_1 > 0$. The distance d between a tetrahedron and the surface is defined as the smallest distance between the tetrahedron's centroid and the surface. The weight function κ can be used to consider absolute surface curvature at the surface point closest to a tetrahedron. When using $p = \kappa = 1$, tetrahedral volumes increase in a linear fashion with increasing surface distance.

The overall goal is to minimize the difference between actual and desired tetrahedral volumes $V(d)$ (see [18]). Figure 3.6 shows a slice of an unstructured-tetrahedral grid generated by this method. Three tetrahedra are shown.

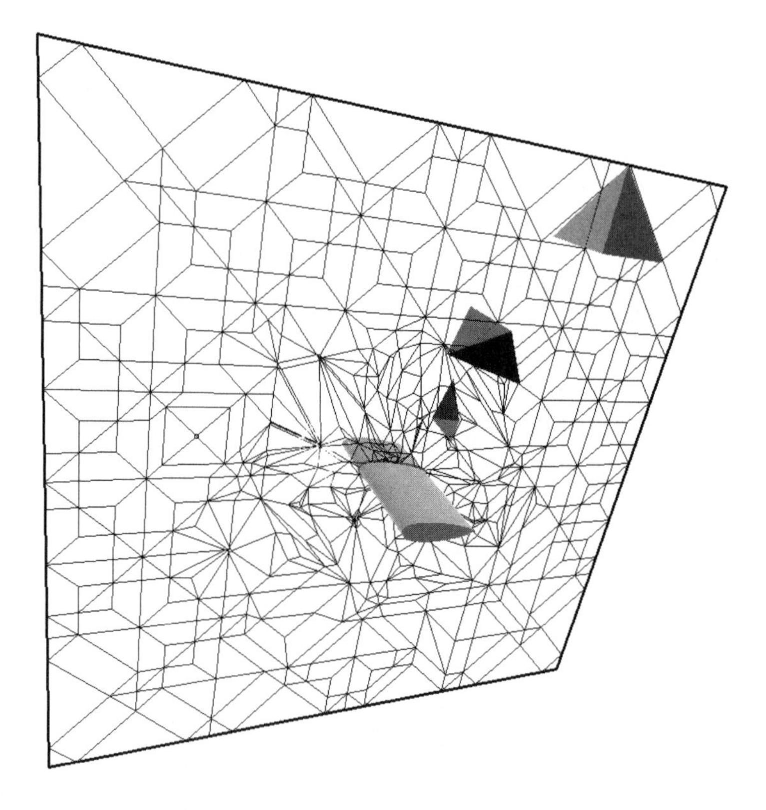

Figure 3.6: Slice of unstructured-tetrahedral grid around wing.

3.2.5 Unstructured Grid Generation Based on the Advancing Front Method

The advancing front method generates unstructured grids in a completely different way. The input—for the generation of a triangular (tetrahedral) grid—is a set of oriented line segments (oriented triangles) discretizing the boundary curves (surfaces) of some geometry. Triangles (tetrahedra) are then constructed by advancing a *front* into the interior of the field until it is completely filled with elements. Depending on local edge and angle configurations of the current front, new elements are created by connecting existing points or by inserting new points—according to some distribution function—and connecting them with existing ones (see [29]). The approach is similar to hyperbolic grid generation.

The desired point distribution is typically defined by a set of points with associated spacing parameters (background grid). Interpolating these spacing parameters yields the desired spacing at any point in the field. Points are inserted such that the desired spacing is optimally satisfied. The advancing front strategy is also used for the generation of unstructured grids consisting of quadrilaterals (hexahedra). Figure 3.7 shows various stages of the advancing front algorithm.

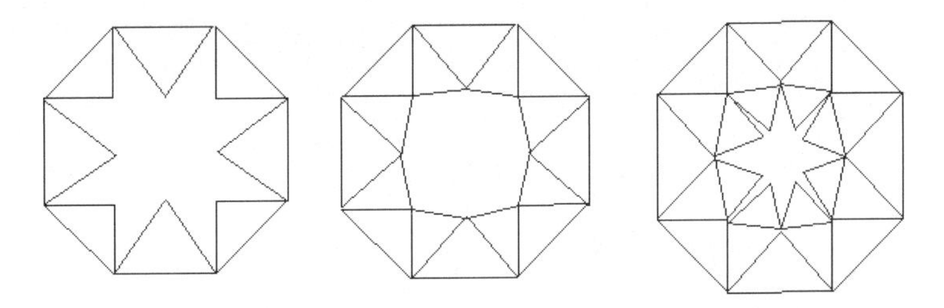

Figure 3.7: Different stages of the advancing front algorithm for 2D example.

3.2.6 Chimera Grids

Chimera grids (overlapping structured grids) are a collection of partially overlapping structured "component grids." Chimera grids are produced by generating a separate structured grid around or above each component of a geometry. The component grids are overlapping, and information is passed from one component grid to the other via an appropriate interpolation scheme. Often, Chimera grids are generated with a hyperbolic grid generation method and "growing" each component grid from the associated geometry component. The approach is described in detail in [3] and [8]. Figure 3.8 shows an example.

3.2.7 Hybrid Grids

Hybrid grids are the combination of structured and unstructured grids. When generating a hybrid grid, a field is discretized by structured grid topologies wherever possible, and regions between the different structured grids are discretized by unstructured grids. Thus, this approach leads to grids that combine the strength of structured grids (computational

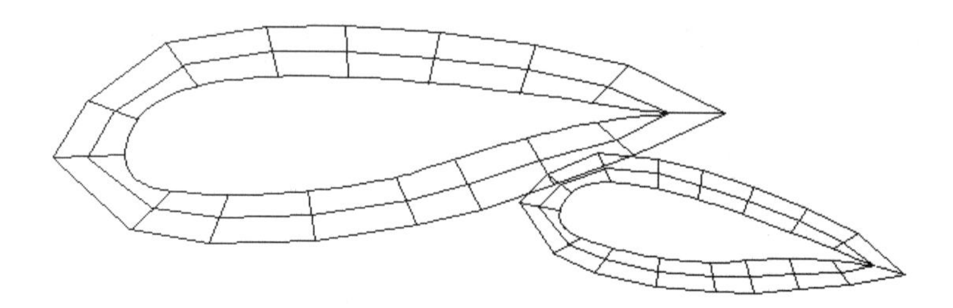

Figure 3.8: Chimera grid for 2D wing configuration.

efficiency and high-aspect ratio elements) and unstructured grids (flexibility). Figure 3.9 shows a hybrid grid.

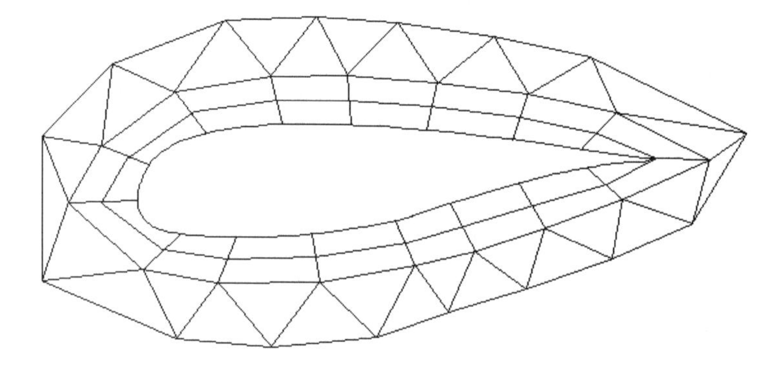

Figure 3.9: Hybrid grid around 2D wing configuration.

3.2.8 Hierarchical Grid Generation Based on Quadtrees and Octrees

By applying a subdivision technique based on quadtrees (octrees) to certain elements of a grid, multiple levels of different resolution are defined. This approach leads to so-called hierarchical (embedded) grids. Depending on the local complexity of the geometry or the local complexity of an intermediate field solution, more elements are adaptively created by splitting parent elements.

In the structured case, a quadrilateral (hexahedron) is subdivided into four (eight) quadrilaterals (hexahedra) which are defined by splitting the edges at their midpoints (see [33, 45]). The subdivision algorithm must ensure that no "cracks" are introduced in physical xyz space when splitting edges in the parameter space of a parametric surface. As a result of the subdivision process, nodes are created that lie in the interior of edges (faces). They are usually called "hanging nodes."

Hierarchical subdivision schemes based on quadtrees (2D) and octrees (3D) are also used for the automatic generation of entire unstructured grids inside simply connected regions. For this purpose, the boundary of the region to be discretized must be given by a set of line segments (triangles), and the bounding box of the region is recursively subdivided until no cuboid contains more than one of the points used in the boundary discretization. The cuboids lying in the interior of the region are then subdivided into triangles (tetrahedra). Special care is necessary to preserve the region's boundary. This approach is discussed in [61, 62, 63]. Figure 3.10 shows the quadtree implied by the boundary of a simply connected region.

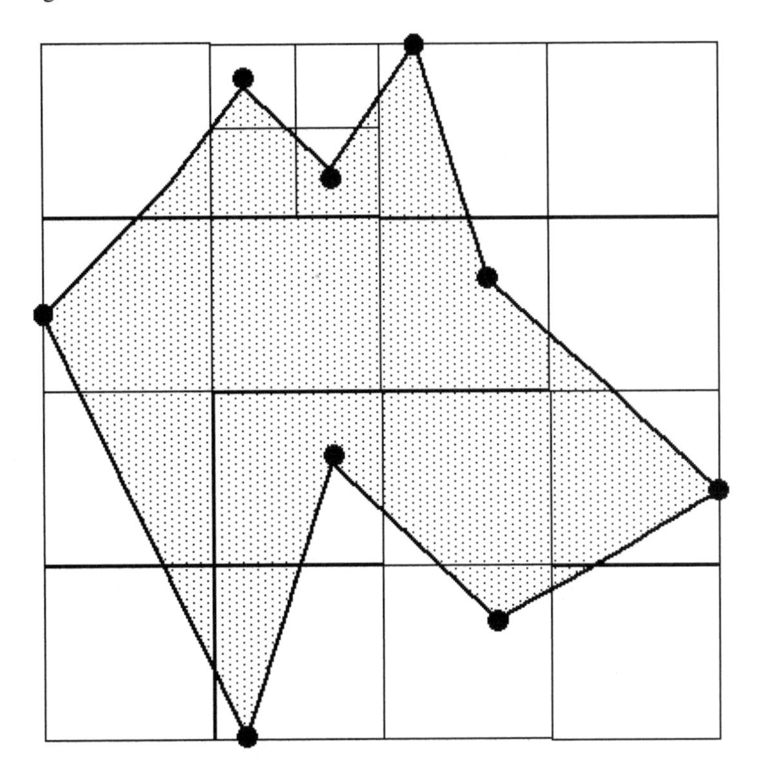

Figure 3.10: Quadtree implied by boundary line segments of 2D region—interior (shaded) must be triangulated.

3.3 The National Grid Project (NGP)

3.3.1 Survey

The National Grid Project (NGP) started in 1991. The purpose of this project is the development of a next-generation, interactive grid generation system for a consortium of U.S. industrial and governmental laboratories. This effort has led to a grid generation system that combines geometric modeling (CAD), structured grid generation, and unstructured grid generation functionality. The interdisciplinary character of the ERC-CFS has made it possible to design and develop this general grid generation system.

Nonuniform rational B-splines (NURBS) are used throughout the system for the representation of geometries. The user interface is based on MOTIF and X, and Iris GL is used for graphics. The system interfaces with CAD systems by converting given CAD data (IGES format) to NURBS and vice versa. NURBS approximations of original data are generated only if an exact conversion is impossible. The NGP system provides various CAD functions for the creation and manipulation of geometries, which can also be used for the definition of field boundary curves and surfaces. In addition, the CAD system provides a method for the continuous approximation of geometries with discontinuities.

CAD functionality that is helpful for the block definition and grid generation process is continuously added to the system. Currently, most CAD operations can be performed only on a single NURBS surface, that is, it is not possible to perform all CAD operations across multiple NURBS surfaces with different parameter spaces. Algorithms that allow certain operations across surface boundaries will be developed. The NGP system allows the user to

- interface with CAD systems via IGES
- create geometries via a NURBS-based CAD subsystem
- perform geometry-processing operations, such as surface-surface intersection, trimming, and surface subdivision
- correct geometries with discontinuities ("gaps," overlapping surfaces, or intersecting surfaces)
- automatically generate the topology information of the boundary representation (B-rep) of a CAD model, that is, generate the connectivity information for curves and surfaces
- construct unstructured grids (triangular, quadrilateral, and tetrahedral elements) and multiblock structured grids (quadrilateral and hexahedral elements), including Chimera and hybrid grids
- analyze—according to various criteria—and visualize the quality of 2D and 3D grids (see [41])
- save the state of the system at any time and provide a script language that can be edited.

Figure 3.11 shows an example of an aircraft geometry and the far-field boundary created with the NGP CAD system. It also shows the general layout of the NGP interface.

Most of the requirements that were defined by the NGP consortium are satisfied at this point, three years after the start of the project. Currently, the system can handle only four-sided surfaces connected in a full-face interface fashion, that is, surfaces must share entire boundary edges and not parts thereof. Partial edge (and partial face) matching will be added to the NGP system. When dealing with trimmed surfaces (that is, surfaces containing holes defined by so-called trimming curves), the user must generate a set of new four-sided surfaces from each trimmed surface. Thus, each trimmed surface is replaced by a set of new surfaces without trimming curves (holes). This process will eventually be automated. The current capabilities of the NGP system are described in more detail in [43].

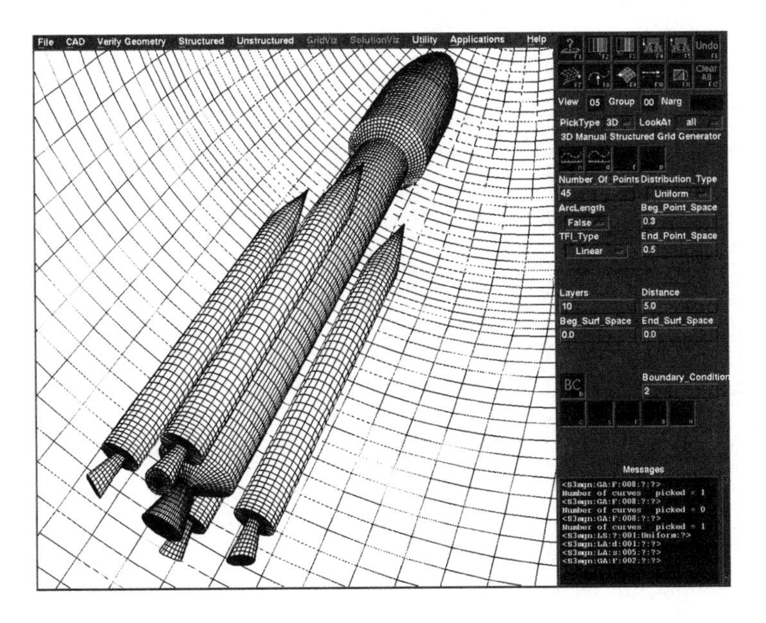

Figure 3.11: Geometry and far-field boundary created with the NGP CAD system.

3.3.2 Approximation of Geometries with Discontinuities

For grid generation purposes, a geometry that is to be discretized must be error-free, that is, it must not contain discontinuities (overlapping surfaces, "gaps" between surfaces, or surface intersections). Unfortunately, CAD data imported from other CAD systems very often contains such errors.

The NGP system provides an interactive geometry correction technique that locally approximates a given geometry by a B-spline surface. The technique must be applied to all regions containing discontinuities. Eventually, a continuous geometry is obtained that consists partially of original NURBS surfaces and partially of B-spline approximations. The continuous geometry is then used for grid generation.

The geometry correction technique is based on constructing an initial local surface approximation (a bilinearly blended Coons patch) which is projected onto the given surfaces. The user must ensure that the B-spline surface approximations are connected with each other and with certain given NURBS surfaces in a full-face interface fashion. This can done with the CAD system. Generating a local surface approximation requires these steps:

1. Definition of four surface boundary curves
2. Generation of $N \times N$ points on the bilinearly blended Coons patch defined by the four surface boundary curves
3. Projection of the $N \times N$ points onto the given surfaces
4. Generation of "artificial projections" whenever certain points of the Coons patch cannot be projected onto any original surface

When trying to project a point of the Coons patch onto the given surfaces, a projection might or might not be found. If one or more projections are found within a small distance to the Coons patch, the one closest to the Coons patch is chosen. If no projection is found, an "artificial projection" is approximated by applying a scattered data interpolation scheme to the projections that have been found.

5. Interpolation of the points resulting from steps (3) and (4) by a bicubic B-spline surface

Points on the Coons patch are denoted by $\mathbf{x}_{i,j}$ and are obtained by applying the bilinear blending procedure to a finite set of points on the four specified boundary curves (see Section 3.2.2, Equation (3.1)). The points $\mathbf{x}_{i,j}$ are then projected onto the given surfaces in the normal direction $\mathbf{n}_{i,j}$ of the Coons patch, which is approximated by

$$\mathbf{n}_{i,j} = \frac{(\mathbf{x}_{i+1,j} - \mathbf{x}_{i-1,j}) \times (\mathbf{x}_{i,j+1} - \mathbf{x}_{i,j-1})}{||\ (\mathbf{x}_{i+1,j} - \mathbf{x}_{i-1,j}) \times (\mathbf{x}_{i,j+1} - \mathbf{x}_{i,j-1})\ ||}, \tag{3.5}$$

where $||\ ||$ is the Euclidean norm. A family of line segments, defined by the point pairs in

$$\{\mathbf{a}_{i,j} = \mathbf{x}_{i,j} - d\mathbf{n}_{i,j},\ \mathbf{b}_{i,j} = \mathbf{x}_{i,j} + d\mathbf{n}_{i,j} \mid d > 0,\ i, j = 1, \ldots, N\} \tag{3.6}$$

is intersected with the given geometry. If a line segment $\overline{\mathbf{a}_{i,j}\mathbf{b}_{i,j}}$ has multiple intersections with the given geometry, the one closest to $\mathbf{x}_{i,j}$ is chosen as the projection of $\mathbf{x}_{i,j}$.

The original surfaces might be discontinuous, and, in this case, certain points $\mathbf{x}_{i,j}$ cannot be projected onto an original surface. Each projection $\mathbf{p}_{i,j}$ that has been found can be represented as the linear combination $\mathbf{p}_{i,j} = (1-t_{i,j})\mathbf{a}_{i,j} + t_{i,j}\mathbf{b}_{i,j}$. Thus, the computation of "artificial projections" reduces to the construction of a scattered data interpolant. Hardy's reciprocal multiquadric method is used for interpolating the known values $t_{i,j}$ (see [11]). The system of linear equations to be solved is

$$t_{i,j} = \sum_{I,J \in \{1,\ldots,N\}} c_{I,J} \left((u_{I,J} - u_{i,j})^2 + (v_{I,J} - v_{i,j})^2 + R \right)^{-0.5}, \quad i, j \in \{1, \ldots, N\}, \tag{3.7}$$

where all values $t_{i,j}$, $u_{i,j}$, and $v_{i,j}$ are considered for which a projection has been found. The value for $R > 0$ is chosen according to the point spacing on the Coons patch. The NGP system uses a localized version of this method. The solution of the system (3.7) is used to approximate the required "artificial projections." The approximation of "artificial projections" from the ones that can be found is illustrated in Figure 3.12.

The resulting $N \times N$ points (projections and "artificial projections") are interpolated by a C^1 continuous, bicubic B-spline surface. An error estimate is computed for each local surface approximant. Curves on the original geometry, for example, surface boundary curves or trimming curves, can be preserved by this method.

The algorithm is described in detail in [17, 24, 50]. The underlying concepts are well known in geometric modeling and can be found in [10] and [22]. Figure 3.13 shows the original CAD data of a car body geometry with "holes" (upper two images) and its continuous approximation (lower two images).

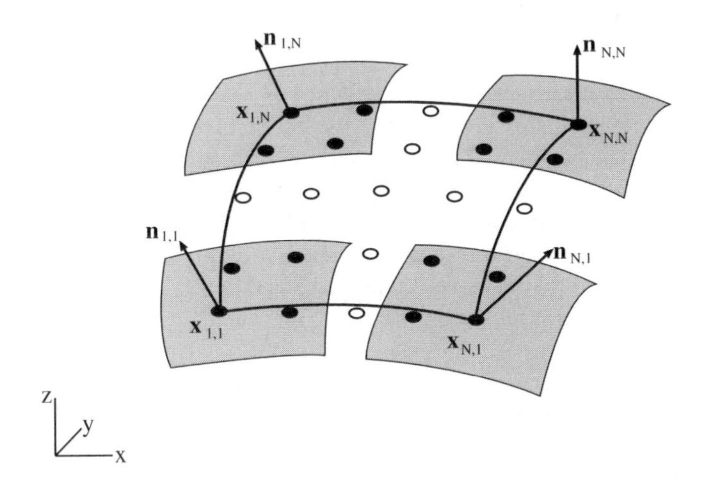

Figure 3.12: Projections (bullets) and estimated "artificial projections" (circles).

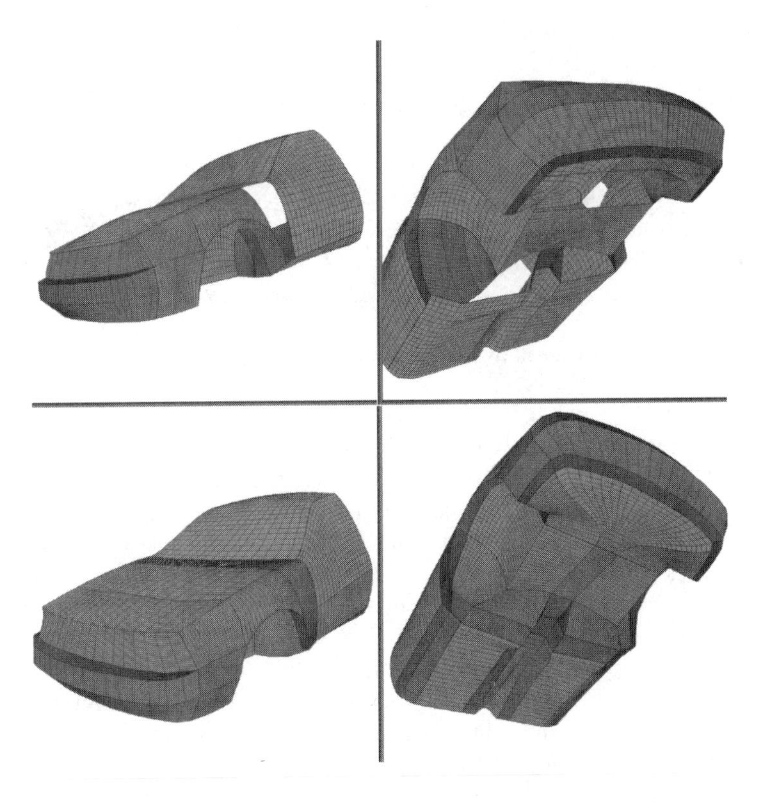

Figure 3.13: Car body with "holes" *(top)* and its approximation *(bottom)*.

3.3.3 Grid Generation with the NGP System

The structured and unstructured grid generation modules of the NGP system are based on NURBS. The representation of curves and surfaces as NURBS has turned out to be beneficial for the grid generation process. The generation of a structured grid for a single NURBS surface requires two steps. The first step is the generation of boundary curve points considering user-specified distribution functions. The second step is the generation of grid points in the surface's interior. An initial surface grid is computed by applying TFI to the boundary points. This surface grid is then "smoothed" yielding a surface grid with nearly orthogonally intersecting grid lines (see Section 3.2.2 and [53]).

The same principle is used for the generation of 3D structured volume grids. Currently, the user has to define the vertices, edges, and faces of the blocks surrounding a 3D geometry. All blocks must be connected in a full-face interface fashion; partial face matching is not possible at this time. It is planned to use automatic procedures for the block definition phase (see [7]). Figure 3.14 shows the initial TFI surface grid and the smoothed surface grid for the space shuttle.

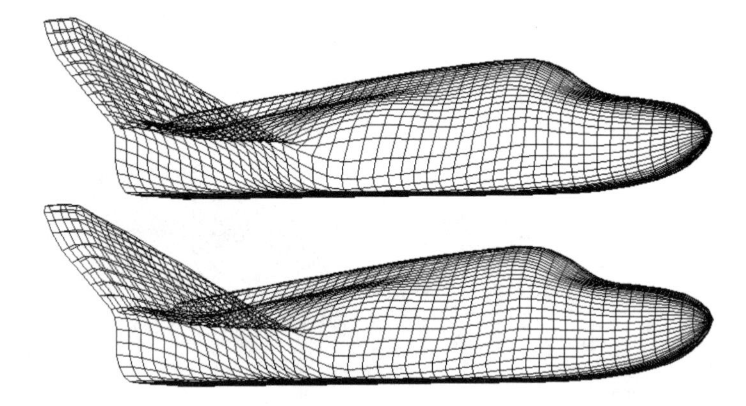

Figure 3.14: Initial grid obtained by transfinite interpolation *(top)* and elliptically smoothed grid *(bottom)* for space shuttle.

The generation of unstructured-triangular surface grids is based on the Delaunay triangulation. Once surface boundary curve discretizations are established, scattered points are generated in a surface's interior, and the Delaunay triangulation is computed. Point insertion and triangulation are performed in a surface's parameter space while considering metric properties of the surface in 3D space. Poorly parameterized surfaces can sometimes lead to poor surface triangulations. Appropriate reparameterization of parametric surfaces is currently being investigated.

The unstructured grid generation module can handle surfaces with any number of closed, nonintersecting trimming curves. Usually, the point distributions on the boundary and trimming curves imply the point distribution in a surface's interior. Local variations in point distribution can be achieved by using point, line, and plane sources. Point densities decrease with increasing distance to a source. Volume grids are computed by the Delaunay triangulation as well. The resulting triangulations are clipped against the boundaries of the field to obtain boundary conformity.

The unstructured grid generation module requires little user input and operates highly automated. The underlying concepts and theories are discussed in [54, 59, 60]). Figure 3.15 shows an unstructured-tetrahedral grid around a car body.

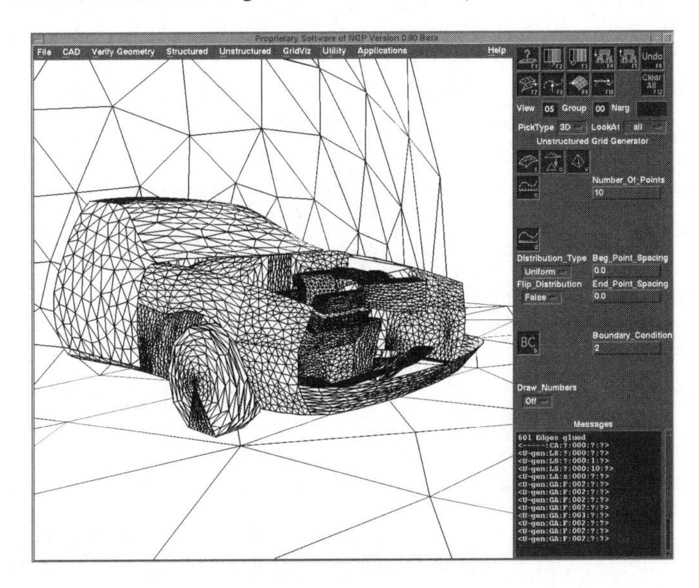

Figure 3.15: Unstructured-tetrahedral grid generated with the NGP system.

The NGP system also allows for hyperbolic grid generation and supports the Chimera approach for multiple blocks. These two approaches are often combined (see [6]).

3.4 Visualization Efforts at the ERC-CFS

3.4.1 The Scientific Visualization Thrust

The Scientific Visualization Thrust at the ERC-CFS provides image-oriented analytical tools to allow scientists and researchers to better understand and explain physical phenomena. To accomplish this goal, both measured and computational data is visualized. The application focus has been diverse and has included oceanographic, aerodynamic, hydrodynamic, meteorological, geophysical, and acoustic work. The work has also spanned the functional gamut—from visualization to allow application scientists and engineers to understand a measured phenomenon or the results of a computational model—to visualization to allow application scientists and engineers to explain their results to their supervisors and funding sources. In some cases, the understanding must be obtained quickly so that the next sample may be taken in an operational scenario; in other cases, the understanding can be obtained only after extensive and rigorous analysis in a laboratory.

Unfortunately, it is still true that many revered and respected scientists and researchers seldom actually create visualizations themselves, but instead direct "visualization scientists and engineers" to accomplish effective visualizations. Thus, although photorealism is usually a secondary goal, since the visualization is not done interactively, high-quality imagery is usually expected if not demanded. For example, Gouraud shading and anti-aliasing is expected, that is, "jaggies" are unacceptable. Even so, interactivity or the ability to explore a data set is very important to the visualization engineer in this case, because often the application scientist is not exactly sure for what he or she is searching and because new or better understanding is often obtained by data exploration. Interactivity is also crucial for "field-deployment" scenarios, in which rapid visualization of measured data is important, and for concurrent visualization, in which the computational scientist seeks to save expensive supercomputer time by looking for mistakes in an evolving simulation.

The quantity and quality of the visualization technology is another metric that can be applied to the scenarios for which we develop visualization techniques and toolkits. Within the ERC-CFS, the personnel have access to state-of-the-art equipment such as:

- Powerful visualization computers
- Equipment to convert computer-generated imagery to video format
- Various video recording equipment (for example, video laser disks and Betacam SP, U-matic SP, S-VHS, and VHS VTRs)
- Immersive display devices

However, the visualization technology available to users in other environments constrains much of the research and development. For example, as will be described in the section on SCIRT below, a rapid data approximation and visualization scheme had to be created to allow engineers in the field to decide where to take the next sample. This software had to function on a lower-end graphics workstation in the field, yet allow engineers in the lab to postprocess the data to quantify more precisely the measured environment. In another scenario described below (one of the AGP environments), there is only one visualization workstation in a lab with multiple desktop workstations and an FDDI-attached supercomputer for computational field simulations. In this environment, the scientist's visualizations typically consist of lower-end, image-processing (2D)-type analysis and batch-mode-oriented, high-quality, video-based animations. Minimal expenditure has been made to allow the scientist to interactively visualize the results of the simulation of time-varying phenomena in an animated fashion on a workstation that supports high-spatial, high-intensity, and high-temporal resolution.

Thus the research and development efforts have spanned the gamut from high-end, nonreal-time imagery production to computationally efficient, interactive, near-real-time visualization. In the next few sections, some of that work is described in hopes of showing the diversity and value of analytical scientific visualization.

3.4.2 The Oceanographic Visualization Interactive Research Tool (OVIRT)

In 1992 and 1993, an interactive 3D scalar field visualization system, called the Oceanographic Visualization Interactive Research Tool (OVIRT), was developed at the ERC-CFS for the Naval Oceanographic Office (NAVO), Stennis Space Center, Mississippi. The purpose of this project was the development of effective visualization tools for oceanographic sound speed data.

The scalar field visualization, scattered data modeling, and isosurface generation techniques on which OVIRT is based are discussed in [16, 38, 39, 40]. OVIRT is described in more detail in [34].

OVIRT is used for visualizing environmental ocean data. It extends some of the classical 2D oceanographic displays into a 3D visualization environment. Its five major visualization tools are *cutting planes*, *minicubes*, *isosurfaces*, *sonic surfaces*, and *direct volume rendering* (DVR).

The cutting planes tool provides three orthogonal planes which can be independently and interactively moved through the volume. The minicubes routine renders small cubes inside the given volume. Their faces are shaded according to function value, for example, sound speed, pressure, temperature, or salinity. The isosurface tool allows the generation of constant value surfaces (see [38]). The DVR techniques give a global, translucent view of the data.

The sonic surfaces are an extension of 2D surfaces which have been classically used to display acoustic propagation paths and inflection lines within the ocean. The surfaces delineate the extent and axis of the sound channels. Other features in OVIRT include the ability to overlay the shoreline and inlay the bathymetry. Traditional analysis of sound velocity in the ocean involves graphical depictions of horizontal and vertical slices of the ocean. The limitation of those techniques is that the analyst views only a limited portion at once. To achieve a total view requires assimilating multiple executions of the display routine and mentally composing the results. The principle function of OVIRT is to visualize a 3D ocean sound speed field as a volume.

The user interface is based on X and MOTIF, and rendering is done with Iris GL using mixed-mode programming. The data is given on a rectilinear grid which is uniformly spaced in the x and y directions (latitude and longitude) and nonuniformly in the z direction (depth). In Figure 3.16, sound speed is rendered by the minicubes method ($10 \times 10 \times 10$ resolution). Some of the cubes are clipped against the bathymetry (wire frame). Trilinear interpolation is used for generating the function values at the cubes' vertices.

In Figure 3.17, the original nonuniform spacing in the z direction is used to define the vertices of the small cubes, creating cubes which increase in height as the depth increases, but which reflect the original sampling grid.

In an attempt to extend the visualization techniques with which the analyst is familiar, the 2D diagrams of sound speed versus depth (sound speed profiles) were extended to create 3D visualizations in which all the sound speed profiles for a particular latitude or longitude can be simultaneously viewed. The technique has been named *marching wiggles*. Traditionally, the analyst could not see the sonic profiles or sonic surfaces inlaid in the ocean. The sonic surfaces of particular interest to an oceanographer are (from the surface down):

- the *shallow sound channel* (SSC) *axis*
- the *sonic layer depth* (SLD)
- the *deep sound channel* (DSC) *axis*
- the *critical depth* (CD)

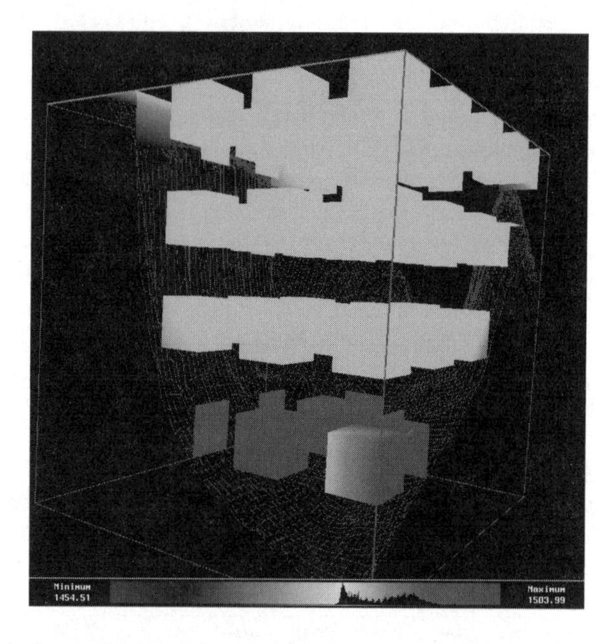

Figure 3.16: Minicubes method used for rendering sound speed with OVIRT system (uniform spacing and uniform cubes). See Color Plate 13.

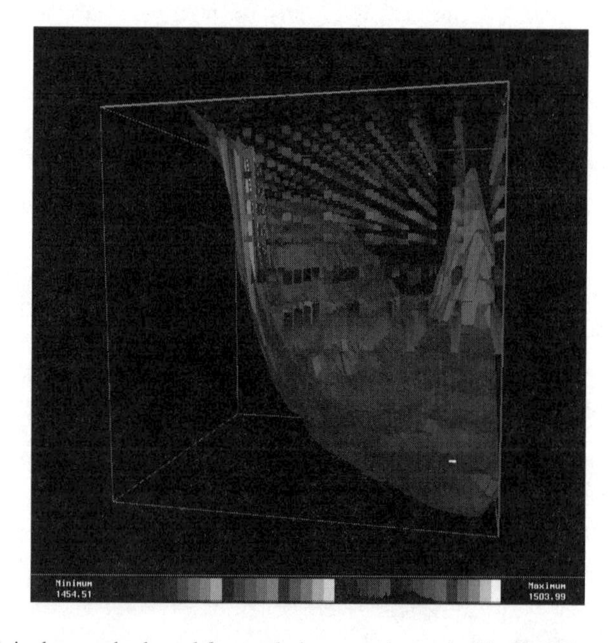

Figure 3.17: Minicubes method used for rendering sound speed with OVIRT system (nonuniform spacing and nonuniform cubes). See Color Plate 14.

These surfaces can be extracted from the collection of sound speed profiles. In general, they bound or are the axis of a sound channel. Figure 3.18 illustrates the surfaces for a 2D longitude slice.

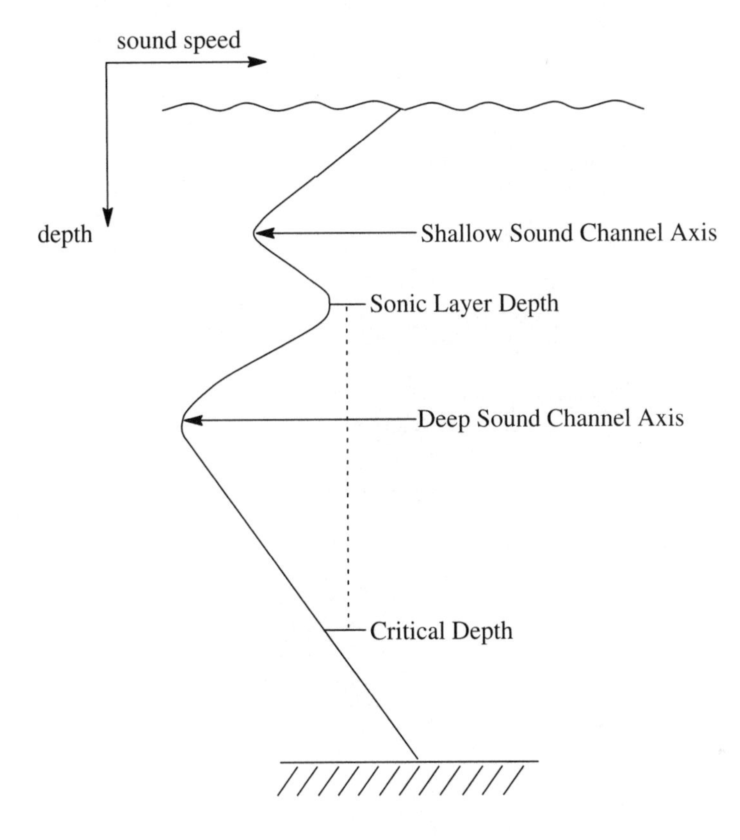

Figure 3.18: Shallow sound channel axis, sonic layer depth, deep sound channel axis, and critical depth.

Considering just a single sound speed profile (sound speed versus depth), the SSC is the region between the sea surface and the depth of the first sound speed minimum. The SLD is defined as the depth at which the first relative maximum of sound speed occurs in the upper levels of the ocean. In general, the next relative maximum is the CD (the sound speed is the same at the SLD and the CD). Sound rays will bounce between the SLD and the CD when emitted from a source between these two surfaces. If the acoustic source is above the SLD (below the CD), the sound rays will bend upward (downward). Figure 3.19 shows the SLD (red), the DSC axis (green), and the CD (blue).

The isosurface generation module is based on the algorithms given in [30] and [40]. Two DVR techniques are included in OVIRT, based on the methods in [28] and [44]. It has been found that DVR techniques do not convey adequate information regarding the location of sonic surfaces in the ocean. Better opacity mapping functions combined with robust classification techniques are necessary (see [9]). OVIRT has the capability to display

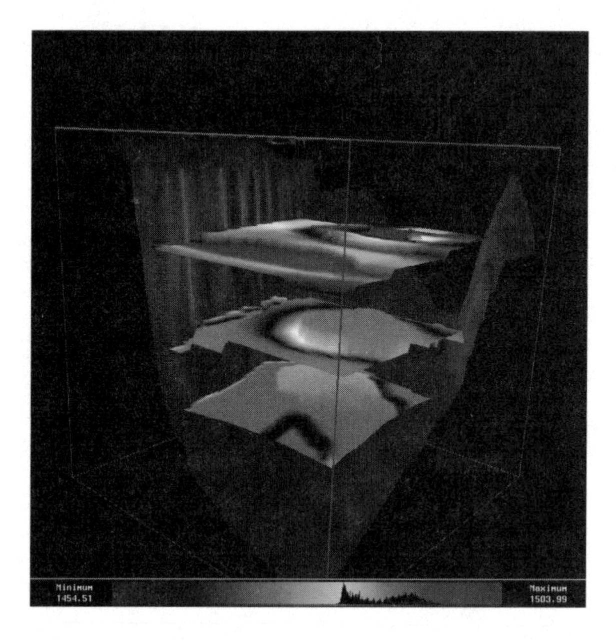

Figure 3.19: Sonic layer depth, deep sound channel axis, and critical depth generated and rendered with OVIRT system. See Color Plate 15.

animations of DVR imagery since, in general, DVR is not an interactive technique, whereas the other techniques are.

OVIRT provides a macro capability which allows the user to create and execute macro commands. Applications can be executed remotely, yet visualization can be done on a local workstation. Computationally intensive modules have been ported so that they can be executed remotely on SGI and Sun machines, as well as the Cray YMP. Other features include

- script and journal capabilities
- interactive query capability of both location and function value via picking
- rendering modes for combined rendering of bathymetry, cutting planes, and sonic surfaces

3.4.3 The Site Characterization Interactive Research Toolkit (SCIRT)

In 1993 and 1994, a scattered data interpolation and visualization system, called the Site Characterization Interactive Research Toolkit (SCIRT), was developed at the ERC-CFS for the U.S. Army Corps of Engineers, Waterways Experiment Station, Vicksburg, Mississippi. The purpose of the project was the development of a visualization tool for scattered data collected by the Site Characterization and Analysis Penetrometer System (SCAPS).

At each point, the soil classification number (SCN) and a concentration value are given. The data is usually collected by probing vertically into the ground. SCIRT provides techniques for the visualization of the SCN and concentration value. The system provides various scattered data interpolation techniques, clipping against the convex hull of the given 3D data points, and the scalar field visualization methods provided in OVIRT with extensions, for example, the visualization of concentration isosurfaces inside transparently rendered soil types and the indication of probe locations.

Since the visualization algorithms provided by OVIRT require a rectilinear grid, the given SCN and concentration data sets cannot be rendered directly. Scattered data interpolation and evaluation on a rectilinear grid are done in a preprocessing step. The system contains these modules:

- Convex hull and triangulation algorithms for scattered 3D points
- Global and local scattered data interpolation methods
- Overlay and transparency methods for the visualization of concentration isosurfaces inside surrounding soil types
- Scalar field visualization techniques supported by OVIRT with the capability of combining various scalar field visualizations into one image
- Local-viewing feature by applying visualization techniques to a user-specified box inside the data set

The visualization results depend greatly on the particular scattered data interpolation method used for the creation of the required rectilinear data set. The problem is to find a trivariate function $f(\mathbf{x}) = f(x, y, z)$ which interpolates the given values f_i at the sites $\mathbf{x}_i$, $i = 1, \ldots, N$. Several techniques have been developed over the last three decades for the interpolation (and approximation) of scattered data (see [11]).

Some scattered data interpolation techniques do not require a grid for the given data. Examples are Shepard's and Hardy's multiquadric methods. Shepard's global scattered data interpolant is the function

$$f(\mathbf{x}) = \frac{\sum_{i=1}^{N} \frac{f_i}{d_i^2}}{\sum_{i=1}^{N} \frac{1}{d_i^2}}, \tag{3.8}$$

where $f(\mathbf{x}_i) = f_i$ and $d_i^2 = (x - x_i)^2 + (y - y_i)^2 + (z - z_i)^2$ is the square of the Euclidean distance between $\mathbf{x}$ and a data point $\mathbf{x}_i$. Hardy's multiquadric method is based on constructing the function

$$f(\mathbf{x}) = \sum_{i=1}^{N} c_i \left((x - x_i)^2 + (y - y_i)^2 + (z - z_i)^2 + R\right)^{0.5}, \tag{3.9}$$

$R > 0$, by solving the $N \times N$ system of equations given by the condition $f(\mathbf{x}_i) = f_i$. Usually, Hardy's multiquadric method yields much better results.

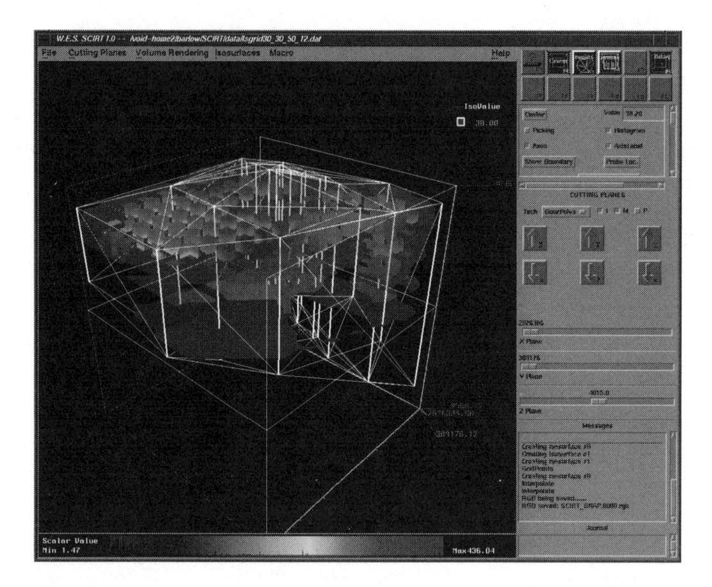

Figure 3.20: Transparent isosurfaces, minicubes, and cutting planes (clipped against convex hull) in a single image generated with SCIRT. See Color Plate 16.

Some techniques require a triangulation of the data sites $\mathbf{x}_i$. Triangulation-based techniques can be used to construct interpolants for each tetrahedron with a specified degree of continuity at the interfaces. Performing piecewise linear interpolation leads to a C^0 continuous function, and its value inside a particular tetrahedron is given by

$$f(\mathbf{x}) = \sum_{j=1}^{4} u_j f_j, \tag{3.10}$$

where u_1, u_2, u_3, and u_4 are the barycentric coordinates of the point $\mathbf{x}$ inside the tetrahedron with vertices $\mathbf{x}_j$, $j = 1, 2, 3, 4$, and function values f_j at its vertices. Higher-degree continuity for triangulation-based methods is achieved by considering derivative estimates at the given data points. Interpolation methods for C^1 and C^2 continuous interpolation will be added to SCIRT.

The SCIRT system supports Shepard's method, Hardy's multiquadric method, a localized version of Hardy's multiquadric method, and the triangulation-based piecewise linear method. The interpolants are evaluated on a rectilinear grid and are then visualized using the techniques provided by OVIRT. Examples are shown in Figure 3.20.

A future research issue is the reduction of the given scattered data sets. Typically, the number of holes used for probing is rather small, and the number of measurements in depth is extremely large.

3.4.4 The ATOC-GAMOT Project (AGP)

In late 1993, work was initiated on a time-varying volume visualization system in support of a large environmental project attempting to measure global warming by using acoustic tomography. The underlying physics is that the speed of sound in water is proportional to the temperature, pressure, and salinity. If the pressure and salinity are known and the transmission time is measured with sufficient accuracy, the temperature can be determined with similar accuracy. For example, travel time variations of 0.2 sec over 3300 km paths indicate temperature changes of 0.1° Celsius. By utilizing many intersecting transmission paths, tomography can be used to determine the temperature throughout the ocean volume, much as computer-aided tomography (CAT) is used in medicine. The visualization task is twofold: visualize time-varying acoustic propagation in the ocean and visualize ocean model data. The ocean model is used to forecast the state of the ocean through which the acoustic signal will pass. The acoustic signal is the basis of the whole experiment.

Visualization of the acoustic propagation has required a number of different visualization approaches. One area of interest has been which eigenrays, that is, which acoustic projections, arrive at the receiver, with how much intensity, and at what time. Not only do the conditions vary within the sound channel over time, but the rays that deviate the most from the sound channel axis travel the fastest. Thus the signal does not arrive all at once, but over a finite time interval (for example, 3.5 sec) that varies diurnally. The computational model generates depth, distance, and time-traveled triples for each step of every ray that is determined to strike the receiver. The environmental context (bathymetry, shoreline, and so on) and conditions (salinity, temperature, flow regime, and such) are all important information. To visualize this data, a system called Acoustic Ray Visualizer (ARV) has been developed. ARV automatically determines the appropriate section of bathymetry and topology to use from the data output by the computational model. The user options include pausing the propagation, moving with the propagating ray, varying the number of time steps between each image, visualizing only some of the eigenrays, and recording the image corresponding to each time step to create a movie. A snapshot of the propagating rays is shown in Figure 3.21. The top 2D window gives a side view of the propagating rays, so that a more precise visual representation of the different propagation rates can be seen. The large 3D window shows the rays moving over the bathymetry.

Another interest is the variance of the acoustic path due to varying ocean conditions over transoceanic paths. Even with fixed source and receiver locations, these acoustic paths are known to deviate by 10 to 15 km due to seasonal variations. This is another time-varying 3D problem, but with much longer time scales. Associated with this interest is a desire to simply see the time-varying sound speed anomalies, which are derived by perturbing climatological data sets with dynamic ocean models. A very flexible program, ViewPerturb, has been developed to visualize this type of data. The interface allows the user to design his or her own color map based on the HSV color model, to clip the range of data mapped into the color map, and to create a movie from the rendered images. A number of different readers have been attached to the system, so that other time-varying 2D fields, for example, vorticity magnitude, sea-surface height, or salinity, can be visualized. An example of a relative vorticity field visualized using ViewPerturb is shown in Figure 3.22.

Visualization of the dynamic ocean models requires a significant understanding of computational fluid dynamics. The biggest hindrance to effective visualization is data overload. The ocean model (see [23]) has a 989×657 time-invariant uniform rectilinear grid in the horizontal with a latitude/longitude spacing of 0.125/0.176 degrees and six time-varying isopycnal layers. The internal time step of the model is on the order of 24 minutes, but data is only saved every 3.05 model days. This generates about 3GB of data per model year and

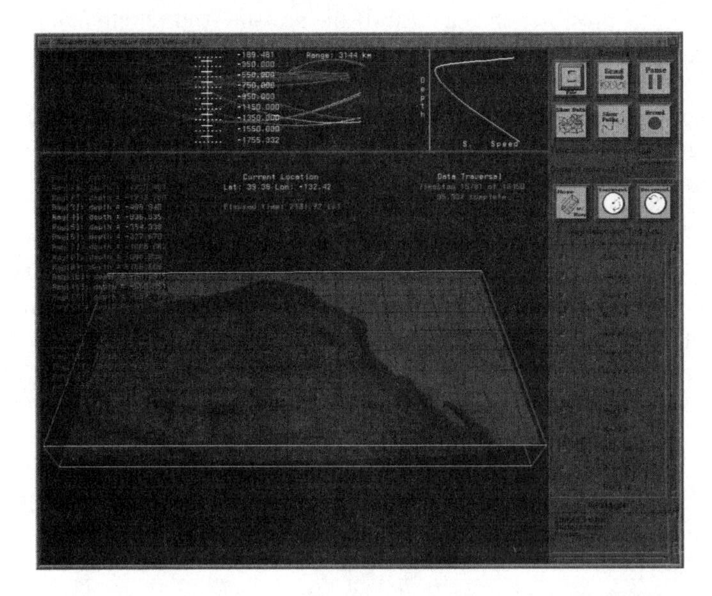

Figure 3.21: Snapshot of Acoustic Ray Visualizer (ARV). See Color Plate 17.

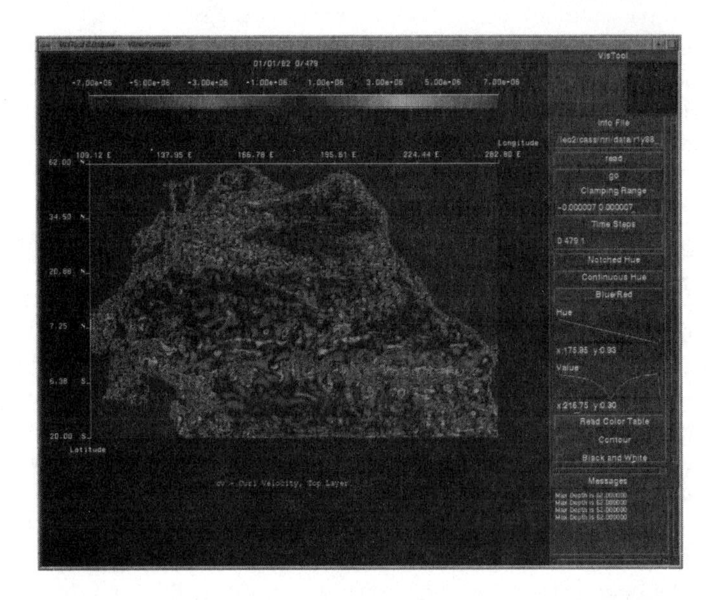

Figure 3.22: Relative vorticity for Pacific Ocean on January 1, 1982. See Color Plate 18.

at present, the model has been run for 13.5 model years. The visualizations that are needed are time-varying features such as waves, currents, and eddies. The geographic context must be given, since the flow physics is a function of the geographical location. The flow can be represented using moving arrows or a partially redundant mapping into the HSV color space. An example of this color wheel visualization scheme is shown in Figure 3.23. The direction of flow is mapped to a hue and the magnitude is mapped to both saturation and value (brightness). Thus high-magnitude southerly flows are bright red and small northerly flows are black with a slight tint of cyan.

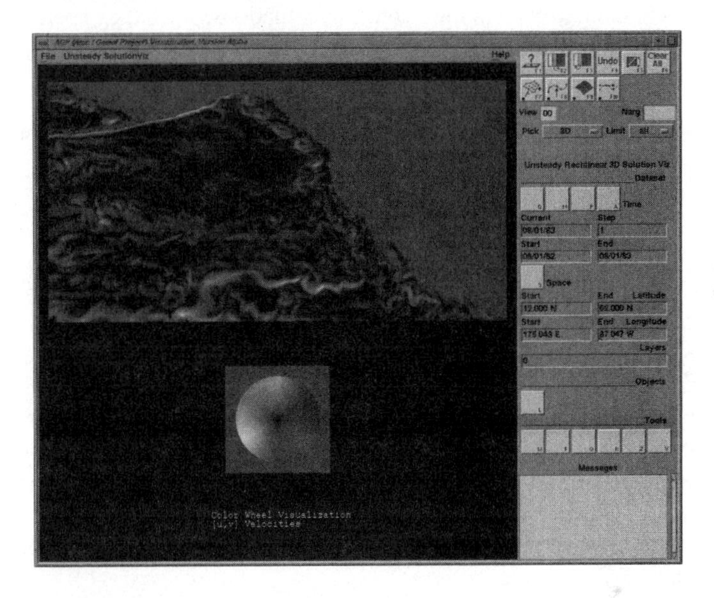

Figure 3.23: Flow in NE Pacific on August 1, 1983; direction of flow is mapped to a hue and magnitude to saturation and value; "rotated" color wheels close to coastline representing eddies. See Color Plate 19.

3.4.5 Feature Extraction from Oceanographic Data Sets

The ability to automatically recognize and track underlying features such as eddies and fronts from oceanographic data sets is an excellent tool, especially if important features can be detected automatically during the visualization process. The existence and therefore the locations of features are sometimes unknown even to a knowledgeable user at the beginning of the analysis phase. Without an automatic mechanism, intensive and time-consuming searches must be performed, otherwise features may not be revealed. In the analysis of the evolution and interaction of features, their temporal motions and fluctuations and their cross-correlations over multiparameter fields are best observed through animation. This again complicates the searching (via data classification) problem. In order to best characterize the features, locally optimized data classification must be achieved at each time step, at each depth level, and for each parameter. This is a difficult, if not impossible, task with traditional data classification methods.

A requirement of the feature extraction algorithm is the ability to operate effectively in noisy data sets. Although there is no commonly accepted measurement, the fundamental criterion is that the "signal-to-noise ratio" should be maximized at the output of the feature extractor (see [14]). The feature extraction methods for volumetric data sets can be considered an extension of those for computer vision and image processing. They can be classified into region-based and edge-based approaches. Region-based methods deal only with functional values. Examples can be found in [47, 48, 46]. The entire process usually contains extrema detection, thresholding, and feature isolation algorithms. Region-based methods are reliable even in low "contrast" situations, since they operate only on the original functional values. They are good if the possible features have unique and known functional values, especially when the unique values are extrema.

Edge-based methods attempt to detect features directly by tracing the boundary of possible features. Edge-based methods assume that features have some values that are significantly different from those outside the feature; therefore features must exhibit obvious spatial edges along the boundaries. Edges can be detected by computing the gradients. Edge-based methods are the most intuitive way to track features that can be best profiled by their edges. However, little work had been done in this area. The probable reasons include: 1) unlike images, data volumes are not necessarily regularly spaced, or even rectilinear, making the edge computation more complicated; and 2) 3D feature recognition algorithms are much more complicated. In [35], the 3D image edge operator was first extended and made applicable to nonregular 3D data volumes. That 3D image edge operator handles rectilinear data sets with irregular grid intervals in one direction, as are often used in environmental simulations. The edge operator was developed for locating, contouring, and tracking oceanic features, such as eddies (see [65]). It was later applied to cyclone and jet stream detection in meteorological data. The algorithms track features over space and time (see [64]).

Figure 3.24 shows the difference in simply detecting the edges and applying the filtering techniques based on parameters such as eccentricity and shape presented in [35]. Figure 3.25 shows eddies propagating through the Gulf of Mexico.

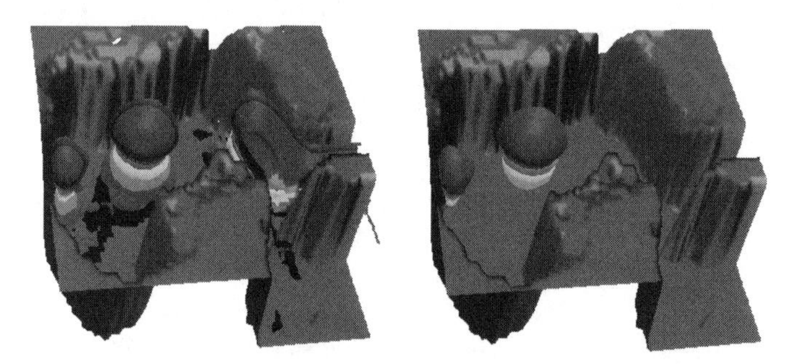

Figure 3.24: Difference in straightforward isosurfacing and enhanced edge-detection scheme. See Color Plate 20.

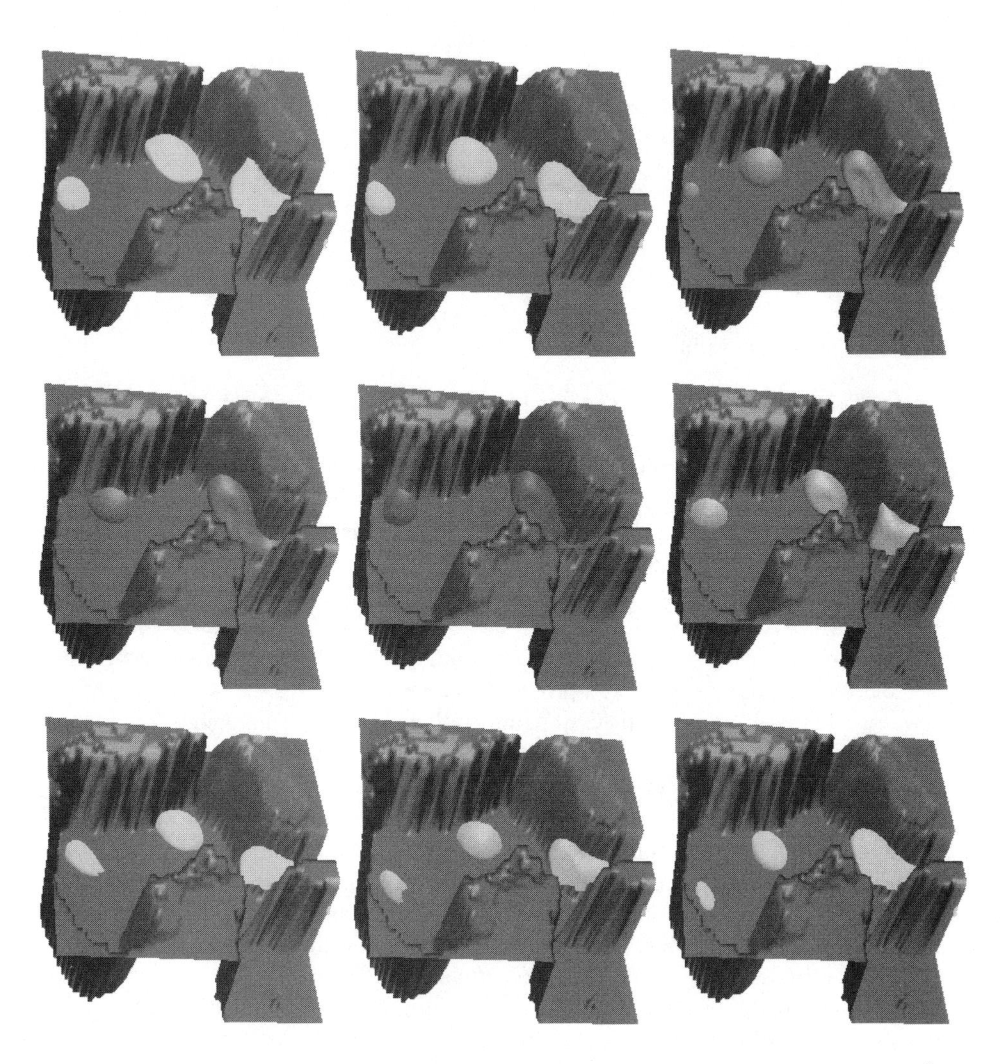

Figure 3.25: Example of eddies propagating through the Gulf of Mexico; sequence progresses from left to right and from top to bottom. See Color Plate 21.

3.4.6 Data Compression for Oceanographic Data Sets

As mentioned above, many real data sets, whether created by measurement or simulation, are "too big to visualize interactively." Thus an important issue in scientific visualization systems is to reduce the storage space and the access time of these data sets in order to speed up the visualization process. Progressive transmission offers a solution. The principle is to transmit the least amount of data necessary to generate a usable approximation of the original data. If the user requires further refinement, more data is fetched into the memory and a higher-resolution data set is reconstructed.

A progressive refinement scheme consists of four steps:

1. data encoding
2. data decoding
3. a refinement strategy
4. the rendering algorithm

State-of-the-art data encoding/decoding schemes consist of a wavelet transform (WT) technique and entropy coding. Because of its multiresolutional nature, the WT is very suitable for hierarchical manipulations such as progressive data transmission and tree-structured graphics rendering. Sophisticated methods for entropy coding and refinement control have been previously developed (see [55, 4, 31]). The need was for a better WT, which was more computationally efficient and produced more precise reconstruction of functional values in continuous space.

Previous work (see [36]) applied the truncated version of Battle-Lemarié (BL) wavelets to volume data and proposed a fast superposition algorithm for reconstruction of functional values in continuous space as a progressive transmission scheme. However, the BL wavelets are not perfect. First, they are computationally slow. This greatly limits their value in applications with strict speed requirements. Second, in most visualization techniques such as isosurface rendering and fluid topology extraction, the functional values in the continuous space must be calculated conveniently. By using an infinitely supported function—as is necessary with the BL wavelets—for the superposition basis, a complex approximation algorithm has to be utilized. A third drawback is that lossless data compression is difficult because the denominators of the coefficients are not in the form of 2^n.

A new progressive transmission scheme using spline biorthogonal wavelet bases was developed (see [51]). Because all the filter coefficients in the biorthogonal spline WT are dyadic rationals, multiplication and division operations can be simplified to addition and subtraction of the exponent for floating-point numbers, or to shifting of the exponent for integer numbers. This makes the transform very fast. This property also makes lossless coding of floating-point numbers possible. Symmetry is more easily achieved, resulting in a better handling of the data boundary and halving the number of multiplications required.

To combine a progressive transmission scheme with a visualization algorithm such as isosurface generation or ray casting, the functional values in the continuous volume must be approximated efficiently from the transform coefficients. Because the transform basis functions are compactly supported and can be explicitly formulated in a polynomial form, the reconstruction of functional values from the WT coefficients is straightforward.

With the given technique, the exact functional value at any real x can be computed through linear interpolation of the function values at the two discrete neighbors which surround x. In this case, the resulting $\overline{f}(x)$ is C^0 continuous. By increasing the filter length to k, $\overline{f}(x)$ will be C^{k-2} continuous. If the tensor product of the one-dimensional bases is used as the basis for the multidimensional WT, the same conclusion can be derived.

The scheme has been applied to data from an ocean model (see [23]). Model output from a 337×468 horizontal section of the Northeast (NE) Pacific was used. Layer thickness and current data in floating-point format—"prognostic variables" from the model—were visualized. Each of the six time-varying layers were coded independently.

To apply the biorthogonal WT to the data set, the boundary conditions had to be considered. In each frame (time step) of ocean model data, valid functional values are available only within the ocean area of the rectangular region (Figure 3.26a). This area is constant over time. To apply a 2D WT the ocean area must be approximated using square blocks. The block size used was $2^n \times 2^n$, where n was larger than the level of the WT in each data frame. The approximated ocean area had to cover the entire original ocean area (Figure 3.26b). For grid points in the approximated ocean area but not in the original area, interpolation and extrapolation techniques are used to yield function values. This introduces some discontinuity at the boundary. In the implementation, second-order Lagrange interpolation is used. Since all the WT bases are symmetric, the function values outside the boundary are achieved simply by reflection.

As shown in Figure 3.27, after applying the transform twice in each spatial dimension in a frame and once in the temporal dimension, only 1/32 of the transform coefficients are transmitted to the decoder. The reconstructed data has good quality and meets the requirements of further scientific visualization processes. By using the refinement control strategy proposed by Blandford (see [4]), only a small amount of data is needed for a better reconstruction.

In Figure 3.28, layer thickness compressed at the rate of 50:1 is shown. For a 2D vector field, the two components are encoded independently. Streamlines are generated both for the original field and the field reconstructed using only 1/32 of the WT coefficients; the results are shown in Figure 3.29. The additional compression (from 32:1 to 50:1) is obtained by using entropy encoding.

Applying a topology extraction algorithm (see [20]) to this vector field, the global and the stable topology structures remained relatively unchanged, even at a 16:1 compression ratio. With progressive refinement, more and more local and unstable features can be extracted (Figure 3.30). This is the low-pass effect of the WT.

Lossless compression was obtained by recording the necessary extra bits in an associated octree. The lossless compression ratio is about 1.5:1. The decoding speed, an important factor in interactive visualization system, is about 10 frames/sec on an SGI $Indigo^2$ machine. This satisfies the need of fast volume rendering in the application.

In summary, the WT scheme discussed above has several features that make it attractive. The transform itself is fast, and high-compression ratios can be achieved. The reconstructed data is of good quality and can be refined with a small amount of additional data. Boundaries can be handled gracefully. The reconstruction of functional values in continuous space from the WT coefficients requires only a simple polynomial, which is especially useful for scientific visualization applications.

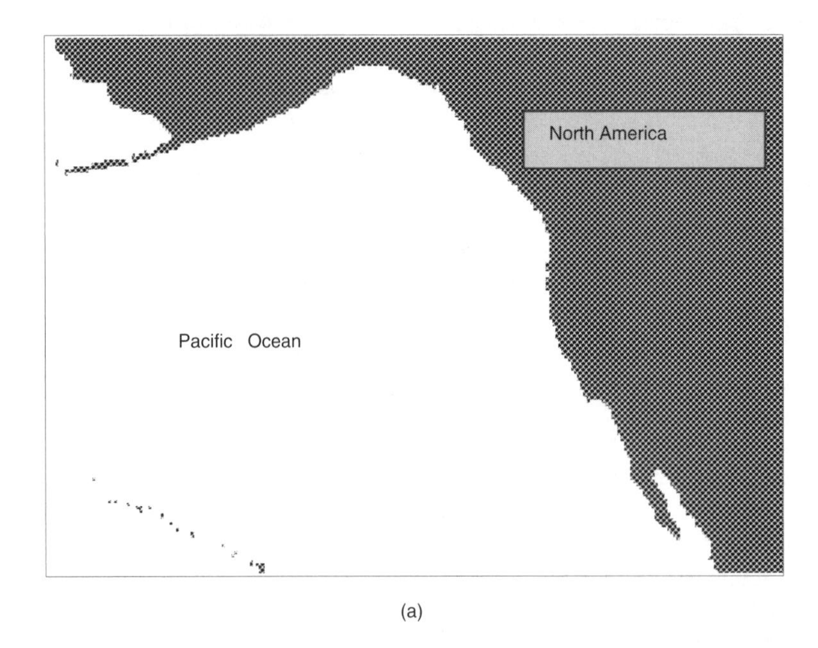

(a)

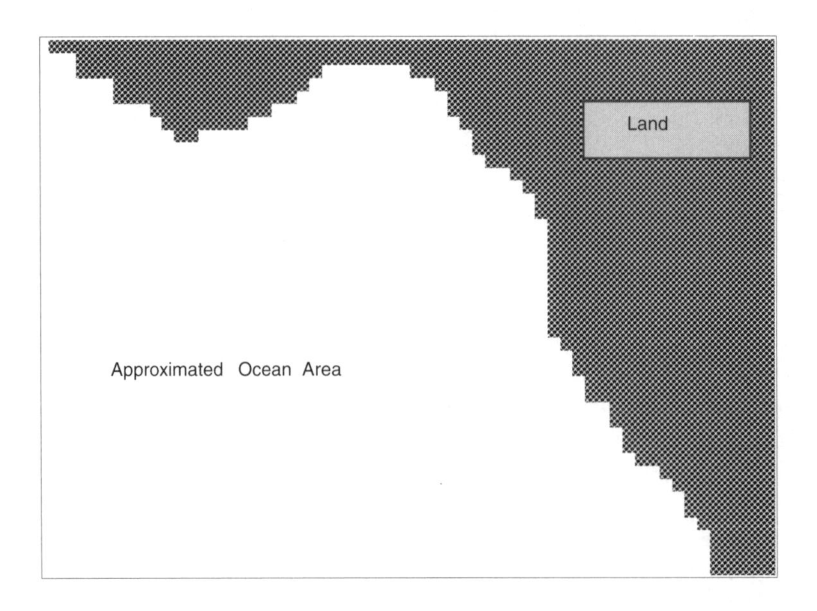

(b)

Figure 3.26: Land and ocean areas in each data frame. *(a)* Ocean area in original data (white area); *(b)* Approximation of ocean area using 8×8 blocks. Approximated version covers original ocean area entirely.

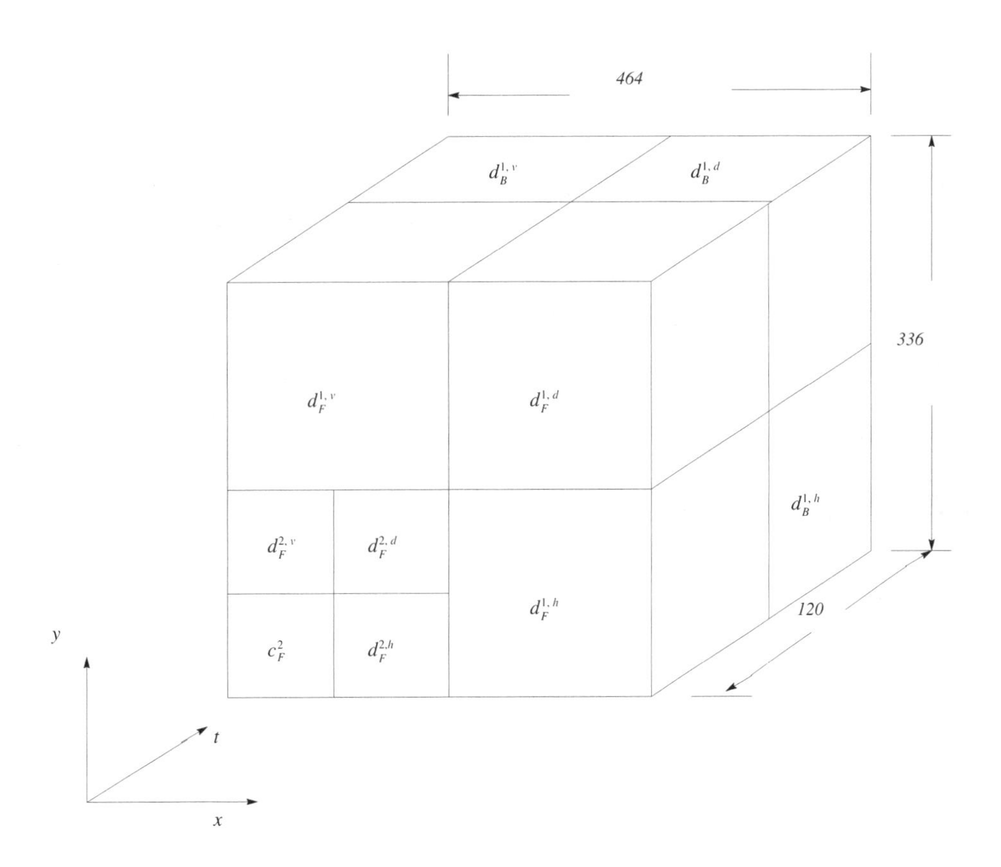

Figure 3.27: Wavelet transform coefficients of 3D ocean model data.

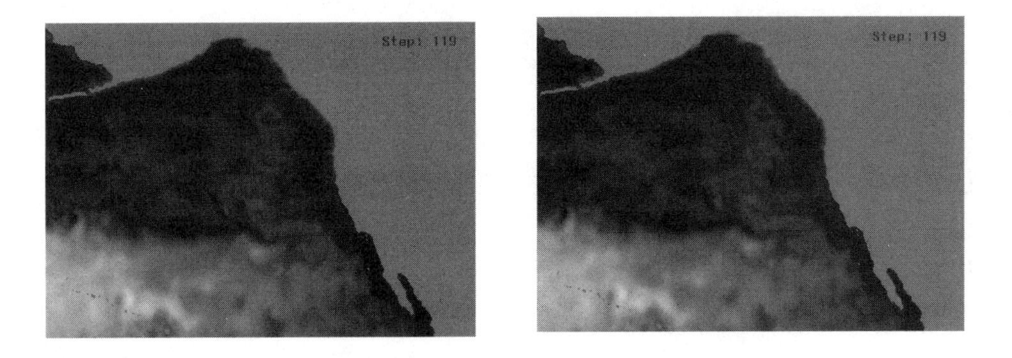

Figure 3.28: Layer thickness for NE Pacific Ocean. *(left)* original data; *(right)* reconstructed data (compression ratio 50:1). See Color Plate 22.

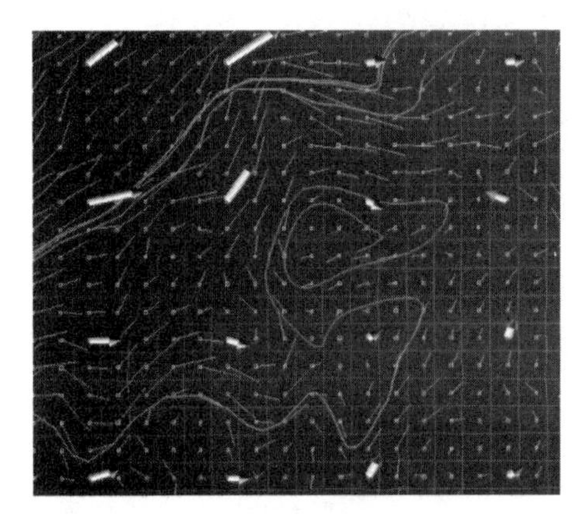

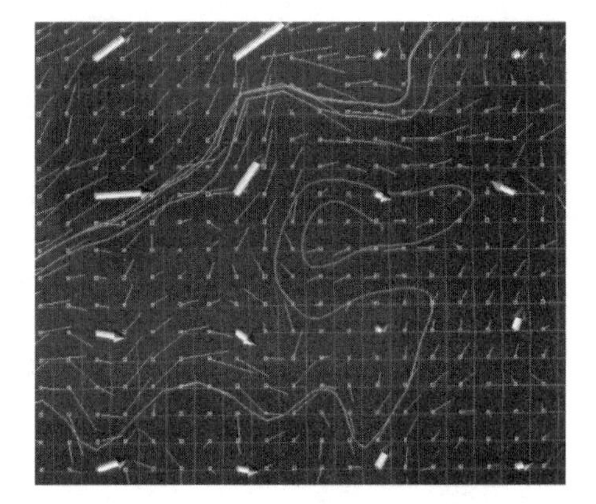

Figure 3.29: Velocity data for NE Pacific Ocean. *(left)* original data; *(right)* reconstructed data (compression ratio 50:1). See Color Plate 23.

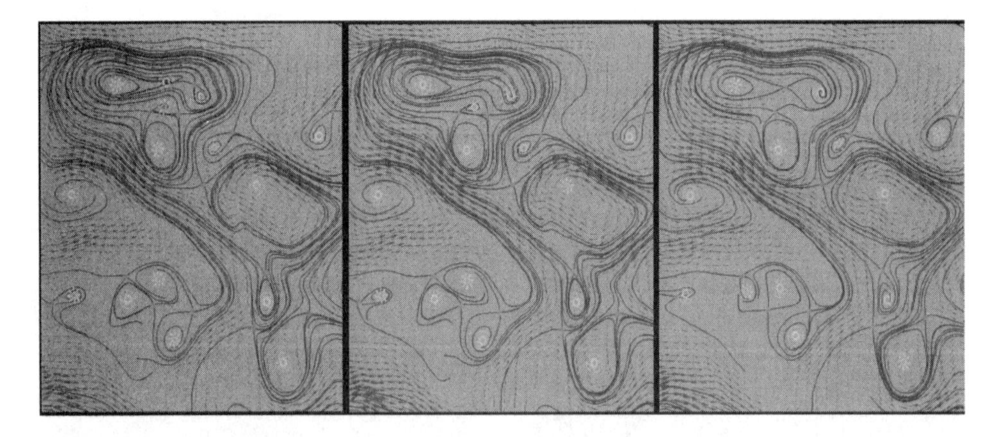

Figure 3.30: Vector field topology extraction on reconstructed data. *(left)* original data; *(middle)* reconstructed data using 1/4 of the WT coefficients; *(right)* reconstructed data using 1/16 of the WT coefficients. See Color Plate 24.

3.4.7 Concurrent Visualization

Computational field simulation visualization has typically been a postprocess. However, concurrent visualization allows the user to more effectively monitor the evolving simulation. The problem has typically been that the creator of the computational field simulation program does not know graphics programming or have access to adequate and/or easy-to-use visualization libraries. A secondary problem is that there is always a need for more compute power or more memory on the machine on which the simulation is being run. In an attempt to develop a concurrent visualization scheme, FAST (see [1] and [56]) was modified to take as input a script from a remote process. FAST was chosen because it was designed to deal with aerodynamic data on 3D curvilinear grids, the source program was available, it has a powerful and extensive scripting facility, and it had been used in previous visualization development work at the ERC-CFS.

Concurrent visualization can be described as either integrated or distributed (see [27]). Integrated systems combine the CFS solution program(s) and visualization modules into one program. This allows a shared memory space and simplified synchronization of the visualization and solution routines. Distributed systems separate the solution program(s) and visualization modules into different programs which communicate with each other. Since there is no longer shared memory, a large amount of communication and data transfer must take place; however, a distributed system offers the opportunity to exploit powerful computational engines connected via high-speed networks with high-performance graphics workstations.

The need to visualize time-varying data was the motivation for this work. Since FAST Version 1.1a did not support time-dependent solutions directly, a concurrent visualization system was developed which visualized the data as it was generated. The system is a master-slave design where the simulation software controls FAST via network communication. The network communication is based on the DTM library (see [37]). DTM handles the message processing required when communicating between dissimilar architectures. The simulation software is part of a family of programs known as UNCLE (see [52]). The original program, written in FORTRAN, was augmented with several C routines allowing the software to indirectly control the full functionality of FAST. Thus, instead of simply generating large simulation solution files which are cumbersome to visualize, the software has the ability to generate potentially smaller visualization files and the user has the ability to watch simulations develop.

Control of FAST was obtained via its scripting facility and named pipes (see [49]). FAST reads from the pipe and waits for data until an EOF signal is read. In order to keep the pipe open, another program, FASTd (FAST daemon), is started which receives messages from the network and translates them into script commands which are written to the named pipe. When a special message is received, signaling that the simulation is complete, FASTd closes the pipe and exits. The visualization entities created by FAST can be ARCGraph files (see [21]), which, as visualization objects (see [13]) and not images, allow the viewing angle and color map to be modified in postproduction.

This approach has shown great promise for utilizing distributed computing techniques to interactively visualize simulation results. The greatest hindrance to concurrent visualization is the lack of standard file and data formats for time-varying scientific data.

3.5 Conclusions

Various recent and ongoing research projects conducted in grid generation and scientific visualization at the ERC-CFS have been discussed. The most common grid generation techniques have been reviewed due to their importance for the development of visualization techniques for different grid types. The survey attempts to create more interest in grid generation aspects among visualization experts.

The basic principles of structured and unstructured grid generation have been pointed out. The NGP system is an example of a state-of-the-art system incorporating a wide variety of geometric modeling and grid generation methods to solve complex, real-world CFS problems.

The survey also provides insight into many scientific visualization projects, the most prominent one being the ATOC-GAMOT project dealing with the visualization of ocean currents and ocean sound propagation.

Acknowledgments

This work was supported by the National Science Foundation (NSF) under contract EEC-8907070 (ERC-CFS) and ASC-9210439 (Research Initiation Award), the Strategic Environmental Research and Development Program (SERDP) and the Advanced Research Projects Agency (ARPA) via contract EEC-8907070 (ERC-CFS), the National Grid Project (NGP) consortium, the Office of Naval Research (ONR) under research grant N00014-92-J-4109, the Naval Oceanographic Office (NAVO) via contract NAS13-330, and the U.S. Army Corps of Engineers under contract DACA 39-93-K-0055.

Special thanks go to the members of the Grid Generation and Scientific Visualization thrusts and the research and development team of the NGP system at the ERC-CFS.

Bibliography

[1] G.V. Bancroft, F.J. Merritt, T.C. Plessel, P.G. Kelaita, R.K. McCabe, and A. Globus, "FAST: A Multiprocessing Environment for Visualization of CFD," *Visualization '90*, A. Kaufman, editor, IEEE Computer Society Press, Los Alamitos, Calif., 1990, pp. 14–27.

[2] R.E. Barnhill, G. Farin, and B. Hamann, "NURBS and Grid Generation," *Modeling, Mesh Generation, and Adaptive Numerical Methods for Partial Differential Equations*, IMA Volumes in Mathematics and its Applications 75, I. Babuska, J.E. Flaherty, W.D. Henshaw, J.E. Hopcroft, J.E. Oliger, and T. Tezduyar, editors, Springer-Verlag, New York, N.Y., 1995, pages 1–22.

[3] J.A. Benek, P.G. Buning, and J.L. Steer, *A 3-D Chimera Grid Embedding Technique*, AIAA paper 85-1523, AIAA, 1985.

[4] R.P. Blandford, "Wavelet Encoding and Variable Resolution Progression Transmission," *NASA Space and Earth Science Data Compression Workshop Proceedings*, 1993, pp. 25–34.

[5] J.E. Castillo, *Mathematical Aspects of Numerical Grid Generation*, SIAM, Philadelphia, Pa., 1991.

[6] W.M. Chan and J.L. Steger, "Enhancements of a Three-Dimensional Hyperbolic Grid Generation Scheme," *Applied Mathematics and Computation*, Vol. 51, 1992, pp. 181–205.

[7] J.F. Dannenhoffer, *A New Method for Creating Grid Abstractions for Complex Configurations*, AIAA paper 93-0428, AIAA, 1993.

[8] F.C. Dougherty, J.A. Benek, and J.L. Steger, *On the Application of a Chimera Grid Scheme to Store Separation*, NASA paper TM-88193, NASA, 1985.

[9] T. Elvins, "A Survey of Algorithms for Volume Visualization. *Computer Graphics*, Vol. 26, No. 3, 1992, pp. 194–201.

[10] G. Farin, *Curves and Surfaces for Computer Aided Geometric Design*, third edition, Academic Press, San Diego, Calif., 1993.

[11] R. Franke, "Scattered Data Interpolation: Tests of Some Methods," *Math. Comp.*, Vol. 38, 1982, pp. 181–200.

[12] P.L. George, *Automatic Mesh Generation*, Wiley & Sons, New York, N.Y., 1991.

[13] A. Globus, *A Software Model for Visualization of Time Dependent 3-D Computational Fluid Dynamic Results*, Technical Report RNR-92-031, NASA Ames Research Center, 1992.

[14] R.C. Gonzalez and R.E. Woods, *Digital Image Processing*, Addison-Wesley, New York, N.Y., 1992.

[15] W.J. Gordon, "Blending-Function Methods of Bivariate and Multivariate Interpolation and Approximation," *SIAM Journal of Numerical Analysis*, Vol. 8, No. 1, 1971, pp. 158–177.

[16] B. Hamann, "Modeling Contours of Trivariate Data," *Modélisation Mathématique et Analysis Numérique [Mathematical Modelling and Numerical Analysis]*, Vol. 26, No. 1, 1992, pp. 51–75.

[17] B. Hamann, "Construction of B-Spline Approximations for Use in Numerical Grid Generation," *Applied Mathematics and Computation*, Vol. 65, Nos. 1 & 2, special issue, 1994, pp. 295–314.

[18] B. Hamann, J.L. Chen, and G. Hong, "Automatic Generation of Unstructured Grids for Volumes Outside or Inside Closed Surfaces," *Numerical Grid Generation in Computational Fluid Dynamics and Related Fields*, N.P. Weatherill, P.R. Eiseman, J. Häuser, and J.F. Thompson, editors, Pineridge Press Ltd., Swansea, UK, 1994, pp. 187–197.

[19] B. Hamann and R.F. Sarraga, "Finite Elements, Grid Generation, and Geometric Design," *Computer Aided Geometric Design*, Vol. 12, No. 7, special issue, Elsevier Science Publishing Co. Inc., New York, N.Y., 1995, pp. 647–784.

[20] J.L. Helman and L. Hesselink, "Representation and Display of Vector Field Topology in Fluid Flow Data Sets," *Computer*, Vol. 22, No. 8, 1989, pp. 27–36.

[21] E.A. Hibbard and G. Makatura, *ARCGraph System User's Manual, Version 7.0*, NASA Ames Research Center, 1988.

[22] J. Hoschek and D. Lasser, *Fundamentals of Computer Aided Geometric Design*, A K Peters, Ltd., Wellesley, Mass., 1993.

[23] H.E. Hurlburt, A.J. Wallcraft, Z. Sirkes, and E.J. Metzger, "Modeling of the Global and Pacific Oceans: On the Path to Eddy-Resolving Ocean Prediction," *Oceanography*, Vol. 5, No. 1, 1992, pp. 9–18.

[24] B.A. Jean and B. Hamann, "Interactive Techniques for Correcting CAD/CAM Data," *Numerical Grid Generation in Computational Fluid Dynamics and Related Fields*, N.P. Weatherill, P.R. Eiseman, J. Häuser, and J.F. Thompson, editors, Pineridge Press Ltd., Swansea, UK, 1994, pp. 317–328.

[25] A. Khamayseh and B. Hamann, "Elliptic Grid Generation Using NURBS Surfaces," *Computer Aided Geometric Design*, Vol. 13, No. 4, 1996. pp. 369–386.

[26] P.M. Knupp and S. Steinberg, *Fundamentals of Grid Generation*, CRC Press, Inc., Boca Raton, Fla., 1993.

[27] J.C. Lakey and R.J. Moorhead, "Concurrent Visualization of Time-Varying CFD Simulations," *Visual Data Exploration and Analysis*, Vol. 2178, R.J. Moorhead, D.E. Silver, and S.P. Uselton, editors, SPIE—The International Society for Optical Engineering, Bellingham, Wash., 1994, pp. 123–126.

[28] M. Levoy, "Display of Surfaces from Volume Data," *IEEE Computer Graphics and Applications*, Vol. 8, No. 3, 1988, pp. 29–37.

[29] R. Löhner and P. Parikh, "Three-Dimensional Grid Generation by the Advancing Front Method," *International Journal for Numerical Methods in Fluids*, Vol. 8, 1988, pp. 1135–1149.

[30] W.E. Lorensen and H.E. Cline, "Marching Cubes: A High Resolution 3D Surface Construction Algorithm," *Computer Graphics*, Vol. 21, No. 4, 1987, pp. 163–169.

[31] S.G. Mallat, "A Theory for Multiresolution Signal Decomposition: The Wavelet Representation," *IEEE Transactions on Pattern Analysis and Machine Intelligence*, Vol. 11, No. 7, 1989, pp. 674–693.

[32] C.W. Mastin and J.F. Thompson, "Quasiconformal Mappings and Grid Generation," *SIAM Journal on Scientific and Statistical Computing*, Vol. 5, 1984, pp. 305–310.

[33] D. Meagher, "Geometric Modelling Using Octree Encoding," *Comp. Graph. Image Proc.*, Vol. 19, 1982, pp. 129–147.

[34] R.J. Moorhead, B. Hamann, C. Everitt, S.C. Jones, J. McAllister, and J.H. Barlow, "Oceanographic Visualization Interactive Research Tool (OVIRT)," *Visual Data Exploration and Analysis*, Vol. 2178, R.J. Moorhead, D.E. Silver, and S.P. Uselton, editors, SPIE—The International Society for Optical Engineering, Bellingham, Wash., 1994, pp. 24–30.

[35] R.J. Moorhead and Z. Zhu, "Feature Extraction for Oceanographic Data Using a 3D Edge Operator," *Visualization '93*, G.M. Nielson and D. Bergeron, editors, IEEE Computer Society Press, Los Alamitos, Calif., 1993, pp. 402–405.

[36] S. Muraki, "Volume Data and Wavelet Transforms," *IEEE Computer Graphics and Applications*, Vol. 13, No. 4, 1993, pp. 50–56.

[37] *Data Transfer Mechanism Programming Manual, Version 2.3*, National Center for Supercomputing Applications, University of Illinois at Urbana-Champaign, 1992.

[38] G.M. Nielson, T.A. Foley, B. Hamann, and D.A. Lane, "Visualizing and Modeling Scattered Multivariate Data," *IEEE Computer Graphics and Applications*, Vol. 11, No. 3, 1991, pp. 47–55.

[39] G.M. Nielson and B. Hamann, "Techniques for the Interactive Visualization of Volumetric Data," *Visualization '90*, A. Kaufman, editor, IEEE Computer Society Press, Los Alamitos, Calif., 1990, pp. 45–50.

[40] G.M. Nielson and B. Hamann, "The Asymptotic Decider: Resolving the Ambiguity in Marching Cubes," *Visualization '91*, G.M. Nielson and L.J. Rosenblum, editors, IEEE Computer Society Press, Los Alamitos, Calif., 1991, pp. 83–91.

[41] K.L. Parmley, J.F. Dannenhoffer, and N.P. Weatherill, *Techniques for the Visual Evaluation of Computational Grids*, AIAA paper 93-3353, AIAA, 1993.

[42] F.P. Preparata and M.I. Shamos, *Computational Geometry*, third printing, Springer-Verlag, New York, N.Y., 1990.

[43] M.G. Remotigue, "The National Grid Project: Making Dreams into Reality," *Numerical Grid Generation in Computational Fluid Dynamics and Related Fields*, N.P. Weatherill, P.R. Eiseman, J. Häuser, and J.F. Thompson, editors, Pineridge Press Ltd., Swansea, UK, 1994, pp. 429–439.

[44] P. Sabella, "A Rendering Algorithm for Visualizing 3D Scalar Fields," *Computer Graphics*, Vol. 22, No. 4, 1988, pp. 51–58.

[45] H. Samet, "The Quadtree and Related Hierarchical Data Structures," *Computing Surveys*, Vol. 16, No. 2, 1984, pp. 187–285.

[46] R. Samtaney, D.E. Silver, N.J. Zabusky, and J. Cao, "Visualizing Features and Tracking their Evolution," *Computer*, Vol. 27, No. 7, 1994, pp. 20–27.

[47] D.E. Silver and N.J. Zabusky, "3D Visualization and Quantification of Evolving Amorphous Objects," *Extracting Meaning from Complex Data: Processing, Display, Interaction II*, Vol. 1459, E.J. Farrell, editor, SPIE—The International Society for Optical Engineering, Bellingham, Wash., 1991, pp. 97–108.

[48] D.E. Silver and N.J. Zabusky, "Quantifying Visualizations for Reduced Modeling in Nonlinear Science: Extracting Structures from Data Sets," *Journal of Visual Communications and Image Representation*, Vol. 4, No. 1, 1993, pp. 46–61.

[49] M. Sobell, *A Practical Guide to the Unix System*, Benjamin-Cummings, 1989.

[50] B.K. Soni and B. Hamann, "Computational Geometry Tools in Grid Generation," *Advances in Hydro-Science & Engineering*, Volume I, Part B, S.S.Y. Wang, editor, 1993, pp. 2004–2009.

[51] H. Tao and R.J. Moorhead, "Progressive Transmission of Scientific Data Using a Biorthogonal Wavelet Transform," *Visualization '94*, R.D. Bergeron and A. Kaufman, editors, IEEE Computer Society Press, Los Alamitos, Calif., 1994.

[52] L.K. Taylor and D.L. Whitfield, *Unsteady Three-Dimensional Incompressible Euler and Navier-Stokes Solver for Stationary and Dynamic Grids*, AIAA paper 91-1650, AIAA, 1991.

[53] J.F. Thompson, Z.U.A. Warsi, and C.W. Mastin, *Numerical Grid Generation*, North-Holland, New York, N.Y., 1985.

[54] J.F. Thompson and N.P. Weatherill, *Aspects of Numerical Grid Generation: Current Science and Art*, AIAA paper 93-3539-CP, AIAA, 1993.

[55] K.H. Tzou, "Progressive Image Transmission: A Review and Comparison of Techniques," *Optical Engineering*, Vol. 26, No. 7, 1987, pp. 581–589.

[56] P. Walatka, J. Clucas, K. McCabe, T. Plessel, and R. Potter, *FAST User's Guide, FAST 1.1a*, NASA Ames Research Center, NAS Division, RND Branch, 1994.

[57] N.P. Weatherill, *The Generation of Unstructured Grids Using Dirichlet Tessellations*, MAE Report 1715, Princeton University, 1985.

[58] N.P. Weatherill, "A Method for Generating Irregular Computational Grids in Multiply Connected Planar Domains," *International Journal for Numerical Methods in Fluids*, Vol. 8, 1988, pp. 181–197.

[59] N.P. Weatherill, "The Delaunay Triangulation in CFD," *Computers and Mathematics with Applications*, Vol. 24, 1992, pp. 129–150.

[60] N.P. Weatherill, O. Hassan, D.L. Marcum, and M.J. Marchant, "Grid Generation by the Delaunay Triangulation," *von Karman Institute for Fluid Dynamics 1993-1994 Lecture Series*, 1994.

[61] M.A. Yerri and M.S. Shephard, "A Modified Quadtree Approach to Finite Element Mesh Generation," *IEEE Computer Graphics and Applications*, Vol. 3, 1983.

[62] M.A. Yerri and M.S. Shephard, "Automatic 3D Mesh Generation by the Modified-Octree Technique," *Int. J. Num. Meth. Eng.*, Vol. 20, 1984, pp. 1965–1990.

[63] M.A. Yerri and M.S. Shephard, "Automatic Mesh Generation for Three-Dimensional Solids," *Comp. Struct.*, Vol. 20, 1985.

[64] Z. Zhu and R.J. Moorhead, "Exploring Feature Detection Techniques for Time-Varying Volumetric Data," *IEEE Workshop on Visualization and Machine Vision*, Seattle, Wash., 1994, pp. 45–54.

[65] Z. Zhu, R.J. Moorhead, H. Anand, and L.R. Raju, "Feature Extraction and Tracking in Oceanographic Visualization," *Visual Data Exploration and Analysis*, Vol. 2178, R.J. Moorhead, D.E. Silver, and S.P. Uselton, editors, SPIE—The International Society for Optical Engineering, Bellingham, Wash., 1994, pp. 31–39.

Chapter 4

An Environment for Computational Steering

Jarke J. van Wijk and Robert van Liere

Abstract. *Computational steering is the ultimate goal of interactive simulation: researchers change parameters of their simulation and immediately receive feedback on the effect. We present a general and flexible environment for computational steering. Within this environment a researcher can easily develop user interfaces and 2D visualizations of the simulation. Direct manipulation is supported, and the required changes of the simulation are minimal. The architecture of the environment is based on a Data Manager that takes care of centralized data storage and event notification, and satellites that produce and visualize data. One of these satellites is a graphics tool to define a user interface interactively and to visualize the data. The central concept here is the Parameterized Graphics Object: an interface is built up from graphics objects whose properties are functions of data in the Data Manager. The scope of these tools is not limited to computational steering, but extends to many other application domains.*

Keywords: visualization, computational steering, interaction, direct manipulation.

4.1 Introduction

4.1.1 Computational Steering

Scientific visualization has become a major research area since 1987, when the influential report of the National Science Foundation (NSF) [5] was published. In recent years many new methods, techniques, and packages have been developed. Most of these developments are limited to postprocessing of data sets. Usually the assumption is made that all data is generated first and that next the researcher iterates through the remaining steps of the visualization pipeline (selection, filtering, mapping, and rendering) to achieve insight in the generated data. Hence, the interaction with the simulation is limited.

Marshall et al. [4] distinguish two alternatives to this postprocessing approach to visualization. The first step to more interaction with the simulation is with *tracking*. After each time step of the simulation, the resulting data for that time step is sent into the visualization pipeline and can be inspected. If the researcher considers the results invalid, then the simulation can be stopped at an early stage, and restarted with a different set of input parameters. The next step, *computational steering*, goes a lot further, and can be considered as the ultimate goal of interactive computing. Computational steering enables the researcher to change parameters of the simulation while the simulation is progressing. According to Marshall et al. [4]: "Interaction with the computational model and the resulting graphics display is fundamental in scientific visualization. Steering enhances productivity by greatly reducing the time between changes to model parameters and the viewing of the results."

Our aim is to provide researchers with a Computational Steering Environment (CSE) that encourages exploratory investigation of a simulation by the researcher. In the following subsections we sum up the requirements, discuss existing solutions, and give an overview of the remainder of this chapter, in which we present our solution.

4.1.2 Requirements

The data flow between the researcher and the simulation via a CSE is shown in Figure 4.1. The researcher can enter new values for parameters and view visualizations of the resulting data. Hence, input widgets such as text fields, sliders, and buttons must be provided, as well as a variety of visualization methods, such as graphs, text, graphics objects, and so on. With such objects, input and output are separated. For more direct control, objects must be provided that allow for two-way communication: both input and output. It must be possible to select and drag visualization objects, thereby directly controlling parameters of the simulation. In other words, direct manipulation must be provided.

The simulation receives new parameter values from the CSE and sends newly calculated results to the CSE. We assume that the simulation can handle changes of parameters on-the-fly, and that it can provide meaningful intermediate results within a time interval that is acceptable to the researcher. The concept of direct manipulation has a counterpart in the interface between CSE and simulation. Some variables, typically state variables, are continuously updated by the simulation, but can also be changed from outside of the simulation.

As an example of an interface with a high degree of interaction, consider Figure 4.2. This shows an interactive graphics interface to a simulation of a set of bouncing balls in two dimensions. The balls are depicted as circles, each with a small red line that indicates the direction and magnitude of its velocity. For every time step the simulation calculates the position and velocity of all balls. Various control parameters such as the size of the balls, damping, attraction, and a constant field force, can be set. This can be done via sliders, by typing, or by dragging the arrows. The positions of the balls (state variables) are continuously updated by the simulation, but can also be changed by the researcher by dragging their graphical representations to other positions. Figure 4.2 is a screen dump of a result of the CSE presented in this chapter. In Section 4.4.1 we come back to this example and show how it was defined.

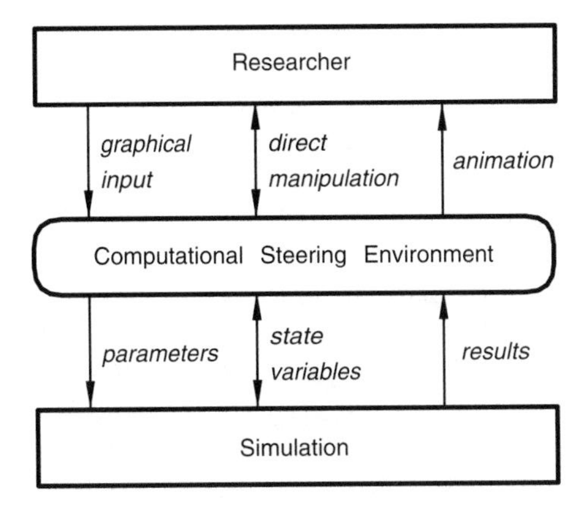

Figure 4.1: Data flow between researcher, CSE, and simulation.

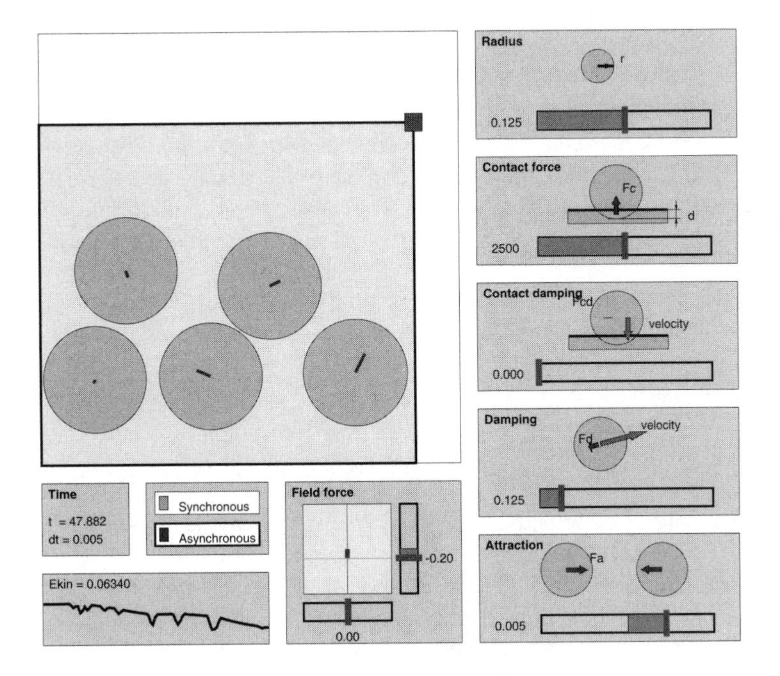

Figure 4.2: Simulation of bouncing balls. See Color Plate 25.

The researcher must be enabled to create and refine the interface to the simulation easily and incrementally. The process of achieving insight via simulation is an incremental process. For all stages of the visualization pipeline (from simulation to rendering), the cycle *specification, implementation, application* is continuously reiterated. In the process of gaining insight via computational steering, a researcher typically wants to look at and to control other, possibly new, variables, and to visualize them in different ways.

The CSE must be able to deal with multiple processes simultaneously. There are three reasons for this: First, it must be possible to integrate existing tools, such as special purpose packages for grid-editing or visualization, into the CSE. Second, a simulation is often built up of several processes running on distant computer servers. Third, it would be convenient if several researchers could view and control the data simultaneously.

The final requirement concerns the underlying data model and the amount of data within the CSE. The type of data to be handled depends very much on the type of simulation, and therefore can vary from simple scalar data to large, three-dimensional, time-dependent vector and tensor field data sets. The underlying data model must be flexible enough to support a wide range of data types and data quantities.

4.1.3 Related Work

An optimal result for the researcher can be achieved if we build a complete system, incorporating handling of input and output, as well as the simulation, using basic graphics libraries, such as IRIS GL or PHIGS. Typical examples are flight-simulators and packages for mechanical engineering. This approach requires a large effort and considerable experience, and often leads to an inflexible system with fixed functionality. Thus, the requirement for easy modification by the researcher is violated. However, if the application is time-critical and has wide-spread use, the gain can be worth the cost.

Given the set of requirements, how can we realize a more general solution, that is, an environment rather than a turnkey system? Basically, three approaches can be taken:

- Consider the design and implementation of a CSE as a graphics problem, and extend a basic graphics library
- Consider it as a user interface problem and extend the functionality of a User Interface Management System (UIMS)
- Consider it primarily as a visualization problem, and extend an application builder

All these approaches have been pursued. An exhaustive treatment would require far more space than is available here, because very many related approaches, techniques, and concepts exist. We limit ourselves, therefore, to some relevant examples.

Graphics libraries typically offer only low-level functionality. To simplify the definition of interactive applications with direct manipulation, graphics libraries can be enhanced with higher-level interactive objects. With this approach, a tight coupling of application objects and interaction objects can be ensured. For example, the Inventor toolkit [11] provides an object-oriented library which simplifies the development of interactive graphics applications. Van Dam et al. [14] take this a step further. They describe an extensible system that primarily aims at the integration of a variety of simulation and animation concepts in the graphics toolkit. Objects may have geometric, algorithmic, or interactive properties. Objects may send messages to each other which, after being stored in the object, can be edited. The authors show how collision detection [17], constraints, and deformation [10] are handled within this framework.

Various UIMSs have been described that provide support for coupling the application (represented by a set of application objects) and the user interface [6, 8, 9]. User interface and application designers can develop their parts independently and have the UIMS manage the dialogue layer to integrate the two parts. Some UIMSs allow users to specify direct manipulation interfaces through WYSIWYG editors. An example is the Peridot system [7], which allows nonprogrammers to create sophisticated interaction techniques. With the help of graphical constraints, Peridot allows a user to draw graphical objects of the user interface. The user provides the behavior of these graphical objects with a technique called *programming by example*. This results in Peridot generating parameterized procedures which can, in turn, be linked in or interpreted by the application program.

Application builders have emerged as a flexible solution for scientific visualization. The users are provided with a set of modules which can be connected and extended to rapidly prototype new applications and reconfigure existing ones. Some researchers have discussed the use of computational steering in this context [13, 12, 4]. With the current generations of such systems, the user can define user interface panels. The simulation has to be included as a module. The extension of such systems with direct manipulation has been discussed in [3]. In general, the implementation of direct manipulation is cumbersome, because in dataflow environments the relation between the original data and the geometric objects in the visualization is not known.

Finally, a novel approach to exploratory visualization has been taken in VIEW [1]. The key idea of VIEW is a very tight coupling of geometry with an underlying database. The VIEW system allows researchers to interact directly with the visualization. Researchers can select tools for the visualization of their data. This allows and encourages researchers to experiment and explore the underlying data spaces. Scripting languages are used for defining new tools.

All approaches discussed have value in their own right. However, we feel that none satisfies all CSE requirements simultaneously. The extensions of basic graphics libraries discussed are very convenient for graphics application programmers, but are not directly suitable for use by researchers. The visualization of data is outside the scope of current UIMSs. Application building environments do not provide direct manipulation, which we consider an important issue for computational steering. In the VIEW system, new visualization tools must be specified via a specialized scripting language. Here the graphics objects, and the relations between graphics objects, data, and user actions are specified.

4.1.4 Overview

In the following sections we present an environment for computational steering that does satisfy our requirements. The space available here does not allow for a detailed treatment; a more extensive description can be found in [15]. In Section 4.2 we give an overview of the architecture. The central component is a Data Manager, which is responsible for managing data storage and process communication. In Section 4.3 a general tool for graphical input and output is described. We have taken the approach of extending low-level graphics.

We use the notion of a Parameterized Graphics Object (PGO) as the main concept, and a graphics editor as a metaphor for the design of the interface. Examples of applications are given in Section 4.4, followed by a discussion (Section 4.5) and conclusions (Section 8.7).

4.2 Architecture

Flexibility implies that the functionality is spread over several separate processes, for simulation, input, output, as well as auxiliary operations on data. These processes must have access to the same set of data. We therefore use the architecture shown in Figure 4.3. The central process is the *Data Manager*, the other processes we call *satellites*. Satellites can connect to and communicate with the Data Manager.

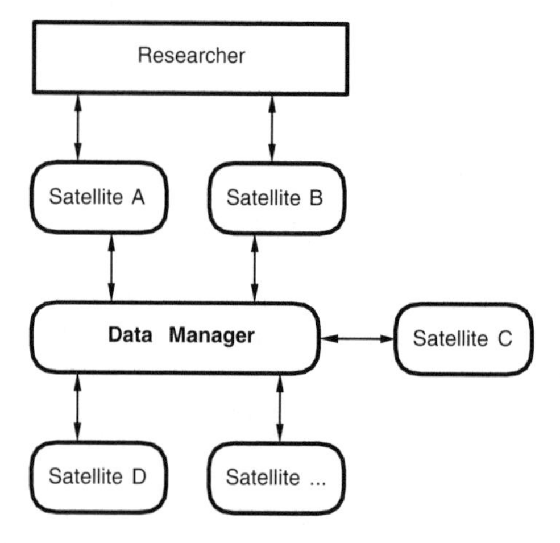

Figure 4.3: Data Manager and satellites.

The purpose of the Data Manager is twofold:

- Manage a database of variables. Satellites can create and do read/write operations on variables. For each variable its name, its type (*floating point, integer, string*), and its current value is stored and managed. Variables can be scalar variables or arrays, in which case the number of dimensions and size of the dimensions is also stored. These sizes can change dynamically.

- Act as an event notification manager. Satellites can subscribe to a set of predefined events. For example, if a satellite subscribes to mutation events on a particular variable, the Data Manager will send a notification to that satellite whenever the variable is updated.

The functionality of the Data Manager is purposely limited: it can be used by satellites for communication, but it does not control these satellites themselves. This is in contrast to dataflow-oriented application builders, in which the underlying execution models dictate when modules will be fired [2, 13, 16]. A number of small but useful general purpose satellites have been developed. With these satellites, variables in the Data Manager can be updated. Some examples are:

- *dmdump* and *dmrestore* dump and restore the values of the variables to and from files
- *dmslice* selects subsets of arrays
- *dmlog* maintains a log of the last N values of a variable
- *dmcalc* evaluates arithmetic expressions on variables
- *dmscheme* is a Scheme interpreter which is extended with the API to the Data Manager
- *ReadDM* and *WriteDM* are two IRIS Explorer modules which translate Data Manager variables into the Explorer data types *cxParameter* or *cxLattice* and back

Communication of a satellite with the Data Manager is done via a small Application Programmers Interface (API). The abstractions used by this low-level API are similar to standard UNIX I/O file handling, with variables instead of files. Satellites use handles to read, sample, and write to variables. A sample call returns immediately, a read call waits until the value of a variable changes. Events are used to indicate state changes in the Data Manager. Various routines are provided to query the status of variables and the Data Manager.

The functionality provided by this low-level API is compact, terse, and complete, but not simple to use. Therefore, on top of this API a Data I/O library was defined, which is tuned to the needs of researchers that want to integrate their simulations within the CSE. The two main calls are `dioConnect` to register a variable whose value must be read and/or written, and `dioUpdate`, which call takes care of updating the variables in the simulation. Entry points in existing code where these calls must be added are generally easy to locate: connection of variables just before the main loop, and update of variables at the end of the main loop, just after the new results of the time step have been calculated. The application programmer does not have to change the control flow of his simulation, and can use a procedural programming language.

4.3 Parameterized Graphics Objects

4.3.1 Overview

In this section a tool for the graphical interaction of a researcher with a simulation is described. As stated before, an important requirement for such a tool is that the visualization of the data, the handling of user input, as well as direct manipulation are provided. One way to solve this is to consider these aspects as disjoint, and to provide ingenious, but unrelated solutions. We have chosen a different solution: look for the greatest common divisor of these aspects, and provide a homogeneous solution. The greatest common divisor of user input widgets and visualization tools is simply graphics. Buttons, sliders, graphs, and histograms all boil down to collections of graphics objects. Therefore, we use Parameterized Graphics Objects (PGOs) as the main concept, and the graphics editor as a metaphor for the design of the interface.

The graphics editor has two modes: specification and application, or shorter, *edit* and *run*. In edit-mode, the researcher can create and edit graphics objects much like in MacDraw-like drawing editors. The properties of those objects can be parameterized to values of variables in the Data Manager. Hence, the researcher draws a specification of the interface. In run-mode, a two-way communication is established between the researcher and the simulation by binding these properties to variables. Data is retrieved and mapped onto the properties of the graphics objects. The researcher can enter text, drag, and pick objects, which is translated into changes of the values of variables. What you draw is what you control. The working method of a researcher is thus:

1. Decide which parameters are important for control and visualization
2. Adapt the simulation to connect those parameters with the Data Manager
3. Edit an interface
4. Run the interface: view and control the simulation
5. Analyze the results and go back to one of the previous steps

In addition, standard satellites can be used or new satellites can be developed. The graphics editor itself is just a satellite, as shown in Figure 4.4. When the researcher interacts with PGOs, data will be written to the Data Manager. Similarly, writes by other satellites to variables will trigger the graphics editor and result in visualizations of the data.

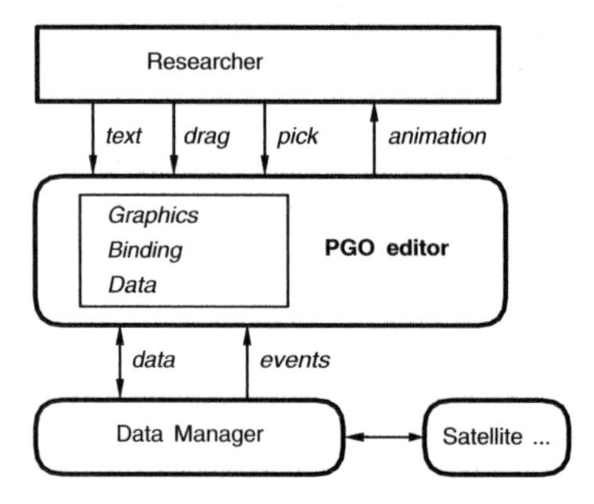

Figure 4.4: The PGO editor in the CSE.

Figure 4.5 shows that the researcher is free to design representations that convey the semantics of the underlying variables: nine different ways to visualize two scalar variables x and y are shown and both the edit- and the run-mode versions of the interface are given. The variables can be presented via standard UI-widgets (a) and (b), or in a business graphics style (c). The parameters can also refer to a position (d), a range (e), or have a mechanical (f) or sensual (g) interpretation. Typical computer graphics interpretations are given in (h) and (i). Many more representations can be conceived.

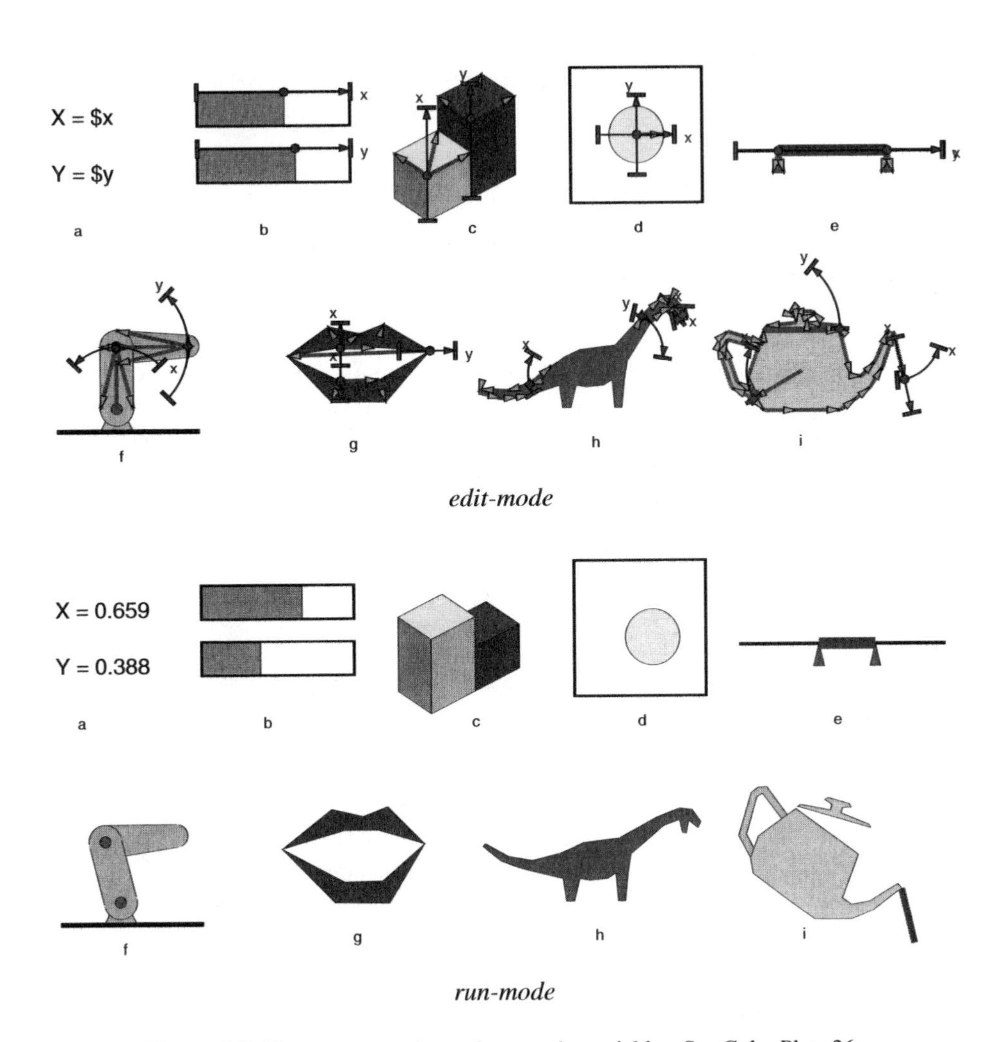

Figure 4.5: Nine representations of two scalar variables. See Color Plate 26.

Figure 4.6 provides an overview of the main objects within the PGO editor. For each object one or more examples of their presentation to the researcher is shown. Various PGOs are shown in the top row of the figure. The geometry of each of these objects is defined by points, the nongeometric properties are defined by various attributes. In the bottom row of the figure the objects for local data management are shown. The object responsible for the binding between graphics and data is the Degree of Freedom, shown in the middle row.

In the following subsections we first describe each object in more detail. After that, we will describe how they interact when the editor is switched to run-mode. Finally, the handling of arrays is discussed. Applications are given in the next section.

4.3.2 Graphics Objects

The PGO editor offers a set of standard graphics objects: fill-area, polyline, rectangle, circle, arc, and text. The geometry of the objects is defined by one or more points. These points are independent of the graphics objects themselves, so that one point can be shared by various graphics objects.

The researcher can define relationships between points (Figure 4.7). Any point can have another point as a reference point or parent point. These relations are shown as gray lines with yellow arrows. Cycles are not allowed, thus, the structure is a forest of trees, with points as nodes and leaves. These relations are used when points are moved. How this is done depends on the type of the point. Two types of points can be used:

- Hinge points (depicted as circles, and labeled H). When a hinge point is moved, the same translation is applied to its child points.

- Fixed points (depicted as diamonds, and labeled F). If a fixed point is moved, then its children are rotated such that the angles between the points and their distances to the parent point remain fixed.

Next these transformations are recursively applied to the children of the transformed points. In Figure 4.7 we show the effect of moving a point in run-mode along a Degree of Freedom.

Graphics objects have four attributes: the hue, saturation, and value of its color, and the linewidth used for the object or its outline. In text objects references to Mapped Variables can be made by using a $, followed by a Mapped Variable name. In run-mode this reference is replaced by the value of the corresponding variable.

4.3.3 Degrees of Freedom

Degrees of Freedom (DOFs) are used to parameterize points and attributes as a function of variables. DOFs have a standard visual representation which is shown in Figure 4.6. Each DOF has a minimum, a maximum, a current value, and possibly a Mapped Variable that is bound to the DOF. In edit-mode all aspects of the DOFs can be changed interactively via dragging and text-editing, in run-mode only the current value can be changed. Seven options are available for assigning DOFs to a point: no DOFs, Cartesian DOFs in either x, y, or both, and polar DOFs in *radius*, *angle*, or both (see Figure 4.8). Polar DOFs can be used only for points with parent points.

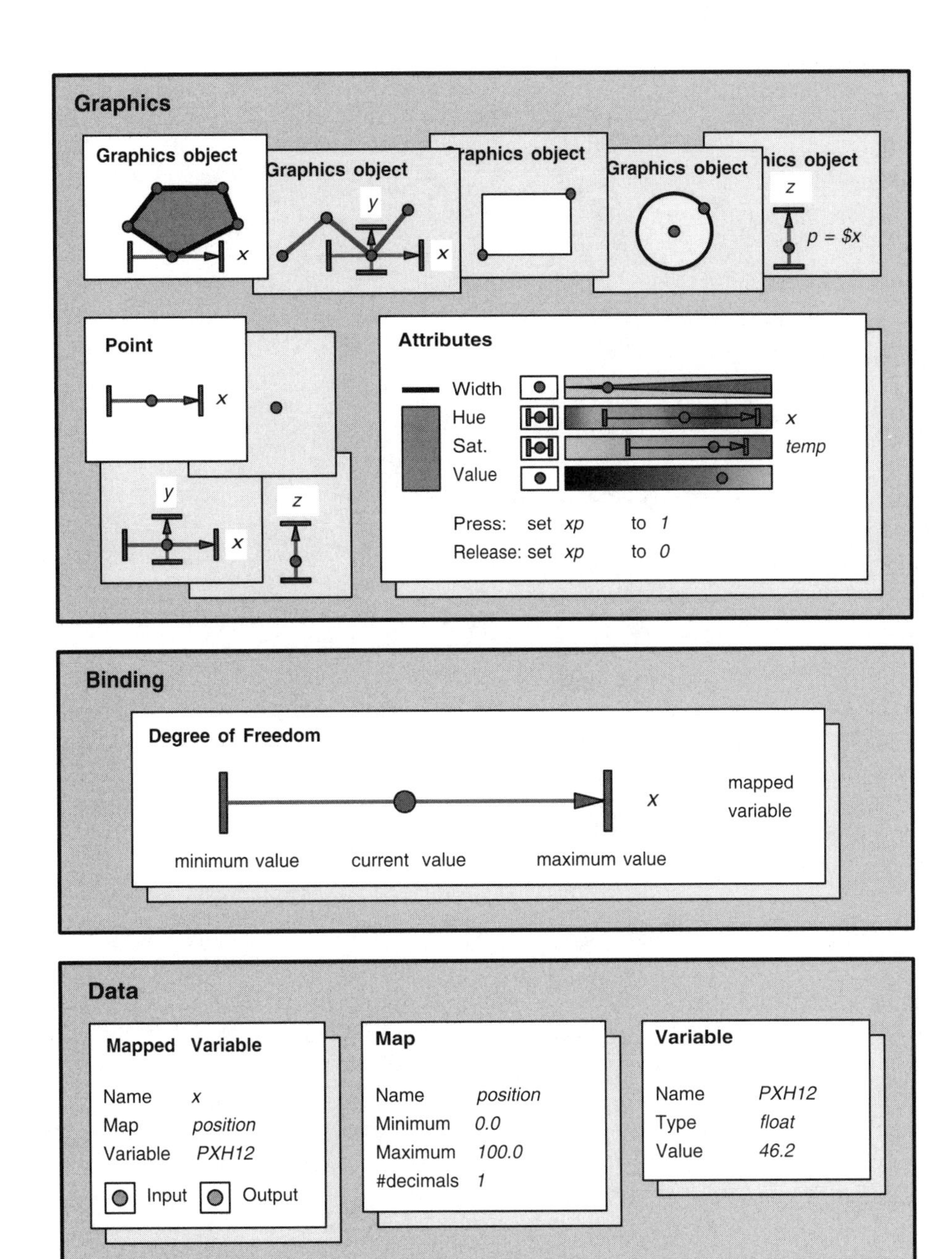

Figure 4.6: Objects in the PGO editor. See Color Plate 27.

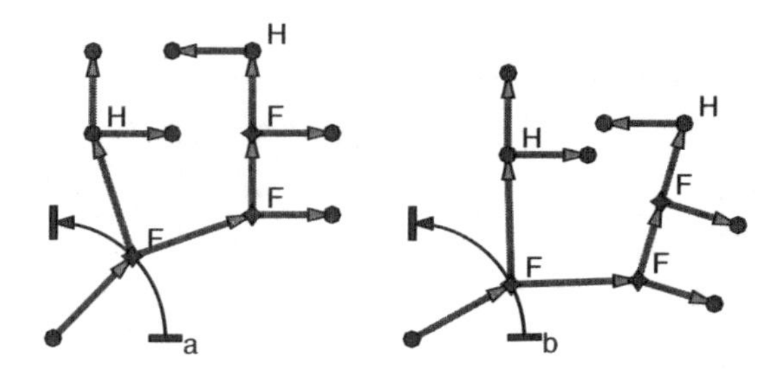

Figure 4.7: Points and point structures.

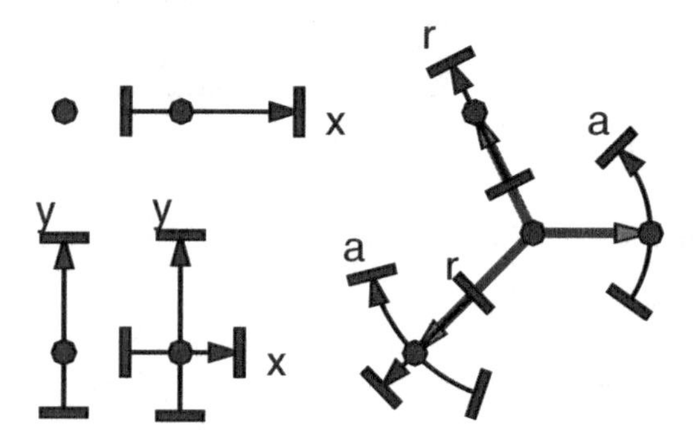

Figure 4.8: Degrees of freedom for points.

With these options in combination with the relations between points, a wide variety of coordinate frames can be defined: relative, absolute, Cartesian, polar, and hierarchical, although the concept of a coordinate frame itself is not provided. This is a typical example of the main principle that has guided us: provide a set of simple primitives and useful operations to combine them, so that the user can easily construct a wide range of higher order concepts.

DOFs can also be used for the attributes' hue, saturation, value, and linewidth, in the same way that DOFs were used to associate a mapped variable with geometry. These attributes are presented to the researcher via a separate attribute window (Figure 4.6). With those DOFs, at most four variable values can be visualized simultaneously.

4.3.4 Mapped Variables

The name that a researcher can specify for a DOF refers to a Mapped Variable (MVAR). The data of an MVAR are references to a Map and a Variable, and two on/off switches for input and output. With these switches, the researcher can select if input information (from the researcher to the Data Manager) and output information (from the Data Manager to the researcher) can pass or not.

The Map contains a specification for how values must be mapped in the communication with the Data Manager. The current implementation is simple: only linear mappings are supported, hence the specification of a minimum and maximum value suffices. Furthermore, the Map contains a specification of the format for textual output. A Variable is a local copy of the information in the Data Manager. Here some bookkeeping information, such as the type, size, name and Data Manager id, and the current value(s) are stored.

A simpler implementation would be to use just one object that includes all information. However, the indirection via the MVARs is very useful. Several Variables can share the same Maps, thus, to change the mapping of those variables only a single map has to be changed. As an example, suppose that a simulation calculates a variety of water temperatures ($T1, T2, ...$), all within the same range (say, $0°$ to $100°$ Celsius). Here we can use the same map ($temperature$) for all variables. Also, one variable can have several associated mappings. For instance, a wide mapping to select a variety of values, and a narrow one to visualize if a value is above or below a threshold value. As an example, to visualize if water is boiling or not, we can use an additional map $boiltemperature$ from $99.5°$ to $100.5°$ Celsius.

4.3.5 Output

In the previous sections we have described the objects in the system. Now we can describe how they interact in run-mode. We start with visualization: the route from data in the Data Manager to images and animations.

Upon switching from edit- to run-mode, the PGO editor first makes a connection to the Data Manager. For each MVAR that is used by a DOF (or that is referenced in a text object), and for which the switch *output* is on, the PGO editor expresses interest in mutation events for the associated variables. The algorithm for updating the image is as follows:

1. When a notification event arrives for a variable, its value (say f) is sampled and stored locally.

2. Via the map (say m) a normalized value f_n is calculated:

$$f_n = \begin{cases} 0 & f \leq m_{min} \\ (f - m_{min})/(m_{max} - m_{min}) & m_{min} < f \leq m_{max} \\ 1 & f > m_{max} \end{cases}$$

   If f is outside the range of the map, f_n is clamped.

3. For all DOFs bound to MVARs the current value d_c is updated via:

$$d_c = (1 - f_n)d_{min} + f_n d_{max}.$$

4. The corresponding objects are updated. Colors are recalculated: mapped from HSV- to RGB-space. The coordinates of points and all their children are updated according to the type of DOF.

5. Finally, the graphics objects are redrawn, using the changed points and the new values for attributes, and thereby displaying the changes in the data. References to MVARs in text objects are replaced by the formatted value of the variable.

Redrawing the image for each change of the data of the Data Manager is expensive, and can lead to erroneous animations. As an example, consider an object with coordinates x and y. Typically, a simulation will change both x and y in a single time step. If the image is redrawn for each change in the value of one of the variables, the observed trajectory of the object will have a staircase form. Therefore, the researcher can specify a trigger variable. If a trigger variable (say t) has been specified, then the PGO editor only awaits changes in t. When t changes, all relevant variables in the Data Manager are sampled, and the image is redrawn.

4.3.6 Input

The processing of user input in run-mode is very similar to the handling of output, except that the direction of the data flow is reversed. Points can be dragged, but now, in contrast to edit-mode, only along the direction(s) specified by the associated DOFs. Changes in the corresponding MVARs are dealt with in the reverse way, as described in the previous subsection. Child points move along just as they would in edit-mode: the DOFs move along with the points.

The researcher can also change attribute values of graphics objects in the attribute window, provided that DOFs have been defined for them. This is done by dragging the current value of the DOF in the attribute window to a new value.

Picking is a useful interaction primitive for the selection of objects and for invoking actions. We have defined picking in line with the data-driven concept of the system: a pick results in a change of data. One standard action is predefined: Set the value of a *variable* to a *value*. For each graphics object the researcher can specify which variable has to be assigned what value, both for press-events (a mouse-button is pressed with the pointer inside the graphics objects) and for release-events (a mouse-button is released).

Text objects can be changed in run-mode by the researcher if the text object contains one or more references to MVARs, and if the 'input'-switches of the MVARs are on. The researcher can click on such an object, the string is replaced by an input box, and the researcher can enter new text. Next, this text is scanned, the value of the variable associated with the MVAR is updated, and this new value is sent to the Data Manager. Also, the image is redrawn to show updates in other graphics objects that depend on the same variable.

4.3.7 Arrays

Often the major part of data to be visualized will be arrays. The manual specification of a large number of similar PGOs, with DOFs parameterized as indexed elements of the array, is a tedious process. Therefore, we automated it. Each graphics object in edit-mode is considered as a template, from which instances are generated after binding with the Data Manager. Multiple instances can be generated from all graphics objects. In addition, polylines and fill-area points can be expanded in-line within a single object (see Figure 4.9).

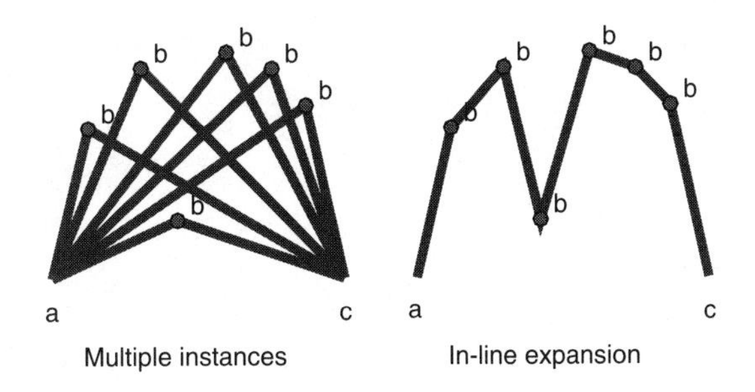

Figure 4.9: Two different expansions of arrays for polyline object.

4.4 Applications

4.4.1 Balls

In Figure 4.10 an edit-mode version of Figure 4.2 is shown. We used array expansion for the balls. Only a single circle with a single velocity line is defined. Each of the variables px, py, vx, and vy is an array. The mapped variable rc is bound to the variable r, with the input-switch to off. Thus, when a ball is dragged, the radius is not influenced. The variable r is scalar. If r were an array (each ball has a different size), the same interface could still be used. The color of the enclosing box is parameterized to the damping factor of the velocity.

The standard satellite *dmlog*, in combination with array expansion, was used to make the graph of the total kinetic energy in the lower-left corner of Figure 4.10. A polyline for a single point with a suitable parameterization was defined. If logging is active, the point is expanded in a series of points, and hence a sliding graph is displayed. This example shows that standard scientific graphics primitives can easily be defined.

4.4.2 Cars

Figure 4.11 shows that the CSE can be used for multidimensional visualization, and as a front end to databases. We used our CSE to visualize a table of 400 different types of cars, where for each car a number of attributes such as displacement, horsepower, and acceleration are given[1]. We developed a small satellite that reads in the data, writes these data to the Data Manager, and can be used to make selections of the data. The data are visualized via three scatterplots. Each item in the scatterplot represents a car type. For each item we used color to visualize the country of origin, and two lines to indicate miles per gallon and weight. Each axis of the scatterplot has an associated range slider to make selections of the cars. Further, the user can click at each item, and the properties of this car type are shown in the lower-right corner.

[1] The data set has been provided by the Committee on Statistical Graphics of the American Statistical Association for its Second (1983) Exposition of Statistical Graphics Technology.

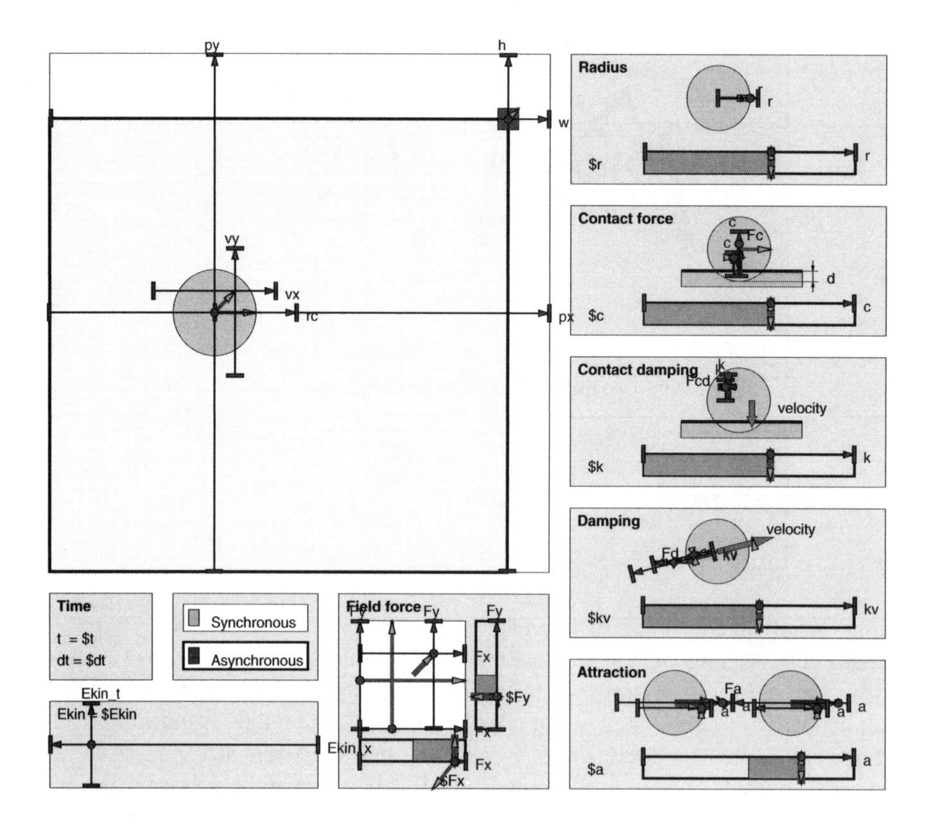

Figure 4.10: Simulation of bouncing balls, edit-mode. See Color Plate 28.

4.4.3 Atmospheric Simulation

We have applied our techniques to a model for smog prediction over Europe. The full-blown model forecasts the levels of air pollution, characterized by approximately 104 reactions between ca. 70 species. For example, the concentrations of ozone (O_3), sulphur dioxide (SO_2), and sulphate aerosol (SO_4) are calculated. The vertical stratification is modeled by four layers: the surface layer, the mixing layer, the reservoir layer, and the upper layer. The physical and chemical model is described by a set of partial differential equations that describe advection, diffusion, emission, wet and dry deposition, fumigation, and chemical reactions.

An important numerical utility to solve these equations is local grid refinement. This technique is used to improve the quality of the model calculations in areas with large spatial gradients (for example in regions with strong emissions) and in areas of interest, specified by the researcher. The tradeoff to be made in local grid refinement is calculation accuracy versus computation speed.

We have used our environment to steer various aspects of the local grid refinement technique. We name only a few of these aspects:

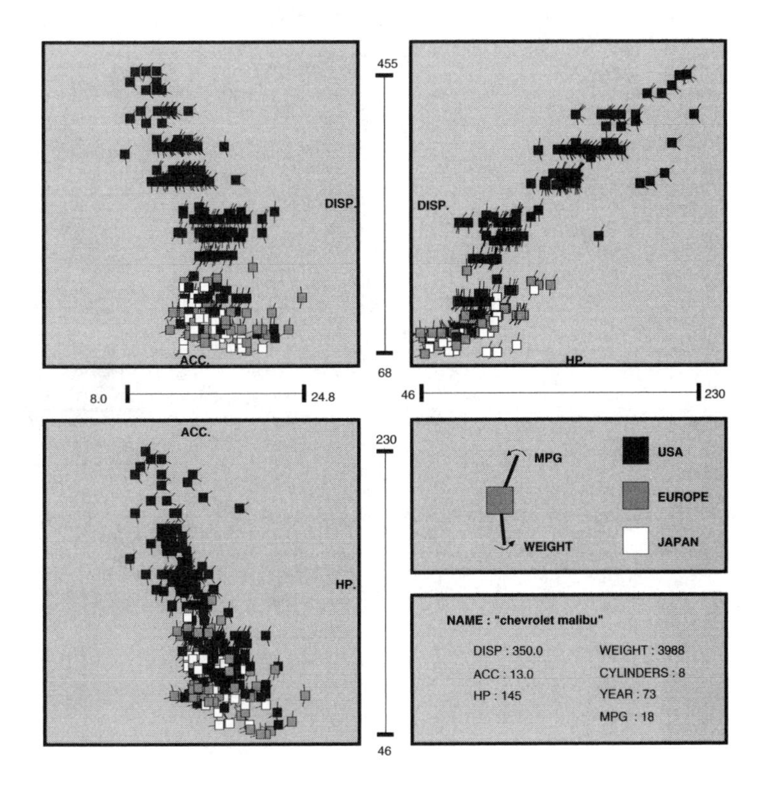

Figure 4.11: Visual front end to a multidimensional database. See Color Plate 29.

- the researcher may select the species and the layer to which grid refinement will be applied
- the researcher may define the tolerance value which determines where refinement is necessary
- the researcher may define a bounding box as a region of interest to drive the grid refinement
- the researcher has interactive control over simulation time

Figure 4.12 shows a snapshot of the ongoing simulation. On the left, in the PGO editor, only the refined grid within a region of interest is shown. Visualized is one species (in this case, ozone) at one layer and the wind field. On the right, in the IRIS Explorer Render module, all four layers of the same species are shown. We used the satellite $ReadDM$ to send data to IRIS Explorer. The $dmSlice$ satellite is shown on the lower-right.

The amount of data produced per time step of the simulation is substantial. For each grid point the concentration of each species is calculated. Depending on the grid refinement

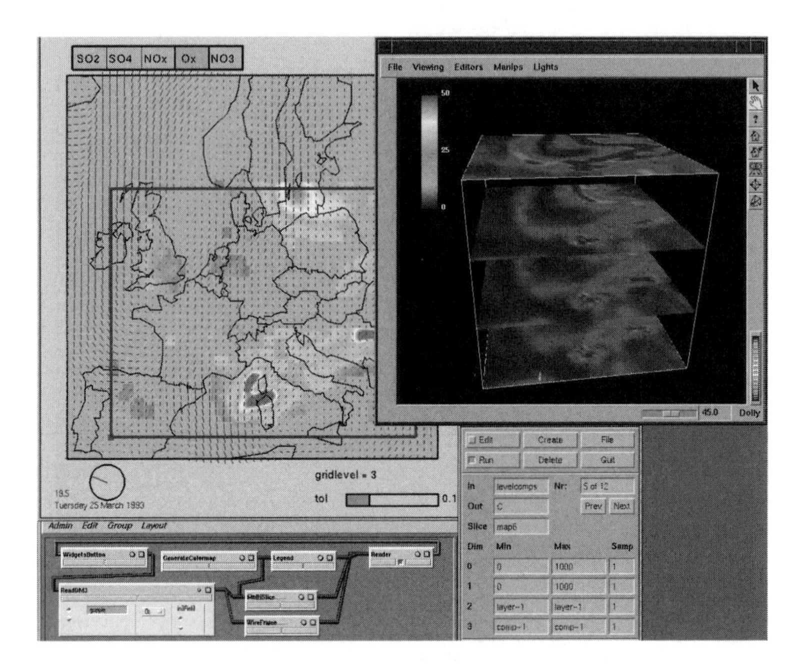

Figure 4.12: Visualization of a smog prediction model with the PGO editor and IRIS Explorer. See Color Plate 30.

level, the amount of data can vary between one to four megabytes per time step. There are approximately 500 time steps in our test simulations.

The geometry of the grid cells and all concentrations are stored as variables in the Data Manager. The *dmslice* satellite is used to interactively select a particular concentration. Concentrations are visualized via specification of a single rectangle that is parameterized with these variables. The wind field is visualized via specification of a single polyline. Both the rectangle and polyline specifications are expanded through the array expansion mechanism.

4.5 Discussion

In the previous section we presented applications. In this section we step back to a higher level of abstraction. We discuss the underlying concepts of the developed CSE, compare our work with related work, and consider future extensions and the scope of the methods and techniques described.

4.5.1 Concepts

The presented CSE has two major components: the Data Manager and the PGO editor. Several concepts are shared by both components:

- The use of *low-level primitives*: a simple data model and graphics objects. The interfaces to these primitives are familiar to the end user: a simple I/O library for data manipulation and a graphics editor for graphics.

- *No higher-level semantics* are defined for the graphics and the data. As a result, the environment is general and flexible. The researcher can easily add meaning to the graphics and data. With the PGO editor the end user can edit and combine PGOs to realize meaningful widgets. Changes to existing simulation software are minimal. The locations where hook functions must be added are easy to locate, data structures and program structures do not have to be changed. To summarize, the CSE itself is not a tool for, say, physically based modeling and animation, but it could be a great environment for the development of such a system.

- Both the Data Manager and the PGO editor rely on *late binding* of names. As a result, it is possible to define new visualizations of the data output by the simulation, while the simulation continues to run.

- All operations in both the Data Manager and the PGO editor are based entirely on *data*. Dragging, picking, and text input are translated into changes of data. The main type of event within the system is the data mutation. Process control (firing algorithms or execution models) can be implemented on top of those events by the satellites.

4.5.2 Comparison

In Section 4.1.3 we considered three approaches to realize an environment for computational steering. Our approach falls in the class of extension of graphics toolkits. Most related to our work are the Inventor system and the 3D interactive system being developed at Brown University. However, there are also important differences. An obvious difference is that those two systems are 3D. We will discuss the extension of our system from 2D to 3D in the next section.

Our starting point was the researcher as an end user: how can we offer an environment that suits his needs and is easy to use? This is reflected in the contents of the chapter. We extensively discussed how the environment looks from the end user point of view, and paid less attention to the computer science methods and techniques used. This was a deliberate choice—for us this was the most important issue. The approach taken by Inventor and at Brown University is primarily of interest for application programmers and not for end users.

Another interesting difference is the handling of user input and hence, direct manipulation. In both Inventor and at Brown University, separate objects are used for this—a distinction is made between geometric objects and manipulators. Relations between those two categories must be specified via text. For the PGOs this distinction was not made. In their terms: each graphics object can serve as a manipulator. Feedback via the manipulated object is always provided. If such a manipulation must have an effect on other objects, then the use of the same names in DOFs suffices for linear mappings. If more complex behavior is required, this can be added as a separate satellite.

4.5.3 Future Work

Concerning the Data Manager, we will consider the use of shared memory and point-to-point communication schemes for more efficient interprocess communication. In the current implementation a write will always transport data to the Data Manager which, in turn, will transport the data to all subscribing satellites. This indirect method of communication can be prohibitively expensive for large data sets. This problem in the current implementation will be overcome by allowing point-to-point channels between satellites. In this way, the Data Manager will act only as a router for large data sets.

In the current environment all graphics are two-dimensional. This allowed a rapid development of the environment, and an early validation of the underlying concepts. Two-dimensional PGOs do not render the CSE useless. Many real-time simulations are two- or even one-dimensional, so a 2D PGO editor is more than justified. Further, the results of three-dimensional simulations can be mapped on two-dimensional data (slices), or data of lower dimensionality. Finally, we have shown that we can include existing tools for 3D visualization in our environment. Nevertheless, a 3D version of the PGOs would be welcome. This would mean that we have to implement solutions for viewing, direct manipulation, and cursor positioning in 3D. But that is primarily a problem related to 3D, and not to our environment. We expect that the basic concepts, and especially the one-dimensional DOFs, can be transposed to 3D without major problems.

We have tailored our environment to computational steering (see Figure 4.1). It is an interesting exercise to replace *researcher* by *user* and *simulation* by *application*. We expect that the environment and its underlying concepts will be useful for many other application areas. We hope to investigate this in the future. Some examples of possible applications for our visual specification techniques are custom widgets in User Interface Management Systems, interactive animation and simulation for Computer Aided Education and hypermedia, and template-driven graphics editors.

4.6 Conclusions

We have shown that the combined use of:

- an architecture with a central Data Manager
- parameterized graphics objects

results in an environment that provides

- easy integration of different processes
- easy definition of visualizations
- easy definition of customized user-interface widgets
- direct manipulation by default

The primary target application for the environment is computational steering, but the same environment, or at least the same concepts, can probably be applied in many other application areas. In short, we have developed a system that makes it easy for an end user to specify and use interactive computer graphics.

Bibliography

[1] L. Bergman, J. Richardson, D. Richardson, and F. Brooks Jr., "VIEW—An Exploratory Molecular Visualization System with User-Definable Interaction Sequences," *Computer Graphics*, Vol. 27, No. 6 (SIGGRAPH '93), 1993, pp. 117–126.

[2] IRIS Explorer Development Team, *IRIS Explorer^{TM} Module Writer's Guide*, Technical Report 007-1369-020, Silicon Graphics Computer Systems, Mountain View, Calif., 1993.

[3] W. Felger and F. Schroder, "The Visualization Input Pipeline—Enabling Semantic Interaction in Scientific Visualization," *Computer Graphics Forum*, Vol. II, No. 3, 1992, pp. 139–152.

[4] R.E. Marshall, J.L. Kempf, D. Scott Dyer, and C-C Yen, "Visualization Methods and Simulation Steering a 3D Turbulence Model of Lake Erie," *1990 Symp. on Interactive 3D Graphics, Computer Graphics*, Vol. 24, No. 2, 1990, pp. 89–97.

[5] B. McCormick, T. Defanti, and M. Brown, "Visualization in Scientific Computing," *Computer Graphics*, Vol. 22, No. 6 (SIGGRAPH '88), 1988, pp. 103–111.

[6] B. Myers, "User Interface Tools: Introduction and Survey," *IEEE Software*, Vol. 6, No. 1, 1989, pp. 15–23.

[7] B. Myers, "Creating User Interfaces Using Programming by Example, Visual Programming, and Constraints," *ACM Transactions on Programming Languages and Systems*, Vol. 12, No. 2, 1990, pp. 143–177.

[8] D. Olsen Jr., "Workshop on Software Tools for User Interface Management," *Computer Graphics*, Vol. 21, No. 2, 1987, pp. 71–147.

[9] G. Pfaff, editor, *User Interface Management Systems*, Springer-Verlag, 1985.

[10] S. Snibbe, K. Herndon, D. Robbins, D. Brookshire Conner, and A. van Dam, "Using Deformations to Explore 3D Widget Design," *Computer Graphics*, Vol. 26, No. 6 (SIGGRAPH '92), 1992, pp. 351–353.

[11] P.S. Strauss and R. Carey, "An Object-Oriented 3D Graphics Toolkit," *Computer Graphics*, Vol. 26, No. 6 (SIGGRAPH '92), 1992, pp. 341–351.

[12] C. Upson, "Volumetric Visualization Techniques," *State of the Art in Computer Graphics*, D.F. Rogers and R.A Earnshaw, editors, Springer-Verlag, 1991, pp. 313–350.

[13] C. Upson, T. Faulhaber, D. Kamins, D. Laidlaw, D. Schelgel, J. Vroom, R. Gurwitz, and A. van Dam, "The Application Visualization System : A Computational Environment for Scientific Visualization," *IEEE Computer Graphics and Applications*, Vol. 9, No. 7, 1989, pp. 30–42.

[14] A. van Dam, "VR as a Forcing Function: Software Implications of a New Paradigm," *IEEE Virtual Reality Symposium Proceedings*, 1993, pp. 6–9.

[15] J.J. van Wijk and R. van Liere, *An Environment for Computational Steering*, Technical Report CS-R9450, Centre for Mathematics and Informatics, Amsterdam, The Netherlands, 1993.

[16] C. Williams, J. Rasure, and C. Hansen, "The State of the Art of Visual Languages for Visualization," *Proceedings Visualization '92*, 1992, pp. 202–209.

[17] R. Zeleznik, D. Brookshire Conner, M. Wloka, D. Aliaga, N. Huang, P. Hubbard, B. Knep, H. Kaufman, J. Hughes, and A. van Dam, "An Object-Oriented Framework for the Integration of Interactive Animation Techniques," *Computer Graphics*, Vol. 25, No. 4 (SIGGRAPH '91), 1991, pp. 105–111.

Chapter 5

Scientific Visualization of Large-Scale Unsteady Fluid Flows

David A. Lane

Abstract. *In a numerical flow simulation, it is common to generate several thousand time steps of unsteady (time-dependent) flow data. Each time step may require tens to hundreds of megabytes of disk storage, and the entire unsteady flow data set may be hundreds of gigabytes. Interactive visualization of unsteady flow data of this magnitude is impossible with the current hardware capabilities. Particle tracing is an effective technique to visualize unsteady flows. Streaklines, which are computed by tracking particles in unsteady flows, depict time-varying phenomena that are sometimes difficult or impossible to see with other flow visualization techniques. This chapter provides a tutorial for particle tracing in steady and unsteady flows. First, the life cycle of a typical numerical flow simulation is outlined. The current approaches for visualizing unsteady flows are then described. Many systems exist for flow visualization, some of which are discussed. A tutorial for particle tracing is then given. A particle tracing system called UFAT has been developed for unsteady flow visualization. The features and implementation of UFAT are described. The steps for computing streaklines are summarized. Several unsteady flow data sets were visualized using UFAT, and the results are presented.*

5.1 Introduction

The life cycle of a numerical flow simulation in Computational Fluid Dynamics (CFD) generally consists of three phases: grid generation, flow calculation, and visualization. First, a numerical grid is constructed to enclose the boundaries of the object in the flow using a grid generation tool. The grid may be rectilinear, curvilinear (structured), block structured, unstructured, or a mix of the structured and unstructured (hybrid). In CFD, structured and unstructured grids are commonly used. A multizoned curvilinear grid consists of one or more structured grids (sometimes referred to as blocks or zones). Some blocks may overlap other blocks. Each cell in a 3D curvilinear grid has a hexahedron-like configuration (warped bricks); the faces of the cell will not generally be planar. Figure 5.1 shows a multizoned curvilinear grid. An unstructured grid consists of cells that are not necessarily hexahedra. For example, the cells may be tetrahedra, prisms, or pyramids. For a good discussion of the types of grids used in scientific visualization, see Speray and Kennon [29]. In the second phase of the flow simulation, a system of Navier-Stokes equations that simulate the flow condition is solved. Euler equations are commonly used. The solution yields momentum and other flow quantities such as density and stagnation energy. From these quantities, velocity can be derived. Several methods can be used to set up the flow equations. Using a finite difference method, the flow quantities are solved at the grid points. A finite volume method solves the flow quantities, usually at the grid cell centers instead of at the grid points. Using a finite element method, the flow quantities in each cell are represented by a set of basis functions. The calculation of the flow solution can be computationally expensive, depending on the grid size. In the last phase, the velocity field and other flow quantities computed from the flow calculation are analyzed using a number of visualization techniques, including contour plots, particle traces, isosurfaces, and volume rendering.

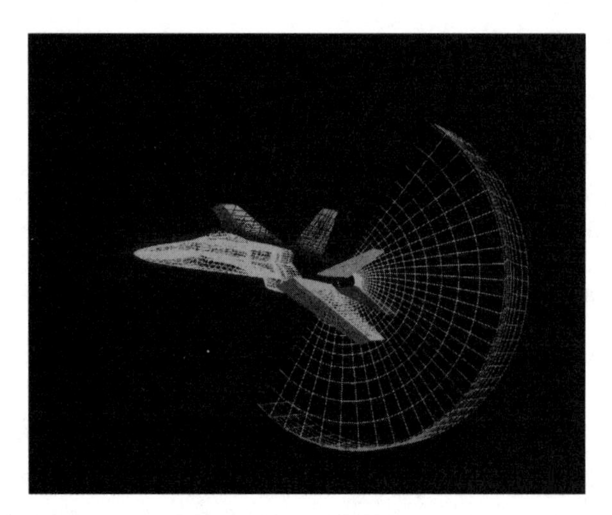

Figure 5.1: A multizoned curvilinear grid surrounding a F18 aircraft, generated by Yehia Rizk and Ken Gee, NASA Ames Research Center. Each sub-block is distinguished by color. See Color Plate 31.

With the increase in computing power over the past decade, numerical simulations of 3D unsteady flow fields are becoming more common. The disk requirements for 3D unsteady flow simulations have also increased dramatically. The data generated from an unsteady flow simulation is usually several orders of magnitude larger than the data generated from a steady flow simulation. Some unsteady flow data sets require hundreds of gigabytes of disk space. Visualizing data of this magnitude is a Grand Challenge problem in scientific visualization.

5.2 Problem

At NASA Ames Research Center, CFD scientists are performing complex 3D unsteady flow simulations using the supercomputers located at the Numerical Aerodynamic Simulation (NAS) facility. Table 5.1 shows five grids that were recently used in several simulations.

Grid	No. of Grid Points	No. of Blocks
Clipped Delta Wing	250×10^3	7
Descending Delta Wing	900×10^3	4
V-22 Tilt Rotor	1.3×10^6	25
Harrier Jet	2.8×10^6	18
SOFIA Airplane	3.2×10^6	35

Table 5.1: Five multizoned curvilinear grids used for unsteady flow simulation.

Two types of files are usually generated in a simulation: the grid file contains the physical coordinates of the grid points and the solution file contains the momentum and other flow quantities computed in the flow calculation. The size of the grid file depends on the number of grid points and the size of the solution file depends on the flow quantities saved from the flow calculation. If the grid consists of multiple blocks, then the grid file also contains an integer code for each grid point that indicates if the point is in an overlapped region or near a wall boundary (see "Particle Tracing in Multizonal Grids" below). In unsteady flows, there is one solution file per time step. If the grid moves in time, then there is also one grid file per time step. Table 5.2 shows the disk requirements for storing the grid and solution files per time step for the grids given in Table 5.1. In Table 5.2, the solution files assume that five flow quantities are saved: stagnation energy, density, and the x, y, z components of momentum.

Grid	Grid File	Solution File	Total Per Time Step
Clipped Delta Wing	4 MB	5 MB	9 MB
Descending Delta Wing	16 MB	20 MB	36 MB
V-22 Tilt Rotor	23 MB	29 MB	52 MB
Harrier Jet	45 MB	56 MB	101 MB
SOFIA Airplane	53 MB	66 MB	119 MB

Table 5.2: The disk space, in megabytes (MB), required for each time step.

It is common to have thousands of time steps in an unsteady flow simulation. Table 5.3 shows the disk requirement for storing the data generated in one flow simulation for each of the data sets shown in Table 5.2. For example, the descending delta wing consists of 90,000 time steps. The disk space required to save all of these time steps is 3,240 GB (90,000$\times$36 MB). Clearly, it is impossible to interactively visualize data of this magnitude with the current hardware capability.

Grid	Number of Time Steps	Total Per Simulation
Clipped Delta Wing	5,000 per cycle for 3 cycles	135 GB
Descending Delta Wing	90,000	3,240 GB
V-22 Tilt Rotor	1,450 per cycle for 3 cycles	226 GB
Harrier Jet	1,000	101 GB
SOFIA Airplane	10,000	1,190 GB

Table 5.3: The disk space, in gigabytes (GB), required per flow simulation.

5.3 Current Approaches

For interactive visualization, it is ideal to store all time steps of the data in the physical memory of the system so that the scientist can loop through the time steps interactively. However, as shown in Table 5.3, some 3D unsteady flow data sets are too large to be visualized interactively. Even the simple task of storing a few time steps of the flow data becomes a problem due to insufficient disk space. Therefore, postvisualization is necessary. For postvisualization, scientists often save their data during the flow calculation at fixed time intervals. The flow data set that is extracted this way will be referred to as the *saved data set* hereafter. For example, every 50th time step of the descending delta wing data (see Table 5.2) was saved for postvisualization. The disk requirement for the saved data set is 64.8 GB (90,000/50$\times$36 MB). Currently, it is difficult to find a system with tens of gigabytes of memory for interactive visualization.

Two approaches for postvisualization are sometimes used. The first approach is to store as many time steps of the saved data as possible in memory. If the size of the flow data is small (that is, on the order of hundreds of megabytes), then this approach is attractive. However, for large-scale data sets such as the V-22 and SOFIA data sets listed in Table 5.3, this approach would allow only a few time steps to be visualized interactively. This is unacceptable for most simulations. The second approach is to subsample the flow data at a lower grid resolution so that more time steps can fit in memory. However, this is generally not a good approach because the resolution of the flow is lost. For example, the presence of a vortex may not be detected. Furthermore, additional preprocessing time is required for subsampling. Both of these two approaches only allow the scientist to visualize a subset of the saved data, and important characteristics of the flow may not be displayed.

A more effective approach is to use every time step of the saved data without subsampling. A requirement is that the system has enough memory to store at least a few time steps of the data. The flow data are visualized by loading one time step of data into memory and then applying the desired visualization techniques one time step at a time. For example, compute color contours on a grid surface at each time step for all time steps. Visualizing flow data at instants in time is sometimes referred to as *instantaneous flow visualization*. A good survey of several instantaneous flow visualization techniques can be found in Post and van Wijk [23].

The interactive rate of instantaneous flow visualization is limited by the disk I/O performance of the system. If the flow data at each time step is hundreds of megabytes, then the time required to read each time step's data could be a few seconds, depending on the system's disk I/O performance. For example, the Harrier jet requires 101 MB per time step for both the grid and solution data files (see Table 5.2). It would take approximately one to two seconds to read the grid and solution files at each time step if the disk I/O rate ranges from 50 MB to 100 MB per second. In addition to the read time, computation time is required by the visualization technique used; for example, particle tracing and isosurface calculation. If there are thousands of particles, then it may take several more seconds or minutes for the computation. For this reason, the interactive rate of instantaneous flow visualization of large-scale flow data can be very slow. Although instantaneous flow visualization can be effective (though slow) for some analysis, it may not reveal time-varying phenomena in the flow. Effective unsteady flow visualization techniques should consider time as a parameter in the calculation to depict flow phenomena that evolve in time. Time-dependent particle tracing, which is described in Section 5.5, is an example of an unsteady flow visualization technique.

It is also possible to incorporate visualization into the flow calculation phase. This method generates the graphics objects (for example, particle traces) while the solutions are being computed. The method would also reduce disk requirements because the solution files do not need to be saved for postvisualization; however, when visualization parameters (for example, seed locations) need to be changed, the flow would need to be recalculated, which is computationally expensive.

5.4 Flow Visualization Systems

There are many existing systems that allow instantaneous flow visualization. PLOT3D (Buning and Steger [6]) is a popular CFD visualization tool that provides several instantaneous visualization techniques. FAST (Bancroft et al. [1]) consists of a set of modules that support PLOT3D functionalities and several other new features like topology, function calculator, and animation. COMADI (Vollmers [31]), HIGHEND (Pagendarm [21]), Visual3 (Haimes and Giles [12]), and many other visualization systems also provide several instantaneous visualization techniques. Most of these systems allow unsteady flow visualization by looping through the time steps and visualizing the flow data one time step at a time; however, many of them do not consider time as a parameter in the calculation.

Because simulations of large-scale 3D unsteady flows have only become possible within the past decade, there are currently few systems developed specifically for unsteady flow visualization. PLOT4D and Streaker developed by Smith et al. [28] perform unsteady flow visualization by visualizing trivariate functions of two spatial variables and one time variable. For example, $f(x, y, t)$ or $f(x, z, t)$. Jespersen and Levit [14] and Vaziri et al. [30] have developed unsteady flow visualization systems using the CM5 for the calculation and the results are sent over the network to a local graphics workstation for display. The Virtual Wind Tunnel (Bryson and Levit [3]) performs distributed unsteady flow visualization using a Convex C3240 system and an SGI graphics workstation. The particle traces are computed on the Convex system and then sent over the network to the graphics workstation which displays the traces and provides local view manipulation. pV3 (Haimes [11]) distributes the computation to one or more servers using the PVM message-passing library. A nice feature of pV3 is that it allows the user to view the solution while it is being computed. This "plug-into/unplug" feature allows the user to monitor the flow calculation and to terminate the calculation when necessary. Only a few of the systems mentioned above compute streaklines, pathlines, and timelines.

I have developed a particle tracing system called the Unsteady Flow Analysis Toolkit (UFAT) which computes particle traces in unsteady flow [17, 18]. UFAT is unique from existing systems in that it computes particle traces from a large number of time steps, performs particle tracing in unsteady flow with moving curvilinear grids, supports a save/restore option, and allows playback (see Section 5.6).

5.5 Particle Tracing

An effective way to visualize unsteady flow is to compute particle traces. In steady flows, a *streamline* is a field line tangent to the velocity field at an instant in time. For unsteady flows, three types of particle traces can be computed: pathlines, streaklines, and timelines. A *pathline* shows the trajectory of a single particle released from one fixed location, called the *seed location*. A *streakline* is a line joining the positions, at an instant in time, of all particles that have been previously released from a seed location. Streaklines can be simulated by releasing particles continuously from the seed locations at each time step. In hydrodynamics, streaklines can be simulated by releasing small hydrogen bubbles rapidly from the seed locations. For streakline calculation, particles are released from specified seed locations and tracked through the given time steps. A *timeline* is a line connecting particles that have been released at the same time. In an instantaneous flow field, streamlines, pathlines, and streaklines are identical (Schlichting [26]). In experimental flow visualization, timelines are simulated by releasing a line of particles simultaneously at some time interval. Figures 5.2 and 5.3 show streamlines, pathlines, streaklines, and timelines computed near an oscillating 2D airfoil. Streaklines can reveal very different information than those shown with instantaneous streamlines. The vortices behind the airfoil are visible with the streaklines shown in Figure 5.3, but the streamlines shown in Figure 5.2 do not reveal them clearly. Furthermore, animated streaklines effectively show the development of flow phenomena in time. There are many research papers on the subject of particle tracing in steady flows; however, few of them discuss time-dependent particle tracing. This section provides a tutorial for particle tracing in steady and unsteady flows.

5.5.1 Numerical Models

Most existing methods for particle tracing are based on numerical integration schemes. There are implicit methods for computing streamlines using stream functions (see Kenwright and Mallinson [16]); however, most of these methods have not been extended for unsteady flows. In the next section, a multistage method based on the fourth-order Runge-Kutta (RK4) integration is described for particle tracing in steady flows. In the following section, the method is then extended for unsteady flows.

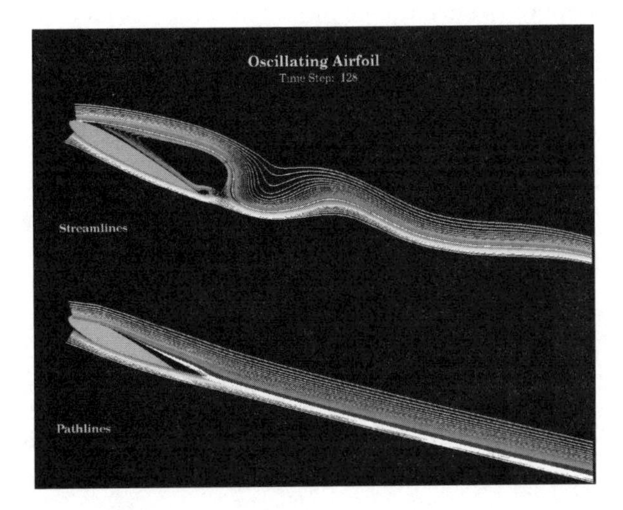

Figure 5.2: Streamlines and pathlines pass through an oscillating airfoil. Particle traces are colored by release location. See Color Plate 32.

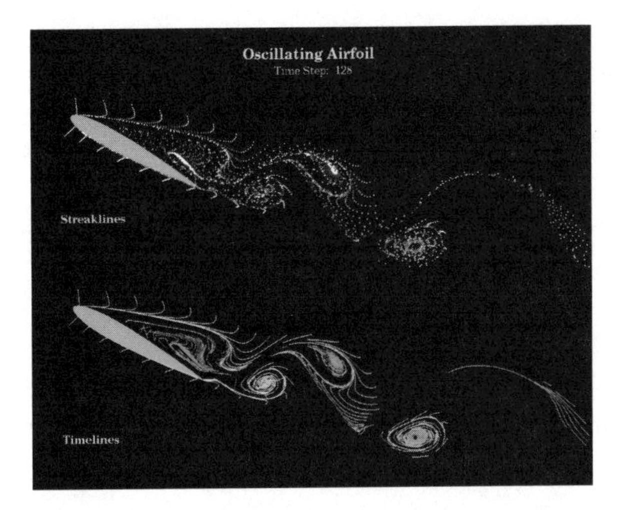

Figure 5.3: Streaklines and timelines pass through an oscillating airfoil. Streaklines are colored by release location and timelines are colored by release time. See Color Plate 33.

Particle Tracing in Steady Flows

Given a vector function $\mathbf{v}(\mathbf{p})$ defined for all $\mathbf{p}$ in the grid domain G, the path of a massless particle at position $\mathbf{p}$ can be determined by solving the following ordinary differential equation:

$$\frac{d\mathbf{p}}{dt} = \mathbf{v}(\mathbf{p}(t)). \tag{5.1}$$

Function $\mathbf{p}(t)$, which represents the position of the particle at time t, can be computed by integrating (5.1). Hence, the particle position after time Δt is as follows:

$$\mathbf{p}(t + \Delta t) = \mathbf{p}(t) + \int_t^{t+\Delta t} \mathbf{v}(\mathbf{p}(t))dt. \tag{5.2}$$

Equation (5.2) can be evaluated using a numerical integration scheme. A common scheme is the RK4 method. Let $k = 0$ and $\mathbf{p}_k$ be the current position of the particle. Then, for $\mathbf{p}_k \in G$, let

$$\begin{aligned} \mathbf{a} = h\mathbf{v}(\mathbf{p}_k),\ \mathbf{b} = h\mathbf{v}(\mathbf{p}_k + \mathbf{a}/2),\ \mathbf{c} = h\mathbf{v}(\mathbf{p}_k + \mathbf{b}/2),\ \mathbf{d} = h\mathbf{v}(\mathbf{p}_k + \mathbf{c}), \\ \mathbf{p}_{k+1} = \mathbf{p}_k + (\mathbf{a} + 2\mathbf{b} + 2\mathbf{c} + \mathbf{d})/6, \quad \text{and} \quad k = k + 1, \end{aligned} \tag{5.3}$$

where $h = \Delta t$ is the step size. To obtain the path of the particle at $\mathbf{p}_k$, (5.3) is evaluated repeatedly until $\mathbf{p}_k$ leaves the grid domain G.

Particle Tracing in Unsteady Flows

In unsteady flows, velocity changes in time, so $\mathbf{v}$ is a function of space and time. Suppose the vector function $\mathbf{v}(\mathbf{p}, t)$ is defined for all $\mathbf{p} \in G$ and $t \in [t_1, t_n]$, where n is the number of time steps in the unsteady flow. The path of a particle at position $\mathbf{p}$ can be computed by solving the following equation:

$$\frac{d\mathbf{p}}{dt} = \mathbf{v}(\mathbf{p}(t), t). \tag{5.4}$$

Integrating (5.4) yields:

$$\mathbf{p}(t + \Delta t) = \mathbf{p}(t) + \int_t^{t+\Delta t} \mathbf{v}(\mathbf{p}(t), t)dt. \tag{5.5}$$

As with steady flows, we can integrate (5.5) numerically using the RK4 method. Let $t = t_1$, $k = 0$, and $\mathbf{p}_k$ be the current position of the particle. Then, for $t < t_n$ and $\mathbf{p}_k \in G$, let

$$\begin{aligned} \mathbf{a} = h\mathbf{v}(\mathbf{p}_k, t), \quad \mathbf{b} = h\mathbf{v}(\mathbf{p}_k + \mathbf{a}/2, t + h/2), \\ \mathbf{c} = h\mathbf{v}(\mathbf{p}_k + \mathbf{b}/2, t + h/2), \quad \mathbf{d} = h\mathbf{v}(\mathbf{p}_k + \mathbf{c}, t + h), \\ \mathbf{p}_{k+1} = \mathbf{p}_k + (\mathbf{a} + 2\mathbf{b} + 2\mathbf{c} + \mathbf{d})/6, \quad t = t + h, \quad k = k + 1. \end{aligned} \tag{5.6}$$

Equation (5.6) is evaluated repeatedly until $\mathbf{p}_k$ leaves the grid domain G or $t \geq t_n$.

5.5.2 Particle Tracing Issues

There are several fundamental issues in particle tracing: physical versus computational space tracing, cell search, step size selection, and velocity interpolation; see Buning [4, 5] and Post and van Walsum [22]. All of these are concerned with the accuracy and speed of particle tracing. In Murman and Powell [20], a first-order integration scheme was shown to produce inaccurate particle traces. They found that using large step size in the integration is undesirable and that the step size should be a fraction of the cell size. Shirayama [27] also concluded that the first-order integration scheme could lead to erroneous results and suggested that a higher-order integration be used. Sadarjoen et al. [25] compared the accuracy and speed of physical space and computational space tracing methods. Trilinear interpolation, inverse distance weighting, and volume weighting schemes for velocity interpolation were evaluated. In Darmofal and Haimes [9], several multistage and multistep integration schemes were compared, and the accuracy and memory requirements of the two schemes were discussed. They concluded that the step size should be adaptive and that a fourth-order interpolation and integration scheme should be used. In the following sections, I review these issues and discuss additional considerations for unsteady flows.

Physical Space Versus Computational Space Tracing

In physical space, the grid is curvilinear and the grid cells are warped bricks. The curvilinear grid is defined by a set of points: $\{(x_{ijk}, y_{ijk}, z_{ijk}), i = 1, \ldots, n_x, j = 1, \ldots, n_y, k = 1, \ldots, n_z\}$. In flow calculation, the grid is transformed to computational space, where each grid cell is a unit cube. The transformed grid consists of the following set of points: $\{(i, j, k), i = 1, \ldots, n_x, j = 1, \ldots, n_y, k = 1, \ldots, n_z\}$. During particle integration, it is necessary to determine the grid cell that a particle currently lies in. This requires cell search (also referred to as point location). In computational space, the grid is uniform and the cell in which the particle currently lies can be easily determined. For example, suppose the computational coordinates of the particle's position are (ξ, η, ζ), then the particle lies in grid cell $(\texttt{int}(\xi), \texttt{int}(\eta), \texttt{int}(\zeta))$. Although cell search is fast and simple in computational space, there are some disadvantages for tracing in computational space. Firstly, the velocity is usually given in physical space. In order to perform integration in computational space, the velocity needs to be converted into computational space. This requires additional calculation time for the velocity transformation. Secondly, during the transformation, accuracy may be lost due to the transformation scheme used. Lastly, if the grid cell is badly deformed (irregularities exist), then the transformed velocity may be infinite [5]. For these reasons, particle tracing is commonly performed in physical space.

The main reason for tracing in physical space is accuracy. The disadvantage is that cell search is more time consuming in physical space than in computational space. From simple code profiling, Kenwright and Lane found the time spent in cell search could require more than 25 percent of the particle tracing time [15]. Cell search is required whenever the particle moves to a new position. For a multistage integration method such as the RK4 method described earlier, cell search is required at the intermediate stages of the integration. For example, cell search is required for $\mathbf{p}_k + \mathbf{a}/2$, $\mathbf{p}_k + \mathbf{b}/2$, and $\mathbf{p}_k + \mathbf{c}$ in Equations (5.3) and (5.6). If the step size h is relatively small, then the particle is likely to move within the current cell or jump no more than one cell. Hence, a local cell search can be performed to find the new position. If the grid is multizoned, then a global cell search is required when the particle moves to a new block (referred to as grid jumping). Because global cell search is more computationally expensive than local cell search, grid jumping can increase the particle tracing time considerably. In Hultquist [13], techniques were suggested for speeding up particle tracing in physical space. These include cell caching, extrapolation in cell search, and cell tagging for grid jumping. Recently, Kenwright and Lane [15] were able to improve the speed of particle tracing in physical space by several factors. By decomposing the grid cell into tetrahedra, cell search time was reduced. Furthermore, particle integration, velocity interpolation, and step size control were performed in physical space.

Cell Search

Given $\mathbf{p}$ in the physical space, the problem is to find the corresponding point $\mathbf{c}$ in the computational space. The first step is to find the grid cell that $\mathbf{p}$ lies in. An intuitive method would be to search for the closest point in the grid using all points. However, this could be expensive if the grid consists of millions of points. Buning [5] suggested searching along edges of the grid cells to find the closest grid point, and then using a "stencil walk" approach to find the exact offset of the particle inside the grid cell. The stencil walk approach, which is based on the Newton-Raphson approach, is summarized below:

1. Select the center of the grid cell (i, j, k) as the initial guess of c, where (i, j, k) is the closest grid point to $\mathbf{p}$. Thus, let $\mathbf{c} = (\xi, \eta, \zeta)$, where $\xi = i + 0.5$, $\eta = j + 0.5$, and $\zeta = k + 0.5$.

2. Convert (ξ, η, ζ) to its corresponding physical point $\mathbf{p}(\xi, \eta, \zeta)$ using trilinear interpolation.

$$\begin{aligned}\mathbf{p}(\xi, \eta, \zeta) \quad = \quad & [(\mathbf{p}_{i,j,k}(1-\alpha) + \mathbf{p}_{i+1,j,k}\alpha)(1-\beta) \\ & +(\mathbf{p}_{i,j+1,k}(1-\alpha) + \mathbf{p}_{i+1,j+1,k}\alpha)\beta](1-\gamma) \\ & +[(\mathbf{p}_{i,j,k+1}(1-\alpha) + \mathbf{p}_{i+1,j,k+1}\alpha)(1-\beta) \\ & +(\mathbf{p}_{i,j+1,k+1}(1-\alpha) + \mathbf{p}_{i+1,j+1,k+1}\alpha)\beta]\gamma, \qquad (5.7)\end{aligned}$$

where $\alpha = \xi - i$, $\beta = \eta - j$, and $\gamma = \zeta - k$.

3. Compute the difference vector $\Delta\mathbf{p}$, where $\Delta\mathbf{p} = \mathbf{p} - \mathbf{p}(\xi, \eta, \zeta)$. This vector indicates how close $\mathbf{p}(\xi, \eta, \zeta)$ is to $\mathbf{p}$ in physical space.

4. Convert $\Delta\mathbf{p}$ to $\Delta\mathbf{c}$, where $\Delta\mathbf{c}$ is the difference vector mapped into computational space. Let $\Delta\mathbf{p} = (\Delta x, \Delta y, \Delta z)$ and $\Delta\mathbf{c} = (\Delta\alpha, \Delta\beta, \Delta\gamma)$. Then,

$$\begin{bmatrix} \Delta\alpha \\ \Delta\beta \\ \Delta\gamma \end{bmatrix} = J^{-1} \begin{bmatrix} \Delta x \\ \Delta y \\ \Delta z \end{bmatrix}, \quad \text{where}$$

$$J = \begin{bmatrix} x_\xi & x_\eta & x_\zeta \\ y_\xi & y_\eta & y_\zeta \\ z_\xi & z_\eta & z_\zeta \end{bmatrix} \quad \text{and} \quad J^{-1} = \begin{bmatrix} \xi_x & \xi_y & \xi_z \\ \eta_x & \eta_y & \eta_z \\ \zeta_x & \zeta_y & \zeta_z \end{bmatrix}. \qquad (5.8)$$

The metric terms in J^{-1} are

$$\xi_x = D(y_\eta z_\zeta - y_\zeta z_\eta),\ \xi_y = D(x_\zeta z_\eta - x_\eta z_\zeta),\ \xi_z = D(x_\eta y_\zeta - x_\zeta y_\eta),$$
$$\eta_x = D(y_\zeta z_\xi - y_\xi z_\zeta),\ \eta_y = D(x_\xi z_\zeta - x_\zeta z_\xi),\ \eta_z = D(x_\zeta y_\xi - x_\xi y_\zeta),$$
$$\zeta_x = D(y_\xi z_\eta - y_\eta z_\xi),\ \zeta_y = D(x_\eta z_\xi - x_\xi z_\eta),\ \zeta_z = D(x_\xi y_\eta - x_\eta y_\xi), \quad (5.9)$$

where D is the determinant of the Jacobian matrix J and

$$D = 1/(x_\xi y_\eta z_\zeta - x_\xi y_\zeta z_\eta - y_\xi x_\eta z_\zeta + x_\zeta y_\xi z_\eta + x_\eta y_\zeta z_\xi - x_\zeta y_\eta z_\xi). \quad (5.10)$$

The partial derivatives in the Jacobian matrix J are the partial derivatives of (5.7), where $\mathbf{p} = (p_x, p_y, p_z)$. For example, $x_\xi = \partial p_x/\partial\xi$, $y_\xi = \partial p_y/\partial\xi$, $z_\xi = \partial p_z/\partial\xi$, and so on.

5. Let $\alpha = \alpha + \Delta\alpha$, $\beta = \beta + \Delta\beta$, and $\gamma = \gamma + \Delta\gamma$. If $\alpha, \beta, \gamma \notin$ [0,1], then $\mathbf{p}$ is outside the current cell. Increase i by 1 if $\alpha > 1$ or decrease i by 1 if $\alpha < 0$. Update j and k according to β and γ, respectively, then go to step 1.

6. Let $\xi = \xi + \Delta\alpha$, $\eta = \eta + \Delta\beta$, $\zeta = \zeta + \Delta\gamma$. If $|\Delta c| < \epsilon$, where ϵ is the chosen tolerance, then $\mathbf{p}(\xi, \eta, \zeta)$ is close enough to $\mathbf{p}$ and its corresponding point (ξ, η, ζ) in computational space has been found. Otherwise, go to step 2.

Step Size Selection

The distance that the particle traverses at each integration step is based on the step size h and the velocity at $\mathbf{p}$. The larger h is, the further $\mathbf{p}$ traverses. If h is too large, then the resulting particle trace can be inaccurate because the particle may have missed important flow features. This is especially true if the flow changes direction rapidly. Likewise, if h is too small, then particles may unnecessarily take too many steps to traverse the grid, which would increase the computation time. A good rule for selecting h is based on the velocity at the current grid cell: If the velocity is large, then h should be small. Buning [5] suggested letting $h = c/max(|U|, |V|, |W|)$, where (U, V, W) represent the computational velocity at $\mathbf{p}$ and

$$\begin{bmatrix} U \\ V \\ W \end{bmatrix} = J^{-1} \begin{bmatrix} u \\ v \\ w \end{bmatrix}. \quad (5.11)$$

Matrix J^{-1} is given in (5.8). The computational velocity is used so that the number of steps in each cell is consistent. For example, if $c = 0.2$, then the particle will traverse no more than one-fifth of a computational cell at each step. Small c yields small steps. A common scheme for adaptively setting h is step doubling, which successively reduces h until some desired accuracy is obtained (see Press et al. [10]). Another scheme for determining h is to base it on the curvature of the particle trace. If the curvature is high, then h should be small.

Velocity Interpolation

In particle tracing, the velocity at the current position of the particle is required to advance the particle. Velocity interpolation is performed at each stage of the RK4 integration. The velocity at $\mathbf{p}$ can be interpolated using the velocities at the corners of the grid cell that contains $\mathbf{p}$. A fast and simple scheme is trilinear interpolation. If $\mathbf{p}$ is in grid cell (i, j, k) and $\mathbf{p}$ has the fractional offsets (α, β, γ) from the grid point at (i, j, k), then

$$\begin{aligned} \mathbf{v}(i+\alpha, j+\beta, k+\gamma) &= [(\mathbf{v}_{i,j,k}(1-\alpha) + \mathbf{v}_{i+1,j,k}\alpha)(1-\beta) \\ &\quad +(\mathbf{v}_{i,j+1,k}(1-\alpha) + \mathbf{v}_{i+1,j+1,k}\alpha)\beta](1-\gamma) \\ &\quad +[(\mathbf{v}_{i,j,k+1}(1-\alpha) + \mathbf{v}_{i+1,j,k+1}\alpha)(1-\beta) \\ &\quad +(\mathbf{v}_{i,j+1,k+1}(1-\alpha) + \mathbf{v}_{i+1,j+1,k+1}\alpha)\beta]\gamma, \quad (5.12) \end{aligned}$$

where $\mathbf{v}_{i,j,k}$ is the velocity at grid point (i, j, k). The trilinear interpolant assumes that the velocity varies linearly across the edges of the cell. Though trilinear interpolation is simple, accuracy may be lost if the grid cell is deformed. In Yeung and Pope [32], higher-order interpolations were compared to linear interpolation. They showed that cubic spline interpolation gives more accurate results than linear interpolation in turbulent flows, and that inaccuracy due to interpolation was greater than that due to integration when small step size is used. The disadvantage of using a cubic spline interpolant is that it is more expensive and requires velocities from 64 nodes versus eight nodes.

Steady flows do not change in time and we can assume that the number of time steps is infinite. Thus, the velocity function is defined for any t. However, unsteady flows vary in time and the velocity function is known only at time steps $t_1, \ldots, t_n$. At time t, if $t \neq t_l$ for $l = 1, \ldots, n$, then a temporal interpolation of velocity is performed prior to the spatial interpolation given in (5.12). If $t_l \leq t \leq t_{l+1}$, then let

$$\mathbf{v}_{i,j,k}(\delta) = (1-\delta)\mathbf{v}_{i,j,k}^{t_l} + \delta\mathbf{v}_{i,j,k}^{t_{l+1}} \qquad \text{and} \qquad \delta = (t - t_l)/(t_{l+1} - t_l),$$

where $\mathbf{v}_{i,j,k}^{t_l}$ and $\mathbf{v}_{i,j,k}^{t_{l+1}}$ are the velocities at grid point (i, j, k) at time t_l and t_{l+1}, respectively. After the above temporal interpolation of velocity, Equation (5.12) can then be evaluated by letting $\mathbf{v}_{i,j,k} = \mathbf{v}_{i,j,k}(\delta)$. The above interpolation assumes that the flow varies linearly between the time steps and that it is only second-order accurate in time over a complete RK4 integration. Darmofal and Haimes [9] suggested a double time-step RK4 method, which is fourth-order accurate in time and does not require temporal interpolation of velocity. The method uses the velocities at the given time steps for the intermediate stages of the RK4 integration, and h is the delta time between two consecutive time steps. Thus, Equation (5.6) becomes

$$\mathbf{a} = 2h\mathbf{v}(\mathbf{p}_k, t), \quad \mathbf{b} = 2h\mathbf{v}(\mathbf{p}_k + \mathbf{a}/2, t+h),$$
$$\mathbf{c} = 2h\mathbf{v}(\mathbf{p}_k + \mathbf{b}/2, t+h), \quad \mathbf{d} = 2h\mathbf{v}(\mathbf{p}_k + \mathbf{c}, t+2h),$$
$$\mathbf{p}_{k+1} = \mathbf{p}_k + (\mathbf{a} + 2\mathbf{b} + 2\mathbf{c} + \mathbf{d})/6, \quad t = t + 2h, \quad k = k+1.$$

Although this method is attractive, the step size is uniform and the accuracy of the integration is dictated by the temporal resolution of the saved time steps. For large unsteady flow data sets, the data are usually saved at some fixed time interval (such as every 50th time step) which may not have sufficient resolution for this method. Furthermore, particles are saved only at every other time step.

Particle Tracing in Multizoned Grids

When tracing particles in a multizoned grid, special consideration is required to track the particles from one block to another. A common grid generation technique used for multi-zoned grids is the 3D Chimera grid-embedding scheme [2]. The output grid generated by this scheme is used by PLOT3D and FAST. The Chimera scheme stores an integer code called *iblank* at each grid point. The iblank of a grid point indicates whether the grid point lies in more than one block. The iblank may also indicate whether the grid point is near a wall boundary or that the velocity at a grid point is invalid. When a particle leaves the current block, the iblanks at the corners of the grid cell are checked to determine if the particle can continue to another block. If this is possible, then a global cell search is performed to find the grid cell in the new block where the particle lies.

Particle Tracing in Moving Grids

For unsteady flow simulations, the grid may move in time. Particle tracing in moving grids requires additional interpolations. In cell search, to find the current grid cell containing $\mathbf{p}$ at time t, an interpolated grid cell is generated. The grid cell is simply a linear interpolation of the grid cells at t_l and t_{l+1} if $t_l \leq t \leq t_{l+1}$. It is not necessary to interpolate the entire grid, only the current grid cell in which $\mathbf{p}$ lies. Thus, at each intermediate stage of the RK4 integration, an interpolated grid cell is computed for velocity interpolation and cell search.

To transform the velocity from physical space to computational space in unsteady flows with moving grids, Equations (5.8) and (5.11) need to be modified to consider t [24] and the grid velocity (x_τ, y_τ, z_τ). Let

$$\begin{bmatrix} U \\ V \\ W \\ T \end{bmatrix} = J^{-1} \begin{bmatrix} u \\ v \\ w \\ t \end{bmatrix},$$

$$J = \begin{bmatrix} x_\xi & x_\eta & x_\zeta & x_\tau \\ y_\xi & y_\eta & y_\zeta & y_\tau \\ z_\xi & z_\eta & z_\zeta & z_\tau \\ t_\xi & t_\eta & t_\zeta & t_\tau \end{bmatrix} \quad \text{and} \quad J^{-1} = \begin{bmatrix} \xi_x & \xi_y & \xi_z & \xi_t \\ \eta_x & \eta_y & \eta_z & \eta_t \\ \zeta_x & \zeta_y & \zeta_z & \zeta_t \\ \tau_x & \tau_y & \tau_z & \tau_t \end{bmatrix}.$$

The partial derivatives t_ξ, t_η, t_ζ are all zero, $t_\tau = t_{l+1} - t_l$, and

$$(x_\tau, y_\tau, z_\tau) = \mathbf{p}^{t_{l+1}}(\xi, \eta, \zeta) - \mathbf{p}^{t_l}(\xi, \eta, \zeta),$$

where $\mathbf{p}^{t_l}(\xi, \eta, \zeta)$ and $\mathbf{p}^{t_{l+1}}(\xi, \eta, \zeta)$ are $\mathbf{p}(\xi, \eta, \zeta)$ at time t_l and t_{l+1} respectively. The nine metric terms in the upper-left corner of J^{-1} are the same as those given in (5.9), and the remaining terms are

$$\xi_t = (-x_\tau \xi_x - y_\tau \xi_y - z_\tau \xi_z)/t_\tau D, \qquad \eta_t = (-x_\tau \eta_x - y_\tau \eta_y - z_\tau \eta_z)/t_\tau D,$$
$$\zeta_t = (-x_\tau \zeta_x - y_\tau \zeta_y - z_\tau \zeta_z)/t_\tau D, \quad \tau_x = 0, \quad \tau_y = 0, \quad \tau_z = 0, \quad \text{and} \quad \tau_t = 1/t_\tau,$$

where the determinant D is given in (5.10).

5.6 Unsteady Flow Analysis Toolkit

A particle tracing system called Unsteady Flow Analysis Toolkit (UFAT) has been developed to compute particle traces in unsteady flows [18]. UFAT has been used to generate streaklines from several large 3D unsteady flows with multizoned curvilinear grids. Some of these grids move in time. UFAT was developed specifically for unsteady flows with a large number of time steps. The major features in UFAT are listed below:

- Computes streamlines, streaklines, pathlines, and timelines
- Performs particle tracing in multizoned curvilinear grids with rigid-body motion
- Allows particle tracing of unsteady flows with hundreds of time steps
- Assigns color to particles based on position, time, or a scalar quantity
- Provides RK2 and RK4 integration schemes with adaptive step size
- Allows particle tracing restricted to a grid surface (oil flow)
- Saves particle traces to a graphics metafile for playback
- Provides save-and-restore option for nonconsecutive run sessions. This feature allows particle traces to be computed from many time steps without requiring all time steps of the flow data to be on line at the same time.

5.6.1 Implementation

Because it is usually impossible to keep all time steps of an unsteady flow data set in memory, UFAT stores only two consecutive time steps of the data in memory. The flow data at time steps t_l and t_{l+1} are used to integrate particles from t_l to t_{l+1}. UFAT uses PLOT3D's particle tracing library for cell search and grid jumping algorithms. I have modified the library to support these and other algorithms in unsteady flows.

UFAT releases particles from the user-specified seed locations at a given time-step interval and tracks the particles. For streaklines, particles are released from the seed locations at each time step. For streamlines, particle traces are computed for the specified seed locations at the given time steps; thus, a set of streamlines is computed for each time step. For pathlines, particles are released from the seed locations at the first time step only, and are then tracked through the given time steps. Pathlines show the trajectories of particles released from the seed locations. For timelines, particles are tracked in the same manner as streaklines, however, they are represented in a different form. For timelines, particles that are released simultaneously are connected (see Figure 5.3).

UFAT currently runs on the Cray, Convex, and SGI systems. The output of UFAT is a graphics metafile. This metafile can then be rendered using a graphics program. Currently, the metafile is written in a format that can be displayed by FAST.

5.6.2 Algorithms

Below is a basic procedure for tracing a given particle $\mathbf{p}$ in the grid domain G and the steady velocity field $\mathbf{v}$.

```
Procedure Trace_Particle( p, G, v )
c = Search_Cell( p, G )
While p ∈ G do
      v_p = Interpolate( p, c, G, v )
      h = Compute_Step_Size( p, c, G, v_p )
      p = Advance( p, h, v_p, G)
      c = Search_Cell( p, G )
End While
```

Search_Cell() searches for the grid cell that $\mathbf{p}$ lies in. Interpolate() interpolates the velocity at $\mathbf{p}$. Based on the interpolated velocity at $\mathbf{p}$, Compute_Step_Size() computes the step size h. Advance() advances particle $\mathbf{p}$ using the RK4 method described in Section 5.5.1. For time-dependent particle tracing in a static grid, the above procedure needs to be modified to consider velocity as a function of time. Let $\Omega = \{(t_l, \mathbf{v}_l), l = 1, \ldots, n\}$, where $\mathbf{v}_l$ is the velocity field at time t_l and n is the number of time steps in the unsteady flow. The procedure for tracing particle $\mathbf{p}$ from time t_1 to t_n is given below.

```
Procedure Trace_Time_Dependent_Particle(p, G, Ω)
t = t1
l = 1
c = Search_Cell( p, G )
While (p ∈ G) and (t < tn) do
      v_p = Interpolate(p, c, G, t, tl, tl+1, vl, vl+1 )
      h = Compute_Step_Size( p, c, G, v_p )
      p = Advance( p, h, v_p, G, t, tl, tl+1, vl, vl+1 )
      c = Search_Cell( p, G )
      t = t + h
      if (t > tl+1) l = l + 1
End While
```

In Interpolate(), the velocity is interpolated in time and space. If the grid moves in time (unsteady grid), then it is also necessary to interpolate the grid cell at each t. This implies that Search_Cell() also needs to interpolate the grid cell in time.

5.6.3 Summary of Steps for Streaklines

Below is a summary of the steps for computing streaklines from an unsteady flow data set.

1. Release one particle from every seed location.

2. Read the first time step's grid and solution files.

3. For the remaining time steps, do the following:

3.1 Read the current time step's grid (if unsteady grid) and solution files.

3.2 Advance all particles from the previous time step to the current time step.

3.3 Release one particle from every seed location.

3.4 Compute color values of all particles based on their positions, time at release, or a specified scalar quantity.

3.5 Save all active particles and their color values to the graphics metafile.

5.7 Examples

This section depicts streaklines computed from several unsteady flow data sets given in Section 5.2. The flow calculations were performed on Cray YMP and C90 supercomputers. The streaklines were computed using UFAT on a Convex C3240 system and animated using FAST on a Silicon Graphics Reality Engine. The streaklines shown in this section are snapshots from the animation, which is an effective way to visualize streaklines. The streaklines are represented by individual particles instead of "connected" particles, because when adjacent particles are too far apart, streaklines can become jagged. Figure 5.4 shows streaklines computed from a simulation of the Harrier jet at time step 106. Particles, which are released from the two exhaust pipes of the jet, are colored by time at release. Blue represents the earliest time and magenta represents the most recent time.

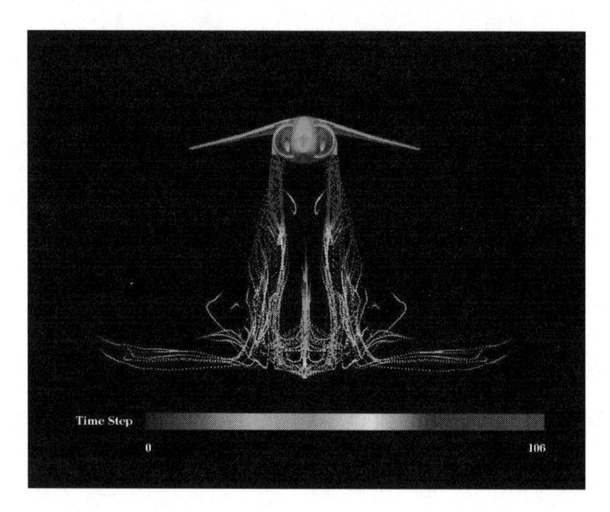

Figure 5.4: Streaklines released from two jet exits of the Harrier jet. Particles are colored by time at release. See Color Plate 34.

The second example is a delta wing in descent. For this simulation, the effects of two thrust-reverser jets in slow-speed flight near the ground are studied (Chawla and van Dalsem [8]). Figure 5.5 shows streaklines surrounding the delta wing at time step 135. The particles are colored by time at release. At each time step, particles are released near the ground and the two jet exits. In the animation, the interactions of the particles released from the jet exits and those released near the ground are clearly visible.

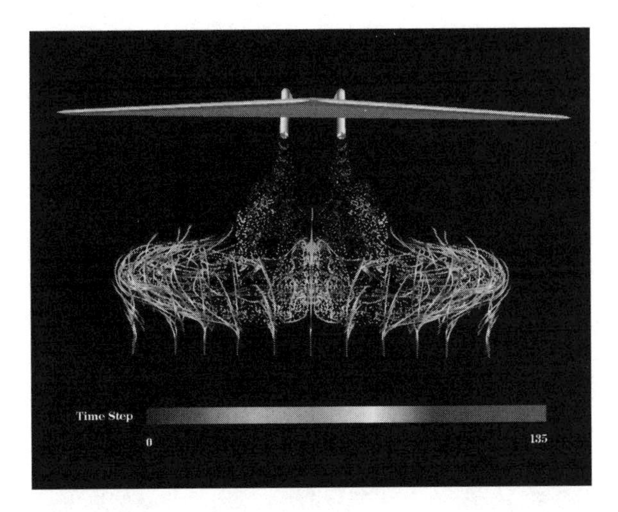

Figure 5.5: Interactions of particles released from the two jets of a descending delta wing and those released near the surface of the ground are shown. Particles are colored by time at release. See Color Plate 35.

Figure 5.6 shows streaklines surrounding the V-22 tilt rotor aircraft after three blade revolutions. The V-22 aircraft has two propellers that rotate in opposite directions. The simulation studies the use of Chimera overset grid methods for computing viscous flow about a complete tilt rotor aircraft (Meakin [19]). In the simulation, there are 1,450 time steps in one blade revolution and each time step requires 52 MB for the grid and solution files (see Table 5.2). Thus, a total of 75.4 GB is required per revolution. Three revolutions are calculated in the simulation. Because there was not enough disk space to store 226.2 GB (3×75.4 GB) of flow data, the data were saved at every 15th time step during the flow calculation. This results in 97 time steps per revolution and 15 GB ($3\times97\times52$ MB) of data were used for the visualization. Particles are released near a rotor blade and the nacelle. In Figure 5.6, the surface of the V-22 aircraft is colored by pressure and the particles are colored by time at release. At each time step, 400 particles are released.

Figure 5.7 shows particles released near the noise of a delta wing with wing rock motion. Particles are represented by spheres to give a better depth perception. The surface of the delta wing is colored by pressure. The simulation is for the development of an experimentally validated CFD tool, which will be used to predict and analyze high-angle-of-attack maneuver aerodynamics (Chaderjian and Schiff [7]).

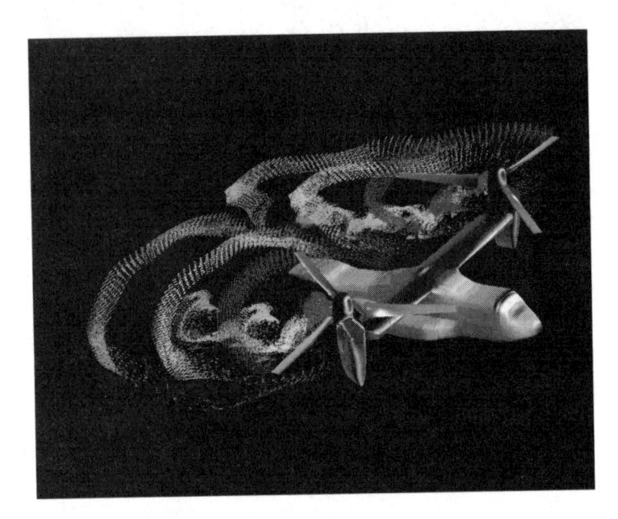

Figure 5.6: Streaklines surrounding the V-22 tilt rotor aircraft after three blade revolutions. The V-22 is colored by pressure and the particles are colored by time at release. See Color Plate 36.

Figure 5.7: Particles released near the noise of a delta wing with wing rock motion. Spiral flow is evident above the left wing and vortex breakdown is visible above the right wing. See Color Plate 37.

5.8 Conclusions

Unsteady flow visualization is a relatively new problem in scientific visualization. The time-dependent nature and the large-scale magnitude of the data make unsteady flow visualization challenging and interesting. Because most existing flow visualization techniques were developed for steady flow data, there is a current need for unsteady flow visualization techniques. The technique described in this chapter uses streaklines to depict time-varying phenomena in unsteady flows. Presently, unsteady flow visualization is likely to rely on scripting and subsampling approaches because of limited hardware capabilities. The size of unsteady flow data sets will continue to increase in the future. There is a continuing need to increase the storage, networking, and computing capabilities as numerical flow simulations become more complex.

Acknowledgments

This work was performed in the Numerical Aerodynamic Simulation Systems Division at NASA Ames Research Center under contract NAS 2-12961. I thank Jill Dunbar, David Kenwright, Gregory Nielson, and the reviewers for their helpful comments. I also thank many colleagues at the NAS division for their support. I thank the following CFD scientists for providing their data sets: Neal Chaderjian and Lewis Schiff (rolling delta wing), Kalpana Chawla (descending delta wing), Sungho Ko (oscillating airfoil), Robert Meakin (V-22 aircraft), and Merritt Smith (Harrier jet). The geometry of the V-22 aircraft was provided by Bell Helicopter Textron Inc. and Boeing Helicopters.

Bibliography

[1] G. Bancroft, F. Merritt, T. Plessel, P. Kelaita, K. McCabe, and A. Globus, "FAST: A multi-processed environment for visualization of computational fluid dynamics," *Proceedings of Visualization '90*, A. Kaufman, editor, San Francisco, Calif., Oct. 1990, pp. 14–27.

[2] J. Benek, P. Buning, and J. Steger, "A 3-D Chimera Grid Embedding Technique," *7th Computational Fluid Dynamics Conference*, AIAA 85-1523, Cincinnati, Ohio, July 1985.

[3] S. Bryson and C. Levit, "The Virtual Wind Tunnel," *IEEE Computer Graphics and Applications*, Vol. 12, No. 4, July 1992, pp. 25–34.

[4] P. Buning, "Sources of Error in the Graphical Analysis of CFD Results," *Journal of Scientific Computing*, Vol. 3, No. 2, 1988.pp. 149–164.

[5] P. Buning, "Numerical Algorithms in CFD Post-Processing, Computer Graphics and Flow Visualization in Computational Fluid Dynamics," *von Karman Institute for Fluid Dynamics Lecture Series 1989-07*, 1989.

[6] P. Buning and J. Steger, "Graphics and Flow Visualization in Computational Fluid Dynamics," *7th Computational Fluid Dynamics Conference*, AIAA 85-1507, Cincinnati, Ohio, July 1985.

[7] N. Chaderjian and L. Schiff, *Navier-Stokes Prediction of Large-Amplitude Forced and Free-to-Roll Delta-Wing Oscillations*, AIAA 94-1884-cp, June 1994.

[8] K. Chawla and W. van Dalsem, "Numerical Simulation of STOL Operations Using Thrust-Vectoring," *AIAA Aircraft Design Systems Meeting*, AIAA 92-4254, Hilton Head, S.C., Aug. 1992.

[9] D. Darmofal and R. Haimes, "An Analysis of 3-D Particle Path Integration Algorithms," *12th Computational Fluid Dynamics Conference*, AIAA 95-1713, San Diego, Calif., June 1995.

[10] W. Press et al., *Numerical Recipes*, Cambridge University Press, 1986.

[11] R. Haimes, *pV3: A Distributed System for Large-Scale Unsteady CFD Visualization*, AIAA 94-0321, 32nd AIAA Aerospace Sciences Meeting and Exhibit, Reno, Nev., Jan. 1994.

[12] R. Haimes and M. Giles, *VISUAL3: Interactive Unsteady Unstructured 3-D Visualization*, AIAA 91-0794, 29th AIAA Aerospace Sciences Meeting and Exhibit, Reno, Nev., Jan. 1991.

[13] J. Hultquist, *Improving the Performance of Particle Tracing in Curvilinear Grids*, AIAA 94-0324, 32nd AIAA Aerospace Sciences Meeting and Exhibit, Reno, Nev., Jan. 1994.

[14] D. Jespersen and C. Levit, *Numerical Simulation of Flow Past a Tapered Cylinder*, AIAA 91-0801, 29th AIAA Aerospace Sciences Meeting and Exhibit, Reno, Nev., Jan. 1991.

[15] D. Kenwright and D. Lane, "Optimization of Time-Dependent Particle Tracing Using Tetrahedral Decomposition," *Proceedings of Visualization '95*, G. Nielson and D. Silver, editors, Atlanta, Ga., Oct. 1995, pp. 321–328.

[16] D. Kenwright and G. Mallinson, "A 3-D Streamline Tracking Algorithm Using Dual Stream Functions," *Proceedings of Visualization '92*, A. Kaufman and G. Nielson, editors, Boston, Mass., Oct. 1992, pp. 62–68.

[17] D. Lane, "Visualization of Time-Dependent Flow Fields," *Proceedings of Visualization '93*, G. Nielson and D. Bergeron, editors, San Jose, Calif., Oct. 1993, pp. 32–38.

[18] D. Lane, "UFAT—A Particle Tracer for Time-Dependent Flow Fields," *Proceedings of Visualization '94*, D. Bergeron and A. Kaufman, editors, Washington, D.C., Oct. 1994, pp. 257–264.

[19] R. Meakin, "Moving Body Overset Grid Methods for Complete Aircraft Tiltrotor Simulations," *11th AIAA Computational Fluid Dynamics Conference*, Orlando, Fla., July 1993.

[20] E. Murman and K. Powell, "Trajectory Integration in Vortical Flows," *AIAA Journal*, Vol. 27, No. 7, July 1989, pp. 982–984.

[21] H. Pagendarm, "HIGHEND, A Visualization System for 3-D Data with Special Support for Postprocessing of Fluid Dynamics Data," *Visualization in Scientific Computing*, M. Grave, Y. LeLous, and W. Hewitt, editors, Springer-Verlag, Heidelberg, 1993.

[22] F. Post and T. van Walsum, "Fluid Flow Visualization," *Focus on Scientific Visualization*, H. Hagen, H. Mueller, and G. Nielson, editors, Springer, Berlin, 1993, pp. 1–40.

[23] F. Post and J. van Wijk, "Visual Representation of Vector Fields: Recent Developments and Research Directions," *Scientific Visualization: Advances and Challenges*, L. Rosenblum et al., editors, Academic Press, San Diego, Calif., 1993, pp. 367–390.

[24] T. Pulliam, "Efficient Solution Methods for Navier-Stokes Equations," *von Karman Institute for Fluid Dynamics Lecture Series*, 1986.

[25] A. Sadarjoen, T. van Walsum, A. Hin, and F. Post, "Particle Tracing Algorithms for 3D Curvilinear Grids," *5th Eurographics Workshop on Visualization in Scientific Computing*, Rostock, Germany, May 1994.

[26] H. Schlichting, *Boundary Layer-Theory*, McGraw Hill, New York, 1979.

[27] S. Shirayama, "Processing of Computed Vector Fields for Visualization. *Journal of Computational Physics*, Vol. 106, 1993, pp. 30–41.

[28] M. Smith, W. van Dalsem, F. Dougherty, and P. Buning, *Analysis and Visualization of Complex Unsteady Three-Dimensional Flows*, AIAA 89-0139, 27th AIAA Aerospace Sciences Meeting and Exhibit, Reno, Nev., Jan. 1989.

[29] D. Speray and S. Kennon, "Volume Probes," *Proceedings of San Diego Workshop on Volume Visualization, Computer Graphics*, Vol. 24, No. 5, Nov. 1990, pp. 5–12.

[30] A. Vaziri, M. Kremenetsky, M. Fitzgibbon, and C. Levit, *Experiences with CM/AVS to Visualize and Compute Simulation Data on the CM-5*, NAS Applied Research Technical Report RNR 94-005, NASA Ames Research Center, 1994.

[31] H. Vollmers, "The Recovering of Flow Features from Large Numerical Data Bases," *VKI Lecture Series on Computer Graphics and Flow Visualization in Computational Fluid Dynamics*, von Karman Institute for Fluid Dynamics Lecture Series, 1991.

[32] P. Yeung and S. Pope, "An Algorithm for Tracking Fluid Particles in Numerical Simulations of Homogeneous Turbulence," *Journal of Computational Physics*, Vol. 79, 1988, pp. 373–416.

Part II

Frameworks and Methodologies

Chapter 6

Integrated Volume Rendering and Data Analysis in Wavelet Space

M. H. Gross

Abstract. *The following chapter describes a framework for volume data analysis and visualization using wavelet transforms (WTs). It is based on the idea that WTs have proved to provide powerful features for various applications in the field of data analysis. Due to the basic properties of this transform, such as local support and orientation selectivity, many researchers tried to exploit the WT for the extraction of local data features. In particular, in image texture analysis, wavelet-based feature extractors accomplished highly accurate segmentation results that can be extended straightforwardly to volumetric data sets.*

In addition, the compact coding properties of orthonormal wavelets allow us to decompose any finite energy function and to approximate it from its bases. Therefore, we can develop rendering methods that provide an approximate solution of the low albedo volume rendering equation. Since generally, the wavelets are not given in a closed form, the approach reported in this chapter is based on a piecewise polynomial representation. Isosurfaces can easily be obtained from the data, either by ray tracing of the bases or by simple marching cubes techniques.

Hence, the WT provides a uniform data representation that features both data analysis and data visualization. This chapter first introduces the mathematical foundations of the WT, it briefly reviews different types of basis functions, and it stresses implementation details using iterated QMF-pair filters. Furthermore, separable extensions to multiple dimensions are explained and methods for deriving local data features from the wavelet pyramid are elucidated. For data analysis purposes, a newly developed image texture analysis pipeline contains a WT, a principal component analysis, normalization procedures, and a neural network. Volume rendering is accomplished by projecting the 3D wavelets onto the viewing ray and by piecewise analytic integration of the rendering equation. The methods reported here are illustrated by various examples.

6.1 Introduction

The wavelet transform has gained much attention for providing innovative solutions to various technical problems in a broad range of applications. Following the mathematical formulation of the wavelet transform [12, 5, 21], a great deal of work has been done to employ wavelets for hierarchical data coding and representation. This chapter, however, reports on a framework for integrated volume rendering and data analysis in wavelet space. Here, the properties of the WT are exploited to derive local data features that allow us to accomplish segmentation and classification tasks. Here, the WT provides a unique data representation that enables us to incorporate data analysis as an important preprocessing step in rendering. This new concept is based on the following observations: In image processing, research has been focused on the investigation of multiresolution analysis techniques for texture feature extraction. [3] used Gabor functions of different spectral ranges and orientations to derive a multiscale sight onto the texture. [26] employed a set of local linear transforms as an initial step in a texture recognition pipeline and combined them with dimensionality reduction techniques. Orthogonal wavelets and wavelet packages for texture analysis have been introduced by [17] and [4]. Obviously, most approaches derive global wavelet features in terms of means from a specific type of texture and they transform each subset separately into wavelet space. This procedure renders high classification rates but it neither takes into account texture boundaries nor does it take advantage of the spatial localization of the wavelet transform. Consequently, for many practical applications, for example, remote sensing or medical imaging, difficulties will arise in applying these techniques because they do not provide large spatially coherent textures. [9] and [8] describe a different approach for deriving features from wavelet transforms using the spatial and frequency localization of the WT. Here, the segmentation pipeline consists of the WT, principal component analysis, normalization, and a neural classifier. This method can easily be extended to volume textures.

In volume rendering, initial work has also been done to employ wavelets for data representation. [22] describes a method to obtain isosurfaces from volume data by computing a continuous approximation with wavelet bases. [13] used wavelets to elegantly control volume morphing algorithms and [19] considers line integration as a texture splatting problem. [11] describes a new method to approximate the volume rendering equation using wavelet transforms. For this purpose, the initial volume data set is transformed into wavelet space using separable 3D extensions of orthonormal wavelet types. Since some wavelets, such as the Daubechies, are not given in a closed form solution, they approximate the basis functions with piecewise polynomial splines. This allows a continuous hierarchical approximation of the 3D data set. Due to the local support of the wavelet bases, local level-of-detail can be controlled efficiently. This enables us to emphasize locally interesting features. Once this continuous piecewise polynomial approximation is computed, the volume intensity function along the ray can be easily formulated and a linear approximation of the exponential absorption term provides a polynomial approximation of the entire rendering integral. This finally leads towards analytic solutions.

In order to define an integrated framework, this chapter reviews and unifies the initial concepts provided by [10] and [9]. The organization of the chapter is as follows: First of all, the mathematical basics of the wavelet transform are briefly elucidated and how to obtain separable 2D and 3D extensions is shown. Different wavelet types are discussed and compared to each other in terms of smoothness and compact support. The second section introduces the concepts of integrating volume rendering and data analysis in wavelet space. The data analysis pipeline is illuminated in detail in Section 6.3 and examples are given for the segmentation performance on image textures. Section 6.4 sheds light on the continuous approximation of the data with wavelet bases and provides an approximation of the volume rendering integral for the ray-casting process. A method for computing isosurfaces in wavelet space is illustrated using examples from laser range data sets. We compare results obtained from Kalra's [16] method to those of the marching cubes [20] reconstruction technique.

6.2 Mathematical Foundations

6.2.1 The Wavelet Transform

The 1D wavelet transform (WT) is an integral transform of any finite energy function $f(x) \in L^2(\mathbf{R})$ using a set of self-similar basis functions $\psi_{ab}(x)$. Its generic continuous form description for real functions is provided as the following inner product:

$$WT_{f,\psi}(a,b) = \langle f, \psi_{ab} \rangle = \int_{-\infty}^{\infty} \psi_{ab}(x) f(x) dx \quad a, b \in \mathbf{R}. \tag{6.1}$$

$L^2(\mathbf{R})$ denotes the Hilbert space of square integrable functions.

The individual basis functions on the real axis are derived from each other by scaling and shifting one prototype function ψ controlled by the parameters a and b, respectively [12].

$$\psi_{ab}(x) = \frac{1}{\sqrt{|a|}} \psi \left(\frac{x-b}{a} \right). \tag{6.2}$$

One required property of the orthonormal bases is their band-pass behavior which is defined as

$$\Psi_{ab}(0) = 0, \; \Psi_{ab}(\omega): \quad \textit{Fourier Transform of } \psi_{ab}(x) \tag{6.3}$$

Like any other type of linear transform, the WT enables the decomposition and the expansion of the initial function $f(x)$ by linear combinations of the basis functions.

In order to handle this method with a computer, it is necessary to set up a discrete version. A dyadic scaling of the bases with $a = 2^m$ and a unit shift $b = p2^m$ yields:

$$\psi_{mp}(x) := 2^{-\frac{m}{2}} \psi(2^{-m}x - p), \quad (\psi_{mp})_{p \in \mathbf{Z}} \; \textit{basis of vectorspace } U_m \tag{6.4}$$

where $m: \; 1, .., M$ denotes the depth of the iteration.

In many construction schemes, the bases are furthermore supposed to be orthonormal to each other [5].

$$\begin{aligned} \langle \psi_{mp}, \psi_{\tilde{m}\tilde{p}} \rangle &= \int_{-\infty}^{\infty} \psi_{mp}(x) \psi_{\tilde{m}\tilde{p}}(x) dx = \delta_{m\tilde{m}} \delta_{p\tilde{p}} \\ \text{with } \delta_{ij} &:= \begin{cases} 1 & \text{if } i = j \\ 0 & \text{else} \end{cases} \qquad (\textit{Kronecker-Delta-function}) \end{aligned} \tag{6.5}$$

Thus, the transform with discrete orthonormal wavelets can be formalized as

$$DWT_{f,\psi}(m,p) = 2^{-\frac{m}{2}} \int_{-\infty}^{\infty} \psi(2^{-m}x - p) f(x) dx \qquad m, p \in \mathbf{Z} \tag{6.6}$$

Mallat [21] stresses this concept by defining a set of multiresolution function spaces V_m that provide approximations of all $f(x) \in L^2(\mathbf{R})$. The space of resolution m is derived from the higher resolution space by adding the orthogonal complement space U_m.

$$V_{m-1} = V_m \oplus U_m \tag{6.7}$$

where $\oplus$: *direct sum.*

It can be proved that the so-called scaling functions $\phi \in L^2(\mathbf{R})$ with $\phi_{mp}(x) = 2^{-\frac{m}{2}}\phi(2^{-m}x - p)$ provide orthonormal bases of the vectorspaces V_m in each resolution step. The wavelet $\psi \in L^2(\mathbf{R})$ with $\psi_{mp}(x) = 2^{-\frac{m}{2}}\psi(2^{-m}x - p)$, however, is proven to be a basis of the orthogonal complement space U_m.

The statements explained above offer an iterated decomposition scheme, where an initial discrete function can be successively approximated for a given iteration depth M using the scaling function and wavelets of each orthogonal complement space U_m.

$$V_0 = V_M \oplus U_M \oplus \cdots \oplus U_1 \tag{6.8}$$

Hence, we obtain the following approximation:

$$\begin{aligned} f(x) := \sum_p c_p^0 \phi_{0p}(x) &= \sum_p c_p^1 \phi_{1p}(x) + \sum_p d_p^1 \psi_{1p}(x) \qquad f(x) \in V_0 \\ &= \sum_p c_p^M \phi_{Mp}(x) + \sum_{m=1}^{M} \sum_p d_p^m \psi_{mp}(x) \end{aligned} \tag{6.9}$$

where c_p^M and d_p^m denote the coefficients of the transform.

6.2.2 Extensions to Multiple Dimensions

So far, the definitions have been restricted to the one-dimensional case. For multidimensional signal processing, as image analysis or volume rendering, it is necessary to extend the method to multiple dimensions. Besides the nontrivial nonseparable case [24], there is a straightforward way to accomplish this by means of nonstandard tensor product extensions of the one-dimensional representatives of the function spaces and of their bases. Once V_m is given, we can define a 2D version V_m^2 by

$$\begin{aligned} V_m^2 &= V_m \otimes V_m \\ V_m^2 &= (V_{m+1} \oplus U_{m+1}) \otimes (V_{m+1} \oplus U_{m+1}) \\ V_m^2 &= V_{m+1}^2 \oplus U_{m+1}^{2,1} \oplus U_{m+1}^{2,2} \oplus U_{m+1}^{2,3} \end{aligned} \tag{6.10}$$

where $\otimes$: *Tensor product operator.*

In Equation (6.10) the initial space is broken up into three spaces that account for the differences in the signal and into a low-pass part. The resulting 2D versions of the respective wavelet bases yield:

$$\begin{aligned}
\phi^2_{mpq}(x,y) &:= 2^{-m}\phi(2^{-m}x-p)\phi(2^{-m}y-q) \\
\psi^{2,1}_{mpq}(x,y) &:= 2^{-m}\psi(2^{-m}x-p)\phi(2^{-m}y-q) \\
\psi^{2,2}_{mpq}(x,y) &:= 2^{-m}\phi(2^{-m}x-p)\psi(2^{-m}y-q) \\
\psi^{2,3}_{mpq}(x,y) &:= 2^{-m}\psi(2^{-m}x-p)\psi(2^{-m}y-q)
\end{aligned} \tag{6.11}$$

Unfortunately, some wavelet types are not defined in a closed form solution and consequently the convolution products of Equation (6.1) cannot be explicitly computed. The implementation of orthogonal wavelet transforms often employs so-called quadrature mirror pair filters (QMFs). The basic scheme of this filter bank is illustrated in Figure 6.1a for a decomposition of 2D data sets. The initial data set is filtered along the x- and y-axis and subsampled by the factor 2 using the two filters $H'(\omega) = \sum_{n=-\infty}^{\infty} h(-n)e^{in\omega}$ and $G'(\omega) = e^{-i\omega}\overline{H'(\omega+\pi)}$ with $h(n) := 2^{-\frac{1}{2}}\int_{-\infty}^{\infty}\phi\left(\frac{x}{2}\right)\phi(x-n)dx$, respectively. The results of this process are three detail signals $D^1_{2^m}f, \ldots, D^3_{2^m}f$ that account for the oriented wavelets in this channel and a low-pass signal $A_{2^m}f$ that is further decomposed. This iterated scheme corresponds to a dyadic subband coding of the data [2] and is illustrated in Figure 6.1b. Note that the filters can be considered operators projecting from one function space into another.

As stated earlier, the WT is located both in the spatial and in the frequency plane within the boundaries of the Heisenberg principle. This property allows us to locally adapt the level-of-detail (lod) of any reconstruction to interesting features and to control the approximation, neglecting unimportant or low-energy coefficients of the transform. Figure 6.2 illustrates the effects of local lod-filtering for different wavelet types (see also Section 6.2.3). The pyramid shows especially the orientation selectivity of the different frequency channels.

The QMF method features critical sampling of the spectrum and the WT can be formulated as a filter design problem for $H(\omega)$ where the conditions for orthogonal wavelets collapse in Fourier space to [14]:

$$\begin{aligned}
|H(\omega)|^2 + |H(\omega+\pi)|^2 &= 1 \\
|H(0)| &= 0
\end{aligned} \tag{6.12}$$

For volumetric data sets we can straightforwardly set up a 3D orthogonal wavelet basis using the same tensor product technique. We obtain the function space V^3_m:

$$\begin{aligned}
V^3_m &= V_m \otimes V_m \otimes V_m \\
V^3_m &= (V_{m+1} \oplus U_{m+1}) \otimes (V_{m+1} \oplus U_{m+1}) \otimes (V_{m+1} \oplus U_{m+1}) \\
V^3_m &= V^3_{m+1} \oplus U^{3,1}_{m+1} \oplus U^{3,2}_{m+1} \oplus \cdots \oplus U^{3,7}_{m+1}
\end{aligned} \tag{6.13}$$

Equation (6.13) shows, that at each decomposition step, the space is broken up into seven orthogonal complements that account for the principal orientations of the data, respectively. The corresponding 3D versions of the wavelets and of the scaling function can

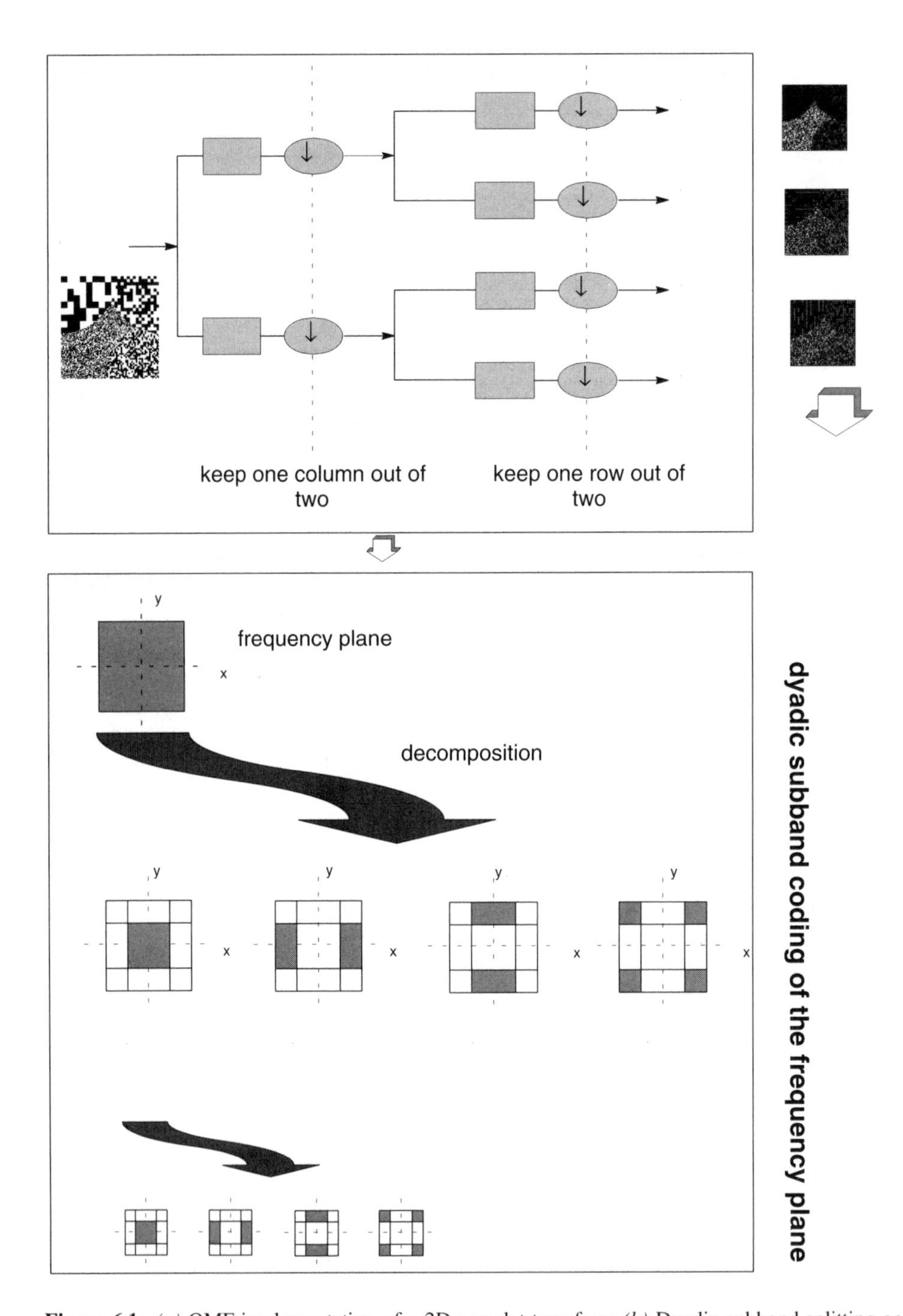

Figure 6.1: *(a)* QMF implementation of a 2D wavelet transform *(b)* Dyadic subband splitting according to [2].

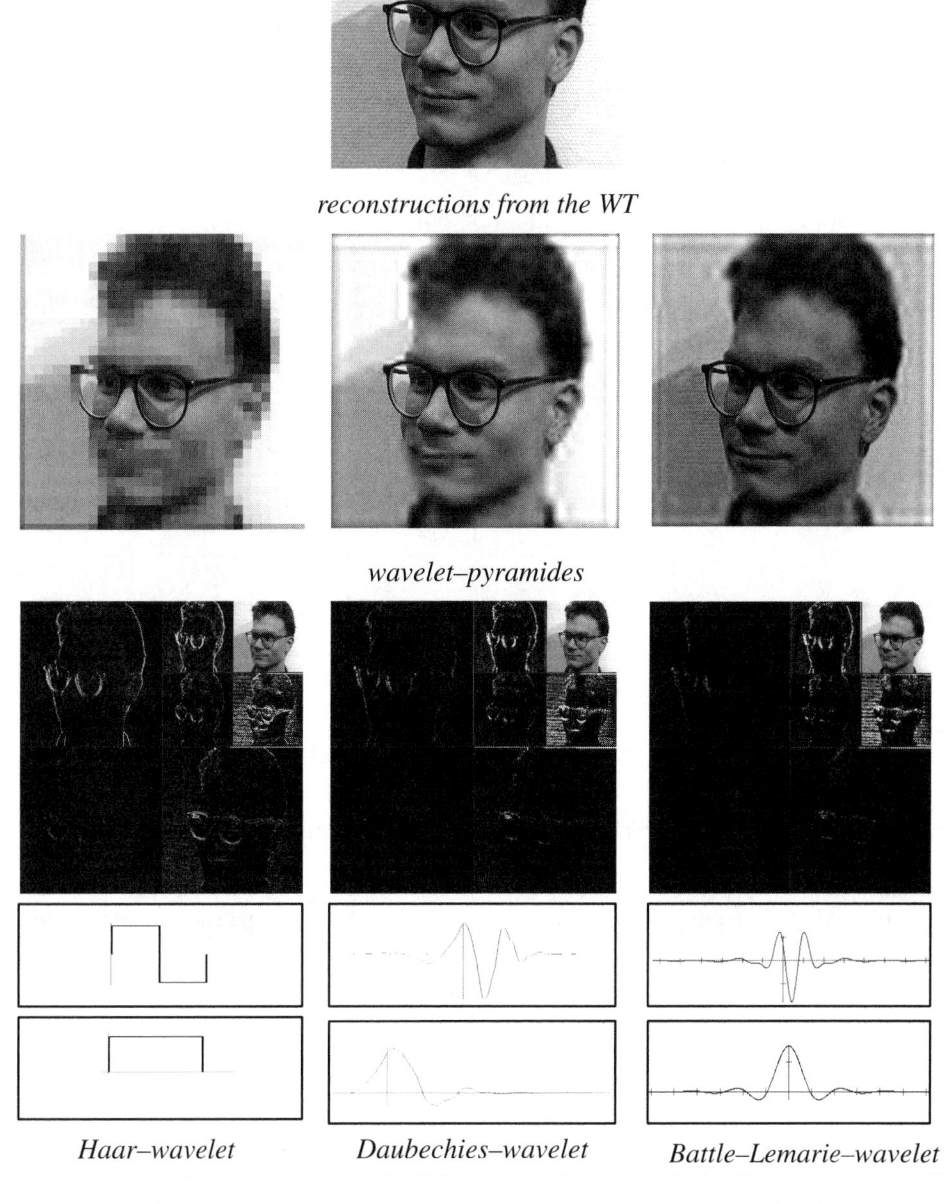

Figure 6.2: Initial image, its wavelet transforms and reconstructions with local variations of the level-of-detail for different wavelet types.

easily be derived from their one-dimensional relatives, as

$$
\begin{aligned}
\phi^3_{mpqr}(x,y,z) &:= 2^{-\frac{3m}{2}}\phi(2^{-m}x-p)\phi(2^{-m}y-q)\phi(2^{-m}z-r) \\
\psi^{3,1}_{mpqr}(x,y,z) &:= 2^{-\frac{3m}{2}}\phi(2^{-m}x-p)\phi(2^{-m}y-q)\psi(2^{-m}z-r) \qquad (6.14)\\
\psi^{3,2}_{mpqr}(x,y,z) &:= 2^{-\frac{3m}{2}}\phi(2^{-m}x-p)\psi(2^{-m}y-q)\phi(2^{-m}z-r) \\
\psi^{3,3}_{mpqr}(x,y,z) &:= 2^{-\frac{3m}{2}}\phi(2^{-m}x-p)\psi(2^{-m}y-q)\psi(2^{-m}z-r) \\
\psi^{3,4}_{mpqr}(x,y,z) &:= 2^{-\frac{3m}{2}}\psi(2^{-m}x-p)\phi(2^{-m}y-q)\phi(2^{-m}z-r) \\
\psi^{3,5}_{mpqr}(x,y,z) &:= 2^{-\frac{3m}{2}}\psi(2^{-m}x-p)\phi(2^{-m}y-q)\psi(2^{-m}z-r) \\
\psi^{3,6}_{mpqr}(x,y,z) &:= 2^{-\frac{3m}{2}}\psi(2^{-m}x-p)\psi(2^{-m}y-q)\phi(2^{-m}z-r) \\
\psi^{3,7}_{mpqr}(x,y,z) &:= 2^{-\frac{3m}{2}}\psi(2^{-m}x-p)\psi(2^{-m}y-q)\psi(2^{-m}z-r)
\end{aligned}
$$

The resulting data pyramid is illustrated in Figure 6.3. In each branch of this tree the different signal components are emphasized, being extracted from the corresponding oriented wavelet function.

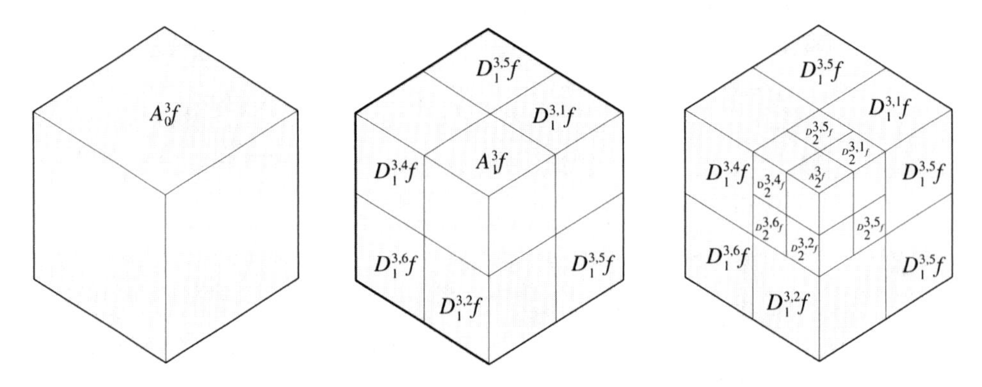

Figure 6.3: Iterated decomposition of the initial volume data by the wavelet transform.

Let $\tilde{c}_j \in \{c^m_{pqr}, d^{m,1}_{pqr}, d^{m,2}_{pqr}, d^{m,3}_{pqr}, d^{m,4}_{pqr}, d^{m,5}_{pqr}, d^{m,6}_{pqr}, d^{m,7}_{pqr}\}$ and $w_j(x,y,z)$ be the wavelet $\psi^{3,1}_{mpqr}$ or scaling basis function ϕ^3_{mpqr}. We write the expansion of the volume function $f(x,y,z)$:

$$
\begin{aligned}
f(x,y,z) &= \sum_r\sum_q\sum_p c^M_{pqr}\phi^M_{pqr} + \Bigl(\sum_{m=1}^{M}\bigl(d^{m,1}_{pqr}\psi^{m,1}_{pqr} + d^{m,2}_{pqr}\psi^{m,2}_{pqr} + d^{m,3}_{pqr}\psi^{m,3}_{pqr} \\
&+ d^{m,4}_{pqr}\psi^{m,4}_{pqr} + d^{m,5}_{pqr}\psi^{m,5}_{pqr} + d^{m,6}_{pqr}\psi^{m,6}_{pqr} + d^{m,7}_{pqr}\psi^{m,7}_{pqr}\bigr)\Bigr) \\
&= \sum_{j=1}^{N}\tilde{c}_j \cdot w_j(x,y,z) \qquad (6.15)
\end{aligned}
$$

where N is the number of basis functions in finite dimensional projections.

Due to the immense number of basis functions (for $128^3 > 2$ Mio.), we are forced to define a significance measure, a so-called oracle, to reject unimportant coefficients. A generic significance for orthogonal settings can be derived by computing the local signal energy E_j from $|| \cdot ||_{L^2}$. It is obtained for each basis as

$$E_j = \int_{-\infty}^{\infty} \int_{-\infty}^{\infty} \int_{-\infty}^{\infty} |\tilde{c}_j w_j(x,y,z)|^2 \, dxdydz = | \tilde{c}_j | \cdot || w_j(x,y,z) ||_{L^2}^2 . \tag{6.16}$$

Supposing the basis functions to hold

$$\langle w_i, w_j \rangle = \delta_{ij}, \tag{6.17}$$

we obtain the local energy by the square of the corresponding coefficients

$$E_j = (\tilde{c}_j)^2 . \tag{6.18}$$

The total energy E_{tot} of a 3D signal of size N is obtained by summing up all squared coefficients of the WT:

$$E_{tot} = \sum_{j=1}^{N^3} | \tilde{c}_j |^2 \tag{6.19}$$

Obviously, a good L^2 oracle is given by the sorting and rejection of coefficients according to their magnitude.

In addition, these equations allow us to estimate the error bounds of our approximation when rejecting unimportant coefficients. Once the coefficients are filtered, we obtain a residual approximation as

$$\hat{f}(x,y,z) = \sum_{j=1}^{K} \hat{c}_j \cdot \hat{w}_j(x,y,z) \tag{6.20}$$

where $K \leq N$. The same conditions hold for the 2D case.

6.2.3 Wavelet Bases for Data Representation

The elucidations above do not further restrict the mathematical properties of the wavelet bases and there have been different construction schemes proposed in the literature depending on smoothness, strict local support, and other criteria. This section briefly introduces the most important orthogonal wavelets (see also [9]).

Haar Wavelets

A very simple, but discontinuous basis is given with the Haar wavelet, whose scaling function and corresponding wavelet is defined by

$$\phi(x) := \begin{cases} 1 & \text{for } 0 \leq x \leq 1 \\ 0 & \text{otherwise} \end{cases} \qquad \psi(x) := \begin{cases} 1 & \text{for } 0 \leq x \leq \frac{1}{2} \\ -1 & \text{for } \frac{1}{2} \leq x < 1 \\ 0 & \text{otherwise} \end{cases} \tag{6.21}$$

The Fourier transforms of these functions (Equation (6.22)) show that the expressions have optimal localization properties in the spatial domain but a weak localization in frequency.

$$\Phi(2\omega) = e^{-i\omega}\frac{\sin(\omega)}{\omega} \qquad \Psi(2\omega) = ie^{-i\omega}\frac{\sin^2(\frac{\omega}{2})}{\frac{\omega}{2}} \tag{6.22}$$

where i is a complex operator.

Daubechies Wavelets

In order to obtain better localization in frequency along with a minimal local support in space, and smoother basis functions, Daubechies [5] proposes wavelet types as follows: The smoothness can be measured by the regularity R, (of any function $\phi(x)$) which is defined as the maximum of R such that

$$\mid \Phi(\omega) \mid \leq \frac{c}{(1+ \mid \omega \mid)^{R+1}} \qquad c \in \mathbf{R}^+ \tag{6.23}$$

This relation also describes the continuity of $\phi(x)$, where $\phi(x) \in C^i(\mathbf{R})$ with $i \leq R$. The regularity of the Daubechies wavelets is proportional to the number of vanishing moments V, defined by

$$\int_{-\infty}^{\infty} x^n \psi(x) dx = 0 \qquad n \in \{0, \ldots, V-1\} \tag{6.24}$$

The function ϕ is only constrained by Equation (6.25).

$$\int_{-\infty}^{\infty} \phi(x) dx = 1 \tag{6.25}$$

Note that these wavelets are implemented by their corresponding QMF pairs.

Coiflet Bases

Further restricting the scaling function to a fixed number of vanishing moments following Equation (6.26),

$$\int_{-\infty}^{\infty} x^n \phi(x) dx = 0 \quad n \in \{1, \ldots, V'-1\} \tag{6.26}$$

results in a different wavelet type, the so-called Coiflet wavelet [1]. Besides the Haar wavelet, strict finite support comes along in the orthonormal case with a lack of symmetry. However there is a relationship between the number of vanishing moments, stated earlier, and the symmetry of the function. Thus Coiflet bases appear to be "more symmetric" than Daubechies ones.

Battle-Lemarie Wavelets

Orthogonal spline wavelets with an infinite support can be approximated in the frequency domain. An example with four vanishing moments is given by Equations (6.27) through (6.30) (see [21]).

$$\hat{\phi}(\omega) := \frac{1}{\omega^4 \sqrt{\Sigma_8(\omega)}} \tag{6.27}$$

where

$$\Sigma_8(\omega) := \frac{N_1(\omega) + N_2(\omega)}{105 \left(\sin \frac{\omega}{2}\right)^8} \tag{6.28}$$

with

$$N_1(\omega) := 5 + 30 \left(\cos \frac{\omega}{2}\right)^2 + 30 \left(\sin \frac{\omega}{2}\right)^2 \left(\cos \frac{\omega}{2}\right)^2 \tag{6.29}$$

and

$$N_2(\omega) := 2 \left(\sin \frac{\omega}{2}\right)^4 \left(\cos \frac{\omega}{2}\right)^2 + 70 \left(\cos \frac{\omega}{2}\right)^4 + \frac{2}{3} \left(\sin \frac{\omega}{2}\right)^6 . \tag{6.30}$$

The resulting wavelet type is often referred to in the literature as the *Battle-Lemarie* wavelet.

Figures 6.4 through 6.6 again give a graphical representation of the shape of some 2D wavelet basis functions. We can clearly distinguish between the smooth shape of the Battle-Lemarie wavelet on the one hand and the strict local support of the Daubechies wavelet on the other hand. It is clear to see where the name *"wavelet"* comes from.

We should also note that for continuous approximations we face the competing criteria of providing strict compact support and smooth symmetric shapes along with orthonormality in order to achieve a perfect reconstruction and compact coding. A good introduction to wavelet theory is given in [14].

6.3 A Framework for Integrated Data Analysis and Visualization

As stated earlier, the WT has been successfully used for data feature extraction [26, 17, 4], as well as for volume and isosurface rendering [22, 13, 10]. Particularly in the field of image processing, several approaches to texture analysis have been proposed so far. Most of these techniques use a local WT on single coherent texture samples. In order to overcome this restriction, [9] proposed a new concept for texture feature extraction in images based on a global WT. The features are derived from the local wavelet coefficients and take advantage of the localization properties of the WT. These results motivated us to extend the segmentation scheme to 3D and to embed it into the rendering process—all that in the underlying data space of the WT. Figure 6.7 again illustrates the pipeline, where one global WT is first performed on the initial data set.

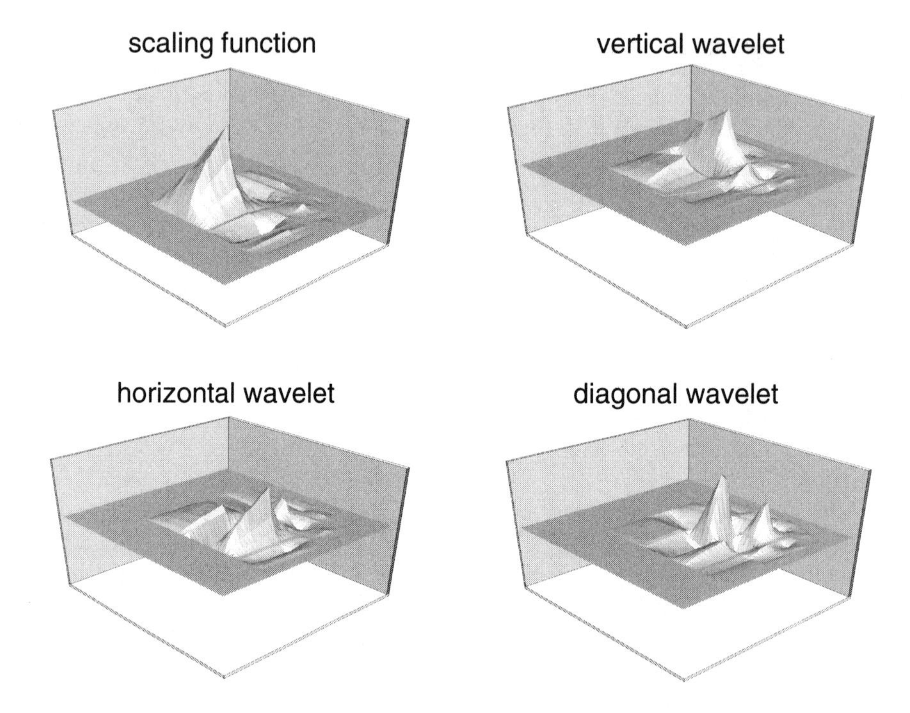

Figure 6.4: 2D Daubechies 4-tap wavelets and scaling function.

Once the data are transformed, the pipeline associates a segmentation result to all voxels in the data set. Therefore, we first extract a wavelet stream of coefficients from the wavelet pyramid for each voxel. This vector is taken as the feature vector describing the local properties of the data surrounding the voxel. These data must be decorrelated because the extraction scheme renders slightly correlated features, in spite of the orthogonality of the transform. After normalizing the features, they are fed into a neural network [11] that accomplishes clustering and classification. The result of this process is a segmentation map that can be imported into the volume renderer. The following two sections shed light on both data analysis and volume rendering [10] based on WTs.

6.4 Data Analysis in Wavelet Space

6.4.1 Feature Extraction in Images and Volumes

As stated earlier, the result of the wavelet transform is a pyramidal representation of the image as illustrated in Figure 6.8b. The initial image is separated into a dyadic set of frequency channels where horizontal, vertical, and diagonal components are split by the corresponding wavelets. Due to the local properties of the wavelet transform, the idea of [9] was to derive local texture features in the adjacency of any pixel (x, y) of the initial image by a set of respective coefficients from the pyramid. Supposing a depth of M in the pyramid, the set $\mathrm{g}(x, y) = \{g_i(x, y)\} = \{g_0, \ldots, g_{4M-1}\}$ renders a feature vector for

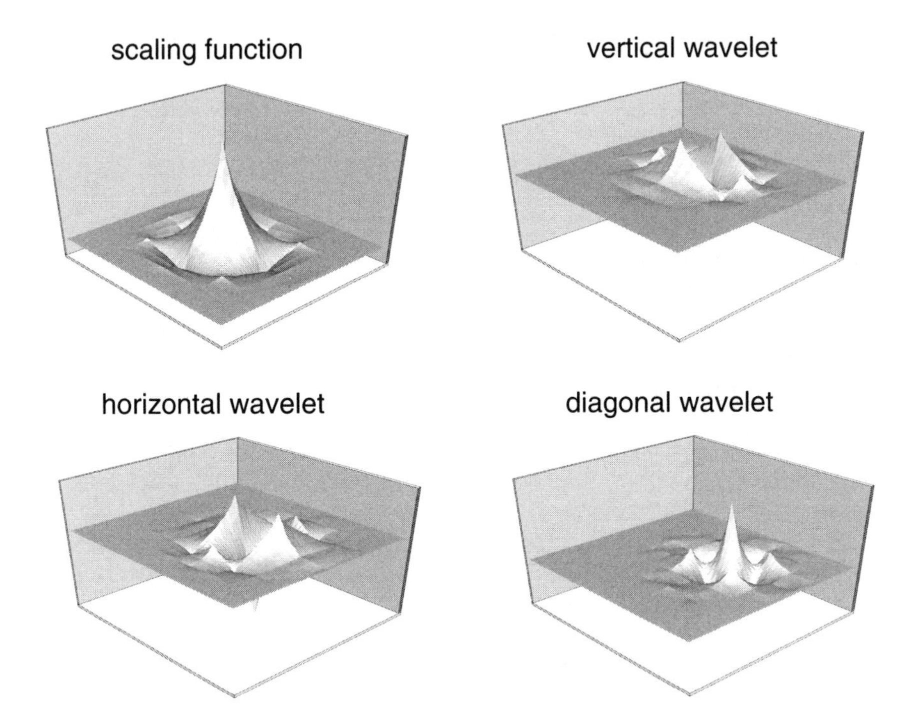

Figure 6.5: 2D Coiflet wavelets and scaling function.

further analysis:

$$g_i(x,y) = \begin{cases} \{d^{m,l}_{x,y} | l = 1 + (i \bmod 4)\} & \text{if } (i \bmod 4 \leq 2) \\ \{c^m_{x,y}\} & \text{else} \end{cases}$$

$$m = 1 + \lfloor \frac{i}{4} \rfloor$$

Since the spatial localization of the WT decreases with lower frequencies, according to the Heisenberg principle, we have to perform a bilinear interpolation between adjacent coefficients in every channel to approximate the contribution of the respective wavelet for a spatial coordinate (x, y) in the image.

Figure 6.8 illustrates these properties of the WT. The initial image consisting of different types of textures is transformed into the wavelet space. The result is a four-level pyramid that separates horizontal, vertical, and diagonal subbands of different spectral ranges. The local texture properties of the image can be described by the respective wavelet coefficients of the pyramid, since they provide a local spectral estimate. The contribution of the corresponding sets $\mathrm{g}(x, y)$ can be demonstrated by the reconstruction of the image, using only coefficients of the scaling functions and of the wavelets inside the marked region. The resulting image in Figure 6.8c features local lod, since the high frequency information is only provided locally.

It should be noted that the coefficients of the scaling function, c^m_{xy}, of each resolution step m, are added into our feature vector. They represent the low-pass parts of the signal

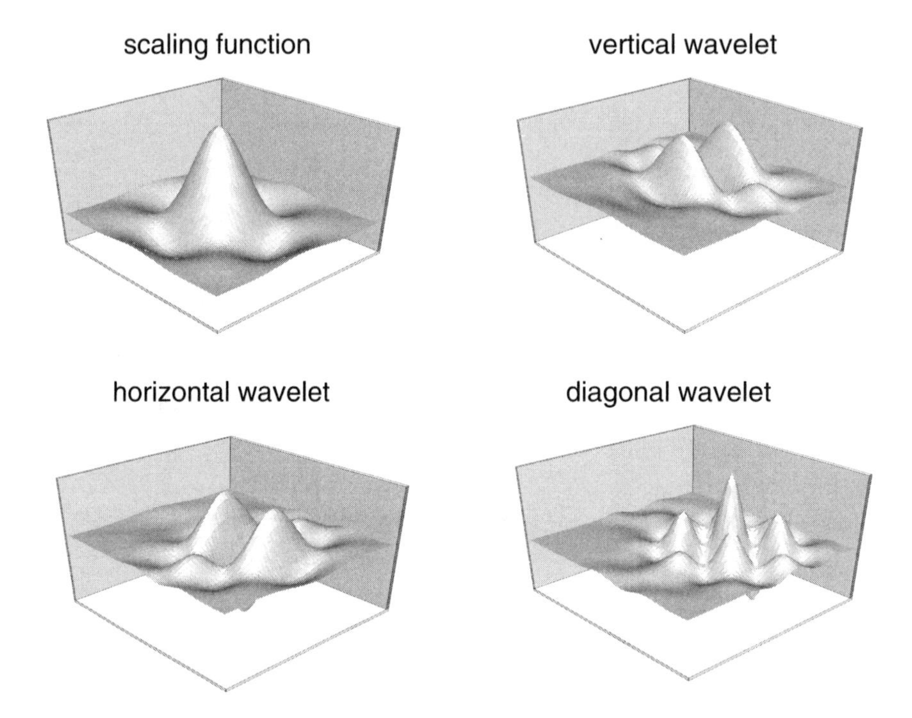

Figure 6.6: 2D Battle-Lemarie wavelets and scaling function.

and are further decomposed. This gives rise to a correlation of the individual features, which turned out to provide better results. For this end, decorrelation is a meaningful step in the pipeline, even when employing orthonormal wavelet types.

This scheme can be extended into 3D as shown in Figure 6.9. The local volume texture properties surrounding a voxel (x, y, z) are described by their respective set of coefficients $\mathrm{g}(x, y, z)$ from the volume pyramid.

6.4.2 A Pipeline for Data Analysis

Figure 6.10 again shows the pipeline for data analysis as first applied in [9] and later extended in [8]. After an initial WT, decorrelation of the coefficients is required to provide an optimal representation of the extracted features. There are many different approaches in mathematical statistics to perform this. The method proposed here employs PCA [7]. Since the amplitude distribution of the data is unknown, the normalization step supposes a uniform distribution and scales the data straightforwardly to its minimum and maximum.

The required cluster analysis of the texture features, as well as the supervised classification step, is accomplished with an extension of the Kohonen Feature Map proposed by [11]. The basic topology of the network is illustrated in Figure 6.11. The competitive layer is extended to 3D and each neuron j is assigned to a specific color triplet (R_j, G_j, B_j) in the RGB color space. During the nonsupervised organization process, the network performs a C-means clustering of the incoming data, where each weight vector $\mathbf{w}_j$ represents the po-

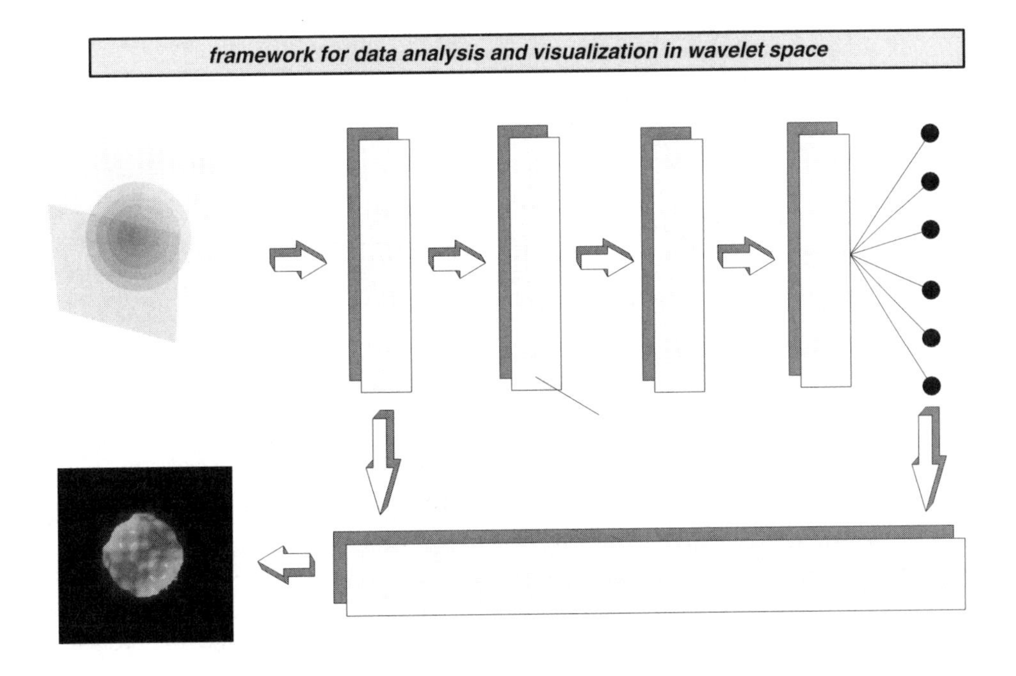

Figure 6.7: Framework for integrated volume rendering and data analysis in wavelet space.

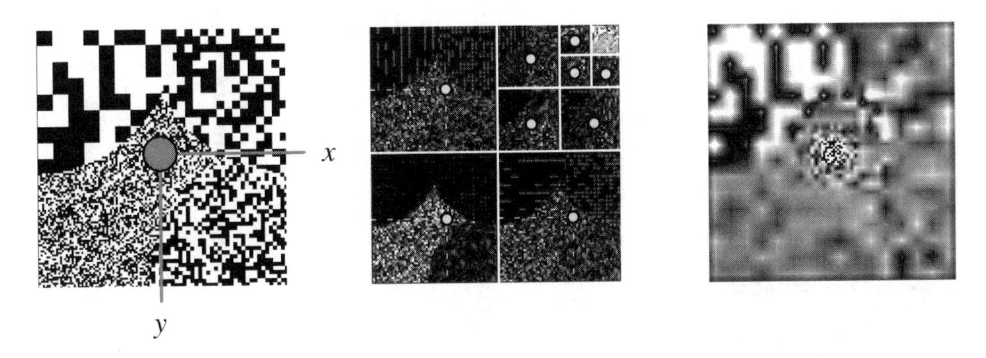

Figure 6.8: *(a)* Initial test image consisting of four different texture types and nonlinear texture boundaries *(b)* Subband representation by an orthogonal WT filter. *(c)* Reconstruction of the image with a spatially varying resolution.

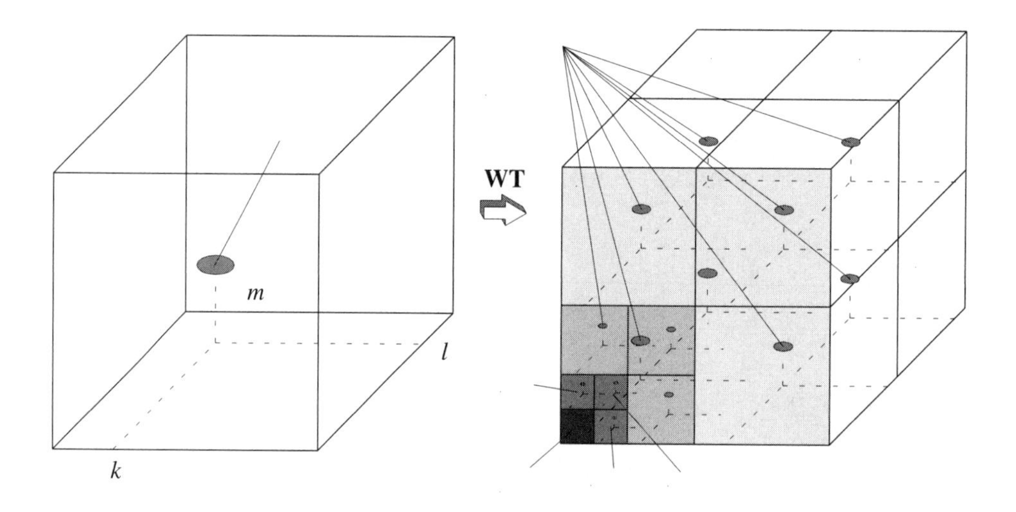

Figure 6.9: Illustration of the extraction of local volume texture features from the volume pyramid.

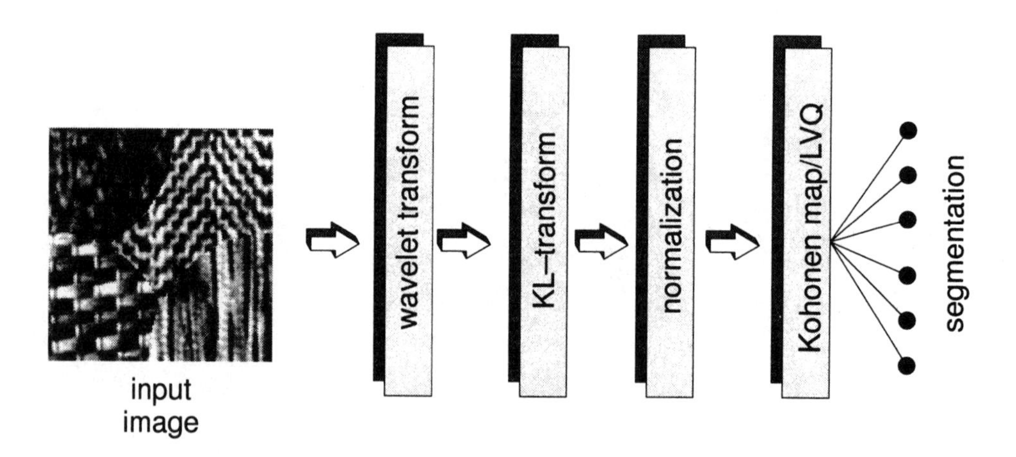

Figure 6.10: Wavelet based data analysis pipeline.

sition of a centroid in feature space. Since the network tries to preserve the data topology, neighbored neurons in the output layer are mapped to neighbored cluster centers in feature space. Visualizing these cluster centers in terms of color results in a mapping of a multi-dimensional data distribution to color similarities in RGB. The updating of the weights $\mathbf{w}_j$ for each neuron j during learning follows the basic rules of Kohonen:

$$\Delta w_{ij} = \alpha(t) \cdot (g_i(x,y) - w_{ij}), \quad \text{if neuron } j \in N_j(t) \tag{6.31}$$

where $\alpha(t)$ is the time-dependent learning rate and g_i is the feature coordinate i. $N_j(t)$ represents the local neighborhood of neuron j with the minimum Euclidian distance to the feature vector $\mathrm{g}(x, y)$.

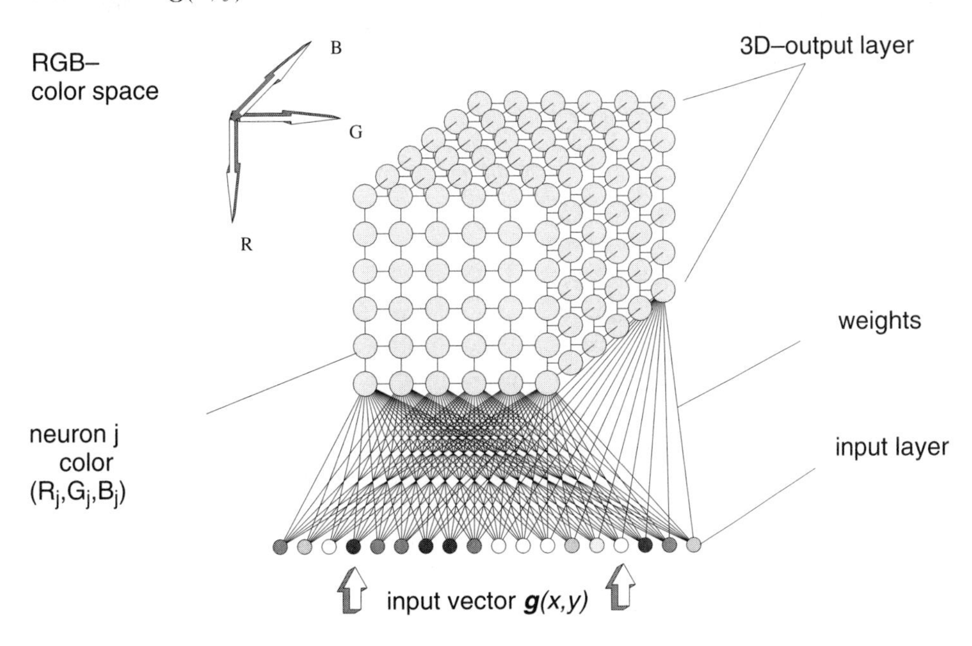

Figure 6.11: 3D Kohonen map and color assignment.

6.4.3 Some Results on Texture Analysis

Figure 6.12 (upper row) shows four different images generated to investigate the capabilities of the approach. The task was to perform a classification of the textures in the different images. The following wavelets were employed:

- Gabor wavelets (eight spatial orientations) (see also [8] for details)
- Daubechies wavelets (4-tap)
- Haar wavelets
- Battle-Lemarie wavelets

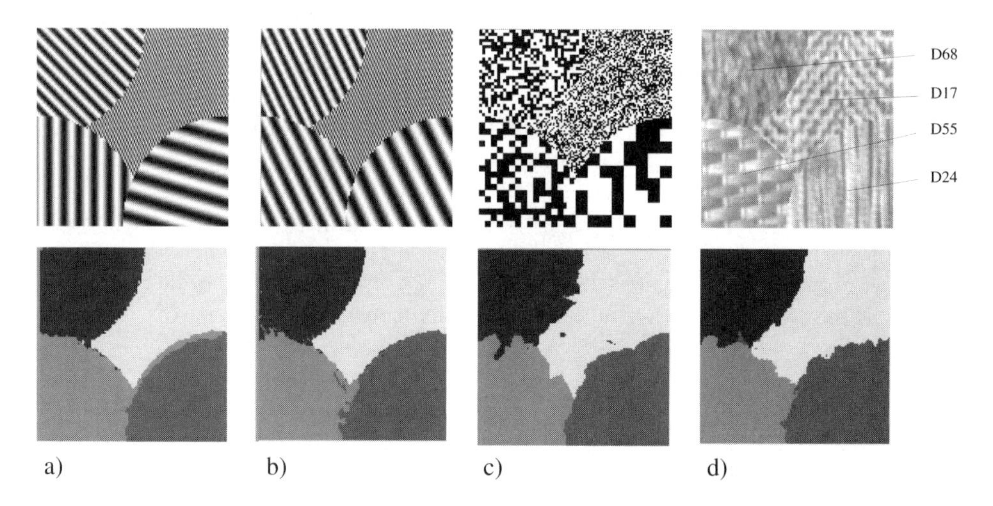

Figure 6.12: Different texture compositions and results: *(a)* Sine1: Sine gratings of different frequencies and orientations *(b)* Sine2: Sine gratings of different frequencies *(c)* Random dots *(d)* Brodatz textures: D17, D24, D55, D68.

Figure 6.13 illustrates the error we obtained for different images and wavelets as a function of the iteration depth. The results show that the accuracy depends on the texture type. It is interesting to compare the homogeneity of the results within the texture regions. The errors recorded appear mostly at the boundaries.

A detailed analysis of the performance of this algorithm for image texture analysis is provided in [9]. The following section of the chapter elaborates on a volume visualization method based on wavelets according to [10].

6.5 Volume Rendering in Wavelet Space

6.5.1 Preliminary Remarks

The basic goal of volume rendering is to find a good approximation of the low albedo volume rendering integral [2, 15], which expresses the relation between the volume intensity and opacity functions and the intensity in the image plane. Most of the standard volume rendering algorithms, therefore, approximate an integral equation with the ray parameter t of the type

$$I = \int_{t_1}^{t_2} C(t)_{\alpha(t)}\, e^{-\int_{t_1}^{t} \alpha(s)ds} dt, \tag{6.32}$$

where $C(t)$ stands for a volume intensity function and includes emitted, scattered, and reflected light. $\alpha(s)$ denotes the opacity function of the data and can be used to encode data features to be enhanced in the final images. Hence, the inner integral includes the self-occlusion of the volume.

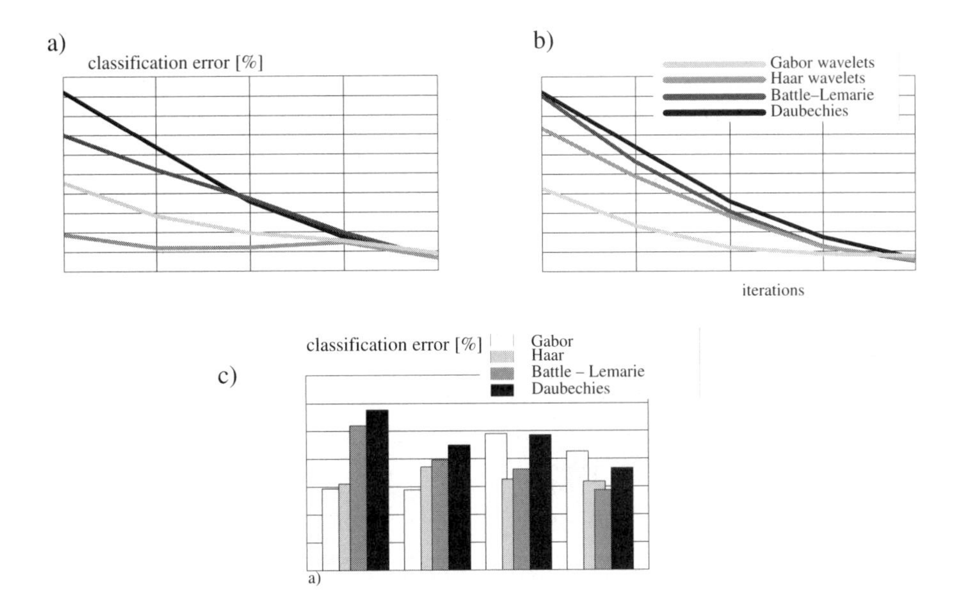

Figure 6.13: Results for *(a)* RANDOT *(b)* BRODATZ *(c)* Best matches for all samples.

The most common way to obtain a numeric solution of Equation (6.32) employs a first-order quadrature of the inner integral along with a linearization of the exponential. The outer integral is also solved by a finite sum of uniform samples. We yield

$$I = \sum_{k=1}^{M} C_k \alpha_k \prod_{i=1}^{k-1} (1 - \alpha_i) \tag{6.33}$$

where α_k are the opacity samples along the ray and C_k are the local color values derived from the illumination model. Note that I must be computed for each spectral sample λ, that is, at least in R,G,B.

A good mathematical analysis of the problem and error bounds of numeric quadrature is provided in [23].

6.5.2 Approximate Solutions of the Volume Rendering Equation

In Section 6.2 we elaborated that some wavelet types are not given in a closed form solution and that properties of smoothness, symmetry, and the keeping of the orthonormality can be achieved only by infinite support. That is, explicit approximations are vulnerable to truncation errors. In order to compare different wavelet types and to be independent of the basis function, we require a generic continuous representation scheme. Furthermore, the final goal is to find an approximate and analytic representation of the ray intensity function. Thus, the discrete values of the wavelet functions obtained by feeding Dirac-pulses into the filter bank are interpolated by piecewise polynomials using cubic splines in each interval.

Note that this approach provides a good framework for fundamental studies and investigations, however, it is not suited for building fast rendering schemes, such as in [19].

From the sections above, we know that a 3D wavelet decomposition of a volume data set accomplishes an expansion of the volume function $f(x, y, z)$ according to Equation (6.15).

The separability of the bases $\hat{w}_j$ allows us to write Equation (6.20) as

$$\hat{f}(x, y, z) = \sum_{j=1}^{K} \hat{c}_j \, b_j^1(x) \, b_j^2(y) \, b_j^3(z) \tag{6.34}$$

where the b_j represent the x, y, and z components of the tensor product basis functions.

The image generation with ray-casting starts from a parameterization of the ray as

$$\begin{pmatrix} x \\ y \\ z \end{pmatrix} = \begin{pmatrix} \alpha_x t + \beta_x \\ \alpha_y t + \beta_y \\ \alpha_z t + \beta_z \end{pmatrix} \quad \boldsymbol{\alpha} = [\alpha_x, \alpha_y, \alpha_z]^T \;, \; \boldsymbol{\beta} = [\beta_x, \beta_y, \beta_z]^T, \tag{6.35}$$

where $\boldsymbol{\alpha}$ and $\boldsymbol{\beta}$ are the viewing direction and the eyepoint, respectively.

We obtain the intensity function along the ray with

$$\hat{f}(t) = \hat{f}\begin{pmatrix} \alpha_x t + \beta_x \\ \alpha_y t + \beta_y \\ \alpha_z t + \beta_z \end{pmatrix} = \sum_{j=1}^{K} \hat{c}_j \, b_j^1(\alpha_x t + \beta_x) \, b_j^2(\alpha_y t + \beta_y) \, b_j^3(\alpha_z t + \beta_z) \tag{6.36}$$

This scheme is illustrated in Figure 6.14. The ray intensity function is provided by projecting the individual basis functions onto the ray and by superimposing them. This is accomplished by scaling with $\alpha = \{\alpha_x, \alpha_y, \alpha_z\}$ and by translating with $\beta = \{\beta_x, \beta_y, \beta_z\}$. Due to the piecewise spline interpolation, we now get a continuous approximation of the ray intensity function. In each spline interval $\left[t_n^j, t_{n+1}^j\right]$ and for each component of the resulting vector, the cubic polynomials, $b_j^{i,n}(t)$, are given for each wavelet w_j, as monomials

$$b_j^{i,n}(t) = \sum_{k=0}^{3} \tilde{a}_{k,j}^{i,n} t^k \qquad i = 1, 2, 3 \tag{6.37}$$

and their coefficients $\tilde{a}_{k,j}^{i,n}$ as

$$\tilde{a}_{k,j}^{i,n} = \sum_{l=k}^{3} \begin{pmatrix} l \\ l-k \end{pmatrix} a_{l,j}^{i,n} \alpha^k \beta^{l-k} \tag{6.38}$$

where $a_{l,j}^{i,n}$ are spline coefficients in interval $\left[t_n^j, t_{n+1}^j\right]$. Thus, we can write

$$b_j^{i,n}(t) = \sum_{k=0}^{3} \left(\sum_{l=k}^{3} \begin{pmatrix} l \\ l-k \end{pmatrix} a_{l,j}^{i,n} \alpha^k \beta^{l-k} \right) t^k \quad with\; t \in [t_n^j, t_{n+1}^j] \tag{6.39}$$

The final expression for $\hat{f}(t)$ is obtained by

$$\begin{aligned}\hat{f}(t) = \sum_{j=1}^{K} \hat{c}_j \quad & \left(\sum_{k=0}^{3}\left(\sum_{l=k}^{3}\begin{pmatrix} l \\ l-k \end{pmatrix} a_{l,j}^{1,n}\alpha_x^k\beta_x^{1-k}\right)t^k\right) \\ \cdot & \left(\sum_{k=0}^{3}\left(\sum_{l=k}^{3}\begin{pmatrix} l \\ l-k \end{pmatrix} a_{l,j}^{2,n}\alpha_y^k\beta_y^{1-k}\right)t^k\right) \qquad t \in [t_n^j, t_{n+1}^j] \\ \cdot & \left(\sum_{k=0}^{3}\left(\sum_{l=k}^{3}\begin{pmatrix} l \\ l-k \end{pmatrix} a_{l,j}^{3,n}\alpha_z^k\beta_z^{1-k}\right)t^k\right) \end{aligned} \tag{6.40}$$

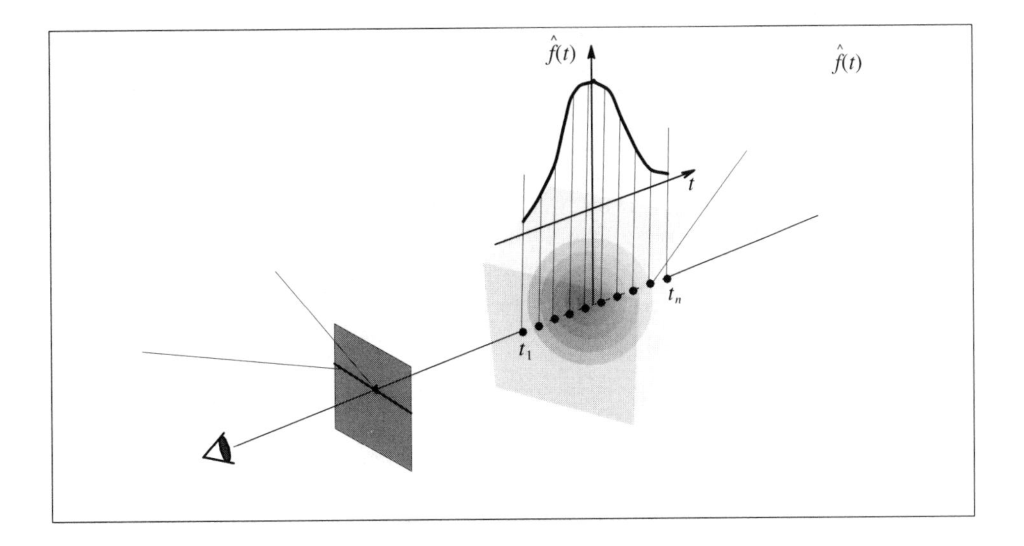

Figure 6.14: Illustration of the rendering process.

Note that Equation (6.40) finally provides an approximation with cubic volume splines that can be straightforwardly integrated. Unfortunately, however, the rendering equation incorporates an exponential absorption term.

Now, depending on the application, there are different ways to compute the image intensity I. In the simple, but important [25] case of constant transfer functions, Equation (6.32) simplifies to

$$I = \int_{t_1}^{t_2} f(t)dt \tag{6.41}$$

We obtain a piecewise analytic solution for each interval $[t_n^j, t_{n+1}^j]$ by the piecewise primitive functions $W_j^n(t) = \int b_j^1(t) \cdot b_j^2(t) \cdot b_j^3(t)\, dt$, with

$$I = \sum_{j=1}^{K} I_j = \sum_{j=1}^{K} \hat{c}_j \sum_{n=0}^{L^j-1} \left[W_j^n(t_{n+1}^j) - W_j^n(t_n^j)\right] \tag{6.42}$$

where L^j is the number of spline intervals. Note that the size and number of the intervals L^j depends both on how the viewing ray intersects the wavelet and on the iteration m. In the case of a closed form representation of the $W_j(t)$, such as with B-spline wavelets, the relations collapse to:

$$I = \sum_{j=1}^{K} I_j = \sum_{j=1}^{K} \hat{c}_j \left[W_j(t_2) - W_j(t_1)\right] \approx \sum_{j=1}^{K} \hat{c}_j \int_{-\infty}^{\infty} \tilde{w}_j(t)dt \tag{6.43}$$

The upper integral can be effectively computed using Fourier projection slicing [19].

In order to include a self-occlusion term and to evaluate the inner integral, we expand the function $\alpha(s)$ by our bases and obtain the following expression:

$$e^{-\int_{t_1}^{t} \alpha(s)ds} = e^{-\sum_{j=1}^{K} \alpha_j \sum_{n=0}^{L^j(t)-1}\left[W_j^n(t_{n+1}^j) - W_j^n(t_n^j)\right]} \qquad \forall j\ ,\ t_0^j = t_1 \quad t_{L^j(t)}^j = t \tag{6.44}$$

where α_j is the wavelet coefficient for $\alpha(x, y, z)$. Due to the local support of the WT, we have only to account for a subset of wavelets along the ray. A linear approximation of the exponential function aligned to the spline intervals yields

$$e^{-\int_{t_1}^{t} \alpha(s)ds} \approx \prod_{j=1}^{K} \alpha_j \prod_{n=0}^{L^j(t)-1} \left(1 - \left[W_j^n(t_{n+1}^j) - W_j^n(t_n^j)\right]\right) \tag{6.45}$$

Again, it should be noted that a closed form representation provides a compact approximation, but worse error bounds.

$$e^{-\int_{t_1}^{t} \alpha(s)ds} \approx \prod_{j=1}^{K} \alpha_j \left(1 - \left[W_j(t) - W_j(t_1)\right]\right) \tag{6.46}$$

The final discrete solution of the rendering equation depends on the shading model. The gradient $\nabla f(x, y, z)$ that is required for shading can easily be computed from

$$\nabla \hat{f}\begin{pmatrix} x \\ y \\ z \end{pmatrix} = \sum_{j=1}^{K} \hat{c}_j \left(\frac{d}{dx} b_j^1(x)\, b_j^2(y)\, b_j^3(z),\, b_j^1(x)\, \frac{d}{dy} b_j^2(y)\, b_j^3(z),\, b_j^1(x)\, b_j^2(y)\, \frac{d}{dz} b_j^3(z)\right) \tag{6.47}$$

We should mention here that the implementation of these equations is not straightforward. Although the piecewise spline interpolation of the basis functions only has to be performed for the 1D prototypes of ψ and ϕ, the calculation of the spline intervals in t is computationally very expensive.

The problem of getting isosurfaces in the images will be treated separately and discussed in the next section. The pictures in this chapter are based on Equation (6.41).

6.5.3 Examples

In Figure 6.15 a Gaussian density distribution was voxelized at a resolution of 32^3 and rendered with different numbers of coefficients and iteration depths. The isosurface was set to $\tau = 0.5$—the intensities below that threshold are represented as a bluish translucency. We should note that symmetry and shape of the wavelet strongly influence the shape of the isosurface. Asymmetric and fractallike functions, for example, the Daubechies wavelet, generate artifacts, such as rips or modulations of the translucency. It is interesting to compare them to the shapes obtained by a standard marching cubes on the initial volume data set.

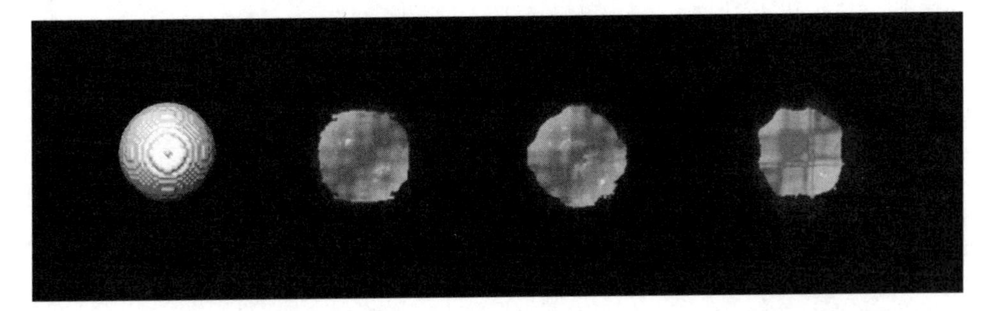

Figure 6.15: Images obtained from a Gaussian density distribution of 32^3 voxels: *(a)* Isosurface at $\tau = 0.5$ with a marching cubes algorithm. *(b)* Isosurfaces and translucent hull obtained from a Battle-Lemarie wavelet with 961 coefficients. *(c)* Isosurfaces and translucent hull obtained from a Coiflet wavelet with 1006 coefficients. *(d)* Isosurfaces and translucent hull obtained from a Daubechies wavelet with 1154 coefficients. See Color Plate 38.

The pictures were generated with our hybrid renderer that composes the volumetric intensities with the isosurfaces. In these examples, the isosurfaces were rendered using the implicit function approach of [16]. The coefficients were filtered according to the significance measures in Section 6.2.2.

6.6 Isosurface Reconstruction

6.6.1 Implicit Function Approach versus Marching Cubes

The wavelet decomposition offers different ways to obtain isosurfaces from volume data. Direct volume integration to obtain opaque surfaces, as proposed by [18] or [6], is usually based on shading models that estimate the surface normal by means of the volume gradient ∇f. Another way to solve the isosurface problem is proposed by [22]. It is based on the idea that isosurfaces of a threshold τ satisfy

$$\hat{f}(x, y, z) = \tau \Rightarrow \hat{f}(x, y, z) - \tau = 0 \tag{6.48}$$

in the continuous approximation of Equation (6.20).

Equation (6.48) leads to an implicit function that can be rendered using methods given in [16]. Detailed descriptions of how to compute the Lipschitz condition are given in [10].

Unfortunately, there are several shortcomings to this approach: Due to the huge number of basis functions, Kalra's method becomes extremely time-consuming. Furthermore, the appearance of the isosurface is strongly influenced by the shape of the wavelet, which must be represented in a continuous form. Hence, it becomes interesting to apply a simple marching cubes technique [20] to the expanded data set and to compose the polygons with the volume data during the rendering process. In these cases, the isosurface generation is no longer performed in wavelet space, but the local data quality is still controlled by the wavelets.

6.6.2 Examples

Figure 6.16a shows the bust of Johann Strauss, as it is derived from a 3D laserscanner. This model was illuminated and voxelized with a resolution of $32^2 \times 64$ voxels. The isosurfaces obtained from a marching cubes are presented for a flat shaded reconstruction in Figure 6.16b. Figures 6.17a,b,c show the isosurface reconstruction with Kalra's method [16], which is immediately accomplished by ray-tracing the basis functions in wavelet space. In Figure 6.18a,b,c the same reconstructions are presented after an inverse transform of the filtered data and with a marching cubes algorithm. In both cases, an increasing number of coefficients was employed to encode the data. We can clearly recognize that the level of detail increases as the number of wavelets is raised.

Figure 6.16: *(a)* Range data of a bust of Johann Strauss, voxelized at a resolution $32^2 \times 64$. *(b)* Isosurface reconstruction using a marching cubes algorithm (range data provided courtesy of the ZGDV, Darmstadt, Germany).

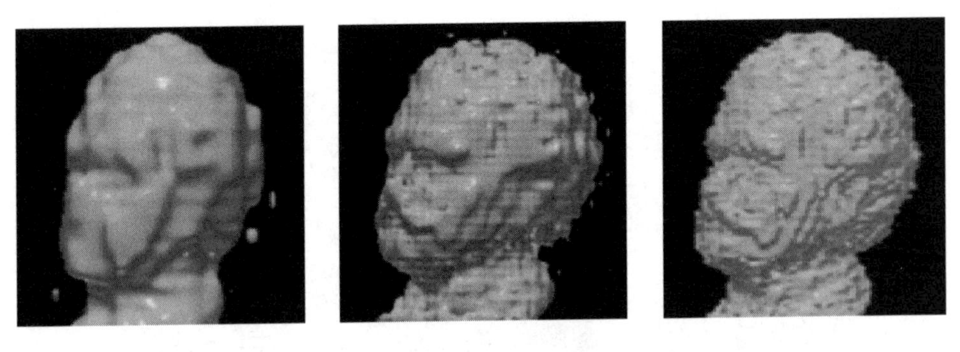

Figure 6.17: Isosurface reconstruction using Kalra's method and 3D wavelets with $\tau = 0.5$: *(a)* 1180 coefficients *(b)* 2832 coefficients *(c)* 9601 coefficients. See Color Plate 39.

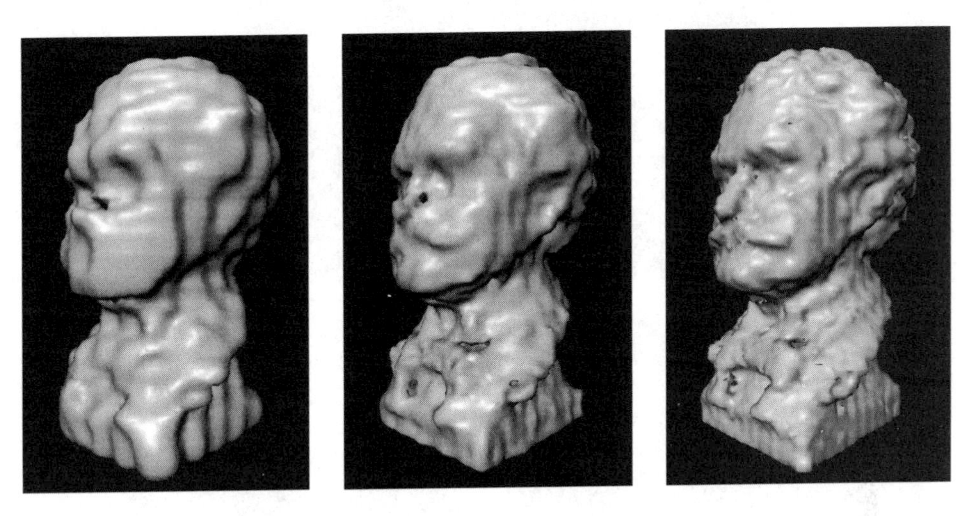

Figure 6.18: Isosurface reconstruction using marching cubes methods with $\tau = 0.5$ and Battle-Lemarie wavelets: *(a)* 1180 coefficients *(b)* 2832 coefficients *(c)* 9601 coefficients. See Color Plate 40.

In Figure 6.19 a series of isosurface reconstructions of a human scull is depicted with an increasing number of coefficients. Due to the Haar wavelets employed for the decomposition, the surface becomes more or less "box-like." Figure 6.20 illustrates the localization properties of the WT. Here, a 3D Gaussian weighting function (transparent ellipsoid in Figure 6.20b) was applied in wavelet space to enhance the approximation. Obviously, we end up with a perfect reconstruction of the isosurface in those spatial regions affected by the Gaussian.

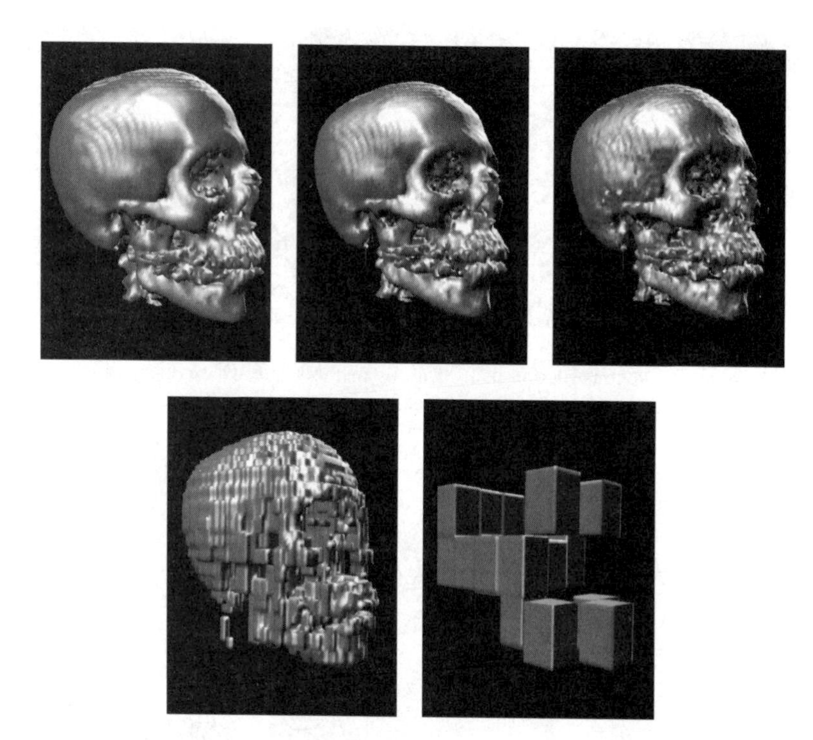

Figure 6.19: Isosurfaces from a human scull obtained by Haar decompositions of the data with different levels of approximation: *(a)* 100% coefficients *(b)* 15.5% coefficients *(c)* 6.8% coefficients *(d)* 0.3% coefficients *(e)* 0.02% coefficients. See Color Plate 41.

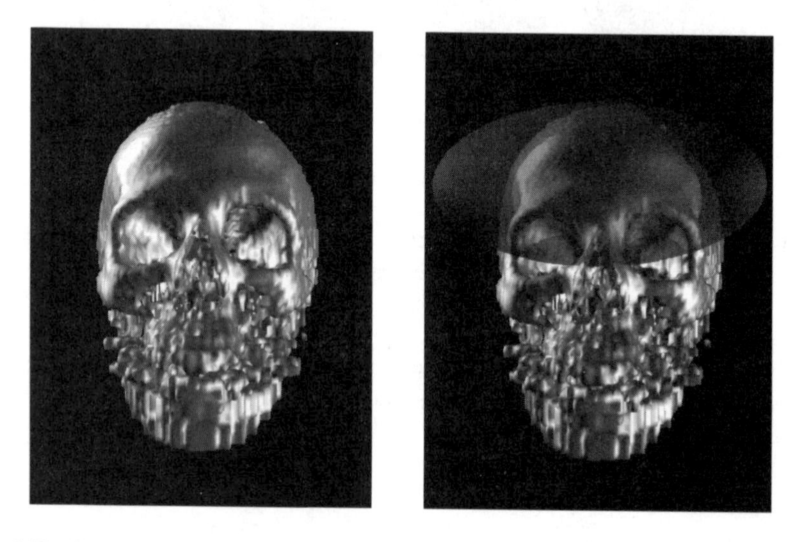

Figure 6.20: Controlling the local level-of-detail of the data by Gaussian weighting functions: *(a)* Result obtained with Haar wavelets. *(b)* Illustration of the weighting function (red ellipsoid). See Color Plate 42.

6.7 Conclusions and Future Research

In this chapter, a new framework for integrated volume rendering and data analysis in wavelet space was elaborated. It was shown that the WT is well suited for the extraction of local data features in images. Future research activities should focus on testing the method on volume data sets and combining it with current imaging techniques, such as morphological processing. Furthermore, the proposed rendering method provides piecewise analytic solutions of the intensity integral because the underlying volume is approximated continuously by polynomials. The quality of the results depends strongly on the type of selected basis function. Yet, it turns out that the projection of the bases onto the ray is computationally expensive. Hence, we must find a wavelet that is smooth, of strict finite support, orthonormal, and that provides a closed-form integral in t. These properties account for both rendering and data analysis, but cannot be satisfied in common. Thus, a compromise must be found in terms of biorthogonal wavelets.

Acknowledgment

Most of this research was done at the Computer Graphics Center in Darmstadt, Germany. It was supported in part by the German Telekom within the KAMEDIN project. The author thanks Lars Lippert, Rolf Koch, Oliver Staadt, and Andy Dreger for implementing parts of the presented methods during their Diploma theses.

Bibliography

[1] A.N. Akansu and R.A. Haddad, *Multiresolution Signal Decomposition*, Academic Press, Inc., 1992.

[2] J.F. Blinn, "Light Reflection Functions for Simulation of Clouds and Dusty Surface," *Computer Graphics*, Vol. 16, No. 3, 1993, pp. 116–123.

[3] A.C. Bovik, M. Clark, and W. Geisler, "Multichannel Texture Analysis Using Localized Spatial Filters," *IEEE Transactions on Pattern Analysis and Machine Intelligence*, Vol. 12, No. 1, 1990, pp. 55–73.

[4] T. Chang and C. Kuo, "Texture Analysis and Classification with Tree-Structured Wavelet Transform," *IEEE Transactions on Image Processing*, Vol. 2, No. 4, 1993, pp. 429–441.

[5] I. Daubechies, "The Wavelet Transform, Time-Frequency Localization and Signal Analysis," *IEEE Transactions on Information Theory*, Vol. 36, 1990, pp. 961–1005.

[6] R.A. Drebin, L. Carpenter, and P. Hanrahan, "Volume Rendering," *Computer Graphics*, Vol. 22, No. 4, 1988, pp. 125–134.

[7] M. Gross, *Visual Computing*, Springer-Verlag, 1994.

[8] M. Gross and R. Koch, "Visualization of Multidimensional Shape and Texture Features in Laser Range Data Using Complex-Valued Gabor Wavelets," *IEEE Transactions on Visualization and Computer Graphics*, Vol. 1, No. 1, 1995, pp. 44–59.

[9] M. Gross, R. Koch, L. Lippert, and A. Dreger, "Multiscale Image Texture Analysis in Wavelet Space," *Proceedings of the IEEE-ICIP, CSPRESS*, Vol. III, 1994, pp. 412–416.

[10] M. Gross, L. Lippert, A. Dreger, and R. Koch, "A New Method to Approximate the Volume Rendering Equation Using Wavelets and Piecewise Polynomials," *Computers & Graphics*, Vol. 19, No. 1, 1995, pp. 47–62.

[11] M. Gross and F. Seibert, "Visualization of Multidimensional Image Data Sets Using a Neural Network," *The Visual Computer*, Vol. 10, 1993, pp. 145–159.

[12] A. Grossmann and J. Morlet, "Decomposition of Hardy Functions into Square Integrable Wavelets of Constant Shape," *SIAM Journal of Mathematical Analysis*, Vol. 15, 1984, pp. 723–736.

[13] Tao He, S. Wang, and A. Kaufman, "Wavelet-Based Volume Morphing," *ACM Volume Rendering Symposium, Proceedings of the IEEE Visualization '94*, 1994, pp. 85–92.

[14] B. Jawerth and W. Sweldens, *An Overview of Wavelet Based Multiresolution Analyses*, internal report, University of South Carolina, Department of Mathematics.

[15] J.T. Kajiya and B.P. Von Herzen, "Ray Tracing Volume Densities," *SIGGRAPH '84*, 1984, pp. 165–174.

[16] D. Kalra and A.H. Barr, "Guaranteed Ray Intersection with Implicit Surfaces," *Computer Graphics (Proc. SIGGRAPH '89)*, Vol. 23, 1989, pp. 297–306.

[17] A. Laine and J. Fan, "Texture Classification by Wavelet Packet Signatures," *IEEE Transactions on Pattern Analysis and Machine Intelligence*, Vol. 15, No. 11, 1993, pp. 1186–1191.

[18] M. Levoy, "Display of Surfaces from volume Data," *IEEE Computer Graphics and Applications*, Vol. 8, No. 5, 1988, pp. 29–37.

[19] L. Lippert and M. Gross, "Fast Wavelet-Based Volume Rendering by Accumulation of Transparent Texture Maps," *Eurographics 95 Proceedings, Computer Graphics Forum*, Vol. 14, No. 3, 1995, pp. 431–443.

[20] W.E. Lorensen and H.E. Cline, "Marching Cubes: A High Resolution 3D Surface Construction Algorithm," *Computer Graphics*, Vol. 21, 1987, pp. 163–196.

[21] S. Mallat, "A Theory for Multiresolution Signal Decomposition: The Wavelet Representation," *IEEE Transactions on Pattern Analysis and Machine Intelligence*, Vol. 11, No. 7, July 1989, pp. 674–693.

[22] S. Muraki, "Volumetric Shape Description of Range Data Using 'Blobby Model'," *Computer Graphics*, Vol. 25, No. 4, 1991, pp. 227–235.

[23] K. Novins and J. Arvo, "Controlled Precision Volume Integration," *ACM Workshop on Volume Visualization*, 1992, pp. 83–89.

[24] E.P. Simoncelli and E.H. Adelson, "Non-Separable Extensions of Quadrature Mirror Filters to Multiple Dimensions," *Proceedings of the IEEE*, Vol. 78, No. 4, 1990, pp. 652–664.

[25] T. Totsuka and M. Levoy, "Frequency Domain Volume Rendering," *Computer Graphics Proceedings, Annual Conference Series*, 1993, pp. 271–278.

[26] M. Unser and M. Eden, "Multiresolution Feature Extraction and Selection for Texture Segmentation," *IEEE Transactions on Pattern Analysis and Machine Intelligence*, Vol. 11, No. 7, 1989, pp. 717–728.

Chapter 7

An Approach to Intelligent Design of Color Visualizations

Philip K. Robertson and Matthew A. Hutchins

Abstract. *We propose an approach to intelligent design of visualizations to allow flexibility between full user control and full automation. The approach uses metavisualizations—visualizations of metadata associated with the data and its mapping into visual attributes of a display—to provide the user with a clear path to exercising or relinquishing control over the choice of data mappings in an adaptive manner. We illustrate this approach with a color management system that allows the mapping of data into perceptual dimensions of a uniform color space, and provides the tools for user control over mappings in a manner only constrained by the user interface to the metavisualizations.*

7.1 Introduction

7.1.1 The Need for Intelligence

Embedding intelligence in visualization systems can help users make appropriate decisions to map data into visual attributes of a data display [8, 9, 18, 17, 3, 6, 1, 4, 15]. Automated display of data, including design of visualizations, is one end of the spectrum; full user control over display design and data mappings is the other end of the spectrum. In this work we take one subset of visualization design—that of choosing appropriate color representations for data—in order to investigate a systematic approach to introducing intelligence into designing visualizations.

A wide range of color representation tools are available to applications specialists using visualization software. There is a clear need for guidance in the choice of color sequences or ranges that are suitable for representing various data types to meet specified interpretation aims. Misinterpretations due to inappropriate choice of color sequences, poor use of devices with differing color characteristics, perceptual artifacts, and ambiguities are evident in many visual presentations of data. When directed interpretation is required, these problems can make the visualization less than convincing.

Guidance can come from user knowledge or system intelligence. Domain-specific constraints, requirements, or practices are often available only in example form or in the experience of specialists. Domain-independent guidance can be drawn from understanding of visual perception. Specific device behavior information is necessary to make effective use of various output systems and attain compatibility between screens and hard copy.

7.1.2 An Approach to Introducing Intelligence

In practice, the design of color visualizations is a typical design problem, characterized by multiple alternatives, conflicting constraints, high levels of complexity, and the need for creativity and imagination. Problems of this sort are not easily solved by direct application of computer "intelligence." Instead, it is more fruitful to concentrate on enhancing the capabilities of the human operator. The system intelligence can be directed towards providing guidance for the human visualization designer, and carrying out low-level tasks such as providing defaults and maintaining consistency, to reduce the complexity of the task and leave the designer free to concentrate on the more difficult problems.

We propose an approach to the design of intelligent visualization systems that allows flexibility between full user control and full automation. The approach is to initially design the system as if it were to be operated fully under control of the human operator, without any system guidance, and then introduce intelligence in the form of autonomous agents that can make sensible decisions based on stored knowledge and underlying reference models. Each agent can take over the operation of a particular part of the system, performing the actions that would normally be performed by the human user. The choices of the intelligent agent are clearly reflected in the user interface of the visualization system, and thus serve both as useful defaults and advice. Centering the activity of the agents around a graphical representation of the operation to be performed—a "metavisualization"—cements the intuitive link between the user, the intelligence being applied, and the task to be performed. The user can override any or all of the choices made by an agent.

The advantage of this approach is that the user's task is changed from designing an entirely new visualization, to modifying a sensible default visualization until it meets certain requirements. Users with varying levels of knowledge can choose to exercise different levels of control over the system. Novice users may choose to make only minor modifications to the system's suggestions, but at the same time gain an understanding of the constraints and possibilities offered by the system. More advanced users are able to exercise more control. However, the clear and unambiguous display of all the system's choices at the user interface gives the advanced user increased confidence in the system's intelligence, so that more choices can be comfortably left to the system.

This approach also provides a clear path for the evolution and adaptation of existing software systems, without necessitating a complete design and rebuild. Intelligent agents can be introduced into such systems to lift some of the burden of work from the user, providing a gradual, flexible path from a user-controlled to a fully automatic system.

7.1.3 A Framework for Demonstration

This investigation is based on a system for mapping data into perceptually uniform color spaces, exploiting graphical gamut representations to help users to make appropriate choices of data mappings [13, 7, 2, 14, 10, 11]. The system was developed as a flexible and extensible system for color management and applications support. The system is not an outcome of the approach described here. Rather, it is an example of a system designed to be operated under full user control, to which intelligent agents of the form described above could be added. The system does provide reasonable defaults in most cases, with user control over the chosen mappings. Underpinning the interactive facilities are models of output devices to ensure consistency and device-independent color [2]. The system includes high-quality color report generation facilities. Its description in this paper is sufficient only to illustrate how best to introduce flexible degrees of guidance into choice of data mappings.

In the following section, we look at how metavisualizations can be used as the basis for designing a fully user controlled interface, and offer examples from the color mapping system. Section 7.3 looks at how the use of metavisualizations provides a natural focus for the introduction of intelligent agents. We also look at the models that are needed as a basis for decision making.

7.2 Designing Interfaces for Full Control

7.2.1 Metadata and Metavisualizations

Most scientific data of the sort used as input to a visualization system has associated with it some form of metadata—additional data *about* the data. Common types of metadata are historical data, describing the original source of the data and the operations that have been performed on it, derived data, such as histograms and other statistics, and descriptive data, such as the structure and type of the data, units for the data values, coordinate system details, and other information that helps to determine how the data should be interpreted. Depending on the sophistication of the data model used by the visualization system, the metadata may be stored entirely as knowledge in the head of the specialist, recorded in a notebook or on a disk label, or it may be stored as readily accessible, structured data along with the data that it refers to.

Metavisualizations are visualizations of the metadata, or visualizations of the process used to visualize the data. Visualizations of derived metadata can be designed to be directly interpreted. Other metavisualizations are typically designed as an aid to interpreting the original visualizations of the data. In common practice, these types of metavisualizations are at best designed and added to a visualization as an afterthought.

7.2.2 Using Metavisualizations for Direct Manipulation

Direct manipulation techniques can be difficult to apply to the process of data visualization. That is, it is difficult to design an interface that supports the design of visualizations through direct manipulation of the data that is to be visualized. Primarily, this is because the data must be visualized in some way *before* it can be manipulated—the visualization needs for manipulation conflict with those for interpretation. Also, particularly in the case of choosing colorings, there are few real-world metaphors that can be easily and naturally applied. Objects that change their colors through manipulation are rare. The alternative approach, taken by most visualization systems, is to allow direct manipulation of the parameters of the visualization process—in other words, direct manipulation of metavisualizations. The metavisualizations are interactive visual representations of the control parameters of the system. The interface design task then becomes one of designing appropriate interactive metavisualizations.

7.2.3 Examples from the Color Mapping System

Metavisualizations used in the color mapping system include interactive color gamut views, axial alignments of data variables within the color gamuts, scatterplots, and return-to-gamut graphical depictions [13, 11]. The metavisualizations provide a means of specifying the rules or constraints applied to generating color mappings for standard data types or interpretation aims in a graphical manner, and a means for interactively defining and preserving such mappings. Core to the system is the display to the user of the bounded perceptual gamut corresponding to the output device being used. The gamut can be shown as a 3D wireframe for overall shape appreciation, and as sets of cross sections through planes of constant hue and constant lightness. Figure 7.1 shows a typical gamut explorer metavisualization. The user can adjust the chosen planes or wireframe view.

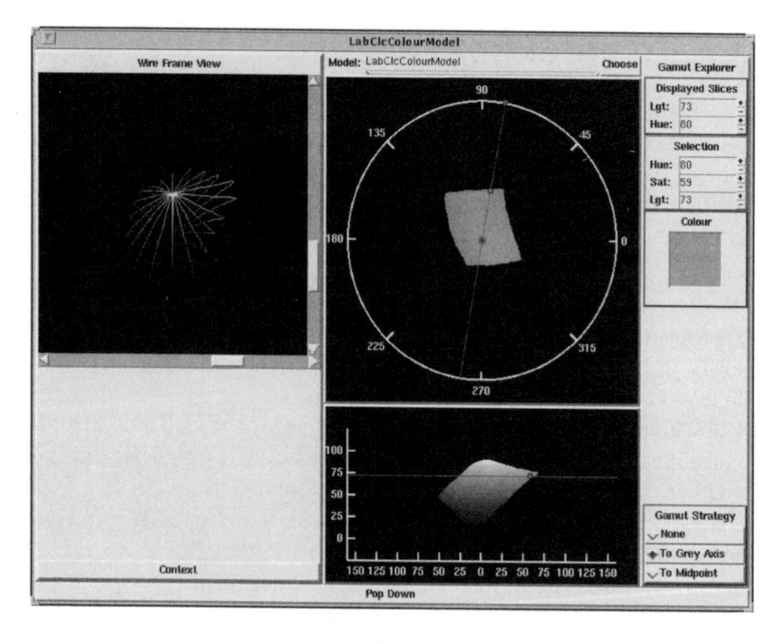

Figure 7.1: Gamut explorer. See Color Plate 43.

The data model used in the color mapping system actively supports the storage of metadata along with all images. The metadata typically describes the structure and "meaning" of the image, separated into meanings for the actual pixel values, and meanings for each of the dimensions. The meaning stored with an image determines how it is displayed, and can be shown in textual form along with the image view. Histograms can also be calculated for any image and displayed as metavisualizations, as illustrated in Figure 7.2.

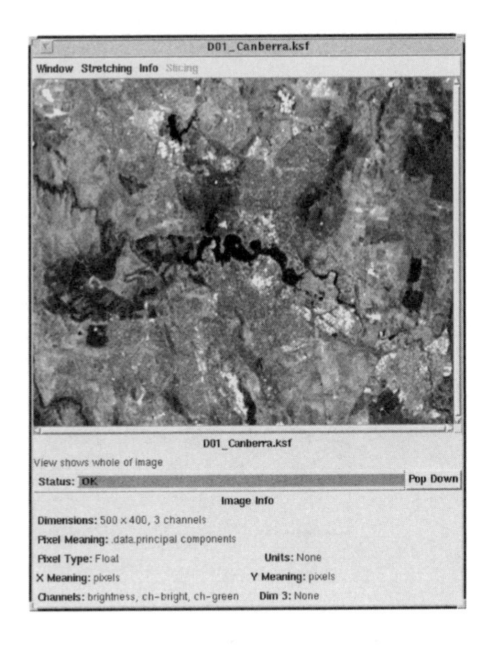

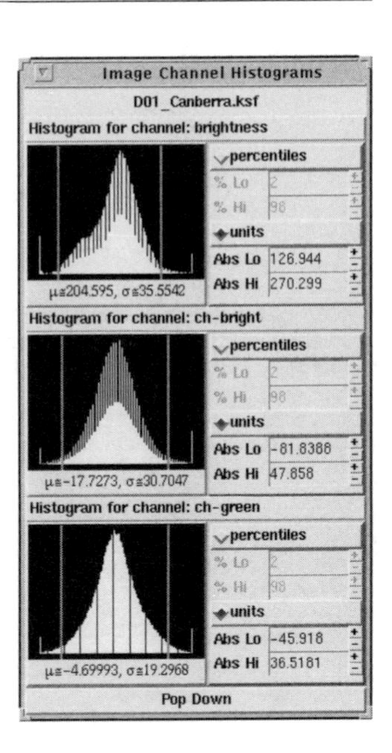

Figure 7.2: Image viewer with channel histograms. See Color Plate 44.

The process of choosing the data mapping is based on defining axes for alignment of the data within the perceptual color space. Predetermined guidance for particular data meanings can be chosen as defaults. The user can adjust the endpoint limits (shown by bars) on the data histograms, and on the chosen color space alignments (also shown by enclosing bars). The result of the mapping on the data can be viewed simultaneously, within performance limitations of the workstation, or on a reduced size data icon for immediacy of feedback. Figure 7.3 shows the result of applying a typical data mapping, including the metavisualizations of the data mapper.

Data points that are mapped outside the gamut under a chosen data mapping, or device change, must be portrayed in some color. The color most appropriate may depend on the application, the user's preference, the device limitations, and visual perception issues. Most important is that the user be able to see the method chosen, and modify it if required, through the metavisualization. Figure 7.4 shows metavisualizations of the effects of a return-to-gamut strategy.

Of course, the use of metavisualizations is not limited to the user interface of the visualization system. One often overlooked aspect of choosing data mappings is the need for an encapsulated representation of the mapping chosen to accompany the data display. Although the full gamut mapping metavisualization set may be required for proper interpretation, often just the key aspects of that mapping are enough for an experienced interpreter to use, or explain to others, the color representation. Figure 7.5 shows the encapsulated metavisualizations generated by the system for a typical example. These encapsulated representations, plus accompanying text, are generated automatically and included as determined by the user or application.

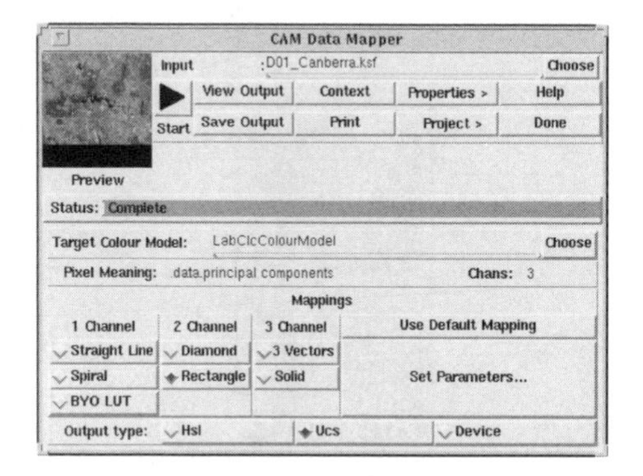

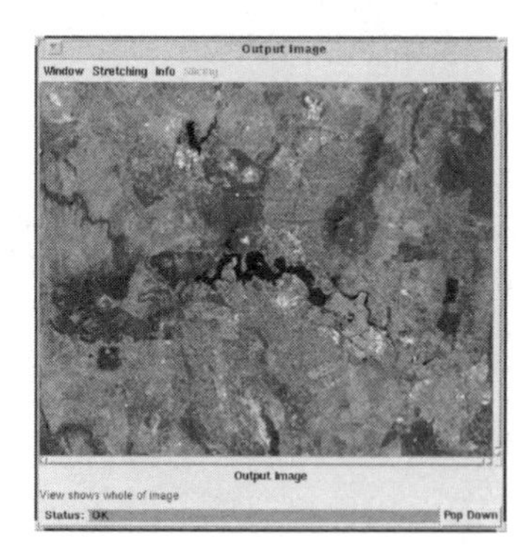

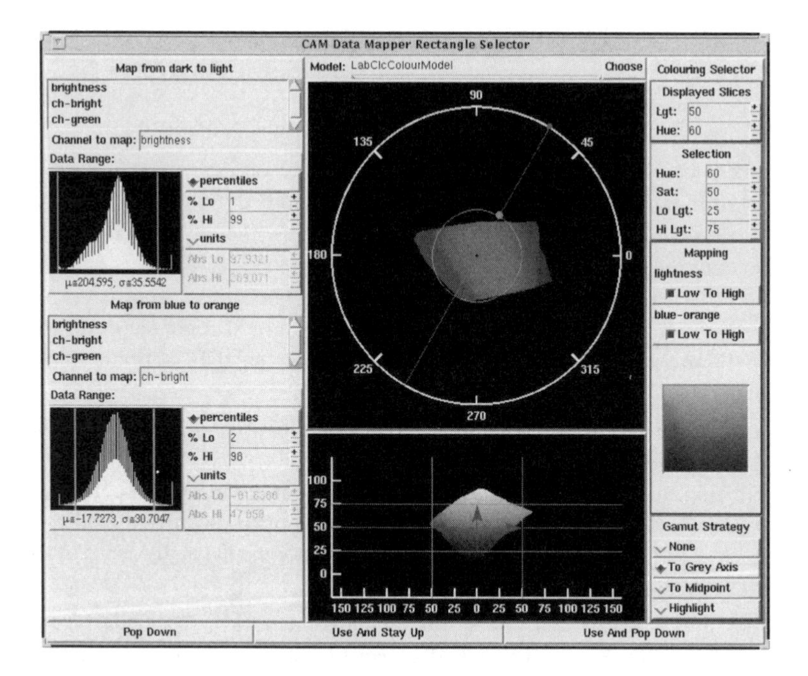

Figure 7.3: Data mapper, output image view, and coloring selector. See Color Plate 45.

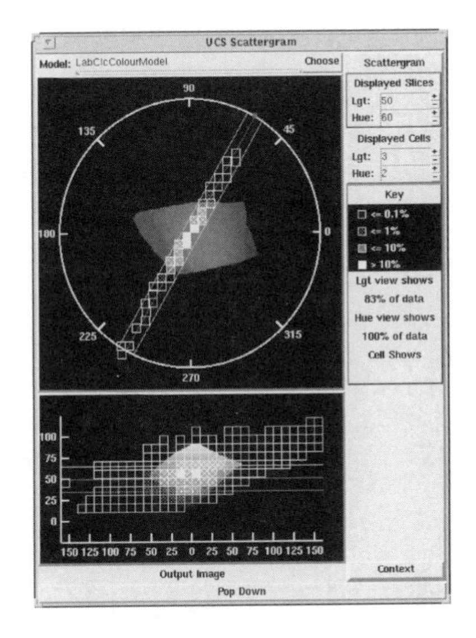

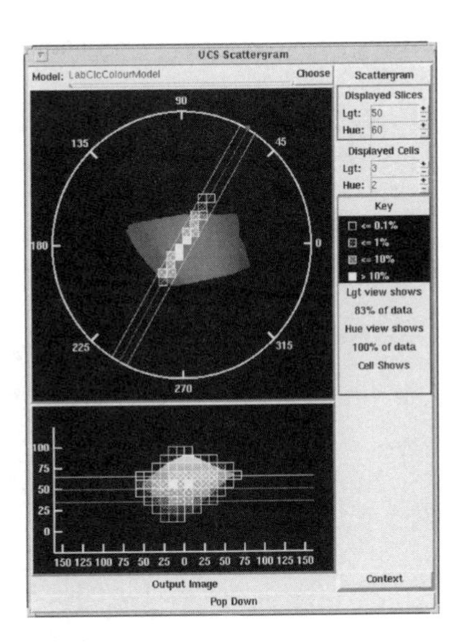

Figure 7.4: Color scattergrams, before and after return-to-gamut. See Color Plate 46.

Many work sessions end in report generation. The system provides a compositing capability, together with layout support, to generate reports. Figure 7.6 shows a typical page from a report, composed from the metavisualizations generated by the system and additional textual information provided by the user.

7.3 Incorporating Intelligent Guidance

7.3.1 Observations

An interface based on interactive metavisualizations is well suited for the introduction of intelligent agents. A well-designed metavisualization will capture exactly the information that must be communicated to the user of the system in order for an intelligent choice to be made about parameters for the visualization process. That is, in order to design the interface, the designer must develop a complete model of the process, and the knowledge that is required at each step in the process, in order to make sensible decisions. This knowledge will then be presented to the user in the form of an appropriate metavisualization. The same knowledge will be required by an intelligent agent designed to assist in the user's task. In other words, the automation component can be fully encapsulated by the functional descriptions of the role of the metavisualization in the interface. The task of designing the embedding of intelligence is therefore effectively subsumed within the task of designing a good user interface to allow user control over the data mapping task.

From a system design and development point of view, then, the intelligent agents have well-defined functions that are easily implemented and validated, by reference to the metavisualizations that they are associated with. From the user's point of view, the metavisualizations serve as a clear illustration of the action of the system. Like the primary

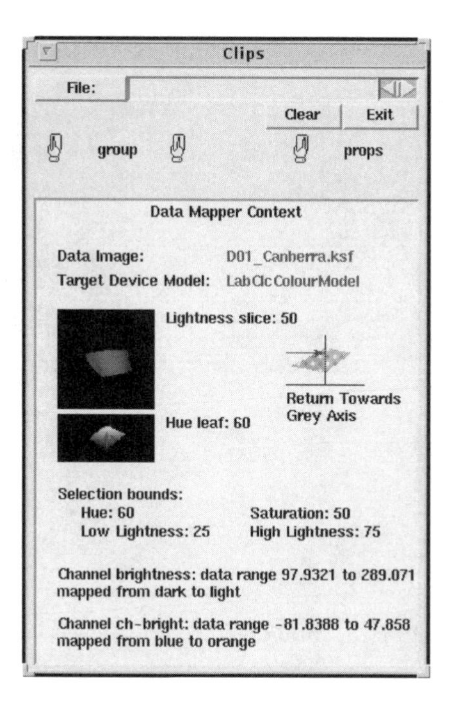

Figure 7.5: Encapsulated metavisualizations as context. See Color Plate 47.

visualization they are helping to design, the metavisualizations will exploit the human perceptual system to simply and quickly convey complex information, in this case information about the constraints, possibilities, and relationships of the various parts and parameters of the visualization process.

7.3.2 Models for Guidance

In order to provide intelligent guidance, the agents must have access to underlying knowledge and reference models. Earlier work has focused on the need for a reference model of the visualization process, and well-defined models for key components of the process [4, 5, 12, 16]. The details of the models used in this work are not critical; what is important is that they are sufficiently rich to allow intelligent agents to extract information to meet the needs of each of the metavisualizations. To successfully introduce intelligence into the color mapping system, the following models are needed within an overall reference framework:

- a data model, storing sufficient metadata to determine the display of the appropriate metavisualizations;

- an output device model, describing the characteristics and limitations of the chosen output device;

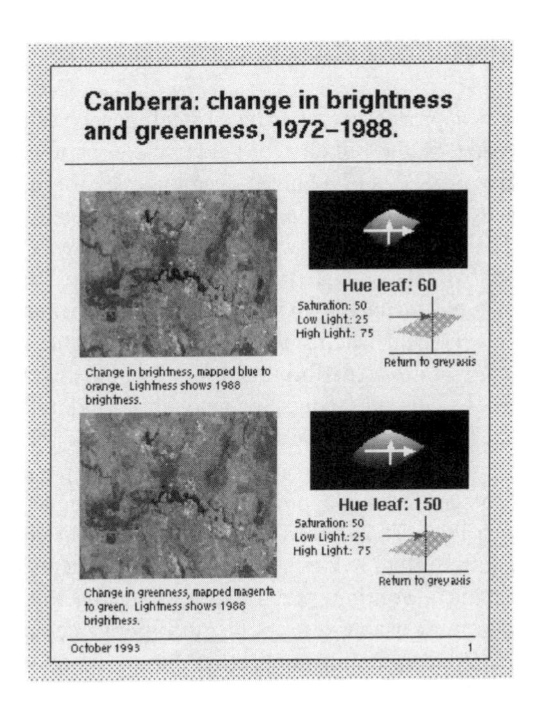

Figure 7.6: Integration of data and metavisualization into a report. See Color Plate 48.

- a user-perception model, containing generic interpretation capabilities of the human visual system;
- a domain model, storing coloring conventions and preferences of an application area;
- a user profile, storing the specific preferences of a particular user;
- a model of the user's interpretation aims for the visualization.

The color mapping system already makes use of some of these models to support its metavisualizations. For example, Figure 7.3 shows how information from the data model and the device model are presented to the user for fully controlled choice of data mapping. The system uses this information directly, along with an implicit perception model built into a case-based decision system, to choose sensible default mappings for any input image.

Ideally, software systems should be designed with a complete knowledge of the intended application domain and task structure. In practice, this is constrained by the complex, changing nature of users' tasks, and because software typically has to be designed generically, to suit a wide range of users. The danger in designing a system to suit all users is that it will end up being tailored for none. The intelligent use of an explicit, user-modifiable domain model and user profile can help to overcome this problem, by providing a system that can adapt to, or be adapted for, individual users.

7.4 Conclusion

The basic premise of our approach is that artificial intelligence need not be designed to replace human intelligence. Rather, intelligent guidance can be used to directly support the creative and cognitive processes of a human operator. We propose the introduction of intelligence through the use of autonomous agents that act on the components of the system in the same manner as the human operator, with the results of those actions reflected in the state of the user interface. An interface based on metavisualizations is an ideal platform for introducing intelligence in this manner, both for the system designer and the user. From the designer's point of view, the metavisualizations precisely define the scope of each agent, and the knowledge that it requires to perform its tasks. From the user's point of view, the action of the agents serves to directly reduce the amount of work that must be done, in a manner that is immediately comprehensible, and completely reversible.

Although not yet implemented as a working system, our experiences with the color mapping system have shown that this approach, linking intelligent agents to interactive metavisualizations, is feasible. They have also shown which aspects of the models and intelligent processes require further research, and which components are sufficiently well designed to be integrated into a working system. We conclude that this approach provides a robust framework for designing usable systems that incorporate complex capabilities, and an incremental approach to incorporating intelligence into the design of visualizations.

Acknowledgments

This work has utilized a color mapping system as its basis. The authors acknowledge the input and advice of the developers of the system, Stephen Barrass, Don Bone, Chris Gunn, Dione Smith, and Duncan Stevenson, all in the Visualisation Systems Group at the CSIRO Division of Information Technology.

Bibliography

[1] C. Beshers, "Automated Design of Virtual Worlds for Visualising Multivariate Relations," *Proceedings of IEEE Visualization '92*, Boston, Mass., Oct. 1992, pp. 283–290.

[2] D. Bone, "Adaptive Colour-Printer Modelling Using Regularized Linear Splines," *Proc. Symp. on Device Independent Color Imaging, IS&T, SPIE Symp. on Electronic Imaging Science & Technology*, (CSIRO Div. of Information Technology Tech. Rep. TR-HJ-92-19), Feb. 1993.

[3] S. Casner, "A Task-Analytic Approach to the Automated Design of Graphic Presentations," *ACM Transactions on Graphics*, Vol. 10, No. 2, 1991, pp. 111–151.

[4] L. De Ferrari and P.K. Robertson, "Automating the Design of Visualisations—Is a Reference Model Needed," *Workshop on Automated Design of Visualisation, IEEE Visualisation '92*, Boston, Mass., Oct. 1992.

[5] R.B. Haber and D. McNabb, "Visualization Idioms: A Conceptual Model for Scientific Visualization Systems," *Visualisation in Scientific Computing*, G. Neilson and B. Shriver, editors, IEEE Computer Society Press, 1990, pp. 74–92.

[6] W. Hibbard, C. Dyer, and B. Paul, "Display of Scientific Data Structures for Algorithm Visualisation," *Proceedings of IEEE Visualization '92*, Boston, Mass., Oct. 1992, pp. 139–146,

[7] M. Hutchins, D. Stevenson, and S. Barrass, "Effective Colour Visualisations on Paper," *Proceedings of Digital Image Computing, Technologies and Applications DICTA '93*, Sydney, Australia, Dec. 8–10, 1993.

[8] J. Mackinlay, "Automating the Design of Graphical Presentations of Relational Information," *ACM Transactions on Graphics*, Vol. 5, No. 2, Apr. 1986, pp. 110–141.

[9] P. Robertson, "A Methodology for Choosing Data Representations," *IEEE Computer Graphics and Applications*, Vol. 11, No. 3, May 1991, pp. 56–67.

[10] P.K. Robertson, "Perceptual Framework for Color Image Display," *Proceedings of SPIE Symposium on Visual Communications and Image Processing II*, Boston, Mass., Oct. 1987, pp. 147–150.

[11] P.K. Robertson, "Visualising Colour Gamuts: A User-Interface for the Effective Use of Perceptual Colour Spaces in Data Displays," *IEEE Computer Graphics and Applications*, Vol. 8, No. 5, Sep. 1988, pp. 50–64.

[12] P.K. Robertson and L. De Ferrari, "Systematic Approaches to Visualisation: Is a Reference Model Needed?," Proc. ONR Workshop on Data Visualisation, Darmstadt, July 1993, *Scientific Visualization: Advances and Challenges*, Academic Press, 1994, pp. 287–305.

[13] P.K. Robertson, M. Hutchins, D.R. Stevenson, S. Barrass, C. Gunn, and D. Smith, "Mapping Data into Colour Gamuts: Using Interaction to Increase Usability and Reduce Complexity," *Computers and Graphics*, Vol. 18, No. 5, 1994.

[14] P.K. Robertson and J.F. O'Callaghan, "The Generation of Color Sequences for Univariate and Bivariate Mapping," *IEEE Computer Graphics and Applications*, Vol. 6, No. 2, Feb. 1986, pp. 24–32.

[15] B.E. Rogowitz and L.A. Treinish, "An Architecture for Rule-Based Visualization," *Proceedings of IEEE Visualization '93*, San Jose, Calif., Oct. 1993, pp. 236–244.

[16] B.E. Rogowitz and L.A. Treinish, "Data Structures and Perceptual Structures," *Proc. SPIE Symp. Human Vision, Visual Processing and Digital Display IV*, Vol. 1913, 1993.

[17] S. Roth and J. Mattis, "Data Characterization for Intelligent Graphics Presentation," *Proceedings of CHI*, 1990.

[18] H. Senay and E. Ignatius, "Compositional Analysis and Synthesis of Scientific Data Visualization Techniques," *Scientific Visualization of Physical Phenomena*, N.M. Patrikalakis, editor, Springer-Verlag, 1991, pp. 269–281.

Chapter 8

Engineering Perceptually Effective Visualizations for Abstract Data

Stephen G. Eick

Abstract. *Visualization is an emerging discipline practiced by skilled artisans who are handcrafting the current systems. In the process of constructing many displays of abstract corporate data, we have distilled some aspects of the successful displays into a set of guidelines for creating perceptually effective visualizations. By describing our guidelines, illustrating their usage with several novel examples, and discussing our software infrastructure, we hope to further the progress of visualization from a craft into a production technology.*

8.1 Introduction

Visualization is a powerful link between the two most powerful information processing systems—the human mind and the modern computer. It is a key technology for extracting information, and therefore it is becoming more and more necessary in our increasingly information-rich society. Visualization techniques can enable us to navigate and explore the fast growing number of networked databases far more easily and to discover far more rapidly the information hidden in the ever increasing volumes of data available.

As a discipline, visualization is still emerging, tracking the revolution in computing and computer graphics. It is being practiced by researchers and scientists looking for better display techniques. Skilled practitioners are handcrafting the current systems. Much of the research is exploratory, and producing novel and creative visualizations is as much art as science.

Visualization simply means presenting information in pictorial form. However, the process by which people extract knowledge from pictures is not well understood. There are some limited perceptual theories in particular domains, but much of the work is based on heuristics and trial-and-error.

An emerging discipline progresses through four stages. It starts as a craft, and is practiced by skilled artisans using heuristic methods. Later, researchers formulate scientific principles and theories to gain insights about the processes. Eventually, engineers refine these principles and insights to determine production rules. Finally, the technology becomes widely available.

In handcrafting some two dozen visualization systems at AT&T Bell Laboratories over the last several years and applying them in many case studies to analyze particular data sets, my colleagues and I have formulated some design guidelines for building novel, effective, information-conveying visualizations of abstract data. These heuristic design guidelines are firmly rooted in practical experience and they have enabled us to create many novel and effective displays of data. By presenting our principles and some of the innovative visualizations that have resulted from them, describing our motivating data sets, and relating our techniques to perception, we hope to further the progress of visualization from a craft to a production technology. In addition, perhaps promoting our design rules will stimulate the application of the scientific method in an area of great importance.

Section 8.2 of this chapter describes our focus on abstract corporate databases and introduces the examples we will consider. Section 8.3 presents the visualization guidelines. Sections 8.3.1 through 8.3.8 each describe one of the guidelines in detail and relate it to engineering and perception. Although there are no guidelines for the use of 3D and sound as yet, these are promising areas for research, and they are considered in Sections 8.4.1 and 8.4.2. Techniques for implementing interactive visualization systems, and examples of visualizations created using our guidelines are shown in Section 8.5 and related research is discussed in Section 8.6.

8.2 Focus

Our focus is on visualizing moderately-sized corporate databases containing abstract data.

Abstract data, unlike physical data, has no natural physical representation. Examples of abstract data include traffic flows through the internet, statements in a computer program, and purchases at a grocery store, none of which has an obvious shape. Historically, work on visualization has concentrated on scientific data, such as medical images, fluid flows, or Landsat images, each of which has an obvious physical representation. Breakthroughs in 3D computer graphics have enabled this work to flourish [14, 17, 23]. However, it is hard to take advantage of these breakthroughs to present data that does not have an obvious 3D representation. The research challenge, then, is to discover useful 2D and 3D representations of abstract data.

Moderately-sized means that the database has between 1,000 and 1,000,000 multidimensional observations, possibly with a time component. Larger databases can be studied by subsetting, or by using scientific visualization techniques such as contours, surfaces, and smoothing [24, 23]. The idea is that the cardinality of entities being visualized is of the same order as the number of pixels on a high-resolution workstation monitor. For smaller databases, say between 100 and 1,000 observations, statistical graphics are an ideal tool [33, 9, 8]. For less than 100 observations, spreadsheet graphics are useful.

Developing general visualization guidelines for arbitrary-sized, abstract databases is likely an unsolvable problem (or at least a rich research area), but much progress can be made by partitioning a difficult problem and focusing on particular pieces of it. Databases with 10,000 to 1,000,000 observations are commonplace and many widely-used and important data sets fall within this range. Typical examples include the number of

- retail customers for a small corporation
- employees in a large corporation
- lines of code in a typical computer program
- messages sent between processors on a massively parallel computer
- movies available for purchase or rent
- cities in the United States
- links in a large computer network
- items in the federal budget
- articles published in the statistical literature
- entities in a project management database for a large software system
- items put on sale during the year at a hardware chain
- TV shows that will be available per week when 500 channel capacity cable systems are widely deployed

For each of these databases my colleagues and I have developed working visualizations.

8.2.1 Example Data Sets

Four example data sets are used for illustration:

1. **The lines of code in a large computer system.** The source code database contains the history of a 10,000,000-line switching system developed over the last two decades. For production systems such as this switching system, version control databases containing the complete history of every change made to the source code are necessary for managing the software project and maintaining code integrity. Associated with every line, there are numeric variables—such as the date it was last modified and the number of times it was executed in the regression test—and categorical variables—such as the name of the programmer last modifying it or the software feature that it supports. The code visualizations (Figures 8.1, 8.2, and 8.3) are generated by the SeeSoft[1] software visualization system. They focus on visualizing software modules containing from tens of thousands up to a few hundred thousand lines of code. A detailed description of several case studies analyzing this data set may be found in [11] and [2] and will not be repeated here. However, the interesting analysis tasks involve understanding the age of the code, frequency of change, locations of the bugs, the programmers writing the code, the functionality of the modules, and the active areas of development.

[1] SeeSoft is a trademark of AT&T Bell Laboratories

Figure 8.1: SeeSoft reduced representation overview showing 13,589 lines of code, color coded by age. The newest lines are in red and the oldest in blue with a color spectrum in between. Lines in the same hue were written at approximately the same time and are likely related. See Color Plate 49.

2. **National telecommunications calling patterns.** This data set contains the node-to-node call attempts, switch loads, and their associated geographic locations for Christmas 1994, traditionally one of the heaviest calling days of the year. (See Figures 8.4 and 8.5.) A related data set, calling volume after the 1989 California earthquake, is analyzed as a case study in [4]. The important questions involve the magnitudes of the flows, the capacities of the links, link and node utilization, and variations through time.

3. **Local telecommunications traffic.** This data set came from a local telephone traffic study of all phone calls made during a four-hour period. There is no geographic information, but the data set contains the complete calling history for a few thousand subscribers making several thousand calls during the four-hour period. (See Figures 8.6, 8.7, and 8.8.) The key task for this data set is to identify the communities of interest from the calling patterns. (For more detail see [12].)

4. **Retail sales promotional history.** This data set contains the weekly sales history for a 147-store hardware chain and related advertizing information. This includes sales information and unit counts for each item that was advertized in weekly newspaper inserts over a 47-week period. (See Figures 8.9 and 8.10.) The analysis challenge for this data set is to identify and track the promotional effectiveness for the items advertized in the chain-wide weekly fliers.

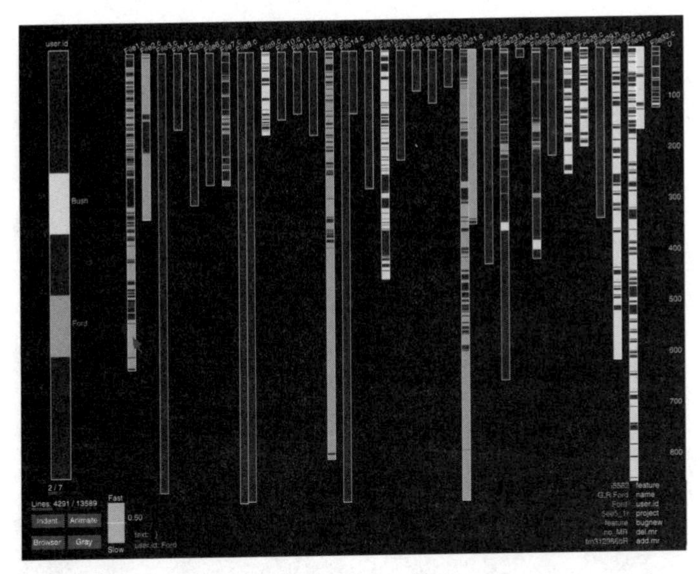

Figure 8.2: SeeSoft line indentation has been turned off and here the lines are color coded to show the `user_id` of the author. `User_id` *Ford* wrote the green lines and *Bush* wrote the yellow lines. See Color Plate 50.

8.3 Engineering Design Guidelines

Inventing a pleasing and informative visualization of abstract data is a difficult task. In the process of developing many special-purpose visualization systems for nongeometric data sets, we have arrived at a set of general guidelines to aid us in engineering effective systems. It is extremely difficult to invent general guideliness that might be expected to remain valid under all conditions and for all interpretation tasks. Ours certainly do not and there are exceptions, even some in our own figures. We believe, however, that for analyzing nongeometric data sets with the analysis tasks associated with corporate information systems, some of our guidelines are likely to be sustaining. In addition, we hope that our guidelines provide a foundation for further progress as visualization moves from a craft to an engineering discipline.

8.3.1 Guideline One: Task-Specific

Understand the data analysis task and thereby ensure that the visualization is focused on the user's needs.

Since the analysis needs of each data set are often unique, some of the best visualizations are task-oriented. These visualizations help frame interesting questions as well as answer them. As with all engineering, the best systems are designed to meet the user's requirements. By focusing on the specific analysis needs and targeting the user's tasks, we ensure a thorough understanding of the system requirements and can then engineer displays and representations to meet these needs. For data analysis, a visual display is useful if it leads to insights and understanding. The best visualizations are built with this goal in mind.

Figure 8.3: An experimental version of SeeSoft showing three views of code. The filled columns represent each line of code using a few pixels positioned horizontally along each row, the indented columns use the traditional SeeSoft representation with one line per row, and the browser window combines both views. By representing more than one line in a row, more code can be shown on a single view. See Color Plate 51.

8.3.2 Guideline Two: Reduced Representation

Use a reduced representation overview to display the entire database on a single screen.

Items in the database are represented by tiny glyphs positioned on the screen in a pleasing, informative, and context-preserving arrangement. The best arrangements are perceptually linked to an important characteristic in the database, thereby facilitating the rapid transfer of information.

For a reduced representation to be effective, the glyphs encoding data must be carefully chosen. A poor choice of symbols or positioning will produce a display that is jumbled because of overplotting and clutter. To fit a large database onto a single screen the area devoted to any particular item may have to be quite small, and therefore choosing and placing the glyphs is critical. The glyphs should be small, so that as much information as possible can be packed on the screen, and they must overplot gracefully. For efficiency, they must be quickly and easily rendered, and therefore simple geometric shapes work well; possibilities include lines, segments, tick marks, points, circles, rectangles, spheres, and so on.

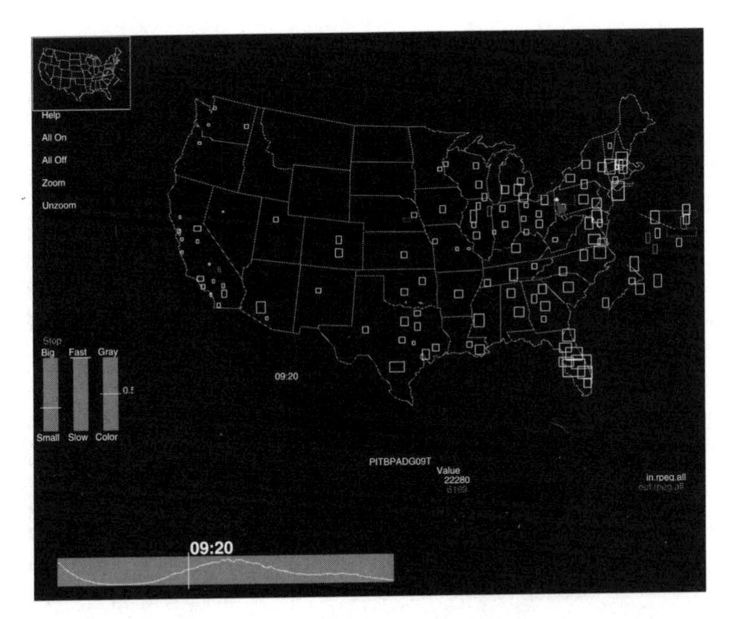

Figure 8.4: SeeNet showing AT&T calling volume by switch Christmas morning, 1994. The size of each rectangle shows the volume with the x-extent coding the inbound volume, the y-extent coding the outbound volume, and color showing direction. Red rectangles indicate sinks and green rectangles indicate source. In this time period there is heavy calling volume both into and out of Florida and also heavy volume into the five red gateway switches destined for the international network. See Color Plate 52.

For perceptual effectiveness, glyphs should be positioned in a way that seems natural for the data set under study. There are a number of possibilities including

- in a grid (Figure 8.9)
- as a scatterplot
- on a map (Figures 8.4 and 8.5)
- along a time axis
- on a surface
- in space using 3D
- with position showing internal relationships (Figure 8.6)

The positioning should convey some essential and interesting aspect of the data set, such as time, spatial information, or structure. A scatterplot is useful for data sets with two independent dimensions, a map works well for spatial data, and temporal data can be plotted along a time axis. The position can encode independent dimensions of the data set, as with longitude and latitude on a map, or it can be a redundant cue. The glyphs and the positioning for the examples in this chapter include rows in columns (Figures 8.1 and 8.2),

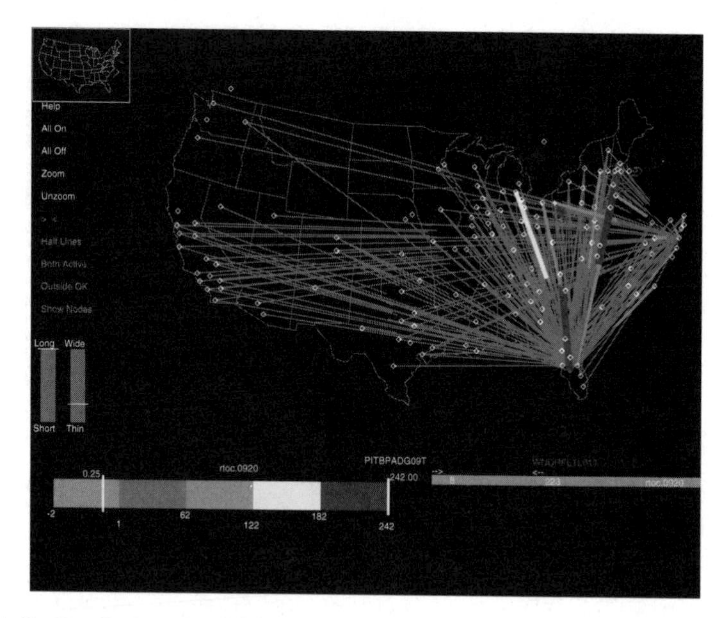

Figure 8.5: SeeNet display of the long distance telephone network showing overload between the nodes. See Color Plate 53.

marks positioned in a grid, rectangles positioned on a map (Figure 8.4), and circles and lines positioned in space showing logical coupling (Figure 8.6).

To display lines of source code in files, glyphs in the form of thin rows may represent the lines of code and glyphs in the form of columns may represent the files. Tying the row length and indentation to the actual code text and positioning the rows within the columns preserves context and creates a natural mapping that is perceptually effective. Programmers can immediately recognize features in the code, for example, loops from the indentation, comments at the tops of the files, and `cases` in `switch` statements from the sawtooth patterns (see Figure 8.1).

The reduced representation overview should attempt to use every pixel to encode information. The detail—when layered properly on the display so it does not obscure the overall pattern—produces images with a high information density [32]. Displays, particularly maps, with a high information density are visually engaging and interesting, and they invite close study.

By showing the complete database in a single view, it is possible to discover database-wide patterns. The most effective representations are, of course, those that make the patterns most obvious and perceptually salient. The reduced representation overview unites and coordinates the other views. Unification may be accomplished through a common use of color and a common, overall perspective.

Often, navigating within a database is a frustrating problem for visualization users. In many systems it is easy for users to get lost. By showing the entire database on a single screen, the reduced representation overview can serve as a navigation framework for the other views. There is no possibility of getting lost because the entire database is visible.

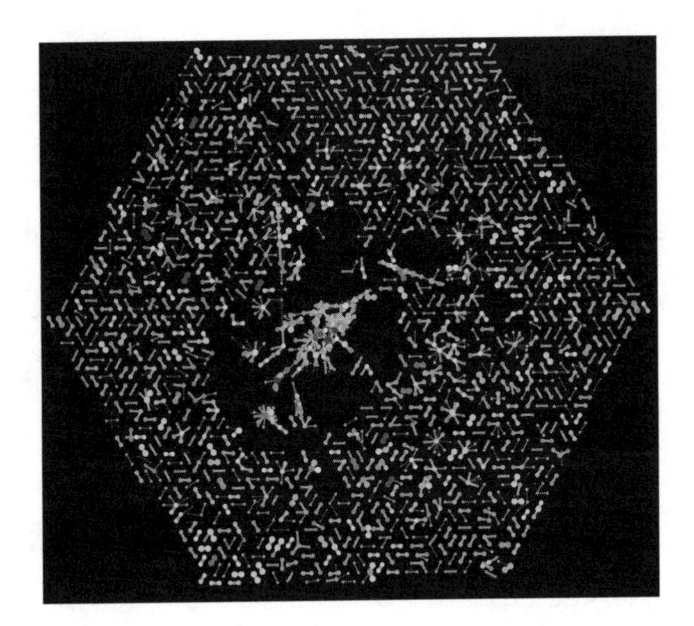

Figure 8.6: NicheWorks visualization of local exchange calling patterns. Each subscriber making a call is represented by a node with lines drawn between the nodes showing individual conversations. The nodes are positioned to show the calling patterns. The cluster of nodes in the center of the diagram represents callers who all talk to each other. The time spent talking is coded by the node colors and size. The link colors code the length of individual conversations. See Color Plate 54.

Figures 8.1 and 8.3 show examples of the overview uniting and navigating. In Figure 8.1 the yellow rectangle in the middle of the fifth file, the red rectangle on the eighth, and the green rectangle on the thirteenth file are linked to browser windows (a browser window is visible in Figure 8.3) showing the code text corresponding to the rows inside the rectangles. The windows are linked to the rectangles through the color and the user may scroll the browser windows by moving the rectangles on the overview.

8.3.3 Guideline Three: Data Encoding

Use color and other visual characteristics to encode the data.

Tying the color and other visual attributes of the glyphs to code characteristics shows the distribution of the statistics in the database. For coding information there are several visual attributes that are effective:

1. position
2. size (area, length, height)
3. shape (orientation)
4. color
5. motion (blinking)

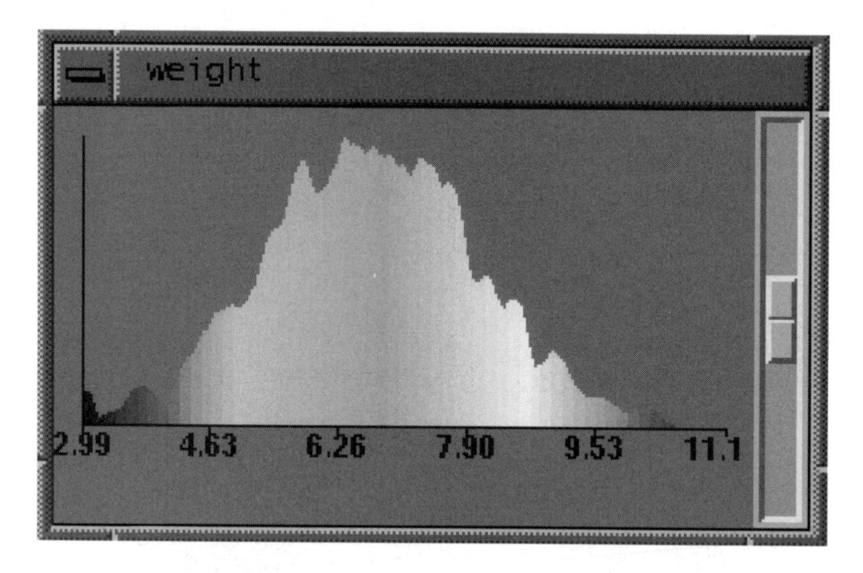

Figure 8.7: Histograms show the distribution of the statistics; users may interactively filter out nodes or links that fall in uninteresting portions of the distribution. The histogram also maps colors from the main display to values. See Color Plate 55.

Thus, for example, the colors of the rows in Figure 8.1 encode the age of each line, and in Figure 8.2 the colors encode the authors. The colors may be tied to any line-oriented statistics, such as the functionality of each line, the number of times it was executed in a regression test, and so on. The sizes of the nodes and thicknesses of the links in Figure 8.6 might encode total subscriber usage and usages between two subscribers, respectively. If the glyphs are rectangles, the x- and y-extents might encode two different, but related, statistics, for example, the inbound and outbound overload in Figure 8.4.

For quantitative comparisons, the most effective perceptual data encoding variable is size [5, page 69], actually length or height [10]. Lengths and heights are more easily and accurately compared than are areas, particularly for proportionally shaped glyphs. Shape is useful for visual segmentation, although it is less effective for small glyphs.

In practice, color, particularly when linked to interactive scales, works well with small glyphs, provided the color scale is carefully chosen. The color scale standard is the RGB rainbow spectrum, which unfortunately has several undesirable properties: the rainbow scale is not naturally ordered; the hue and brightness interact; and the scale stresses the greens. For maximal perceptual effectiveness Levokwitz et al. [19, 20] recommend perceptually uniform color scales, where increments in colors are perceived to have equal increments in the data range.

Cleveland [8] recommends a two hue color scale for encoding data with both positive and negative values. On such a scale, white is tied to the middle of the data range and the full saturation of the two colors encodes, respectively, the largest positive and negative data items.

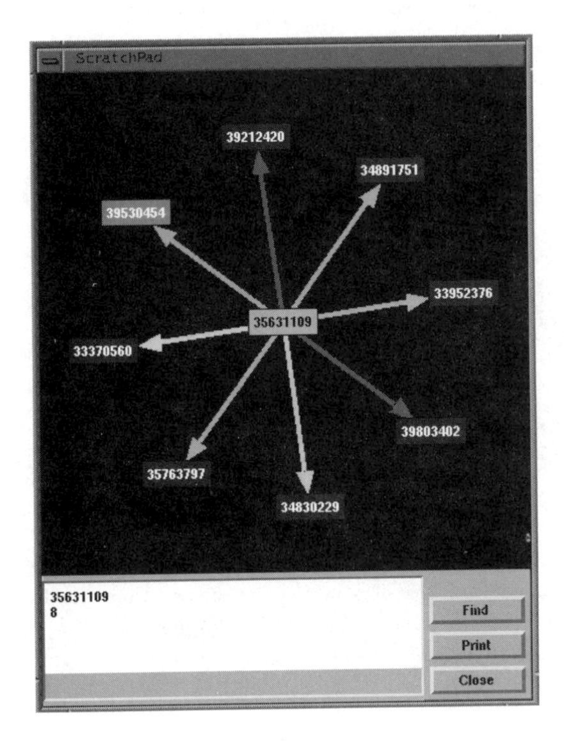

Figure 8.8: A `scratch pad` shows the details for any node. See Color Plate 56.

8.3.4 Guideline Four: Filtering

Use interactive filters to focus the display.

With many data items, the reduced representation overview may become too busy to interpret. Interactive filters can reduce the visual complexity by turning off (not displaying) particular data items, thereby focusing on the interesting and informative patterns in the data. In Figure 8.2 the data displayed in Figure 8.1 has been recolored to highlight those portions of the code written by two of the seven authors.

The interactive color scale (Figure 8.1), a linked histogram (Figure 8.7), and double-edged sliders (Figure 8.5) are all effective filters. Each is mouse-sensitive, enabling the user to interactively turn on or off regions of interest. In Figure 8.2 a range of the color scale (red, orange, blue, violet) has been deactivated to highlight only those portions of the code written by each of two authors (shown in yellow and green). In Figure 8.5 the user has adjusted the lower threshold on the doubled-edged slider, thereby turning off some of the links with low statistic values.

In short, since the display is parameterized, the user can adjust the parameters to filter out data that is not of immediate interest, focus on that which is, and look for underlying patterns. As with focusing a camera, humans effortlessly solve computationally complex problems that are involved in determining when a display is interpretable.

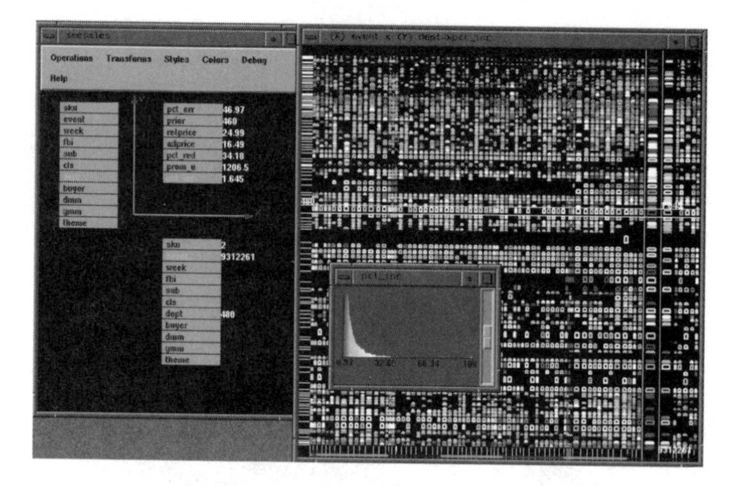

Figure 8.9: SeeSales showing retail sales data. Items for viewing are selected using the control panel (left) and represented as a color-coded rectangle showing on an item-by-week grid. "Hotter" colors (pinks and whites) represent large increases and "colder" colors (blues, greens) represent smaller increases. The two widest columns on the right are the Thanksgiving weekend and day after Christmas sales. The histogram shows the distribution of the percentage increases in sales. See Color Plate 57.

8.3.5 Guideline Five: Drill Down

Use the "drill down" technique to obtain details about particular items.

Users, upon discovering interesting patterns, need access to the actual data values. There are two mechanisms for providing detailed data. When the user touches any glyph with the mouse, the data values for that item are displayed. Detailed information should be available by pointing the mouse—mouse clicks should be unnecessary. For example, in the SeeSoft system, as the user touches any row with the mouse, the corresponding line of code and abstract data associated with the line (author, date, release, and so on) are displayed (see Figure 8.1).

The second mechanism for providing details is with specific views such as the NicheWorks[2] scratch pad (see Figure 8.8) or the SeeSoft browser windows. Browser windows—special purpose linked views coordinated by the overview—will show details on demand. In SeeSoft, the browsers show the program text using a standard readable font.

8.3.6 Guideline Six: Multiple Linked Views

Use multiple linked views to answer specific questions.

For important data sets, one view is often not sufficient to answer all interesting questions. Many views, each answering separate, but related questions, may work together to provide insight. Each view should show one aspect of the data or its distribution. The views should be tightly linked so that operations in one view, such as color scale manipulations,

[2]NicheWorks is a trademark of AT&T Bell Laboratories

Figure 8.10: An experimental 3D representation with a cutting plane showing all sales increases above a threshold. See Color Plate 58.

propagate instantly to the other views. Together, the combination of several simple views is much more powerful than the sum of the individual views taken one at a time.

The reduced representation overview serves as a navigation guide and a coordination mechanism for other linked views. Good mechanisms for coordinating multiple views involve color-coding navigation cursors, marking particular parts of the screen, or leaving other visual clues. For example, in the SeeSoft system, the browser windows showing the code text are linked to the overview through the use of color-coded rectangles. The frame color on each browser window matches its corresponding rectangle. Users may scroll the browsers either by moving the rectangles on the overview or by using the scroll bar. Color manipulations affect both views of the data simultaneously.

8.3.7 Guideline Seven: User Interface

Provide a direct manipulation user interface to facilitate interaction.

A high-interaction user interface increases the effectiveness of visualizations by allowing the user to manipulate any item on the screen. The computer should respond within 50ms or at most 100ms. The user perceives this as an instantaneous response and feels as if he or she is causing the action as the display parameters are manipulated. For some operations point-and-click interfaces fail because the computer response is too slow and jerky. The best interfaces provide the user with constant and continuous feedback, as when driving a car.

In SeeSoft, for example, filtering, focusing, and linking all occur in real time, following mouse movements. By using continuous, simple, reversible mouse movements, the user can probe the display by wandering around it with the mouse.

Developing visualization systems with high-interaction user interfaces requires careful software design. With the moderately priced machines currently available, we find that software implementing multiple windows with tight linking between them must run as a single process to ensure adequate performance. Our experience has been that it is difficult to provide an instantaneous response to user commands with available standard IPC (interprocess communication) mechanisms, even on fast workstations.

The requirement for near-instantaneous responses restricts the types of manipulations and the complexity of displays. As computers and networks become faster, these restrictions will become less onerous, but there will always be tension between visual detail and appropriate interaction techniques.

8.3.8 Guideline Eight: Animation and Motion

Use animation to show the evolution of temporally oriented data.

Many commonly encountered data sets have a temporal aspect. Animation, with each frame representing a single time period, is an ideal tool for analyzing large time-oriented data sets.

In general, it is possible to animate over any variable indexing the frames. However, for animation to be effective, the frames must evolve smoothly and continuously as with pictures in a movie. Big or unexpected changes are jarring and stand out perceptually.

8.4 Techniques Under Development

We are currently investigating the use of 3D and sound in our visualizations of abstract data. Although we do not yet understand their use well enough to codify it in guidelines, we have some preliminary insights.

8.4.1 3D

For abstract data visualization, 3D has been underexploited, but it is an active research area. The depth dimension can be used as a natural way to pack more information onto the screen without overloading it, and eventually, after 3D representations of abstract data are fully explored, it may be possible to increase the information content, the ratio of information to pixels[3], by a factor of 10. However, it is difficult to find natural embeddings of nongeometric data in 3D that are more interpretable than their 2D counterparts.

Perceptually, it is possible for 3D displays to increase the information density beyond that possible with 2D displays by enabling our minds to create *virtual pixels*. For example, the exact position of a pedestrian walking behind a tall picket fence is clear even though his or her position can only be seen between the fence posts. The reason is that from the fragmentary input and motion our minds create an exact position. Applying this idea to network visualization, for example, complex 2D node and link displays become visually

[3] Tufte [31, page 162].

confused with too many line crossings. For the same network visualized using a 3D representation, however, the links may no longer cross. In a 3D representation it is easy to see when one link passes behind or in front of another.

3D representations are closely tied to motion because motion and depth cues are often necessary to perceive objects in 3D on a 2D computer monitor. Thus in the statistics literature, rotating 3D point clouds have long been used to generalize to the 2D scatterplot [13, 9]. Recently, several novel uses of 3D for showing the structure of data sets have appeared [27] in the human factors literature.

Figure 8.10 shows an experimental cityscape 3D visualization of retail sales data [16]. The height of each column codes the increase in sales above the cutting plane. The two rows with huge columns are the Thanksgiving weekend and day-after-Christmas sales.

8.4.2 Sound

Sound is another underexploited medium for encoding data [22]. There are many aspects to sound—pitch, timbre, loudness, and so on—and many ways that sound can encode data. Sound is fundamentally different from a visual display in four ways:

- it arrives through an independent channel
- its bandwidth is lower than vision
- it is immediate instead of persistent as with visual representations
- it is serial in time

Sound arrives at the brain through the ears, a different path from the eyes, and is processed by different parts of the brain. It has a lower bandwidth than vision. For example, it is generally faster to read text than to listen to it being spoken. It is immediate and serial in time, which makes repetition difficult.

For abstract data visualization, audible alarms may be used to alert the user to unusual events, tones can encode data, and voice messages can convey detailed information. In our experience, using tones to encode data is often perceptually challenging. Other researchers have had more successes than we have achieved. In general, because sound has low bandwidth, using sound to encode complicated data sets will be difficult.

Sound works very well for alerting, particularly for monitoring tasks, enabling it to cut through visual clutter. Since much communication is via speech, this suggests that using voice to convey detail is natural. In animations, researchers have used voice to announce the passage of time. This was effective because it enables the users to concentrate on the data display and not on the time slider.

8.5 Implementation Techniques

Developing new techniques suitable for nontraditional, irregular, or large data sets is complex and time consuming, even for experts. To support our research, we have developed an infrastructure for producing highly interactive data visualizations using an object-oriented

C++ class library. Vz, our cross-platform library (X11, OpenGL, and MS Windows), embodies linked views, direct manipulation, and data abstractions in a selective manner. Using Vz, it is possible to create novel, production-quality visualization systems more quickly and easily than with alternative environments, and yet maintain the runtime performance of custom C code. Vz focuses on providing a set of tailorable components suited to interaction graphics and not "canned" classes, such as charts and graphs.

To simplify coding complexity, Vz factors out common windowing code, so application developers can focus on visualization aspects of the implementation. Vz uses C++'s multiple inheritance to enable applications to incrementally add new functionality by inheriting. For developing graphics, object-orientation is clearly a method of choice.

In interactive visualization systems, speed is critical, particularly on lower end machines, and so three aspects of Vz's design ensure optimal performance:

- Applications run as a single process in one address space
- C++ is compiled and executes at C speeds
- Vz supports and encourages native access to the graphics hardware and the use of special purpose graphics libraries

Running the applications in a single address spaces avoids the context-switch overhead of IPC and also facilitates porting to single process operating systems such as MS Windows. One of the advantages of partitioning code into separate processes connected by IPC is that it simplifies the software development by ensuring clean interfaces between the processes. Vz captures this advantage by using C++'s object-oriented capabilities and strong typing. An advantage of C++ over interpreted languages is its execution speed.

Other techniques for speed involve precalculating, incrementally rendering at runtime, and color map animation. On all modern workstations, color map manipulations are done in hardware and synchronized with the vertical refreshes of the monitor. This means that operations implemented with color map manipulations, such as deactivating a range of colors on the color scale, involve no rerendering. Vz supports multiple linked windows, using a *publish and subscribe* metaphor. Each view contains a local copy of the variables that it is interested in and *subscribes* to the global copy. If any view changes the global copy, it *publishes* the change and all subscribers are notified and can update their local copies and redraw as necessary. This simple linking mechanism coordinates all of the views through a straightforward interface, *publish and subscribe*, and allows for high-bandwidth communication between the views because all share the same address space.

8.6 Related Research

Because of the importance of visualization there is a rich history of related work. Since it is impossible to provide an exhaustive review, we have selected some interesting papers that have influenced our formulation of the guidelines.

The task guideline is motivated by Bertin's classic work on graphics constructions [5] and some more research by several authors. Using his semiological approach, Bertin carefully explores the use of visual attributes to encode data based on perceptual considerations. He also introduces the idea of interactively manipulating graphical displays using specially built manual tools. Some of Bertin's ideas on perception and visual attributes are captured in guidelines one and three.

More recently, several researchers have investigated automated, task-based methods for visualizing specific classes of data. See, for example, Mackinlay for relational information [21] and Casner for networks [7]. In clearly related work, Robertson proposed a construction methodology for matching the data characteristics and the interpretation aims using a natural scene paradigm [28]. Other authors have proposed knowledge-based systems for designing visualizations [29] and [15].

The idea of using glyphs to encode abstract, nongeometric data is well established (for a recent paper describing the state of the art in glyph-based visualization see [26]). Our reduced representation guideline builds on traditional glyph approaches by suggesting small, very compact, information-rich glyphs. For a recent, interesting example of this guideline, Keim, Kriegel, and Seidl show an entire database using a pixel-like representation [18]. They position many small glyphs, each representing an item in the database, in a grid according to their relevance to a database query. The iconographic displays of Pickett and Grinstein are another example of our reduced representation guideline [25].

Becker, Cleveland, and Wilks's survey paper on dynamic graphics [3] describes several early cases of our interactive and filtering guidelines. Their paper discusses a suite of interactive techniques for analyzing multidimensional data, often shown in scatterplots, with particular emphasis on conditioning. Our interactive techniques, particularly color manipulations, are generalizations of their ideas (for another paper showing the utility of this guideline see [6]).

Ahlberg and Shneiderman describe an example of our interactive guideline in their dynamic scatterplot displays, called starfields [1]. Recently, in closely related work, Shneiderman describes several visualization systems using "Dynamic Queries" [30]. His double-edged sliders and dynamic queries are a special case of our filtering guideline. In general, effective filters may be visual displays that function both to visualize results and also as control panels.

The animation and motion guideline is motivated by Becker, Eick, and Wilks's research in network visualization [4]. They describe techniques for analyzing large temporally oriented networks using interactive techniques and animation.

In early work, Fisherkeller, Friedman, and Tukey describe an interactive analysis system for multidimensional data [13]. This work is particularly interesting because it is one of the earliest examples of interactive 3D computer graphics showing data. Their PRIM-9 system featured a set of dynamic tools for projecting, rotating, isolating, and masking multidimensional data in up to nine dimensions. 3D rotations were central to their methods.

Robertson, Card, and Mackinlay describe some innovative techniques to visualize 3D structures [27] using distortion techniques. 3D representations are promising and an active research area.

8.7 Conclusions

We have presented a set of guidelines for engineering perceptually effective visualizations of abstract data and illustrated their use with several examples. These guidelines have been distilled from our experience in constructing visualizations over the last several years. We have focused our guidelines on systems for visualizing corporated-sized databases.

By following these guidelines, engineers can build informative visualizations of abstract data. These visualizations will enable users to see interesting patterns in the database, find trends, discover additional knowledge, and extract information from data.

Our aim in presenting these guidelines is to further the transition of visualization from a craft to an engineering discipline. We would like to stimulate more careful thought about how information should be encoded visually for the most effective transfer to humans.

Acknowledgments

I would like to gratefully acknowledge the contributions of my colleagues in creating our visualization systems, software infrastructure, and suite of applications. Specifically, Joe Steffen and Eric E. Sumner collaborated on the original SeeSoft system, shown in Figures 8.1 and 8.2. Thomas Ball created the views of software shown in Figure 8.3. Richard Becker and Allan Wilks helped to write the SeeNet system, shown in Figures 8.4 and 8.5. Dave Atkins, Paul Lucas, and Graham Wills are the authors of the Vz C++ library. Besides working on Vz, Graham Wills also developed the NicheWorks system shown in Figure 8.6. Jackie Antis and John Pyrce are the authors of the SeeSales shown in Figures 8.9 and 8.10. Brian Johnson and Anselm Spoerri carefully proofread drafts of this chapter.

Bibliography

[1] Christopher Ahlberg and Ben Shneiderman, "Visual Information Seeking: Tight Coupling of Dynamic Query Filters with Starfield Displays," *CHI '94 Conference Proceedings*, 1994, pp. 313–317.

[2] Marla J. Baker and Stephen G. Eick, "Space-Filling Software Displays," *Journal of Visual Languages and Computing*, Vol. 6, No. 2, June 1995. To appear.

[3] Richard A. Becker, William S. Cleveland, and Allan R. Wilks, "Dynamic Graphics for Data Analysis," *Statistical Science*, Vol. 2, 1987, pp. 355–395.

[4] Richard A. Becker, Stephen G. Eick, and Allan R. Wilks, "Visualizing Network Data," *IEEE Transactions on Visualization and Graphics*, 1995. To appear.

[5] Jacques Bertin, *Semiology Of Graphics*, University of Wisconsin Press, Ltd., London, England, 1983.

[6] A. Buja, J. A. McDonald, J. Michalak, and W. Stuetzle, "Interactive Data Visualization Using Focusing and Linking," *IEEE Visualization '91 Conference Proceedings*, San Diego, Calif., Oct. 1991, pp. 156–163.

[7] Stephen M. Casner, "A Task-Analytic Approach to the Automated Design of Graphics Presentations," *ACM Transactions on Graphics*, Vol. 10, No. 2, 1991, pp. 111–151.

[8] William S. Cleveland, *Visualizing Data*, Hobart Press, Summit, N.J., 1993.

[9] William S. Cleveland and Marylyn E. McGill, editors. *Dynamic Graphics for Statistics*. Wadsworth & Brooks/Cole, Pacific Grove, Calif., 1988.

[10] William S. Cleveland and Robert McGill, "Graphical Perception: Theory, Experimentation and Application to the Development of Graphical Methods," *Journal of the American Statistical Association*, Vol. 79, 1984, pp. 531–554.

[11] Stephen G. Eick, Joseph L. Steffen, and Jr. Eric E. Sumner, "Seesoft^{TM}—A Tool for Visualizing Line Oriented Software Statistics," *IEEE Transactions on Software Engineering*, Vol. 18, No. 11, Nov. 1992, pp. 957–968.

[12] Stephen G. Eick and Graham J. Wills, "Navigating Large Networks with Hierarchies," *Visualization '93 Conference Proceedings*, San Jose, Calif., 25–29 Oct. 1993, pp. 204–210.

[13] Mary Anne Fisherkeller, Jerome H. Friedman, and John W. Tukey, "PRIM-9: An Interactive Multidimensional Data Display and Analysis System," *Data: Its Use, Organization, and Management*, The Association for Computing Machinery, New York, N.Y., 1975, pp. 140–145.

[14] James D. Foley, Andries van Dam, Steven K. Feiner, and John F. Hughes, *Computer Graphics Principles And Practice*, Addison-Wesley, Reading, Mass., 1990.

[15] J. Goldstein, S. F. Roth, J. Kolojejchick, and J. Mattis, "A Framework for Knowledge-Based Interactive Data Exploration," *Journal of Visual Languages and Computing*, Vol. 5, Dec. 1992, pp. 339–363.

[16] William C. Hill and James D. Hollan, "Deixis and the Future of Visualization Excellence," *IEEE Visualization '91 Conference Proceedings*, San Diego, Calif., Oct. 1991, pp. 314–320.

[17] Arie Kaufman, *Volume Visualization*, IEEE Computer Society Press, Los Alamitos, Calif., 1991.

[18] Daniel A. Keim and Hans-Peter Kriegel, "Visdb: Database Exploration Using Multidimensional Visualization," *IEEE Computer Graphics and Applications*, Vol. 14, No. 5, Sep. 1994, pp. 40–49.

[19] Haim Levkowitz and G. T. Herman, "Color Scales for Image Data," *IEEE Computer Graphics and Applications*, Vol. 12, No. 1, 1992, pp. 78–80.

[20] Haim Levkowitz, Richard A. Holub, Gary W. Meyer, and Philip K. Robertson, "Color vs. Black and White in Visualization," *IEEE Computer Graphics and Applications*, Vol. 12, No. 4, 1992, pp. 20–22.

[21] Jock D. Mackinlay, "Automating the Design of Graphical Presentations of Relational Information," *ACM Transactions on Graphics*, Vol. 5, No. 2, Apr. 1991, pp. 110–141.

[22] Tara M. Madhyastha and Daniel A. Reed, "Data Sonification: Do You See What I Hear?" *IEEE Software*, Vol. 12, No. 2, Mar. 1995, pp. 45–56.

[23] Gregory M. Nielson, Thomas A. Foley, Bernd Hamann, and David Lane, "Visualizing and Modeling Scattered Multivariate Data," *IEEE Computer Graphics And Applications*, Vol. 11, No. 3, 1991, pp. 47–54.

[24] Gregory M. Nielson, Bruce Shriver, and Lawrence J. Rosenblum, editors, *Visualization in Scientific Computing*, IEEE Computer Society Press, Los Alamitos, Calif., 1989.

[25] R. M. Pickett and G. G. Grinstein, "Iconographic Displays for Visualizing Multidimensional Data," *IEEE Conference on Systems, Man and Cybernetics*, IEEE Press, Piscataway, N.J., 1988, pp. 514–519.

[26] Willian Ribarsky, Eric Ayers, John Eble, and Sougata Mukherjea, "Glyphmaker: Creating Customized Visualizations of Complex Data," *IEEE Computer*, Vol. 27, No. 7, July 1994, pp. 57–64.

[27] George G. Robertson, Stuard K. Card, and Jock D. Mackinlay, "Information Visualization Using 3D Interactive Animation," *Journal of the ACM*, Vol. 36, No. 4, 1993, pp. 56–71.

[28] Philip K. Robertson, "A Methodology for Choosing Data Representations," *IEEE Computer Graphics and Applications*, Vol. 11, No. 3, May 1991, pp. 56–67.

[29] Hikmet Senay and Eve Ignatius, "A Knowledge-Based System for Visualization Design," *IEEE Computer Graphics and Applications*, Vol. 14, No. 6, Nov. 1994, pp. 36–47.

[30] Ben Shneiderman, "Dynamic Queries for Visual Information Seeking," *IEEE Software*, Vol. 11, No. 6, Nov. 1994, pp. 70–77.

[31] Edward R. Tufte, *The Visual Display of Quantitative Information*, Graphics Press, Cheshire, Connecticut, 1983.

[32] Edward R. Tufte. *Envisioning Information*. Graphics Press, Cheshire, Conn., 1990.

[33] John W. Tukey, *Exploratory Data Analysis*, Addison-Wesley, Reading, Mass., 1977.

Chapter 9

Studies in Comparative Visualization of Flow Features

Hans-Georg Pagendarm and Frits H. Post

Abstract. *This chapter introduces two important concepts: feature visualization and comparative visualization. Features are patterns or structures hidden in complex data describing physical phenomena. Comparative visualization is concerned with the analysis of differences and sources of error in a simulation or visualization process. Using examples from fluid dynamics, different types of comparative visualization of flow features are illustrated and analysed. Feature extraction techniques for vortices, shock waves, and skin friction are described, and comparative analysis shows their strengths and weaknesses. Some important lessons may be learned from these examples. Comparative visualization is identified as an important tool for data exploration, which provides incentives for explanations and further investigations, and increases awareness of possible problems within the visualization process itself.*

9.1 Introduction

Comparing results or methods is common practice in scientific research. An absolute measure for the accuracy of a data production or sampling process is not often available from any accepted and practically usable theory. In such cases an estimate of the accuracy or the validity of data may be obtained from a comparison with data of different origin but describing the same physics.

The general idea of comparative visualization [8] is that data from two or more different sources are visualized with the intention to show similarities and differences. Differences in visual appearance can be caused by many factors; we list only a few here:

- different physical phenomena
- different experimental or numerical conditions

- measurement artifacts: noise, sampling resolution, interference with the phenomenon, and so on
- numerical inaccuracies
- different mathematics or logic
- the visualization process

Note that each of these sources of difference can be the goal of a comparative study. We will focus here on comparisons of data from two different sources providing information on the same physical phenomenon, and also on comparison of different visualization techniques providing different views of the same data. Using these scenarios, we will illustrate various techniques and applications of comparison.

Comparison is an established way to raise awareness of users and researchers in visualization towards certain difficulties or sources of error. To illustrate this, we will present three cases of comparative analysis in this chapter. The first two cases are concerned with different visualization techniques for the same data: two techniques for visualization of vortices, and two techniques for detecting shock waves. For these two cases, the same data will be used from a numerical flow simulation for a blunt fin/wedge configuration [2] (see Figure 9.1). The data are a result of a Navier-Stokes solution in a hypersonic flow. In Section 9.2, two techniques for vortex visualization are compared, which produce different results from the same data.

Another significant phenomenon in the same flow field, discussed in Section 9.3, is a pattern of shock waves. The flow is studied to learn more about the interaction of shock waves and complex three-dimensional boundary layers or viscous flows in general. Even though the geometry appears to be simple, an extremely complex three-dimensional flow field is generated.

In a wind tunnel experiment, the vortices of Figure 9.1 can produce characteristic traces on the wall of the fin. These patterns have been visualized in the numerical data as well as in a wind tunnel experiment. The third case, discussed in Section 9.4, is a direct comparison of these numerical and experimental data, showing the relations between these two types of visualization. The differences found in the initial visual comparison are explained, and a new visualization technique is suggested to verify this.

The purpose of this chapter is to show the role that comparative visualization can play in a critical evaluation of different visualization techniques, and how comparative visualization can help to clarify the relation between experiment and simulation.

9.2 Visualization of Vortices

This first example was chosen to illustrate the influence of the scale of an extracted feature with respect to the scale of the discrete resolution of the data. Two alternative methods to visualize vortices in vector fields are compared and show significantly different results. This makes one of the two methods superior for a given data set. Which method to choose may well vary within a single data set depending on the position within the data. More importantly, this behavior is usually not expected by the user of the visualization method.

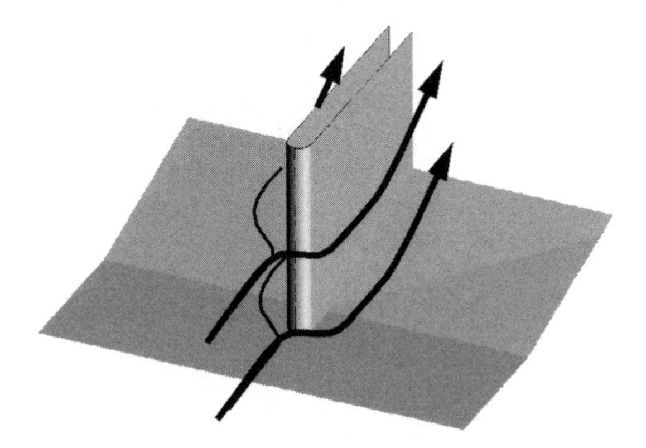

Figure 9.1: Geometry of the flat plate, wedge, and fin configuration placed in a hypersonic flow field (schematic view).

9.2.1 Streamlines and Streamribbons

There is a large number of techniques to analyse and visualize vortical behavior in vector fields (see Banks and Singer [1] for a brief review). In particular in experimental flow visualization, there is a long tradition of visualizing material transport by injection of dye into a water flow [4]. The numerical counterpart of this technique is the calculation of particle paths. Techniques for generating streamlines are well established; for a survey of algorithms, see Kenwright [3]; for implementation issues on streamline integration, see Murman and Powell [5]; for a comparative analysis, see Sadarjoen et al. [10].

However, streamlines can give a poor representation of a complex velocity field, even if they are generated from carefully selected seed points. Figure 9.2 shows a set of streamlines (red and black) which are well placed in order to illustrate the flow characteristics. Exactly the same lines will be used for further analysis in the following images. In spite of the fact that a number of additional yellow and blue lines are used to enhance the image, no deep insight in the flow field is provided. Insertion of even more lines would tend to clutter the image. Streamlines lack a direct representation of swirl, or the streamwise rotational motion of the fluid. This is best represented by vorticity, defined as the curl of velocity, and thus derived from the velocity gradient. Thus, a high density of adjacent streamlines is needed to visualize vorticity.

Consequently, some researchers suggested the use of streamribbons to illustrate this important property of flows. Volpe [13] suggested filling the gap between two streamlines which run more or less in parallel by creating a bundle of lines to give the impression of a solid ribbon. Another way of creating streamribbons is obtained by constructing a mesh of polygons between two adjacent streamlines. We will denote this technique as 'adjacent streamlines' (ASL for short). A simple method to create such a polygon mesh is by the use of a marching triangulation algorithm as described by Pagendarm [6], which demonstrates the problems of such an approach in diverging flow fields.

Divergence of the vector field becomes visible by the increasing width of the ribbons. However, if the two limiting streamlines of the ribbon diverge, they are no longer subject to

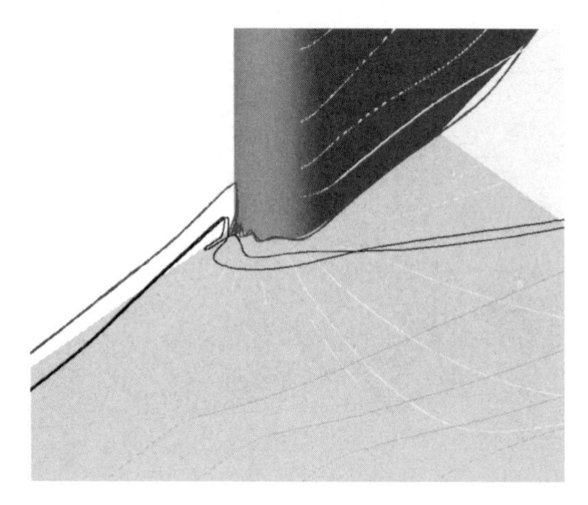

Figure 9.2: Poor representation of a vector field by streamlines. See Color Plate 59.

Figure 9.3: Two TSR streamribbons visualize the same data as in Figure 9.2 and clearly depict the axis of two vortices as well as the swirl in the flow. The scale is identical to Figure 9.2. See Color Plate 60.

the same amount of vorticity. They behave quite independently. Since this should not occur for physical reasons, such an event clearly demonstrates the limitations of the method.

These limitations may be overcome by evaluating the data at the location of a single streamline and constructing the ribbon from the angular rotation of the vector field around this line. One way to achieve this is calculating two adjacent streamlines and keeping their distance constant. Such an algorithm may be extracted from Van Wijk [12]. Streamribbons may also be considered as a simplification of a more complex analysis of the stress tensor [11]. For direct comparison to the ASL technique, a straightforward implementation calculating the angular velocity from the curl of the velocity vector field was used [9]. The ribbon is a surface of constant width, centered around a single streamline. The amount of twist of the ribbon is directly linked to angular velocity at the streamline. We will call this the 'twisting ribbon around a single streamline' technique here, or TSR for short.

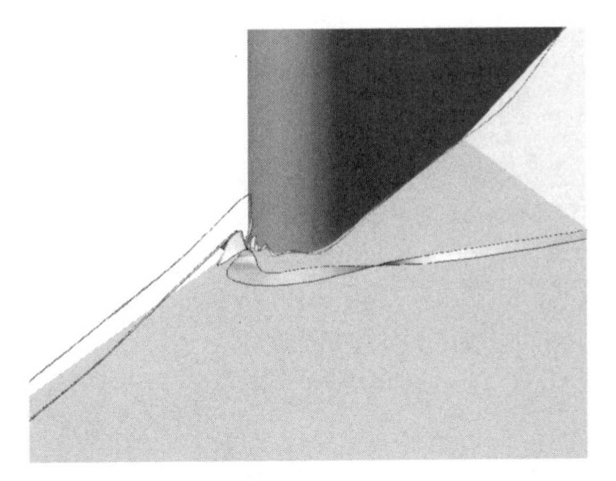

Figure 9.4: The situation of Figure 9.3 visualized by constructing a ribbon from adjacent streamlines, or an ASL ribbon. See Color Plate 61.

Figure 9.3 demonstrates, in comparison to Figure 9.2, how the two dominant vortices in the flow field may be intuitively visualized using the TSR method. One would expect that similar information may be extracted from an image constructed using the ASL method (Figure 9.4).

9.2.2 Evaluation of the Two Techniques

Close examination reveals that the ASL method fails because the two streamlines used for constructing the ribbon do not follow a pure vortical motion, due to interference with other features of the flow. On the other hand, the streamlines converge, which makes it difficult to visualize the vortices without enlargement of the image. Note that the streamlines shown in the image are not incorrect, they just fail to clearly show the desired feature.

The TSR method visualizes pure rotational motion along a single streamline, which is not easily affected by other flow features. It behaves much like an icon depicting the position and swirl of vortices. The scale of the icon (the width of the ribbon) may be adjusted in a wide range to meet the image scale. The ASL method does not allow independent control of the ribbon width, as the width is defined by the paths of the streamlines.

A surface, as defined by two adjacent streamlines connected by a polygon mesh, suggests a linear variation between the streamlines, which may not be warranted at the given distance. The TSR method shows velocity direction and axial rotation, where the latter is a gradient quantity. Approximation of the gradient using finite differences is always linked to local cell size.

In the case of Figure 9.5, a strong enlargement by a factor of 20 is required to make this method useful. The image is expected to show a certain behavior of streamribbons in case vortices are present. Due to the fact that the feature "vortex" is hidden in discrete data, this behavior of ribbons may not occur if an unsuitable method is chosen. In the worst case, the feature is not detectable at all with one of the alternative methods. Note that the method of choice may well be different depending on the data or simply depending on the position within the same data.

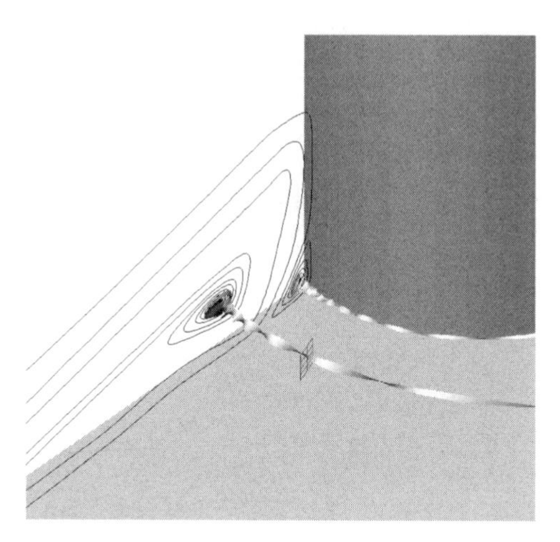

Figure 9.5: ASL streamribbons provide useful visualization if they are used within the proximity of the vortex core where the flow is clearly dominated by vortical effects. Some cells of the numerical grid are shown in the image to emphasize that the relevant region is restricted to a very small scale of a few cells in these data. See Color Plate 62.

In this example, the failure of the ASL method was caused by the need to use streamlines outside the close proximity of the vortex core in order to meet the requirements of the image scale and make features visible from a distant view. When visualized at the same scale level as the feature, both methods work equally well, however, the TSR method would need further effort to provide additional information such as convergence or divergence in the vector field. The ASL method provides such information implicitly, even down to the size of numerical grid cells.

Going down to very small scales in the range of one or two grid cells, one may find that ribbons constructed from two adjacent streamlines may still pick up small but significant differences in the velocity field. Figure 9.6 illustrates this effect by connecting a number of such ribbons combined to a stream surface which eventually forms an "S"-shape. However, the TSR method does not respond to variation in the vector field at this scale because the evaluation of the curl uses the data values at the vertices of eight adjacent cells, which has a smoothing effect.

For the example of visualizing vortical behavior by streamribbons, one could conclude that the TSR method is superior for small-scale features in large images. When the size of the image is of the same order as the size of the vortex, both methods compared perform well. However, the ASL method can provide additional information about divergence or convergence of the vector field. At scales on the order of grid cells, the TSR method would be preferred to detect physical features, while the ASL method would provide information about the numerical smoothness of the data. The important message is that there is no general rule to select one of the two methods in question and that conditions may well vary within a single data set.

In cases where a large number of distributed vortices need to be visualized, there would be a need for more complex visualization techniques. For example, Banks and Singer [1]

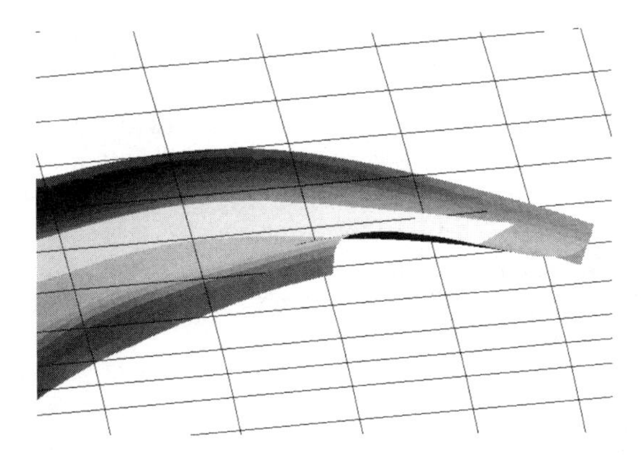

Figure 9.6: The blue stream surface is constructed from 30 adjacent streamlines. While the red ribbon picks up the vortical motion of the flow correctly, the green ribbon fails because of the limited numerical resolution. See Color Plate 63.

suggested a method for vortex extraction from complex flow fields such as turbulent shear layers.

9.2.3 Techniques of Comparative Visualization

Comparative visualization may be used to uncover hidden discrepancies between different data intended to describe the same physical phenomena [8]. This approach assumes that the visualization methods used do not introduce significant errors or discrepancies which might affect the resulting comparative analysis. However, it is important to note that various visualization methods which can, in principle, visualize the same phenomenon in the data may well give different results. This effect may also be studied using comparative visualization techniques (see Figure 9.7).

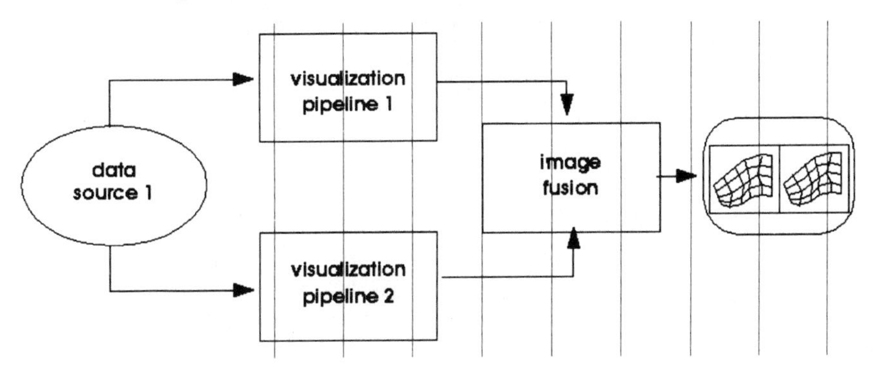

Figure 9.7: Comparing visualization methods.

The comparison may be done by presenting images side-by-side, such as Figures 9.3 and 9.4. If the visualization system allows combination of the resulting representations at the level of 3D graphical objects, the object created by the visualization pipeline may be placed within the same image, as was done in Figure 9.6.

The example in the next section will illustrate how two different methods to extract and visualize a three-dimensional shock wave produce a significantly different result. Again, the resulting shock waves may be compared by putting images side-by-side, which already reveals major differences between the two methods used. However, integrating both resulting graphical objects into a single visualization illustrates the significance of the differences. Further discussions explain the reason for those differences, as well as their significance for investigation of the underlying physics.

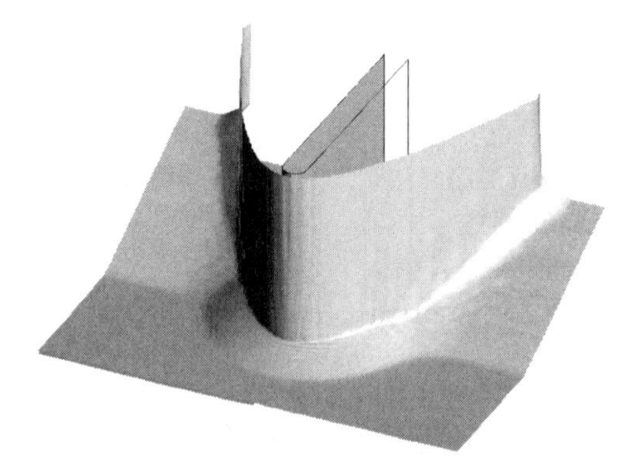

Figure 9.8: Front shock wave visualized using an isosurface close to the free-stream Mach-number. The flow field is the same as in Figures 9.2–9.6. See Color Plate 64.

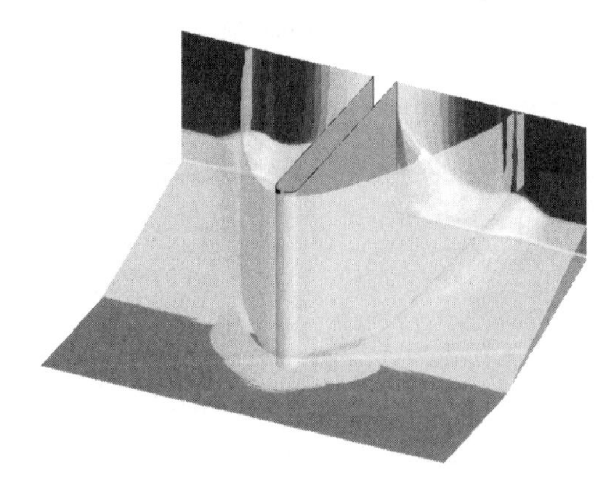

Figure 9.9: Shock wave computed from the location of the maximum local gradient. This method allows us to visualize a secondary shock wave which is visible behind the transparent front shock. See Color Plate 65.

9.3 Visualization of Shock Waves

Returning to the same flow field as used for the vortex visualization, a second dominant feature, a pattern of shock waves in the same data will be visualized.

Since one of the vortex visualization techniques (Section 9.2), is capable of visualizing the vortex seen from greater distances, one might eventually combine the vortex visualization with the visualization of a large-scale shock wave. We will concentrate on the less complex shock visualization here.

The flow in the example approaches from the left with a supersonic speed equivalent to a Mach-number of five. It then hits the wedge at the bottom as well as the blunt fin. Both obstacles cause a complex pattern of shock waves in the flow field. Physically, a shock is marked by an abrupt change in Mach-number. Ahead of the shock, the Mach-number is essentially undisturbed and equal to the free-stream value of five. This is valid for the first or front shock wave in the field. A useful method to find the position of the shock wave in the data is to calculate an isosurface for a Mach-number slightly below the free-stream value (Figure 9.8).

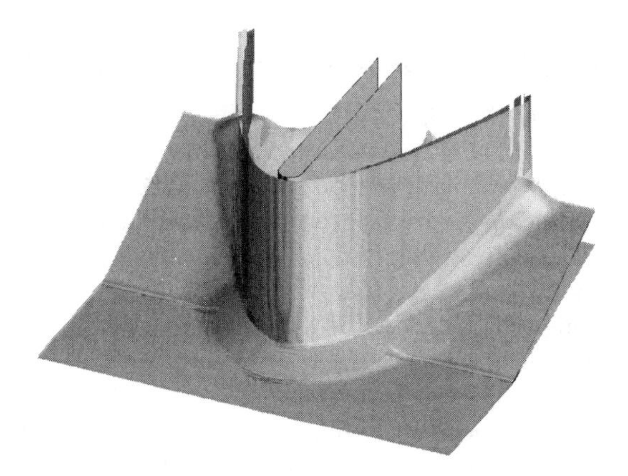

Figure 9.10: Direct comparison of the shock visualization resulting from two alternative methods shows significant spatial displacement of the extracted feature. The blue transparent surface is equivalent to the one in Figure 9.8, the surface behind it matches the golden transparent surface shown in Figure 9.8, and is pseudocolored with Mach-number. See Color Plate 66.

Obviously, in a discrete representation of the flow field, a shock wave is never sharply defined. The method, therefore, visualizes the onset of a transition to a lower Mach-number, which may be spread over several grid cells in streamwise direction. This method can only find the first shock wave at the front, where the deviation from the constant free-stream value occurs, and the Mach-number at the shock is known. The surface is then visualized as a Mach-isosurface. At secondary shocks, the Mach-number is unknown, and isosurface extraction is not possible.

A second shock visualization method [7] locates the local maxima of the gradient of the Mach-number and constructs a surface at these positions. This method is not restricted to locating the first shock wave, but it is able to find any shock in the field. Note that if shocks were represented by sharp discontinuities, both methods would show the same surface.

Again, both alternative methods considered separately (Figures 9.8 and 9.9) provide an acceptable visualization of the shock feature. Once both surfaces are plotted in a single image, a significant spatial displacement becomes visible (Figure 9.10).

There is no physical or mathematical reason to consider either of the two methods incorrect. It is only the discrete representation of the phenomenon in the data that leads to the spatial displacement depending on the algorithm used. Users of visualization software typically are unaware of this type of mismatch between the continuous physical reality and the discrete representation of features in their data. This occurs because the effects are unexpectedly larger than the numerical inaccuracy or noise in the data which is quantified by order-of-accuracy evaluations or noise estimates.

In many cases, comparison of data may only be possible at the image level, that is, by putting images side-by-side (Figure 9.11a). However, as the previous example clearly showed, more detailed information may be obtained when two representations are fed into a common visualization pipeline (Figure 9.11b). In this way we can ensure that the differences we see are not caused by errors in the visualization process. A framework of comparative visualization was given by Pagendarm and Post [8].

9.4 Visualization of Near-Wall Flow Fields: Simulation and Experiment

In this section we will show examples of comparative visualization in fluid dynamics research using data from different sources. Such sources could be two numerical simulations of the same physics using different algorithms. Obviously, this can be treated similarly to the comparison of visualization algorithms shown above, since data-level comparison is relatively easy to achieve in many cases when the data originates from numerical simulations. Comparative visualization of data from experiments with numerical simulation usually requires a larger effort.

Pagendarm and Walter [9] demonstrated the comparison of near-wall flow fields using oil-flow visualization techniques in a wind tunnel and wall friction lines resulting from a numerical simulation. The use of oil-flow visualization allows global acquisition of near-wall-velocity directional information. The technique employs a dye dispersed in a special oil. This oil is sprayed on the solid walls of the model in the wind tunnel. Due to the viscous action of the flow close to the wall, the oil moves slowly in the local flow direction. When the oil evaporates, it leaves behind a trace of dye, which marks the local flow direction.

In order to compare such experiments with numerical flow simulations, we must visualize a near-wall flow field in a way that is visually comparable to images of the experiment. The direction of a flow velocity field is often visualized using streamlines. In simulations of viscous flows, the velocity of the flow at a solid wall is zero by definition. This prevents the calculation of streamlines directly on the walls. We would like to find the limiting streamline at locations where the velocity goes to zero, while the direction of the velocity vector is determined by the direction of the velocity near the wall. Experienced aerodynamicists will be able to imagine, even qualitatively, the overall three-dimensional flow pattern in the flow field from these wall patterns.

In particular, skin-friction lines show the location of separation and reattachment of the flow at the wall. As mentioned, earlier experiments were performed in a wind tunnel to provide measurements and visualization to match the numerical flow simulation. Oil-flow patterns were recorded photographically. Due to the limited accessibility of the wind tunnel during the experiments, the photograph shows a perspective view of the fin taken from a side of the wind tunnel.

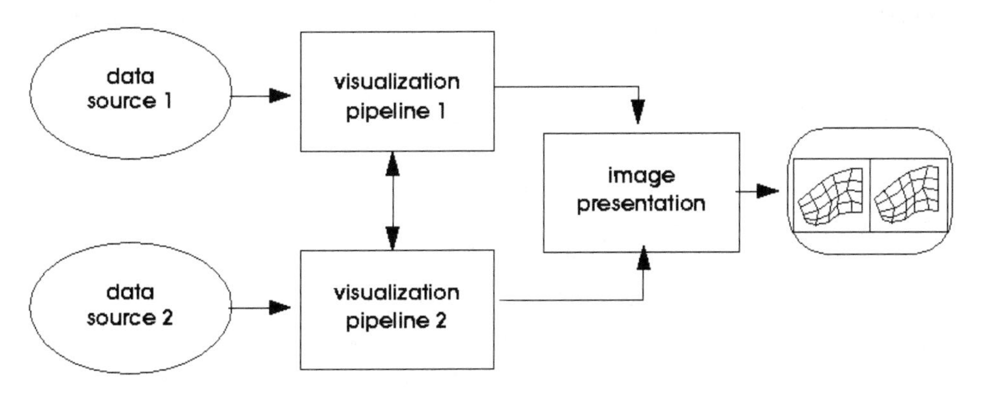

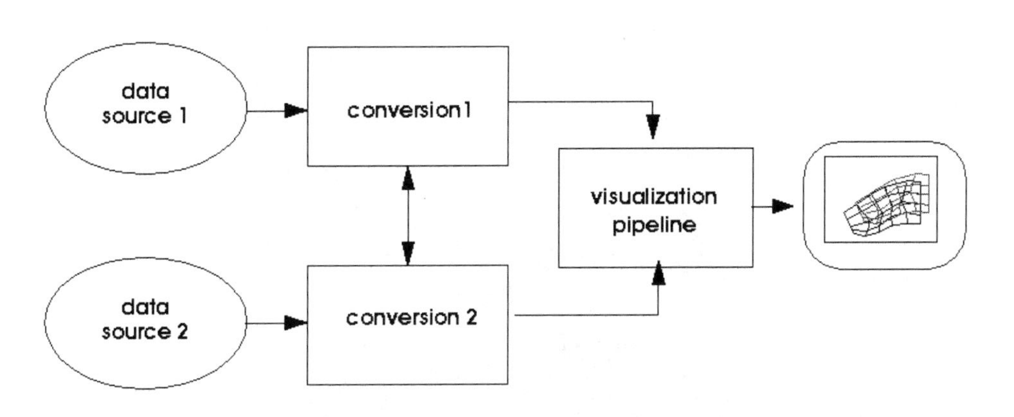

Figure 9.11: Two approaches to comparative visualization [8].

Unfortunately, some of the details of the positions and the equipment used when recording the experiments were no longer available. Therefore, for direct comparison of significant lines and oil-flow pattern, the perspective had to be reconstructed from the edges of the model visible in the image.

After careful matching of the viewing conditions, the skin-friction lines could be projected into the image showing the oil-flow pattern in the wind tunnel experiment. The resulting combined image (Figure 9.13) increases confidence in the overall correctness of the numerical simulation.

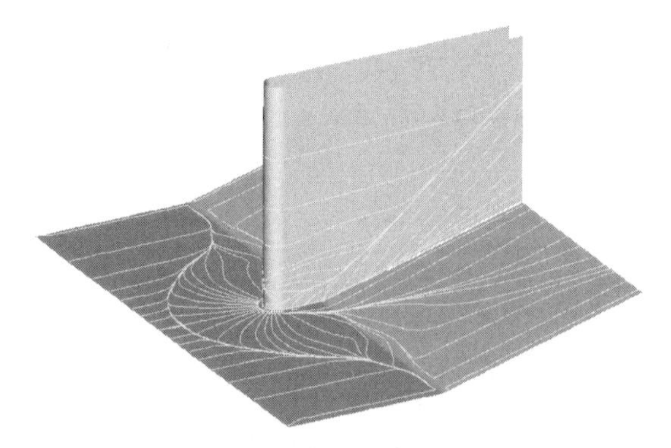

Figure 9.12: Skin-friction lines on walls of the blunt fin and wedge. See Color Plate 67.

Figure 9.13: Skin-friction lines on walls of the blunt fin and wedge. See Color Plate 68.

Some local discrepancies could be explained by a low grid resolution in the numerical simulation. However, the experiment clearly shows a second weak separation line in the lower-side wall of the fin, which does not show up in the numerically generated skin-friction lines (Figure 9.14). The experiment suggests that there is a vortex at some distance from the wall (see Figure 9.3), which is the reason for this pattern.

Figure 9.14: The oil-flow pattern shows a second weak separation trace s_2, which is not visible in the numerical data. See Color Plate 69.

The flow shows a horseshoe type of vortex. The simulation also reveals a smaller secondary vortex that rotates in the opposite direction. In particular, in the vicinity of the stagnation region ahead of the fin, the vortices are accelerated when they pass the fin. This is advantageous for the visualization since a streamline released in this region is sucked into the vortex and remains very close to the central core for a long time.

Therefore, streamlines were carefully selected to hit the vortex close to the plane of symmetry, then follow the center of the vortex past the fin. The TSR ribbons were calculated for two streamlines, one for each of the vortices.

To find streamlines that stay within the vortex core, the starting point for the streamline must be placed accurately at the center of the vortex. This may be done interactively, or using vortex core detection algorithms (see Banks and Singer [1], and further references in their paper).

The full path of the line consists of two parts: one part integrated forward through the converging vortex core, and a second part that approaches the fin from the inflow boundary, which was integrated backwards to meet the vortex exactly. Careful selection of streamlines and calculation of ribbons allows representation of the vortex pattern with effective and easily perceived visual objects that do not clutter the image (Figure 9.15).

In the case discussed here, the horseshoe vortex passes the fin with increasing distance, while the secondary vortex stays very close to the side wall of the fin. When visualized in combination with the oil-flow pattern obtained in the wind tunnel experiment, the position of this secondary vortex nicely explains the weak trace of a separating flow halfway up the fin. Figure 9.16 shows the vortex core close to the fin, slightly below the lower separation trace. The vortex must be below this trace to explain the topology of the pattern on the fin. As expected, this vortex is only weakly represented, due to low grid resolution in the simulation. This may explain why the skin friction lines as they are shown in Figures 9.12 and 9.13 do not show this separation.

Figure 9.15: Ribbons represent vortices in the flow field. The image allows the examination of three-dimensional phenomena with respect to their traces on the solid walls. See Color Plate 70.

Figure 9.16: Three-dimensional vortex cores from the numerical simulation visualized in combination with oil-flow traces from wind tunnel experiments. See Color Plate 71.

9.5 Discussion and Conclusions

In the three cases described above, a variety of issues in scientific visualization were discussed. First, the problem of evaluating different visualization techniques was discussed. It turns out that this is not just a matter of accuracy; other issues are involved, such as the choice of parameters to be visualized, and the relation between grid resolution and the physical size of features. All of the visualization techniques involved—streamribbons, shock wave surfaces, and skin friction lines—are physically correct and accurate representations, but each technique has its own functionality, and its own conditions of applicability.

A set of adjacent streamlines is a good representation of flow direction, but it does not directly show certain derived quantities, such as rotation or divergence. Using the velocity gradient tensor to derive good approximations of these quantities gives a clear result, and is always linked to grid resolution. However, the gradient quantities are less intuitive and must be visualized separately, whereas the streamlines are a more intuitive and integrated representation. Eventually, a combination of both techniques may be desirable.

The scale of physical features in relation with the size of the visual representation is another interesting issue. Commonly, a trilinear field is assumed within each grid cell, so very complex curves cannot be expected inside a single cell. But, in order to make good use of computed data, at least one sample should be taken in each cell. So a visualization technique should somehow be linked to the local grid resolution, which is not the case with the adjacent streamlines technique. In the case of the shock waves, the limited resolution of the data smoothes down the important discontinuities, and this introduces a significant difference in detected position, depending on the technique employed. The use of the approximated second derivatives is more general, as it detects all shock waves, but the smoothing effect of the gradient calculation places the shock surface at the center of the transition zone.

The role of comparative visualization is mainly limited to presentation techniques, but the shock wave example in particular shows the value of simultaneous presentation in a single image. Similar techniques such as image fusion and differential display could also be applied for this purpose.

One important lesson is that users should be made more aware of the problems in choosing visualization techniques in a particular case. This is not merely a matter of good or bad, but also of the right conditions of applicability. The example of the near-surface flow suggests that comparing experiment and simulation does not directly lead to a validation of the simulation model. The comparison is mainly performed at the level of features, and the numerical data can be further explored to look for hidden features that do not show up at the first try. This shows that comparative visualization is not a one-time event, but that it initiates an iterative process of matching images, explaining differences, and testing possible explanations using other visualizations. Perhaps the incentive to explain, to generate and test hypotheses is the most important effect of comparison, which lies at the heart of the process of scientific research. Reliable visualization tools and techniques for comparative visualization can help users to concentrate on the underlying physics and simulation models, rather than on the internal problems of the visualization process.

Acknowledgments

This research is a result of the informal cooperation between the authors at DLR in Göttingen and at TU Delft on comparative visualization in CFD research. The data of the blunt fin/wedge configuration were kindly supplied by T. Gerhold [2]. Birgit Walter contributed wall-shear and vortex visualization.

Bibliography

[1] D.C. Banks and B.A. Singer, "Vortex Tubes in Turbulent Flows: Identification, Representation, Reconstruction," *Proceedings Visualization '94*, R.D. Bergeron and A.E. Kaufman, editors, IEEE Computer Science Press, 1994, pp. 132–139.

[2] T. Gerhold, *Numerische Simulation und Analyse der turbulenten Hyperschallströmung um einen stumpfen Fin mit Rampe*, Technical Report DLR-FB 94–19, DLR, 1994.

[3] D.N. Kenwright, *Dual Stream Function Methods for Generating Three-Dimensional Stream Lines*, Ph.D. thesis, University of Auckland, Department of Mechanical Engineering, Aug. 1993.

[4] W. Merzkirch, *Flow Visualization*, 2nd edition, Academic Press, New York, 1987.

[5] E. Murman and K. Powell, "Trajectory Integration in Vortical Flow," *AIAA Journal*, Vol. 27, No. 7, July 1989, pp. 982–984.

[6] H.-G. Pagendarm, "Flow Visualization Techniques in Computer Graphics," *Computer Graphics and Flow Visualization in Computational Fluid Dynamics*, VKI Lecture Series Monograph 1991-07, Von Karman Institute for Fluid Dynamics, Rhode-St.-Genese, Belgium, 1991.

[7] H.-G. Pagendarm and B. Seitz, "An Algorithm for Detection and Visualization of Discontinuities in Scientific Data Fields Applied to Flow Data with Shock Waves," *Scientific Visualization—Advanced Software Techniques*, P. Palamidese, editor, Ellis Horwood Ltd., Chichester, UK, 1993, pp. 161–177.

[8] H.-G. Pagendarm and F.H. Post, "Comparative Visualization—Approaches and Examples," *Visualization in Scientific Computing*, M. Göbel, H. Müller and B. Urban, editors, Springer, Wien, 1995, pp. 95–108.

[9] H.-G. Pagendarm and B. Walter, "Feature Detection from Vector Quantities in a Numerically Simulated Hypersonic Flow Field in Combination with Experimental Flow Visualization," *Proceedings Visualization '94*, R.D. Bergeron and A.E. Kaufman, editors, IEEE Computer Science Press, 1994, pp. 117–123.

[10] A. Sadarjoen, T. van Walsum, A.J.S Hin, and F.H. Post, "Particle Tracing Algorithms for 3D Curvilinear Grids," *Scientific Visualization: Overviews, Methodologies, and Techniques*, G. Nielson, H. Müller and H. Hagen, editors, IEEE Computer Science Press, 1994, (Chapter 14 in this volume).

[11] W.J. Schroeder, C.R. Volpe, and W.E. Lorensen, "The Stream Polygon: A Technique for 3D Vector Visualization," *Proceedings Visualization '91*, G.M. Nielson and L. Rosenblum, editors, IEEE Computer Society Press, 1991, pp. 126–132.

[12] J.J. van Wijk, "Flow Visualization with Surface Particles," *IEEE Computer Graphics and Applications*, Vol. 13, No. 4, 1993, pp. 18–24.

[13] G. Volpe, *Streamlines and Streamribbons in Aerodynamics*, AIAA Paper 89-0140, 27th Aerospace Science Meeting, Reno, Nev., Jan 9–12, 1989.

Chapter 10

Toward a Systematic Analysis for Designing Visualizations

William L. Hibbard, Charles R. Dyer, and Brian E. Paul

10.1 Introduction

Scientific visualization is a computational process that transforms invisible data inside of computers into visible images on a display device. The field of scientific visualization started as a set of ad hoc techniques for meeting particular needs, but is now evolving into a systematic study. There have been workshops focused on defining systematic models for the data that are the inputs to the visualization process [9, 24], there have been efforts to define models of the displays that are the output of the visualization process [1, 11], and a variety of formalisms have been developed for expressing the process itself. For example, the visualization process can be expressed as a rendering pipeline [14, 15], as a flow diagram [6, 25], or as an object-oriented program [8, 12, 20]. Visualization processes are quite complex, with many different choices of ways to display the same data. Thus, there is also considerable interest in automating the design of displays [2, 11, 17, 23] using expert systems techniques. That is, users define a set of visualization goals, and these techniques apply a set of visualization rules to design a visualization process.

The purpose of this chapter is to explore the possibilities for systematic mathematical analyses of the visualization process that can be used as the basis for formal rather than heuristic techniques for designing the visualization process. We let U denote a set of data objects, V denote a set of displays, and $D : U \rightarrow V$ denote a visualization mapping from data to displays. Our approach is to define analytic conditions on the function $D : U \rightarrow V$ that express various visualization goals and to study the classes of functions satisfying those conditions. We will particularly focus on defining various mathematical structures on U and V that can be used as the basis for defining analytic conditions on D. These include order structures, topological structures, metric structures, algebraic structures and symmetry structures. Given similar mathematical structures on U and V it is natural to define conditions on D requiring that it preserve the structure in the mapping from U to V (that is, that D is some sort of isomorphism of the structure).

We note that a variety of computing problems first attacked with heuristic techniques are now attacked systematically. Examples include the problems of automating symbolic integration and of automating theorem proving. We believe that there is sufficient mathematical structure in the problem of scientific visualization to enable a systematic attack. Even principles of human perception may be expressed in terms of mathematical structures on the display model V.

In the next section we present two examples of analytic conditions on visualization mappings in order to illustrate the basic approach. Then in Sections 10.3, 10.4, and 10.5 we describe the basic mathematical structures of data and display models. Finally in Section 10.6 we discuss the meanings and implications of analytic conditions on visualization mappings based on various kinds of mathematical structures.

10.2 Conditions on Visualization Mappings - Part I

There is an enormous variety of ways to define requirements on visualization functions. For example, one of the simplest and most obvious requirements is that users be able to distinguish different data objects from their displays (that is, that different data objects have different displays). This requires that the function $D : U \to V$ be one-to-one, which can be stated as the following condition, true for all $u, u' \in U$,

Condition 1: $u = u' \Leftrightarrow D(u) = D(u')$

We can think of this as saying that the mapping D defines an isomorphism between equality on U and equality on V. This is a weak condition, satisfied by many functions that do a poor job of visually communicating data. Thus we need to define other conditions.

Our lattice model for visualization [7] provides a more complex example of formally defining requirements on visualization functions. A lattice is an ordered set L such that every pair of elements $l, l' \in L$ has a least upper bound (denoted by $l \vee l'$) and a greatest lower bound (denoted by $l \wedge l'$) [4, 21, 22]. Mathematical objects and idealized displays contain infinite precision real numbers and functions with infinite domains, but computer data objects and real displays contain only finite amounts of information and must therefore generally be approximations to mathematical objects and ideal displays. Let M be a set of mathematical objects (including real numbers, finite vectors of real numbers, functions of real variables, vectors of functions, and so on) and let U be a set of data objects. Since data objects are finite strings over finite alphabets, U must be countable. However, M is uncountable so, in general, each data object $u \in U$ represents a large set $math(u) \subseteq M$ of mathematical objects. We can define an order relation on U by:

$$u \leq u' \Leftrightarrow math(u') \subseteq math(u) \tag{10.1}$$

This order relation is based on precision: u' is more precise than u because it represents a more restricted set of mathematical objects. We can define a similar order relation on V. The precision of displays is based on the finite spatial resolution of their pixels or voxels, on their finite color resolution, on their finite time resolution, and so on. Figures 10.1

through 10.4 provide an example of an ordered chain of four image data objects. Each of these images is an approximate representation of a continuous radiance field, going from the least precise in Figure 10.1 to the most precise in Figure 10.4.

Figure 10.1: Least precise image in sequence of four. See Color Plate 72.

In [7] we assumed that U and V are lattices under this order relation and we provided concrete examples of data and display models with lattice structures. We drew on Mackinlay's expressiveness criteria [11], which require that displays express all the facts about data objects, and only those facts. We interpreted facts about data objects as monotone predicates of the form $P : U \to \{undefined, true\}$. [To say that P is monotone means that $u \leq u' \Rightarrow P(u) \leq P(u')$. Furthermore, we assumed that $undefined < true$. This interpretation is based on the assumption that facts about data objects represent facts about mathematical objects.] Then we interpreted the expressiveness criteria by Requirements 1 and 2, as illustrated in Figure 10.5.

Requirement 1: For every monotone predicate $P : U \to \{undefined, true\}$, there is a monotone predicate $Q : V \to \{undefined, true\}$ such that $P(u) = Q(D(u))$ for each $u \in U$.

Requirement 2: For every monotone predicate $Q : D(U) \to \{undefined, true\}$, there is a monotone predicate $P : U \to \{undefined, true\}$ such that $Q(v) = P(D^{-1}(v))$ for each $v \in D(U)$. [Here we use $D(U)$ in place of V so we do not assume that D is onto.]

We showed that $D : U \to V$ satisfies these requirements if and only if it is a *lattice isomorphism* from U to V. That is, for all u, u', $u'' \in U$,

Condition 2: $u'' = u \vee u' \Leftrightarrow D(u'') = D(u) \vee D(u')$

Condition 3: $u'' = u \wedge u' \Leftrightarrow D(u'') = D(u) \wedge D(u')$

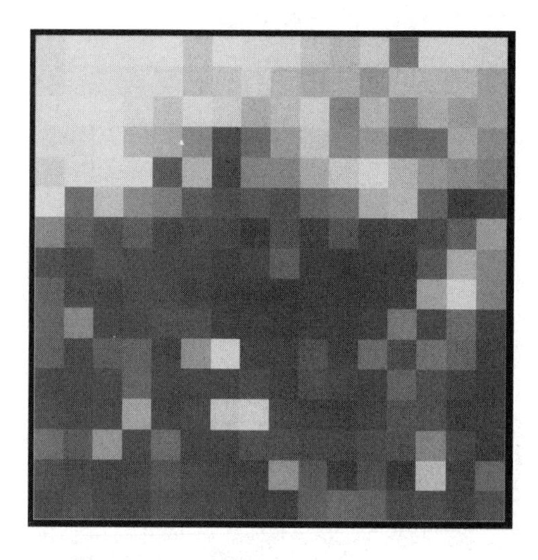

Figure 10.2: Second image in sequence of four, ordered by precision. See Color Plate 73.

Figure 10.3: Third image in sequence of four, ordered by precision. See Color Plate 74.

Figure 10.4: Most precise image in sequence of four. See Color Plate 75.

Condition 4: D is a bijection (that is, one-to-one and onto).

Our lattice results are quite complex, and have only been sketched out here. However, the important point for the current discussion is the similarity in form between Condition 1, which expresses the one-to-one condition on D, and Conditions 2, 3, and 4, which express Requirements 1 and 2. In both of these examples we define structures on U and V, and define conditions on $D : U \to V$ that require that D define an isomorphism between the structure on U and the structure on V. In Condition 1, D defines an isomorphism between the structures of equality on U and V, and in Conditions 2, 3, and 4, D defines an isomorphism between lattice structures on U and V.

In this chapter we will look at other kinds of mathematical structures on the sets U and V, and discuss visualization functions that preserve these structures. Structures on U and V include:

1. Lattice structures (the ordering of data based on precision)
2. Topological structures
3. Metric structures
4. Algebraic structures (for example, involving the operations $+$ and $\times$)
5. Symmetry structures

By defining predicates for equality and for the lattice operations, we can restate Conditions 1, 2, and 3 in a way that makes their similarity even more apparent. Define predicates $EQUAL_U(u, u') \equiv (u = u')$ [that is, $EQUAL_U(u, u') = true$ if $u = u'$ and $EQUAL_U(u, u') = false$ if $u \neq u'$] and $EQUAL_V(v, v') \equiv (v = v')$. Then Condition 1 becomes:

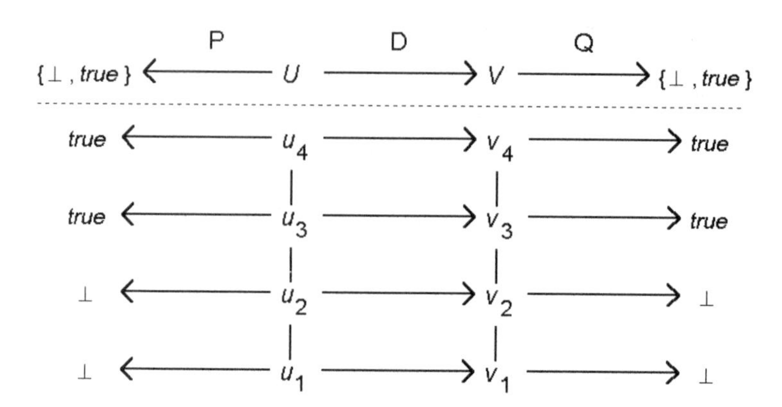

Figure 10.5: The expressiveness conditions specify that $D : U \rightarrow V$ defines a correspondence between monotone predicates on U and V.

Condition 1′: $EQUAL_U(u, u') \Leftrightarrow EQUAL_V(D(u), D(u'))$

By defining predicates $LUB_U(u, u', u'') \equiv (u'' = u \vee u')$ and $GLB_U(u, u', u'') \equiv (u'' = u \vee u')$ [LUB_V and GLB_V are defined similarly], Conditions 2 and 3 become:

Condition 2′: $LUB_U(u, u', u'') \Leftrightarrow LUB_V(D(u), D(u'), D(u''))$

Condition 3′: $GLB_U(u, u', u'') \Leftrightarrow GLB_V(D(u), D(u'), D(u''))$

While Conditions 1′, 2′, and 3′ all have the form of logical equivalencies between predicates on U and V, it is also useful to consider conditions that take the form of one-way implications. If R_U and R_V are similar predicates on U and V, then this distinction is illustrated by the difference between:

$$R_U(u, u', u'') \Leftrightarrow R_V(D(u), D(u'), D(u'')) \tag{10.2}$$

$$R_U(u, u', u'') \Rightarrow R_V(D(u), D(u'), D(u'')) \tag{10.3}$$

$$R_U(u, u', u'') \Leftarrow R_V(D(u), D(u'), D(u'')) \tag{10.4}$$

For example, in Condition 1 the $\Rightarrow$ direction is implicit in the fact that D is a function, so the one-to-one condition could be expressed by the $\Leftarrow$ direction of Condition 1. However, since the $\Rightarrow$ direction is always true, we have stated Condition 1 as an implication both ways (that is, $\Leftrightarrow$).

Conditions on the function $D : U \rightarrow V$ must be defined in terms of structures on U and V. Thus, in the next three sections we describe the basic mathematical structures of data and display models.

10.3 Data Models

Scientists observe nature, formulate laws to fit those observations, and predict future observations in order to test their laws. All of these activities are carried out in terms of mathematical models. The primitive elements of mathematical models are variables such as *temperature, pressure, latitude, time*, and so on. These primitives are aggregated into complex mathematical objects. For example, the function

$$\textit{temperature} = \textit{temperature-field(latitude, longitude, altitude)} \tag{10.5}$$

represents temperatures over the atmosphere. The vector

$$\begin{aligned}\textit{state} \quad = \quad & \{\textit{temperature-field, pressure-field, humidity-field,} \\ & \textit{velocity-x-field, velocity-y-field, velocity-z-field}\}\end{aligned} \tag{10.6}$$

represents the state of the atmosphere (where *pressure-field*, and so forth, are functions similar to *temperature-field*). And the function

$$\textit{state} = \textit{state-history(time)} \tag{10.7}$$

represents a possible history of the atmosphere.

Scientific data objects are finite representations of such mathematical objects. A data model defines a set U of data objects and also defines mathematical structures on U. Scientific data models primarily need to address three kinds of issues:

1. The types of primitive data values used to represent mathematical variables. A primitive type defines a set of primitive values. It may also define an order relation, basic operations (for example, addition, negation, string concatenation), a topology (such as the discrete topology of integers or the continuous topology of real numbers), and a metric (the distances between pairs of values) on the set of values.

2. The ways that primitive values are aggregated into data objects. Vectors of data objects may represent vectors of mathematical objects. Mathematical functions may be represented by arrays containing finite sets of samples of function values. Data aggregates may also include complex networks of values. Some systems support application-specific types of aggregates, such as two-dimensional images and time series.

3. Metadata about the relation between data and the things that they represent. For example, given a meteorological temperature, metadata includes the fact that it is a temperature, its units (for example, Fahrenheit or Kelvin), the location of the temperature and whether it is a point sample or volume average, the time of the temperature, an estimate of its accuracy, how it was produced (such as by a simulation, by direct observation, or deduced from a satellite radiance), and whether the value is missing (such as in case of a sensor or computational failure).

Primitive data types represent mathematical variables. For example, floating point numbers are a primitive data type used to represent real variables, and fixed length (for example, 32-bit) integers represent integer variables. Floating point numbers and fixed length integers define finite samplings of $\mathbb{R}$ (the real numbers) and of $\mathbb{Z}$ (the integers).

Information about the way that primitive data values sample mathematical values is a form of metadata. For example, the information that specifies how 8-bit pixel intensities in a satellite image can be converted to physical radiances (that is, how 8-bit codes sample real radiance values) is a form of metadata called *satellite calibration*. Primitive data types may be enhanced to include other kinds of metadata. For example, a missing data indicator may be added to any set of primitive values. Accuracy metadata may be integrated into a data model by using intervals to represent real numbers (that is, lower and upper bounds). We can also attach names (such as *temperature, pressure, latitude, time*) and units (such as *meters, seconds, mps*) to primitive values.

Although mathematical values in $\mathbb{R}^2$ and $\mathbb{R}^3$ are aggregates, it is often useful to regard them as primitives because of the special role they play in so much scientific data and because this is convenient for managing certain forms of metadata such as locations and accuracies in two and three dimensions.

Aggregate data types represent complex mathematical objects such as vectors and functions. Vectors are represented by tuple data types (often called *records* or *structures*) and functions are represented by arrays containing finite samplings of function values (of course, other representations of functions are possible, such as by Fourier coefficients). Thus the vectors and functions described above could be represented by the following data types:

type *temperature* = real;
. . .
type *latitude* = real;
type *longitude* = real;
type *altitude* = real;
type *time* = real;
type *temperature-field* = array [*latitude*] of array [*longitude*]
    of array [*altitude*] of *temperature*;
. . .
type *state* =
    structure {*temperature-field; pressure-field; water-concentration-field;*
        wind-velocity-x-field; wind-velocity-y-field; wind-velocity-z-field;}
type *state-history* = array [*time*] of *state*;

Alternatively, we could define a primitive data type *lat_lon_alt* to represent three-dimensional locations. Its values would come from some sampling of $\mathbb{R}^3$, and it could be used to redefine the *temperature-field* type as follows:

type *lat_lon_alt* = real3d;
type *temperature-field* = array [*lat_lon_alt*] of temperature;

Our goal is to define analytic conditions on visualization mappings in terms of mathematical structures defined on the set U of data objects (and on the set V of displays).

Aggregate data objects are organized collections of values of primitive data objects, and we can often extend the natural mathematical structures defined on primitive types to aggregate types. For example, if P is a set of pixel values then we can regard $P^{1024 \times 1024}$ (that is, this is the cross product of 1024×1024 copies of P) as the set of all images of size 1024×1024. There are natural ways to extend algebraic, topology, metrics, symmetries, and other mathematical structures to this cross product. However, a data model may define images of many different sizes, and may include many different data types. In this case we must treat the set U of data objects as the union of many cross products. For example, can define the set of 2D images of size $N \times J$ as:

$$IMAGE2D_{N,J} = P^{N \times J} \tag{10.8}$$

and define the set of all 2D images as the union:

$$IMAGE2D = \bigcup \{IMAGE2D_{N,J} | 1 \leq N \ \& \ 1 \leq J\} \tag{10.9}$$

We can make similar definitions for 3D images:

$$IMAGE3D_{N,J,K} = P^{N \times J \times K} \tag{10.10}$$

$$IMAGE3D = \bigcup \{IMAGE3D_{N,J,K} | 1 \leq N \ \& \ 1 \leq J \ \& \ 1 \leq K\}. \tag{10.11}$$

These 2D and 3D images correspond to the data types:

> type *image2d* = array [*row*] of array [*column*] of *pixel*;
> type *image3d* = array [*row*] of array [*column*] of array [*level*] of *pixel*;

where sets of images of various sizes are subtypes of *image2d* and *image3d*.

In general, we assume that a data model defines a set U of data objects partitioned among a set T of different data types. Let $U(t)$ denote the set of data objects of a type $t \in T$. Then $U = \bigcup\{U(t) | t \in T\}$. Every type $t \in T$ is further partitioned into a set subtypes $ST(t)$ such that all objects of subtype $s \in ST(t)$ have the same size. Let $U(s)$ denote the set of objects of subtype $s \in ST(t)$. Then $U(t) = \bigcup\{U(s) | s \in ST(t)\}$. Define I_s as a set of indices for the primitive values occurring in data objects of subtype s, and let $\{u_i | i \in I_s\}$ denote the set of primitive values occurring in a data object $u \in U(s)$. This can be used to extend predicates from sets of primitive values to sets of aggregates of the same subtype. Given a predicate $P(a, a')$ defined for pairs of primitive values, given a subtype s, and given $u, u' \in U(s)$ we can extend the predicate P to aggregate objects of subtype s as follows:

$$P(u, u') \equiv \forall i \in I_s.\ P(u_i, u'_i) \tag{10.12}$$

where i is an index of the primitive values occurring in data objects of subtype s. Furthermore, for any pair u and u' of data objects not of the same subtype, we can define $P(u, u') \equiv false$.

Equation (10.12) is inappropriate for some structures. For example, a metric structure defines distances between data objects and these should be additive in the distances between primitive components of data objects. Given a predicate $METRIC(u_i, u'_i, r_i)$ defined for pairs of primitive values (where $r_i \in \mathbb{R}$ is the distance between u_i and u'_i), given a subtype s, and given $u, u' \in U(s)$ we can extend $METRIC$ to aggregate objects of subtype s as a Cartesian distance summed over primitives, as follows:

$$METRIC(u, u', r) \equiv [r = sqrt(\Sigma\{r_i^2 | METRIC(u_i, u'_i, r_i) \; \& \; i \in I_s\})] \qquad (10.13)$$

If u and u' are data objects of different subtypes then we define $METRIC(u, u', r) \equiv false$ for all r (that is, no distance is defined between objects u and u' with different subtypes).

10.4 Display Models

Since computers generate displays as data objects, display models are similar to data models. A display model defines a set of displays and defines mathematical structures on this set. Bertin identified key elements of display models in his study of 2D static displays [1]. He defined a display as an aggregate of graphical marks, and he identified eight primitive variables of a graphical mark: two spatial coordinates, size, value, texture, color, orientation, and shape.

Bertin's display model corresponds to what can be physically displayed on a two-dimensional screen at one time. However, computer-generated displays generate the illusion of three dimensions and show motion by changing screen contents at short intervals. We can even regard various forms of user interaction as an integral part of the display. Thus we distinguish between physical and logical display models. We let V' denote the set of physical displays, which are two-dimensional and static, and we let V denote the set of logical displays, which are three-dimensional, animated, and interactive. We define a mapping $RENDER : V \to V'$ that includes traditional graphics operations such as isosurface generation, volume rendering, projection from three to two dimensions (rotate, zoom, and translate), clipping, hidden surface removal, shading, compositing, and animation. A changing set of mappings $RENDER : V \to V'$ expresses the three-dimensional, animated, interactive nature of displays in V.

We can extend Bertin's ideas to define a logical display model V in terms of a variety of different graphical primitive types. For example, we can model displays as animated sequences of 3D arrays of voxels with the following types:

type *color* = real3d;
type *transparency* = real;
type *reflectivity* = real;
type *voxel* = structure {*color; transparency; reflectivity;*}

type *location* = real3d;

type *volume* = array [*location*] of *voxel*;

type *animation_step* = real;
type *display* = array [*animation_step*] of *volume*;

The *transparency* and *reflectivity* values are interpreted by volume rendering techniques. We could add scalar and vector elements to the *voxel* tuple that would be interpreted by contour and flow rendering techniques.

Display models can also be defined for higher-level graphical constructs. For example, we can define a graphical primitive type that is an index into a set of icons (this is essentially an index into a set of shapes of graphical marks), or we can define graphical primitives that are parameters of geometrical shapes (such as the major and minor axes and orientations of elliptical graphical marks). We can also define graphical aggregates for various types of diagrams and charts.

Graphical primitives and aggregates can be defined to model a variety of user interaction techniques. For example, a 3D array of voxels models the user's control over 3D to 2D projection (that is, to rotate, pan, and zoom in 3D). We can also define a graphical primitive that enables users to interactively select subsets of graphical marks (that is, the user defines a range of values for this primitive and only those graphical marks whose primitive values fall in the selected range are displayed). We can define graphical primitives for nested sets of graphical coordinates to be implemented by Beshers and Feiner's "world-within-worlds" technique [2] (that is, the user interactively moves a small coordinate system around inside a larger coordinate system). And, we can define primitives for hypertext links (or hypermedia links) so that when a user picks a graphical mark, the display changes to whatever is referenced by the link value of the mark (such links in display aggregates are analogous to pointers in data aggregates).

Because of the similarity of display models to data models, we can define predicates on sets of displays similar to those we define on sets of data objects. Because the primitive variables of a display model are often real numbers, we can define arithmetic operations, order relations, topology, and metrics on them. We can use the technique of Equations (10.12) and (10.13) to extend these structures to displays defined as aggregates of primitive values.

We have been defining display models in terms of the way that displays are generated. However, it is also possible to define models of the way that displays are perceived. For example, perceptual color spaces [3, 16, 18] are mathematical descriptions of the way that we perceive color. They map colors into $\mathbb{R}^3$ in a way that the Cartesian metric on $\mathbb{R}^3$ models perceptual distances between colors. A similar Cartesian metric can be defined for texture perception [10]. A metric is an operation $DIST(v, v')$ that produces a real number distance between v and v'. As in Section 10.3, we restate this as a predicate $METRIC_V(v, v', r)$ where v and v' are displays in V and $r = DIST(v, v')$. Psychology experiments can determine minimal perceptible differences between colors, and these are used to define a metric on color space according to how many "just perceptible" steps exist between each pair of colors. If this approach is used to define a predicate $METRIC_V$ then we can express the requirement that perceived color differences correspond with quantitative differences between data objects by the condition:

Condition 5: $METRIC_U(u, u', r) \Rightarrow METRIC_V(D(u), D(u'), K_D * r)$

Note that K_D is a constant of proportionality and is included because there is no reasonable way to compare absolute scales of distance between perception and data.

We have used a one-way implication in Condition 5 because the definition of $METRIC_U(u, u', r)$ in Equation (10.13) allows there to be pairs of data objects u and u' for which no distance is defined, whereas distances may be defined for every pair of displays (for example, if a display model contains display objects of only a single type and size).

The way that displays are perceived can also be expressed in terms of symmetry structures. Specifically, symmetry structures on display models can express certain invariants of perception to changes in displays. For example, translating an animation sequence forward or backward in time (that is, starting the movie a few minutes earlier or later) does not change the way that we perceive it. Similarly, translating a display to a different spatial position on a workstation screen does not change the way we perceive it. These invariancies can be expressed in terms of symmetry groups on a display model V.

10.5 Scientific Data and Display Models Based on Lattices

Scientific data objects are finite representations of mathematical objects such as real variables, vectors, and functions. We let U be a set of data objects and M be a set of mathematical objects. As we described in Section 10.2, each data object $u \in U$ represents a set of mathematical objects $math(u)$, and Equation (10.1) defines an order relation on U by $u \leq u' \Leftrightarrow math(u') \subseteq math(u)$ (this says that u' is more precise than u because it represents a more restricted set of mathematical objects).

In [7] we used this order relation to define lattice-structured data and display models. We showed how data objects of many different data types could be embedded in a single lattice, and showed that very similar lattice structures can be applied for data and display models. Common forms of scientific metadata are integrated into data objects in this lattice-structured data model. For example, real numbers are represented by real intervals. The interval $[a, b]$ represents any real number $x \in [a, b]$. Thus, by Equation (10.1), real intervals are ordered by the inverse of containment (that is, smaller intervals are more precise). Intervals integrate metadata about precision. Functions are represented by finite sets of samples of their values, such as the way a satellite image is a finite sampling of a continuous radiance field. Figures 10.1 through 10.4 illustrate the order relations between four such samplings. Information about sampling integrates metadata like satellite navigation (the assignment of latitudes and longitudes to pixels) and satellite calibration (the assignment of real radiances to coded pixel values). Any value may be marked as missing—another common form of metadata. The *missing* data value can represent any mathematical value so $math(missing) = M$ and therefore *missing* is the least element of a data lattice. All of these forms of metadata fit naturally into the lattice structure of the data model since they document how data objects approximate mathematical objects.

Under certain circumstances arithmetical operations on a data lattice U can be defined in terms of arithmetical operations on the set M of mathematical objects that they represent. Given $A \subseteq U$, Equation (10.1) [that is, $u \leq u' \Leftrightarrow math(u') \subseteq math(u)$] implies that:

$$math(\bigvee A) \subseteq \bigcap\{math(u) | u \in A\} \tag{10.14}$$

where $\bigvee A$ denotes the least upper bound of the members of A. However, if U is a complete lattice (an assumption satisfied by the data model defined in [7]) and if Equation (10.14) can be replaced by the equality:

$$math(\bigvee A) = \bigcap\{math(u) | u \in A\} \tag{10.15}$$

then we can define arithmetic operations on the data model U in terms of arithmetical operations on the set of mathematical objects M. For example, if addition is defined on M then addition can be defined on U as follows:

$$u_1 + u_2 = \bigvee\{u | \forall u_1' \in math(u_1).\ \forall u_2' \in math(u_2).\ u_1' + u_2' \in math(u)\} \tag{10.16}$$

That is, the sum $u_1 + u_2$ is the least upper bound (which exists if U is complete) of all u that contain the sums of all members of $math(u_1)$ and $math(u_2)$. Subtraction and other mathematical operations can be defined by equations similar to Equation (10.16). If arithmetical operations are not defined between mathematical objects of different types we need to restate Equation (10.16) as:

$$\begin{aligned} u_1 + u_2 = \bigvee\{u | \forall u_1' \in math(u_1).\ \forall u_2' \in math(u_2). \\ \exists u_3' \in math(u).\ SUM_M(u_1', u_2', u_3')\} \end{aligned} \tag{10.17}$$

where the predicate SUM_M is defined by $SUM_M(u_1', u_2', u_3') \equiv (u_3' = u_1' + u_2')$.

As an example, Equation (10.16) can be used to define interval arithmetic [13] in a lattice consisting of all real intervals where $math([a, b]) = \{x \in \mathbb{R} | a \leq x \leq b\}$ and including the empty interval (greatest element of the lattice) and the infinite interval (least element of the lattice). Equation (10.15) holds in this lattice so Equation (10.16) can be used to derive the usual definitions for addition and subtraction of intervals:

$$[a, b] + [c, d] = [a + c, b + d] \tag{10.18}$$

$$[a, b] - [c, d] = [a - d, b - c] \tag{10.19}$$

Equations (10.16) and (10.17) can be applied to define arithmetical operations on a wide variety of other lattice-structured data models.

10.5.1 Lattice Structured Data and Display Models Based on Logic

The basic idea of lattice-structured data models is that data objects provide finite amounts of information that define subsets of the uncountable set of mathematical objects. We can think of each bit in a data object as providing a statement that restricts the choice of mathematical objects. This view can be formalized as a logical lattice that illustrates the duality between the intensive and extensive natures of data.

Specifically, let M be a set of mathematical objects and let X be a countable set of logical statements about a mathematical object $o \in M$ (that is, o is a free variable in each statement $x \in X$). The statements in X may be the implicit meanings of bits in data object representations, they may be statements about o in the SQL language, or they may be statements about o in any appropriate logical theory. Let $math(x)$ be the set of objects in M that satisfy the statement x. For any set $T \subseteq X$ define:

$$math(T) = \bigcap\{math(x)|x \in T\} \tag{10.20}$$

Thus, an object satisfies T if it satisfies all the statements in T. Then M is a model for the theory X and logical implication can be expressed as:

$$S \Rightarrow T \equiv math(S) \subseteq math(T) \tag{10.21}$$

so by Equation (10.1), implication is the basis for the order relation:

$$T \leq S \Leftrightarrow S \Rightarrow T \tag{10.22}$$

on subsets of X. However, rather than taking the set of all subsets of X as our data model, we define the set of closed sets by:

$$CL(X) = \{T \subseteq X | \forall x \in X. \, [(T \Rightarrow x) \Rightarrow x \in T]\} \tag{10.23}$$

That is, a set of statements is in $CL(X)$ if it is closed under logical implication. Given $T \subseteq X$, define:

$$CLOSURE(T) = \{x \in X | T \Rightarrow x\} \tag{10.24}$$

Then $CLOSURE(T) \in CL(X)$ and $CLOSURE(T)$ is the least member of $CL(X)$ containing T. We further note that for all $T \subseteq X$, $math(T) = math(CLOSURE(T))$. Thus we take $CL(X)$ as our data model. A set of statements $T \subseteq CL(X)$ is the intensive representation of a data object and the set $math(T)$ is its extensive representation.

$CL(X)$ is a lattice. Given $S, T \in CL(X)$, we can define the lattice operations by $S \vee T = CLOSURE(S \cup T)$ and $S \wedge T = CLOSURE(S \cap T)$. In fact, $CL(X)$ is a complete lattice. Note that $\phi \in CL(X)$ (that is, the empty set) and $X \in CL(X)$, and that $math(\phi) = M$ and $math(X) = \phi$ (assuming that X contains some statement x and its negation $\neg x$).

Computer data objects are finite, so we further define:

$$FIN(X) = \{S \in CL(X) | \exists T \subseteq S. \, T \text{ finite} \,\&\, \forall x \subseteq S. \, (T \Rightarrow x)\} \tag{10.25}$$

That is, any set of statements in $FIN(X)$ is logically equivalent to a finite set of statements. Thus $FIN(X)$ is a model for the set of all finite data objects.

Sets of logical statements are plausibly the most general possible model for data objects and displays. The fact that logical implication can be used to define a lattice structure on this model illustrates the generality of lattices as models for data and displays.

10.5.2 Order Structures on Data and Display Models Based on Probability

The order relation on data and display models defined by Equation (10.1) expresses the notion of precision, assuming that each data object represents a set of mathematical objects. However, there are scientific data that are more properly interpreted as representing probability distributions over a set of mathematical objects. Equation (10.1) cannot be applied in these situations.

As in Section 10.2, let M be a set of mathematical objects and let U be a set of data objects. We assume that a measure is defined on M and identify each data object $u \in U$ with a function $p(u, m)$ on M (where m varies over M) such that $\forall u \in U.\ \forall m \in M.\ 0.0 \leq p(u, m)$ and $\forall u \in U.\ \int p(u, m)dm = 1$. That is, $p(u, .)$ is a probability distribution on M.

The entropy of a distribution $p(u, .)$ is defined by:

$$E(p(u, .)) = -\int p(u, m) \log(p(u, m))dm \tag{10.26}$$

Then we define an order relation on data objects in terms of entropy, as follows:

$$\begin{aligned} u < u' \equiv \text{there exists a function } f\colon [0,1] \times M \to [0, \infty) \text{ such that:} \quad (10.27) \\ \forall m \;\in\; & M.\ f(0, m) = p(u, m) \\ \forall m \;\in\; & M.\ f(1, m) = p(u', m) \\ \forall m \;\in\; & M.\ f(z, m) \text{ is continuous in} z \text{for } z \in [0, 1] \\ \forall z \;\in\; & [0, 1].\ \int f(z, m)dm = 1 \\ & [\text{that is, } f(z, .) \text{ is a probability distribution over } M] \\ \forall z, z' \;\in\; & [0, 1].\ z < z' \Rightarrow E(f(z, .)) > E(f(z', .)) \end{aligned}$$

That is, $u < u'$ if there is a continuous deformation of $p(u, .)$ into $p(u', .)$ with monotonically decreasing entropy. It is easy to check that Equation (10.27) does define an order relation, and decreasing entropy does express the intuition of increasing precision.

However, data objects do not form a lattice under this order relation, as can easily be seen over a domain M containing three members (the whole space of distributions over M can be represented by a two-dimensional triangle and simple graphical reasoning can be used to see that the order relation does not form a lattice). Furthermore, there is no obvious way to define an order relation on probability distributions that:

1. Expresses the intuition of precision, and

2. Forms a lattice.

Thus, unfortunately, there is no obvious way to apply the results of [7] to data models based on probability distributions.

10.6 Conditions on Visualization Mappings - Part II

Given a data model U and a display model V, both with complex structure, we can define conditions on visualization functions $D : U \to V$ in terms of those structures. We express structures on data and display models as predicates and define conditions on D in the form of Equations (10.2) through (10.4). We will discuss a variety of different kinds of structures on U and V. We will ask what requirements on visualization are expressed by conditions relating to each kind of structure, and we will ask what functions satisfy those conditions.

10.6.1 Conditions on the Structure of Equality

Condition 1 requires that D map different data objects to different displays. This natural requirement enables users to distinguish between different data objects based on their displays. However, it allows an enormous variety of visualization functions. Many of those functions will generate displays that are not useful (that is, they do not help users to understand data objects) as illustrated by the following proposition.

Proposition 1: If D satisfies Condition 1, then D' satisfies Condition 1 if and only if there is a permutation $F : V \to V$ such that $D' = F \circ D$. [A permutation is a one-to-one function from a set onto itself.]

Because of the large number of visualization functions satisfying Condition 1, we must define other conditions that can be combined with Condition 1.

10.6.2 Conditions on Lattice Structures

Conditions 2, 3, and 4 express Requirements 1 and 2 as described in Section 10.2. These requirements are an interpretation, in a lattice context, of expressiveness criteria defined by others [11]. These are intuitive, although strict, requirements.

In [7] we developed a lattice-structured scientific data model that included primitive types suitable for approximate representations of real numbers, integers, and text strings, and included aggregates suitable for approximate representations of mathematical vectors and functions. We also developed a lattice-structured display model that included interactive, animated, three-dimensional displays. In the context of these data and display models we showed the following:

Proposition 2: If D satisfies Conditions 2, 3, and 4, then D can be factored into continuous mappings from primitive data types to primitive display types.

That is, mappings from data aggregates to display aggregates can be factored into mappings from data primitives to display primitives. While this has been accepted as intuitive in the past, we have shown that it completely characterizes all visualization mappings that satisfy the expressiveness requirements. Figure 10.6 illustrates how the mapping from a data aggregate to a display aggregate is factored into mappings from data primitives to display primitives. Figure 10.7 illustrates how a data aggregate can be displayed according to four different ways of factoring aggregate mappings into primitive mappings.

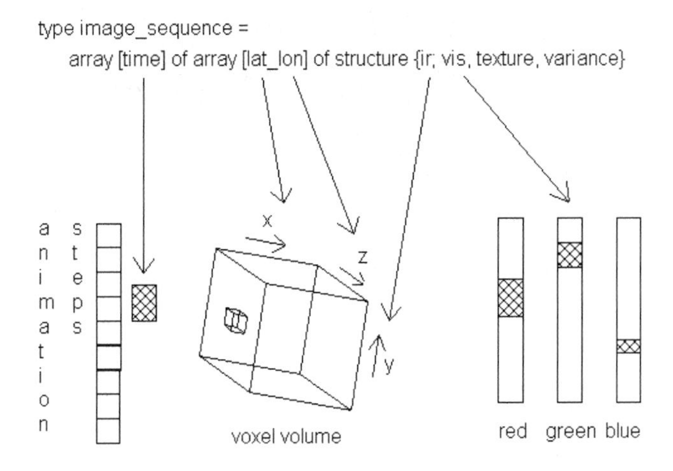

Figure 10.6: Mappings from data primitives to display primitives.

The lattice model developed in [7] was appropriate for a simple type hierarchy. However, this lattice model may be extended to include aggregates defined using *references*. In a data model, references may be implemented as pointers, and in a display model, references may be implemented as hypermedia links. Pointers and hypermedia links can be used to construct very complex data and display structures. In the study of programming language semantics, data objects defined with references are modeled by *recursively defined data types*. Sets of data objects of recursively defined types are defined as infinite limits [19] and large classes of recursively defined data types are embedded in lattices called *universal domains* [5, 22]. These universal domains can be interpreted as the lattices of Conditions 2, 3, and 4 and thus may provide a basis for analyzing the display of complex data structures.

10.6.3 Conditions on Topology Structures

Conditions based on topology structures on U and V can be used to express continuity requirements on visualization functions. Intuitively, a visualization function $D : U \to V$ is continuous if it maps neighboring data objects to neighboring displays. A topology is defined as a collection of open sets. Thus we define a predicate $OPEN_U(A)$ on subsets $A \subseteq U$ which is true if A is open, and a predicate $OPEN_U(B)$ on subsets $B \subseteq V$ which is

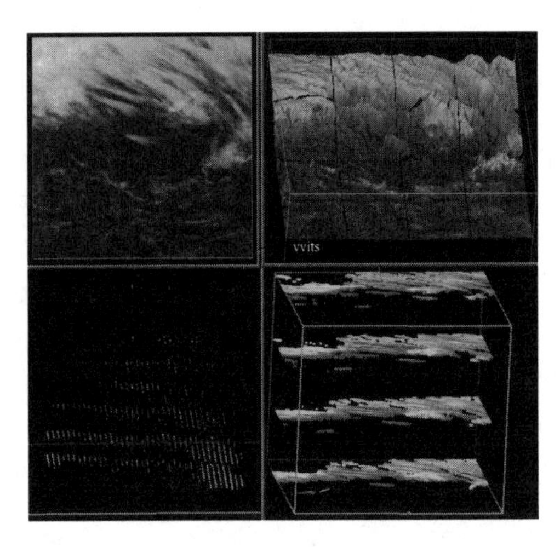

Figure 10.7: A time sequence of multivariate images displayed according to four different sets of mappings. The top-right window uses the mappings shown in Figure 10.6, the top-left maps *ir* (red) and *vis* (blue-green) to *color*, the bottom-right maps *ir* to *selector* (only pixels with selected *ir* radiances are displayed) and *time* to the y axis, and the bottom-left maps *ir, vis,* and *variance* to the x, y, and z axes, maps *texture* to *color* (*variance* and *texture* are pixel fields derived from *ir* radiance). See Color Plate 76.

true if B is open. We also define the notational convention $D(A) = \{D(a) | a \in A\}$. Then we can define a condition:

Condition 6: $OPEN_U(A) \Leftarrow OPEN_V(D(A))$

This condition requires that D be continuous. If we require implication both ways, as in:

Condition 7: $OPEN_U(A) \Leftrightarrow OPEN_V(D(A))$

then we require that both D and D^{-1} be continuous. Intuitively, this adds the requirement that neighboring displays are the displays of neighboring data objects. Condition 7 requires that U be *homeomorphic* with $D(U)$ (that is, that U and $D(U)$ have the same topology).

As in Section 10.3 assume that the data model U is defined as various types of aggregates of primitive values. Then we can define a topology on U in terms of the natural topologies on the sets $\mathbb{Z}$, $\mathbb{R}$, $\mathbb{R}^2$ and $\mathbb{R}^3$ (and possibly other sets) used for primitive values. Extending these topologies to a topology on U is a bit more complex than applying Equation (10.12). First, U is a disjoint union:

$$U = \bigcup \{U(s) | t \in T \;\&\; s \in ST(t)\}. \tag{10.28}$$

As in Section 10.3, let I_s be the set of indices of the primitive values of objects in $U(s)$ for a subtype s. Also, for $i \in Is$, let U_i be the set of values of the i-th primitive. Then we can interpret $U(s)$ as the cross product $\times\{U_i | i \in I_s\}$ and define the topology of $U(s)$ as product topology on this cross product. We can further define the topology on U whose connected components are the $U(s)$. A similar topology can be defined on V.

Conditions 6 and 7 provide no information about the relation between values of D on different connected components of U. However, Condition 7 requires a one-to-one correspondence between connected components of U and connected components of V. This may be impossible if U is partitioned into many different types and V consists of a single type.

10.6.4 Conditions on Metric Structures

We can also generate a topology using the $METRIC$ predicate defined in Equation (10.13). Because the definition of $METRIC_U(u, u', r)$ in Equation (10.13) allows there to be pairs u and u' for which no distance is defined, it is not a proper metric. Nevertheless, it can still generate a topology whose basis is the set of open balls around every point $u \in U$ (that is, the ball of radius R around u is $\{u' \in U | \exists r < R.\ METRIC_U(u, u', r)\}$). According to this topology, any two data objects u and u' of unlike subtype are in disconnected components of U. In fact, this is the same topology that we just defined in Section 10.6.3.

Condition 5 about metric structure, as defined in Section 10.4, is much stronger than Conditions 6 and 7 about topology. As we described in Section 10.4, Condition 5 can be used to express requirements relating perceptual distance (for example, the perceptual distance between two colors) to quantitative differences between data objects.

To understand what set of visualization functions satisfy Condition 5, consider that metric preserving mappings between finite-dimensional Cartesian spaces are restricted to combinations of rotations, translations, and reflections. The definition of the $METRIC$ predicate in Equation (10.13) partitions U into $U(s)$ for each type $t \in T$ and subtype $s \in ST(t)$. As in Section 10.6.3, $U(s)$ is a cross product of primitive sets and, given metrics on the primitive value sets, we can define the Cartesian metric on $U(s)$. Thus we have the following:

Proposition 3: If D satisfies Condition 5, if F_s is a rigid rotation/translation/reflection on the finite Cartesian space $U(s)$, if Z_s is a pure scaling by a factor of K on the finite Cartesian space $U(s)$, and if $F : U \rightarrow U$ is defined by $F(u) = Z_s(F_s(u))$ for $u \in U(s)$, then $D \circ F$ satisfies Condition 4 with $K_{D \circ F} = K * K_D$.

It is interesting to compare Condition 5 with Conditions 2, 3, and 4. Proposition 2 says that, for visualization functions satisfying Conditions 2, 3, and 4, the values of a particular primitive display type are functions of just one primitive data type, although that may be any continuous function. In contrast, the rotations of Proposition 3 allow values from different primitive data types to be mixed together in determining the values of primitive display types, but only according to linear functions. Taken together, Conditions 2 through 5 are a very strong constraint on visualization mappings.

10.6.5 Conditions on Algebraic Structures

There are a wide variety of operations on the primitive values of data objects and displays, such as addition, multiplication, string concatenation, and so on. These operations can be expressed as predicates, such as $SUM(u, u', u'') \equiv (u'' = u + u')$, and these predicates can be extended to all data objects and displays by the technique of Equation (10.12). Under the proper circumstances, the technique of Equation (10.16) can also be used to define arithmetical operations on data objects in terms of arithmetical operations on the mathematical objects that they represent. Given either of these ways of defining arithmetical operations on data and display models, we can define visualization conditions such as:

Condition 8: $SUM_U(u, u', u'') \Rightarrow SUM_V(D(u), D(u'), D(u''))$

We can also define arithmetical operations between a scalar field (such as the real numbers $\mathbb{R}$) and members of data and display models U and V. In particular, we can define scalar multiplication $r \times u$, and the condition:

Condition 9: $u' = r \times u \Rightarrow D(u') = r \times D(u)$

If the operator $D : U \to V$ satisfies Conditions 8 and 9, we say it is *linear*.

10.6.6 Conditions on Symmetry Structures

As we discussed at the end of Section 10.4, symmetry structures may express perceptual invariance to certain changes in displays, such as translations of images in time and space. Symmetries are generally defined by groups of operations, so we define a group G_V acting on the display model V. That is, every $g \in G_V$ is a bijection $g : V \to V$.

The intuition that perception is invariant to various symmetries cannot be expressed in the forms of Equations (10.2) through (10.4). In order to express this intuition, we need to define a condition on the set of displays rather than a condition on individual displays. Let S be a set of "acceptable" displays, perhaps satisfying one or more of Conditions 1 through 9. Then the idea that perception is invariant to symmetry can be stated as follows:

Condition 10: If $D \in S$ then $g \circ D \in S$ for all $g \in G_V$

It is interesting to note that Condition 10 is similar in form to the conclusions of Propositions 1 and 3. In fact, the conclusion of Proposition 1 is exactly Condition 10 with G_V equal to the group of all permutations of V. In the conclusion of Proposition 3, the group acts on U rather than V, and is a group of linear transformations acting on the subtypes of U. These similarities suggest that such conditions on the set S of acceptable displays may play a natural role in analyzing visualization problems.

We may also define conditions that relate symmetries of data to symmetries of displays. In order to do this, assume symmetry groups G_U and G_V acting on both U and V. Then define the condition:

Condition 11: There is a group isomorphism $I_D : G_U \leftrightarrow G_V$ such that $\forall g \in G_U.$ $D \circ g = I_D(g) \circ D$

Condition 11 says that the visualization function D maps symmetries of U to symmetries of V, according to the isomorphism I_D. Note that while Conditions 1 through 9 define various kinds of isomorphisms between structures on U and V, Condition 11 defines an isomorphism between groups of operations on U and V. By defining different symmetry groups G_U and G_V, Condition 11 can be used to express a variety of visualization requirements.

10.7 Conclusions

In this chapter we have discussed a few possibilities for defining a mathematical basis for analyzing visualization. We can express various properties of data and display models in terms of mathematical structures defined on those models. We can use these structures to define conditions on visualization mappings, as summarized in the following table:

structure on U and V	mapping $D : U \to V$
lattice	isomorphism
metric	isometric
topology	continuous, homeomorphism
arithmetic	linear
symmetry	set of valid D invariant to composition with symmetries

In particular, we are interested in exploring alternatives to heuristic algorithms for automating the design of displays.

Acknowledgments

This work was supported by NASA grant NAG8-828, and by the National Science Foundation and the Defense Advanced Research Projects Agency under Cooperative Agreement NCR-8919038 with the Corporation for National Research Initiatives. We would like to thank John Benson and Robert Krauss for their helpful comments.

Bibliography

[1] J. Bertin, *Semiology of Graphics*, W.J. Berg, tr., University of Wisconsin Press, 1983.

[2] C. Beshers and S. Feiner, "Automated Design of Virtual Worlds for Visualizing Multivariate Relations," *Proc. Visualization '92*, 1992, pp. 283–290.

[3] *CIE Recommendations on Uniform Color Spaces—Color Difference Equations, Psychometric Color Terms*, CIE Publication No. 15(E-13.1) 1971/(TC-1.3) 1978, Supplement No. 2, 9–12, Bureau Central de la CIE, Paris, France, 1978.

[4] B.A. Davey and H.A. Priestly, *Introduction to Lattices and Order*, Cambridge University Press, 1990.

[5] C.A. Gunter and D.S. Scott, "Semantic Domains," *Handbook of Theoretical Computer Science, Volume B*, J. van Leeuwen, editor, The MIT Press/Elsevier, 1990, pp. 633–674.

[6] P. Haberli' "ConMan: A Visual Programming Language for Interactive Graphics," *Computer Graphics*, Vol. 22, No. 4, 1988, pp. 103–111.

[7] W. Hibbard, C. Dyer, and B. Paul, "A Lattice Model for Data Display," *Proceedings of IEEE Visualization '94*, 1994, pp. 310–317.

[8] J.P.M. Hultquist and E.L. Raible, "SuperGlue: A Programming Environment for Scientific Visualization," *Proceedings of Visualization '92*, 1992, pp. 243–250.

[9] J.P. Lee and G.G. Grinstein, editors, *Database Issues for Data Visualization, Lecture Notes in Computer Science*, Number 871. Springer-Verlag, 1994.

[10] R. Li and P.K. Robertson, "Towards Perceptual Control of Markov Random Field Textures," *Perceptual Issues in Visualization*, G. Grinstein and H. Levkowitz, editors, Springer-Verlag, 1995, pp. 83–94.

[11] J. Mackinlay, "Automating the Design of Graphical Presentations of Relational Information," *ACM Transactions on Graphics*, Vol. 5, No. 2, 1986, pp. 110–141.

[12] C. McConnell and D. Lawton, "IU Software Environments," *Proceedings of IUW*, 1988, pp. 666–677.

[13] R.E. Moore, *Interval Analysis*, Prentice Hall, 1966.

[14] T. Nadas and A. Fournier, "GRAPE: An Environment to Build Display Processes," *Computer Graphics*, Vol. 21. No. 4, 1987, pp. 103–111.

[15] M. Potmesil and E. Hoffert, "FRAMES: Software Tools for Modeling, Animation and Rendering of 3D Scenes," *Computer Graphics*, Vol. 21, No. 4, 1987, pp. 75–84.

[16] P.K. Robertson, "Visualizing Color Gamuts: A User Interface for the Effective Use of Perceptual Color Spaces in Data Analysis," *IEEE Computer Graphics and Applications*, Vol. 8, No. 5, 1988, pp. 50–64.

[17] P.K. Robertson, "A Methodology for Choosing Data Representations," *IEEE Computer Graphics and Applications*, Vol. 11, No. 3, 1991, pp. 56–67.

[18] P.K. Robertson and J.F. O'Callaghan, "The Application of Perceptual Color Spaces to the Display of Remotely Sensed Imagery," *IEEE Transactions on Geoscience and Remote Sensing*, Vol. 26, No. 1, 1988, pp. 49–59.

[19] D.A. Schmidt, *Denotational Semantics*, William C. Brown, 1986.

[20] W.J. Schroeder, W.E. Lorenson, G.D. Montanaro, and C.R. Volpe, "VISAGE: An Object-Oriented Scientific Visualization System," *Proceedings of Visualization '92*, 1992, pp. 219–226.

[21] D.S. Scott, "The Lattice of Flow Diagrams," *Symposium on Semantics of Algorithmic Languages*, E. Engler, editor, Springer-Verlag, 1971, pp. 311–366.

[22] D.S. Scott, "Data Types as Lattices," *Siam J. Comput.*, Vol. 5, No. 3, 1976, pp. 522–587.

[23] H. Senay and E. Ignatius, "Compositional Analysis and Synthesis of Scientific Data Visualization Techniques," *Scientific Visualization of Physical Phenomena*, N.M. Patrikalakis, editor, Springer-Verlag, 1991, pp. 269–281.

[24] L.A. Treinish, "SIGGRAPH '90 Workshop Report: Data Structure and Access Software for Scientific Visualization," *Computer Graphics*, Vol. 25, No. 2, 1991, pp. 104–118.

[25] C. Upson, Jr. T. Faulhaber, D. Kamins, D. Laidlaw, D. Schlegel, J. Vroom, R. Gurwitz, and A. van Dam, "The Application Visualization System: A Computational Environment for Scientific Visualization," *Computer Graphics and Applications*, Vol. 9, No. 4, 1989, pp. 30–42.

Chapter 11

Controlled Interpolation for Scientific Visualization

Ken Brodlie and Petros Mashwama

Abstract. *Interpolation is a fundamental process in scientific visualization. It is the process by which we create an empirical model of the phenomenon we wish to visualize, from a set of data samples. In this chapter, we show how the scientist can control the interpolation process according to properties he may wish the model to have. The properties we consider are:*

- *satisfying simple bounds—such as positivity*
- *continuity of derivatives*
- *interpolation of derivative values*

in different combinations. We look at interpolation in 1D, 2D, and 3D; and we look at linear and cubic approaches. The cubic approach will generally provide the two derivative properties mentioned above, but not the satisfaction of simple constraints. We show how cubic methods can be easily modified to satisfy bounds—provided one is prepared to forego interpolation of derivatives. If derivatives must be honored, we show how bounds can be achieved by, for 1D using additional knots, and for 2D and 3D, a modification of the blending function method. Indeed blending functions offer a useful tool for volume visualization where one wants greater accuracy than trilinear interpolation, without the complexity of tricubic.

11.1 Introduction

Scientific visualization helps us to understand some physical phenomenon about which we have only partial knowledge. The visualization process takes some data—either measured as in medical imaging, or calculated as in computational science—and creates a view of that data so that we can gain insight into the underlying phenomenon.

There is an important first step in the visualization process. The data we are given are only a sample, and so it is necessary to create some empirical model of the underlying entity we are trying to visualize. When the given data are relatively sparsely distributed, we require *interpolation* to create this empirical model. So, in reality, we visualize not the data, but our empirical model. Thus interpolation is quite fundamental to scientific visualization.

In mathematical terms, if we suppose the underlying entity is a scalar function f of a number of variables $\mathbf{x}$, and if we suppose we are given data samples with values f_i at points $\mathbf{x}_i$, for $i = 1, 2, \ldots n$, then the interpolation process constructs an empirical model $F(\mathbf{x})$ which matches the data samples, and is an approximation to the unknown reality, $f(\mathbf{x})$. We visualize $F(\mathbf{x})$ and thereby hope to gain insight about $f(\mathbf{x})$.

It is clear that interpolation is a very ill-posed problem—there are an infinity of functions $F(\mathbf{x})$ which will fit any given data set. We need to allow the scientist to incorporate additional knowledge about the underlying entity so that the most suitable $F(\mathbf{x})$ is selected.

The aim of this chapter is to show how knowledge of different properties of the entity—for example, upper or lower bounds—can be used by the scientist to *control* the interpolation process. We shall look at 1D, 2D, and 3D: curves, surfaces, and volumes.

11.2 A Simple Example

First we look at a very simple example which will motivate the work. Coal is burnt in a furnace, and an instrument records the percentage of oxygen in the flue gas as time proceeds. Initially, there is 20.8 percent of oxygen as normal. When the coal burns, so the oxygen is used up, and so the percentage in the flue gas drops close to zero before gradually increasing again. Seven data samples are taken, and the values are shown in Table 11.1.

Time (mins)	0	2	4	10	28	30	32
Oxygen (%)	20.8	8.8	4.2	0.5	3.9	6.2	9.6

Table 11.1: Levels of oxygen in flue gas.

This could hardly be a simpler visualization problem, yet it illustrates the key role of interpolation.

Suppose we decide *not* to interpolate, but look only at the data. This 'visualization' is shown in Figure 11.1, and is not very useful. We want to know what happens between data points. That is, we want to see the empirical model of the relationship between oxygen level and time.

Linear interpolation is a simple option, and Figure 11.2 shows the visualization. It is unconvincing, because we know the underlying entity, the oxygen level, should not have a slope discontinuity at the sample points.

This leads us to try piecewise cubic interpolation, for example, cubic splines, where the interpolation scheme provides slope continuity (indeed both first and second derivative continuity in the case of cubic splines). This visualization is shown in Figure 11.3. But look again—we know that the oxygen levels never fall below zero, so this is equally unsatisfactory.

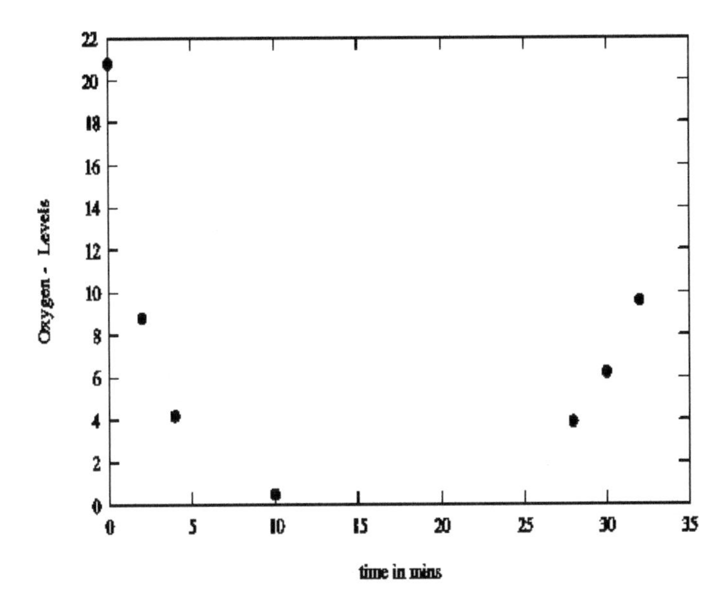

Figure 11.1: Visualization of the data itself.

It is not that these methods are bad—it is just that we have two pieces of knowledge about the underlying entity (continuity of first derivative; lower bound of zero) that we want to use to control the interpolation. Linear interpolation gives positivity but not slope continuity; cubic interpolation gives slope continuity, but not necessarily positivity. It is possible to achieve both, as Figure 11.4 shows, and as will be explained later in the chapter.

11.3 Controlling the Interpolation

The above example sets the scene for the rest of this chapter. We look at controlling the interpolation process for one-, two-, and three-dimensional data so as to provide a number of different properties:

Property 1: Satisfying upper and lower bounds.

This is a common property in scientific applications. For example, if the quantity to be visualized is a percentage as was the case in Section 11.2, then the model must lie between 0 and 100; if we are visualizing the density of some material, we wish the model to be everywhere positive. The 'overshoots' of cubic spline interpolation shown in Figure 11.3 are a frequent annoyance when visualizing such data.

Linear methods tend to satisfy this property, but in an extreme way which is often not what we want: the extremes of the model match the extremes of the data. Thus we distinguish:

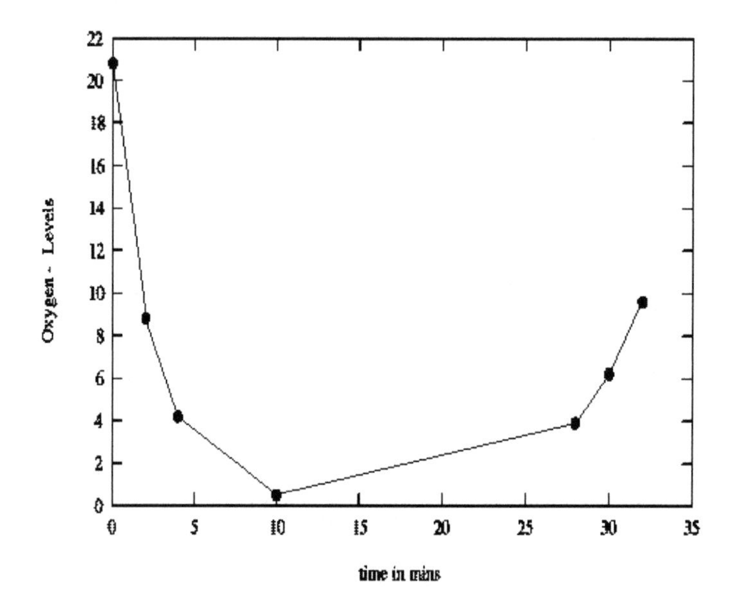

Figure 11.2: Visualization using linear interpolation.

- *Property 1: Type A.* Model not constrained by extremes of data, but can be controlled to lie between specified bounds.
- *Property 1: Type B.* Model constrained by extremes of data.

Property 2: Continuity of derivatives.

Continuity of derivatives is often inherent—and of course, important for a good visual appearance, particularly if the data is specified on a coarse grid.

Property 3: Interpolation of specified derivative values at data points.

Preservation of derivative data is important in computational applications, where typically quantities are specified through a set of differential equations—if the function value is known, the derivative is given from the differential equation, and it makes sense to use it in the visualization. A simple 1D application of this is in chemical kinetics: the concentrations of species are defined over a period of time in a reaction by a set of ordinary differential equations. As we solve the equations over time, we get both the concentration value and its rate of change.

We look at each property individually, but the combination of Property 1 with the two derivative cases is of special interest—so we have in addition:

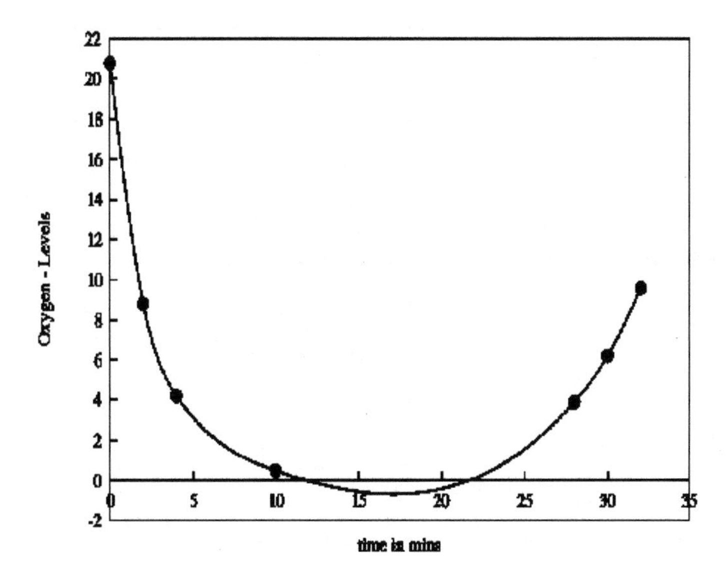

Figure 11.3: Visualization using cubic spline interpolation.

Property 4: Satisfying bounds *and* continuity of derivatives (that is, both 1 and 2).

Property 5: Satisfying bounds *and* interpolation of derivative values (that is, both 1 and 3).

The chapter is structured so as to discuss curves, surfaces, and volumes in turn, and the concluding section gives an overview across the different dimensions.

11.4 One-Dimensional Interpolation

11.4.1 Data and Notation

In the one-dimensional case, we are given data samples with values f_i at points x_i, for $i = 1, 2, \ldots n$. We shall use the notation d_i to denote the first derivative at each point: in some cases, this will be given as data (as in the differential equations applications mentioned earlier); in other cases, it will be useful to calculate d_i as an estimate of the first derivative, as part of the interpolation process.

Section 11.2 provided a simple example of this one-dimensional interpolation case.

11.4.2 Piecewise Linear Interpolation

Piecewise linear interpolation (Figure 11.2) creates a model function $F(x)$ as follows:

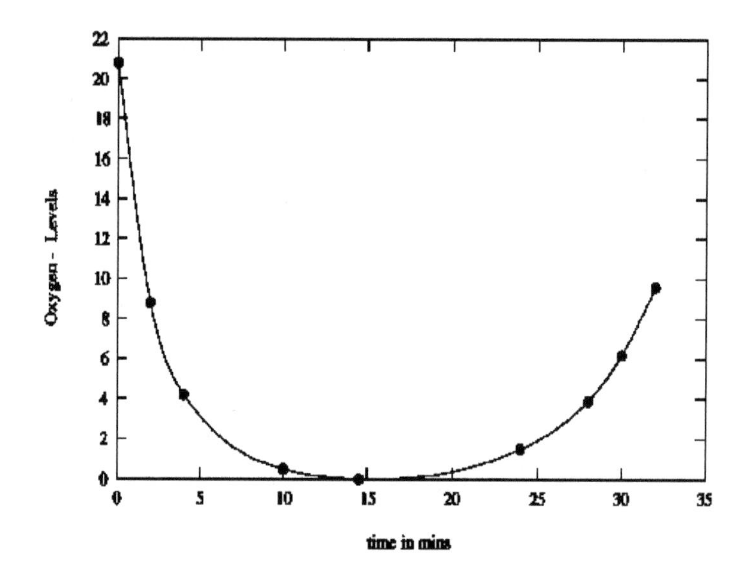

Figure 11.4: Visualization with a controlled interpolant which preserves positivity.

$$F(x) = f_i + t(f_{i+1} - f_i),$$

where

$$t = (x - x_i)/(x_{i+1} - x_i), x \in [x_i, x_{i+1}]$$

This is very simple (one multiplication and two additions to obtain t, followed by the same to evaluate $F(x)$). In terms of the properties we are interested in, the model is bounded by the extremes of the data. Thus Property 1: Type B is achieved—this is usually more restrictive than one would wish.

None of the other properties are provided by this method—it has only C^0 continuity.

11.4.3 Piecewise Cubic Interpolation

Suppose now that in addition to knowing the function values f_i at the sample points, we also know the derivative values, d_i. Then we can uniquely define a piecewise cubic polynomial $F(x)$ which matches values and derivatives as:

$$F(x) = f_i H_1(x) + f_{i+1} H_2(x) + d_i H_3(x) + d_{i+1} H_4(x), x \in [x_i, x_{i+1}]$$

where the $H_j(x)$ are the cubic Hermite basis functions for the interval $[x_i, x_{i+1}]$:

$$H_1(x) = \phi((x_{i+1} - x)/h_i),$$

$$H_2(x) = \phi((x - x_i)/h_i),$$

$$H_3(x) = -h_i\psi((x_{i+1} - x)/h_i),$$

$$H_4(x) = h_i\psi((x - x_i)/h_i),$$

where $h_i = x_{i+1} - x_i, \ \ \phi(t) = 3t^2 - 2t^3, \ \ \psi(t) = t^3 - t^2$.

Property 2, continuity of first derivatives, is satisfied by construction, because adjacent cubic pieces share a common slope at each data point. Property 3 is also satisfied, because known derivative values d_i are directly incorporated in the scheme.

If the slopes d_i are not known, there is a variety of techniques for estimating them. A common technique is cubic spline interpolation (see Chapter 4 of the book by Lancaster and Salkauskas [8]), where the slopes are chosen to give a stronger continuity property, namely C^2, or second derivative continuity. Another, simpler method is the osculatory method, where the slope at a data point is taken as the slope of the unique quadratic function which interpolates the sample point and its two immediate neighbors. For a full review of slope estimation methods, see the papers by Brodlie [2, 3].

Property 1, and hence Properties 4 and 5, are not satisfied in general, as one suspects from Figure 11.3, and this is the subject of the next section.

11.4.4 Bounded Piecewise Cubic Interpolation

In the simple example earlier, there was a need to impose a lower bound on the model, namely that $F(x)$ be bounded below by zero. Piecewise cubic interpolation may not yield a positive function $F(x)$, even though the data values f_i are themselves positive—hence the disaster in Figure 11.3.

The conditions for positivity of a cubic polynomial have been studied by Schmidt and Hess [12], and by Carlson and Fritsch [5].

Theorem 11.4.1 *Consider a single cubic piece, $U(x)$ say, on the interval $[x_1, x_2]$, with $f_1 \geq 0$ and $f_2 \geq 0$. A sufficient condition for $U(x)$ to be non-negative on $[x_1, x_2]$ is that:*

$$(d_1, d_2) \in S$$

where

$$S = \{(a, b) : a \geq -3f_1/h, b \leq 3f_2/h\}$$

Essentially, the theorem says that positivity is lost when the slope at either endpoint is too steep.

Consider now piecewise cubic interpolation to a set of n data points, where the data values $f_1, f_2, \ldots f_n$ are all non-negative.

Theorem 11.4.2 *A sufficient condition for the piecewise cubic interpolant $F(x)$ to be non-negative is that:*

$$d_i \in [-3f_i/(x_{i+1} - x_i), 3f_i/(x_i - x_{i-1})], i = 2, 3, \dots n - 1$$

with $d_1 \geq -3f_1/(x_2 - x_1)$ and $d_n \leq 3f_n/(x_n - x_{n-1})$.

This theoretical result allows us to propose modified piecewise cubic interpolants which satisfy either Property 4 (bounds and slope continuity) or the stronger Property 5 (bounds and slope interpolation).

We continue just to think of the bound $F(x) \geq 0$ for simplicity, and look at each case in turn.

Property 4: Satisfying bounds by projecting derivative values.

If the slope values have only been estimated rather than specified, it seems not unreasonable to modify them to fit within the valid range given in Theorem 11.4.2. A simple strategy is to project each derivative value d_i onto the valid interval—that is, any value outside the interval is mapped to the nearest endpoint.

Property 5: Satisfying bounds and interpolating derivative values by inserting additional knots.

If the slope values are to be honored, then Butt and Brodlie [4] show how it is possible to replace any offending cubic piece by two, or sometimes three, smaller cubic pieces by inserting additional knots in the interval between data points. This is the strategy that was adopted to produce the visualization shown in Figure 11.4.

The above discussion has been in terms of positivity, but it is easy to generalise to the case where $F(x)$ is to be bounded above and/or below by a linear function of x. The corresponding theorem is:

Theorem 11.4.3 *The cubic polynomial $U(x)$ satisfies $U(x) \geq mx + c$, for all $x \in [x_1, x_2]$ if*

$$(d_1, d_2) \in W$$

where

$$W = \{(a, b) : a \geq -\frac{3f_1 - 3(mx_1 + c) - mh}{h}, b \leq \frac{3f_2 - 3(mx_2 + c) + mh}{h}\}$$

The results for a single interval can be applied successively over the intervals $[x_1, x_2]$, $[x_2, x_3], \dots, [x_{n-1}, x_n]$—to give conditions for a piecewise cubic interpolant lying above a linear function. There is an extension to cater for the case of lying below a linear function.

Again it is possible to use slope projection or knot insertion to control the boundedness—see again Butt and Brodlie [4].

11.5 Two-Dimensional Interpolation

11.5.1 Data Arrangement and an Example

We now turn to 2D, and we assume that the data samples are taken on a rectangular mesh, $D = [x_1, x_2, \ldots x_m] \times [y_1, y_2, \ldots y_n]$, with data values $f_{ij}, i = 1, 2, \ldots m, j = 1, 2, \ldots n$.

As an illustration, we shall also use a worked example throughout the section. This is a test function used by Lancaster and Salkauskas [8], and defined as:

$$f = \begin{cases} 1, & \text{if } y - x \geq \frac{1}{2} \\ 2(y - x), & \text{if } 0 \leq y - x \leq \frac{1}{2} \\ \frac{1}{2}\{\cos(4\pi[(x - \frac{3}{2})^2 & \\ +(y - \frac{1}{2})^2]^{\frac{1}{2}}) + 1\}, & \text{if } [(x - \frac{3}{2})^2 + (y - \frac{1}{2})^2] \leq \frac{1}{16} \\ 0, & \text{otherwise} \end{cases}$$

$f(x, y)$ is defined over the rectangle with vertices $(0, 0), (2, 0), (2, 1), (0, 1)$. The function is illustrated in Figure 11.5: notice that the function is bounded by 0 and 1, and that there are slope discontinuities at the top and bottom of the ramp.

To provide test data for the methods described below, the function f was evaluated on a 11 by 6 grid over the rectangle $[0, 2] \times [0, 1]$, that is, at grid spacing of 0.2.

11.5.2 Piecewise Bilinear Interpolation

This is a simple approach where a bilinear model function

$$F(x, y) = a + bx + cy + dxy \tag{11.1}$$

is fitted within each grid rectangle.

In terms of our properties, the situation is similar to the 1D case. Property 1: Type B is satisfied in the sense that the interpolant lies within the bounds of the data, but this may be more severe than one wants in practice. The remaining properties are not satisfied, because there is a slope discontinuity across all mesh lines. Figure 11.6 shows a visualization created by evaluating the bilinear interpolant on a fine mesh over the rectangle.

In practice, it is important to execute the interpolation process efficiently, and this is discussed by Hill [7]. As with all interpolation of this type, it is efficient to transform the domain first to a unit interval, and then evaluate the function. Thus we shall assume our bilinear interpolant is to be defined within the unit square $[0, 1] \times [0, 1]$. The expression for $F(x, y)$ in terms of basis functions:

$$F(x, y) = (1 - x)(1 - y)f_{00} + x(1 - y)f_{10} + (1 - x)yf_{01} + xyf_{11}$$

is inefficient (eight multiplications, five additions), and Hill shows that it is better to carry out two 1D interpolations in x, say, to get $F(x, 0)$ and $F(x, 1)$, and a further 1D interpolation in y, to get $F(x, y)$. This will involve three multiplications and six additions.

It is worth noting, however, that if a, b, c, d in Equation (11.1) are known, then evaluation of $F(x, y)$ by:

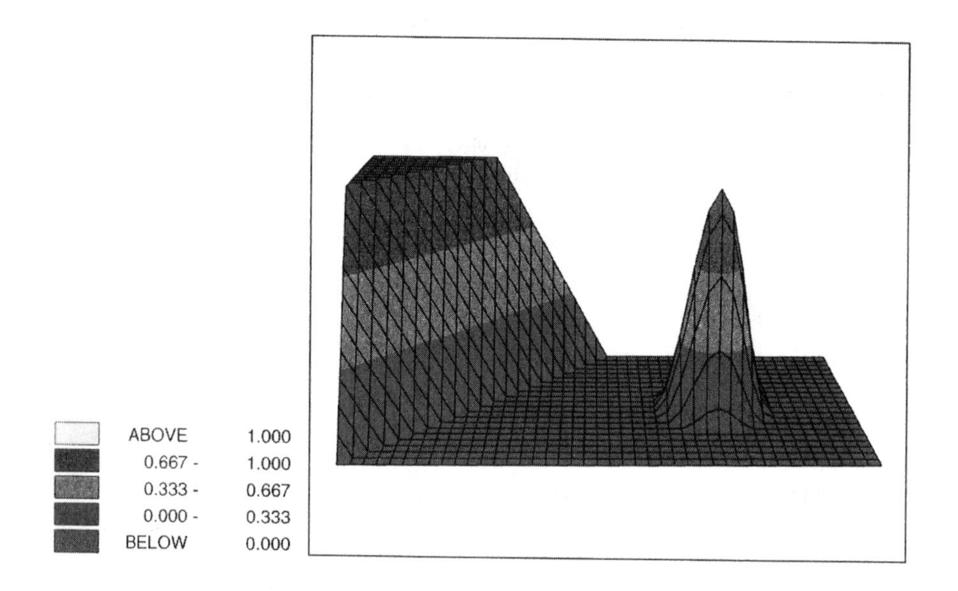

Figure 11.5: The test function. See Color Plate 77.

$$\begin{aligned} F(x, y) &= a + bx + cy + dxy \\ &= a + x(b + dy) + cy \end{aligned}$$

only requires three multiplications and three additions—plus the start-up cost of finding a, b, c, d. These constants can be found by solving the four equations:

$$F(x, y) = f_{xy}, x = 0, 1; y = 0, 1$$

which is surprisingly easy: just four additions! An advantage is that the partial derivatives F_x and F_y can also be found at little additional cost:

$$\begin{aligned} F_x(x, y) &= b + dy \\ F_y(x, y) &= c + dx \end{aligned}$$

11.5.3 Piecewise Bicubic Interpolation

To address Property 2, and achieve slope continuity, we need to move to piecewise bicubic interpolation. This assumes we are either given, or can estimate, first partial derivatives and the mixed partial derivative at each grid point.

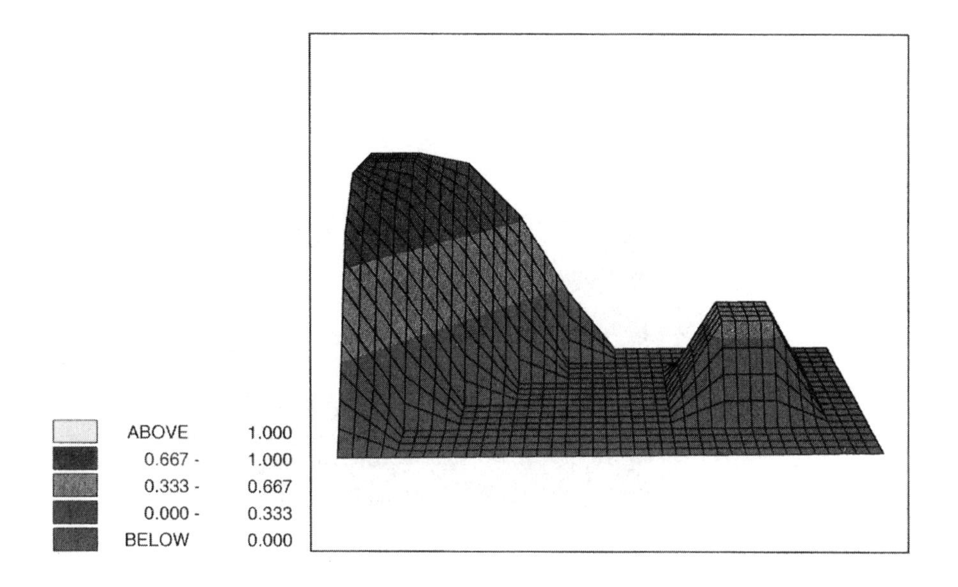

Figure 11.6: Piecewise bilinear interpolant. See Color Plate 78.

A bicubic $U(x, y)$ over a single rectangular data element $D_e = [a, b] \times [c, d]$ may be written as follows in terms of values and derivatives u, u_x, u_y, u_{xy} at the corner points:

$$\begin{aligned} U(x,y) \quad = \quad & u(a,c)H_1(x)G_1(y) + u(b,c)H_2(x)G_1(y) + \\ & u(a,d)H_1(x)G_2(y) + u(b,d)H_2(x)G_2(y) + \\ & u_x(a,c)H_3(x)G_1(y) + u_x(b,c)H_4(x)G_1(y) + \\ & u_x(a,d)H_3(x)G_2(y) + u_x(b,d)H_4(x)G_2(y) + \\ & u_y(a,c)H_1(x)G_3(y) + u_y(b,c)H_2(x)G_3(y) + \\ & u_y(a,d)H_1(x)G_4(y) + u_y(b,d)H_2(x)G_4(y) + \\ & u_{xy}(a,c)H_3(x)G_3(y) + u_{xy}(b,c)H_4(x)G_3(y) + \\ & u_{xy}(a,d)H_3(x)G_4(y) + u_{xy}(b,d)H_4(x)G_4(y) \end{aligned}$$

where the H_i and G_j are the cubic Hermite basis functions (see [8] for example).

A piecewise bicubic $F(x, y)$ can then be defined over the rectangular mesh. Property 2 is satisfied by construction: adjacent pieces share derivatives, so the interpolant is C^1. Property 3 is likewise satisfied: the scheme uses directly known derivative values.

If derivatives are not known, then they can be estimated as in the 1D case.

Figure 11.7 shows a visualization created from piecewise bicubic interpolation: the first partial derivatives in x and y are estimated from the data using the osculatory method; a similar estimation gives the mixed partial derivatives. Property 1, and hence Properties 4 and 5, are not achieved. We know that the function $f(x, y)$ lies within the range $[0, 1]$ and would like to impose that constraint on $F(x, y)$. This is not possible in general, and one can see in Figure 11.7 areas at the top and bottom of the ramp, and at the base of the circular peak, where these bounds have been exceeded.

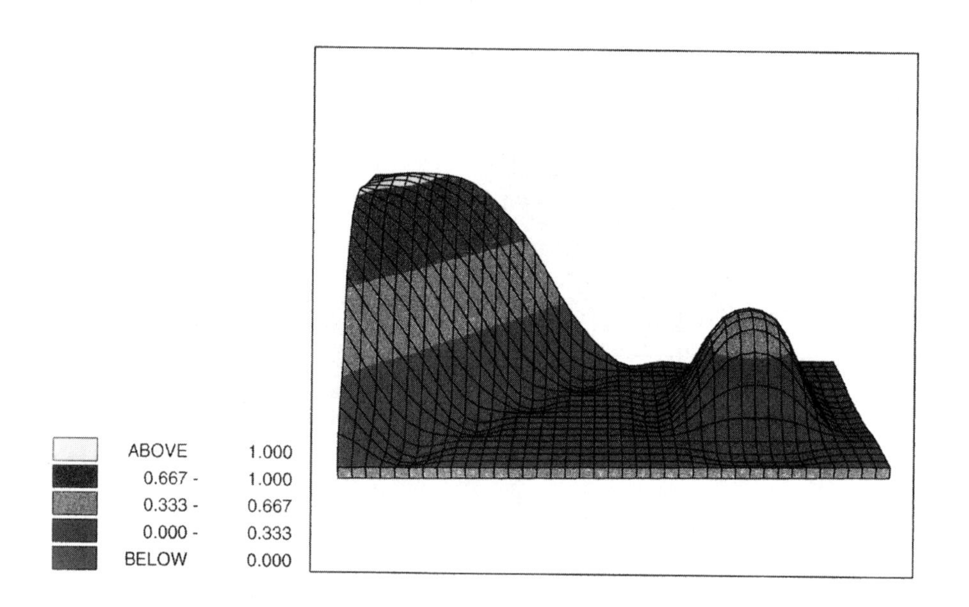

Figure 11.7: Piecewise bicubic interpolant. See Color Plate 79.

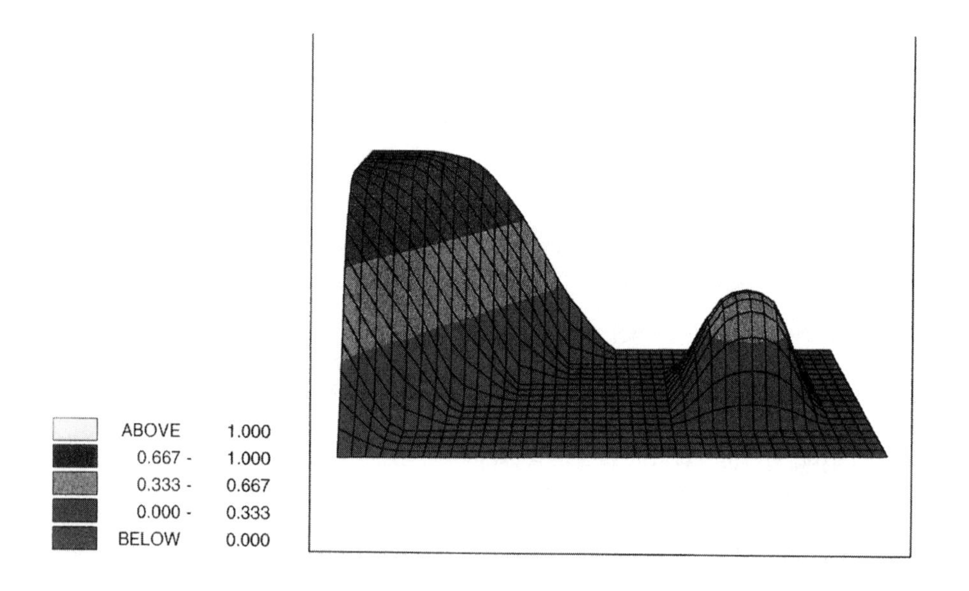

Figure 11.8: Bounded piecewise bicubic interpolant. See Color Plate 80.

11.5.4 Bounded Piecewise Bicubic Interpolation

The theoretical work of Schmidt and Hess on positivity of 1D cubics has been extended to 2D bicubics by Brodlie, Butt, and Mashwama [1]. Positivity is achieved provided the derivative values lie within a certain range. They prove the following theorem:

Theorem 11.5.1 *Let the piecewise bicubic interpolant $F(x, y)$ be defined over the rectangular mesh $D = [x_1, x_m] \times [y_1, y_n]$. Suppose that the given data $f_{ij} = f(x_i, y_j)$ is positive. Then $F(x, y)$ will be positive over the mesh D if the following sufficient conditions hold:*

$$f_x(x_i, y_j) \in \left[-\frac{3f(x_i, y_j)}{x_{i+1} - x_i}, \frac{3f(x_i, y_j)}{x_i - x_{i-1}}\right]$$

$$f_y(x_i, y_j) \in \left[-\frac{3f(x_i, y_j)}{y_{j+1} - y_j}, \frac{3f(x_i, y_j)}{y_j - y_{j-1}}\right]$$

$$\begin{aligned} f_{xy}(x_i, y_j) \geq \; & max\left[-\frac{3f_y(x_i, y_j)}{x_{i+1} - x_i} - \frac{3}{y_{j+1} - y_j}\left\{f_x(x_i, y_j) + \frac{3f(x_i, y_j)}{x_{i+1} - x_i}\right\},\right. \\ & \left.\frac{3f_y(x_i, y_j)}{x_i - x_{i-1}} + \frac{3}{y_j - y_{j-1}}\left\{f_x(x_i, y_j) - \frac{3f(x_i, y_j)}{x_i - x_{i-1}}\right\}\right] \end{aligned}$$

$$\begin{aligned} f_{xy}(x_i, y_j) \leq \; & min\left[-\frac{3f_y(x_i, y_j)}{x_{i+1} - x_i} + \frac{3}{y_j - y_{j-1}}\left\{f_x(x_i, y_j) + \frac{3f(x_i, y_j)}{x_{i+1} - x_i}\right\},\right. \\ & \left.\frac{3f_y(x_i, y_j)}{x_i - x_{i-1}} - \frac{3}{y_{j+1} - y_j}\left\{f_x(x_i, y_j) - \frac{3f(x_i, y_j)}{x_i - x_{i-1}}\right\}\right] \end{aligned}$$

for $i = 2, 3, \ldots, m-1; j = 2, 3, \ldots, n-1$, with equivalent, one-sided conditions for x_1, x_m, y_1, y_n.

This theorem shows the conditions for Property 1: Type A, for the simple case of a lower bound of 0. It is straightforward (though the algebra is lengthy!) to extend this to conditions for $F(x, y)$ to lie between two planes, and the details are given in [1].

Now we look separately at practical ways of achieving Properties 4 and 5.

Property 4: Satisfying bounds by projecting derivative values.

First we consider the case where the derivatives have been estimated, and there is no problem in modifying them. The derivative projection idea extends easily, with the values being moved to within the valid ranges as defined in Theorem 11.5.1.

Figure 11.8 shows a visualization created by bounded piecewise bicubic interpolation: the derivatives estimated as for visualization in Figure 11.7 are projected onto a valid range to preserve a lower bound of 0, and upper bound of 1.

Property 5: Satisfying bounds and interpolating derivative values—can it be done?

In the 1D case, it was possible to do this by inserting additional knots so that two, or three, cubic pieces were used. Unfortunately this idea does not extend in any obvious way to 2D, and indeed we are not aware of any easy means of modifying piecewise bicubic interpolation to achieve Property 5. In the next section, we look at a different approach.

11.5.5 Blending Functions

We revisit a technique which was popular some years ago, but which now is perhaps less well known. This is the technique of *blending functions*, which are described by Gordon [6] and by Lancaster and Salkauskas [8]. The idea is essentially the same as the Coons patch—see, for example, Rogers and Adams [11]—for parametric surface fitting, but here applied to nonparametric surfaces. The method assumes that not only are data values at the grid points known, but also that we know the cross sections along grid lines. These cross sections are then blended, again in a piecewise fashion, to give a continuous interpolant across the entire mesh.

As before, it is simplest to reduce the domain of a grid rectangle to the unit square $[0,1] \times [0,1]$. We suppose the cross sections parallel to the x-axis are prescribed as $C_{x0}(x)$ and $C_{x1}(x)$ at $y = 0, 1$; and the cross sections parallel to the y-axis are prescribed as $C_{0y}(y)$ and $C_{1y}(y)$ at $x = 0, 1$. Using the Lagrange cardinal functions,

$$\phi_0(x) = 1 - x,$$

$$\phi_1(x) = x,$$

$$\psi_0(y) = 1 - y,$$

$$\psi_1(y) = y,$$

it is easy to blend by linear interpolation in x and in y to create two separate surfaces that will each match a pair of opposite curves:

$$B^x(x,y) = C_{x0}(x)\psi_0(y) + C_{x1}(x)\psi_1(y)$$

$$B^y(x,y) = C_{0y}(y)\phi_0(x) + C_{1y}(y)\phi_1(x)$$

Gordon showed that a single surface which matches *all four* boundary curves can be found by adding these two surfaces, and subtracting the bilinear interpolant, $P(x,y)$ say. Thus the blended interpolant is defined as:

$$F(x,y) = B^x(x,y) + B^y(x,y) - P(x,y)$$

Note that $P(x, y)$ can be written as follows:

$$P(x,y) = \sum_{i,j=0}^{1,1} f_{ij}\phi_i(x)\psi_j(y)$$

It is easy to extend this to a complete mesh of grid rectangles.

It is unusual in visualization applications for the cross sections to be known—typically only the grid-point values and perhaps derivatives are known. However as explained earlier, there is a range of 1D interpolation techniques which will construct a curve through discrete points. Thus the cross sections can easily be created from the data using, say, piecewise cubic interpolation, and then blending applied.

Property 2 is only partly satisfied. If the cross sections are continuous in first derivative, then this property will be inherited by the blended surface—but only along the grid lines. There will be a discontinuity of slope across the grid lines, except of course at the mesh points themselves. However, the fact that continuity is maintained at the mesh points encourages a belief that the discontinuities at other points are not too severe.

Property 3 is satisfied if the derivative information is included in the construction of the cross sections.

Figure 11.9 shows a visualization of our test case using a blending function. Osculatory interpolation was used to create the curves along the grid lines, which were then blended to give the interpolant shown.

Property 1 is not in general satisfied: again we would like to keep the surface in the range $[0, 1]$ but this has clearly not been achieved. However, can we use the control that is possible in the construction of 1D interpolation, to create cross sections that will blend to create surfaces within prescribed bounds? This would enable us to achieve the elusive Property 5 which we failed to achieve with piecewise bicubic interpolation. This is the subject of the next section.

11.5.6 Bounded Blending Functions

Again for simplicity, we consider just the case of a lower bound of 0, and we work within the unit square $[0, 1] \times [0, 1]$.

The blended interpolant can be written as:

$$\begin{aligned} F(x,y) &= B^x(x,y) + B^y(x,y) - P(x,y) \\ &= (B^x(x,y) - \frac{1}{2}P(x,y)) + (B^y(x,y) - \frac{1}{2}P(x,y)), \\ &= F_1(x,y) + F_2(x,y) \end{aligned}$$

Thus it is sufficient if we can construct F_1 and F_2 separately to be positive.

Consider first:

$$F_1(x,y) = B^x(x,y) - \frac{1}{2}P(x,y)$$

For $F_1(x,y)$ to be positive, it is required that at least $F_1(x_i,y_j) \geq 0$—which is true as:

$$\begin{aligned} F_1(x_i,y_j) &= B^x(x_i,y_j) - \frac{1}{2}P(x_i,y_j) \\ &= f_{ij} - \frac{1}{2}f_{ij} \\ &= \frac{1}{2}f_{ij} \end{aligned}$$

Expanding $F_1(x,y)$ we get

$$(C_{x0}(x)-\frac{1}{2}(f_{00}\phi_0(x)+f_{10}\phi_1(x)))\psi_0(y)+(C_{x1}(x)-\frac{1}{2}(f_{01}\phi_0(x)+f_{11}\phi_1(x)))\psi_1(y)$$

But we saw in Section 11.4.4 (Theorem 11.4.3) that it is possible to create a cubic interpolant which lies above a linear function. Thus we can construct $C_{x0}(x)$ such that:

$$(C_{x0}(x) - \frac{1}{2}(f_{00}\phi_0(x) + f_{10}\phi_1(x))) \geq 0$$

and $C_{x1}(x)$ such that:

$$(C_{x1}(x) - \frac{1}{2}(f_{01}\phi_0(x) + f_{11}\phi_1(x))) \geq 0$$

Since $\psi_0(y)$ and $\psi_1(y)$ are both positive, it follows that $F_1(x,y)$ is positive. A similar argument shows that B^y can be constructed so that $F_2(x,y)$ is positive—and hence $F(x,y)$ itself is positive.

Thus we now have a means of achieving Property 5—the positivity bound is obtained yet the derivative values are preserved.

Figure 11.10 shows a visualization created in this way. Notice the interpolant remains positive at the foot of the ramp and the foot of the peak.

We have described only the simple case of a lower bound of 0. It is straightforward to extend the argument to any lower bound, and indeed to include an upper bound. In fact, in Figure 11.10, an upper bound has also been imposed to limit the maximum value to 1.

11.6 Three-Dimensional Interpolation

11.6.1 Data Arrangement and an Example

We turn finally to 3D, and we assume that the data samples are taken on a rectilinear mesh:

$$D = [x_1, x_2, \ldots x_m] \times [y_1, y_2, \ldots y_n] \times [z_1, z_2, \ldots z_p],$$

with data values $f_{ijk}, i = 1, 2, \ldots m, j = 1, 2, \ldots n, k = 1, 2, \ldots p$.

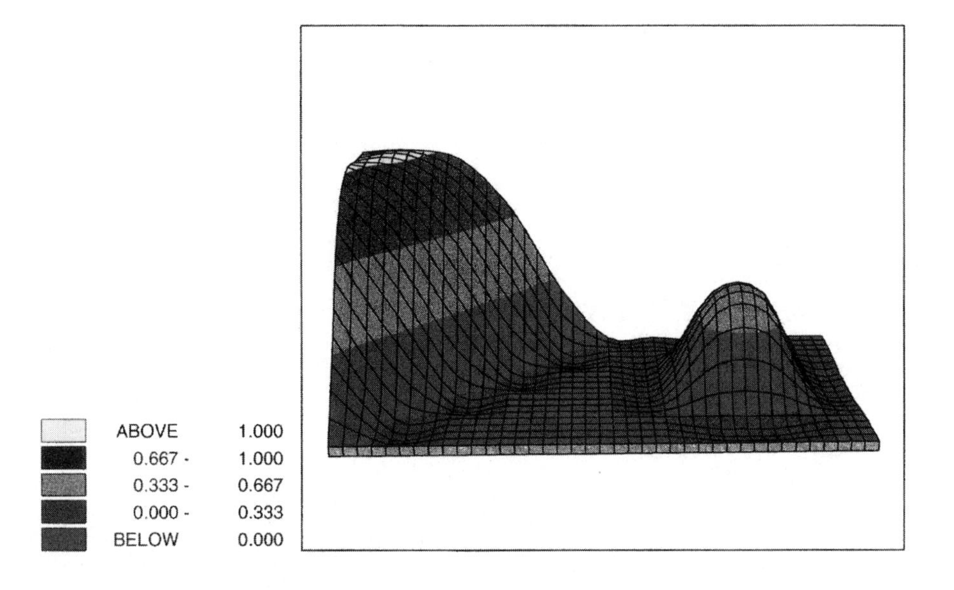

Figure 11.9: Blending function interpolant. See Color Plate 81.

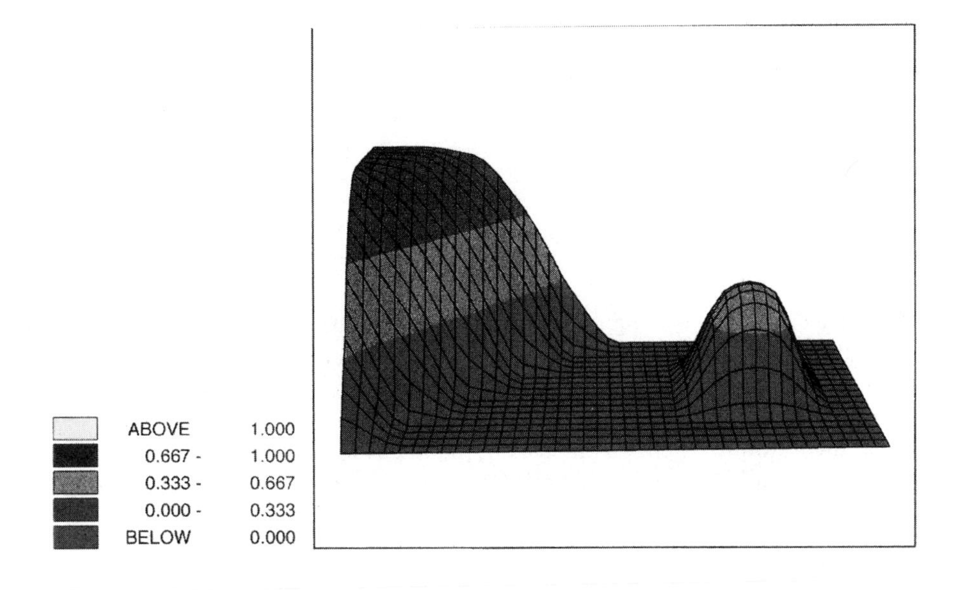

Figure 11.10: Bounded blending function interpolant. See Color Plate 82.

As an illustration, we shall use a worked example which is a simplified extension to 3D of the test function we used for the 2D case. This is defined as:

$$f = \begin{cases} \frac{1}{2}\{\cos(4\pi[(x-\frac{1}{2})^2 \\ +(y-\frac{1}{2})^2 + (z-\frac{1}{2})^2]^{\frac{1}{2}}) + 1\}, & \text{if } [(x-\frac{1}{2})^2 + (y-\frac{1}{2})^2 + (z-\frac{1}{2})^2] \leq \frac{1}{16} \\ 0, & \text{otherwise} \end{cases}$$

To provide test data, the function was evaluated on a $7 \times 11 \times 8$ grid over the region $[0.2, 0.8] \times [0.0, 1.0] \times [0.2, 0.9]$.

11.6.2 Piecewise Trilinear Interpolation

This is by far the most common interpolation method in 3D scientific visualization. A trilinear model function

$$F(x,y,z) = a + bx + cy + dz + exy + fyz + gzx + hxyz \tag{11.2}$$

is fitted within each grid cell.

As far as properties are concerned, Property 1: Type B is satisfied in the severe sense of bounding the interpolant by the data extremes. The other properties are not satisfied, as the interpolant is only C^0 and there are derivative discontinuities across the cell faces.

Thus trilinear interpolation does not score well against our list of properties, but as the dimension increases, so the need for simplicity and efficiency grows—and higher order methods are less competitive. As in 1D and 2D, it is important to carry out the trilinear interpolation as cheaply as possible. If we work again in a unit cell, the trilinear interpolant expressed in terms of basis functions is:

$$\begin{aligned} F(x,y,z) &= (1-x)(1-y)(1-z)f_{000} + x(1-y)(1-z)f_{100} \\ &\quad + (1-x)y(1-z)f_{010} + xy(1-z)f_{110} \\ &\quad + (1-x)(1-y)zf_{001} + x(1-y)zf_{101} \\ &\quad + (1-x)yzf_{011} + xyzf_{111} \end{aligned}$$

As Hill [7] again shows, a straightforward evaluation of this expression is expensive (24 multiplications, 10 additions). He recommends a cascaded approach of four 1D interpolations in x, say, two 1D interpolations in y, and one 1D interpolation in z—taking seven multiplications and 14 additions.

But again it is worth looking at the cost of computing the coefficients in Equation (11.2) directly. Evaluation of the expression itself takes seven multiplications and six additions, if done carefully as:

$$\begin{aligned} F(x,y,z) &= a + bx + cy + dz + exy + fyz + gzx + hxyz \\ &= a + x(b + y(e + hz)) + y(c + fz) + z(d + gx) \end{aligned}$$

The cost of computing the coefficients by solving the set of eight equations:

$$F(x, y, z) = f_{xyz}, x = 0, 1; y = 0, 1; z = 0, 1$$

is again surprisingly easy: just thirteen additions! So there are benefits if more than one interpolation is carried out. Moreover, the gradient of the model function is cheaply evaluated from the partial derivative expressions:

$$F_x = b + ey + z(g + hy)$$

$$F_y = c + fz + x(e + hz)$$

$$F_z = d + gx + y(f + hx)$$

This is particularly useful in visualization where the gradient is needed in the rendering process.

11.6.3 Piecewise Tricubic Interpolation

As computers become more powerful, and the demand for more accurate visualization increases, so one can expect interest to increase in higher-order interpolation for 3D, which will allow Properties 2 and 3 to be achieved.

Piecewise tricubic interpolation will provide this, and is already occasionally used. The model function for one tricubic piece has 64 terms—eight per vertex of the cell. These eight terms involve the value, the three first partial derivatives, the three mixed second partial derivatives, and the single third mixed partial derivative.

As in the lower dimensional cases, Properties 2 and 3 come automatically by construction. Property 1, however, is not attained.

11.6.4 Bounded Piecewise Tricubic Interpolation

We can derive conditions analogous to the 2D case, to establish when the model tricubic will be everywhere positive—see the thesis by Mashwama [9] for details. Then, if the derivatives have only been estimated, it is reasonable to apply the projection method to constrain them to a valid range—and we have a modified tricubic interpolant which preserves a lower bound. Thus Property 4 is achieved.

This leaves Property 5—and we study this in the next two sections.

11.6.5 Blending Functions for 3D

It is natural to seek a solution through blending functions, since they were successful in 2D. Indeed there is a nice extension of blending functions from 2D to 3D. Suppose we know the cross sections of the function along all mesh lines. As usual, it will be convenient to work with a unit cell $[0, 1] \times [0, 1] \times [0, 1]$—see Figure 11.11. The value of the function at

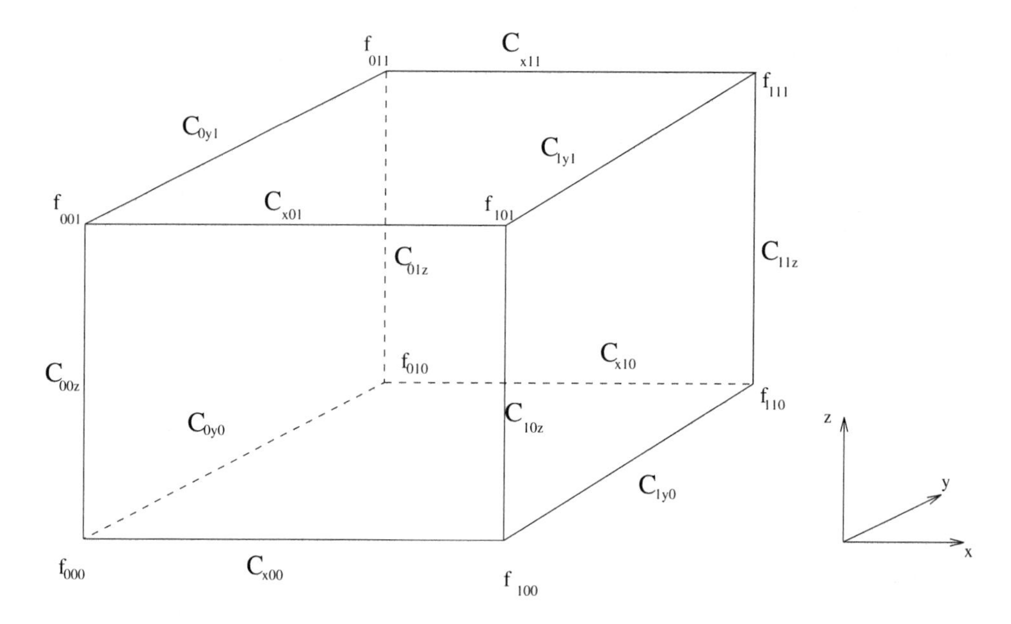

Figure 11.11: Notation for 3D blending functions.

grid point $x = i; y = j; z = k$ is denoted f_{ijk}, for $i = 0, 1; j = 0, 1; k = 0, 1$. The four cross sections parallel to x are denoted C_{xjk} for $j = 0, 1; k = 0, 1$; and similarly for cross sections in y and z directions. There are 12 cross sections in all.

Then within a cell, we can blend the curves together into a 3D interpolant which matches the curves along all the sides of the cube. The interpolant is:

$$F(x, y, z) = B_x(x, y, z) + B_y(x, y, z) + B_z(x, y, z) - 2T(x, y, z)$$

The function $T(x, y, z)$ is the trilinear interpolant. Each $B_j(x, y, z)$ is a bilinear blending of the curves in the direction of the corresponding axis; that is,

$$B_x(x, y, z) = (1 - y)(1 - z)C_{x00} + (1 - y)zC_{x01} + y(1 - z)C_{x10} + yzC_{x11}$$

$$B_y(x, y, z) = (1 - x)(1 - z)C_{0y0} + (1 - x)zC_{0y1} + x(1 - z)C_{1y0} + xzC_{1y1}$$

$$B_z(x, y, z) = (1 - x)(1 - y)C_{00z} + (1 - x)yC_{01z} + x(1 - y)C_{10z} + xyC_{11z}$$

It is easy to extend this to a complete mesh of grid rectangles, and this gives us a novel technique for 3D interpolation.

Of course it is unlikely in a visualization application that we would know the cross-section curves C explicitly. However, we can easily generate these using 1D interpolation methods, giving us an interpolation scheme somewhat simpler than piecewise tricubic.

There are essentially 32 terms—four per vertex of the cell, rather than eight in the tricubic case, because mixed partial derivatives are not included.

Property 2 is only partly satisfied. There is derivative continuity along all mesh lines, and within cells, but not across cell boundaries. Again we hope the pinning of derivative continuity at the grid points will make the discontinuities elsewhere relatively small. Property 3 is satisfied if the derivative information is included in the construction of the cross sections.

Property 1, however, is not attained, since, of course, any of the curves C may fall below zero.

Can we use our control over the generation of curves to achieve control over the bounds of the 3D blended interpolant, and thus attain Property 5? We look at this in the next section.

11.6.6 Bounded Blending Functions

Again for simplicity, we consider just the case of a lower bound of 0, and we work within the unit cell $[0, 1] \times [0, 1] \times [0, 1]$.

The blending interpolant can be written as:

$$\begin{aligned} F(x, y, z) &= B_x(x, y, z) + B_y(x, y, z) + B_z(x, y, z) - 2T(x, y, z) \\ &= (B_x(x, y, z) - \frac{2}{3}T(x, y, z)) + (B_y(x, y, z) - \frac{2}{3}T(x, y, z)) \\ &\quad + (B_z(x, y, z) - \frac{2}{3}T(x, y, z)) \\ &= F_1(x, y, z) + F_2(x, y, z) + F_3(x, y, z) \end{aligned}$$

It is sufficient to prove that each function $F_i(x, y, z)$ is positive. Consider $F_1(x, y, z)$ which we can write as:

$$\begin{aligned} F_1(x, y, z) &= (1 - y)(1 - z)(C_{x00} - \frac{2}{3}((1 - x)f_{000} + xf_{100})) \\ &\quad + (1 - y)z(C_{x01} - \frac{2}{3}((1 - x)f_{001} + xf_{101})) \\ &\quad + y(1 - z)(C_{x10} - \frac{2}{3}((1 - x)f_{010} + xf_{110})) \\ &\quad + yz(C_{x11} - \frac{2}{3}((1 - x)f_{011} + xf_{111})) \end{aligned}$$

We can keep each of these four terms positive by ensuring that each curve C_{xjk} lies above a linear function, which is two-thirds of the straight line joining the endpoints. We saw the conditions for this in Section 11.4.4 (Theorem 11.4.3), and by knot insertion we can construct the required curves.

Thus we have achieved Property 5—the bound is satisfied, and the derivative values preserved.

11.7 Conclusions and Future Work

We have shown how it is possible to control interpolation in scientific visualization, so as to achieve certain properties which the scientist may wish to associate with the empirical model being created.

Piecewise linear interpolation (bilinear in 2D, trilinear in 3D) will satisfy bounds—but in a severe way so that the bounds of the interpolant are forced to be constrained by the extremes of the data. Linear interpolation has neither of the derivative properties—continuity or interpolation—that we might wish to achieve.

Piecewise cubic interpolation (bicubic in 2D, tricubic in 3D) will provide the continuity and interpolation of derivative values—but will not satisfy bounds on the interpolant. However, we do have theoretical results which tell us conditions on the (partial) derivatives of the cubic pieces at the data points that will ensure bounds are satisfied. Thus, if we need to satisfy bounds, we can simply modify the derivative values so that the conditions are met. This will lose the property of interpolating derivative values. If this property is important, then we can, in 1D, use extra knots in 'difficult' regions; or in 2D and 3D, use the modified blending methods described in Sections 11.5 and 11.6—but these latter methods do not provide a surface or volume which is C^1 everywhere.

We summarize this work in the following table. We show which properties are attained, and give some idea of the complexity by identifying the number of terms in a single piece of interpolant.

	Method	Prop 1	Prop 2	Prop 3	Prop 4	Prop 5	Terms
Curves	Piecewise linear	B	No	No	No	No	2
	Piecewise cubic	No	Yes	Yes	No	No	4
	P'wise cubic - proj deriv	A	Yes	No	Yes	No	4
	P'wise cubic - add knots	A	Yes	Yes	Yes	Yes	4+
Surfaces	Piecewise bilinear	B	No	No	No	No	4
	Piecewise bicubic	No	Yes	Yes	No	No	16
	P'wise bicubic - proj deriv	A	Yes	No	Yes	No	16
	Blending function	No	Yes	Yes	No	No	12
	Bounded blending	A	No	Yes	No	Yes	12+
Volumes	Piecewise trilinear	B	No	No	No	No	8
	Piecewise tricubic	No	Yes	Yes	No	No	64
	P'wise bicubic - proj deriv	A	Yes	No	Yes	No	64
	Blending function	No	Yes	Yes	No	No	32
	Bounded blending	A	No	Yes	No	Yes	32+

Table 11.2: Comparison of approaches.

Finally there remain two points of interest for further work. Firstly, we note that there seems no easy way of achieving Properties 1, 2, *and* 3, together, except for 1D. Some higher-order interpolant might be needed.

Secondly, there is a quite different approach which we are currently exploring. Suppose we wish to create a positive interpolant. We first *transform* the sample values by applying logarithms; that is,

$$f_i^* = \log(f_i)$$

Note that f_i^* may be negative. Now use *any* interpolation scheme to create a model function F^*. Then the function:

$$F = \exp(F^*)$$

interpolates the data samples, and is positive everywhere! Notice this technique can be used with any interpolation scheme, and with any arrangement of data. It can thus be used in conjunction with scattered data interpolation schemes, such as multiquadric interpolation and modified Shepard methods—see Nielson [10].

Acknowledgments

We would like to thank Dr. Sohail Butt who helped develop the knot insertion method for positive preserving 1D interpolation. We would also like to thank Professor John Nelder, who pointed out that statisticians frequently apply the log transformation to maintain positivity.

Bibliography

[1] Ken Brodlie, Sohail Butt, and Petros Mashwama, "Visualization of Surface Data to Preserve Positivity and Other Simple Constraints," *submitted to Computers and Graphics*, 1994.

[2] K.W. Brodlie, "A Review of Methods for Curve and Function Drawing," *Mathematical Methods in Computer Graphics and Design*, K.W. Brodlie, editor, Academic Press, New York and London, 1980, pp. 1–37.

[3] K.W. Brodlie, "Methods for Drawing Curves," *Fundamental Algorithms for Computer Graphics*, R.A. Earnshaw, editor, Springer-Verlag, 1985, pp. 303–324.

[4] S. Butt and K.W. Brodlie, "Preserving Positivity Using Piecewise Cubic Interpolation," *Computer and Graphics*, Vol. 17, No. 1, 1993, pp. 55–64.

[5] R.E. Carlson and F.N. Fritsch, "Monotone Piecewise Bicubic Interpolation," *SIAM J. of Numerical Analysis*, Vol. 22, No. 2, 1985, pp. 386–400.

[6] W.J. Gordon, "Distributive Lattices and the Approximation of Multivariate Functions," *Approximations with Special Emphasis on Spline Functions*. I.J. Schoenberg, editor, Academic Press, 1969.

[7] S. Hill, "Tri-Linear Interpolation," *Graphics Gems IV*, P.S. Heckbert, editor, AP Professional, 1994, pp. 521–525.

[8] P. Lancaster and K. Salkauskas, *Curve and Surface Fitting: An Introduction*, Academic Press, London, 1986.

[9] P. Mashwama, *Shape Preserving Interpolation for Scientific Visualization*, Ph.D. thesis, University of Leeds, 1994, in preparation.

[10] G.M. Nielson, "Scattered Data Modelling," *IEEE Computer Graphics and Applications*, Vol. 13, No. 1, 1993, pp. 60–70.

[11] D.F. Rogers and J.A. Adams, *Mathematical Elements for Computer Graphics*, second edition, McGraw Hill, New York, 1990.

[12] J.W. Schmidt and W. Hess, "Positivity of Cubic Polynomial on Intervals and Positive Spline Interpolation," *BIT*, Vol. 28, 1988, pp. 340–352.

Part III

Techniques and Algorithms

Chapter 12

Feature Visualization

D. Silver

Abstract. *Identifying and isolating features is an important part of visualization and a crucial step for the analysis and understanding of large time-dependent data sets (either from observation or simulation). In this chapter, we address this concern, namely the investigation and implementation of basic 2D and 3D feature based methods to enhance current visualization techniques and provide the building blocks for automatic feature extraction, tracking, and analysis. These methods incorporate ideas from scientific visualization, computer vision, image processing, and mathematical morphology. Our focus is in the area of computational fluid dynamics, and we show the applicability of these methods to the quantification and tracking of three-dimensional vortex and turbulence bursts.*

Keywords: visualization, feature extraction, feature tracking, computer vision, CFD.

12.1 Introduction

The aim of the visualization of massive scientific data sets is to devise algorithms and methods that transform numerical data into pictures and other graphic representations, thereby facilitating comprehension and interpretation. By coloring and connecting different segments of data (based upon a user-defined color map and threshold), the algorithms highlight regions of activity. If a particular region is of interest, the scientist attempts to identify and quantify the region: what is it, what is its cause, how does it evolve, how long does it persist, and so on. The aim in many disciplines is to study the evolution and essential dynamics of these amorphous regions or features and describe them for modified time periods, thus obtaining a deeper understanding of the observed phenomena or a reduced-and-simpler model of the original set of complex equations. For example, one tracks the progression of a storm for weather prediction, the change in the ozone "hole" for knowledge about the environment, or the movement of air over an aircraft or automobile for better design.

The main concentration of the visualization community has been on rendering algorithms for 3D data sets. Feature based methods are not as common (that is, integrating scientific visualization and computer vision), mainly because they are domain-specific. However, there are some generic techniques which are highlighted in this chapter.

Features are fundamental to the analysis/visualization process for many reasons:

Reduction of Visual Clutter. With the advent of faster parallel supercomputers and more sophisticated sensing equipment, larger data sets are being generated. These data sets may contain hundreds of small-scale structures which cannot be distinguished from each other with traditional visualization techniques (see for example [26]). Methods to isolate distinct regions will provide better visual cues and enable more detailed study of the phenomena being investigated.

Localized Measurements. Visualization is an aid in the analysis of physical phenomena and is only part of the scientific process. Full study requires measurements and quantification of the observed data fields. Some of these quantifications can be incorporated into visualization routines, enhancing the utility of the resulting pictures and, in many instances, improving the visualization. For example, isosurfaces can be colored by the values within their domains, such as the local maxima, highlighting the more "active" or intense regions of the data set. Other examples are presented in later sections.

Tracking. The reason for collecting or generating time-dependent data sets is to study the evolution of different parts of the field. Tracking routines can be incorporated as a regular part of the visualization process to display regions individually over time and highlight interesting interactions.

Juxtaposition. Simulations are often verified and compared to similar experiments or other simulations. Some comparisons can be performed accurately while others require a more qualitative approach. Juxtaposition of major features and their topologies is an essential part of this process.

Data Reduction. Focusing on distinct regions and only displaying or saving them performs a type of *data reduction* which can reduce storage space and postprocessing time. Hierarchies of different data *abstractions* can also be used for data reduction.

Database Management. Databases of features can be built, and sophisticated queries can be supported once features are identified and *classified*. Storage and retrieval of data sets based upon content would be a big boost to massive data analysis and dissemination.

Related Work

Feature based tools are typical in 2D image processing and computer vision [1, 10]. There has been some work in feature extraction in 3D especially for medical imaging. Many of the techniques used are applicable to other scientific domains. However, the majority of the literature in scientific visualization in concerned with producing better images instead of with producing information for further analysis. For research on feature extraction in particular domains see [5, 12, 15, 11, 23, 20, 18, 14, 2].

Tracking objects in a series of two-dimensional images is widely dealt with in computer vision [1] (*motion tracking* and *optical flow*). The major issue is to find a particular feature in a series of consecutive frames. The process of matching an object in one frame to one in another frame is called the *correspondence problem*. The objects are generally matched using a range of attributes such as centroids, moments, gradients, geometry, pixel values, and so forth.

Tracking objects is common in the field of oceanography, where remote observations are continually being generated. For example, in [6, 9], clouds are tracked by calculating attributes of the clouds (such as centroid and area) and searching for matches in the next data set. In fluid dynamics, tracking has been performed on vortex tubes by using skeletons as in [24]. Regions were located from one data set to the next by searching about windows in the direction of the object's velocity.

In this chapter, we describe our motivation and efforts to extract, quantify, and track coherent amorphous regions from three-dimensional scalar (and vector) fields [17]. The data sets used for demonstrations are from ongoing research in computational fluid dynamics. The domain is described in the next section. However, the results and methods are general and can be applied to many other disciplines with continuous scalar and vector fields such as oceanography, chemistry, materials science, and meteorology.

12.2 Example Domain

The data sets are obtained from ongoing studies of 3D *vortex dynamics*[1]. The generation, interaction, and dispersal or mixing of vorticity[2] plays a profound role in a wide class of practical and fundamental fluid flows. "Vortex dynamics" is the motion of fluids at high Reynolds number that are dominated by localized concentrations of vorticity or *coherent vortex structures* (CVS) embedded in a nearly passive/incoherent background (or sea) of intermixed and distributed vorticity. This is important in aircraft wake control, combustion, chemical species mixing, atmospheric/oceanographic species dispersal (hurricanes, tornadoes, polar vortex or ozone holes), combustion, process control, and so on.

One of the fundamental interactions in compressible hydrodynamics is between a shock wave and a density inhomogeneity. The physical situation may be characterized by a shock wave propagating through a fluid of density, ρ_1, striking a contact interface and passing into a region of density, ρ_2 (see Figure 12.1). The governing equations are the compressible Euler equations. The physical processes can be divided into two phases: a rapid vorticity deposition phase due to the baroclinicity, and a vorticity evolution phase during which the interface is characterized by the presence of CVS. Typically, the late time evolution of the fluid interface is in the turbulent regime. Quantification of properties (such as circulation, area, and centroid) of the CVS and their interactions with each other are essential for a physical understanding and development of reduced models of the ensuing turbulent mixing phase. (For further details about this problem see [16].)

An interface separates two gases (Air and R22 in this case) which is accelerated by a shock of strength $M = 2.0$. In this case, the grid was uniform $254 \times 62 \times 62$ ($\Delta x = \Delta y =$

[1] These simulations were performed by R. Samtaney and N. Zabusky, Rutgers University.

[2] Vorticity is defined as $\omega = \nabla \times u$, where u is the velocity field.

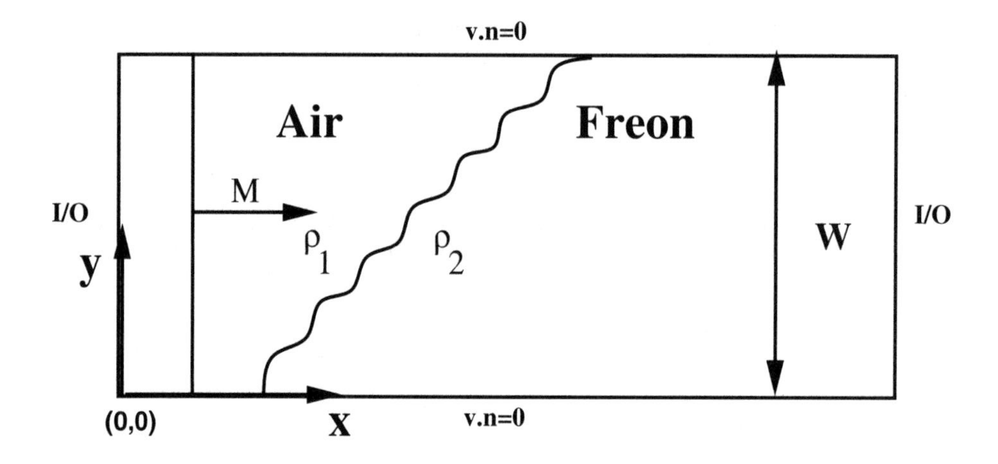

Figure 12.1: Schematic of the shock-interface problem

$\Delta z = 1$) (the simulation parameters are $x_0 = 60.5$, $\alpha = 45^o$, $A = 10.0$, and $\lambda = 62$). The simulation was terminated at $t_{end} = 300$ (that is, for 300 time steps).

In Figure 12.2, vorticity isosurface (magnitude $\omega_{th} = 0.15$) is shown for times *t= (a) 40, (b) 100, (c) 150, (d) 200, (e) 250, (f) 300.* At $t = 40$, the shock has just traversed the interface. The dominant vorticity on the interface is negative in the z direction and positive (negative) in the y direction [16].

In Figure 12.4, a 128^3 data set from a pseudospectral simulation of coherent turbulent vortex structures is shown. The variable being visualized is vorticity magnitude. The simulation involved 100 timesteps. (The simulation is courtesy of V. Fernandez, Rutgers University.)

12.3 Feature Extraction

Figure 12.2 demonstrates the need for feature extraction and tracking. Many of the *interesting* regions (such as the ring vortex) are occluded by other regions. Furthermore, the visualization alone does not provide any measurements, such as volume and mass. When analyzing 300 of such data sets the problem grows and it is very hard to even see and track visually distinct regions (this is especially true when the number of regions is very large as in [26]). It would be helpful to focus on the evolution of particular regions (with their changing volume and strength). An example of this is shown in Figure 12.3. Two views (top-XZ plane and front-XY plane) of the lower wall vortex are shown. This region was extracted from the series using our feature extraction routines and then displayed in a timeline.

The regions were extracted using a *region growing* algorithm to segment the entire data set based upon a particular threshold value and a connectivity. A more detailed description is presented in [20]. Multidimensional thresholding can also be used, that is, segmenting the data set based upon more than one variable, such as the direction of the vector field and the scalar value. After segmentation, the regions above the threshold (which are stored as a set of connected nodes) are *quantified*, and the attributes of these regions are computed, including, mass, moments, volume, and so on. Bounding surfaces are also computed. These will resemble the isosurface when multidimensional thresholding is not performed. Note, the segmentation effectively separates the different connected components of the "isosurface." Each separate region can be considered an *object* or feature and can be colored by any one of the quantifications computed.

In Figure 12.3, the lower wall vortex exhibits an interesting shape. The upstream part of this vortex is in the form of a ring and there is a stem attached to this ring which arises at the $z = \lambda$ plane, curves upstream, runs through the ring, and finally curves downstream and is attached to the $z = 0$ plane. At $t = 150$, observe that the vortex ring develops a "notch" due to the strain induced by the stem vortex. This results in an antiparallel configuration of vortex lines. The stems of the vortex ring are nearly aligned in the x direction. If the tails of the vortex are ignored, the "hairpin-like" shape of the vortex is clear at $t = 200$. "Hairpin-like" coherent structures are of importance in turbulent flows. Isolating this coherent feature from the entire flow field helps us better understand topological aspects and the evolution of these types of coherent structures.

In these examples, the data sets are rectilinear (regular), so an octree data structure can be used to aid in the segmentation process and enable quick transitions between different threshold values and other criteria [7, 25, 8]. In the case of irregular data sets, the same segmentation can be performed in the computational domain and then the output (that is, bounding surface) can be transformed to the physical domain. During segmentation, extra connectivity criteria are needed to account for periodic boundary conditions. The octree can be updated to reflect this so that a correct region total and segmentation results. The case for unstructured data is more complex [3].

12.4 Abstract Representations

While bounding surfaces are an effective method to represent regions, there are other types of *abstractions* which may be beneficial. An abstraction can be thought of as a simple representation of a feature, much like an icon. For example, the simplest representation of a region in a data set is a point, which could be either the weighted centroid of the region or a local maxima. (Both of these values can be computed during the segmentation process [20].) Abstractions are useful for both data compaction and data analysis.

Because of the massive amount of data being generated and visualized, data compression techniques are becoming increasingly important to reduce the local storage requirements and to ease browsing large data sites. Compression is a preanalysis routine and is generally done on a global basis even though much of the field may not be of prime interest. Most compression routines are also hierarchical so that different storage/lossey requirements can be satisfied. The original data set can usually be retrieved from tape.

Abstraction can also be used for data compaction. In this case each region is compressed separately into different hierarchical categories. In the aforementioned example, a point is the extreme abstraction of a feature (a region/feature is just a subset of the data set). A bounding surface of a region can also be thought of as an abstraction because it represents the shape of the outside (but not the distribution of values on the inside). Other

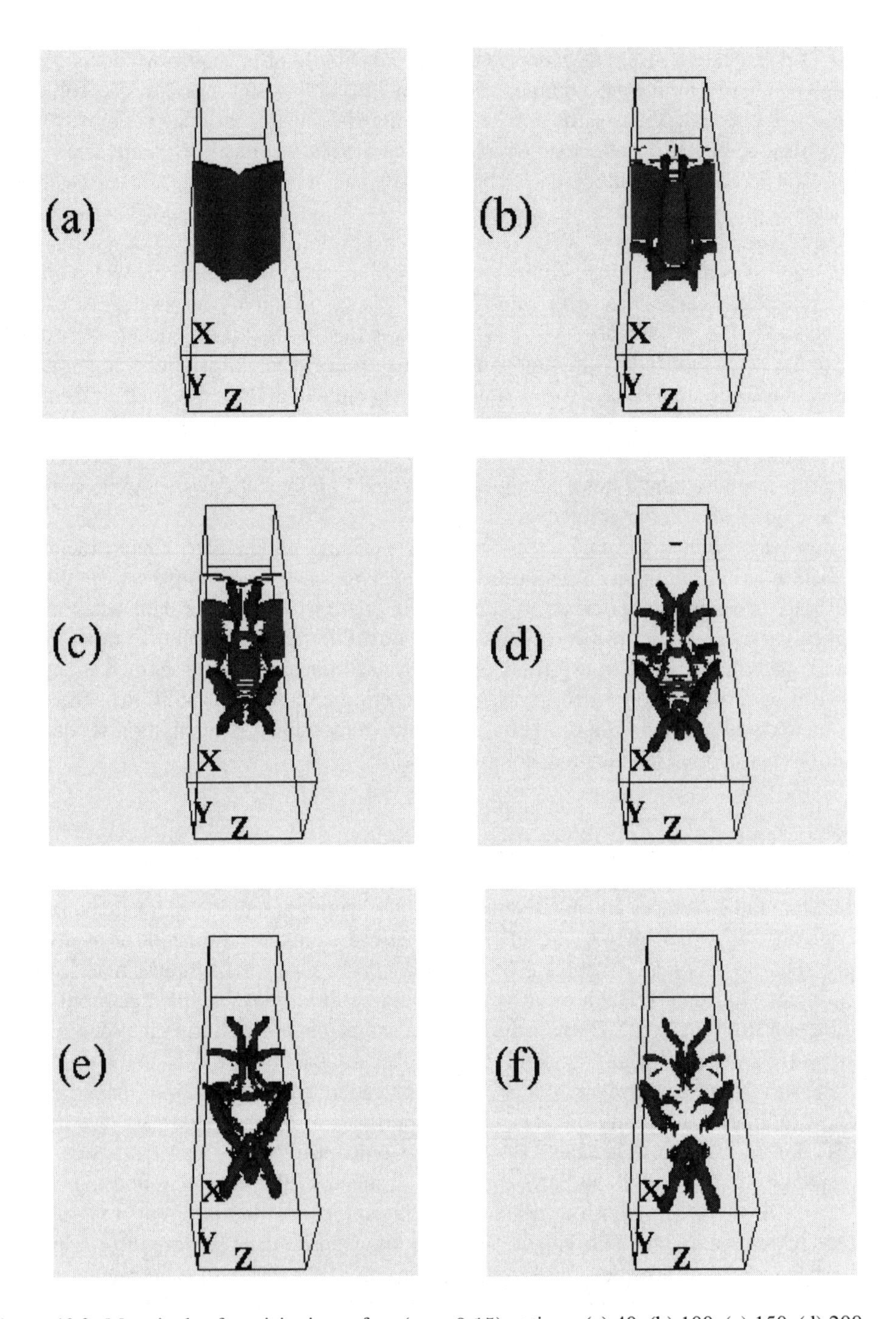

Figure 12.2: Magnitude of vorticity isosurface ($\omega = 0.15$) at times (a) 40, (b) 100, (c) 150, (d) 200, (e) 250 and (f) 300. Air-R22 interface, $M = 2.0$, $\alpha = 45^{\circ}$, $A/\lambda = 10/62$.

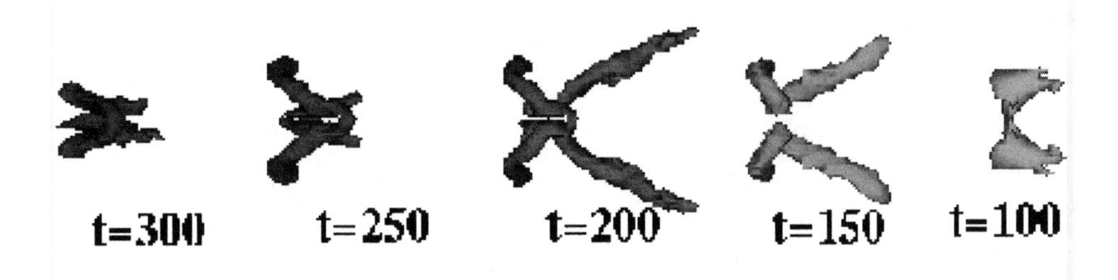

Figure 12.3: Time sequence of lower wall coherent structure, front view (XY plane). Air-R22 interface, $M = 2.0$, $\alpha = 45^o$, $A/\lambda = 10/62$.

abstractions include geometric entities such as ellipsoids, tubes, sheets, skeletal lines, and reduced complexity surfaces [14].

Ellipsoids (or ellipses in 2D) provide a simple representation for blob-like regions, highlighting the central axis and orientation. The ellipsoid's origin is at the centroid of the region ($\overline{x_k} = W^{-1} \int_\Omega w(\mathbf{x})\, x_k \, d\Omega$ for $k = 1, 2, \ldots, d$, and d is the dimension of the field. The mass (or more accurately, integrated sum) is defined as $W = \int_\Omega w(\mathbf{x})\, d\Omega$ where $\mathbf{x} = (x_1, x_2, \ldots, x_d)$, $w(\mathbf{x})$ is the scalar value at the node $\mathbf{x}$, and $\Omega = \{\mathbf{x} | \omega(\mathbf{x}) > T_\omega\}$ for a particular threshold value T_ω). The axes are defined from the eigenvalues and eigenvectors of the tensor of second moments and they are normalized with respect to the volume of the region [21, 14]. In Figure 12.4, ellipsoids are fit to segmented regions from a 128^3 scalar data set of vorticity magnitude. The ellipsoids are fit to regions at a threshold value of 38 percent of the maximum, and they are colored by the interior local maxima within each region [17, 18]. (The isosurfaces are shown on the left side for reference.)

While ellipsoids do not accurately capture the essence of all regions (that is, regions with many branches), they provide essential information concisely. For example, flat sheet-like or pancake-like regions are immediately noticed from the aspect ratio of the ellipsoid axes. Ellipsoids are a very compact representation of a region, requiring only twelve floating-point values, but still containing quantifiable information such as volume and mass.

Another useful abstraction is the skeleton or medial axis. This is especially helpful for abstracting tubes or finger-like regions. The medial axis can be generated from the shape of the region by using dilation and distance transform techniques (from Computer Vision [1]). Also useful are value based skeletal representations, such as "local-maxima" lines or vortex cores [22, 14]. An example is shown in Figure 12.5. The medial lines in this figure were computed by searching for local maxima in a particular direction.

In this abstraction hierarchy, the maxima/centroid value are at the bottom of the tree followed by ellipsoids or other simple shapes, skeletal lines, higher moments, bounding surfaces of varying accuracy (for example, there are different methods available to smooth triangular surfaces and reduce the number of triangles produced by the isosurface routines [13]), followed by the subset of nodes which comprise the regions. The surrounding sea between regions can also be represented hierarchically, either using standard compression techniques or, more quantitatively, using statistical methods. A major advantage is

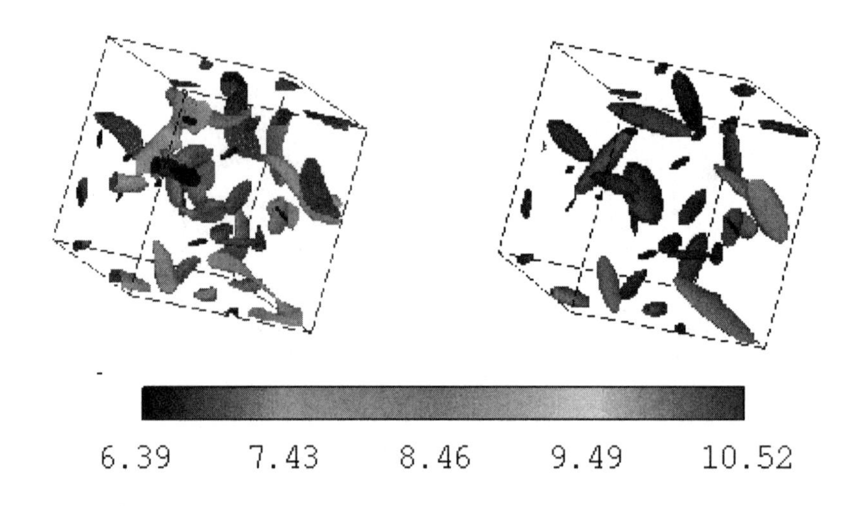

Figure 12.4: Ellipsoid fitting: Ellipsoids are fit to regions segmented from a vorticity magnitude data set. Regions are colored by their interior local maxima. See Color Plate 83.

that information is available at each level for quantitative analysis and not just visual inspection. For example, the ellipsoids provide the orientation of the region from which angles/distances between regions can be measured, skeletons allow distance measure as well as curvature, and so forth. In addition, these abstractions are useful for formulating reduced models: ellipses can be found in the analyses of vortex mergers and binding [4], and skeletal lines are useful in the Biot-Savart integral representation which assumes that all the vorticity is concentrated along a line ("filament" or "string") as opposed to being space filling. Simulations of this reduced model can then be juxtaposed with the more complex model [27].

12.5 Tracking and Correlation

Once objects are extracted they can be tracked. Tracking involves correlating the extracted objects from one data set to the extracted objects in a subsequent data set. Tracking can also be used to help locate a particular object in the next data set by constructing a search window. The tracking algorithm assumes continuation between data sets and small variations in movement and change (sampling frequency).

The evolutionary events of observed phenomena can be characterized in one of the following ways: it may continue with possible shape changes, it may dissipate completely, it may bifurcate into two or more new regions, or it may combine with another region. Furthermore, a new object may be created. These actions are illustrated in Figure 12.6.

The feature extraction and tracking algorithm can be coupled with the simulation in a postprocessing mode. The different modes will affect the amount of storage and the amount

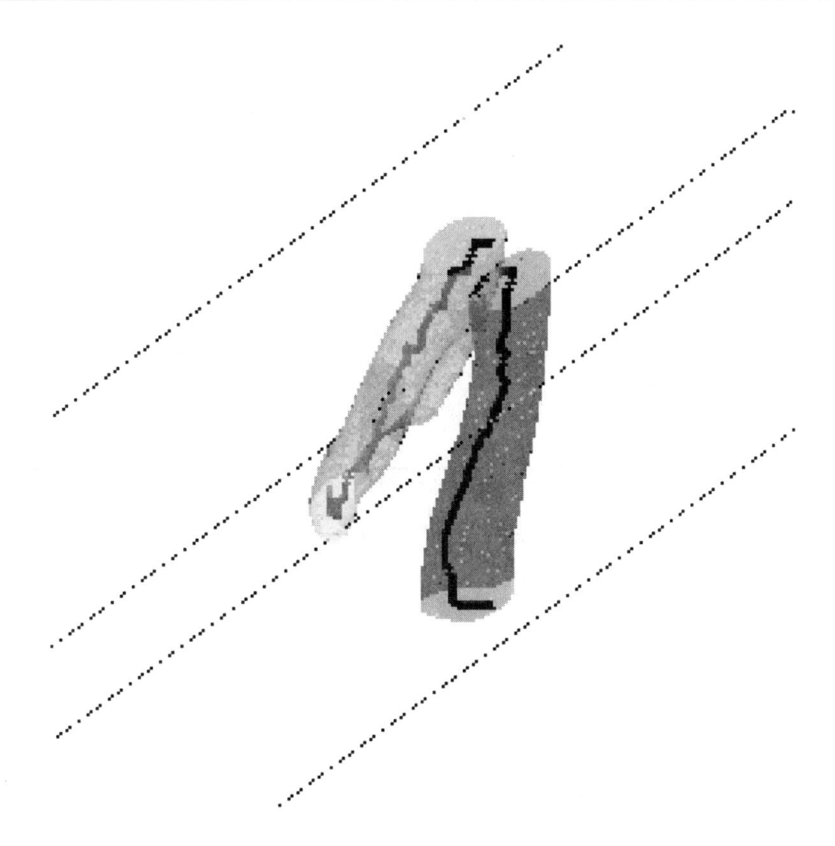

Figure 12.5: Skeleton using local maxima lines and vortex direction. See Color Plate 84.

of information maintained for tracking. Since the actual objects may consume a great deal of storage, it is sometimes more convenient to save only the computed attributes, such as centroid, mass, volume, and so on. These values can accurately capture the essence of the regions extracted.

After the tracking procedure, the history of each object is known. Different representations can be used to display this information. For example, the histories of objects in Figure 12.6 can be characterized as follows (note that creation and dissipation are not included here):

a,b: $1(a) \rightarrow 1(b), 2(a) \rightarrow 2(b) + 3(b), 3(a) + 4(a) \rightarrow 4(b);$

b,c: $2(b) \rightarrow 2(c), 3(b) \rightarrow 1(c), 4(b) \rightarrow 3(c) + 4(c);$

This information can also be represented as a directed acyclic graph, DAG, as shown in Figure 12.7. In both of these methods, a legend of the object names and their attributes is needed for complete understanding (the file used by the tracking program stores the object legends).

Tracking has been performed on the data sets of Figure 12.2 [17]. In Figure 12.8, the "hairpin-like" vortex has been isolated, tracked, and shaded in five time steps between $t = 200, 600$. A recent improved tracking algorithm is described in [19].

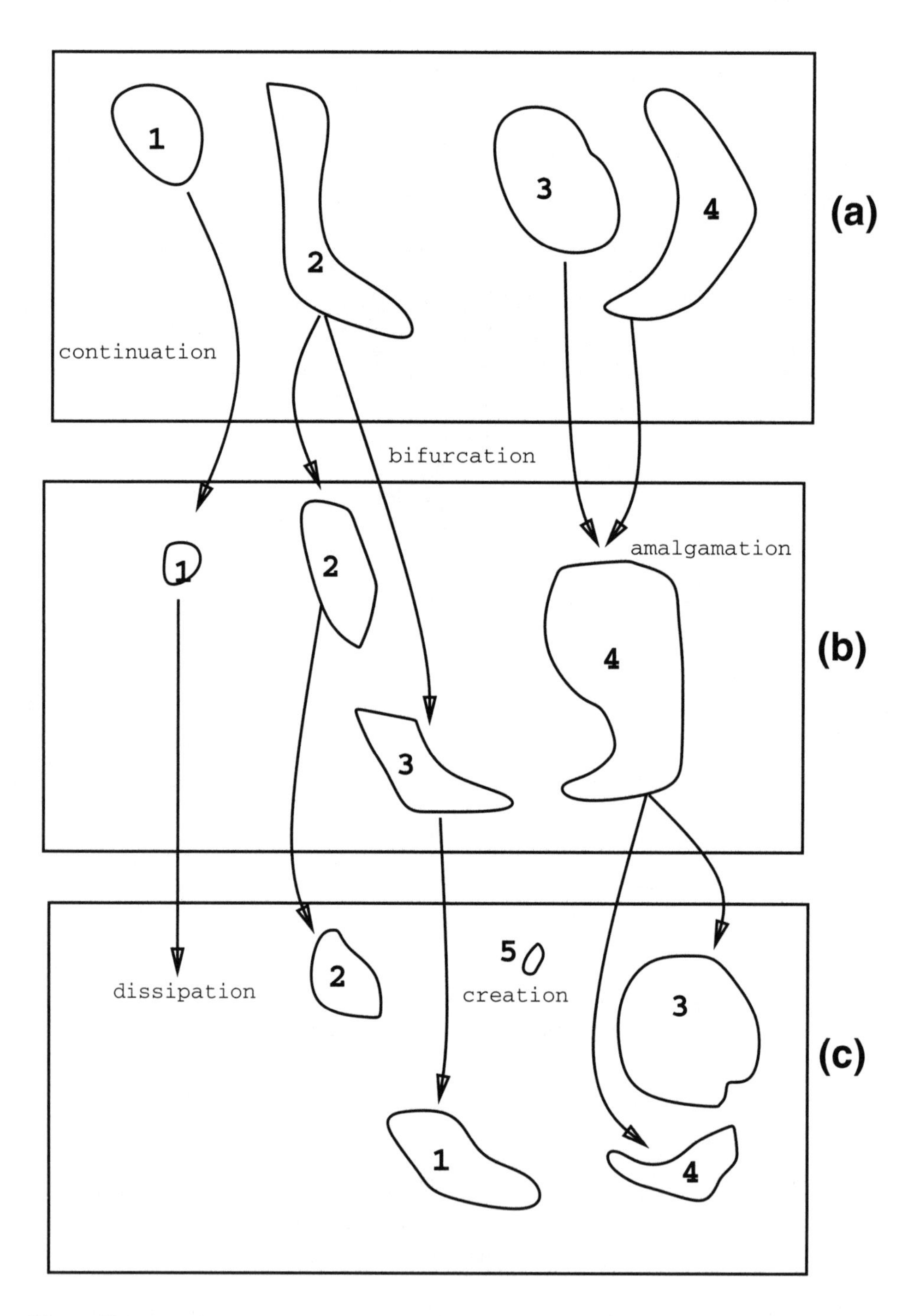

Figure 12.6: Tracking interactions: continuation, creation, dissipation, bifurcation, and amalgamation.

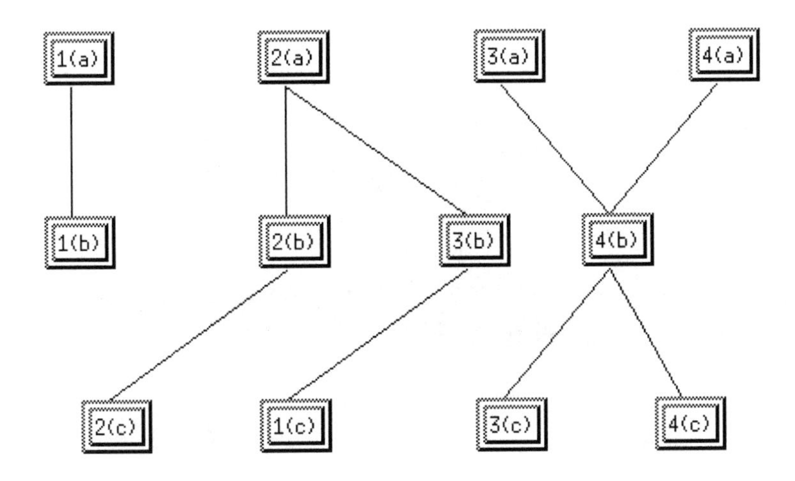

Figure 12.7: Graphical history of the evolution of objects

12.6 Analysis and Future Work

One of our goals in feature extraction and tracking is to classify structures and identify events. Once objects are isolated, they can be catalogued in a database for later use. One can envision a sophisticated database for scientific applications where events found in one simulation can be searched for in others and then automatically rendered. Event searching is particularly useful when simulations involve hundreds or thousands of time steps. One such event we are interested in is *reconnection*. Loosely defined, reconnection is the rapid collapse of antiparallel vortex flux tubes, the dissipation and breaking of vortex lines in each tube, and the subsequent connection of the vortex lines in one tube with those in the other. Vortex reconnection has been connected with intermittency in turbulence, finite-time singularities in Euler equations, noise in jets, and in three-dimensional shock-interface problems. Many reconnection events may be present over a large time-dependent simulation. Tracking scalar reconnection is one step in identifying locations for further investigation (with vortex lines).

Large data sets also demonstrate the need for a more thorough melding of the visualization routines with the ongoing simulation. This would cut down on storage requirements and redundant loading of data into memory, and enable larger and more complex features to be tracked and quantified. There is also the need to parallelize many of the feature extraction/visualization routines so they can easily be called by simulation processes during execution.

The ultimate goal of visualization is to aid in the understanding and analysis of data. In this chapter, we have presented a feature-oriented approach to visualization and have demonstrated basic feature extraction and tracking algorithms. These methods greatly enhance and simplify visualization and provide the tools for the management and understanding of massive data sets.

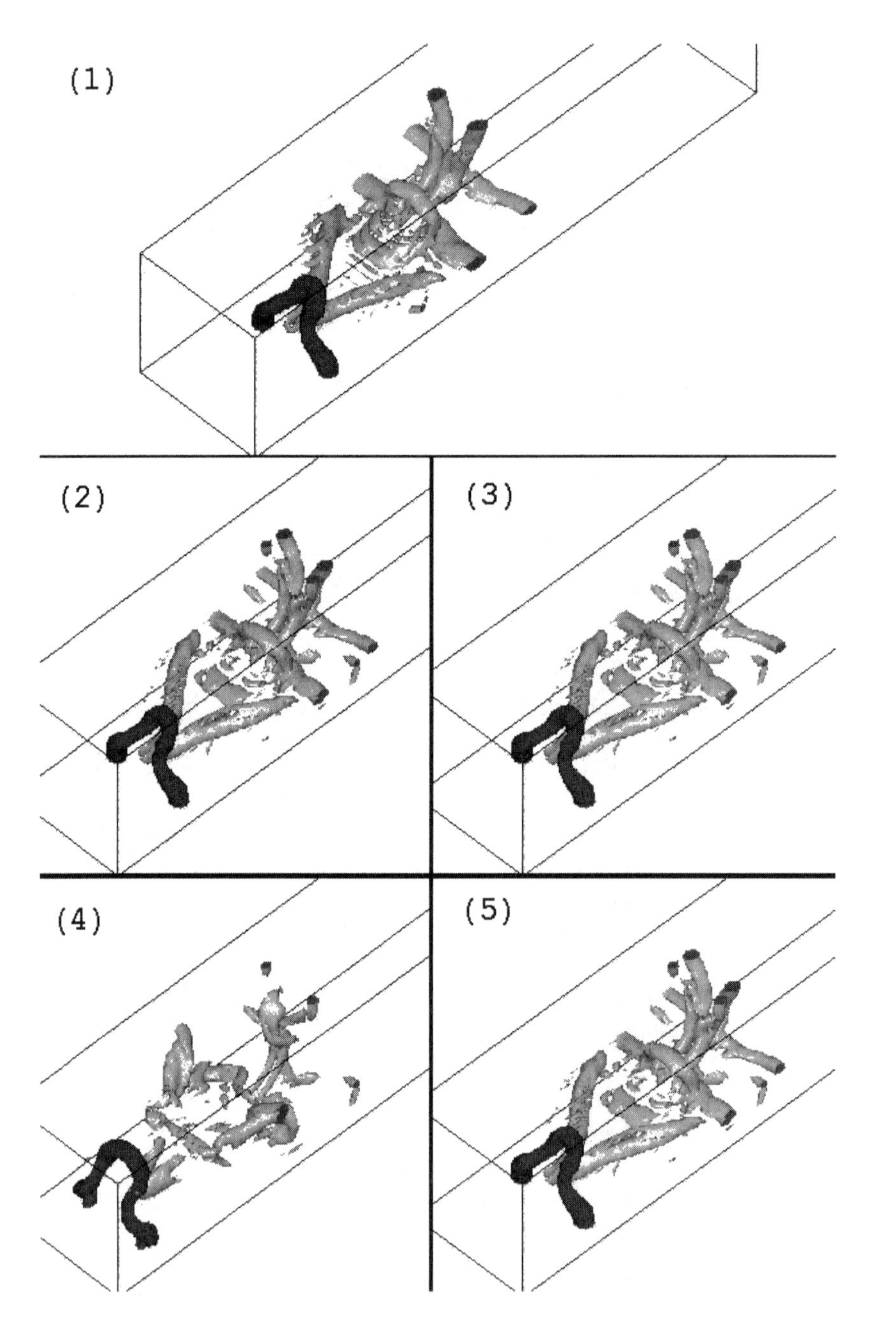

Figure 12.8: The hairpin-like vortex has been isolated, tracked, and shaded to highlight its topology. See Color Plate 85.

Acknowledgments

This work is part of an ongoing research effort in visualization and quantification at the Lab for Visiometrics and Modeling at Rutgers University. Special thanks go to R. Samtaney, V. Fernandez, and N. Zabusky for providing the simulations and some of the renderings. These computations were done on the CM-5 at NCSA [16] and ACL-LANL. The Lab gratefully acknowledges the support of the NASA Ames Research Center (NAG 2-829), ARPA HPCD DABT-63-93-C-0064, DOE DE-FG02-93ER25179.A000, and the CAIP (Computer Aids for Industrial Productivity) Center, Rutgers University. MPEG movies demonstrating some of the concepts in this paper are available over our World Wide Web site: http://www.caip.rutgers.edu/vizlab.html.

Bibliography

[1] D. Ballard and C. Brown, *Computer Vision*, Prentice-Hall, Inc., Englewood Cliffs, N.J., 1982.

[2] David Banks and Bart Singer, "Vortex Tubes in Turbulent Flows: Identification, Representation, Reconstruction," *Proceedings Visualization '94*, IEEE Computer Society Press, 1994, pp. 132–139.

[3] S. Bhat, *Segmentation of Unstructured Datasets*, M.S. thesis, Rutgers University, Dept. of Electrical and Computer Engineering, New Brunswick, N.J., 1996.

[4] J.D. Buntine and D.I. Pullin, "Merger and Cancellation of Strained Vortices," *J. Fluid Mech.*, Vol. 205, 1989, pp. 263–295.

[5] H. Cline, W. Lorensen, R. Herfkens, G. Johnson, and G. Clover, "Vascular Morphology by Three-Dimensional Magnetic Resonance Imaging," *Magnetic Resonance Imaging*, Vol. 1, 1989, pp. 45–54.

[6] M. Desbois, Y. Arnaud, and J. Maizi, "Automatic Tracking and Characterization of African Convective Systems on Meteosat Pictures," *American Meteorological Society*, May 1992.

[7] M. Gao, *Data Extraction And Abstraction In 3D Visualization*, M.S. thesis, Rutgers University, Dept. of Electrical and Computer Engineering, 1992.

[8] A. Globus, "Octree Optimization," *Symposium on Electronic Imaging Science and Technology*, SPIE/SPSE, 1991.

[9] J. Holland and X. Yan, "Ocean Thermal Feature Recognition, Discrimination, and Tracking Using Infrared Satellite Imagery," *IEEE Transactions On Geoscience and Remote Sensing*, Vol. 30, No. 5, Sep. 1992, pp. 1046–1053.

[10] B. Jahne, *Digital Image Processing*, Springer-Verlag, 1991.

[11] K. Ma, M. Cohen, and J. Painter, "Volume Seeds: A Volume Exploration Technique," *The Journal of Visualization and Computer Animation*, Vol. 4, No. 2, 1991, pp. 135–140.

[12] J. Miller, D. Breen, W. Lorensen, R. O'Bara, and M. Wozny, "Geometrically Deformed Models: A Method for Extracting Closed Geometric Models from Volumes," *Computer Graphics*, Vol. 4, July 1991, pp. 217–226.

[13] P. Ning and J. Bloomenthal, "An Evaluation of Implicit Surface Tilers," *Computer Graphics*, Vol. 13, No. 6, 1993, pp. 33–41.

[14] F. Post, T. van Walsum, F.H. Post, and D. Silver, "Iconic Techniques for Feature Visualization," *Proceedings of IEEE Visualization '95*, Atlanta, Ga., Oct. 1995, pp. 288–295.

[15] X. Qu and Davis, "An Extended Cuberille Model for Identification and Display of 3D Objects from 3D Gray Value Data," *Proceedings of Graphics Interface '92*, May 1992.

[16] R. Samtaney, *Vorticity in Shock-Accelerated Density-Stratified Interfaces: An Analytical and Computational Study*, Ph.D. thesis, Rutgers University, 1993.

[17] R. Samtaney, D. Silver, N. Zabusky, and J. Cao, "Visualizing Features and Tracking their Evolution," *IEEE Computer*, Vol. 27, No. 7, July 1994, pp. 20–27.

[18] D. Silver, "Object-Oriented Visualization," *IEEE Computer Graphics and Applications*, Vol. 15, No. 3, May 1995.

[19] D. Silver and X. Wang, "Volume Tracking," to appear in *Proceedings of IEEE Visualization '96*, San Francisco, Calif., Oct. 1996.

[20] D. Silver and N. Zabusky, "Quantifying Visualizations for Reduced Modeling in Nonlinear Science: Extracting Structures from Data Sets," *Journal of Visual Communication and Image Representation*, Vol. 4, No. 1, Mar. 1993, pp. 46–61.

[21] D. Silver, N.J. Zabusky, V. Fernandez, M. Gao, and R. Samtaney, "Ellipsoidal Quantification of Evolving Phenomena," *Visualization of Physical Phenomena. Proceedings of Computer Graphics International '91 Symposium*, N.M. Patrikalakis, editor, Springer-Verlag, June 1991, pp. 573–588.

[22] B. Singer and D. Banks, *Predictor-Corrector Scheme for Vortex Identification*, Technical Report, NASA Langley, Mar. 1993.

[23] J. Udupa, "Applications of Digital Topology in Medical Three-Dimensional Imaging," *Topology and its Applications*, Vol. 46, 1992, pp. 181–197.

[24] J. Villasenor and A. Vincent, "An Algorithm for Space Recognition and Time Tracking of Vorticity Tubes in Turbulence," *CVGIP: Image Understanding*, Vol. 55, No. 1, Jan. 1992, pp. 27–35.

[25] J. Wilhelms and A. Van Gelder, "Octrees for Faster Isosurface Generation," *ACM Transactions on Graphics*, Vol. 11, No. 3, July 1992, pp. 201–227.

[26] P. Woodward, "Interactive Scientific Visualization of Fluid Flow," *IEEE Computer*, Vol. 26, No. 10, Oct. 1993.

[27] N. Zabusky, D. Silver, R. Pelz, and Vizgroup '93, "Visiometrics, Juxtaposition and Modeling," *Physics Today*, Vol. 46, No. 3, Mar. 1993, pp. 24–31.

Chapter 13

Flow Surface Probes for Vector Field Visualization

Cláudio Silva, Lichan Hong, and Arie Kaufman

Abstract. *We present a technique called flow surface probes for interactive visualization of 3D steady vector fields. The user places an arbitrarily-shaped seed surface in the flow field, which is then deformed over time by the local or global 3D vector field, producing an animation sequence. This sequence and individual frames are used to probe the behavior of the flow field.*

13.1 Introduction

In 2D fluid flow visualization, techniques such as stream lines, arrow plots, and the like are commonly used to visualize vector fields with satisfactory results. However, in the 3D case, they usually generate cluttering displays that convey little, if any, useful information. As pointed out by Hesselink, Post, and van Wijk [9], the problem seems to be a fundamental one, as there is no intuitive and psychologically meaningful representation of 3D vectors. Even though we can use arrows to represent single vectors, there is no such simple metaphor for 3D vector fields. Without any obvious visual representation for 3D vector data, vector fields become difficult to understand. Even more difficult is the visualization of tensor fields, which are much more complex and abstract entities.

In some applications, it might be desirable to study the steady vector field defined on an arbitrarily shaped surface. For instance, the user may wish to visualize the wind velocities at a constant distance from a plane wing, that is, the vector field defined on a surface wrapping around the wing. A different kind of visualization occurs when the user wants to observe how points defined over a surface are continuously advected by the vector field. For instance, one may want to see how a flag changes its shape with respect to a 3D vector field defining wind.

In this chapter a new technique called *flow surface probes* is presented to address both of the above problems. Using our technique, the user places an arbitrarily shaped seed surface in the flow field by specifying surface parameter values and locations. Then, the surface can be continuously deformed over time by the vector field defined on the seed surface, which shows the behavior of the local vector field at the seed location. Alternatively, the seed surface can be deformed over time by the vector field as the surface moves through the flow, which provides an understanding of the global structure of the steady field. Therefore, through a sequence of interactive animations, the user can obtain a better understanding of the steady vector field. Our basic assumption is that surfaces are good visualization objects, as field structure can be mapped to shape, and the human eye is well tuned to the task of deriving shape from surface shading as determined by light reflection.

Currently, we have only tried this technique on steady vector fields; probing of unsteady dynamic vector fields should also be possible with this approach. However, for brevity of the description and examples, we have limited this chapter to steady vector fields. In Section 13.2 we review some of the related approaches to 3D vector field visualization. In Section 13.3 we present the definitions and mathematical background, followed by implementation notes in Section 13.4. Then, in Section 13.5 we describe several examples of different flow surface probes.

13.2 Related Work

In order to overcome the problems intrinsic to visualizing 3D vector and tensor fields, many techniques and systems have been developed. The simplest one is the contraction technique [8], which reduces the vector field into some scalar field, such as the vector magnitude, and then renders the scalar field with well-known visualization techniques, such as cutting plane, isosurface, or volume rendering. A problem intrinsic to this approach is that the directional information inherent in the vector field is unavoidably lost. With arrow plots (or hedgehogs) [12], one can represent the direction of a vector by the direction of an arrow, and the magnitude of the vector by the length of the arrow. Unfortunately, this method brings problems such as directional ambiguity and cluttering effects. We classify these techniques as of the *first kind*, that is, they predominantly try to visualize some localized part of the vector field.

Stream lines [2], stream surfaces [10, 15], time surfaces [15], and stream polygons [14] are useful in understanding global properties of the vector field. A stream line is a curve everywhere tangent to the vector field. Usually, the user specifies a seed point somewhere in the flow, and the seed point is advected along the trajectories of the velocity field by solving the following differential equation:

$$\frac{d\vec{x}}{ds} = \vec{v} \quad \text{or} \quad \frac{dx_i}{ds} = v_i(x_1, x_2, x_3, t) \tag{13.1}$$

where $\vec{v}$ is the velocity field, $\vec{x}$ is the position, t is time (for steady fields t is constant), and s is simply a parameter along the streamline. It is important not to confuse s with time. If t is constant, the resulting curves are stream lines at the instant t (see [1] for details).

Similarly, starting with a seed curve $\vec{\alpha}(t_1)$, a stream surface $\vec{r}(t_1, t_2)$ can be generated by solving the stream line equation for each point in the seed curve. Curves $\vec{r}(Constant, t_2)$ are stream lines and $\vec{r}(t_1, Constant)$ are time lines [15]. Also, by sweeping a stream polygon through the flow and changing the shape of the stream polygon with physically meaningful parameters like rigid body rotation and shear strain, one can obtain more information about the flow field [14]. A time surface [15] can be viewed as the distribution of a set of boundary points after being advected for time t, and it can be generated by isosurface rendering. An unfortunate limitation of using isosurfaces to generate time surfaces is speed, making it difficult to obtain real-time animations. We classify these techniques as of the *second kind*, as they try to convey global information of the vector field.

Some other techniques also try to visualize tensor data. For instance, de Leeuw and van Wijk [3] developed an interesting gadget for local flow field visualization. Their method basically transforms the vector field to a local coordinate frame, then decomposes the tensor into information parallel to the flow (acceleration, shear, curvature) and perpendicular to the flow (torsion, convergence). Finally, they map each property to a different graphical representation. The stream polygon technique also makes use of tensor decomposition. By mapping tensor parameters onto a stream polygon when it is swept through the flow, interesting tensor information is conveyed. Other methods also exist; for instance, Delmarcelle and Hesselink [4] use tensor decomposition techniques to visualize symmetric tensor fields.

The flow surface probes technique we present in this chapter can be classified as of both the first and the second kind of visualization techniques. When deformed by the local vector field on the seed surface, the flow surface can be used as a complementary probe to hedgehogs. Alternatively, when deformed by the vector field as the surface moves through the flow field, the flow surface can be used to generate time surfaces interactively, and thus assists the user in probing the global structure of the vector field. Our motivation for developing this tool was to solve the ambiguity and cluttering problems with current techniques of the first kind, and at the same time to allow the user to generate general time surfaces interactively.

13.3 Flow Surface Probes

The concept of flow surface probes is rather intuitive and can be used to visualize a number of important aspects of a vector field. This technique can be viewed as a complementary tool for other vector field visualization methods. In order to help the user understand the behavior of the vector field, we use an arbitrarily shaped (user-defined) surface to extract the section or region of interest, and we deform the surface by the vector field over time. The user uses both single frames and complete animations of the deformation, to augment his knowledge of the vector field.

Our formalization of this technique uses regular curves and surfaces [5]. The main reason for the "regular" requirement is to be able to compute tangents and normals, and to guarantee that the surfaces are continuous and not self-intersecting. We need these conditions not only for shading purposes but also in our advection calculations.

We first define a *seed surface* as a regular parameterized surface $X_0(u, v)$ in R^3, where, for convenience, we use rectangular $I \times J$ parameterization patches for it. A *deformation path* is a regular parameterized curve $\vec{\alpha}(t)$ in R^3. In order to deform the seed surface we need a family of such paths, with the further constraint that each curve "start" on the seed surface. That is, we need a family of $\vec{\alpha}_{(u,v)}(t)$, where for fixed u and v each $\vec{\alpha}_{(u,v)}(t)$ is a regular parameterized curve, such that $\vec{\alpha}_{(u,v)}(0) = X_0(u,v)$, for $\forall u \in I, \forall v \in J$.

The flow surface $X_t(u,v)$ at animation time step $t \in R$ is defined as the "movement" of $X_0(u,v)$ when it is being "dragged" by the family of deformation paths. ("Animation time" is used here to avoid confusion with an unsteady field where the field itself changes over time.) Flow surfaces have the following simple form in terms of deformation paths:

$$X_t(u,v) = \vec{\alpha}_{(u,v)}(t), \quad u \in I, \quad v \in J, \quad t \in R \tag{13.2}$$

Notice that $X_0(u,v)$ is, by definition, the seed surface and our "dragging" operation does not permit slipping. Also note, for $X_t(u,v)$ to be a surface in general, some further constraints would have to be imposed on the deformation paths. Specifically, for $X_t(u,v)$ to have a tangent at every point, $|\frac{\partial X}{\partial u} \times \frac{\partial X}{\partial v}|$ needs to be nonzero. This creates a requirement that $\alpha_{(u,v)}(t)$ not only be a differentiable function on t, but also on u and v.

This simple formulation does not imply that one set of deformation paths may correctly model parts of the field with critical points (because of the regular point requirements on the surfaces and self-intersection problems). We are currently working on extending our method so that critical points can be detected and the regularity conditions can be kept intact, by breaking down the paths and the surfaces in sets of regular surfaces where the break occurs. For implementation purposes, as the deformation paths actually obtained are linearly interpolated, we always get a consistent flow surface of our deformation paths.

In the process of visualizing a vector field, one can specify not only a set of deformation paths but also a mapping from scalar data at positions $X_t(u,v)$ on the flow surface to colors at every time step, to provide more information about the flow field. In our implementation, the deformation paths are defined either by the local vector field on the seed surface or the dynamic vector field on the flow surface as the surface moves through the field, and the mapping is from the vector magnitudes of the points on the surface to colors. As shown in Section 13.5, the mapping demonstrates a nice property of our technique.

13.4 Implementation Issues

The definitions in the previous section are based on continuous 3D space. To implement the technique, we need to discretize the flow surface, that is, to produce a tessellation of the flow surface with desirable resolution, in order to simulate the continuous 3D surface. Therefore, our algorithm starts with a tessellation of the seed surface as the basis for drawing the flow surface. Currently, we assume that the tessellation is composed of patches formed by a set of points $p_{(u,v)}(t)$, and that each patch can be an arbitrary polygon. At every time step, the algorithm advects the points according to their deformation paths to generate new positions for the patches, which are then colored and drawn on the screen. Depending on the application, the user can specify how long the time step is and for how many time steps he wants to observe the flow surface moving over space. Also, the user can specify a deformation path for every tessellation point of the seed surface. The fact that we do not change the tessellation adaptively as the surface is being advected is a shortcoming of our current implementation; in the future we expect to allow new points to be added or old points deleted from the flow surface as it is being advected [16, 10]. Our methods for tessellations are presented in the next section, followed by the deformations paths.

13.4.1 Surface Tessellation

Generally, for the algorithm to produce accurate images, one needs to generate a tessellation of the seed surface that correctly models the surface. At this time we use the very simple approach of trying to produce uniformly spaced points (or at least approximately uniformly spaced). Because of performance considerations, it is not desirable to generate patches formed by a large number of points. Currently, we use quadrilateral and triangular patches in our implementation. The triangular patches are used for spherical probes, while all the other probes we have implemented so far use quadrilateral patches. It may be computationally very expensive to achieve accurate tessellations with a desired resolution, but we have simplified the problem by using simple tessellation strategies.

We have so far implemented two classes of seed surfaces, spherical surfaces and Bézier surfaces (including planes). Others should not be difficult to add. To create a triangular tessellation of a spherical surface, we start with an octahedron and keep subdividing each face of the octahedron into four by introducing vertices at the middle of every edge and normalizing their magnitudes so that they lie on the spherical surface, until we achieve the desired resolution. This method generates a nice triangular mesh approximating the spherical surface.

One fast and easy way to generate a tessellation is to use a mapping $M(u, v)$ to transform uniformly spaced points from a plane patch $I \times J$ to the seed surface. This method does *not* guarantee that the tessellation has uniformly spaced points. It works quite nicely for surfaces that have low gaussian curvatures, but it does not, in general, produce a good mesh. Even in the case of a cylinder, which has zero gaussian curvature and is locally isometric to a plane, points uniformly spaced on the boundary of the interval in the plane are mapped to the same point on the cylinder (see Figure 13.1). Nonetheless, this method usually generates nice looking quadrilateral tessellations and it is fairly easy to implement.

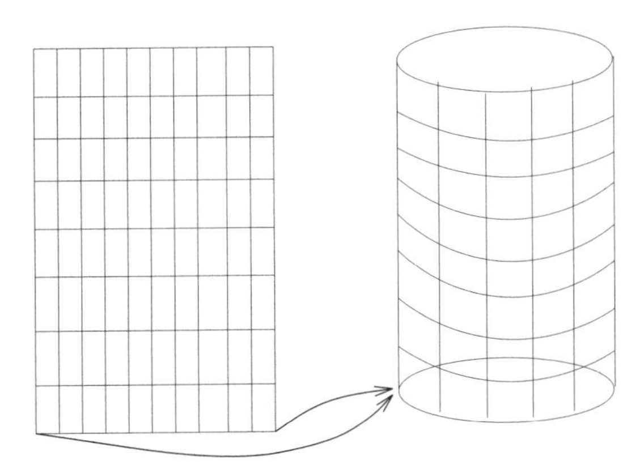

Figure 13.1: Two distinct points in the plane are mapped to the same point of the cylinder

We have used the above method to generate tessellations for tensor product Bézier surfaces [6]. Bézier surfaces are rather general, thus enabling the user to specify a large class of seed surfaces with desired curvatures (for example, one may want to wrap a seed surface around a plane wing). Such surfaces are represented as

$$b^{m,n}(u,v) = \sum_{i=0}^{m}\sum_{j=0}^{n} b_{i,j} B_i^m(u) B_j^n(v) \tag{13.3}$$

where $B_i^m(u) = \binom{n}{i} u^i (1-u)^{n-i}$ are the Bernstein polynomials and $b_{i,j} \in R^3$ are the control points.

13.4.2 Deformation Paths

The deformation paths determine the changes in positions of all the tessellation points on the flow surface. They act on all the points individually, and the surface is constructed by connecting those tessellation points. Theoretically, one can specify a path for each tessellation point of the seed surface. However, for our visualization purpose, we are only interested in two kinds of deformation paths: one is the deformation path defined by the local vector field on the seed surface, the other is the deformation path defined by the global vector field as the surface moves through the flow field.

Our goal in the visualization of a set of local vectors over a surface is to complement the use of hedgehogs. To this end, given any region of interest, the user places a seed surface X_0 in the vector field, and our method will sample the vector field (interpolating when necessary) at the tessellation vertices, that is, find $\vec{v}_{(u,v)}$ at each tessellation point on the seed surface and use $\vec{\alpha}_{(u,v)}(t) = \vec{v}_{(u,v)}t$ as the deformation path. Notice that in this case $\vec{v}$ is sampled only once, at the seed surface. As shown in Section 13.5, this technique does provide more information about the local vector field than the previous hedgehog technique, and it is extremely simple to use.

With this idea, at each time step a new tessellation point is advected by the vector of the corresponding point on the seed surface. That is, in our case, the deformation path of a tessellation point on the seed surface becomes a 3D line segment, which can be defined as:

$$X_t(u,v) = X_0(u,v) + \vec{v}_{(u,v)}t \tag{13.4}$$

where $\vec{v}_{(u,v)}$ is the velocity of point $p_{(u,v)}(0)$ defined on the seed surface and t is the animation time. When a flow surface is advected with this equation we call it a *local probe*.

Moreover, to explore the global property of the vector field, we can apply our technique to construct another version of time surfaces. What we have to modify is our definition of the deformation paths. Instead of being line segments, the deformation paths are defined as the stream lines of the points in the tessellations. In this case, we can take $\vec{\alpha}_{(u,v)}(t)$ as the solution to the following differential equation:

$$\frac{d\vec{\alpha}_{u,v}(t)}{dt} = \vec{v}_{(u,v)} \tag{13.5}$$

where $\vec{v}_{(u,v)}$ depends on the position. To achieve interactive animation, we simply calculate the advection by the first-order Euler integration formula:

$$X_{t+\Delta t}(u, v) = X_t(u, v) + \vec{v}_{(u,v)}\Delta t \tag{13.6}$$

where $\vec{v}_{(u,v)}$ is the velocity of point $p_{(u,v)}(t)$ at the current surface X_t. In this case, as in any other case where the deformation paths depend globally on the field, we call the flow surface a *global probe*. Besides time surfaces, users can employ different kinds of global probing simply by changing the deformation path formulations. This will modify the way flow surfaces get advected at each animation time step.

13.4.3 Choosing and Placing Seed Surfaces

The problem of choosing and placing seed surfaces in the vector field is a hard one and is highly dependent on the application at hand. It seems some kinds of surfaces might be more applicable in general than others. For instance, spherical probes seem to work best when placed around (but not too close to) critical points. This way the behavior of the field can be easily inferred from the change in the flow surface. Spherical seed surfaces are clearly among the easiest to use because of their overall symmetry. Another kind of seed surface that is general and easy to use is planes. Planes can be easily specified because of their concise representations (like the spheres), and also their shape makes it easy to interpret changes.

The placement of seed surfaces can be greatly simplified if there are methods to identify important features (for the particular application) that generate some form of surface. After one has the representation for a seed surface, it is possible to use any of the already available methods to approximate a surface with Bézier patches in order to generate a surface that is suitable for use within our framework.

13.5 Examples of Flow Surface Probes

With the following examples, we show the kind of visual effects that can be generated with the flow surface probes. It should be noted that the still images of flow surface probes shown in this chapter are not as powerful as the interactive animations.

There are two classes of examples, each associated with one type of advection scheme used to change the tessellation points, as described in Section 13.4.2. The first class, called local probing, uses the vector field defined on the tessellation of the seed surface. The second class, time surfaces, uses the vector field defined on the tessellation of the flow surface, which will be changed dynamically as the surface moves through the flow field. In all our examples, the techniques described in Section 13.4 for tessellation generation and deformation path specification are used.

13.5.1 Local Probing of Electrical Charges

First, to give the flavor of our technique, we show three sets of examples using three different kinds of surface probes. In these examples, the vector fields used were analytically calculated by modeling the electric field of a few particles using Coulomb's law and the principle of superposition [7].

Planes

A possible way of visualizing the local vector field defined over a region of interest is by placing a plane somewhere inside the vector field, and using a contraction technique to reduce the vector field defined on the plane into a scalar field. In Figure 13.2, the 3D vectors on the plane are reduced to the vector magnitudes, which are then mapped to color, where red represents large magnitudes, blue represents small magnitudes, and green represents those ranging in the middle. It can be observed that Figure 13.2 unfortunately loses the directional information inherent in the original vector field. In Figure 13.3, hedgehogs are added to Figure 13.2 to show the direction and magnitude of the vectors on the plane. Even though we can rotate the plane and observe it from various directions, cluttering is an annoying, unavoidable problem in Figure 13.3.

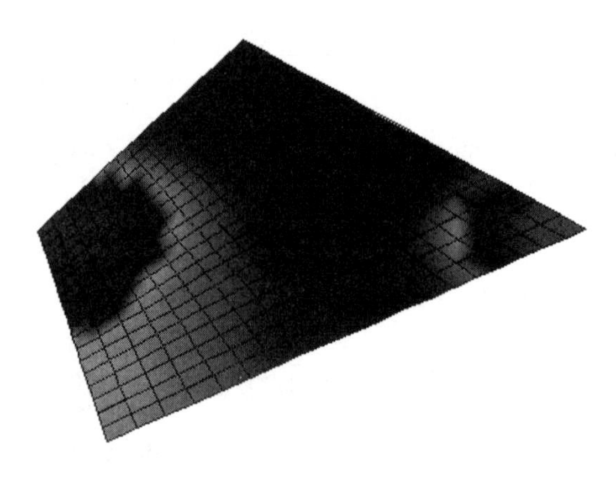

Figure 13.2: A seed plane with vector magnitudes mapped to colors (red represents large magnitudes, blue represents small magnitudes, and green represents those in the middle). See Color Plate 86.

In our approach, instead of showing the hedgehogs, we deform the plane with the local vector field defined on the seed plane. In addition, we map the vector magnitudes onto the plane using colors. An animation sequence of the deformed plane shows how the tessellation points on the seed plane are advected by the vectors, thus providing information helpful in understanding the local vector field defined on the seed plane. Figure 13.4 shows one of the animation frames. Figure 13.5 shows, at a certain animation time step, a deformation of a plane inserted into another vector field, with the seed plane shown at the bottom of the figure in white color. One can also use transparency and different color mappings in order to obtain a better view of the deformation.

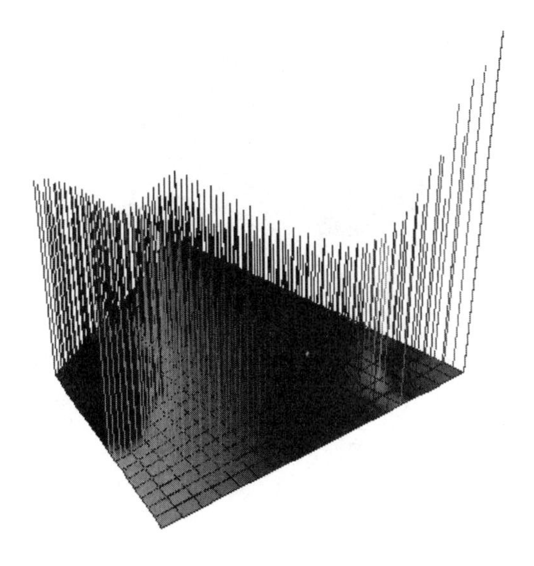

Figure 13.3: Hedgehogs have been added to Figure 13.2. See Color Plate 87.

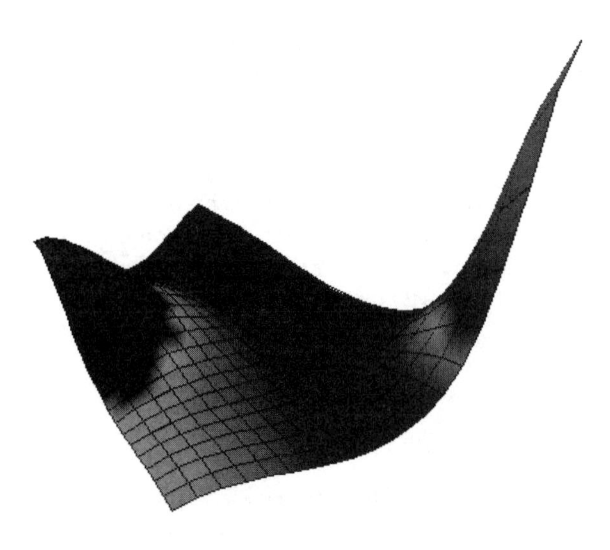

Figure 13.4: An animation frame of a flow surface probe evolving from the planar seed surface shown in Figure 13.2. See Color Plate 88.

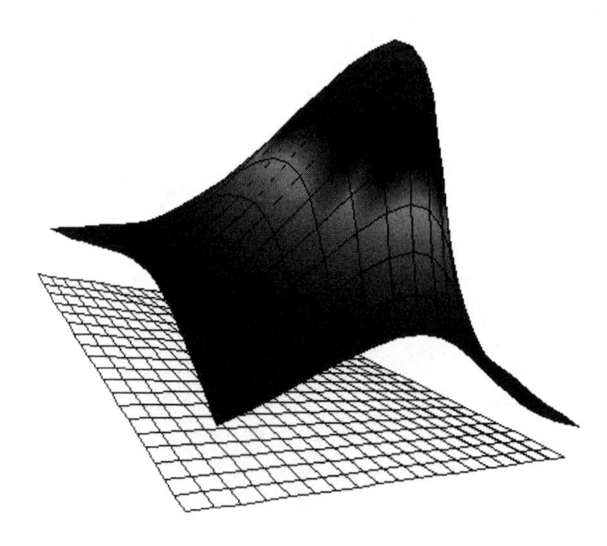

Figure 13.5: An animation frame of a flow surface probe in another vector field (the seed plane is shown at the bottom in white). See Color Plate 89.

Bézier Curved Surfaces

In some applications, it may be desirable to observe the vector field around an embedded object in the flow field—for instance, the vector field at a constant distance from a plane wing. Another method that addresses this problem is described in Max et al. [13]. With our flow surface probe technique, we can instead wrap a curved surface around the object and observe the deformation of the curved surface by the local vector field. In Figure 13.6, a Bézier seed surface wrapped around an object is displayed in white color, and the image shows the curved surface deformed after a certain animation time. Figure 13.7 shows, in another vector field, the deformation of a Bézier surface wrapping around a curved object at a certain animation time step, with the seed surface at the bottom of the figure shown in white color.

Spherical Surfaces

Some applications, such as vector fields generated by electrical charges, suggest that the seed surface be a spherical surface. As described in Section 13.4.1, the method for generating a triangular tessellation of a spherical surface is quite different from that for a Bézier surface. Figure 13.8 shows a seed spherical surface with the vector magnitude mapped to color space. Figure 13.9 shows the deformation of the spherical seed surface (in Figure 13.8) after a certain animation time step. Spherical surfaces may be most useful in visualizing vector fields close to critical points.

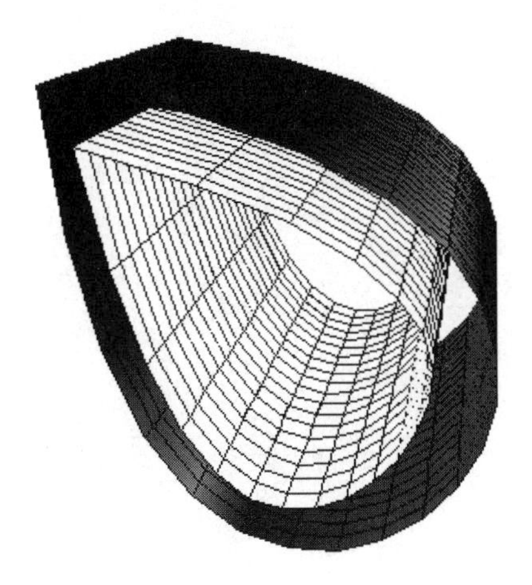

Figure 13.6: An animation frame of a flow surface probe with a seed surface (not shown) as a Bézier surface wrapped around the object shown in white. See Color Plate 90.

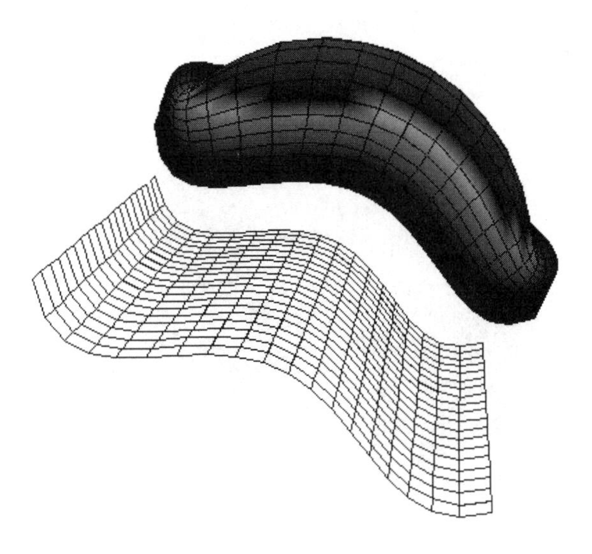

Figure 13.7: An animation frame of a flow surface probe with the seed Bézier surface at the bottom in white. See Color Plate 91.

Figure 13.8: A seed spherical surface. See Color Plate 92.

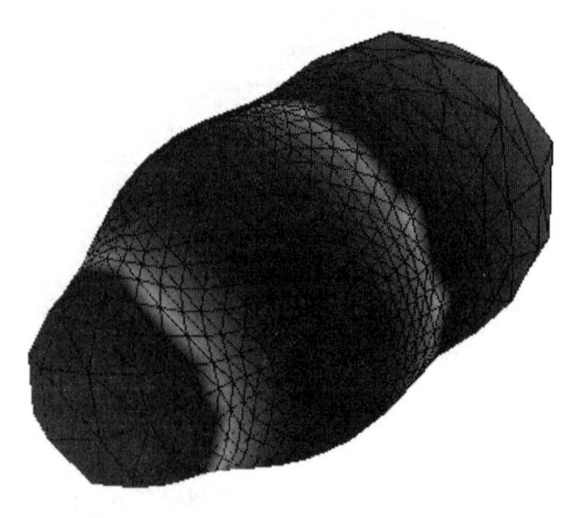

Figure 13.9: An animation frame of a flow surface with the spherical seed surface shown in Figure 13.8. See Color Plate 93.

13.5.2 Local Probing of Blunt Fin

Our technique can also be used to probe vector fields with complex structures, such as the blunt fin data set [11]. The blunt fin data set consists of $39 \times 31 \times 31$ curvilinear cells, and it measures an airflow over a flat plate with a blunt fin rising from the plate. To probe the vector field, the user may insert flow surfaces in areas of interest and watch the animation procedure. What needs to be done in real time is to sample the velocities for the tessellation points of the seed surface, which poses no problem for most graphics workstations. Figure 13.10 shows a sequence of a flow surface at animation times 0, 10, 20, and 40, respectively (the seed surface is a plane), with the grid mesh of the data set shown in white.

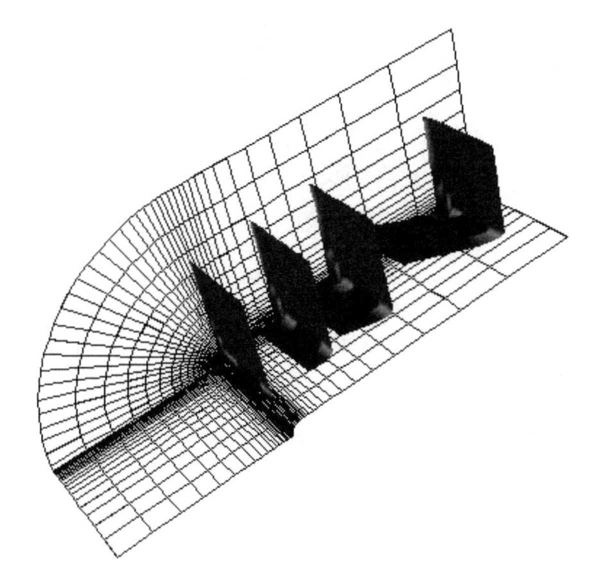

Figure 13.10: A sequence of flow surface at animation times 0, 10, 20, and 40, respectively, with the grid mesh in white. See Color Plate 94.

It is important to realize that even though we show the flow surface over space, at animation time t, the surface is not being advected with respect to vectors on the current position, but only changed linearly with the vectors defined on the seed surface. One can then clearly study how the vectors defined on the seed surface behave. Interesting visualization can be generated by changing the seed surface slightly and producing another animation sequence.

13.5.3 Time Surfaces of Blunt Fin

Now that we have shown how the flow surface probes can be used to probe the local vector field as hedgehogs, we also give an example to show how this technique can be used to create interactive time surfaces. We use the same blunt fin data set as in the last example, but use the first-order Euler integration to advect the tessellation points at every time step instead. Figure 13.11 shows a sequence of time surfaces at animation times 0, 10, 20, 30, and 40, respectively (the seed surface is the same plane as that of Figure 13.10), with the grid mesh shown in white.

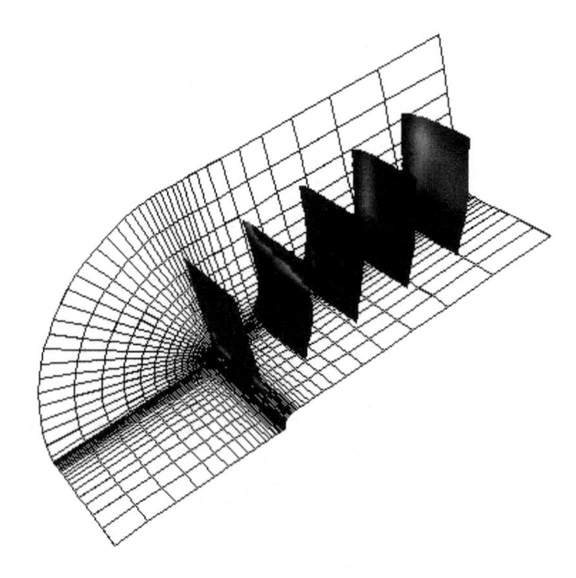

Figure 13.11: A sequence of time surfaces at animation times 0, 10, 20, 30, and 40 respectively, with the grid mesh in white. See Color Plate 95.

Enlightening pictures can be generated by using both time surfaces and local probes. One can first use time surfaces to locate areas of interest, and then probe the local vector field of a particular region with various different probes, as we suggested in Section 13.5.2.

13.6 Conclusions and Future Work

Our beta implementation uses Geomview, from the Geometry Center at the University of Minnesota, as the drawing engine, and for inputting and outputting files as a means to enter vector fields and other necessary data into the system. Even though Geomview saved us a lot of interface design and implementation work, it was not designed to be used for our purposes. We are currently designing another user interface specifically for placing the seed surfaces.

So far we have implemented Bézier surfaces, planes, and spherical surfaces. There should be no difficulty in adding other flow surface probes. A shortcoming of this technique is that it may be hard to specify the desired flow surfaces inside the flow field, and even to specify surfaces of a given form. In the future, we might try to use one of the new free-form surfaces [16] to ease the user interaction.

There are several important concerns with our flow surface implementation. One is the fact that, to achieve real-time rendering, we currently only use a simple Euler integration scheme to calculate the time surfaces, which definitely affects the accuracy of the surfaces. It is fairly easy to use a second-order Runge-Kutta instead, at the expense of more operations. The other issue is how to dynamically change the tessellation as the surface moves in the flow field. In our current implementation, we always keep the same tessellation over animation time. Using the same tessellation eventually gives bad results, especially for strongly rotational and divergent fields. Possibly, we could use the method of Witkin and Heckbert [17], which employs points to model surfaces, coupled with the idea of Hultquist [10], to generate new points when the current points get further apart, and delete points when they get too close. We are currently considering these issues.

We believe that this real-time and easy-to-implement technique is general and can be used to complement existing techniques in assisting scientists and engineers in understanding 3D vector data sets. By choosing a well-known surface, like a plane or a sphere, and deforming it with the vector field, the user obtains a better view of the rate of change of the vector field in different directions than if a complex seed surface is chosen. Using this technique, the user can get a good grasp of the directions and magnitudes of the vector field.

We also believe that in many applications it is very important to provide the user with the capability to "wrap" an object of interest with a flow surface. In this way, the user visualizes a large part of the vector field rather than just a single direction of interest. This makes our tool more useful in visualizing global properties of the field over some large area. It is also interesting to use the tool to find directions of interest. For instance, if the user places a spherical surface around an expanding field, the area of largest rate of growth is clearly shown with the largest deformation of the sphere. This applies similarly to a contracting field.

Acknowledgments

This research has been supported by the National Science Foundation under grant number CCR-9205047 and by the Department of Energy under the PICS grant. Cláudio Silva received partial support from CNPq (Brazilian Research Council). We are grateful to Sam Uselton and NASA for supplying us with the Blunt Fin data set. We would like to thank Eduardo Prado of the Mathematics Department at University of São Paulo, Brazil for enlightening discussions about vector fields and differential geometry.

Bibliography

[1] R. Aris, *Vectors, Tensors and the Basic Equations of Fluid Mechanics*, Dover, N.Y., 1962.

[2] S. Bryson and C. Levit, "The Virtual Windtunnel: An Environment for the Exploration of Three-Dimensional Unsteady Flows," *Proc. Visualization '91*, 1991, pp. 17–24.

[3] W. C. de Leeuw and J. J. van Wijk, "A Probe for Local Flow Field Visualization," *Proc. Visualization '93*, 1993, pp. 39–45.

[4] T. Delmarcelle and L. Hesselink, "Visualizing Second-Order Tensor Fields with Hyperstreamlines," *IEEE Computer Graphics and Applications*, 1993, pp. 25–33.

[5] M. do Carmo, *Differential Geometry of Curves and Surfaces*, Prentice-Hall, Englewood Cliffs, N.J., 1976.

[6] G. Farin, *Curves and Surfaces for CAGD*, Academic Press, New York, N.Y., 1993.

[7] R. Feynman, R. Leighton, and M. Sands, *The Feynman Lectures on Physics (Vol. 2)*, Addison-Wesley, Reading, Mass., 1964.

[8] P. Hall, "Volume Rendering for Vector Fields," *The Visual Computer*, Vol. 10, 1993, pp. 69–78.

[9] L. Hesselink, F. Post, and J. van Wijk, "Research Issues in Vector and Tensor Field Visualization," *IEEE Computer Graphics and Applications*, Vol. 14, No. 2, 1994, pp. 76–79.

[10] J. Hultquist, "Constructing Stream Surfaces in Steady 3D Vector Fields," *Proc. Visualization '92*, 1992, pp. 171–178.

[11] C.-M. Hung and P.G. Bunning, *Simulation of Blunt-Fin Induced Shock Wave and Turbulent Boundary Layer Separation*, AIAA Paper 84-0457, 1984.

[12] R. Klassen and S. Harrington, "Shadowed Hedgehogs: A Technique for Visualizing 2D Slices of 3D Vector Fields," *Proc. Visualization '91*, 1991, pp. 148–153.

[13] N. Max, R. Crawfis, and C. Grant, "Visualizing 3D Velocity Fields near Contour Surfaces," *Proc. Visualization '94*, 1994, pp. 248–255.

[14] W. Schroeder, C. Volpe, and W. Lorensen, "The Stream Polygon: A Technique for 3D Vector Field Visualization," *Proc. Visualization '91*, 1991, pp. 126–132.

[15] J. van Wijk, "Implicit Stream Surface," *Proc. Visualization '93*, 1993, pp. 245–252.

[16] W. Welch and A. Witkin, "Free-Form Shape Design Using Triangulated Surfaces," *Computer Graphics (Proc. SIGGRAPH '94)*, 1994, pp. 247–256.

[17] A. Witkin and P. S. Heckbert, "Using Particles to Sample and Control Implicit Surfaces," *Computer Graphics (Proc. SIGGRAPH '94)*, 1994, pp. 269–277.

Chapter 14

Particle Tracing Algorithms for 3D Curvilinear Grids

I. Ari Sadarjoen, Theo van Walsum, Andrea J.S. Hin, and Frits H. Post

Abstract. *This chapter presents a comparison of several particle tracing algorithms on curvilinear grids. The fundamentals of particle tracing algorithms are described and used to split tracing algorithms into basic components. Based on this decomposition, two different strategies for particle tracing are described in greater detail: tracing in computational space and tracing in physical space. Accuracy and speed tests are performed for both types of algorithms. From these tests it is concluded that particle tracing algorithms in physical space generally perform better than algorithms in computational space.*

Keywords: scientific visualization, vector field visualization, particle tracing, interpolation, grid transformation

14.1 Introduction

Computational Fluid Dynamics (CFD) is concerned with modeling fluid flows. Increasingly sophisticated software is being developed to simulate interesting flow phenomena. Scientists can gain insight into the resulting data by visualization. One method of visualizing a velocity field is particle tracing, which involves releasing particles into a flow and calculating their positions at specific times.

In general, CFD simulations provide a velocity field defined on a discrete grid. The simplest grids are block-shaped with cubical cells. Particle tracing algorithms for such Cartesian grids are investigated in [17, 6]. In CFD practice, the grids are often boundary-fitted and therefore curvilinear, with the purpose of solving flows in complex geometries. Particle tracing algorithms for CFD grids were presented in [2, 1]. Some algorithms transform a curvilinear grid to a Cartesian grid and perform particle tracing in the Cartesian space [13]. A more detailed description of this method was recently given in [12].

In most papers on particle tracing, the scope is limited to a *single* method. If alternatives are given at all, they are not closely examined. Often, not all details of the particle tracing process are given, suggesting that some operations are straightforward. In commercial visualization packages, background information concerning the applied algorithms is seldom given.

The present chapter addresses the above issues. Detailed descriptions are given of the particle tracing process, which is split into distinct components. The strengths and weaknesses of different implementations of these components are studied and visualized. The main aspects considered are accuracy and performance.

After covering the fundamentals of particle tracing in Section 14.2, the details of two different classes of particle tracing algorithms will be described in Sections 14.3 and 14.4. Considerations for the implementation of several algorithms, as well as an overview of test cases, are given in Section 14.5. The test results and a discussion are presented in Section 14.6. Finally, Section 14.7 derives some conclusions.

14.2 Fundamentals of Particle Tracing

Although many terms are in use for grid types, we will reserve the term *Cartesian grids* for grids with straight grid lines and unit cubes as cells, and *curvilinear grids* for grids with curved (or more precisely: piecewise straight) grid lines and a regular topology. We will start by explaining the principles of particle tracing in Cartesian grids.

14.2.1 Cartesian Grids

The computation of a particle path is based on a numerical integration of the ordinary differential equation

$$\frac{d\mathbf{x}}{dt} = \mathbf{v}(\mathbf{x}) \tag{14.1}$$

where t denotes time, $\mathbf{x}$ the position of the particle, and $\mathbf{v}(\mathbf{x})$ the velocity field. The starting position $\mathbf{x}_0$ of the particle provides the initial condition:

$$\mathbf{x}(t_0) = \mathbf{x}_0 \tag{14.2}$$

The solution is a sequence of particle positions $(\mathbf{x}(t_0), \mathbf{x}(t_1), \ldots)$.

A particle tracing algorithm must perform the following steps. First, a search is performed for the cell which contains the initial position of the particle. To determine the velocity at this position, the velocities at the cell corners are interpolated. Then, an integration step calculates the next position of the particle. Again, a search is performed, this time for the cell containing the new position. The process of interpolation, integration, and point location is repeated until the particle leaves the grid. This process can be translated into pseudocode representing the *general structure* of a particle tracing algorithm:

```
find cell containing initial position                 (point location)
while particle in grid
    determine velocity at current position            (interpolation)
    calculate new position                            (integration)
    find cell containing new position                 (point location)
endwhile
```

Note that the above pseudocode is merely intended to show the main algorithm components; in real implementations many refinements and optimizations could be made.

Point Location

Determining which cell contains a specified point is called point location. The coordinates of a point can be divided into their integer and fractional parts: $\mathbf{x} = (x, y, z) = [i, j, k] + (\alpha, \beta, \gamma)$, where i, j, k are integers and $\alpha, \beta, \gamma \in [0, 1]$. In this chapter, we will refer to $[i, j, k]$ as the *indices* and (α, β, γ) as the *offsets*. Point location in a Cartesian grid is as simple as truncating the coordinates to their integer parts. Here, determining the offsets is also considered to be part of the point location operation, but sometimes we will strictly distinguish between point location and offset determination.

Interpolation

To obtain a value of the velocity field at points other than the grid nodes, it is necessary to determine an interpolated value from the nodes surrounding the point. The standard way to do this in cubical cells is trilinear interpolation. This requires the cell indices and the offsets obtained through point location. Let $I, J, K \in \{0, 1\}$ and let the basis functions Ψ be defined as $\Psi_0(\alpha) = 1 - \alpha$ and $\Psi_1(\alpha) = \alpha$. Then, for a point $\mathbf{x}$ in a cell $[i, j, k]$ with offsets (α, β, γ), the function $\mathcal{T}$ determines the interpolated value from the corner velocities $\mathbf{v}_{i,j,k} \ldots \mathbf{v}_{i+1,j+1,k+1}$.

$$\mathbf{v} = \mathcal{T}(\mathbf{v}, \alpha, \beta, \gamma) = \sum_{I,J,K=0}^{1} \mathbf{v}_{i+I,j+J,k+K} \cdot \Psi_I(\alpha)\Psi_J(\beta)\Psi_K(\gamma) \tag{14.3}$$

Integration

Many integration methods are known in the literature, ranging from the simple first-order Euler scheme to the fourth-order Runge-Kutta scheme or even higher-order methods, applied with fixed or variable time steps. A well-known second-order method is Heun's scheme, also known as a second-order Runge-Kutta scheme. Starting from position $\mathbf{x}_n$ at time $t = t_n$, the position $\mathbf{x}_{n+1}$ at time $t = t_{n+1}$ is calculated in two steps:

$$\mathbf{x}^*_{n+1} = \mathbf{x}_n + \Delta t \cdot \mathbf{v}(\mathbf{x}_n) \tag{14.4}$$

$$\mathbf{x}_{n+1} = \mathbf{x}_n + \Delta t \cdot \frac{1}{2}\left\{\mathbf{v}(\mathbf{x}_n) + \mathbf{v}(\mathbf{x}^*_{n+1})\right\} \tag{14.5}$$

14.2.2 Curvilinear Grids

In practice, the grids used in many CFD applications are *not* Cartesian. To handle complex geometries, curvilinear, boundary-fitted grids are used. These grids are also referred to as *structured* grids, because they have a regular topological structure, as opposed to unstructured grids as used in Finite Element Methods.

While this allows for a large variety of geometries, numerical procedures are more difficult in curved grids. This is the reason why in CFD flow solvers, the curvilinear grid in the physical domain is often internally transformed to a Cartesian grid in a new domain. The physical domain will be called *P-space* ($\mathcal{P}$) and the new domain will be called computational space or *C-space* ($\mathcal{C}$). Many calculations are performed more efficiently in the simple Cartesian grid in C-space.

In particle tracing in a curvilinear grid, similar problems arise. Especially point location and interpolation become more complex. This suggests the use of a similar procedure in determining particle paths for visualization purposes as for flow solving. A transformation is defined from physical space to computational space such that the curvilinear grid becomes a Cartesian grid (see Figure 14.1).

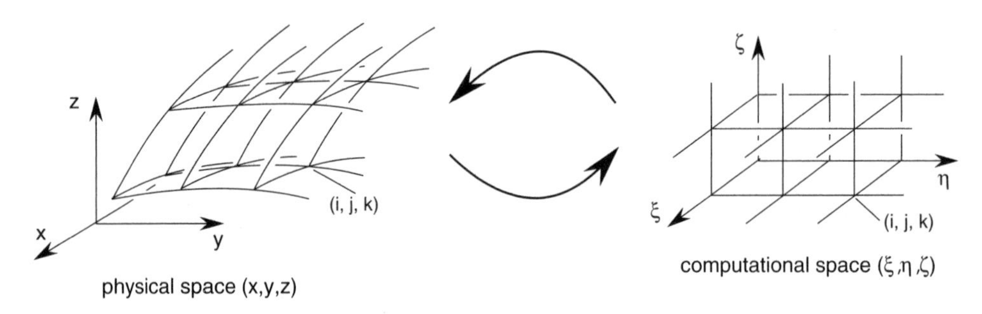

Figure 14.1: Transformation between $\mathcal{P}$ and $\mathcal{C}$.

In a Cartesian grid, point location and interpolation can be carried out as described in the previous subsection. We must transform the positions of the complete path back to the physical domain to be able to visualize it, but the use of the convenient computational domain may increase the efficiency of the algorithm. The details will be discussed in Section 14.3.

An alternative is to calculate the particle path directly in P-space ($\mathcal{P}$). This would involve more complex point location and interpolation, but would avoid grid transformation. This alternative will be covered in Section 14.4.

14.2.3 Restrictions

We will restrict our attention to stepwise integration methods. Also, we will only use the second-order Runge-Kutta method with fixed time steps. For reasons of simplicity, we consider the flow field to be *time-independent*. A paper that discusses visualization of time-dependent flow fields is [7].

14.3 C-Space Algorithms

Particle tracing in computational space proceeds in almost the same way as described in Section 14.2.1, except that the physical velocities $\mathbf{v}$ must be transformed to $\mathbf{u}$ in $\mathcal{C}$ and instead of Equation (14.1), the following differential equation is solved:

$$\frac{d\boldsymbol{\xi}}{dt} = \mathbf{u}(\boldsymbol{\xi}) \tag{14.6}$$

where $\boldsymbol{\xi}$ denotes a position in $\mathcal{C}$. The solution is now a sequence of positions $(\boldsymbol{\xi}(t_0), \boldsymbol{\xi}(t_1), \ldots)$ in computational coordinates. As a consequence, the particle path calculated in $\mathcal{C}$ must be transformed to physical space for visualization purposes. The general form of the algorithm then becomes:

```
find cell containing initial position                (point location)
while particle in grid
     transform corner velocities from P to C         (transform velocities)
     determine C-velocity at current position        (interpolation)
     calculate new position in C-space               (integration)
     transform C-position to P                       (transform position)
     find cell containing new C-position             (point location)
endwhile
```

The point location and interpolation parts in this algorithm are straightforward, because the computational grid is Cartesian. However, from the pseudocode it can be seen that two transformation operations are introduced. First, the physical velocities must be transformed to C-space. Second, the calculated particle positions must be transformed back to P-space. The former operation is not necessary if it is possible to use the velocities in computational space that are used during the flow solving process. However, we also assume that the results from the flow solver are physical velocities. It should be emphasized that in the above algorithm the velocity field is not transformed and stored in its entirety, but only the velocities in the current cell are transformed on-the-fly. In other words, they are *local* transformations, since in most of the cases no single, global grid transformation from $\mathcal{P}$ to $\mathcal{C}$ exists.

14.3.1 Transformation of Positions from $\mathcal{C}$ to $\mathcal{P}$

The transformation of a position from $\mathcal{C}$ to $\mathcal{P}$ is relatively straightforward. What is necessary is a transformation which maps the corner nodes of a cubic cell in computational space to the corner nodes of a cell in physical space. The edges between nodes are assumed to be straight in either space. This leads to a transformation which is equivalent to a trilinear interpolation between the cell corner P-space positions, as in Equation (14.3). The function $\mathcal{T}$ transforms a C-space position $\boldsymbol{\xi} = (\xi, \eta, \zeta)$ consisting of integer parts $[i, j, k]$ and fractional parts (α, β, γ) to a P-space position $\mathbf{x}$ as

$$\mathbf{x} = \mathcal{T}(\mathbf{x}, \alpha, \beta, \gamma) \tag{14.7}$$

14.3.2 Transformation of Velocities from $\mathcal{P}$ to $\mathcal{C}$

A velocity $\mathbf{u}$ in $\mathcal{C}$ is transformed to $\mathbf{v}$ in $\mathcal{P}$ according to:

$$\mathbf{v} = \mathbf{J} \cdot \mathbf{u} \tag{14.8}$$

and similarly, a velocity $\mathbf{v}$ in $\mathcal{P}$ can be transformed to $\mathbf{u}$ in $\mathcal{C}$ with

$$\mathbf{u} = \mathbf{J}^{-1} \cdot \mathbf{v} \tag{14.9}$$

The matrix $\mathbf{J}$ is called the *Jacobian* and contains the partial derivatives:

$$\mathbf{J} = \begin{pmatrix} x_\xi & x_\eta & x_\zeta \\ y_\xi & y_\eta & y_\zeta \\ z_\xi & z_\eta & z_\zeta \end{pmatrix} \tag{14.10}$$

where x_ξ is short for $\frac{\partial x}{\partial \xi}$, and so on.

The matrix $\mathbf{J}$ can be thought of as consisting of the columns $(\mathbf{j_1} \mid \mathbf{j_2} \mid \mathbf{j_3})$. These are in fact the partial derivatives $\frac{\partial \mathbf{x}}{\partial \xi}, \frac{\partial \mathbf{x}}{\partial \eta}, \frac{\partial \mathbf{x}}{\partial \zeta}$. As we are dealing with (discrete) grids, the Jacobian must be calculated with finite differences. There are several methods for doing this. Another aspect is the *number* of Jacobians used for each cell.

Jacobian Approximations

To approximate a Jacobian in a grid point, several types of differencing are available. Given the coordinates of the grid nodes $\mathbf{x}_{i,j,k}$ where $[i, j, k]$ lie within the grid boundaries, there are at least three possibilities:

1. forward differences: $\mathbf{j}_1 = \frac{\Delta \mathbf{x}}{\Delta \xi} = \mathbf{x}_{i+1,j,k} - \mathbf{x}_{i,j,k}$
2. backward differences: $\mathbf{j}_1 = \frac{\Delta \mathbf{x}}{\Delta \xi} = \mathbf{x}_{i,j,k} - \mathbf{x}_{i-1,j,k}$
3. central differences : $\mathbf{j}_1 = \frac{\Delta \mathbf{x}}{\Delta \xi} = (\mathbf{x}_{i+1,j,k} - \mathbf{x}_{i-1,j,k})/2$

Calculating the derivatives $\mathbf{j}_2 = \frac{\Delta \mathbf{x}}{\Delta \eta}$ and $\mathbf{j}_3 = \frac{\Delta \mathbf{x}}{\Delta \zeta}$ proceeds in a similar way.

In a 2D cell, these differences can be combined to the cases in Figure 14.2. In an arbitrary grid node (Figure 14.2a), either forward differences (Figure 14.2b), backward differences (Figure 14.2c), or central differences (Figure 14.2d) can be calculated. Alternatively, mixed differences, such as forward for $\mathbf{j}_1$ and backward for $\mathbf{j}_2$ (Figure 14.2e), or vice versa (Figure 14.2f) can be used.

Number of Jacobians

The other aspect is the *number* of Jacobians calculated in a cell. Basically, there are three options:

1. one for each cell (see Figure 14.3a)
2. one for each node (see Figure 14.3b)

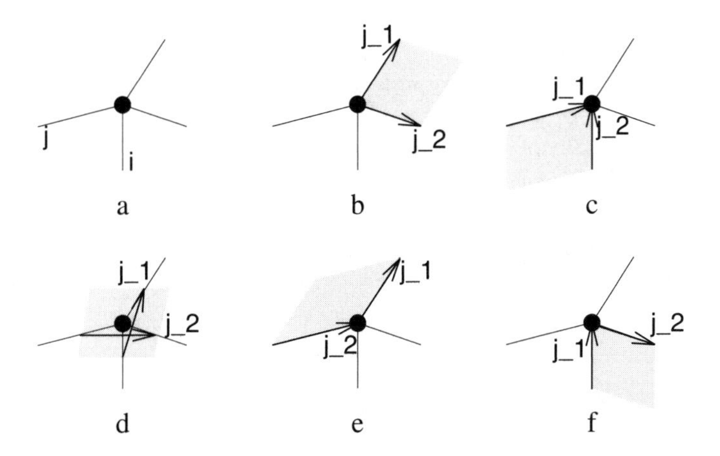

Figure 14.2: Several types of differences for Jacobian calculation.

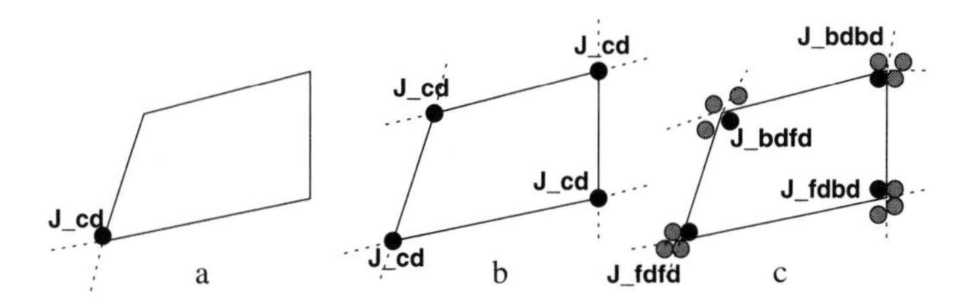

Figure 14.3: Number of Jacobians in a 2D cell.

3. one for each node for each adjacent cell (see Figure 14.3c)

The simplest and fastest method is to calculate only one Jacobian per cell [13]. By computing forward, backward, or central differences in one node, for example in the lower-left node, it is assumed that this Jacobian reasonably represents the deformation throughout the entire cell (see Figure 14.3a).

Another approach is to calculate separate Jacobians *in each grid node* [12]. Now, four different Jacobians have to be calculated in a 2D cell (see Figure 14.3b), or eight in a 3D cell.

Using mixed types of differences leads to one Jacobian per node per adjacent cell. The Jacobian for a node is calculated according to which cell the particle is in. We shall call this *forward/backward differences* (see Figure 14.3c).

This method of calculating Jacobians is mathematically correct when trilinear interpolation is used for point transformation. We give more details on this in the appendix. The results have been confirmed by other researchers who have found similar results [5].

14.4 P-Space Algorithms

The alternative to particle tracing in a simple Cartesian grid in C-space, is particle tracing in a complex curvilinear grid in P-space. This has a considerable impact on the complexity of the algorithm presented in Section 14.2. In particular, point location and interpolation are not straightforward. There is no longer a simple relation between a physical position in space and the grid cell that contains it. Moreover, since the cells in a curvilinear grid are not cubes, it is also not as easy to perform trilinear interpolation because the relative position (fractional offsets) of a position in a cell is hard to determine. The next two subsections will discuss alternative point location and interpolation methods, respectively.

14.4.1 Point Location Methods

We can distinguish between *global* and *incremental* point location. In global point location, a given point in a grid must be found with no previously known cell. In a curvilinear grid this is not an easy task. As with all search algorithms, it is possible to use a simple brute-force algorithm which searches all grid cells one-by-one. Naturally, this is expensive. Auxiliary data structures can be used to perform a smart search [10, 16].

Fortunately, particle tracing is a step-by-step process, in which most of the time, global point location is only necessary to find the cell containing the initial position of a particle. After this, there is always a current starting position and a current starting cell from which the new position is to be found. This occurs at every integration step that starts from the current particle position in some known cell and calculates a new position. The following paragraphs will discuss only incremental point location.

Tetrahedrization

To use tetrahedrization, the curvilinear hexahedral cells are decomposed into tetrahedra. The reasons for this are:

- Tetrahedra are convex, which facilitates testing if a specified point is inside.
- Tetrahedra have planar faces, which facilitates intersection with a line.

Incremental point location can now be performed by drawing a line from the previously known position to the next position [3]. Intersections with the faces of the tetrahedron and containment tests in neighboring cells are used to locate the new point. Tetrahedrization is only performed in the cells along the path of the line and the results do not have to be stored, nor does the tetrahedrization involve insertion of new grid points.

Stencil Walk and Newton-Raphson

Let $\mathbf{P}$ be a point given in physical space that must be found in computational space. In the so-called *stencil walk* method [2], first an initial point $\boldsymbol{\xi}$ in computational space is chosen. This point is transformed to physical space using the transformation $\mathbf{x} = \mathcal{T}(\mathbf{x}, \alpha, \beta, \gamma)$. The difference between the transformed point $\mathbf{x}$ and the target point $\mathbf{P}$ is calculated as $\Delta\mathbf{x} = \mathbf{x} - \mathbf{P}$. This difference vector in physical space is transformed to computational space using $\Delta\boldsymbol{\xi} = \mathbf{J}^{-1}\Delta\mathbf{x}$ and added to the previous point, resulting in a new guess. If one of the elements of $\Delta\boldsymbol{\xi}$ is outside the range $[0, 1]$, the center of the corresponding neighboring cell is the new guess. The iterative process continues until the right cell has been found. Once the correct cell has been found, one can iterate until the value of $\Delta\boldsymbol{\xi}$ is small enough.

We can also use the well-known Newton-Raphson iteration [4, 8] that finds the root to the equation $\mathcal{F}(\boldsymbol{\xi}) = \mathbf{0}$ by the following relation:

$$\boldsymbol{\xi}_{n+1} = \boldsymbol{\xi}_n - \frac{\mathcal{F}(\boldsymbol{\xi})}{\mathcal{F}'(\boldsymbol{\xi})} \tag{14.11}$$

Here, $\mathcal{F}$ is defined as $\mathcal{F}(\boldsymbol{\xi}) = \mathbf{P} - \mathcal{T}(\boldsymbol{\xi})$, $\mathbf{P}$ is again the point in P-space to be found in C-space, and $\mathcal{T}$ is the function that transforms a point from C-space to P-space. Then, the derivative $\mathcal{F}' = -\mathcal{T}'$ is a Jacobian matrix, and the usual 1D division by $\mathcal{F}'$ turns into a matrix inversion in 3D. Experiments have shown that this method converges rapidly [8].

It can be shown that offset calculation using a Newton-Raphson process is identical to offset calculation using a stencil walk process, if the stencil walk algorithm uses interpolated forward/backward Jacobians for the transformation of $\Delta\mathbf{x}$. By substituting the above definition of $\mathcal{F}$ into Equation (14.11), taking into account that $\mathcal{F}' = \mathbf{J}$, the equivalence relation can be easily derived.

14.4.2 Interpolation Methods

Trilinear Interpolation

If the offsets in the cell (α, β, γ) are available, an interpolated velocity can be determined from the data values in the cell corners, as in Equation (14.3). One way to determine (α, β, γ) in a curvilinear grid is by using the Stencil Walk/Newton-Raphson iteration. The following two methods provide alternative ways of interpolation that do not require the offsets.

Inverse Distance Weighting

Let $\mathbf{x}_0, \ldots, \mathbf{x}_7$ be the coordinates of the eight corner nodes of a hexahedral cell, and let $\mathbf{v}_0, \ldots, \mathbf{v}_7$ be the velocities in those nodes. Then, the interpolated value $\mathbf{v}$ in point $\mathbf{X}$ is calculated as a weighted average of the corner values [15, 14]:

$$\mathbf{v} = w_0\mathbf{v}_0 + w_1\mathbf{v}_1 + \cdots + w_7\mathbf{v}_7 \tag{14.12}$$

The weight of each data point is calculated as a function of the Euclidian distance between $\mathbf{x}_i$ and $\mathbf{X}$:

$$d_i = ||\mathbf{x}_i - \mathbf{X}|| \tag{14.13}$$

$$w_i = \frac{\frac{1}{(d_i)^2}}{\sum_{j=0}^{7} \frac{1}{(d_j)^2}} \tag{14.14}$$

In a 2D cell consisting of four nodes, the same principle can be applied (see Figure 14.4).

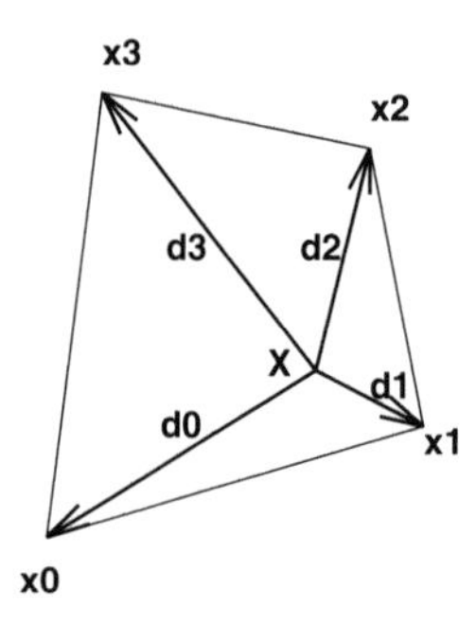

Figure 14.4: Inverse distance weighting.

Should $\mathbf{X}$ be close to a corner node, the distance d_i would be nearly zero and the value w_i would be unpredictable. This is handled in the algorithm by testing whether the distance d_i is close to zero, in which case the weight of that corner is set to one, and the weights of all other nodes are set to zero. The advantage of this method is that it does not require the fractional offsets. The disadvantage is that the interpolated field is discontinuous over cell boundaries.

Volume Weighting

In volume weighting, interpolation is performed within a tetrahedron, based on volume weights. Let $\mathbf{X}$ be a point in a tetrahedron consisting of nodes $\mathbf{x}_0$, $\mathbf{x}_1$, $\mathbf{x}_2$, and $\mathbf{x}_3$. Then, the tetrahedron can be subdivided into four tetrahedra, in all of which $\mathbf{X}$ is a corner node. The weight for each node of the main tetrahedron is the ratio of the volume of the subtetrahedron to the volume of the main tetrahedron [2].

$$w_0 = \frac{V_{123X}}{V_{0123}} \quad w_1 = \frac{V_{023X}}{V_{0123}} \quad w_2 = \frac{V_{013X}}{V_{0123}} \quad w_3 = \frac{V_{012X}}{V_{0123}} \tag{14.15}$$

The volume of an arbitrary tetrahedron ABCD is calculated as

$$V_{ABCD} = \frac{1}{6}|\vec{AB} \cdot (\vec{AC} \times \vec{AD})| \tag{14.16}$$

This interpolation method can take advantage of the information created in decomposing cells into tetrahedra, if tetrahedrization is used as the point location method.

The principle is demonstrated in 2D in Figure 14.5, where the equivalent of tetrahedral volumes are triangular areas. The weight of node $\mathbf{x}_0$ is the ratio of the area A_0 of the opposing subtriangle to the area of the main triangle. Volume weighting is continuous over the cell faces.

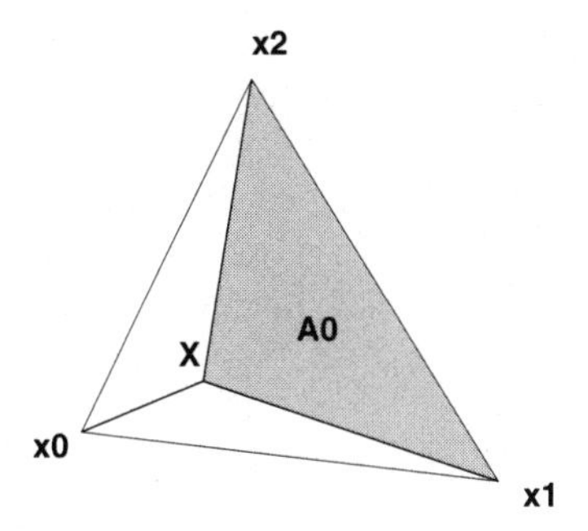

Figure 14.5: Area weighting.

Note that in 3D only four nodes of the cell are used for the interpolation, since only one of the subtetrahedra forming the hexahedral cell is used, whereas in IDW or trilinear interpolation, eight nodes are used.

14.5 Implementation and Test Flows

Based on the descriptions in Section 14.3, the algorithms listed in Tables 14.1 and 14.2 have been implemented.

Algorithm	Transformation
`C-FD1`	1 forward diff. per cell
`C-FD8`	8 forward diff. per cell
`C-CD8`	8 central diff. per cell
`C-FDBD`	forward/backward diff.

Table 14.1: C-space algorithms.

Algorithm	Interpolation
P-T-IDW	Inv. Dist. Weighting
P-T-VOL	Volume Weighting
P-SW-TRI	Stencil Walk + trilin.

Table 14.2: P-space algorithms.

Table 14.1 lists the C-space algorithms. In all of these, point location and interpolation are straightforward as described in Section 14.2. Only the transformation methods vary. In C-FD1, one Jacobian is calculated per cell using forward differences in one node. In C-FD8 and C-CD8, eight Jacobians are calculated per cell using forward and central differences, respectively. In C-FDBD, eight Jacobians are calculated per cell using mixed differences.

Table 14.2 lists the P-space algorithms, each with a code of the form (P-xx-yy). Here, -xx- stands for the point location method, and yy for the interpolation method used. P-T-IDW uses the tetrahedrization (-T-) for point location and inverse distance weighting for interpolation. P-T-VOL also uses tetrahedrization for point location, but uses volume weighting for interpolation. P-SW-TRI uses a Stencil Walk for point location, and trilinear interpolation with fractional offsets obtained by the point location process.

All implemented algorithms employ a second-order Runge-Kutta method as described in Section 14.2. First-order methods (such as Euler) were found to be inadequate by several researchers [1, 12]. Others use higher-order methods as well. But since the integration method was not our main point of interest, we did not implement these.

Test Flows

Ideally, to test the algorithms, flows should be used for which it is possible to analytically calculate the path travelled by a particle. Given a particle starting position, it should be possible to calculate the particle position $\mathbf{x}(t)$ for any given time t. This allows for a comparison of the particle paths computed by the algorithms to the theoretical paths.

Unfortunately, in 3D flows it is seldom possible to determine the analytical solution of a particle trajectory. This makes it more difficult to verify the accuracy of 3D particle paths. Therefore, we used only two-dimensional flows (although they were defined on a 3D grid). Nevertheless, there is no reason to expect that the third dimension should behave differently from the other two. Altogether, in this study we used seven test flows: two 2D theoretical flows, two 2D CFD simulated flows, and three 3D simulated flows [11]. In this chapter we will present only two typical 2D examples, one theoretical and one simulated flow.

Uniform Flow in Straight Pipe with Curved Grid

One kind of flow, where the theoretical particle paths are known, is a uniform flow. This test flow is defined on a $21 \times 41 \times 2$ curvilinear grid, which originates from a practical

CFD application (see Figure 14.6). The grid represents a vertical pipe with an inlet located at the bottom and an outlet at the top.

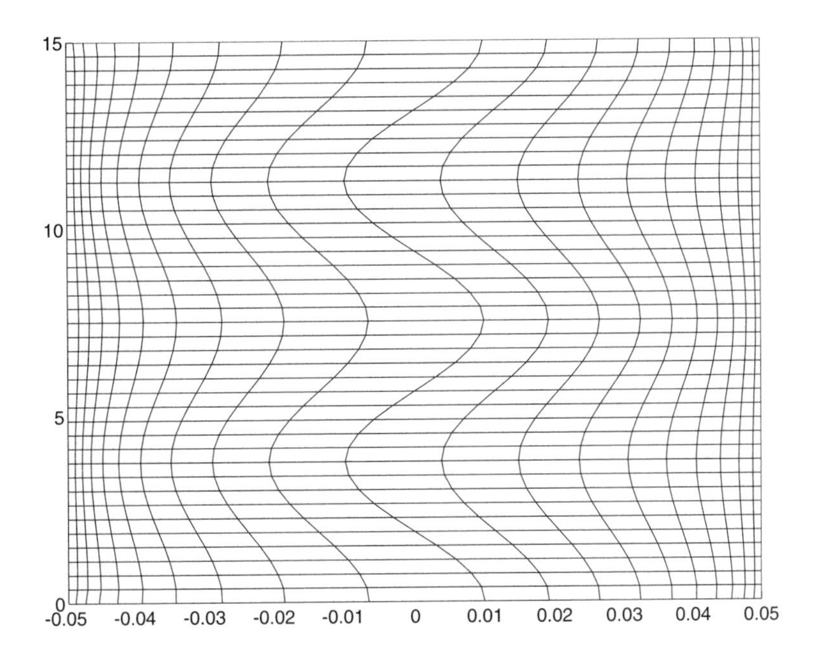

Figure 14.6: Curved grid.

Flow in L-Shaped Pipe

The second test flow uses an L-shaped pipe, in which the flow has been calculated using the `ISNaS` CFD-simulation package [9]. The grid shown in Figure 14.7 consists of $7 \times 23 \times 2$ nodes and contains a sharp discontinuity. The inlet of the pipe is at the right ($x = 2.5$).

14.6 Test Results

14.6.1 Transformation Effect

The two test cases described above were used to test the effect of the transformation component in a C-space particle tracing algorithm. All C-space algorithms (`C-FD1`, `C-FD8`, `C-CD8`, and `C-FDBD`) were applied to these two problems. As a reference, a P-space algorithm (`P-T-IDW`) was used.

Uniform Flow in Straight Pipe with Curved Grid

As the flow field is uniform, the particle paths should be straight vertical lines. All P-space algorithms produce this exact solution. Figure 14.8 shows the resulting particle paths

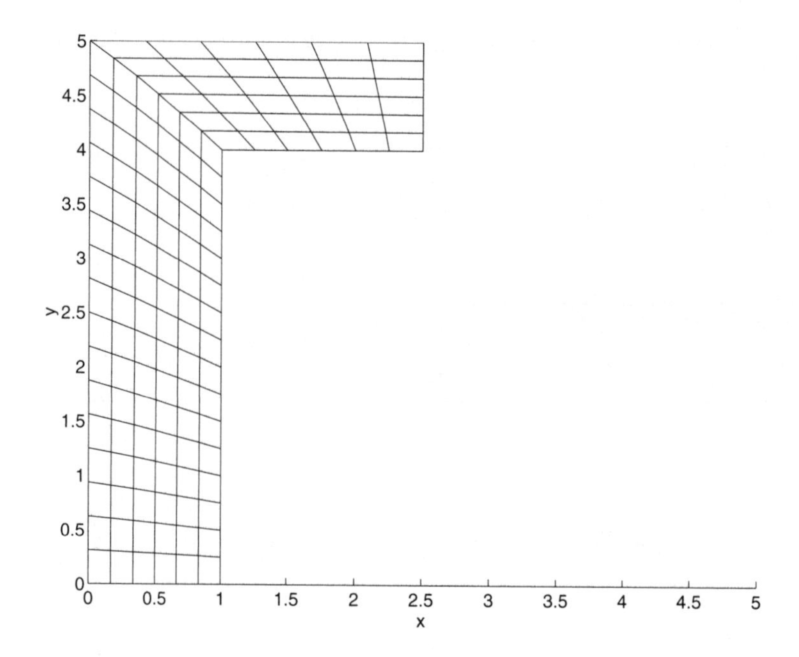

Figure 14.7: L-pipe grid.

calculated by the P-T-IDW algorithm. Since the velocity field in P-space is constant, the P-space algorithms all give the same solution, so any other P-space algorithm could be used instead of P-T-IDW. The particle paths produced by C-FD1 are shown in Figure 14.9; they are not straight. The other C-space algorithms produce similarly inaccurate results.

A closer comparison is possible when particle paths calculated by different algorithms are combined in one figure. In Figure 14.10 the particle paths produced by five algorithms are combined and magnified to highlight the differences. The straight solid line was produced by the P-space algorithm. The other solid line, closest to the first one, but with a few oscillations, was produced by the C-FDBD-algorithm. The other C-space algorithms show greater deviations, in particular C-FD1 and C-FD8. C-CD8 appears to be significantly more accurate, but C-FDBD still gives the best results.

From these results, the conclusion could be drawn that a simple transformation as described in [13] (C-FD1) or [12] (C-FD8) gives inadequate results.

Flow in L-Shaped Pipe

In the previous test, two C-space algorithms were found to be significantly better than the others: the C-CD8 algorithm which uses central differences, and the C-FDBD algorithm which uses forward/backward differences. These two algorithms were applied to the test flow in the L-shaped pipe. The integration time step was set to 0.01, with up to 1000 time steps calculated. Figure 14.11 shows the results of the C-CD8 algorithm, Figure 14.12 shows the results of the C-FDBD algorithm.

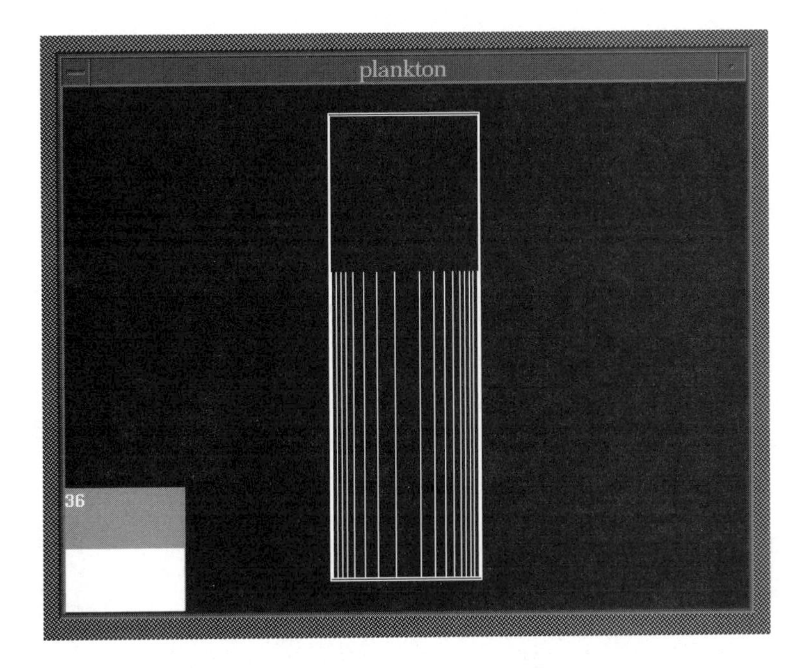

Figure 14.8: Flow in straight pipe with curved grid. Straight particle paths `P-T-IDW`.

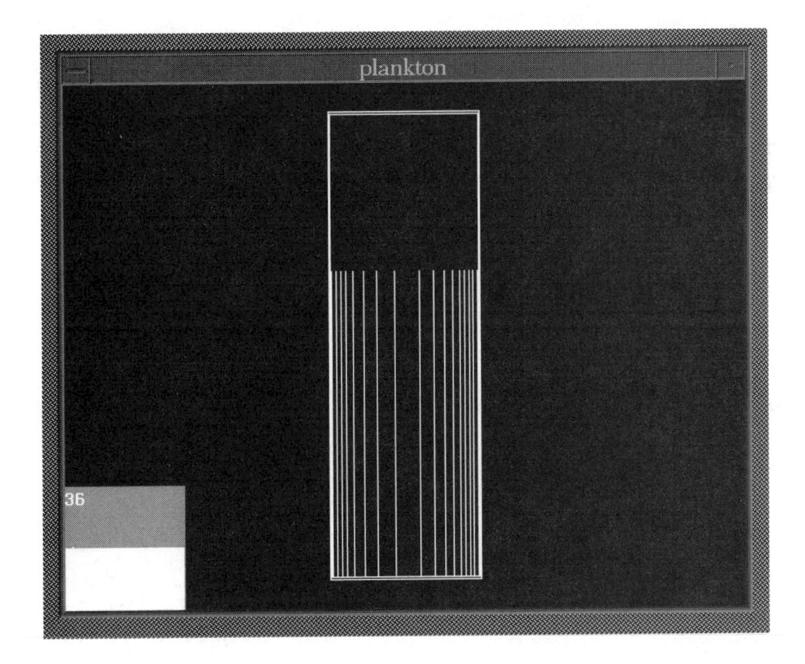

Figure 14.9: Flow in straight pipe with curved grid. Curved particle paths `C-FD1`.

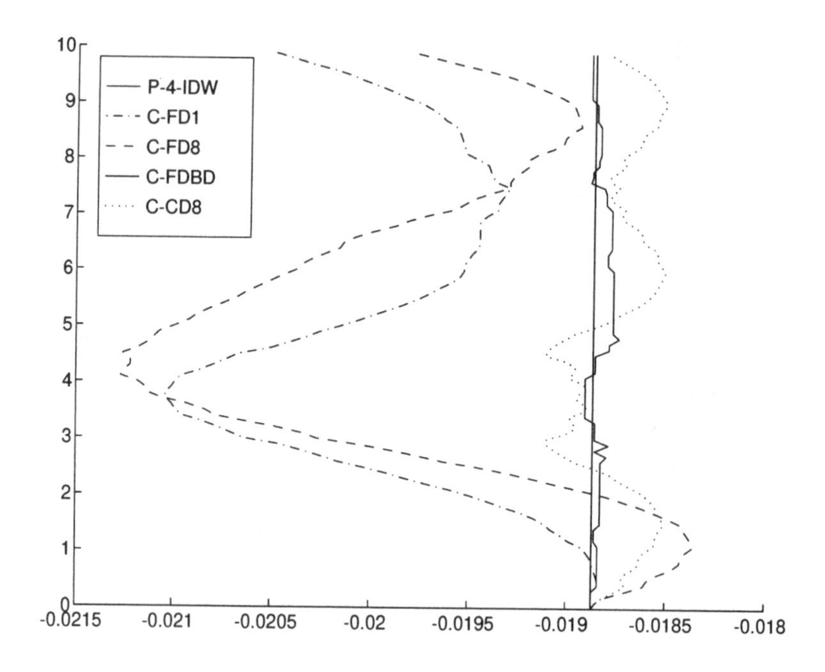

Figure 14.10: Particle paths produced by different algorithms.

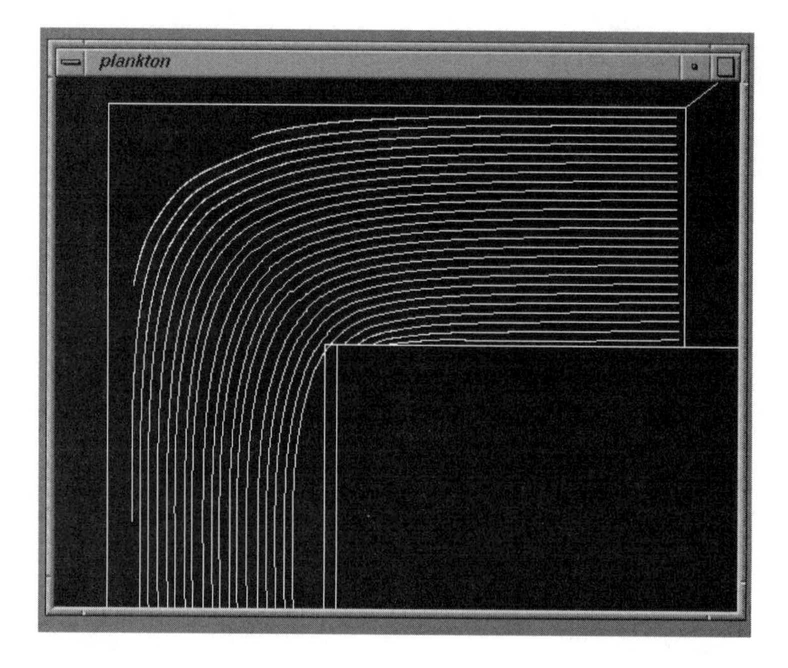

Figure 14.11: Particle paths in L-shaped pipe; Algorithm `C-CD8`.

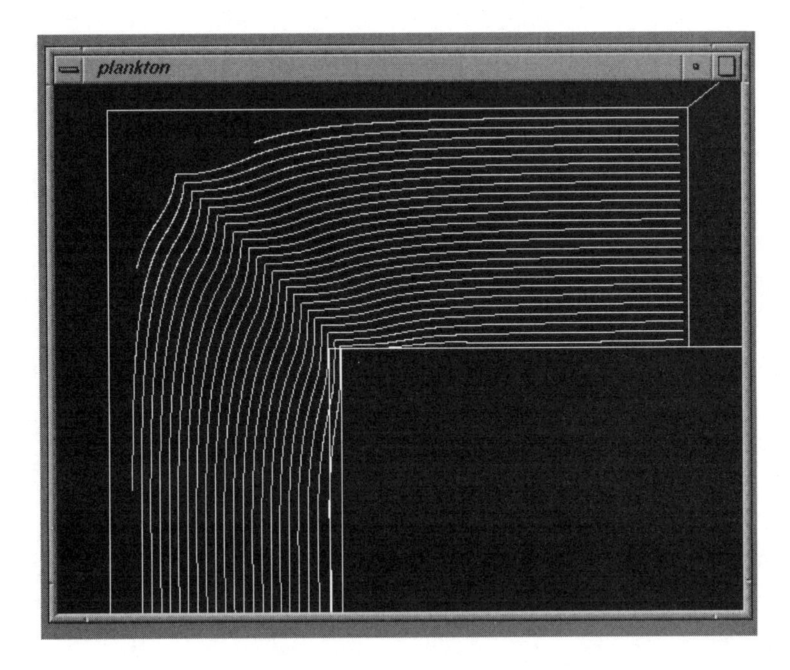

Figure 14.12: Particle paths in L-shaped pipe; Algorithm C-FDBD.

Again, we observe that the results of C-FDBD are better than those of C-CD8, in spite of the fact that C-CD8 is a second-order method and C-FDBD a first-order method. However, the difference between the two methods is that C-CD8 transforms the physical velocity field into a continuous velocity field in C-space, while C-FDBD transforms it into a discontinuous velocity field. Since the grid lines are not differentiable at the grid nodes, the transformed velocity field in C-space must be discontinuous. We will return to this subject in the discussion (Section 14.6.4).

14.6.2 Interpolation Effect

To investigate the effect of the interpolation method on the accuracy of the particle paths, the P-space algorithms (P-T-IDW, P-T-VOL, P-SW-TRI) were examined. These algorithms differ only in their interpolation components. Again, the flow in the L-pipe was used. The same particles as in the previous test were released and their paths traced. Figure 14.13 shows the results for the P-T-VOL algorithm. The results for the other P-space algorithms cannot be visually distinguished and are therefore not included.

14.6.3 Speed Comparison

All algorithms were applied to the flow in the L-shaped pipe and total execution times were measured with timing routines. Table 14.3 lists the results in seconds. The platform used was a SiliconGraphics Iris 4D/310-VGX workstation.

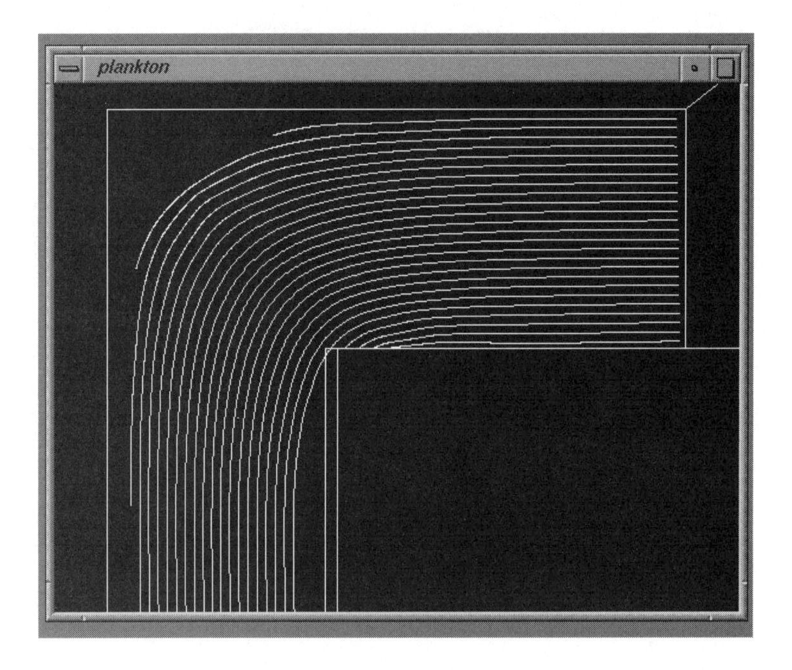

Figure 14.13: Particle paths in L-shaped pipe; Algorithm P-T-VOL.

Algorithm	P-T-IDW	P-T-VOL	P-SW-TRI	
Execution time	56.2	42.4	151.6	
Algorithm	C-FD1	C-FD8	C-CD8	C-FDBD
Execution time	28.3	76.2	76.0	83.1

Table 14.3: Timing results.

14.6.4 Discussion

Accuracy: Interpolation

In general, the differences in accuracy between the P-space algorithms caused by various interpolation methods are marginal. Both the 2D and 3D test cases that we have used have shown this.

Accuracy: Velocity Field Transformation

The accuracy of C-space algorithms is largely determined by the transformation. C-FDBD is the most accurate algorithm. This can be explained from an analysis of the transformation based on a simple test case. Consider the uniform vector field in the four curvilinear cells in Figure 14.14.

When transforming this vector field to computational space, the vectors at the left and right borders ($x = 0$ and $x = 3$) are not a problem because the grid lines are differentiable there. The problem lies in the vectors in the middle of the grid, especially the vector in (2, 1).

Now, assume that particles are released from (1, 0) and (2, 0). In P-space this would result in straight particle paths in $x = 1$ and $x = 2$ (see Figure 14.15). When these particle paths are transformed to C-space, we obtain the curved paths shown in Figure 14.16. Note that here, not vectors are transformed, but *positions* on the paths, using the Newton-Raphson iteration described earlier.

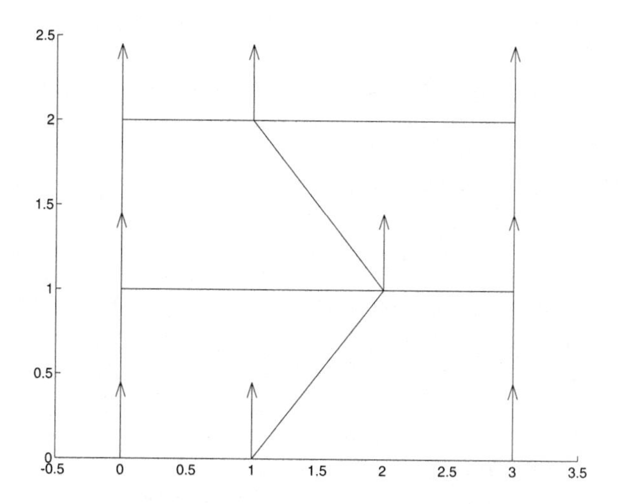

Figure 14.14: Uniform vector field in four cells.

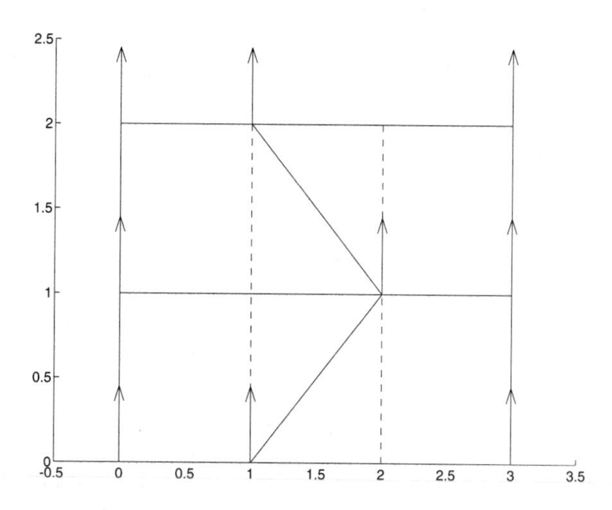

Figure 14.15: Straight particle paths in P-space.

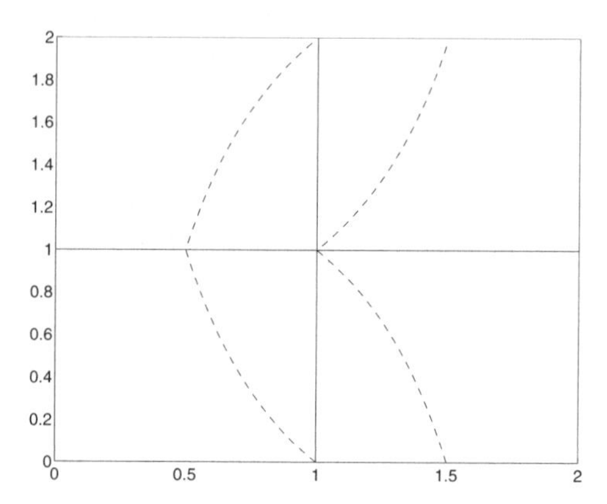

Figure 14.16: Curved particle paths in C-space.

Consider the rightmost path, with a sharp transition at (1, 1). At the cell boundary, the direction of the path changes abruptly. Since the path contains grid node (1, 1) and the velocity in a grid node is defined entirely by the velocity in that node, this means that the velocity vector in (1, 1) must also change in the same way. In other words: the velocity field must be *discontinuous* over cell boundaries. Consequently, the velocity in the corresponding point in P-space, (2, 1), must map to different vectors in C-space, depending on which cell the node belongs to.

The transformation that has this property is `FDBD`, because different Jacobians are used in a node depending on the cell it belongs to. In other words: the continuous velocity field in P-space must be transformed to a discontinuous field in C-space. The reason for this is that the grid lines are also discontinuous, since they are assumed to be straight lines connecting the grid nodes.

This example also shows that the transformations which use either one Jacobian per cell or one Jacobian per node give less accurate results, even if they are second-order, such as the central differencing in the `C-CD8` algorithm. It can also be verified analytically that the tangent vectors of the particle paths are correctly calculated, if all the vectors are transformed with `FDBD`.

Speed

`C-FD1`, which uses one Jacobian per cell, is the fastest algorithm. The other C-space algorithms are roughly six times slower. This can be explained from the fact that eight Jacobians per cell are calculated rather than one.

The P-space algorithms are, in general, faster than the C-space algorithms that use eight Jacobians, although the efficiency of the C-space algorithms could probably be improved upon by applying some optimizations. Among the P-space algorithms, `P-T-VOL` is the fastest. Although the calculation of the weights is more complex in volume weighting than in inverse distance weighting, this disadvantage seems to be outweighed by the fact that only four weights in each cell need to be calculated, rather than eight.

14.7 Conclusions

Particle tracing algorithms have been decomposed into their characteristic components. Various alternatives for these components have been implemented and compared. A distinction has been made between computational space (C-space) and physical space (P-space) algorithms.

The most important component in C-space algorithms is the transformation. Varying this component has a large impact on the accuracy of the results. Varying the interpolation component in P-space algorithms has much less effect on the results.

The accuracy of C-space algorithms depends highly on the deformation of the grid. In Cartesian grids, C-space algorithms are identical to P-space algorithms. If the transformation is not calculated correctly, the accuracy of C-space algorithms decreases rapidly in deformed grids. Among the C-space algorithms, `C-FDBD` has turned out to be the best, both theoretically (see also the appendix) and experimentally.

The reason for considering particle tracing in C-space was efficiency. However, tests have shown that most C-space algorithms are computationally more expensive than P-space algorithms, at least when the data is provided in P-space.

In general, P-space algorithms are more accurate and efficient than C-space algorithms. The effect of varying the interpolation method is small compared to the transformation effect. Therefore, the use of C-space is less useful for particle tracing, but if a C-space algorithm must be used, the `C-FDBD`-algorithm is the best choice.

Acknowledgments

We wish to thank Arthur Mynett of Delft Hydraulics for his help and support and for his comments on earlier versions of this chapter. We thank Jan Mooiman of Delft Hydraulics, for providing interesting data sets and for many useful discussions on fluid dynamics. This work was supported by Delft Hydraulics.

Bibliography

[1] P. Buning, "Sources of Error in the Graphical Analysis of CFD Results," *Journal of Scientific Computing*, Vol. 3, No. 2, 1988, pp. 149–.

[2] P. Buning, "Numerical Algorithms in CFD Post-Processing," *Computer Graphics and Flow Visualization in Computational Fluid Dynamics*, Lecture Series 1989-07, Von Karman Institute for Fluid Dynamics, Brussels, Belgium, 1989.

[3] M. Garrity, "Raytracing Irregular Volume Data," *Computer Graphics*, Vol. 24, No. 5, Nov. 1990, pp. 35–40.

[4] M. Gerritsen, *Geometrical Modelling of 3D Aerodynamic Configurations*, master's thesis, Technische Universiteit Delft, Faculteit Lucht- en Ruimtevaart, 1988.

[5] D.N. Kenwright, Personal Communications, 1994.

[6] K. Kontomaris and T.J. Hanratty, "An Algorithm for Tracking Fluid Particles in a Spectral Simulation of Turbulent Channel Flow," *Journal of Computational Physics*, Vol. 103, 1992, pp. 231–242.

[7] D.A. Lane, "Visualization of Time-Dependent Flow Fields," *Visualization '93*, G.M. Nielson and D. Bergeron, editors, IEEE Computer Society Press, 1993, pp. 32–38.

[8] J. Mooiman, Personal Communications, 1993.

[9] A.E. Mynett, P. Wesseling, A. Segal, and C.G.M. Kassels, "The Isnas Incompressible Navier-Stokes Solver: Invariant Discretization," *Applied Scientific Research*, Vol. 48, 1991, pp. 175–191.

[10] H. Neeman, "A Decomposition Algorithm for Visualizing Irregular Grids," *Computer Graphics*, Vol. 24, No. 5, Nov. 1990, pp. 49–62.

[11] A. Sadarjoen, *Algoritmen voor particle tracing in 3D kromlijnige roosters*, master's thesis, Delft University of Technology, 1994, in Dutch.

[12] S. Shirayama, "Processing of Computed Vector Fields for Visualization," *Journal of Computational Physics*, Vol. 106, 1993, pp. 30–41.

[13] T. Strid, A. Rizzi, and J. Oppelstrup, "Development and Use of some Flow Visualization Algorithms," *Computer Graphics and Flow Visualization in Computational Fluid Dynamics*, Lecture Series 1989-07. Von Karman Institute for Fluid Dynamics, 1989.

[14] D.F. Watson, *Contouring: A Guide to the Analysis and Display of Spatial Data*, Volume 10 of *Computer Methods in the Geosciences*. Pergamon Press, 1992.

[15] J. Wilhelms, J. Challinger, N. Alper, and S. Ramamoorthy, "Direct Volume Rendering of Curvilinear Volumes," *Computer Graphics*, Vol. 24, No. 5, Nov. 1990, pp.41–47.

[16] P. Williams, *Interactive Direct Volume Rendering of Curvilinear and Unstructured Data*, Ph.D. thesis, University of Illinois, 1992.

[17] P.K. Yeung and S.B. Pope, "An Algorithm for Tracking Fluid Particles in Numerical Simulations of Homogeneous Turbulence," *Journal of Computational Physics*, Vol. 79, 1988, pp. 373–416.

A14 Appendix

Calculation of Jacobians

In this appendix, we take a closer look at the transformation of vectors between physical space $\mathcal{P}$ and computational space $\mathcal{C}$. The following is based on the common assumption that points are transformed from $\mathcal{C}$ to $\mathcal{P}$ by trilinear interpolation. In a 2D cell with corner nodes A, B, C, and D (see Figure 14.17), this transformation is defined as

$$\begin{aligned} \mathcal{T}(\alpha,\beta) &= \mathcal{I}(\mathbf{A},\mathbf{B},\mathbf{C},\mathbf{D},\alpha,\beta) \\ &= (1-\alpha)(1-\beta)\mathbf{A} + \alpha(1-\beta)\mathbf{B} + (1-\alpha)\beta\mathbf{C} + \alpha\beta\mathbf{D} \end{aligned} \tag{A14.1}$$

where α and β are normalized coordinates in $\mathcal{C}$.

To transform vectors between $\mathcal{P}$ and $\mathcal{C}$, the transformation Jacobian is used, a matrix with the partial derivatives of the transformation. These can be calculated either continuously or discretely. We will demonstrate that continuous Jacobians are equivalent to the FDBD type of discrete Jacobians, when using trilinear interpolation in a cell.

Continuous Jacobians

In the continuous calculation of a Jacobian, the analytical derivative of the transformation is determined. This derivative is a 2×2 matrix with columns $(\mathbf{j}_1 \mid \mathbf{j}_2) = (\frac{\partial \mathcal{T}}{\partial \alpha} \mid \frac{\partial \mathcal{T}}{\partial \beta})$. With the definition of $\mathcal{T}$ as in Equation (A14.1), we can calculate the first column as:

$$\mathbf{j}_1 = \frac{\partial \mathcal{I}}{\partial \alpha} = -(1-\beta)\mathbf{A} + (1-\beta)\mathbf{B} - \beta\mathbf{C} + \beta\mathbf{D} = (1-\beta)\vec{AB} + \beta\vec{CD}$$

With a similar calculation of $\mathbf{j}_2$, we find

$$\mathbf{J}_{cont}(\alpha,\beta) = \Big((1-\beta)\vec{AB} + \beta\vec{CD} \mid (1-\alpha)\vec{AC} + \alpha\vec{BD}\Big) \tag{A14.2}$$

Discrete Jacobians

In the discrete Jacobian calculation, different Jacobians are calculated in different corners, using finite differences as described in Section 14.3.2. Next, the corner Jacobians are linearly interpolated, in order to correctly represent the cell deformation throughout the entire cell. We will describe how the three types of differencing affect the resulting Jacobians.

Forward Differences: In the case of the C-FD1-algorithm which uses one Jacobian per cell, we have: $\mathbf{J}_A = \mathbf{J}_B = \mathbf{J}_C = \mathbf{J}_D = (\vec{AB} \mid \vec{AC})$. Interpolating does not have any influence in this case, and the result is not equal to $\mathbf{J}_{cont}$ (Equation (A14.2)).

Central Differences: In the approximation with central differences, the Jacobians in each node consist of columns with the averages of the edges connected to that node (see Figure 14.17).

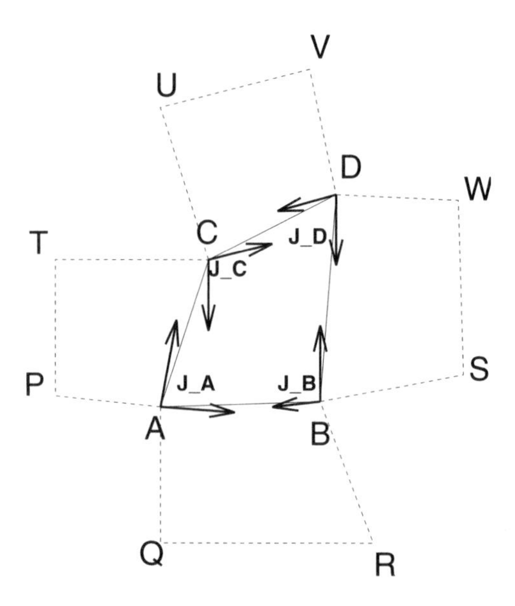

Figure 14.17: Jacobians with central differences.

In this case, the Jacobians in the respective corners are given by:

$$\begin{aligned}
\mathbf{J}_A &= (\frac{1}{2}(\vec{PA} + \vec{AB}) \mid \frac{1}{2}(\vec{QA} + \vec{AC})) \\
\mathbf{J}_B &= (\frac{1}{2}(\vec{AB} + \vec{BS}) \mid \frac{1}{2}(\vec{RB} + \vec{BD})) \\
\mathbf{J}_C &= (\frac{1}{2}(\vec{TC} + \vec{CD}) \mid \frac{1}{2}(\vec{AC} + \vec{CU})) \\
\mathbf{J}_D &= (\frac{1}{2}(\vec{CD} + \vec{DW}) \mid \frac{1}{2}(\vec{BD} + \vec{DV}))
\end{aligned} \tag{A14.3}$$

After interpolation, the first column of the resulting Jacobian is:

$$\begin{aligned}
\mathbf{j}_1 &= \frac{1}{2}(1-\alpha)(1-\beta)(\vec{PA} + \vec{AB}) + \frac{1}{2}\alpha(1-\beta)(\vec{AB} + \vec{BS}) \\
&\quad + \frac{1}{2}(1-\alpha)\beta(\vec{TC} + \vec{CD}) + \frac{1}{2}\alpha\beta(\vec{CD} + \vec{DW}) \\
&= \frac{1}{2}(1-\beta)\left(\vec{AB} + (1-\alpha)\vec{PA} + \alpha\vec{BS}\right) + \frac{1}{2}\beta\left(\vec{CD} + (1-\alpha)\vec{TC} + \alpha\vec{DW}\right)
\end{aligned} \tag{A14.4}$$

This is different from the first column of $\mathbf{J}_{cont}$ (Equation (A14.2)), so again, the discrete Jacobian is not equal to the continuous one.

Forward/Backward Differences: In this case the neighboring cells are not used. However, the approximating Jacobians are better, because *within* the appropriate cell they form a better representation of the cell deformation (see Figure 14.18). Now, the Jacobians

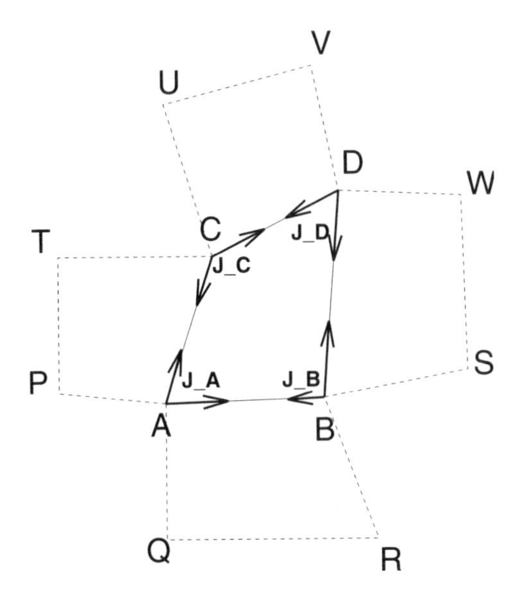

Figure 14.18: Jacobians with FDBD differences.

are given by:

$$\begin{aligned} \mathbf{J}_A &= (\vec{AB} \mid \vec{AC}) \\ \mathbf{J}_B &= (\vec{AB} \mid \vec{BD}) \\ \mathbf{J}_C &= (\vec{CD} \mid \vec{AC}) \\ \mathbf{J}_D &= (\vec{CD} \mid \vec{BD}) \end{aligned} \tag{A14.5}$$

When these are interpolated, the resulting Jacobian is identical to $\mathbf{J}_{cont}$:

$$\begin{aligned} \mathbf{J}_{disc}(\alpha, \beta) &= \mathcal{I}(\mathbf{J}_A, \mathbf{J}_B, \mathbf{J}_C, \mathbf{J}_D, \alpha, \beta) \\ &= (1-\alpha)(1-\beta)\mathbf{J}_A + \alpha(1-\beta)\mathbf{J}_B + (1-\alpha)\beta\mathbf{J}_C + \alpha\beta\mathbf{J}_D \\ &= \Big((1-\beta)\vec{AB} + \beta\vec{CD} \mid (1-\alpha)\vec{AC} + \alpha\vec{BD}\Big) \end{aligned} \tag{A14.6}$$

Hence, in contrast to forward or central differencing, FDBD-differencing gives a correct approximation of the transformation Jacobian within the cell, provided that trilinear interpolation is used for that transformation.

Chapter 15

Variations of the Splitting Box Scheme for Adaptive Generation of Contour Surfaces in Volume Data

Michael Stark, Heinrich Müller, and Ulrike Welsch

Abstract. *A widespread approach to generating isosurfaces or contour surfaces in volume data is the so-called marching cubes algorithm. This algorithm, however, has the disadvantage that the number of polygonal chains generated is considerable. Several attempts to reduce the size of output were undertaken in the past. For instance, we have suggested the splitting box approach. An important advantage against other adaptive solutions is that the problem of cracks inherent to adaptive approximations is solved in a way that neighboring boxes can be processed independently without storing additional context information. In this contribution, a new algorithm is described. It basically follows the splitting box approach, but now bilinear interpolation on grid faces and trilinear interpolation on grid cubes are used. It turns out that the new approach is conceptionally easier to describe and to implement than the original splitting box algorithm. Experimental investigations on several data sets showed, however, that the data reduction is inferior compared with the splitting box algorithm. From an analysis of the causes, an improvement of both algorithms is derived. Further investigations show that the results of the splitting box algorithm are almost optimal.*

15.1 Introduction

The generation of polygonal approximations of contour surfaces in volume data is a widespread approach of scientific visualization in quite different areas of application. In this chapter we assume the volume data to be given by a discrete regular 3D grid of points attributed with values sampled from a scalar function over a finite continuous interval in space.

The contour surfaces are defined by partitioning the grid points in two classes. One example is to distinguish grid points with values above a given threshold from grid points with values below the threshold. For continuous volume data, that is, for a continuous scalar function of the interval observed, the contour surface is just the set of points with values equal to the threshold, that is, an isosurface. However, in the case of noncontinuous volumes as may occur in medical applications, the threshold never needs to be achieved between two separated grid points. For this reason we will prefer the notion of a *contour surface* against the more usual notion of an isosurface.

One approach to calculating contour surfaces in volume data is the *marching cubes algorithm* described in [3]. The marching cubes algorithm examines the cubes defined by eight neighboring grid points. A vertex of a cube is called *black* if its value is greater than or equal to the threshold, and *white* otherwise. The marching cubes algorithm is based on the assumption that the contour surface only has intersections with those edges of a cube which have vertices of different color, and that there is exactly one intersection of the contour surface with an edge of this sort. The intersection point between the edge and the contour surface is estimated by linearly interpolating the values of the vertices. These *contour points* are then connected by polygonal chains separating the black and white vertices of the cube. These *contour chains*, augmented by an inner triangulation, define the contour surface resulting from the marching cubes algorithm.

Since each vertex may be colored either black or white, there are 256 ways of coloring a cube. Using symmetries, the number of really different configurations can be considerably reduced. Care has to be taken in the case of cubes with *ambiguous* faces. These are faces with black vertices on one diagonal and white vertices on the other. One approach to deal with this problem by using bilinear interpolation on the face of a cube is described in [5]. With these basic configurations, a look-up table can be established yielding the contour chain and its triangulation immediately for each cube of the input volume. The marching cubes algorithm scans all cubes of the input grid and reports their triangulated contour chains found in the look-up table.

One disadvantage is the huge amount of contour chains usually generated by the marching cubes algorithm. Since the amount of data has an effect on further modeling and visualization steps, methods using adaptive approximation have been proposed. A topic related to the generation of contour surfaces from volume data is the polygonization of implicit surfaces. Adaptive methods for this problem are described in [1] and [7]. There, a cuboidal part of space is organized in an octree structure. The octree cells are divided until the implicit surface meets certain criteria inside a cell or a maximum number of levels is reached. The surface is approximated in the octree leaves. Since neighboring cells can be of different size, cracks have to be avoided. Both methods use neighborhood information to avoid holes in the approximated surface. The splitting box algorithm described in [4] solves this problem without the use of neighborhood information. To avoid cracks, the approximation of a contour curve depends only on properties of this curve. By forcing the approximation to be the same on neighboring faces even of different size, cracks are avoided.

The remainder of this chapter starts with a new method called the trilinear splitting box algorithm which is conceptually simpler and easier to implement than the splitting box algorithm described in [4]. It basically follows the splitting box approach, but now bilinear interpolation on grid faces and trilinear interpolation on grid cubes are used. The new method is described in Section 15.2. Experimental results are presented in Section 15.3. Unfortunately, it turns out that the number of surface patches needed to approximate the contour surface is higher in comparison with the original splitting box algorithm. From an investigation of the behavior of the trilinear box algorithm, an improvement of both algorithms is derived. This is described in Section 15.4. Finally, the number of surface patches generated by the splitting box algorithm is compared with the number of surface patches generated by a "lower bound" algorithm. This algorithm uses the same recursive subdivision scheme but does not take into account any interpolation conditions. The investigation shows that the splitting box algorithm behaves quite well on realistic data sets.

15.2 Adaptive Trilinear Interpolation

Before starting to describe the new method, some concepts have to be introduced. The method presented deals mainly with cuboidal parts of the grid which are called *boxes* for short. A box consisting of only eight neighboring grid vertices is called *elementary*. The idea of the new method is to search for boxes in the volume data set, which are as big as possible and have the *tl-property*. A box is *tl* if at each grid point in the box the color of the grid point value is equal to the result obtained by trilinear interpolation of the values at the eight vertices of the box. This property guarantees that the topology of the contour surface, which is determined by the black and white grid points, is equal to the result of the marching cubes algorithm as long as no elementary boxes with ambiguous faces are processed. Only the position of a contour surface between two grid points of different color is supposed to change. Inside a box, the contour surface is represented by at most four implicit trilinear surface patches. Therefore, the contour surface is approximated by a piecewise trilinear surface.

15.2.1 The Trilinear Splitting Box Algorithm

Starting with the whole grid, each box is recursively split into two parts. A box is split perpendicular to its four longest edges, see Figure 15.1. If there is not just one such direction, the possible directions are processed in a predetermined order. The recursive subdivision scheme leads to a binary tree of boxes which is traversed in a depth-first order by the algorithm.

During the subdivision process only the edges of a box are considered. An edge is called *valid* if the color of all grid points on the edge is the same as for the result of the linear interpolation of the two values at the vertices of the edge, see Figure 15.2. The values inside a box are checked during the recursive subdivision process when they become part of the edges of a subbox. Furthermore, for each box, two attributes are computed which are related to the validity of the edges. A box is called *valid* if all its edges are *valid*. This attribute is computed before the recursive subdivision of a box. A box which is *valid* is a good candidate for the tl-property. This is checked by recursive subdivision. After returning from the recursion, the attribute concerning the tl-property is computed. An elementary box is always *tl* as long as it has no ambiguous face. An elementary box with ambiguous faces is *tl* if the topology of the contour surface is equal to the elementary box with the original

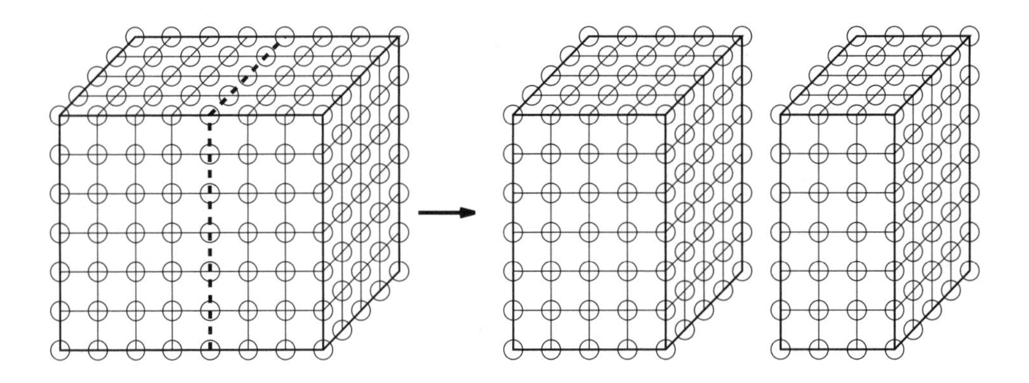

Figure 15.1: Bisecting a box perpendicular to its four longest edges.

grid values. The tl-attribute of a box which is not elementary is computed by combining the tl-attributes of its two subboxes and its own attribute *valid*. A box is *tl* if it is *valid* and its two subboxes are *tl*. Finally, a box is part of the result if it is *tl* and its direct ancestor in the subdivision tree is not.

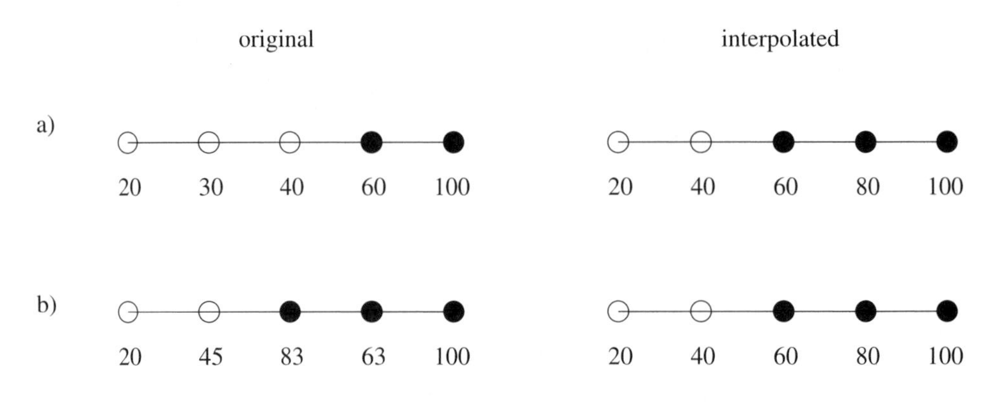

Figure 15.2: a) *invalid* edge b) *valid* edge for threshold 50.

Using trilinear interpolation, all the information about the values inside a box is contained in the values of the eight vertices. Therefore, this information has to be propagated in a useful way to the two subboxes. When a box is split, each of the old vertices is passed on to the respective subbox. The values at the four new vertices have to be computed in a way reflecting the trilinear interpolation scheme. This is included in the processing of the edges. When a box is split into two subboxes, three kinds of edges have to be processed:

1. **Old edges:** Eight edges of a box are not influenced by the splitting process. These edges are passed on together with their attribute *valid* to the respective subbox.

2. **Split edges:** Four edges of a box are affected by the splitting process. If such an edge is *valid*, the attribute *valid* is passed on to the two subedges. Furthermore, the value

at the split point, which is a new vertex of the subboxes, is replaced by the result of the linear interpolation. Otherwise the original grid value is used for the new vertex and the attribute *valid* is computed for the two subedges.

3. **New edges:** The attribute *valid* for the four new edges is computed after the split edges have been processed, so that the values at the new vertices are already updated if necessary.

A pseudocode representation of the trilinear box algorithm is included in the appendix of this paper. The asymptotic runtime of the algorithm is linear in the number of grid points. The additional storage is logarithmic because only one path of the binary box tree is kept at a time. A detailed analysis of runtime and space behavior of the splitting box algorithm is included in [4]. The asymptotical analysis of the trilinear box algorithm is equivalent.

15.2.2 The Problem of Cracks

A typical problem of hierarchical spatial subdivision algorithms is the problem of cracks. Cracks can occur if neighboring boxes of the result are of different size. This can lead to gaps in the approximated contour surface. The trilinear box algorithm avoids this problem by forcing neighboring boxes to have the same values on common edges and faces. If these values are the same, the contour points and contour curves are also the same on common edges and faces. Initially, an edge of grid points not on the border of the grid is common to four boxes. If one of these boxes is split, the edge is always processed in the same way without regard to the rest of the box. Therefore, the effect of the splitting process on this edge is always the same. Similarly, a face of grid points not on the border of the grid is common to two boxes. This face is split in the same way for both boxes. This leads to exactly the same edges in the recursive subdivision process. Since values on common edges are the same, values on common faces are also the same.

15.2.3 Approximating Trilinear Patches by Polygons

A simple approximation of the trilinear patches by polygons for display on screen can be achieved as follows. Polygons are generated inside each elementary box approximating its patches. When the trilinear box algorithm returns from the recursive subdivision, the polygons of the two subboxes can be combined. If both subboxes are *tl* and the box is *valid*, the polygons in the two subboxes can be combined to form an approximation for the patches in the box. The pictures shown in this chapter are obtained by drawing these polygons.

15.3 Experimental Results

In order to compare the number of surface patches generated by the marching cubes algorithm, the splitting box algorithm, and the trilinear box algorithm, the new trilinear box algorithm was implemented in the programming language C. All experimental results were obtained on a Sun SPARC 10-20 workstation. The implementation of the marching cubes algorithm is a variant of the algorithm described in [9]. This has considerable influence on the runtime of the algorithm, which is slower than the look-up table solution originally proposed in [3]. Nevertheless, the number of surface patches generated remains the same as long as the same scheme for dealing with ambiguous faces is used, cf. for instance [5].

The algorithms were applied to three synthetical data sets and two data sets obtained by physical measurements. Figures 15.3 to 15.6 show that the resulting contour surfaces have quite different characteristics. All images of contour surfaces were rendered using flat shading and additional wire frames to visualize the patch structure. They were created on an SGI Indigo using GEOMVIEW [6]. The number of surface patches generated by the three algorithms is compiled in Table 15.1.

The synthetical data sets were obtained by sampling the functions:

$$
\begin{aligned}
f(x,y,z) &= (x-64)^2+(y-64)^2+(z-64)^2, x,y,z \in [0,128] \\
g(x,y,z) &= (x-64)^2-(y-64)^2+(z-64)^2, x,y,z \in [0,128] \\
h(x,y,z) &= (x-64)^2-(y-64)^2-10\cdot(z-64), x,y,z \in [0,128]
\end{aligned}
$$

By variation of the sampling rate, data sets of different sizes were synthesized. The sampled values were quantized to two byte signed integer values. The contour surfaces were defined by the equations:

$$
\begin{aligned}
f(x,y,z) &= 4095.519 \\
g(x,y,z) &= 2047.519 \\
h(x,y,z) &= 0.519
\end{aligned}
$$

The resulting contour surfaces are quadric surfaces. Obviously, the contour surface defined by the first equation is a sphere. The second equation results in a one-connected hyperboloid. The third equation leads to a saddle-like surface, a hyperbolical paraboloid. Table 15.1 shows that the splitting box algorithm works well for these smooth surfaces and leads to a significant reduction in the number of surface patches in comparison with the marching cubes algorithm. The number of surface patches lies in the range of 30 to 40 percent for the sphere, 17 to 21 percent for the hyperboloid, and 16 to 20 percent for the paraboloid. Usually the results get better as the resolution of the data set increases. The trilinear box algorithm generates at least twice as many surface patches as the splitting box algorithm. The results are especially bad for the sphere where the trilinear approximation cannot follow the convex form of the sphere, leading to the wavy look in Figure 15.3. Figures 15.3 and 15.4 show images of the contour surfaces generated by the three algorithms.

The first physical data set is a medical one. It consists of cross-sectional images of the human head. The original data set consists of 128 slices of a resolution of 256×256 at a quantization of 2 bytes per grid point, containing values between 0 and 1,888. In order to have data sets of different sizes for the analysis of runtimes, smaller data sets were generated by averaging neighboring grid points. A simple segmentation with one threshold value was performed. By analyzing the histogram, the threshold was chosen to roughly describe the boundary between air and skin. Since this is a very simple segmentation method, the result not only contains the skin but also several more or less small structures inside the head. Therefore, Table 15.1 includes values for the whole data set as well as for the skin. The surface patches of the skin were extracted from the patches of the whole data

	surface patches				
	MC	SB	SB/MC	TL	TL/MC
Sphere					
017×017×017	1160	464	40.0%	984	84.8%
033×033×033	4760	1912	40.1%	3784	79.5%
065×065×065	19232	6008	31.2%	14488	75.3%
129×129×129	77096	23088	30.0%	51800	67.2%
Hyperboloid					
017×017×017	1016	208	20.5%	448	44.1%
033×033×033	4160	712	17.1%	1808	43.3%
065×065×065	16656	3448	20.7%	8432	50.6%
129×129×129	66752	12808	19.2%	29904	44.8%
Paraboloid					
017×017×017	940	192	20.4%	332	35.3%
033×033×033	3800	640	16.8%	1308	34.4%
065×065×065	15216	2532	16.6%	5840	38.4%
129×129×129	60956	9940	16.3%	22672	37.2%
Head (skin)					
032×032×016	2540	961	37.8%	1981	78.0%
064×064×032	13806	6223	45.1%	8844	67.6%
128×128×064	95195	47175	49.6%	77958	81.9%
256×256×128	622099	387563	62.3%	551764	88.7%
Head (total)					
032×032×016	2992	1300	42.4%	2338	78.1%
064×064×032	17860	9257	51.8%	14945	83.7%
128×128×064	109109	59555	54.6%	91704	84.0%
256×256×128	820628	572167	69.7%	743131	90.6%
Al_2O_3					
031×012×030	4875	2275	46.7%	3173	65.1%
062×025×030	17086	8469	49.6%	11619	68.0%
125×050×030	54468	30712	56.4%	42456	77.9%
250×100×030	163003	101196	62.1%	140472	86.2%

Table 15.1: Number of generated surface patches.

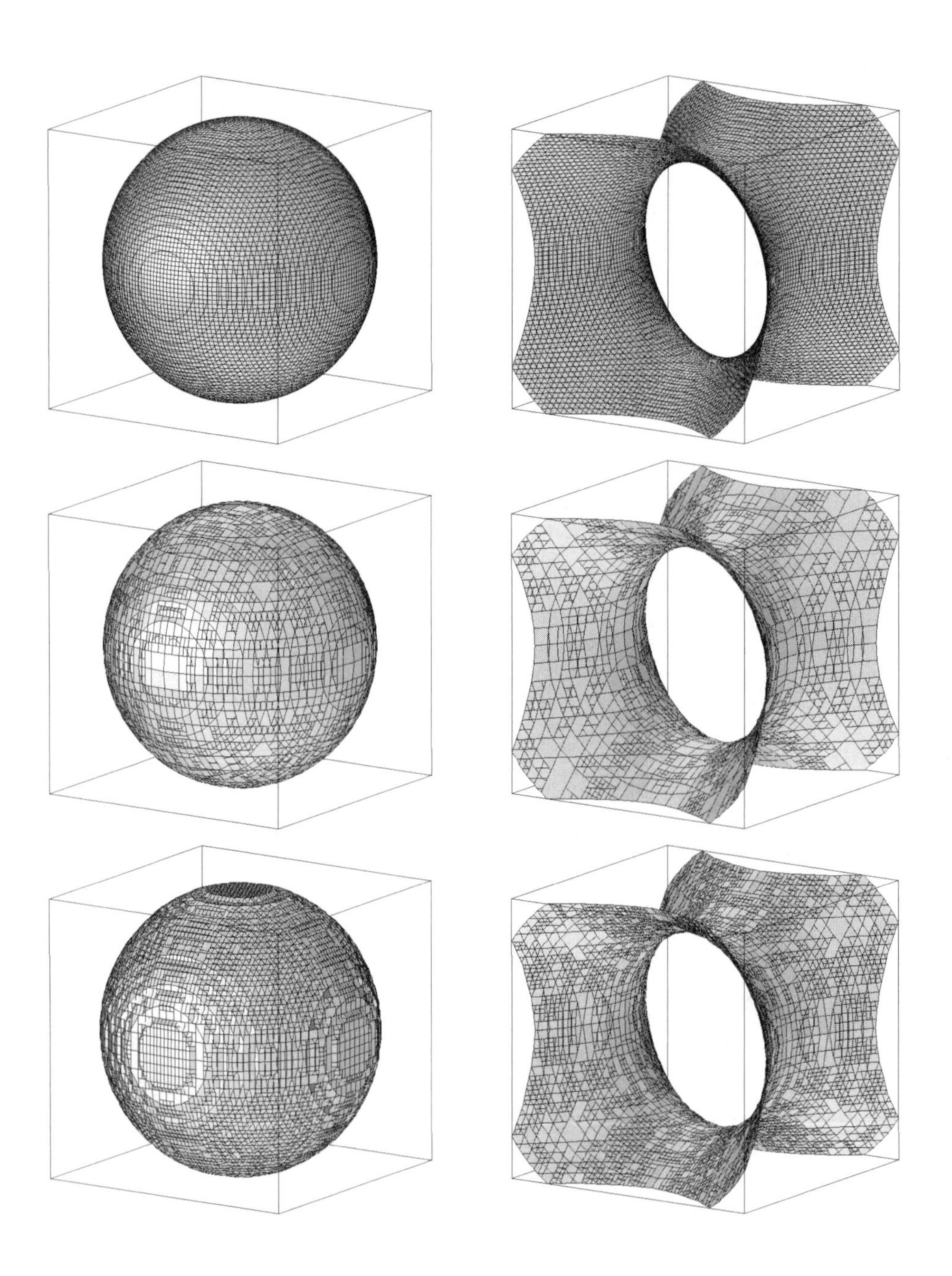

Figure 15.3: Sphere and Hyperboloid generated by the MC, SB, and TL algorithms (from top to bottom) from the data sets of size $65 \times 65 \times 65$.

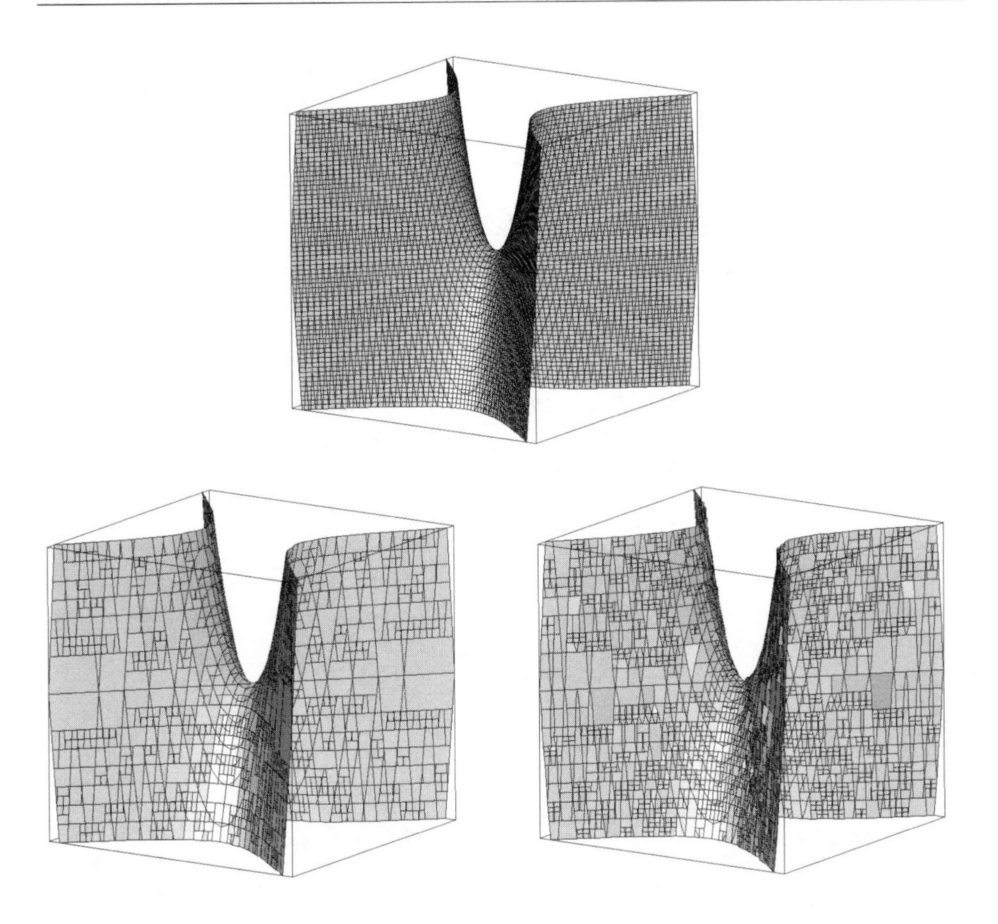

Figure 15.4: Hyperbolical paraboloid generated by the MC *(top)*, SB and TL *(bottom)* algorithms from the data set of size 65 × 65 × 65.

set by computing connected components on this set of patches. Figure 15.5 shows the skin obtained with the different algorithms.

Table 15.1 shows that the percentage of remaining surface patches generated by the splitting box algorithm lies between 38 percent and 62 percent for the skin and between 42 percent and 70 percent for the whole data set. The reductions are not as high as for the quadric surfaces since the data sets contain more detailed and less smooth structures. This is especially true for the results of the whole data set because of the brain structures not visible in Figure 15.5. The gap between the results for the data sets of size 128 × 128 × 64 and 256 × 256 × 128 is caused by noise in the original data set which is filtered out in the smaller data set by the averaging process. As for the synthetic data sets, the reduction in surface patches achieved by the trilinear box algorithm is inferior to the splitting box algorithm. However, the difference is smaller. Also, the visual impression of the contour surface generated by the splitting box algorithm is more pleasant than for the trilinear box algorithm, cf. Figure 15.5.

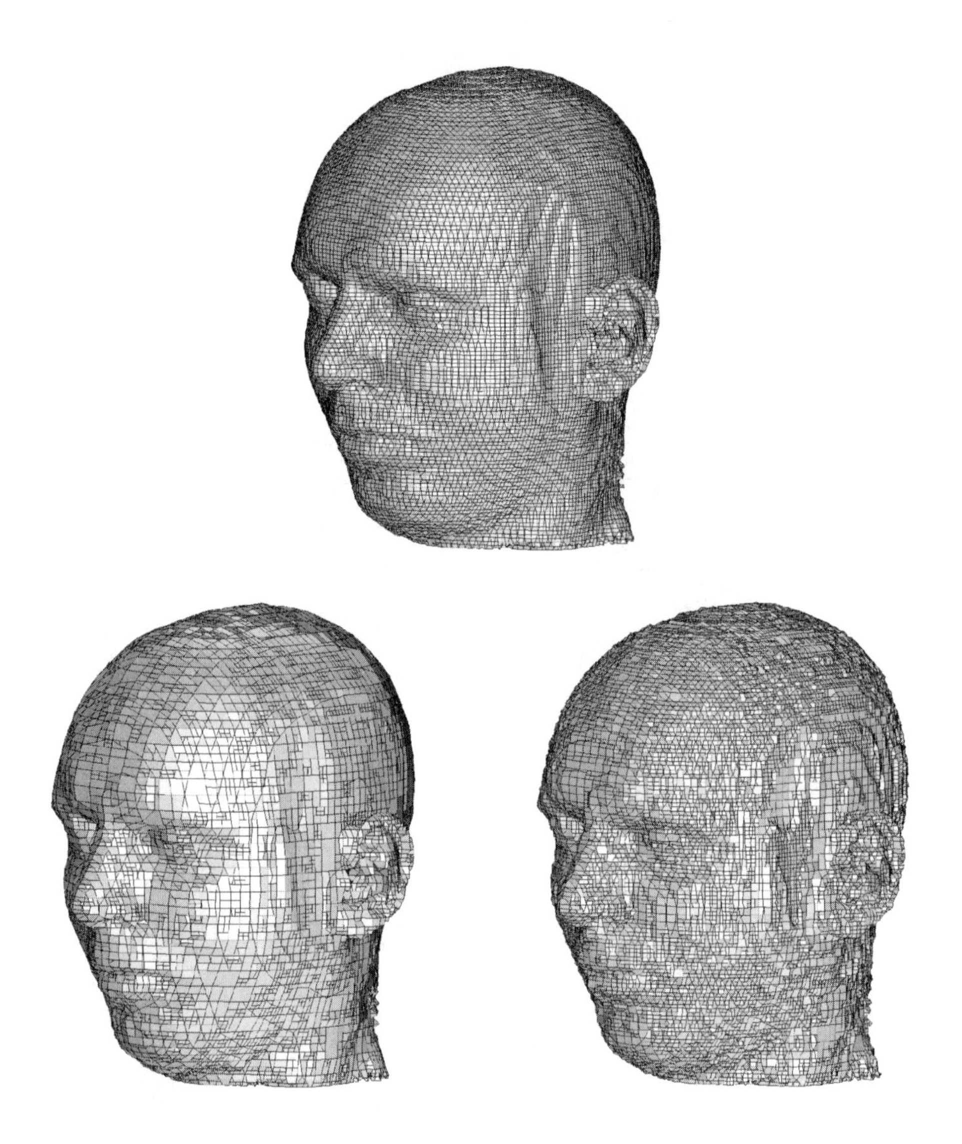

Figure 15.5: Head generated by the MC *(top)*, SB and TL *(bottom, left to right)* algorithms from the data set of size $128 \times 128 \times 64$.

The origin of the second physical data set is materials science. The material investigated is Al_2O_3 contained in a block of plastic. The data set was captured by microcomputer tomography, since the real volume of the investigated object is approximately $9\mathrm{cm}^3$. It consists of 30 slices of resolution 512×512 at a quantization of 1 byte per element. For the experiments, a region of 250×100 was cut out which completely contains the relevant areas. The surface was again determined by a threshold value. The threshold was chosen so that the result shows the boundary of the plastic box as well as the structure inside. Data sets of different size were obtained by averaging within the slices. Figure 15.6 shows contour surfaces generated by the different algorithms for the data set of size $125 \times 50 \times 30$. Table 15.1 shows that the percentage of remaining surface patches decreases for the reduced data sets from 62 percent to 46 percent for the splitting box algorithm and from 86 percent to 65 percent for the trilinear box algorithm. This tendency can be explained with the averaging procedure. The difference between the results of the splitting box and trilinear box algorithms are not as big as for the quadric surfaces. Nevertheless, the trilinear box algorithm produces significantly more patches.

Figure 15.6: Contour surface of Aluminum Oxide generated by the MC, SB, and TL algorithms (from top to bottom) from a data set of size $125 \times 50 \times 30$.

	Splitting-Box	Trilinear-Box
Sphere		
017×017×017	1.2	1.8
033×033×033	6.4	8.8
065×065×065	37.1	47.0
129×129×129	256.7	286.5
Hyperboloid		
017×017×017	0.8	1.2
033×033×033	4.8	6.6
065×065×065	33.3	40.0
129×129×129	241.5	259.8
Paraboloid		
017×017×017	0.8	1.0
033×033×033	4.6	5.9
065×065×065	31.7	36.6
129×129×129	236.3	249.5
Head (total)		
032×032×016	4.7	7.5
064×064×032	35.0	48.4
128×128×064	250.3	326.3
256×256×128	1720.9	1937.8
Al_2O_3		
031×012×030	4.6	6.1
062×025×030	20.0	26.6
125×050×030	80.0	100.9
250×100×030	288.3	353.5

Table 15.2: Runtime in CPU seconds.

Table 15.2 contains the runtimes of the splitting box algorithm and the trilinear box algorithm for data sets of different size. The absolute values are less noteworthy since they depend on the implementation. Rather, they should be regarded as giving the tendency of the runtime for different sizes of input. The important fact is that an increase of the input data by a factor of eight (four for the aluminum oxide) leads to an increase in the runtime smaller than eight (four). This gives evidence for the linear asymptotic run-time behavior of both algorithms. The runtime of the splitting box algorithm is better because the size of the output data is smaller compared with the trilinear box algorithm.

15.4 Improvements

The trilinear box algorithm described so far does not work as well as expected. A detailed analysis of the internal behavior shows that some problems result from interpolation on long edges, although interpolation is only carried out on edges which are *valid*. Interpolation on long edges is of a rather global nature leading to significant changes of values at grid points on the edge. Once the value is changed by interpolation, it is fixed for the rest of the recursive subdivision process. Even if this change leads to subboxes in the subdivision process which are not *valid*, this change cannot be cancelled.

Therefore, it seems useful not to start the algorithm on the whole data set as described before. Instead, the data set is initially partitioned into smaller boxes and the algorithm is started on these boxes. This can be implemented in a simple way by restricting the *valid* property to edges which have a length below a certain value. However, the restriction of the edge length for *valid* edges leads to two conflicting effects on the reduction of surface patches. On one hand, the restriction of the edge length leads to a more local approximation of the data set, increasing the probability that a small *valid* box is really proven to be *tl* by the recursive subdivision process. This leads to a reduction in the number of surface patches. On the other hand, restricting the edge length of *valid* edges and thereby *valid* boxes, also restricts the size of surface patches used for the approximation of the contour surface. So, if there are areas of the contour surface which can be approximated by large patches, this restriction leads to an increase in the number of patches. Therefore, finding the best value for the restriction of edge lengths is an optimization problem. Figure 15.7 shows the number of generated surface patches depending on the edge length. The edge length is defined as the number of grid points on an edge minus 1 so that an elementary box has edges of length 1. The shapes of the plots look like staircases because only certain edge lengths appear during the subdivision process. For the three synthetical data sets, there exist clear global minimums as a result of the two conflicting effects described before. In practice, finding a good value for the edge length is quite easy. For that purpose the algorithm is run without restriction to the edge length. The surface patch of maximum size included in the result indicates the size of the box where it was created. Using edge lengths close to the edge length of this box is a good choice. If much smaller edge lengths are used, the size of resulting surface patches is restricted too much. Much longer edges will lead to the same result as achieved without any edge length restriction.

Table 15.3 shows the number of generated surface patches using an optimal choice for the edge length. The result of the trilinear box algorithm for the three quadric surfaces is better now. Unfortunately, the effect on physical data sets is much smaller.

The idea of restricting the edge length is also applicable to the splitting box algorithm. This leads also to significant improvements leaving the trilinear box algorithm still behind. Figures 15.8 to 15.10 show the generated contour surfaces for the optimal value in Table 15.3.

15.5 A Lower Bound

Both the original splitting box algorithm and the trilinear splitting box algorithm use the same recursive subdivision scheme: searching for boxes which are as big as possible, have the *mc*-property, and satisfy some interpolation condition on the boundaries of the boxes. The interpolation conditions impose a restriction which is particularly responsible for the different behavior of both algorithms with respect to the number and sort—polygonal or hyperbolic—of chains delivered. Discarding the interpolation condition leads to a "lower bound" algorithm giving a number of chains which is the smallest one achievable by the splitting box scheme. This algorithm is of limited practical interest since no approximating

	surface patches						
	MC	length	SB	SB/MC	length	TL	TL/MC
Sphere							
017×017×017	1160	2	272	23.4%	2	760	65.5%
033×033×033	4760	2	1160	24.3%	2	2344	49.2%
065×065×065	19232	2	4760	24.7%	2	7048	36.6%
129×129×129	77096	4	16312	21.2%	2	26544	34.4%
Hyperboloid							
017×017×017	1016	4	208	20.5%	2	400	39.4%
033×033×033	4160	4	624	15.0%	2	1504	36.1%
065×065×065	16656	4	2216	13.3%	2	4832	29.0%
129×129×129	66752	4	6128	9.1%	4	15280	22.9%
Paraboloid							
017×017×017	940	4	128	13.6%	8	236	25.1%
033×033×033	3800	4	456	12.0%	2	1252	32.9%
065×065×065	15216	8	1968	12.9%	4	3824	25.1%
129×129×129	60956	4	5668	9.3%	4	9064	14.9%
Head (skin)							
032×032×016	2540	4	948	37.3%	4	1933	76.5%
064×064×032	13806	7	6054	43.9%	31	8844	67.6%
128×128×064	95195	4	46577	48.9%	15	77917	81.8%
256×256×128	622099	4	387205	62.2%	16	551337	88.6%
Head (total)							
032×032×016	2992	4	1295	43.3%	4	2278	76.1%
064×064×032	17860	7	9088	50.9%	31	14945	83.7%
128×128×064	109109	4	58957	54.0%	15	91698	84.0%
256×256×128	820628	4	571747	69.7%	16	743046	90.5%
Al_2O_3							
031×012×030	4875	4	2265	46.5%	6	3112	63.8%
062×025×030	17086	7	8468	49.6%	61	11619	68.0%
125×050×030	54468	6	30637	56.2%	13	42343	77.7%
250×100×030	163003	4	101164	62.1%	2	139542	85.6%

Table 15.3: Number of generated surface patches using an optimal value for the edge length.

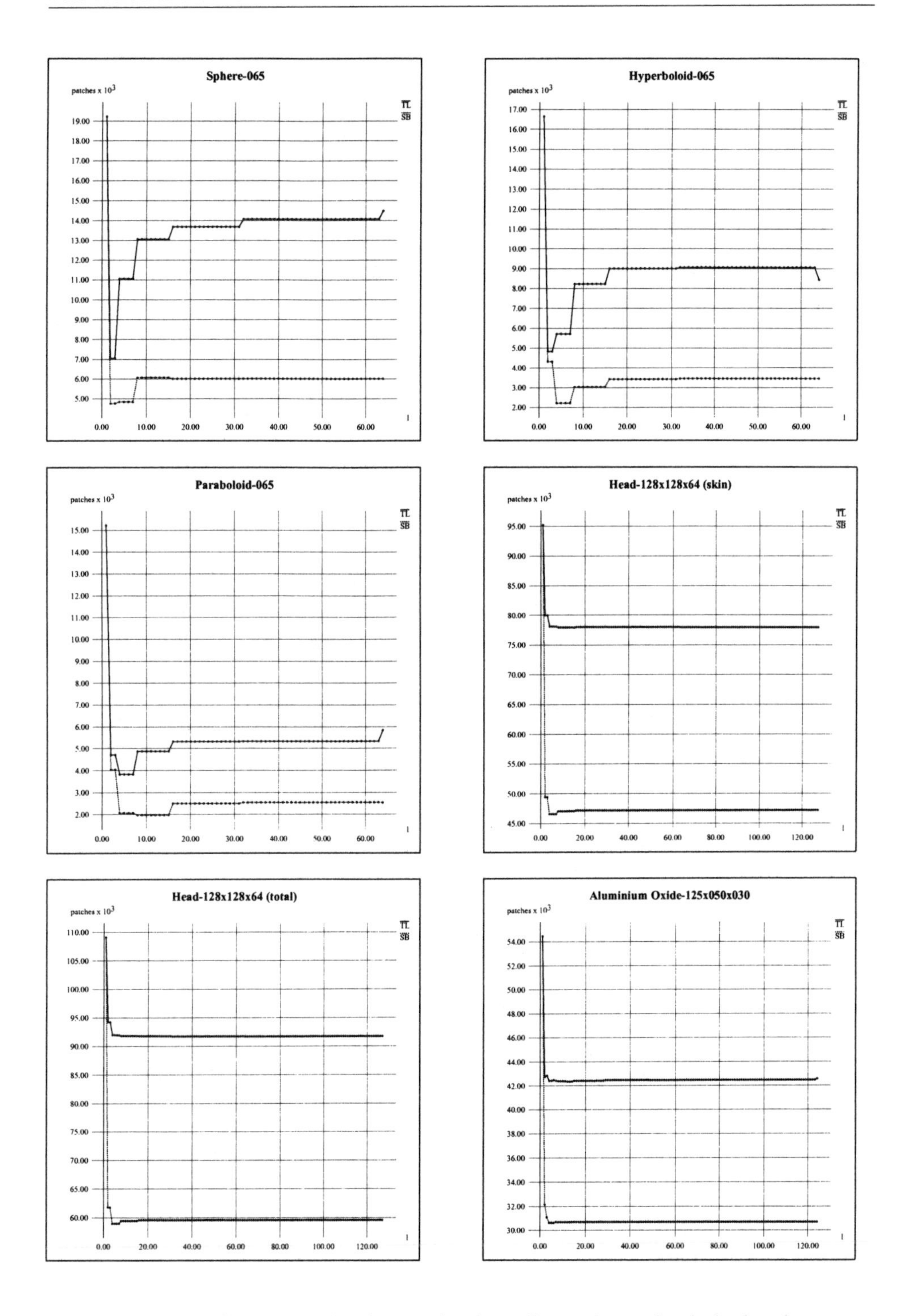

Figure 15.7: Number of surface patches depending on the restricted edge length.

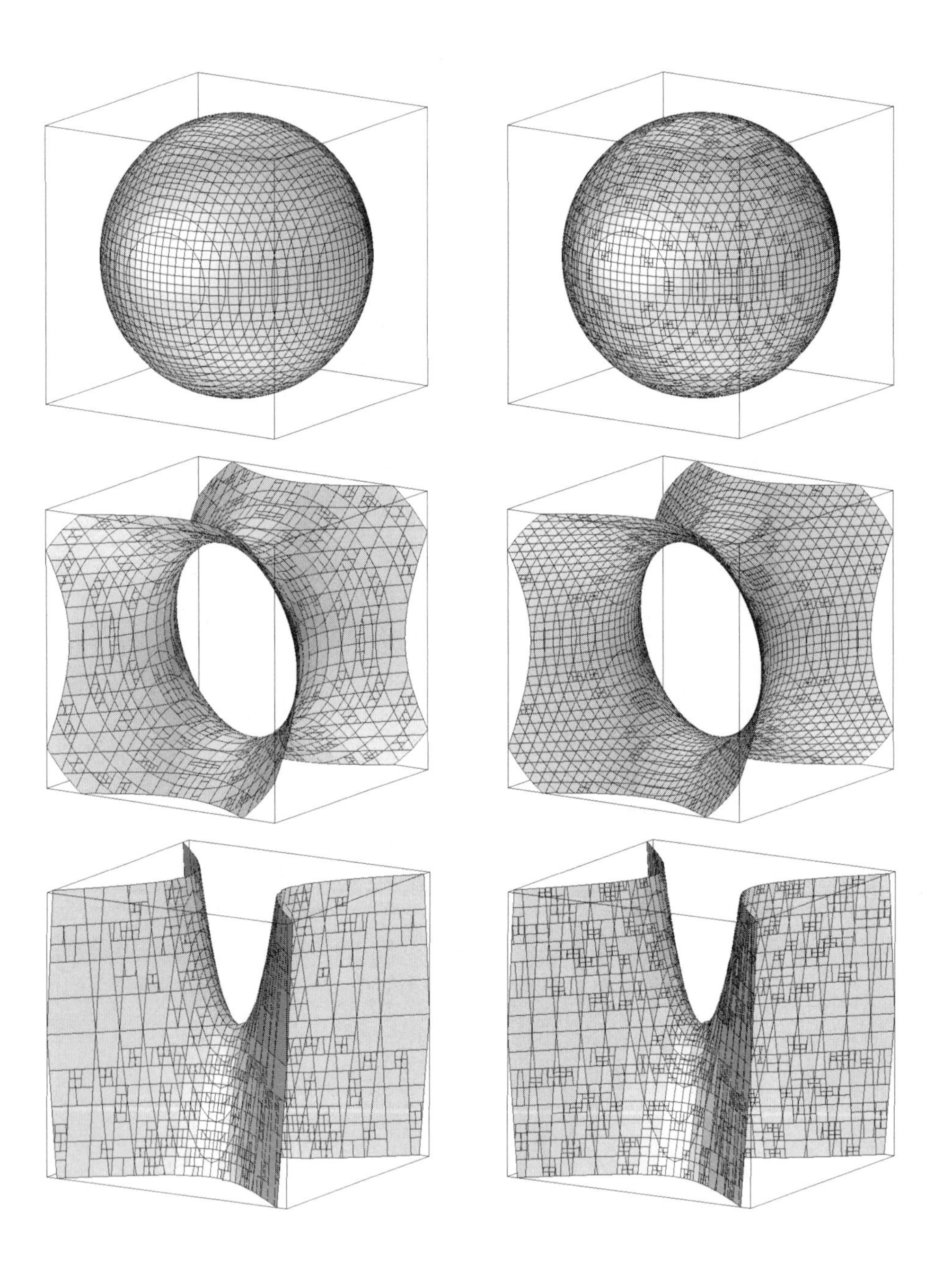

Figure 15.8: Quadric surfaces generated by the SB *(left)* and TL *(right)* algorithms using an optimal edge length.

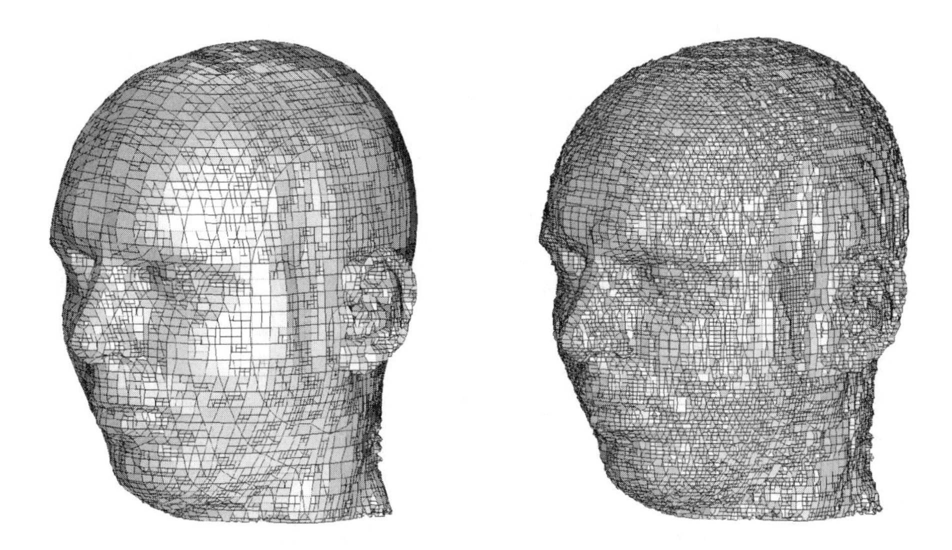

Figure 15.9: Head generated by the SB *(left)* and TL *(right)* algorithms using an optimal edge length.

Figure 15.10: Contour surface of Aluminum Oxide generated by the SB *(top)* and TL *(bottom)* algorithms using an optimal edge length.

patch type is related to the chains obtained. However, it helps to grade the other splitting box algorithms in that it establishes a lower bound on the number of chains.

The algorithm carries out the recursive subdivision until elementary boxes are reached. For each elementary box, contour chains are generated and an attribute called *mc* is set. After returning from the recursion the *mc*-property of a nonelementary box is checked. Such a box is *mc* if all its edges have at most one color change, that is, at most one intersection with the contour surface and its two subboxes are *mc*. In this case, the contour chains of the two subboxes are combined to form bigger contour chains in B.

Analyzing the synthetical data sets in this way leads to eight big contour chains. Table 15.4 shows the number of contour chains generated for the two physical data sets.

	surface patches				
	MC	SB	SB/MC	min	min/MC
Head (skin)					
032×032×016	2540	948	37.3%	509	20.0%
064×064×032	13806	6054	43.9%	4289	31.1%
128×128×064	95195	46577	48.9%	35557	37.4%
256×256×128	622099	387205	62.2%	326807	52.5%
Head (total)					
032×032×016	2992	1295	43.3%	831	27.8%
064×064×032	17860	9088	50.9%	7037	39.4%
128×128×064	109109	58957	54.0%	47418	43.5%
256×256×128	820628	571747	69.7%	505973	61.7%
Al_2O_3					
031×012×030	4875	2265	46.5%	1947	21.9%
062×025×030	17086	8468	49.6%	7317	42.8%
125×050×030	54468	30637	56.2%	25587	47.0%
250×100×030	163003	101164	62.1%	83544	51.3%

Table 15.4: Comparison of the number of surface patches generated by the improved splitting box algorithm and the minimal number of surface patches.

Comparing the number of surface patches generated by the improved splitting box algorithm with the number of surface patches generated by the optimal solution in Table 15.4 shows that the distance between the splitting box algorithm and the optimal solution is quite small.

15.6 Discussion

A different approach to adaptive triangulation is to reduce a given triangular network by retriangulation, cf. for example, [8] and [2]. By these algorithms, impressive reductions are achieved if higher errors of approximation are allowed. The at-first-glance inferior behavior of the trilinear as well as the original splitting box algorithm, in all versions, can be explained by the fact that they perform an approximation within the sampling distance, at least if the respective type of surface patch is used for approximation. Further work should focus on splitting-box-like algorithms allowing higher tolerances. This might be achieved by a combination of the splitting box strategy with a retriangulation algorithm. A first advantage is that the boxes may help to find the patches to be replaced more efficiently. A second advantage might be that the boxes also help to search for possible self-intersections of the coarsened surface more efficiently.

Acknowledgments

Special thanks go to those people who have supported the work by providing test data. Medical data sets were supplied by Dr. B. Kohlberger and J. Burgert from the Universitätsklinik at Freiburg, Germany, in particular, the data of the human head of Kontron GmbH, Germany. Further data sets came from Dr. A. J. Klingert, Fakultät für Informatik, Universität Karlsruhe, for example the aluminum oxide data of the Fraunhofer-Institut für zerstörungsfreie Prüfverfahren (Institute for Nondestructive Test Methods) at Saarbrücken, Germany.

Bibliography

[1] Jules Bloomenthal, "Polygonization of Implicit Surfaces," *Computer Aided Geometric Design*, Vol. 5, No. 4, Nov. 1988, pp. 341–355.

[2] Bernd Hamann, "A Data Reduction Scheme for Triangulated Surfaces," *Computer Aided Geometric Design*, Vol. 11, 1994, pp. 197–214.

[3] William E. Lorensen and Harvey E. Cline, "Marching Cubes: A High Resolution 3D Surface Construction Algorithm," *Computer Graphics*, Vol. 21, No. 4, July 1987, pp. 163–169.

[4] Heinrich Müller and Michael Stark, "Adaptive Generation of Surfaces in Volume Data," *The Visual Computer*, Vol. 9, 1993, pp. 182–199.

[5] Gregory M. Nielson and Bernd Hamann, "The Asymptotic Decider: Resolving the Ambiguity in Marching Cubes," *Visualization 91*, IEEE, 1991, pp. 83–91.

[6] Mark Phillips, *Geomview Manual*, The Geometry Center, University of Minnesota, 1993.

[7] Michael F.W. Schmidt, "Cutting Cubes—Visualizing Implicit Surfaces by Adaptive Polygonization," *The Visual Computer*, Vol. 10, 1993, pp. 101–115.

[8] William J. Schroeder, Jonathan A. Zarge, and William E. Lorensen, "Decimation of Triangle Meshes," *Computer Graphics*, Vol. 26, July 1992, pp. 65–70.

[9] Geoff Wyvill, Craig McPheeters, and Brian Wyvill, "Data Structure for Soft Objects," *The Visual Computer*, Vol. 2, 1986, pp. 227–234.

A15 The Trilinear Box Algorithm

In this appendix, a compact summary of the trilinear box algorithm is given in a Pascal-like notation.

The description of the data structure of a box is reduced to the information necessary for the following presentation. It corresponds to a Pascal record with the following components:

`valid`**:** a boolean variable having the value **true** if the box is valid;

`tl`**:** a boolean variable having the value **true** if the box is tl;

C^***:** a list of at most four contour chains.

```
PROCEDURE Trilinear-Box(B: Box)
BEGIN
  IF box_split(B, B1, B2) THEN
    B1.valid := box_valid(B1);
    B2.valid := box_valid(B2);
    Trilinear-Box(B1);
    Trilinear-Box(B2);
    B.tl := false;
    IF B1.tl
      IF B2.tl
        IF B.valid
          B.tl := true;
          box_combine_polygons(B, B1, B2);
        ELSE
          box_print_polygons(B1);
          box_print_polygons(B2);
          box_remove_polygons(B1);
          box_remove_polygons(B2);
      ELSE
        box_print_polygons(B2);
        box_remove_polygons(B2);
    ELSE
      box_print_polygons(B1);
      box_remove_polygons(B1);
  ELSE
    B.tl := true;
    box_create_polygons(B);
    IF box_ambiguous(B)
      IF not box_check_topology(B)
        box_fix_polygons(B);
        box_print_polygons(B);
        B.tl := false;
END.
```

Chapter 16

Visualization of Deformation Tensor Fields

H. Hagen, S. Hahmann, and H. Weimer

Abstract. *Vector and tensor fields are very important in many application areas. The purpose of this chapter is to present new techniques to visualize deformation tensor fields and infinitesimal bendings.*

16.1 Introduction

Scientific visualization is a new approach in the area of simulation. It allows researchers to observe the results of simulations using complex graphical representations. Visualization provides a method for seeing what is not normally visible, for example, torsion forces inside a body, wind against a wall, heat conduction, flows, plasmas, earthquake mechanisms, molecules, and so on. The purpose of this chapter is to present new visualization techniques for deformation tensor fields. In the second section a mathematical concept for tensor fields is presented. The third section deals with the visualization of deformation tensor fields. In the fourth section infinitesimal deformations are considered and a special vector field is introduced as a "representation or visualization tool" for these special deformations.

16.2 Tensor Fields—A Mathematical Concept

Tensor fields are invariant under parameter transformations and therefore are appropriate "tools" to describe certain "geometric invariant situations."

Definition 16.2.1 $(\mathbf{r}, \mathbf{s})$ **tensor**

(a) V *is an n-dimensional vector space and* V^* *is its dual space. A multilinear map*

$$T : \underbrace{V \times \cdots \times V}_{r} \times \underbrace{V^* \times \cdots \times V^*}_{s} \to \mathbb{R} \tag{16.1}$$

is called an $(\mathbf{r},\mathbf{s})$ **tensor**

(b) $E_1,\ldots,E_n$ *is the basis of* V *and* $E^1,\ldots,E^n$ *is the dual basis of* V^*

$$\begin{aligned} T(A_1,\ldots,A_r,B^1,\ldots,B^s) &= T(a_1^{i_1}E_{i_1},\ldots,a_r^{i_r}E_{i_r},b_{j_1}^1E^{j_1},\ldots,b_{j_s}^sE^{j_s}) \\ &= T(E_{i_1},\ldots,E_{i_r},E^{j_1},\ldots,E^{j_s})a_1^{i_1}\ldots a_r^{i_r}b_{j_1}^1\ldots b_{j_s}^s \\ &=: t_{i_1\ldots i_r}{}^{j_1\ldots j_s}a_1^{i_1}\ldots \end{aligned}$$

The n^{r+s} *real numbers* $t_{i_1\cdots i_r}{}^{j_1\cdots j_s}$ *are called the* **components of the** $(\mathbf{r},\mathbf{s})$ **tensor T**.

Changing basis $\widetilde{E}_i = \alpha_i^j E_j$ *and* $\widetilde{E}^i = \check{\alpha}_j^i E^j$, *an* (r,s) *tensor transforms in this way:*

$$\tilde{t}_{i_1\ldots i_r}{}^{j_1\ldots j_s} = t_{\ell 1\ldots \ell r}{}^{k_1\ldots k_s}\alpha_{i_1}{}^{\ell 1}\ldots\alpha_{i_r}{}^{\ell r}\check{\alpha}_{k_1}{}^{j_1}\ldots\check{\alpha}_{k_s}{}^{j_s}$$

Tensoroperations:

1. Tensors of the same type can be added and scaled like vectors.

2. **Tensorproduct:**

   T – (r,s) tensor and $\widetilde{T}$ – $(\tilde{r},\tilde{s})$ tensor

$$\begin{aligned} &T\circ\widetilde{T}(A_1,\ldots,A_{r+\tilde{r}},B^1,\ldots,B^{s+\tilde{s}}) := \\ &T(A_1,\ldots,A_r,B^1,\ldots,B^s)\cdot\widetilde{T}(A_{r+1},\ldots,A_{r+\tilde{r}},B^{s+1},\ldots,B^{s+\tilde{s}}) \end{aligned}$$

   is called the tensor product of T and $\widetilde{T}$. In components:

$$(t\tilde{t})_{i_1\cdots i_{r+\tilde{r}}}{}^{j_1\cdots j_{s+\tilde{s}}} := t_{i_1\cdots i_r}{}^{j_1\cdots j_s}\tilde{t}_{i_{r+1}\cdots i_{r+\tilde{r}}}{}^{j_{s+1}\cdots j_{s+\tilde{s}}}$$

   $T\circ\widetilde{T}$ is a $(r+\tilde{r},s+\tilde{s})$ tensor

3. **Contraction of a tensor:**

$$T \quad (r,s)\text{tensor} \quad \rightarrow \quad \bar{T} \quad (r-1,s-1)\text{ tensor}$$

$$t_{i\cdots i_r}{}^{j_1\cdots j_s} \quad \rightarrow \quad t_{ii_2\cdots i_r}{}^{ij_2\cdots j_s}$$

   Example: $\{g_{ij}\}$ and $\{h_{ij}\}$ are the components of the first and second fundamental forms of a surface and H is the mean curvature.

$$\begin{aligned} &g^{ij}\,;\,h_{rs} \underset{\text{tensorpr.}}{\longrightarrow} g^{ij}h_{rs} \underset{\text{contr.}}{\longrightarrow} g^{ij}h_{is} \\ &g^{ij}h_{is} \underset{\text{contr.}}{\longrightarrow} g^{ij}h_{ij} = 2H \end{aligned}$$

4. **Inner multiplication with a metric tensor:**

$\{g_{ij}\}$ are the components of a nondegenerated metric tensor

$$t_{i_1\cdots i_r}{}^{j_1\cdots j_s} \qquad g^{ij}t_{j_{i_2}\cdots ir}{}^{j_1\cdots j_s} =: t^{i}_{i_2\cdots i_r}{}^{j_1\cdots j_s}$$

$$t_{i_1\cdots i_r}{}^{j_1\cdots j_s} \qquad g_{ik}t_{i_1\cdots ir}{}^{kj_2\cdots j_s} =: t_{i_1\cdots ir_1}{}^{j_2\cdots js}$$

Example: Weingarten map $h_i{}^j = h_{ir}g^{rj}$

5. **Covariant differentiation:**

The "normal" partial differentiation of an arbitrary tensor is in general not a tensor! Motivated by certain well-known facts from differential geometry [1], we modify the differentiation process:

Theorem and Definition:

A is an (r,s) tensor with components $a_{\ell_1\cdots\ell_r}{}^{q_1\cdots q_s}$.

$$a_{\ell_1\cdots\ell_r}{}^{q_1\cdots q_s}||_i := \frac{\partial a_{\ell_1\cdots\ell_r}{}^{q_1\cdots q_s}}{\partial u^i} - \sum_{m=1}^{r} a_{\ell_1\cdots\ell_{m-1}k\ell_{m+1}\cdots\ell_r}{}^{q_1\cdots q_s}\Gamma^k_{\ell_m i} + \sum_{m=1}^{s} \alpha_{\ell_1\cdots\ell_r}{}^{q_1\cdots q_{m-1}pq_{m+1}\cdots q_s}\Gamma^{q_m}_{pi} \tag{16.2}$$

are the components of an $(r+1,s)$ tensor $A_{||_i}$. This tensor is called the *covariant differentiation* of the (r,s) tensor $a_{\ell_1\cdots\ell_r}{}^{q_1\cdots q_s}$ with respect to the metric tensor g_{ik} (the $\Gamma^k_{i_j}$ are the Christoffelsymbols of the metric tensor g_{ik}).

Application: geodesic curves on a surface

16.3 Visualization of Deformation Tensor Fields

$\{g_{ij}\}$ are the components of the metric tensor of a surface, represented by a regular vector valued map $X : U \to \mathbb{R}^3$, U is a connected domain of $\mathbb{R}^2$. $\{\tilde{g}_{ij}\}$ are the components of the metric tensor of the deformed surface $\tilde{X} : U \to \mathbb{R}^3$.

Then the tensor

$$\gamma_{ik} = \frac{1}{2}(\tilde{g}_{ik} - g_{ik}) \tag{16.3}$$

is called *deformation tensor.*

The purpose of this section is to give the user insight into the deformation tensor field of a surface for a given deformation. We do this by deriving values from the deformation tensor representing, for example, the minimum or maximum deformation at a given surface point and visualizing the distribution of these values on the original surface X. For visualization of the deformation tensor characteristics, we propose two different approaches, one based on generalized focal surfaces, the other using characteristic curves.

γ represents a measurement for deformation on the surface X without referring back to the space $\mathbb{R}^3$ in which the surface lies. This observation implies that an interrogation tool for deformation should preserve this domain independence. The two approaches introduced in this chapter preserve the domain independence. Both are constructed in local coordinate systems only referring back to the point to be interrogated and to the metric tensor of X.

The deformation tensor $\{\gamma_{ij}\}$ is symmetric and real-valued as $\{g_{ij}\}$ and $\{\tilde{g}_{ij}\}$ are symmetric and real-valued. From linear algebra it is well known that a real symmetric matrix always has real eigenvalues and the corresponding eigenvectors are perpendicular (see [6]).

16.3.1 Generalized Focal Surfaces—Basic Concepts

Generalized focal surfaces were introduced in [3] for curvature interrogation of regular surfaces. There they were used to test the convexity of a surface, to pinpoint flat regions and flat points, and to visualize the technical smoothness of surfaces.

In this work we want to apply the concept of generalized focal surfaces for visualization of deformation tensor fields derived from a given deformation of a surface X in three-dimensional space.

A generalized focal surface F for a regular surface X (that is, $X_u \times X_v \neq 0$) and parameter d is the locus of all points

$$F(\mathbf{x}) = X(\mathbf{x}) + a\, d(\mathbf{x})\, N(\mathbf{x}), \quad \mathbf{x} \in U \subset \mathbb{R}^2 \tag{16.4}$$

where X denotes the surface $X : U \to \mathbb{R}^3$ to be interrogated, $d(\mathbf{x})$ is a scalar-valued map $d : U \to \mathbb{R}$, N denotes the field of normals $N : U \to \mathbb{R}^3$ for surface X, that is, $N(\mathbf{x}) \cdot X_u(\mathbf{x}) = 0$, and $N(\mathbf{x}) \cdot X_v(\mathbf{x}) = 0$ for any $\mathbf{x} \in U \subset \mathbb{R}^2$. a is an arbitrary scale factor.

In our case d is derived from the deformation tensor $\{\gamma_{ik}\}$.

16.3.2 Deformation Tensor Interrogation

We already mentioned that the deformation tensor γ can be represented as a real and symmetric matrix $\{\gamma_{ik}\}$. Hence a regular matrix $T \in \mathbb{R}^{2\times 2}$ exists with

$$T^{-1}\gamma T = B, \quad B \in \mathbb{R}^{2\times 2}$$

and B is the diagonal matrix. T is the matrix containing the two perpendicular eigenvectors of γ in its rows and B has the corresponding eigenvalues λ_1 and λ_2 on its diagonal.

To determine the matrices T and B, one solves the characteristic equation

$$det(\gamma - \lambda I) = 0$$

where I denotes the identity matrix and λ is the eigenvalue.

This leads to the two eigenvalues λ_1 and λ_2:

$$\lambda_{1,2} = \frac{\gamma_{11} + \gamma_{22}}{2} \pm \sqrt{\frac{(\gamma_{11} + \gamma_{22})^2}{4} - \gamma_{11}\gamma_{22} + \gamma_{12}\gamma_{21}}.$$

The corresponding eigenvectors $\mathbf{x}_i$, $i = 1, 2$ are nontrivial solutions of the homogeneous system of linear equations

$$A - \lambda_i I = 0.$$

We determined λ_i so that $A - \lambda_i I$ is not regular, that is, the rows of $A - \lambda_i I$ are not linear-independent. Thus, we know that the gaussian algorithm will fail for $A - \lambda_i I$ and we can simply ignore one of the rows of $A - \lambda_i I$. The eigenvector $\mathbf{x}_i$ for λ_i can now be determined as follows:

If $\gamma_{11} - \lambda_i \neq 0$

$$\mathbf{a}_i = \begin{pmatrix} -\frac{\gamma_{12}}{\gamma_{11} - \lambda_i} \\ 1 \end{pmatrix}$$

If $\gamma_{22} - \lambda_i \neq 0$

$$\mathbf{a}_i = \begin{pmatrix} 1 \\ -\frac{\gamma_{21}}{\gamma_{22} - \lambda_i} \end{pmatrix}$$

And

$$\mathbf{x_i} = \frac{1}{||\mathbf{a}_i||}\mathbf{a}_i$$

where $|| \cdot ||$ denotes the Euclidian norm for $i = 1, 2$.

For further explanations and numerically stable algorithms for determination of eigenvalues and eigenvectors see [6].

The directions given by the two perpendicular eigenvectors $\mathbf{x}_1$ and $\mathbf{x}_2$ for deformation tensor γ at an arbitrary surface point $\mathbf{p}$ correspond to the directions of lowest and highest deformation, that appear at $\mathbf{p}$. The corresponding eigenvalues λ_1 and λ_2 are direct measures for the magnitude of deformation appearing in these directions, that is, the minimum of λ_1 and λ_2 is the lowest and their maximum is the highest deformation appearing at point $\mathbf{p}$.

The two eigenvalues λ_1 and λ_2 can now be used to construct a generalized focal surface F for X which can be considered for interrogation of a given deformation of the original surface X. This is done by using either λ_1 or λ_2 as parameter d in Equation (16.4) for construction of the generalized focal surfaces X. Any other value derived from these two eigenvalues can be used similarly; we propose consideration of $\lambda_1\lambda_2$ and $\lambda_1^2 + \lambda_2^2$ which can be used to identify regions where either one or both of the main deformations vanish.

This approach can be combined with grid node selection techniques to pinpoint regions with special characteristics, such as zero eigenvalues, values greater than the mean of all eigenvalues, and so on.

Figure 16.1 shows a convex surface X (red) together with the generalized focal surface $F = X + max(\lambda_1, \lambda_2)\ N$ for the maximum eigenvalue of the deformation tensor. Thus the focal surface visualizes the highest deformation on any surface point.

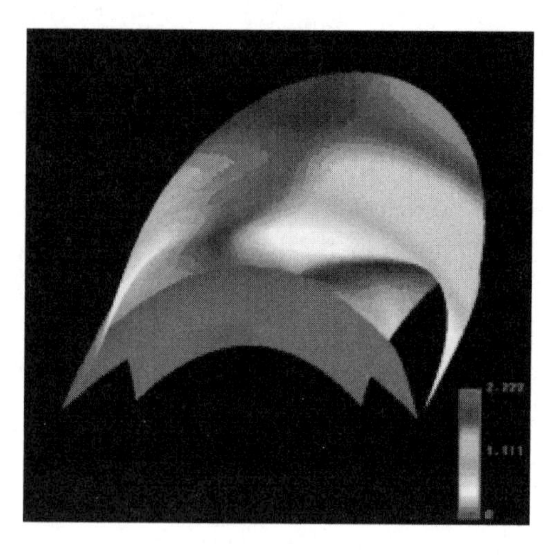

Figure 16.1: Focal surface for maximum eigenvalue. See Color Plate 96.

In Figure 16.2 we zoomed on a detail in the above situation. In this picture the focal surface $F = X + (\lambda_1^2 + \lambda_2^2)N$ is shown. Thus the focal surface can be used to identify regions with zero deformation. The point where the shown generalized focal surface touches the original surface satisfies $\lambda_1^2 + \lambda_2^2 = 0$ and this implies that $\lambda_1 = \lambda_2 = 0$.

Figure 16.2: Point with zero deformation. See Color Plate 97.

16.3.3 Characteristic Curves for Deformation Tensor Visualization

Visualization of a generalized focal surface derived from the eigenvalues λ_1 and λ_2 of the deformation tensor γ omits any directional information contained in the corresponding eigenvectors $\mathbf{x}_1$ and $\mathbf{x}_2$.

To visualize directional information contained in the two eigenvectors $\mathbf{x}_i$, $i = 1, 2$ for deformation tensor γ, as well as the magnitude of the corresponding eigenvalues λ_i, $i = 1, 2$, we consider the characteristic curves

$$\lambda_1 \, x^2 + \lambda_2 \, y^2 = 1, \quad (x, y) \in \mathbb{R}^2 \tag{16.5}$$

for any surface point $\mathbf{p}$ with deformation tensor γ, where λ_i are the eigenvalues of γ. The above equation is the representation of the characteristic curve in the eigenspace with basis $\{\tilde{\mathbf{x}}_1, \tilde{\mathbf{x}}_2\}$ where $\tilde{\mathbf{x}}_i$ denotes the normalized eigenvector $\mathbf{x}_i$ of γ, $i = 1, 2$.

The characteristic curve (16.5) is a closed ellipse for $\lambda_1, \lambda_2 > 0$. It is a parabolic curve for $\lambda_1 \cdot \lambda_2 < 0$. Thus Equation (16.5) can be used to visualize the characteristics of γ.

For evaluation of the implicit representation (16.5) we distinguish five cases:

1. $\lambda_1 > 0$ and $\lambda_2 > 0$:
   The characteristic curve is a closed ellipse for
   $$x \in \left[-\sqrt{\frac{1}{\lambda_1}}, +\sqrt{\frac{1}{\lambda_1}} \right]$$
   and corresponding y-value
   $$y = \pm \sqrt{\frac{1 - \lambda_1 x^2}{\lambda_2}} \, .$$
2. $\lambda_1 > 0$ and $\lambda_2 = 0$:
   Two straight parallel lines, which intersect the x-axis of the local coordinate system at
   $$x = \pm \sqrt{\frac{1}{\lambda_1}} \, .$$
3. $\lambda_1 = 0$ and $\lambda_2 > 0$:
   Two straight parallel lines, which intersect the y-axis of the local coordinate system at
   $$y = \pm \sqrt{\frac{1}{\lambda_2}} \, .$$
4. $\lambda_1 < 0$ and $\lambda_2 > 0$:
   Two open parabolic curves with
   $$y = \pm \sqrt{\frac{1 - \lambda_1 x^2}{\lambda_2}} \quad \text{for } x \in \mathbb{R} \, .$$

5. $\lambda_1 > 0$ and $\lambda_2 < 0$:
   Two open parabolic curves with

$$x = \pm\sqrt{\frac{1 - \lambda_2 y^2}{\lambda_1}} \text{ for } y \in \mathbb{R} .$$

These considerations lead to a set of points for each of the characteristic curves which can be visualized in the tangent plane of the original surface X at point $\mathbf{p}$ and rotated according to the eigenvectors. This can be done by coordinate system transformation which can be considered affine mapping.

$$R\,\mathbf{x} + \mathbf{t} = \mathbf{e}' \tag{16.6}$$

for any point $\mathbf{x} = (x, y)^T$ of characteristic curve (16.5). R represents a rotation matrix and $\mathbf{t}$ a translation.

Translation $\mathbf{t}$ maps the origin of eigenspace $(0, 0)^T$ onto the surface point $\mathbf{p}$, hence

$$\mathbf{t} = \mathbf{p} .$$

To compose matrix R we consider two rotations:

1. Rotation R_1 of the system with eigenspace basis $\{\tilde{\mathbf{x}}_1, \tilde{\mathbf{x}}_2\}$ into the world system with basis $\{(1, 0, 0)^T, (0, 1, 0)^T\}$.

2. Rotation R_2 of the local coordinate system for the tangent plane of surface X at point $\mathbf{p}$ onto the world system $\{(1, 0, 0)^T, (0, 1, 0)^T, (0, 0, 1)^T\}$.

Then R follows as

$$R = R_2^{-1}\, R_1^{-1} .$$

For composition of the two rotation matrices R_1 and R_2, we use the properties of orthogonal matrices. The unit row vectors of orthogonal rotation matrix R_i rotate into the principal axes, and the inverse of R_i is its transposition.

Matrix R_1 should represent a rotation of unit eigenvector $\tilde{\mathbf{x}}_1$ onto the x-axis and of unit eigenvector $\tilde{\mathbf{x}}_2$ onto the y-axis, hence R_1 contains in its first row the components of $\tilde{\mathbf{x}}_1$ and in its second row the components of $\tilde{\mathbf{x}}_2$. As the z-axis should not change, the third row of R_1 has to be the unit vector $(0, 0, 1)^T$.

The second rotation matrix R_2 represents a rotation of the local coordinate system for the tangent plane of surface X at point $\mathbf{p}$ into the world coordinate system. R_2 contains as its first row the first tangent vector X_u of surface X at point $\mathbf{p}$ after normalization. The third row of R_2 contains the vector normal to the tangent plane, which can be calculated as $X_u \times X_v$ and should also be normalized. The second row of R_2 contains a vector, that should be perpendicular to the first and third row, hence the vector product of the first and third row of R_2.

After transposition and multiplication, one receives the final rotation matrix

$$R = R_2^T\, R_1^T .$$

Applying the final system transformation (16.6) on any point calculated from Equation (16.5), one receives an image of the characteristic curve for deformation tensor γ. This image lies in the corresponding tangent plane and is rotated so that the axis of the coordinate system in which Equation (16.5) is represented coincides with the two eigenvectors of γ.

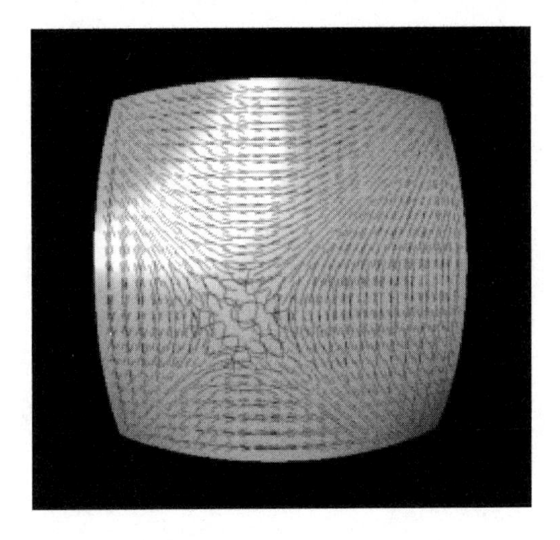

Figure 16.3: Characteristic curves for visualization of a deformation tensor field. See Color Plate 98.

In Figure 16.3 one can see the test surface also used in Figure 16.1. One can clearly identify the point with zero deformation and it is easy to see how the direction of the eigenvectors is distributed over the surface.

16.4 Visual Interrogation of Bendings

In the previous section we dealt with general deformations. The symmetric deformation Tensor $\gamma_{ij} = 1/2(\tilde{g}_{ij} - g_{ij})$ (see Equation (16.3)), based on the metric tensor of the reference surfaces and the deformed one, was used to visualize deformations. Appropriate visualization techniques were presented. A special case of deformations are the so-called infinitesimal bendings. A continuous deformation of a surface is called *(infinitesimal) bending* if the surface is not stretched, that is, the length of any arbitrary surface curve "keeps" unchanged (in first order). The deformation of a paper sheet without stretching and creasing is such a bending.

In this section we present a new concept to interrogate infinitesimal bendings, because the deformation tensor $\{\gamma_{ij}\}$ fails in this case.

A detailed presentation of this concept is given in [4]. This concept mainly asks for stability of a surface with respect to a given deformation and visualizes it with the help of an associate vector field.

Section 16.4.1, that gives the fundamentals of infinitesimal bendings, will point out the relation between the notion of stability and the vector field mentioned above.

Finally in Section 16.4.2 we present a visualization method based on this vector field.

16.4.1 Infinitesimal Bendings—Fundamentals:

We present parametric surfaces in $\mathbb{R}^3$ as vector-valued functions

$$\begin{aligned} X : U &\to \mathbb{R}^3 \\ (u, w) &\mapsto X(u, w) \end{aligned}$$

where U is a connected domain in $\mathbb{R}^2$.

Let $\{X_\varepsilon\}_{\varepsilon \in \mathbb{R}}$ be a continuous one-parameter family of surfaces, where X is contained in this family for $\varepsilon = 0$.

$$\begin{aligned} X_\varepsilon : U \times I &\to \mathbb{R}^3 \\ (u, w; \varepsilon) &\mapsto X_\varepsilon(u, w) \end{aligned}$$

If for each ε, there is a continous mapping between X and X_ε, such that each point P of X corresponds to a unique point P_ε of X_ε, then $\{X_\varepsilon\}$ is called a *deformation* of the surface X. A parametric representation is given by

$$X_\varepsilon(u, w) = X(u, w) + \varepsilon \cdot Z(u, w). \tag{16.7}$$

$Z(u, w)$ is called a *deformation vector field* (see Figure 16.4).

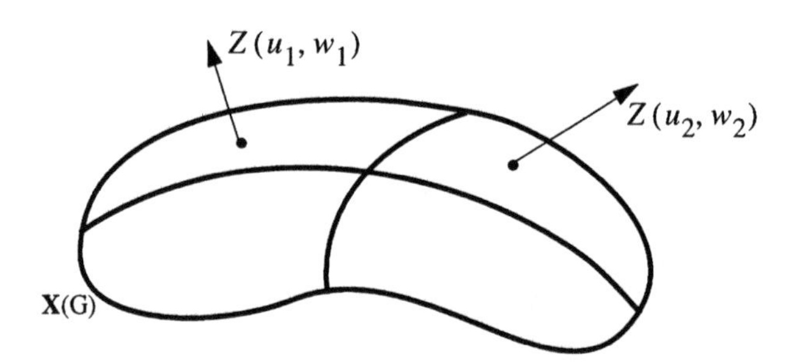

Figure 16.4: Deformation vector field.

A deformation is called an *infinitesimal bending* of first order of the surface X if the length of any arbitrary curve on the surface X keeps unchanged to first order in ε, that is, $L(c_\varepsilon) = L(c) + O(\varepsilon)$, where $c(t) = X(u(t))$ is an arbitrary surface curve. The vector field Z can be seen as the *velocity field* of the surface points in the beginning of the deformation.

Definition 16.4.1 *Let $\{X_t\}_{t \in I}$ ($I \subset \mathbb{R}$) be a deformation. Let $\{f_t\}_{t \in I}$ be a family of functions f_t defined on U such that $\left(\begin{smallmatrix} f \;:\; U \times I \to \mathbb{R} \\ (u,w,t) \mapsto f_t(u,w) \end{smallmatrix} \right)$ is continuous and has continuous partial derivatives.*

The function δf defined on U by

$$\delta f := \frac{\partial f}{\partial t}|_{t=0}$$

is called the **variation of f**.

The following theorem now characterizes the infinitesimal bendings:

Theorem 16.4.2 *Let Z be a deformation vector field of the surface X. Let g_{ij} be the elements of the first fundamental form and denote the first partial derivatives of X and Z by $X_u := \partial X/\partial u$ and $Z_u := \partial Z/\partial u$. The following statements are equivalent:*

$$\begin{array}{ll} i) & Z\ defines\ an\ infinitesimal\ bending\ of\ first\ order\ of\ X \\ ii) & \delta g_{ij} = 0 \qquad (16.8) \\ iii) & < Z_u, X_u >= 0 \\ & < Z_w, X_w >= 0 \qquad (16.9) \\ & < Z_u, X_w > + < Z_w, X_u >= 0 \\ & where < , > denotes\ the\ dot\ product. \end{array}$$

Proof: (see [2]).

Equation (16.8) says that the length of any surface curve does not change in first order in ε during the deformation. This is due to the first fundamental form, which allows us to measure distances on the surface [1]. Now it is easy to see that the deformation tensor $\{\gamma_{ij}\}$ vanishes identically for infinitesimal bendings:

$$\gamma_{ij} = \frac{1}{2}(\tilde{g}_{ij} - g_{ij}) \equiv 0$$

Proof:

$$\begin{aligned} g_{ij} - \tilde{g}_{ij} &\overset{(16.7)}{=} < X_{\varepsilon_i}, X_{\varepsilon_j} > - < X_i, X_j > \\ &= < X_i + \varepsilon Z_i, X_j + \varepsilon Z_j > - < X_i, X_j > \\ &= \varepsilon\,(< X_i, Z_j > + < X_j, Z_i >) + \varepsilon^2 < Z_i, Z_j > \\ &\overset{(16.9)}{=} 0 \end{aligned}$$

because all terms that are of order two or higher in ε are neglected.

The next theorem is at the same time a definition of the rotation vector field and plays an important role in the following.

Theorem 16.4.3 - Existence Theorem - *If the deformation vector field Z verifies the three equations (16.9), then there exists a unique vector field Y with the following properties:*

$$[Y, X_u] = Z_u \ \ and\ \ [Y, X_w] = Z_w. \qquad (16.10)$$

Y is called a **rotation vector field**.

Proof: (see [2]).

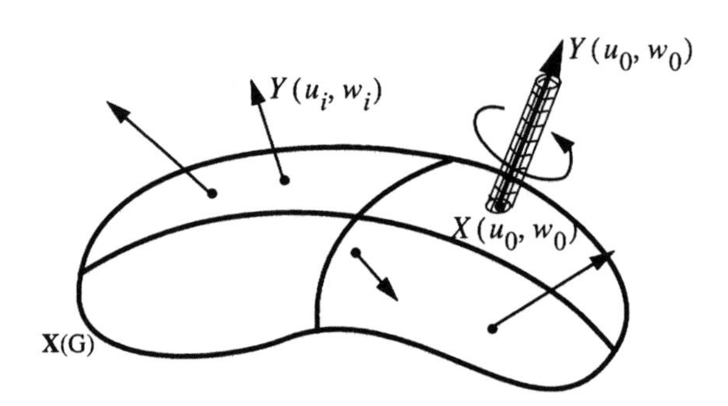

Figure 16.5: Rotation vector field.

To give a geometric interpretation of Y: $Y(u_0, w_0)$ is the rotation vector of a rigid body attached to the surface $X(G)$ at the point $X(u_0, w_0)$ during the deformation (see Figure 16.5).

It is also possible to take Y as a parametric surface. In this case Y is called the *instability surface* of the infinitesimal bending. Later we will see that a stable surface has an instability surface reduced to a single point.

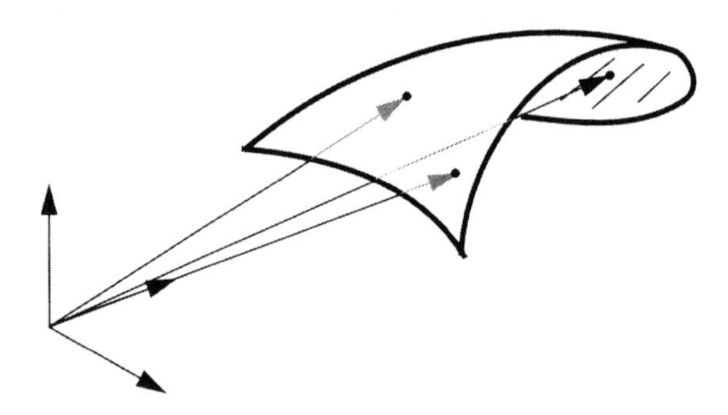

Figure 16.6: Instability surface.

We now take a closer look at a special case of infinitesimal bendings, those with a constant rotation vector field.

Definition 16.4.4 *Infinitesimal bendings are called* **trivial** *(or infinitesimal motion) if and only if*

$$Z(u, w) = [C, X(u, w)] + D \tag{16.11}$$

where C and D are constant vectors.

All bundles of line elements have the same momentary rotation if the rotation vector field is constant. This context leads to the following definition.

Definition 16.4.5 *A surface which allows only trivial infinitesimal bendings is called* **infinitesimal rigid** *under infinitesimal bendings.*

The next theorem characterizes infinitesimal rigid surfaces.

Theorem 16.4.6 *X is infinitesimal rigid if and only if for all deformations which satisfy $\delta g_{ij} = 0$, it also holds that $\delta h_{ij} = 0$. h_{ij} is the second fundamental form.*

Therefore, $\delta g_{ij} = 0$ characterizes the infinitesimal bendings and $\delta h_{ij} = 0$ characterizes the infinitesimal rigid surfaces.

Another way to treat infinitesimal bendings is with the fundamental equations of this theory. The interested reader can find the derivation in [2] and also in [4].

The fundamental equations give the possibility to calculate the rotation vector field in a complicated way.

Another method and a visualization tool is presented in the next section.

16.4.2 Visualizing Bending Stability

The rotation vector field $Y(u, w)$ is not only of central importance for the calculation of the infinitesimal bendings but is also a perfect visualization tool. Because, if the deformation vector field Z is known, then the rotation vector field Y is determined definitely (Theorem 16.4.3). Therefore, Y can be used to visualize the deformation and stability behavior of the surface X. We get the access to the rotation vector field by the existence theorem 16.4.3 in combination with the jerk-free deformation condition

$$Z_{uw} = Z_{wu}.$$

It follows from Equation (16.10) that

$$[Y_u, X_w] = [Y_w, X_u]. \tag{16.12}$$

The problem now states:

Given:	a parametric surface $X : U \to \mathbb{R}^3$
Wanted:	a vector field $Y : U \to \mathbb{R}^3$ such that Equation (16.12) holds.

To solve this system of partial differential equations a finite difference approach or better yet, a B-Spline based approach can be used. All the details can be found in [5].

Parametric surfaces have different bending behaviors: there are surfaces that are more likely to bend and others that are more likely to resist to "pressure." With the help of the rotation vector fields, we want to visualize the bending property. Indeed, the notion of stability is very related to the rotation vector field of infinitesimal bendings: a stable surface in the sense of infinitesimal bendings has a constant rotation vector field. Moreover, the more the rotation vectors vary in their directions, the less the surface is stable. This provides a visual test of the deformation and stability behaviors of the surface as illustrated in the following examples: In Figure 16.7 we see bicubic Bezier surface. The rotation vectors are attached on their corresponding surface points. The different directions of the rotation vectors are easily seen and indicate that they vary more in some regions than in others. In Figure 16.8, we see a special spherical color map from the rotation vectors into the parameter domain of the test surface. Unstable surface regions are emphasized by areas where the colors change a lot.

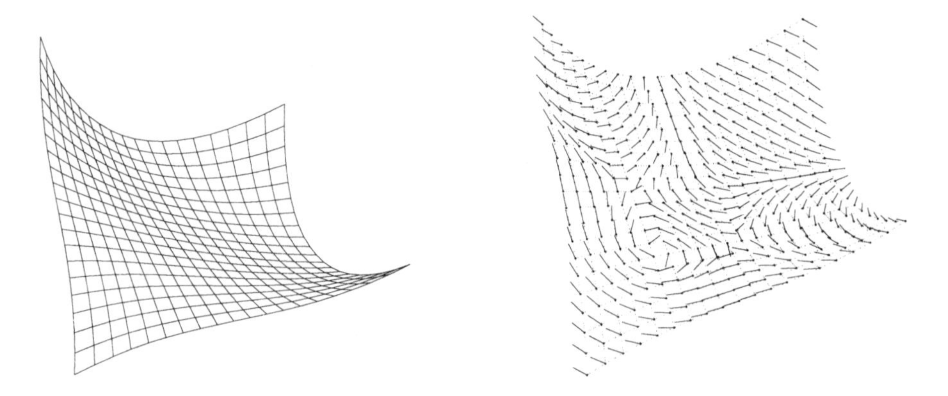

Figure 16.7: Test surface and rotation vectors.

Figure 16.8: Stability analysis. See Color Plate 99.

Bibliography

[1] P. DoCarmo, *Differential Geometry of Curves and Surfaces*, Prentice-Hall Inc., Englewood Cliffs, N.J., 1976.

[2] N.W. Efimov, *Flächenverbiegungen im Großen*, Akademie Verlag, Berlin, 1957.

[3] H. Hagen and St. Hahmann, "Generalized Focal Surfaces: A New Method for Surface Interrogation," *Proceedings of Visualization '92*, A. Kaufmann and G. Nielson, editors, Boston, Mass., 1992, pp. 70–76.

[4] H. Hagen and St. Hahmann, "Stability Concepts for Surfaces," *Computing Suppl.*, Vol. 10, 1995, pp. 189–198.

[5] St. Hahmann and H. Hagen, "Numerical Aspects of Stability Investigations on Surfaces," *Mathematics on Surfaces VI*, Glen Mullineux, editor, 1996, pp. 291–300.

[6] J.H. Wilkinson and C. Reinsch, *Linear Algebra*, Springer-Verlag, Berlin, Heidelberg, New York, 1971.

Chapter 17

An Alchemist's Primer on Reaction-Diffusion Patterns

Alyn Rockwood and Peter Chambers

17.1 Introduction

In 1952, the mathematician Alan Turing introduced the mechanism of reaction-diffusion in a paper entitled "The Chemical Basis of Morphogenesis." This mechanism uses the processes of reaction and diffusion to generate large-scale features in chemical concentrations within an array of reaction sites, or cells. It is worthy of note that Turing's paper [17] is regarded as a completely original work, a rarity in the literature.

Turing was concerned with the emergence of structure from the featureless tissue in the developing organism. He considered the organism as an array of similar cells through which chemicals diffuse according to the local concentration gradients. Within each individual cell, the chemicals react, modifying the local concentrations. Starting with a homogenous system, Turing's nonlinear model is capable of producing patterns in the chemical concentrations as the system evolves.

The ancient craft of alchemy investigated the transmutation of elements. Regarded ambivalently by scientists until the late 19th century, it fell into disrepute due mostly to its stepchild, modern chemistry. It had been an integral part of the philosophical system of Hermeticism which was especially popular in the 17th and 18th centuries as an alternative to the mechanistic philosophies epitomized by Newtonian physics. Hermeticism propounded a tight relationship between all things. Mastery and understanding derived from a knowledge of the strict laws of *sympathy* and *antipathy* that underpinned the universe. It was embraced by many serious intellectuals. Newton dedicated six years of his life to the study and classification of Hermeticism and alchemy.

It is a piece of history with a message. However misguided, these philosophical excursions helped lay the foundations for the modern sciences of pharmacy, chemistry and even physics. After Newton's six year isolation with Hermeticism he returned to the mechanical physics with a new idea about attraction and repulsion of forces which eventually led to his *Principia*. Before that time he had considered the unproductive (for him) idea that particles were connected by invisible "strings."

The reaction-diffusion theory of morphogenesis cannot yet claim scientific provability, although an impressive body of circumstantial evidence exists. We review some of the salient contributions to theory and then contribute our own experience to that body.

It is a tribute to Turing's genius that the salient features of his theory have changed little since its inception and that so much effort has been devoted to apply and verify his work.

17.2 Turing's Work

Turing's model for biological morphogenesis suggests that a system of chemical substances, known as *morphogens*, are able to react together and diffuse throughout the developing tissue. The cells within the tissue act as sites for this diffusion and reaction. The individual cells are not quite identical, having small variations in their properties. These variations are key to the instability of Turing's system.

While the initial state of the system may be homogenous, the nonlinear nature of the reaction-diffusion model permits patterns to develop in the concentrations of the reacting chemicals. The morphogens diffuse at a rate governed by local concentration gradients, and they react together to increase or decrease the amount of morphogen within individual cells. Such patterns, Turing suggested, could act as a basis for large-scale biological features such as mammalian coat patterns, digit development, and other characteristics that possess a sense of regularity. The exact mechanism for the emergence of these patterns is unknown. Turing proposed that as a result of the varying morphogen concentrations, cells would differentiate to select one from a set of possible types. This differentiation might be as basic as selecting hair color, or as complex as determining organ type.

Recently, cellular automata have been used for modeling the dynamics of reaction-diffusion systems. Convenient for simulation on computers, cellular automata provide a basis to investigate certain applications of reaction-diffusion behavior.

Solving the reaction-diffusion equations is best approached numerically. Witkin and Kass state, "The non-linear model largely defies analysis; its behavior must be understood through numerical simulation" [20, page 299]. However, limited analysis is feasible in two circumstances: immediately after the reaction-diffusion mechanism is allowed to occur, when pattern development is highly localized, and when the system has matured to a stable state, and global features have emerged.

Turing, who had no computer, takes an analytical approach and limits the numerical discussion to one-dimensional rings of cells. He developed the patterns on a ring mathematically, which he describes in terms of stationary waves arising from the fundamental instability of the system.

17.3 Cellular Automata

The self-organizing properties of cellular automata have been studied extensively. We extend the original binary nature of cellular automata to multiple scalar variables, increasing the range and variety of pattern types generated.

The basic cellular automaton consists of cells containing Boolean variables, interacting with neighboring cells in arrays of one, two, three, or more dimensions. Periodic conditions solve difficulties at array boundaries. Time is marked in discrete steps; at each step, the value of each cell's variable is updated according to a rule based on the state of the cell's neighbors.

17.3.1 Three-Dimensional Cellular Automata

A cell's three-dimensional *Moore* neighborhood consists of its 26 nearest neighbors: the cells above, below, and to the sides of the cell of interest.

For a simple example of a three-dimensional system's self-organizing abilities, consider the *vote* rule that operates by summing the state values (zero or one) of a Moore neighborhood, and adding the state of the cell itself [16]. The future state of the cell is then determined by the accumulation:

$$0 \text{ if } sum < 14; \; 1 \text{ if } sum \geq 14.$$

Domains of ones and zeroes coalesce, producing stable results within a few iterations.

17.3.2 Visualizing Three-Dimensional Cellular Automata

Figure 17.1(a) shows a three-dimensional system with 64,000 cells after 100 iterations of *vote*. The array is initialized with random values. The result is visualized with a very simple set of rules:

- Each cell containing a one is rendered as a solid, colored cube with Phong lighting.
- Cells containing zero are invisible.
- Perimeter cells are colored red to provide contrast with the interior.

The cellular automata settle into well-delineated domains; global structures form.

17.4 Reaction-Diffusion Systems

The initial state of the reaction-diffusion system's morphogen concentrations may be random or homogenous. After the system has been allowed to evolve, macroscopic concentration patterns appear in the reacting chemicals.

Turing's reaction-diffusion theory has been tested many times and used to explain a variety of different types of pattern development in a diversity of disciplines. This section provides a brief overview.

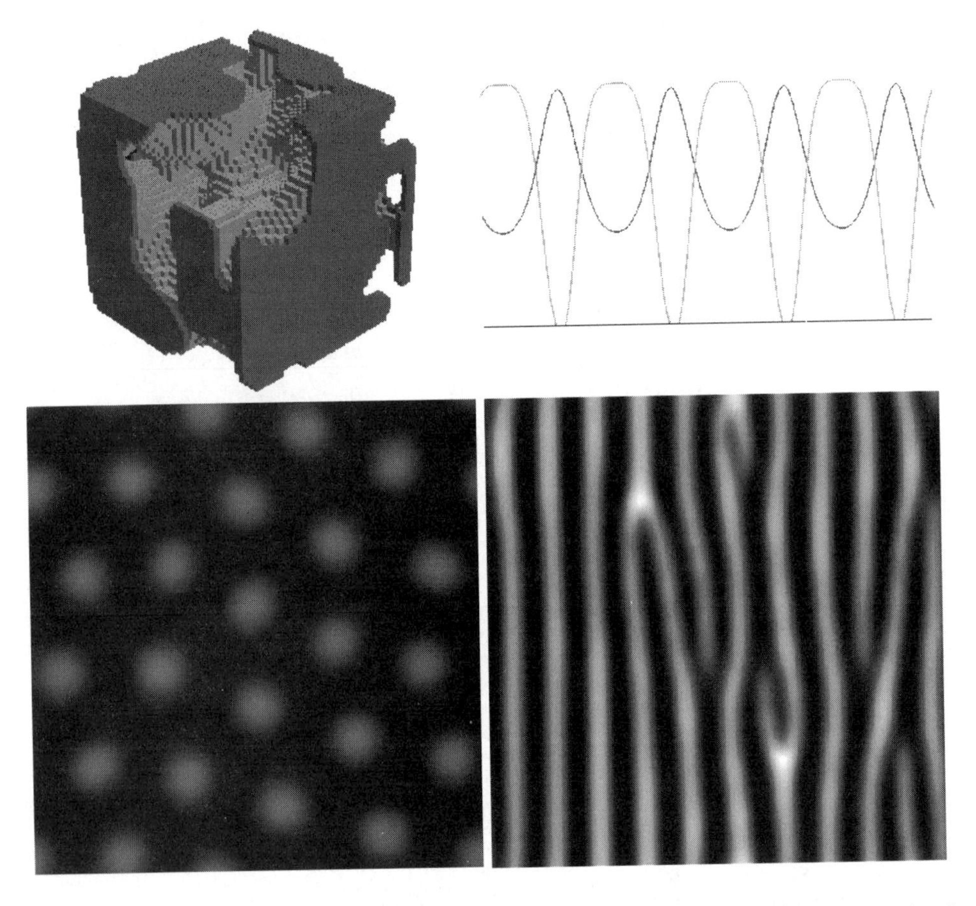

Figure 17.1: *(a)–(d)* (left to right, top to bottom) *(a)* Solid cubes rendering of the three-dimensional cellular automaton data. *(b)* x-y plot of the one-dimensional reaction-diffusion data set. *(c)* Intensity map of the two-dimensional reaction-diffusion data. *(d)* Intensity map of the anisotropic two-dimensional reaction-diffusion data. See Color Plate 100.

17.4.1 Mammalian Coat Patterns

Jonathan Bard applies two-dimensional reaction-diffusion kinetics to the development of coat coloration patterns of zebras, cats, and other mammals [1]. With some fairly small simulation models, Bard demonstrates that Turing's system is capable of generating patterns that indeed bear resemblance to such coat patterns. He extends Turing's model and imposes external (global) parameters to modify the final appearance of the patterns; in this manner he was able to show transitions in the pattern's form from spots to stripes. Bard's conclusion is that the reaction-diffusion system is capable of producing suitable patterns, but verification of the physical existence is lacking.

Jonathan Bard and Ian Lauder discuss reaction-diffusion in a similar context [2].

17.4.2 Waves in Excitable Media

Wave propagation in excitable media provides an excellent example of the emergence of self-organization. Spiral waves in unstirred Belousov-Zhabotinsky reagent are a well-known example of this phenomenon. Dwight Barkley has used a two-variable system of reaction-diffusion equations to model excitable media [3]. His system models the spiral waves very effectively.

Winfree discusses another example of spiral wave propagation with physiological applications [19].

17.4.3 Further Chemical Self-Organization Reactions

French researchers have discovered a physical system that may be generated by the process of reaction-diffusion. In hydrogel-filled spatial reactors, a chlorite-iodide-malonic acid (CIMA) reaction has been observed to produce stationary Turing structures [7]. Research from Texas confirms the Turing patterns in CIMA reactions and refines the results [13].

17.4.4 Drosophilia Morphogenesis

Axel Hunding has explored the application of reaction-diffusion systems to the development of patterns during early embryogenesis in the fruit fly Drosophilia [6]. His model suggests that Turing structures may activate or inhibit certain genes in the organism, thus giving rise to the embryonic patterns. In a rare application of three-dimensional modeling, Hunding's systems appear to be very tightly tied to the gene-activation model described.

17.4.5 Computer Simulations

A great deal of interesting work has focused on the use of computer simulations to investigate the emergence of organization in nonlinear systems. For example, John Pearson uses a two-dimensional system of reaction-diffusion equations to study the manner in which patterns emerge from the underlying kinetics [14]. By carefully exploring the reaction-diffusion system's parameter space, he produces a remarkable variety of elaborate patterns in the reactants.

The work by Witkin and Kass [20] and Turk [18] greatly advanced the graphical application of reaction-diffusion data sets to object textures. Concentrating on surface modeling, these papers develop detailed techniques for mapping the two-dimensional textures, derived from the reaction-diffusion data, onto the visible exterior of an object. These techniques are required to minimize the distortion as the flat texture field is applied to the object.

A number of other works investigate different aspects of the simulation of reaction-diffusion systems on computers [5, 9, 15].

17.5 Reaction-Diffusion Equations

We begin the study with the simplest system that displays the organizational capabilities of reaction-diffusion: the one-dimensional ring.

17.5.1 One-Dimensional Reaction-Diffusion Systems

The equations describing the two-morphogen reaction-diffusion model are given by:

$$\begin{aligned} \frac{\delta a}{\delta t} &= F(a,b) + D_a \nabla^2 a, \\ \frac{\delta b}{\delta t} &= G(a,b) + D_b \nabla^2 b, \end{aligned} \tag{17.1}$$

where a and b are the concentrations of two diffusing morphogens; F and G are functions determining the production rate of a and b; D_a and D_b are diffusion rate constants; and $\nabla^2 a$ and $\nabla^2 b$ are the Laplacians of a and b.

A system to approximate the model is built upon an array of cells through which the morphogens a and b may diffuse. Within each cell, a and b may be created or destroyed according to F and G. Turing proposes a discrete system to solve Equation (17.1):

$$\begin{aligned} \Delta a_i &= s(16 - a_i b_i) + D_a(a_{i+1} + a_{i-1} - 2a_i), \\ \Delta b_i &= s(a_i b_i - b_i - \beta_i) + D_b(b_{i+1} + b_{i-1} - 2b_i), \end{aligned} \tag{17.2}$$

where a_i and b_i are morphogen concentrations in a one-dimensional array of cells; β_i represents the natural variation between individual cells; and s is the reaction rate constant.

17.5.2 Visualizing One-Dimensional Reaction-Diffusion Systems

To visualize one-dimensional reaction-diffusion systems, a simple x-y plot is sufficient. The morphogen concentrations are plotted along the y-axis and the cell positions along the x-axis. Multiple morphogens may be displayed with different colors. Turing's system requires two morphogens, but there exist other systems with as many as five [12].

17.5.3 Evolution of the One-Dimensional Reaction-Diffusion System

Figure 17.1(b) shows the morphogen concentrations for the one-dimensional reaction-diffusion system. Morphogen a is plotted in red, morphogen b is plotted in green. The zero-morphogen baseline is plotted in blue.

For this simulation, the following values are chosen: $D_a = 0.25$, $D_b = 0.0625$, $a_i = 4.0$, $b_i = 4.0$, $\beta_i = 12.0 \pm 0.05$ (randomly selected for each i), $s = 0.00625$, number of cells = 100, iterations = 60,000. This defines a homogenous starting condition. However, after 60,000 iterations, the system acquires a macroscopic organization with symmetric highs and lows of morphogen concentration, as shown in the figure.

Note that the peaks of morphogen a's concentration match the lows of morphogen b's concentration, and vice versa. This is characteristic of a two-morphogen system; high concentration of morphogen a inhibits further similar regions from forming close by.

Figure 17.1(b) is an example of the one-dimensional ring patterns developed by Turing in his original paper.

17.5.4 Two-Dimensional Reaction-Diffusion Systems

The discrete reaction-diffusion equations extend to two dimensions:

$$\begin{aligned}
\Delta a_{i,j} &= s(16 - a_{i,j}b_{i,j}) \\
&+ D_a(a_{i+1,j} + a_{i-1,j}) \\
&+ D_a(a_{i,j+1} + a_{i,j-1} - 4a_{i,j}),
\end{aligned}$$

$$\begin{aligned}
\Delta b_{i,j} &= s(a_{i,j}b_{i,j} - b_{i,j} - \beta_{i,j}) \\
&+ D_b(b_{i+1,j} + b_{i-1,j}) \\
&+ D_b(b_{i,j+1} + b_{i,j-1} - 4b_{i,j}). \qquad (17.3)
\end{aligned}$$

The cells are arranged on a square grid. Each cell diffuses to and from the four neighbors with which it shares an edge, with periodic boundary conditions.

17.5.5 Visualizing Two-Dimensional Reaction-Diffusion Systems

Two methods are used to visualize two-dimensional systems:

1. **Morphogen Intensity Map.** The two-dimensional grid is shaded according to the amount of a particular morphogen that the corresponding cell contains. This may be displayed directly or smoothed with a filter.

2. **Hermite Surface Rendering.** Hermite surface rendering considers the morphogen concentrations as a *height field* above the x-y plane. A smooth surface is passed through each point and generates a picture of a morphogen-concentration "terrain."

The subtle details in the high and low areas are emphasized. The Hermite matrix derivative terms are found by taking finite differences; the twist vectors are set to zero.

17.5.6 Evolution of a Two-Dimensional Reaction-Diffusion System

Initial conditions are the same as the one-dimensional case, except $s = 0.03125$. The system consists of 80×80 cells and is run for 20,000 iterations.

Figure 17.1(c) shows the intensity map visualization of the reaction-diffusion data using a Bartlett filter. The areas of minimum concentration form clearly defined regions with smooth transitions to adjacent cells. It is from patterns of concentration such as this that much of the work on mammalian coat patterns is based.

17.5.7 Anisotropic Two-Dimensional Reaction-Diffusion Systems

If the condition of invariant parameters is relaxed, it is possible to introduce further diversity into the types of patterns produced. Witkin and Kass investigate the behavior of anisotropic reaction-diffusion methods with systems of competing diffusion maps, that is, external diffusion parameter defining procedures [20]. There are a number of ways in which the reaction-diffusion parameters are varied. Perhaps the simplest—and also the most plausible from the biological standpoint—is the imposition of external control upon the diffusion parameters D_a and D_b.

Figure 17.1(d) shows a morphogen intensity map of such a system. The anisotropy biases the system to form stable domains parallel to the direction of the greatest b morphogen diffusion rate. Traditionally, reaction-diffusion systems have been used to model zebra stripes; Meinhardt uses a five-chemical reaction-diffusion system (1982). The results here need only two.

Figure 17.2(a) shows the anisotropic two-dimensional system data rendered with Hermite interpolation. The concentrations of morphogen a are used as the height field, defining a set of points through which the interpolating surface is passed.

The rendering clearly shows the way in which the structures bifurcate and join. From here, the generation of organic surface textures such as tree bark seems a direct step.

This picture has a striking, unexplained similarity to wind-formed sand dunes.

17.5.8 Three-Dimensional Reaction-Diffusion Systems

The now-familiar equations extend easily to three dimensions:

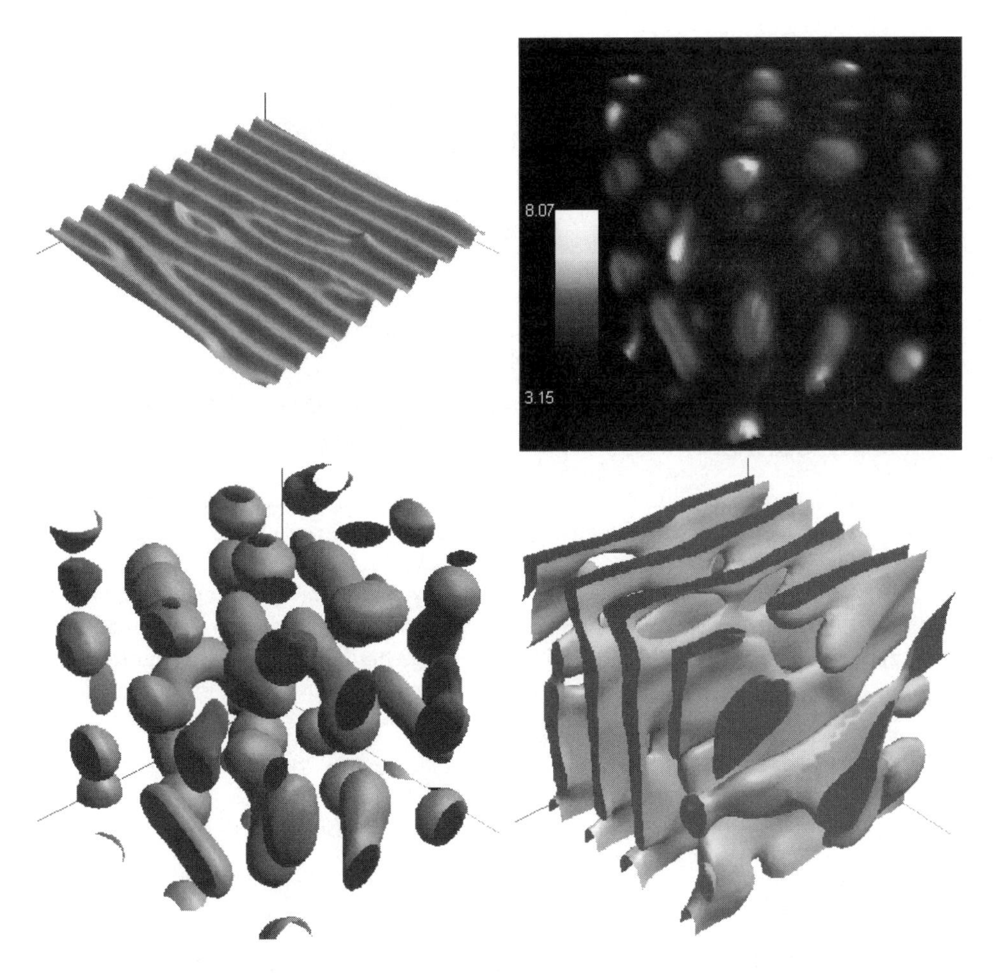

Figure 17.2: *(a)–(d)* (left to right, top to bottom) *(a)* Hermite surface of the anisotropic two-dimensional reaction-diffusion data set. *(b)* Volume rendering of the three-dimensional reaction-diffusion data set. *(c)* Isosurface rendering of the three-dimensional reaction-diffusion data set. *(d)* Isosurface rendering of the anisotropic three-dimensional reaction-diffusion data set. See Color Plate 101.

$$
\begin{aligned}
\Delta a_{i,j,k} &= s(16 - a_{i,j,k} b_{i,j,k}) \\
&+ D_a(a_{i+1,j,k} + a_{i-1,j,k} + a_{i,j+1,k}) \\
&+ D_a(a_{i,j-1,k} + a_{i,j,k+1} - 4a_{i,j,k-1}) \\
&- 6D_a a_{i,j,k}, \\
\\
\Delta b_{i,j,k} &= s(a_{i,j,k} b_{i,j,k} - b_{i,j,k} - \beta_{i,j,k}) \\
&+ D_b(b_{i+1,j,k} + b_{i-1,j,k} + b_{i,j+1,k}) \\
&+ D_b(b_{i,j-1,k} + b_{i,j,k+1} + b_{i,j,k-1}) \\
&- 6D_b b_{i,j,k}. \qquad (17.4)
\end{aligned}
$$

Cells are arranged in a regular cubic grid. Each cell diffuses to and from its six neighbors in the array, with periodic boundary conditions imposed.

17.5.9 Visualizing Three-Dimensional Reaction-Diffusion Systems:

We employ two approaches to the problem of visualizing the internal structure of three-dimensional reaction-diffusion systems:

1. **Isosurface Rendering.** A standard way of visualizing three-dimensional data is by passing an isosurface through the data and then rendering the isosurface. Such an approach has been applied to disciplines such as fluid dynamics, seismography, and medical imaging, and is highly effective at making the data set's form comprehensible.

   Isosurface rendering performs poorly in the presence of noise, but the mature reaction-diffusion data sets may be considered to be noise-free since the scale of the cells is small in comparison with the patterns that arise.

   A straightforward way of rendering an isosurface is with an algorithm known as *marching cubes* [10].

2. **Volume Rendering.** Three-dimensional data sets are common in medical imaging, and much work has gone into producing high-quality images from these data sets. Marc Levoy describes a practical method of rendering three-dimensional data [8].

17.5.10 Example of a Three-Dimensional Reaction-Diffusion System

As with the two-dimensional systems, we consider a basic system with static parameters. This demonstrates clearly the characteristics of a three-dimensional reaction-diffusion system. The parameters are the same as the two-dimensional system. The system is 40 $\times$ 40 $\times$ 40 cells, and is run for 30,000 iterations.

Figure 17.2(b) shows a volume visualization of the three-dimensional data set, enhanced by associating a color map with the morphogen concentration level. When a morphogen concentration sample is taken at a point along the viewing ray through the data set, the value returned is scaled and used as an index into a 256-entry color map.

The color map is shown to the left of Figure 17.2(b), with the low-density colors at the bottom and the high-density colors at the top. Examining the perimeter of the data set shows areas of high-density coloring where the set's boundaries intersect the regions of high morphogen concentration, continuing from one face to the opposite as a result of the periodic boundary conditions. Elsewhere, the blue-colored low- to medium-density regions dominate, reflecting the localized nature of the areas of higher concentration.

Figure 17.2(c) shows an isosurface rendering of the same three-dimensional data set. The isosurface shown is for morphogen a with a threshold value of 5.0. This provides a density rendering that is sparse enough for a considerable amount of internal detail to remain visible, while at the same time showing the overall structure of the data set.

17.5.11 Anisotropic Three-Dimensional Reaction-Diffusion Systems

Figure 17.1(d) introduces anisotropic rate parameters into the three-dimensional reaction-diffusion data; this reaction-diffusion system has clear directional characteristics. There are potential applications to textures that involve layering such as the formation of marble or other sedimentary rock types.

By modifying the reaction parameters, the emergent patterns can be adjusted or given directional characteristics as necessary. In general, the patterns are highly sensitive to the reaction-diffusion parameters; careful experimentation is necessary to achieve the correct balance and stability.

17.6 Characteristics of Solid Texture Functions

A goal of this project—in the inquiring spirit of the alchemist—is to employ the three-dimensional reaction-diffusion data sets as a source of texturing information for graphical objects. These data sets may be considered to be graphics primitives, which we may scale and modify with higher-level operators to create suitable textures. A basic solid texturing function takes as its input a point in space and returns a scalar value. More powerful functions may return a vector; typically this vector is derived from the gradient of the data set at the input point. This gradient may be found by taking partial differences around the point.

17.6.1 Properties of a Solid Texture Function

An ideal solid texture function supports three requirements:

1. **Statistical Invariance under Rotation.** The texture characteristics should be similar, no matter how the object is rotated. Ideally, it should be possible to rotate an object arbitrarily through the solid texture field, and the general appearance of the texture should remain consistent.

2. **Statistical Invariance under Translation.** The texture characteristics should be similar, no matter how the object is translated through the texture field.

3. **A Suitable Bandpass Limit in Frequency.** The features of the texture field should be neither too large nor too small in relation to the object that the texture operates upon. Features that are too large appear incorrect (for example, a wooden object carved out of the wood of a single growth ring rather than displaying grain). Features that are too small cause aliasing artifacts and loss of distinct pattern in the texture.

The property of statistical invariance applies only to rotation and translation; clearly, if an object is scaled, the way in which it intersects the texture field will change.

The three-dimensional reaction-diffusion data sets may be tessellated (stacked in three dimensions) to form a repeated field, since periodic boundary conditions are imposed upon the dynamical systems. This tessellation ensures that the texturing information is continuous between adjacent replications of the field primitive.

17.7 The Rendering Environment

To ensure an orderly, disciplined approach to the study of cellular automata, reaction-diffusion systems, and texture generation, a new operating environment was required. The requirements for the environment were:

1. **Cellular Automata:**
    - Execute and visualize one-, two-, and three-dimensional cellular automata.
    - Permit simple rule selection.
    - Extract texture information from the data.

2. **Reaction-Diffusion Systems:**
    - Execute and visualize one-, two-, and three-dimensional reaction-diffusion systems.
    - Permit simple selection of parameters.
    - Extract texture information from the data.

3. **Texture-Generation Environment:**
    - Support arbitrarily complex objects assembled from basic geometric primitives.
    - Allow easy selection of a variety of texture-mapping methods.
    - Render and manipulate the images in a flexible manner.

This specification demanded a set of features that was satisfied with a formal language, possessing its own syntax and semantics. We created OATS, an acronym for Object And Texture Stream. OATS files permit a variety of different cellular automata, reaction-diffusion, and texture generation systems to be simulated and their results visualized. The language is simple, comprehensible, and forgiving. OATS files support over 250 different commands, each with its own English representation. Many commands take numerical or Boolean arguments, increasing adaptability.

17.7.1 OATS Implementation

An OATS file is a simple text file and is written ostensibly in English. The heart of its structure is the stream: the concept of processing commands serially with minimal need to retain information between one command and the next. This may seem unnecessary, but the design was created out of practicality; some OATS files are in excess of a megabyte in size, and there was simply no room to create data structures to store all this information concurrently.

OATS commands are interpreted since compilation would be of little benefit to the high-level command types. The OATS file is parsed at runtime by a lexical analyzer, which assigns tokens to the various commands from its current dictionary. These tokens are given to the interpreter which then takes the necessary actions.

The use of OATS permits new cellular automata and reaction-diffusion systems to be evaluated with a minimum of effort, necessary for the empirical, alchemist's approach adopted for reaction-diffusion investigation. It also provides an efficient process for documenting the characteristics of such systems.

17.7.2 A Small OATS File

An example will clarify the overall structure and form of an OATS file. This example generates the data and image for the one-dimensional reaction-diffusion system in Figure 17.1(b). The OATS commands are in Courier font with added comments in Times Roman. Additional notes appear after the example.

```
# One-Dimensional Binary CA                  (Comments start with #)
set_window_size 990 660                      (Display window)
start_file                                   (System flags)
start_obj

# Cellular automaton definition:
CA_size 100                                  (Number of automaton cells)
CA_type CA_react_diff                        (Reaction-diffusion system)
CA_dimension CA_one_d                        (One-dimensional)
CA_iterations 60000                          (Loop iterations)
RD_set_react_speed 0.0625                    (Reaction speed s)
CA_init CA_smooth 0.0                        (Initialize the system)

# Run the Cellular Automata ...
 CA_start                                    (This starts the automata)

# Display the results ...
CA_render                                    (Show results to the screen)
image_save one_d_rd                          (Save the results)
end                                          (Completion flag)
```

Notes on the Example OATS File

1. Comments can be placed virtually anywhere, except between a command and its arguments. Comments begin with a # followed by a space. The remainder of the line is ignored and discarded.
2. The `set_window_size` command takes two arguments to define the size of the graphics window on the screen in pixels. This is crucial, since memory must be allocated for pixel data storage.
3. The commands `start_file` and `start_obj` are obligatory statements to define correct syntax. They do not play any role in generating the reaction-diffusion data.
4. The cellular automata definition section is probably clear; the reaction-diffusion parameters are defined here. Parameters not specifically defined always default to valid, typical values which may be correct for many applications. `CA_init` requests memory allocation for the cell data array, and initializes the cell's starting values.
5. `CA_start` actually invokes the reaction-diffusion engine to iterate the system through the requisite number of loops. During this time, calls are made to the multitasking facilities of the operating system to allow other programs to run.
6. `CA_render` paints the image data on the screen, according to the currently selected visualization algorithm. For one-dimensional reaction-diffusion systems, there is only one (the x-y plot) so no choice is made.
7. `image_save` stores the pixel data into a proprietary image data file for later recall, analysis, or photography.
8. Finally, `end` indicates a successful completion to the file. If missing, the system will complain since the file may have been erroneously truncated.

17.7.3 The Flexibility of OATS

The example given above is a very basic OATS file. Generally, OATS files that generate cellular automata or reaction-diffusion data sets contain tens to hundreds of lines, while OATS files that contain polygon data for complex objects may contain tens of thousands of lines. In many cases, the same reaction-diffusion data set will be used to generate texture for several objects. To support this efficiently, the reaction-diffusion data may be stored and recalled without having to be regenerated each time. Facilities are provided to load and store both images and data sets.

A more elaborate example of an OATS file is presented in Appendix A17.

17.8 Conclusions

17.8.1 Cellular Automata

As expected, the simple three-dimensional cellular automata model was well able to produce large-scale order. Furthermore, the visualization method for three-dimensional cellular automata (rendered solid cubes) clearly revealed the internal structure of the automata's data set.

17.8.2 Reaction-Diffusion Systems

The development of one-, two-, and three-dimensional reaction-diffusion models provided a strong foundation for the generation of solid data sets for texture production. For two-dimensional systems, Hermite surface interpolation uses the sense of depth to convey the reaction-diffusion data as a height field. For three-dimensional systems, Levoy's volume rendering method, in combination with a density-indexed color map, achieved a higher comprehension level of the data.

A most effective visualization method applied to the three-dimensional reaction-diffusion data was the rendering of morphogen concentration isosurfaces.

Some examples of reaction-diffusion textures applied to smoothly rendered objects are included in Figure 17.3. The three-dimensional reaction-diffusion data is used to modulate both the color of the objects and the fine surface appearance; several of the examples employ Blinn's bump-mapping [4].

The organic nature of the reaction-diffusion textures underscores the macroscopic patterns developed in the matrix of reacting automata, suggesting a physical basis for the systems. This is in accord with Turing's hypothesis of reaction-diffusion as a mechanism behind morphogenesis.

17.8.3 Texturing with the Reaction-Diffusion Data Sets

We note that the reaction-diffusion data behaves favorably with respect to the three requirements for an effective solid texturing field:

1. **Statistical Invariance under Rotation.** Isotropic reaction-diffusion data comes close to fulfilling this requirement. Textures produced by isotropic systems may vary somewhat under rotation, as a result of the axis alignment of the cubical reacting cells, but this is likely to be minor. Because it is highly directional, anisotropic data does not fulfill this requirement. It nevertheless has interesting qualities that make it appropriate for texture synthesis.

2. **Statistical Invariance under Translation.** When the periodic boundary conditions are applied to the reaction-diffusion system's kinetics, the resulting data set may be tessellated in $\mathbf{R}^3$ to provide continuity of texture. If an object is translated, this tessellation ensures that its texture will be preserved; it will change as the intersection with the texture field changes, but its overall characteristics will remain similar.

3. **A Suitable Bandpass Limit in Frequency.** If the texture field derived from the reaction-diffusion system is sized suitably to the dimensions of the object, then the resulting texture features may acquire the correct scale. The OATS rendering environment provides a number of ways by which the texture field may be scaled appropriately before texture generation.

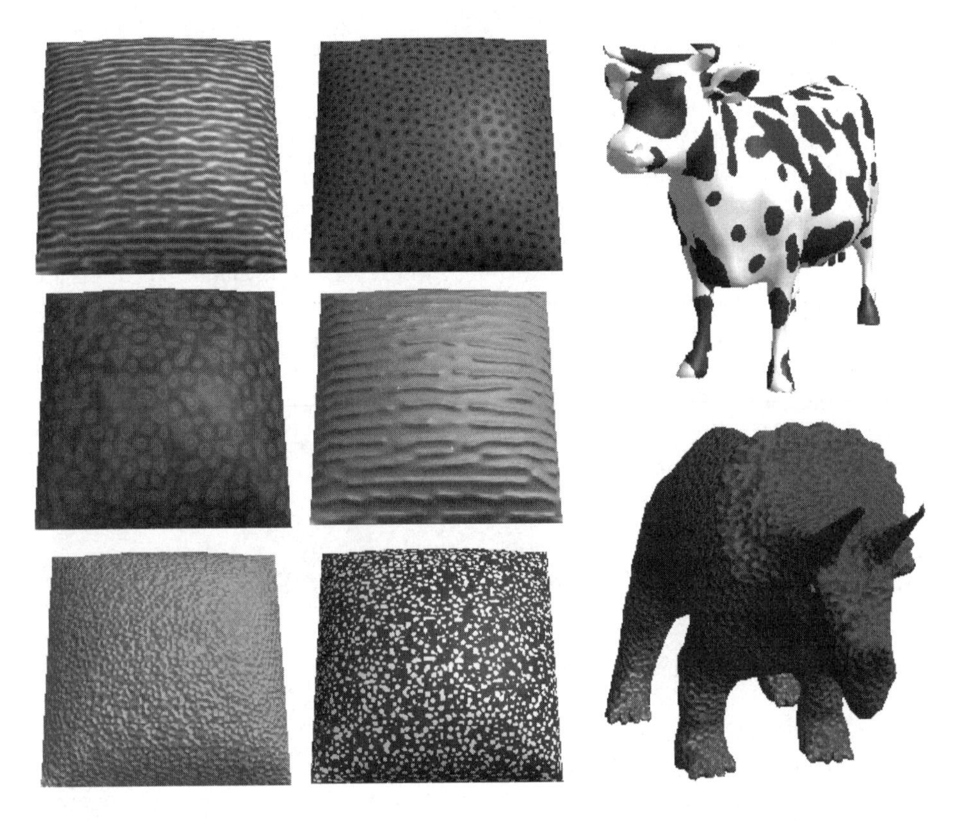

Figure 17.3: Examples of reaction-diffusion textures, (left to right, top to bottom) *Anisotropic color map:* varying morphogen concentrations map to different colors. *Spatial variation of reaction rate:* the reaction rate parameter s is varied, resulting in features of different sizes. *Cow:* Three-dimensional discrete color map. *Interpolated-pair color map:* With a linearly-interpolated color set, smooth transitions occur. *Sand dune pillow:* by applying bump-mapping to anisotropic reaction-diffusion data, the characteristic ridge formation emerges. *Triceratops:* Bump-mapped triceratops, three-dimensional texture. *Jade pillow:* "Dent-mapping," the inverse of bump-mapping, with normalized perturbation vectors. *Discrete color map:* by partitioning morphogen concentrations into two ranges, a pattern defined by the morphogen's isogram appears. See Color Plate 102.

17.8.4 Final Remarks

We can question the foundational processes that generate large-scale order in reaction-diffusion. First, is this process a result of the modeled reaction-diffusion equations, or is it simply the dynamic behavior of cellular automata? Madore and Freedman address this issue [11]. Undoubtedly, the reaction-diffusion system introduces a high degree of refinement to the generated structures. However, we note that Boolean cellular automata, operating with the simplest of rules governing their evolution, are capable of producing structures with a high degree of order (Figure 17.1(a)). Within the constraints of their binary nature, cellular automata generate patterns that possess similar characteristics to the nonlinear reaction-diffusion systems.

Second, there is the issue of numerical accuracy: the nonlinear Turing equations are apt to diverge quickly from the first-order finite difference solutions. A continuous system may well exhibit different patterns from the discretely modeled system; however, Turing's original thesis—morphogens reacting and diffusing between individual cells of finite size—motivates the discrete model.

One might also consider the problem from the point of view of neural networks, which may lend useful insights.

Finally, having provided many indications for Turing's morphogenesis thesis, we conclude by recalling the enthusiasm generated for centuries by alchemists. Whether the answers are positive or negative, one cannot fail to be engaged by the resulting patterns, which will undoubtedly engender continued investigation.

17.9 Future Research

17.9.1 Fractal Characteristics

Both cellular automata and reaction-diffusion systems are capable of exhibiting fractal boundary characteristics related to the modeling of diffusion processes. A study of the form of these boundaries, through the process of subdivision of unstable cells, would be very interesting.

17.9.2 Physical Phenomena

Turing proposes reaction-diffusion as a model of morphogenesis. However, with the extension of the reaction-diffusion system to three dimensions, new fields open up; it is now possible to model solid textures and volume-based properties.

As an example, the anisotropic two- and three-dimensional reaction-diffusion systems produce textures that are readily comparable to tree bark, sand dunes, or sedimentary mineral deposits. Even a cursory examination will reveal curiously similar features in both structures: bifurcations, peaks, and directional shifts.

Although the mechanism of formation may be different, the similarity between the two patterns is undeniable.

Acknowledgments

The cow and triceratops datasets are copyrights of Viewpoint Animation Engineering.

Bibliography

[1] J. Bard, "A Model for Generating Aspects of Zebra and Other Mammalian Coat Patterns," *Journal of Theoretical Biology*, Vol. 93, 1981, pp. 363–385.

[2] J. Bard and I. Lauder, "How Well Does Turing's Theory of Morphogenesis Work?" *Journal of Theoretical Biology*, Vol. 45, 1974, pp. 501–531.

[3] D. Barkley, "A Model for Fast Computer Simulation of Waves in Excitable Media," *Physica D*, Vol. 49, 1991, pp. 61–70.

[4] J.F. Blinn, "Simulation of Wrinkled Surfaces," *Computer Graphics*, Vol. 12, No. 3, 1978, pp. 286–292.

[5] M. Gerhardt and H. Schuster, "A Cellular Automaton Describing the Formation of Spatially Ordered Structures in Chemical Systems," *Physica D*, Vol. 36, 1989, pp. 209–221.

[6] A. Hunding, "Pattern Formation of Reaction-Diffusion Systems in Three Space Coordinates. Supercomputer Simulation of Drosophilia Morphogenesis," *Physica A*, Vol. 188, 1992, pp. 172–177.

[7] P. De Kepper, V. Castets, E. Dulos, and J. Boissonade, "Turing-Type Patterns in Chlorite-Iodide-Malonic Acid Reaction," *Physica D*, Vol. 49, 1991, pp. 161–169.

[8] M. Levoy, "Display of Surfaces from Volume Data," *IEEE Computer Graphics and Applications*, May 1988, pp. 29–37.

[9] W. Li, "Complex Patterns Generated by Next Nearest Neighbors Cellular Automata," *Computers and Graphics*, Vol. 13, No. 4, 1989, pp. 531–537.

[10] W.E. Lorensen and H.E. Cline, "Marching Cubes: A High Resolution 3D Surface Construction Algorithm," *Computer Graphics*, Vol. 21, No. 4, 1987, pp. 163–169.

[11] B.F. Madore and W.L. Freedman, "Self-Organizing Structures," *American Scientist*, Vol. 75, 1987, pp. 252–259.

[12] H. Meinhardt, *Models of Biological Pattern Formation*, Academic Press, New York, 1982.

[13] Q. Ouyang, Z. Noszticzius, and H.L. Swinney, "Spatial Bistability of Two-Dimensional Turing Patterns in a Reaction-Diffusion System," *Journal of Physical Chemistry*, Vol. 96, 1992, pp. 6773–6776.

[14] J. Pearson, "Complex Patterns in a Simple System," *Science*, Vol. 261, 1993, pp. 189–192.

[15] H.E. Schepers and M. Markus, "Two Types of Performance of an Isotropic Cellular Automaton: Stationary (Turing) Patterns and Spiral Waves," *Physica A*, Vol. 188, 1992, pp. 337–343.

[16] T. Toffoli and N. Margolus, *Cellular Automata Machines*, MIT Press, Cambridge, Mass., 1987.

[17] A. Turing, "The Chemical Basis of Morphogenesis," *Philosophical Transactions of the Royal Society (B)*, Vol. 237, 1952, pp. 37–72.

[18] G. Turk, "Generating Textures on Arbitrary Surfaces Using Reaction-Diffusion," *Computer Graphics*, Vol. 25, No. 4, 1991, pp. 289–298.

[19] A. Winfree, *The Three-Dimensional Dynamics of Electrochemical Waves and Cardiac Arrhythmias*, Princeton University Press, Princeton, N.J., 1987.

[20] A. Witkin and M. Kass, "Reaction-Diffusion Textures," *Computer Graphics*, Vol. 25, No. 4, 1991, pp. 299–308.

A17 A Complete Example of an OATS File

The example OATS file given earlier generated and rendered the data set of a one-dimensional reaction-diffusion system. OATS files may be far more elaborate than this, and an example of a more intricate system is given here, with extra comments for clarification on the right-hand side.

The example presented is for the triceratops dinosaur shown in Figure 17.3. The underlying reaction-diffusion system is isotropic and three-dimensional. For reasons of economy, only the first two of the 5,660 polygons are defined in this listing.

```
# Triceratops Dinosaur                    (Heading and comments)
# --------------------

# (c) Viewpoint
# Animation Engineering

# Textured with solid
# reaction-diffusion data

# Converted from .obj format
# Peter Chambers
# Summer 1993

set_window_size 700 700                   (700 by 700 pixel window)
start_file

background plain                          (Set a black background)
0.0 0.0 0.0

start_obj

at -1.442 0.186 0.015                     (Four-parameter
from 22.806 4.351 8.078                   perspective transformation)
up 0.000 1.000 0.000
view_angle 0.400

form solid                                (Filled polygons)
shade Phong                               (Smooth shading)

surface plain                             (A plainly colored surface)
1.0 1.0 1.0                               (Default colors)
0.300 0.600 20                            (Surface reflection constants)
1.500 1.250                               (Illumination intensities)

set_color Brown                           (Set a nice dinosaur color)

faces 5660                                (5,660 triangular polygons)

bounding_box                              (Z-buffer parameters)
```

```
-10.300 -3.692 -2.913
7.416 4.064 2.944

set_noise_type CA_noise              (Solid texture comes from the CA)
perturb_surface bumps                (Texture the surface)
bump_amount -0.5                     (Amount of normal modification)
texture_scale 35.0 35.0 35.0         (Adapt texture to object)
normalize_gradient false             (Do not normalize surface vectors)

CA_size 40                           (40 by 40 by 40 data set)
CA_type CA_react_diff                (Reaction-diffusion kinetics)
CA_dimension CA_three_d              (Three-dimensional system)

CA_iterations 30000                  (Set up 30,000 loops)

CA_noise_type CA_scalar_noise        (Scalar values from the data)
RD_set_react_speed 0.25              (Set the reaction speed s)

CA_init CA_smooth 0.0                (Initialize the data array)

CA_start                             (Run the system to completion)

CA_save tric_rd.dat                  (Save the reaction-diffusion data)

CA_normalize_noise true              (Normalize the data set's range)
CA_set_data_range                    (Scan array for maximum/minimum)
```

(At this point the reaction-diffusion data set is prepared for texturing an object. Since all parameters have been set for the dinosaur, we may now send the polygons to the renderer.)

```
pp 3                                           (Polygon data)
4.394 -0.332 0.134 -0.215 -0.920 0.329
4.494 -0.198 0.299 -0.134 -0.785 0.604
4.085 -0.067 0.445 0.075 -0.792 0.605
```

(Each polygon consists of three world coordinate vertices together with their associated vertex normals.)

```
pp 3
4.394 -0.332 0.134 -0.215 -0.920 0.329
4.085 -0.067 0.445 0.075 -0.792 0.605
4.046 -0.240 0.134 -0.017 -0.962 0.271
```

(5,658 polygons deleted)

```
end                                            (End-of-file marker)
```

Chapter 18

A Visualization Approach to the Study of Singularities on the World Sheets of Open Relativistic Strings

V. V. Dyachin, S. V. Klimenko, and I. N. Nikitin

Abstract. *A visualization approach to the study of singularities on the world sheets of open relativistic strings is presented. World sheets of open relativistic strings in Minkowski space as a central physical object of visual study are considered. Visualization is incorporated for the analysis of the shape of the world sheet and the study of the behavior of the singularities. The results of earlier works on the study of singular world sheets were confirmed and detailed classification of singularities was carried out with the aid of a set of developed visualization tools. The desirable directions for future research using computer animation are discussed.*

18.1 Introduction

A relativistic string is a moving curve in 3D space which spans in its motion some surface in 4D Minkowski space-time. This surface is called the world sheet. The area of the world sheet is proposed to be an action of the string. The minimum action principle gives the equations of motion, which completely define the motion of the string from its initial state.

Historically, string theory arose in high energy theoretical physics in connection with the problems of description of strongly interacting particles (hadrons). Between 1968 and 1975 a number of *dual resonant models* were formulated, which described the mass spectrum of hadrons and amplitudes of their scattering. In 1970 Y. Nambu [11] suggested the possibility of a dual model interpretation in the frame of theory of interacting one-dimensional objects (strings). In this theory the form of scattering amplitude, previously heuristically proposed by G. Veneziano [20], got its natural explanation.

String theory describes hadrons as the system of quarks, connected by a string-like tube filled by a gluonic field. Hadronic strings have a size 10^{-13} cm and a tension of about 10 tons. Breaks of the string look like the decay of a hadron.

In 1974 J. Scherk, J. H. Schwarz and T. Yoneya [19, 21] proposed the string theory of all known interactions: strong, electroweak, and gravitational. In this theory all particles are considered as strings of small size (10^{-33} cm) and great tension. Transformations of particles occur due to fission and fusion of strings.

From the seventies until now many generalizations of string theory were made; they all introduced additional structures on the world sheet, such as rigidness, spin, and group variables. Since 1984 more general many-dimensional objects (membranes) have been considered. Nevertheless, even in the simplest 4D Nambu theory some problems are not solved even now. Modern physics demands the string theory to be quantum, that is, that it must be realized as an algebra of linear self-adjoint operators in some Hilbert space. A geometric formulation of the string theory leads to the fact that all of its results have a clear geometric interpretation. They do not depend on an exact representation of the world sheets, and especially not on their parameterization. The parametric invariance of the theory has not been realized on a quantum level. In addition, no particular parameterization, in which quantum string theory is self-consistent, has been found. This is the main problem in modern string theory.

From a mathematical point of view, classical string theory is a branch of differential geometry and topology, which considers minimal surfaces in d-dimensional Minkowski space. An analogous theory in Euclidean space considers soap films. Topological classification divides the world sheets into open strings (obtained by the mapping of a band into Minkowski space), closed strings (mapping of a cylinder), Y-shaped strings (mapping of three bands, stacked together along one edge), and so on. A finer classification takes into account the smoothness of these mappings. It is known that world sheets can have topologically stable singular points. An important role of singularities in string theory is mentioned by many authors.

The presence of singularities on the string leads to the complicated structure of phase space and hinders the quantization [15, 16]. Linear density of energy-momentum of the string tends to infinity in singular points. Near singular points, string models go beyond the limits of their applicability and string breaking may occur [1]. Singularities on the world sheets behave like point particles that move at the velocity of light, scatter in collisions, and can form bound states [18, 8, 14]. Physical interpretation of singularities on the string as the elements of an exotic hadrons' structure is very attractive, for example, as valent gluons [17].

As far as we know, a detailed classification of singular points on world sheets has not yet been developed. The main goal of this work is the comprehensive study of singularities on the world sheets of open strings. We already have experience in visualization of surfaces in string theory [5]. We have developed a set of visualization tools [13] for drawing complicated three-dimensional surfaces and have used it in this study. We have confirmed the results of earlier investigations of singularities on world sheets [18, 22] and have also observed new phenomena. Below we present a statement of the physical problem and the results we obtained in this study. Detailed proofs can be found in [10].

18.2 The Problem

Let a curve $Q_\mu(\sigma)$ be given in d-dimensional Minkowski space:

$$Q_\mu(\sigma) = (Q_0(\sigma), \vec{Q}(\sigma)),$$

where Q_0 denotes a time component and the vector $\vec{Q}$ denotes $(d-1)$ space components of the d-dimensional vector Q_μ. Let this curve be light-like and periodical:

$$Q'^2(\sigma) = Q_0'^2(\sigma) - \vec{Q}'^2(\sigma) = 0, \qquad Q_\mu(\sigma + 2\pi) = Q_\mu(\sigma) + 2P_\mu.$$

The first condition means that this curve is a trajectory (*world line*) of a particle, moving with light velocity. In the second condition, $2P_\mu$ is a period of the curve. This curve is shown in Figure 18.1. We will call this curve a *supporting curve.*

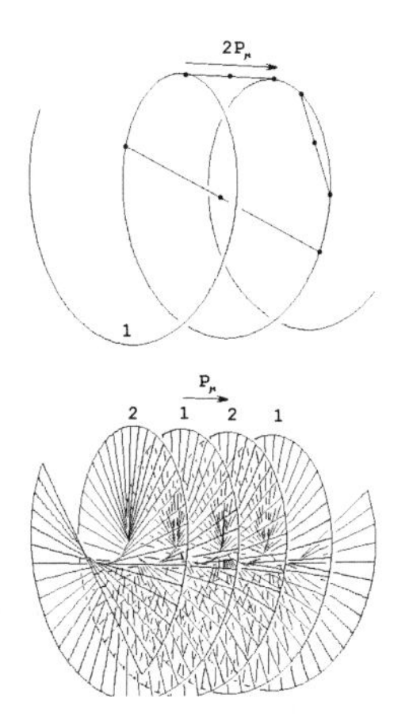

Figure 18.1: A world sheet of the string is constructed as the geometrical locus of middles of segments, connecting all possible pairs of points on the supporting curve. It has two isotropic edges: a supporting curve (1) and its image in translation onto the semiperiod (2). A slice of the world sheet by a plane of constant time gives the string at this instant of time. The time axis is directed from left to right.

A world sheet of open string is constructed as the geometrical locus of middles of segments, connecting all possible pairs of points on a supporting curve [12]

$$x_\mu(\sigma_1, \sigma_2) = \frac{1}{2}\,(Q_\mu(\sigma_1) + Q_\mu(\sigma_2)). \tag{18.1}$$

A surface constructed in such a way has two edges. The first edge coincides with the supporting curve

$$x_\mu(\sigma, \sigma) = Q_\mu(\sigma).$$

The second edge is obtained by translating the supporting curve onto a semiperiod:

$$x_\mu(\sigma + 2\pi, \sigma) = Q_\mu(\sigma) + P_\mu.$$

The world sheet maps onto itself in translation on $2P_\mu$. Being translated on P_μ it also transforms onto itself, but its edges interchange.

Edges of a world sheet are light-like: $Q'^2 = 0$. In addition, any vector from the tangent plane to the world sheet on its edge is orthogonal to Q'_μ. This plane is tangent to the light cone along the direction of Q'_μ. We will call these planes *isotropic*.

Edges of a world sheet are world lines of ends of the string. A slice of a world sheet by a plane of constant time gives the string at this moment in time. It is a curve in $(d-1)$-dimensional space. The tangent to the string l_μ is contained in the tangent plane to the world sheet; that is why on the edge it is orthogonal to the tangent to the edge. In other words, ends of the string move with a velocity of light at a right angle to the direction of the string at this point.

The semiperiod P_μ of the supporting curve is equal to the total momentum of the string. Projection of the supporting curve onto a subspace, orthogonal to P_μ (rest frame of the string) is a closed curve $\vec{Q}(\sigma)$. The supporting curve may be restored from its projection by integration:

$$Q'_0 = |\vec{Q}'| \quad \Rightarrow \quad Q_0(\sigma_1) - Q_0(0) = \int_0^{\sigma_1} |\vec{Q}'(\sigma)| d\sigma = L(\sigma_1), \tag{18.2}$$

where $L(\sigma)$ is a length of the arc of the curve $\vec{Q}(\sigma)$ between the points $\vec{Q}(0)$ and $\vec{Q}(\sigma)$. The total length of the curve $\vec{Q}(\sigma)$ is equal to the double mass of the string: $L(2\pi) = 2\sqrt{P^2}$. One can parameterize the curve $\vec{Q}(\sigma)$ by its length: $\sigma = 2\pi\frac{L}{2\sqrt{P^2}}$, then $Q'_0 = |\vec{Q}'| = \sqrt{P^2}/\pi$. In the rest frame the string moves periodically with period $2P_0$. After time intervals equal to P_0, the string adopts the same shape, only its ends are interchanged.

The world sheet (Equation (18.1)) can have singularities. A point on the surface can be singular, if tangent vectors $\frac{\partial x_\mu}{\partial \sigma_1}$ and $\frac{\partial x_\mu}{\partial \sigma_2}$ are linearly dependent on it:

$$\frac{\partial x_\mu}{\partial \sigma_1} = \frac{1}{2}\, Q'_\mu(\sigma_1) \quad || \quad \frac{\partial x_\mu}{\partial \sigma_2} = \frac{1}{2}\, Q'_\mu(\sigma_2).$$

Therefore, the world sheet has singularities if the supporting curve has two points with parallel tangents (Figure 18.2).

Tangents are parallel at the points of the supporting curve, differed by a period. We found that these points correspond to the edge of the world sheet. The edge is a singular line on the world sheet.

We are mainly interested in points with parallel tangents inside one period of the supporting curve. In the rest frame, this parallel condition takes the form of equality of unit tangent vectors to the projection of the supporting curve $\vec{Q}(\sigma)$:

$$\vec{Q}'(\sigma_1) = \vec{Q}'(\sigma_2) \quad (Q'_0(\sigma_1) = Q'_0(\sigma_2) = \sqrt{P^2}/\pi).$$

NOTE. The construction of a new image by summation of all possible pairs of vectors, contained in two given images, is known in scientific visualization as the *Minkowski sum* of images. This operation is used mainly in two contexts. Surfaces, presented as a Minkowski sum of two given curves, require less memory for storing. Of course, this "squeezing of

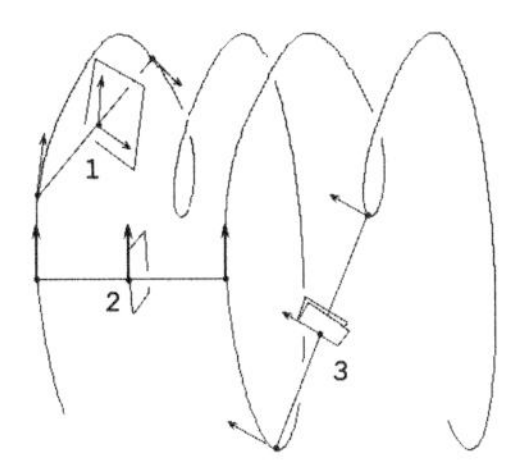

Figure 18.2: A point on a surface is regular if tangent vectors are linearly independent on it, otherwise the point is singular. Numbers on this figure denote: 1: a regular point, 2,3: singular points. Point 2 lies on the edge of the world sheet.

information" occurs due to the fact that such surfaces are special. Not all surfaces can be decomposed into a Minkowski sum of curves. Nevertheless, any object can be decomposed into a Minkowski sum of two objects, where one may have *negative shape*. Axiomatic construction and study of some properties of such nondisplayable objects can be found in [7]. Another application of the Minkowski sum is the interpolation and blending of objects in computer animation [9].

18.3 Visualization Technology

One can see from the previous section that the world sheet of string is a surface with a complicated inner structure. We have developed a set of visualization tools, satisfying the requirements of our study. As was shown above, a world sheet of string is uniquely determined by a supporting curve which, in turn, is defined by its projection onto the rest frame. So we use the projection of the supporting curve as input data for further visualization. The process of the world sheet construction is shown in Figure 18.3. First, the supporting curve was determined using a point-and-click interface by a set of points. Then cubic periodical spline interpolation was used to obtain the representation of the supporting curve projection. The supporting curve could also be specified by an explicit formula placed in a user subroutine. Next, the supporting curve in Minkowski space (the first edge of the world sheet) was derived by integration (Equation (18.2)). The second edge was obtained from the supporting curve by its translation on the semiperiod.

The singular points on the world sheet were obtained using the following algorithm. For each point on the supporting curve $\vec{Q}(\sigma)$, all points with the same unit tangent vector were determined. In the 3D case (2D rest frame), the polar angle $\varphi(\sigma)$ of this vector was calculated: $\vec{Q}'(\sigma)/|\vec{Q}'(\sigma)| = (\cos\varphi(\sigma), \sin\varphi(\sigma)),\ 0 \le \varphi < 2\pi$. All local minima and maxima of the function $\varphi(\sigma)$ were determined, then the equation $\varphi(\sigma_1) = \varphi(\sigma_2)$ was solved for fixed σ_1 and unknown σ_2, applying the dichotomy method in the intervals of monotony of $\varphi(\sigma)$. When σ_1 is varied, the solutions form lines on a parametric plane (σ_1, σ_2); these lines can be mapped in Minkowski space using Equation (18.1). To avoid multiple counting, the singular points were searched in the region $0 \le \sigma_1 < \sigma_2 < 2\pi$, then

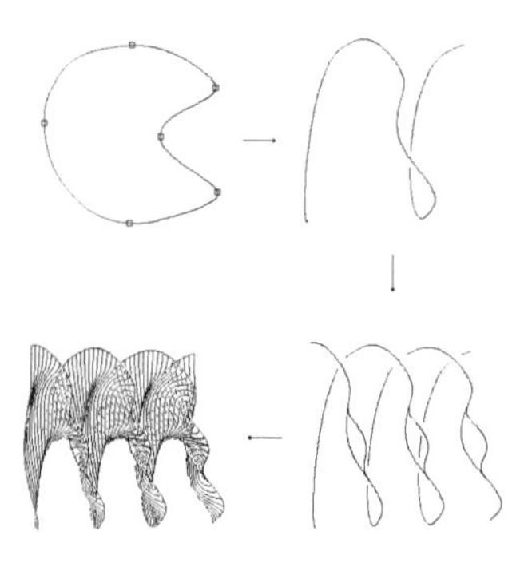

Figure 18.3: World sheet construction.

they could be extended on the whole world sheet by translations on P_μ.

In the 4D case (3D rest frame), the unit tangent vector can be specified by two spherical angles: $\vec{Q}'(\sigma)/|\vec{Q}'(\sigma)| = (\cos\varphi(\sigma)\cos\theta(\sigma), \sin\varphi(\sigma)\cos\theta(\sigma), \sin\theta(\sigma))$. A system of two equations should be solved: $\varphi(\sigma_1) = \varphi(\sigma_2)$ and $\theta(\sigma_1) = \theta(\sigma_2)$. In normal cases, the solutions are located at isolated points on the parametric plane. The solutions of the first equation were found as described above—they form lines on the parametric plane, represented as lists of segments. Then the second equation was solved—searching for segments in which the function $\theta(\sigma_1) - \theta(\sigma_2)$ changes its sign and applying the dichotomy to the segment for refinement of the solution.

The slice of the world sheet by the plane of constant time (the string at this moment in time) can be obtained in the following way. First, we find slices of the edges of the world sheet (positions at the ends of the string), that is, we solve the equations $Q_0(\sigma_1^*) = t$ and $Q_0(\sigma_2^*) + P_0 = t$ in σ_1^* and σ_2^* for given t (the equations have unique solutions because the function $Q_0(\sigma)$ is monotonous for the class of strings we consider). The ends of the string at moment t are placed at the points of the supporting curves $\vec{Q}(\sigma_1^*)$ and $\vec{Q}(\sigma_2^*)$. Then we solve the equation $(Q_0(\sigma_1) + Q_0(\sigma_2))/2 = t$ in σ_2 for fixed σ_1 from the interval $\sigma_2^* < \sigma_1 < \sigma_1^*$. The equation has a unique solution $\sigma_2(\sigma_1)$ in the interval $\sigma_1^* < \sigma_2 < \sigma_2^* + 2\pi$, which can be obtained by the dichotomy method. When σ_1 is varied, the solutions form a line on a parametric plane, which, being mapped to the space by the expression $(\vec{Q}(\sigma_1) + \vec{Q}(\sigma_2(\sigma_1)))/2$, gives the string at moment t.

The image of the world sheet can be constructed using several techniques. For most of figures presented here, all elements mentioned above were put into a temporary buffer of segments and displayed using a halo method. This method produces the wire-frame image of the world sheet, in which all details are clearly visible. Alternatively, the frame of the world sheet could be constructed on a supporting curve as a mesh of points using Equation (18.1). In the 4D case this frame was projected onto a specified 3D space, for example, the rest frame. Then a 3D polygonal surface was constructed using light effects simulation. We used DEC PHIGS 2.1 [3] (the highest available PHIGS release) as a tool for image construction and manipulation.

18.4 Results

Singularities on the world sheet in spaces $d = 3$ and $d = 4$ have different shapes.

When $d = 3$, singularities have the form of kink lines. Near the kink line the world sheet has a fold. The slice of this fold by a constant time plane produces a cusp (kink on a string). Parts of the world sheet, separated by the kink line near the fold, have a common tangent plane. This plane is isotropic, like the tangent plane on its edge. So, the kink moves with light velocity at a right angle to the string direction at the point of kink location. The behavior of kinks on the world sheet is shown in Figures 18.4–18.13. The variety of the processes on the world sheet is complex, but all these processes consist of basic elements shown in these figures.

When $d = 4$, singularities on the world sheet are localized in isolated pinch points. The string undergoes an instantaneous cusp at the moment of passage through the pinch point. In the pinch point the world sheet has no tangent plane. Lines of self-intersection of the world sheet terminate in the pinch points (more precisely, self-intersections of the projection of the world sheet in any 3D space do this). Many interesting properties of pinch points are described in the remarkable book by G. K. Francis [6]. We note here one property: in any projection on a figure plane, the pinch point lies on a contour of the surface, visible or hidden by other parts of surface. World sheets with pinch points are displayed in Figures 18.14–18.17.

If the supporting curve itself has singular points, then corresponding singularities on the world sheet are placed on characteristic lines, passing through these points (Figure 18.18). These lines mark the path of the light signal along the world sheet. Two characteristics pass through each inner point of the world sheet. A single characteristic passes through a point on the edge. Characteristics that pass through kink lines and edges are tangent to these curves. Characteristics, kink lines, and edges exhaust all light-like curves on the world sheet.

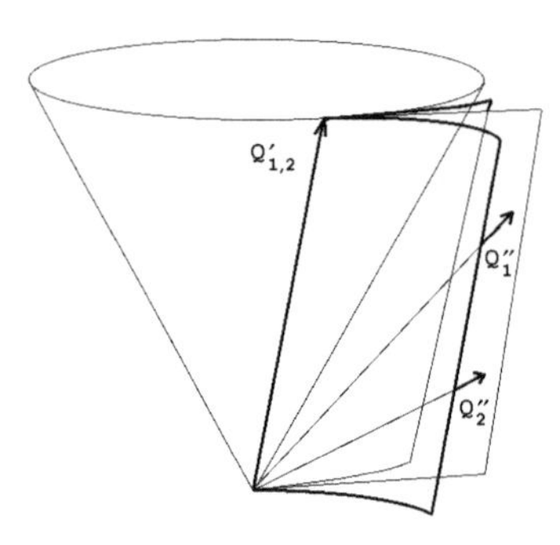

Figure 18.4: The shape of the world sheet near a singular line.

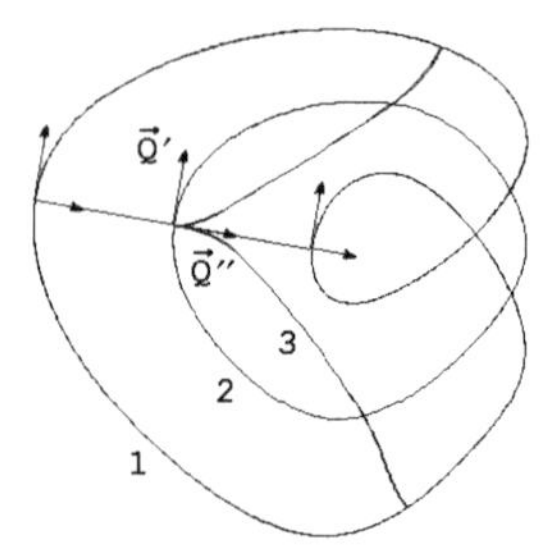

Figure 18.5: In the case $d = 3$ the projection of the supporting curve in the rest frame is a plane closed curve. If this curve is convex (has no inflection points) and its winding number is ν, then the number of kinks on the string is constant in time and equals $\nu - 1$. This figure shows a single kink on the string for the case $\nu = 2$. Numbers on the figure denote: 1: a supporting curve, 2: a trajectory of the kink, 3: the string. The kink moves periodically (with the period P_0) inside the string, never approaching the edges.

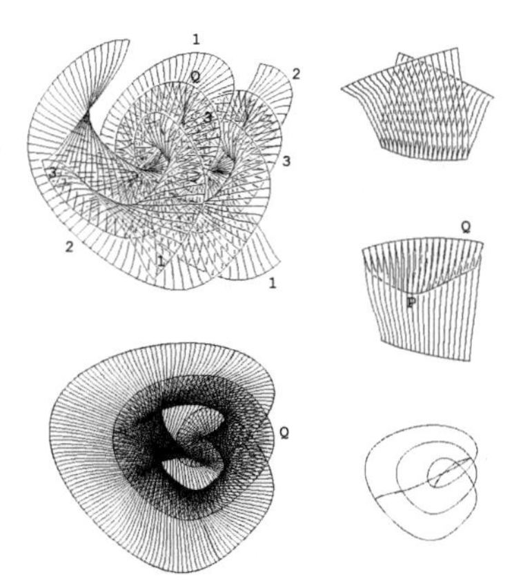

Figure 18.6: The upper-left portion shows the world sheet for the supporting curve shown in the previous figure. The bottom-left portion shows the projection of this world sheet onto the rest frame. Numbers denote: 1,2: the edges of the sheet, 3: a kink line. We note that near the point Q, edge 1 passes first in front of the kink line, then behind it. Near the point Q the world sheet has a self-intersection. The right displays (from top to bottom) are the kink, the line of self-intersection PQ, and the self-intersection on the string.

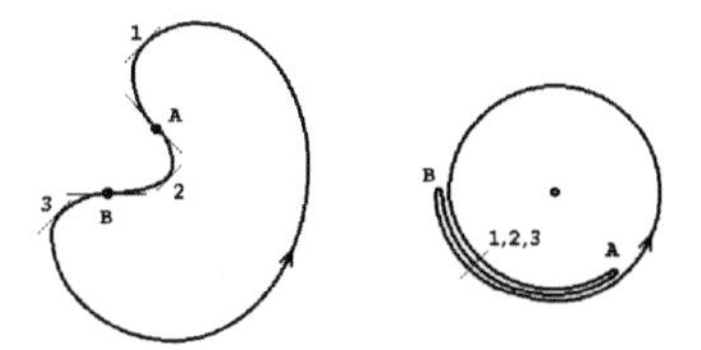

Figure 18.7: A nonconvex supporting curve and its unit tangent vector. A and B are inflection points. Tangents in 1, 2, 3 are collinear, therefore additional kink lines appear on the string.

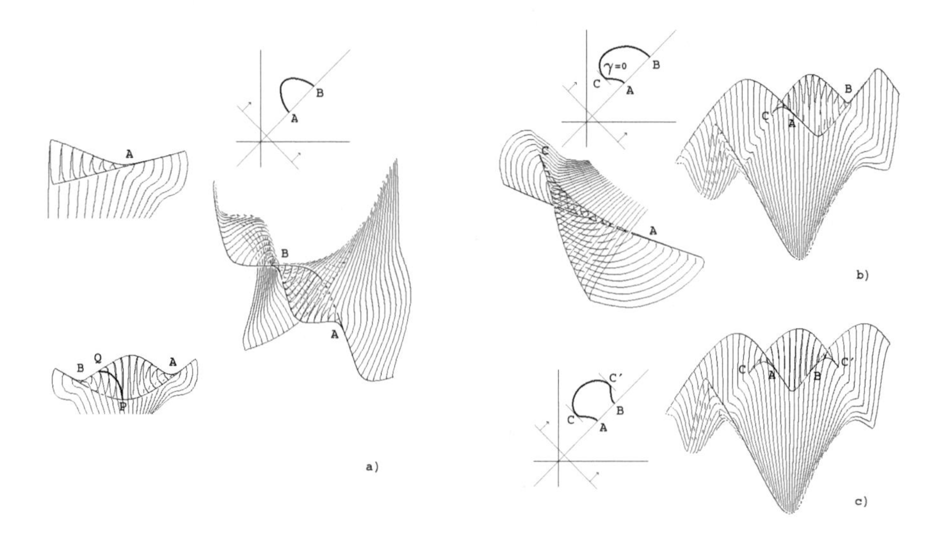

Figure 18.8: World sheets for nonconvex supporting curves. The upper-left parts show the kink line on the plane of parameters (σ_1, σ_2). *(a)* The kink appears on the end of the string at the moment when the end passes through inflection point A. The kink disappears on the same end at the moment of its passage through inflection point B. Between the points A and B the world sheet intersects itself along the line PQ. *(b)* At some moment in point C two kinks appear on the string. One kink disappears on the end of the string at point A, another one disappears at point B. *(c)* At point C a pair of kinks appears. One kink disappears on the end of the string at point A. At point B a new kink appears; it annihilates the second kink in point C$'$.

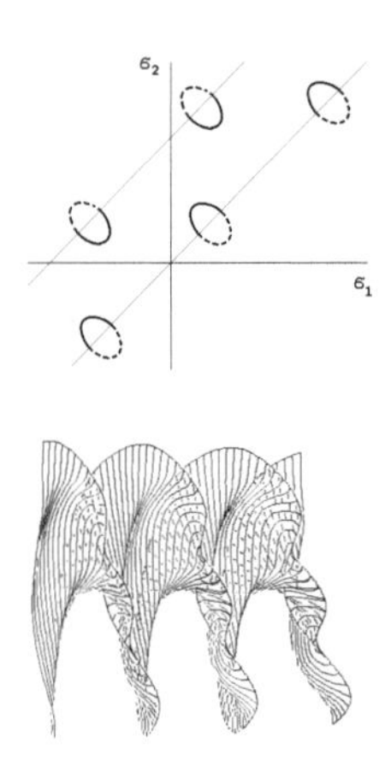

Figure 18.9: Images of the kink line in shifts $\sigma_1 \to \sigma_1 + 2\pi n_1$, $\sigma_2 \to \sigma_2 + 2\pi n_2$, which do not change the parallelism of tangents. Kinks appear and disappear on one end of the string, then after a lapse of time equal to semiperiod P_0, these processes repeat on the other end.

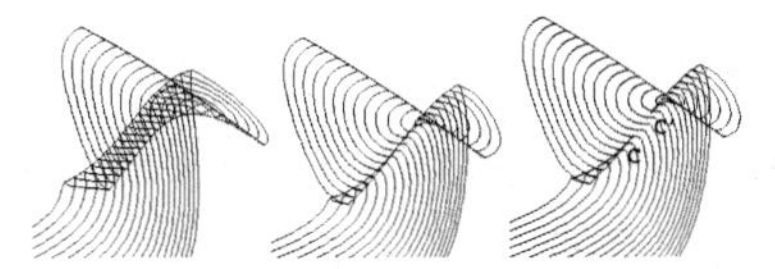

Figure 18.10: Rearrangement of kink lines. In the deformation of the supporting curve the kink lines approach each other, become tangent, then the scattering of the kinks transfers to 'the annihilation channel.'

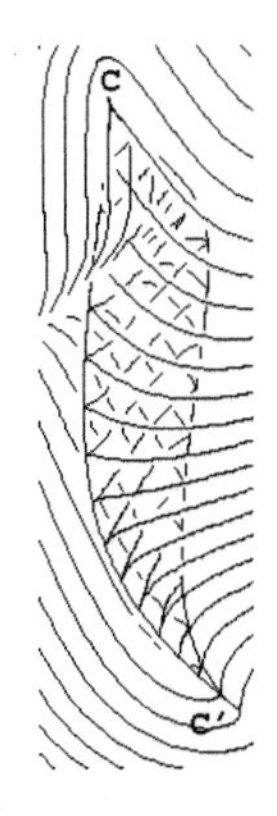

Figure 18.11: Closed kink line on the world sheet. A kink pair appears at point C, disappears at a point C'.

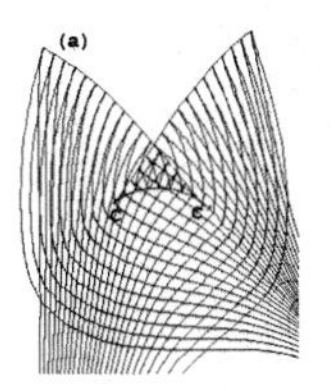

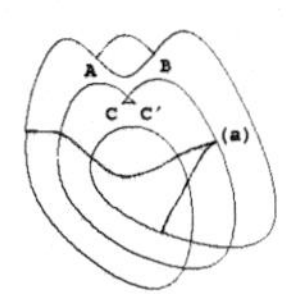

Figure 18.12: The world sheet for a nonconvex supporting curve with $\nu = 2$. One kink (a) is constantly present on the string. At some moment at point C a kink pair appears, one kink of which annihilates a kink (a) at point C'. Another kink continues to move on the string.

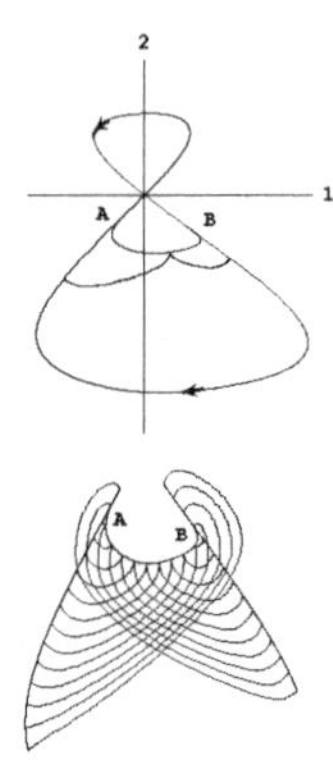

Figure 18.13: The supporting curve with $\nu = 0$. The world sheet and shape of the string. We note that a kink appears at one end of the string and disappears at the other end.

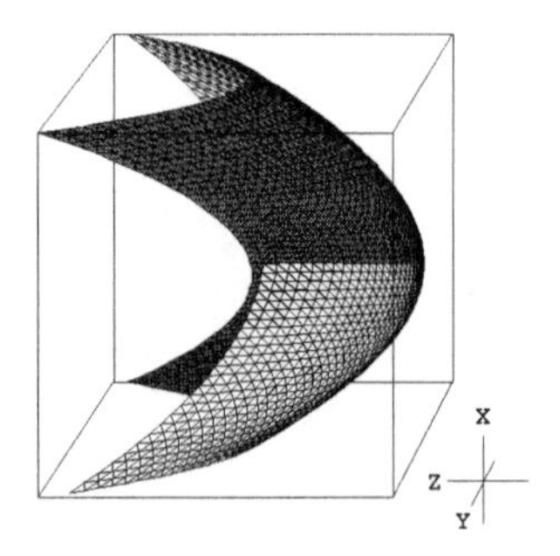

Figure 18.14: The world sheet in the vicinity of a pinch point has a canonical form $(X, Y, Z) = (\xi, \xi\eta, \xi^2 + \eta^2)$. Axes X, Y, Z form a basis in some three-dimensional isotropic space (tangent to the light cone).

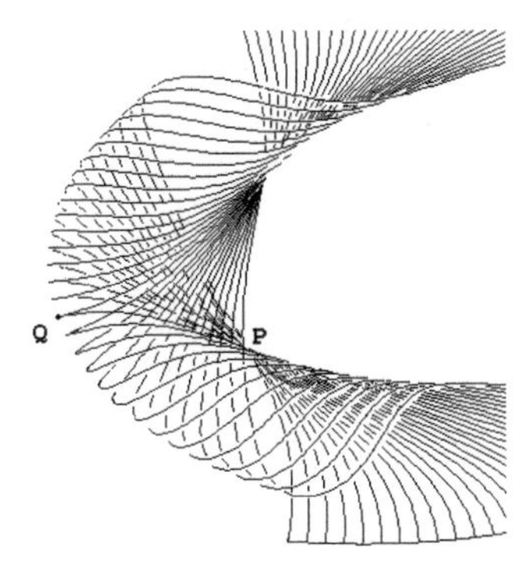

Figure 18.15: Projection of the world sheet onto the rest frame near pinch point Q. This surface intersects itself along the curve PQ, terminating at the pinch point. The string has a cusp at the pinch point.

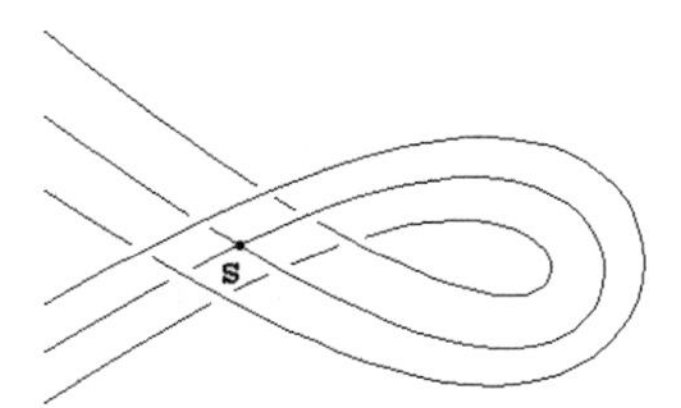

Figure 18.16: The lines of self-intersection present only on projections of the world sheet in 3D space. The world sheet of the previous figure has no self-intersections in 4D space because there are no self-intersections of each string among the strings that form this surface. Stable self-intersections of world sheets in 4D space are localized at isolated points, as shown in this figure.

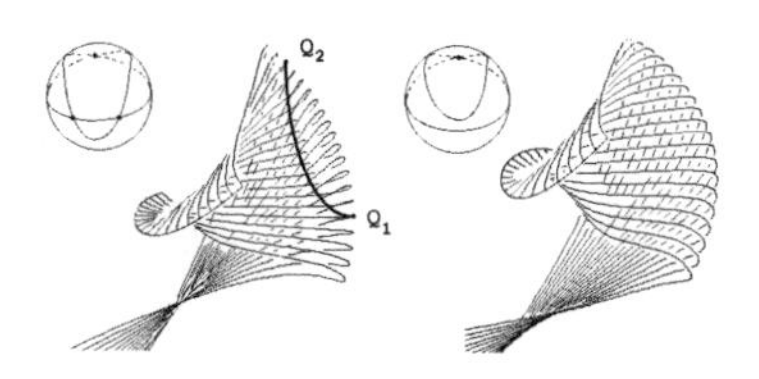

Figure 18.17: In the deformations of the supporting curve the pinch points disappear by pairs, as shown in this figure. (Compare with Figures 9 and 10 in [6].) The upper-left parts show hodographs of unit tangent vector to the supporting curve.

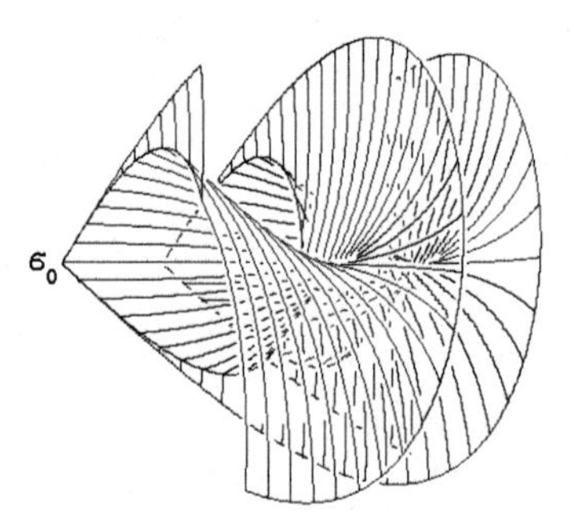

Figure 18.18: Fracture of the world sheet, caused by nonsmoothness of the supporting curve. The fracture line is obtained from the supporting curve by the contraction in two times to point σ_0. Such lines are called characteristics.

18.5 Future Plans

Besides the representation of world sheets with the halo method, we tried some other methods (Figure 18.19). We present these 3D images as stereo-pairs in Figure 18.20. One can view them with the aid of stereoscope or (after some training) without it.

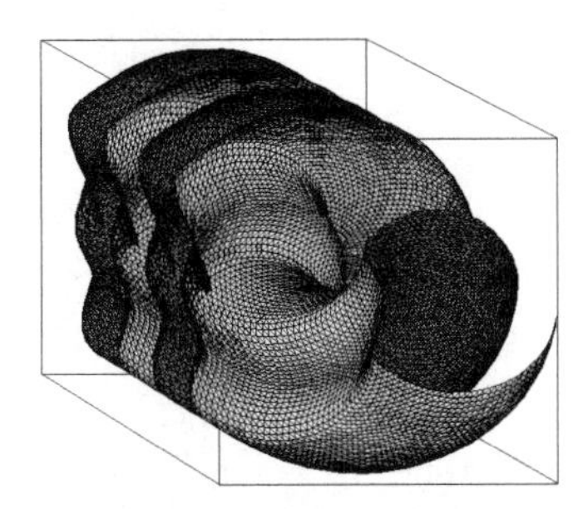

Figure 18.19: The image of the world sheet constructed as a 3D surface.

We also performed the animation of string dynamics. Recall that all surfaces presented here are time evolvents of some physical processes. By creating the sequence of world sheet slices in time like a sequence of film shots, one can observe the motion of a string and the evolution of singularities on it. Such a film as a representation of the world sheet most adequately fits the statement of the physical problem. We have developed a basic tool for string animation, which is applied now for the study of *disconnected strings*—an exotic sector of string theory, correspondent to supporting curves, nonmonotonous in Minkowski time. Typical processes in this sector (periodical creation, annihilation and recombination of strings) are shown in Figure 18.21. A more detailed discussion can be found in [4].

As a next stage we plan to study singular points on strings of other topological types and the processes of string breaking (Figure 18.22). It is also necessary to enhance the programs and tools in use. In particular, we plan to encode the flow of energy on world sheets by color and show the effect of energy concentration in singular points. To create easy to observe stereoimages, we propose employing an autostereogram technique [2]. The further objective is to incorporate all the tools of visualization described and to represent string dynamics as an interactively modifiable stereoscopic movie. The solution of this problem is possible with the help of modern virtual reality systems.

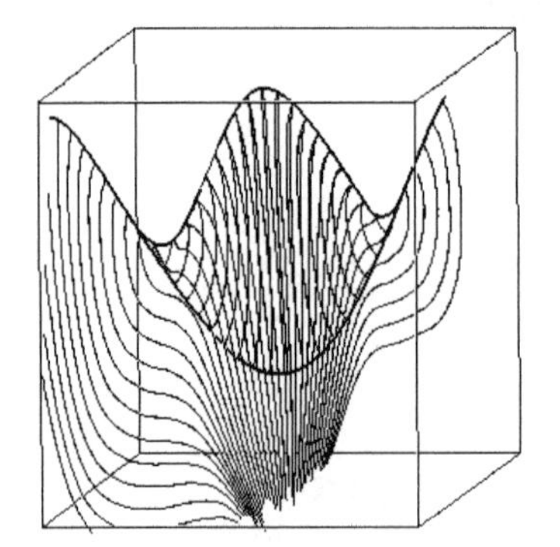

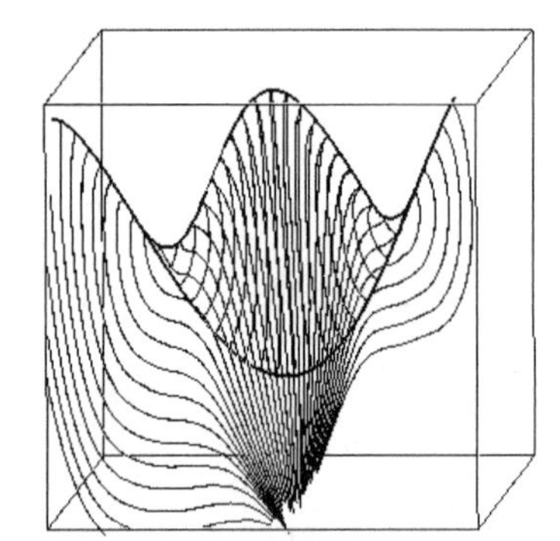

Figure 18.20: Stereo-pair. The left image is destined for right eye, the right image for the left eye. For viewing this image without a stereoscope you should focus on a point midway between your eyes and the page and should adjust your focus in such way that the two images become fused into one. (Note that two other images will appear on each side of the central image; you should ignore them.) The 3D image is located in the middle; it will probably be blurry. If you hold this image for a sufficiently long time, it will become sharp.

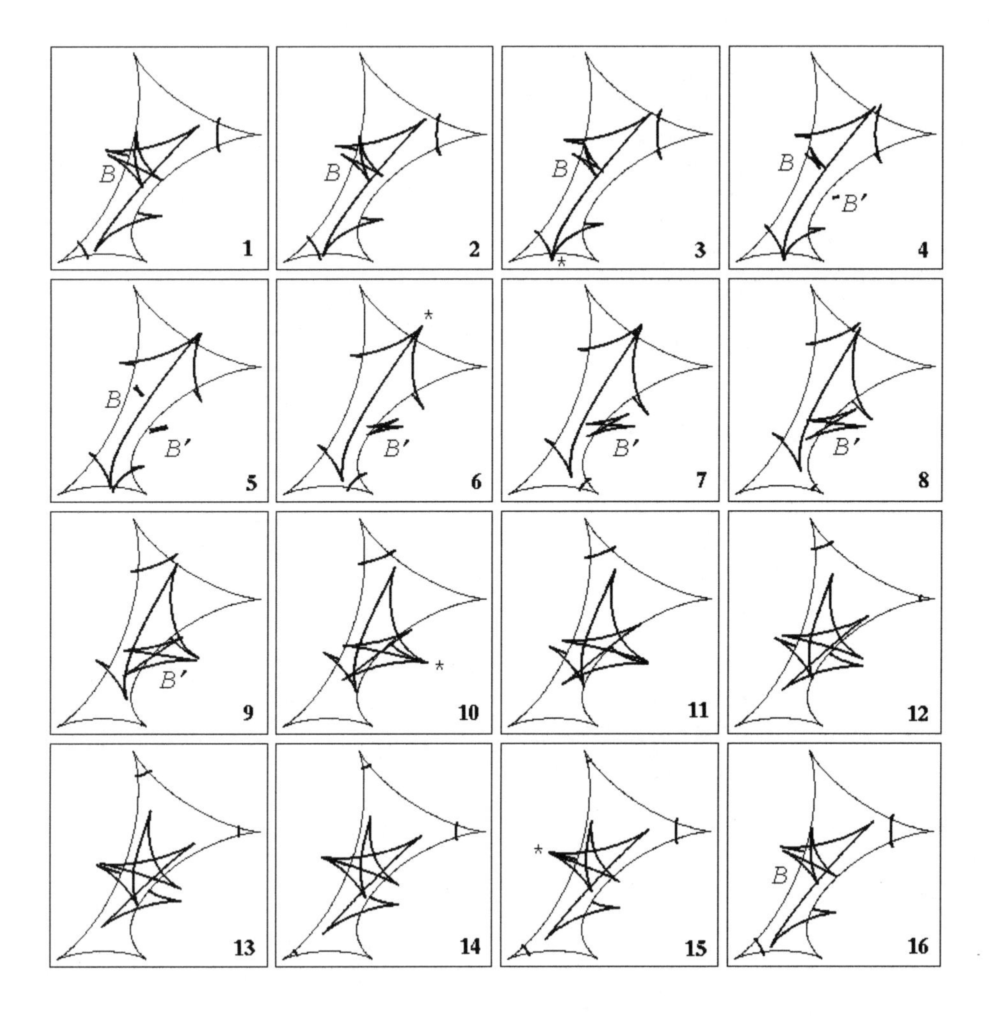

Figure 18.21: Typical processes in the formation of disconnected strings. In the first frame the following four disconnected elements can be found: a Z-shaped long open string in the center, two short open strings in the corners of the supporting curve, and a closed string B, looking like a butterfly. In the process of further evolution, the long string recombines $(*)$ with the short strings, adopting the shape of a reverted Z (frame 8). During this time the butterfly B is decreased and disappears in frame 5. In frame 4 a new butterfly B' appears. In frame 10 it enters into recombination with the long string and is included in it. There are no closed strings in frames 11–14; here the long string in the center is simply connected. In frame 15 the butterfly B is detached from the long string. Frame 16 coincides with frame 1, and the evolution is repeated again.

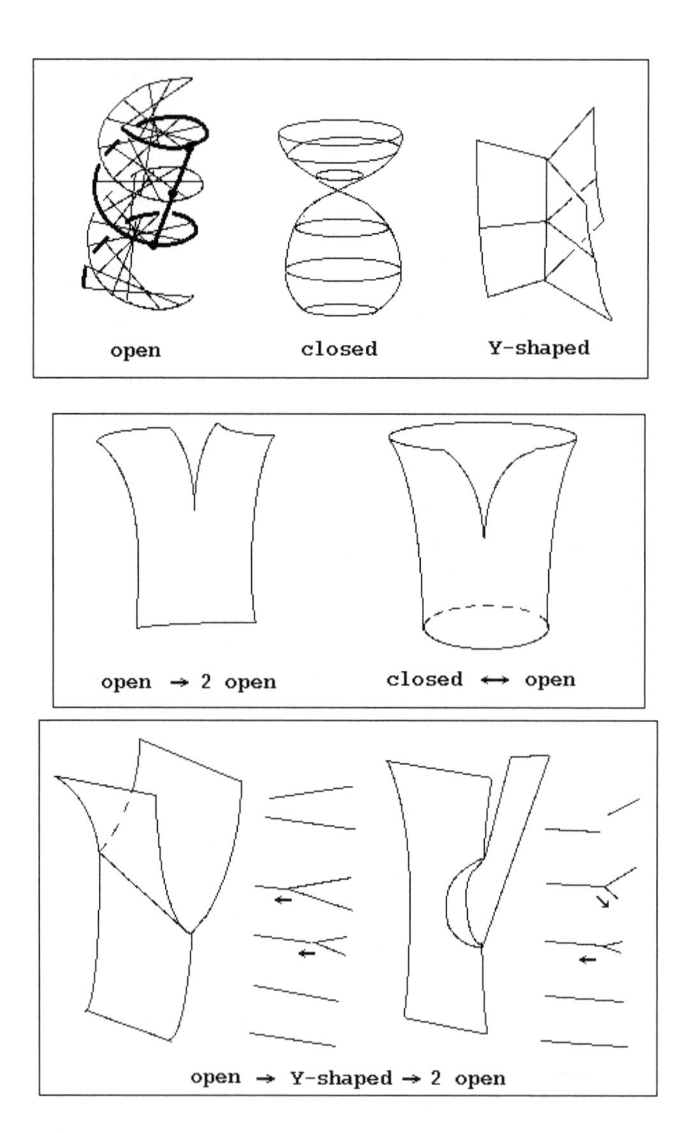

Figure 18.22: Different topological classes of the world sheets. The three upper figures show free strings, the others show various scenarios of string breaking.

Acknowledgments

This work is fulfilled as a part of the international project of the Russian Ministry of Science: "Visualization of Complex Physical Phenomena in Virtual Environment."

Bibliography

[1] X. Artru, *Classical String Phenomenology*, Preprints LPTHE 78/25 (1978), 79/8 (1979), 81/1 (1981), Laboratoire de Physique Théorique et Hautes Energies, Université de Paris-Sua, 91405, Orsay, France, 1978, 1979, 1981.

[2] *Interactive Pictures in 3D*, Benedikt Taschen Verlag GmbH, Köln, 1994.

[3] *DEC PHIGS Reference Manual, Version 2.1*, Feb. 1990.

[4] V.V. Dyachin, S.V. Klimenko, and I.N. Nikitin, "Visualization of Relativistic String Dynamics," submitted to *Programmirovanie [Russian Journal of Computer Science]*.

[5] S.V. Klimenko et al., "Baryon String Model," *Theor. Math. Phys.*, Vol. 64, No. 2, 1985, pp. 245.

[6] G.K. Francis, *A Topological Picturebook*, Springer-Verlag, 1987, 1988.

[7] P.K. Ghosh, "A Unified Computational Framework for Minkowski Operations," *Comp. & Graph.*, Vol. 17, No. 4, 1993, p. 357.

[8] G. P. Jorjadze, A.K. Pogrebkov, and M.K. Polivanov, "Singular Solutions of the Equation $\square\varphi + (m^{(}2/2)\exp\varphi = 0$," *Theor. Math. Phys.*, Vol. 40, 1979, p. 221.

[9] A. Kaul and J. Rossignac, "Solid-Interpolating Deformations: Constructing of Pips," *Proceedings of EUROGRAPHICS '91*, 1991, p. 493.

[10] S.V. Klimenko, I.N. Nikitin, and V.V. Talanov, *Study of Singularities on the World Sheets of Relativistic Strings*, IHEP Preprint 95-7, Protvino, 1995.

[11] Y. Nambu, *Lectures at the Copenhagen Symposium on Symmetries and Quark Models*, Gordon and Breach Book Co., New York, 1970.

[12] I.N. Nikitin, "Relativistic String Configurations with Anomaly Free Quantization," *Nuclear Physics Journal*, Vol. 56, No. 9, 1993, p. 230.

[13] I.N. Nikitin and V.V. Talanov, *On Drawing Complicated 3D Surfaces*, IHEP Preprint 94-23, Protvino, 1994.

[14] A.K. Pogrebkov, "Complete Integrability of Dynamical Systems Generated by the Singular Solutions of Liouville Equation," *Theor. Math. Phys.*, Vol. 45, 1980, p. 161.

[15] G. P. Pron'ko, "Theory of One-Lacuna Configurations of Relativistic String," *Theor. Math. Phys.*, Vol. 59, No. 2, 1984, p. 240.

[16] G. P. Pron'ko, "Hamiltonian Theory of the Relativistic String," *Rev. Math. Phys.*, Vol. 2, No. 3, 1990, p. 355.

[17] G.P. Pron'ko, Private Communication.

[18] G.P. Pron'ko, A.V. Razumov, and L.D. Soloviev, "Classical Dynamics of Relativistic String," *Elementary Particles and Atomic Nuclei Journal*, Vol. 14, No. 3, 1983, p. 558.

[19] J. Scherk and J.H. Schwarz, "Dual Models and Geometry of Space-Time," *Phys. Lett. B*, Vol. 52, No. 3, 1974, p. 347.

[20] G. Veneziano, "Construction of Crossing-Symmetric, Regge-Behaved Amplitudes for Linearly Rising Trajectories," *Nuovo Cim. A*, Vol. 57, No. 1, 1968, p. 190.

[21] T. Yoneya, "Connection of Dual Models of Electrodynamics and Gravidynamics," *Progr. Theor. Phys.*, Vol. 51, No. 11, 1974, p. 1907.

[22] A.A. Zheltuhin, "On Topological Structure of Extrema in the Action Functional for Classical Relativistic String," *Nuclear Physics Journal*, Vol. 34, No. 2, 1981, p. 562.

Chapter 19

Hierarchical Representation of Very Large Data Sets for Visualization Using Wavelets

Pak Chung Wong and R. Daniel Bergeron

Abstract. *The very large data sets that characterize much of today's scientific data bring special challenges to analyzing them. A comprehensive understanding of very large data sets often requires the analysis of measurements obtained at the smallest spatial and temporal resolution possible, yet these measurements must be studied in the context of the global properties of the environment. We present a hierarchical representation model to support progressive refinement data visualizations. It gives scientists an accurate overview of the whole data set at a lower resolution, and zooming capabilities to study the detail once interesting targets are located. We discuss data zooming techniques using statistical measures as well as wavelets. We describe a C++ prototype with a Motif front end that supports progressive refinement.*

19.1 Introduction

Scientific research activities have acquired an almost overwhelming quantity of data and will continue to do so in the future. The need to deal with large amounts of scientific data for visualization is documented in the proceedings of the 1987 National Science Foundation workshop on *Visualization in Scientific Computing* [9]. To quote the report,

> Earth resource satellites sent up years ago are still transmitting data. All scientists can do is warehouse it. The technology does not exist for receiving, analyzing and presenting information in such a way that it is useful to scientists.

The situation has not improved. There is every indication that it is getting worse.

Nevertheless, scientists in different disciplines require scientific data in order to conduct significant research. Major challenges arise when trying to assure maximum scientific gain from this data. It is often the case that scientists are primarily interested in analyzing only small subsets of the data at the highest resolution obtainable. A key requirement, therefore, is to efficiently survey the full data set at a lower resolution and then be able to identify and analyze interesting subsets of the data in greater detail.

The goal of this chapter is to present an efficient and flexible model that allows *rapid analysis* of vast amounts of scientific data in a progressive refinement environment. The analysis of these very large data sets involves a *hierarchy* of data processing stages, each of which are increasingly more coarse approximations of the original data. The model and the prototype we present directly address this *fine to coarse hierarchy* requirement.

19.2 Motivation

Our interest in a progressive refinement environment for scientific research is motivated in part by the very large data obtained from the NASA Global Geospace Science (GGS) Initiative. The primary objective of this mission is to investigate the physical processes controlling the transport of mass, momentum, and energy from the sun and solar wind into the Earth's magnetosphere, ionosphere, and atmosphere. The GGS mission includes two NASA spacecraft, WIND and POLAR, and one spacecraft provided by the Japanese Institute for Space and Astronautical Science, GEOTAIL. The GGS Initiative is part of a broader international effort, the International Solar-Terrestrial Physics (ISTP) program, that includes the United States, Japan, Europe, and Russia.

The design of the GGS mission is based on the need to investigate the physical processes occurring simultaneously in the many diverse regions of near-Earth space. A typical instrument on a spacecraft records data every three seconds, at twenty different energy levels, in six different directions, as it moves through its orbit. This is translated into over five gigabytes of data collected from one instrument on a spacecraft every year. A portion of the data is available to us in Common Data Format (CDF) [12].

After the data is collected, the biggest problem facing scientists is to locate the interesting patterns to study. The sheer quantity of data makes it infeasible to explore the data in its highest resolution. Terabyte-sized data will soon be accumulated for this project. Real-time random access of this data from mass storage (capacity beyond CD-ROM) is still impossible [2]. A scientist needs some mechanism to identify *interesting* regions from a lower resolution representation. The scientific gain from the mission will be greatly enhanced if scientists can access large quantities of data, and efficiently change the data resolution (low resolution displays for surveying purposes and high resolution displays for more detailed studies) throughout the different stages of data analysis.

19.3 Data Resolutions Versus Image Resolutions

There are actually two different stages of hierarchical scaling, as shown in Figure 19.1. Terabyte-sized data is first scaled down to the order of gigabytes, so that it can sit comfort-

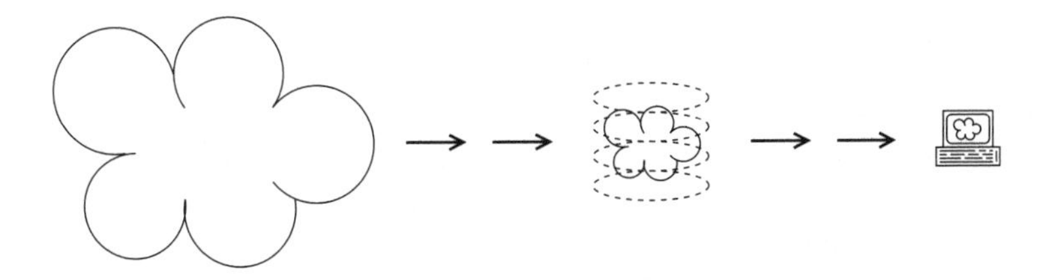

Figure 19.1: Two different stages of scaling.

ably in random access memory such as a CD-ROM. The data is then further scaled down to the order of megabytes, so that it can be visualized on a color workstation display.

The first scaling is important because it may be too expensive to *revisit* the original data frequently. The size of the data may also make it infeasible to keep the data *locally*. On the other hand, zoomed data stored in random access memory such as a CD-ROM can be accessed freely without much memory burden. This version of data might be the best, that is, highest resolution, the scientist ever uses directly. The second scaling is even more important because it is the operation that brings the overview data to the scientist.

19.4 Hierarchical Multiresolution Data Access

The key point of hierarchical multiresolution data access is that data analysis must ultimately be done at the highest *available* spatial and temporal resolution, yet there is far too much data for all of it to be analyzed at this level. Therefore, the analysis process often requires a number of intermediate stages. Figure 19.2 depicts an adaptive zooming hierar-

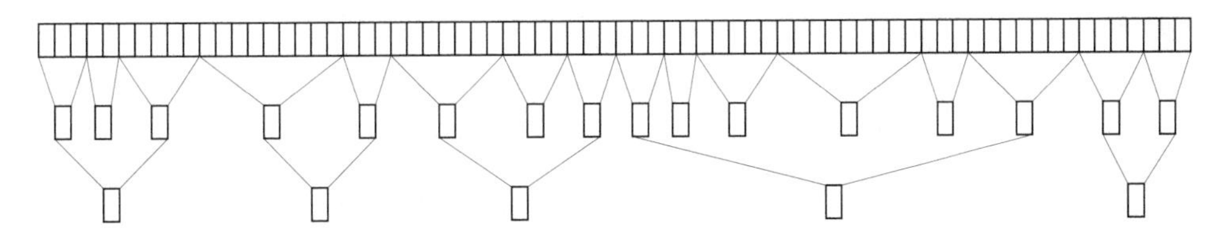

Figure 19.2: A zooming hierarchy of a one-dimensional data stream.

chy of one-dimensional scientific data. In this example, two intermediate stages are created in the hierarchy, and the second stage quickly shrinks to a relatively small size compared to the original. The irregular zooming ranges in intermediate stages allow scientists to emphasize important features and de-emphasize the less interesting areas.

When the size of scientific data gets large enough, the issue of static zooming (precompute and store) versus dynamic zooming (generate on the fly) becomes critical. It is often a case of trading reasonable memory and/or accuracy for reasonable user-response time. A good scientific visualization system should always give users the opportunity to choose the zooming procedures.

For our GGS data, the first analysis stage is the surveying of low resolution data covering a long time period (days, months, or perhaps years) using a subset of the available instruments. At this level a broad overview of the measurements is obtained and interesting subintervals are identified. For example, the magnetopause and the plasma sheet boundary layer are boundary regions that play a critical role in the transport of mass, momentum, and energy from the flowing solar wind into the near-Earth environment. From a survey of many months of data from an orbiting magnetosphere spacecraft, a scientist can identify times when the spacecraft crossed certain boundary regions because the boundary crossing generates distinctive patterns in the data values. The critical requirement is that *the distinctiveness of the pattern must be maintained in the low resolution representations.* This is also the idea behind the irregular zooming range design in Figure 19.2. If such a feature loss occurs, we must find another decomposition approach that preserves the feature.

19.5 Definition of Hierarchy

The goal is to define operations or functions to accurately *zoom down* the size of a very large piece of data into multiple sublevels. Each of the sublevels should contain most, if not all, of the major characteristics of its next higher resolution. Conventional discrete data subsampling methods tend to generalize the data and miss important features such as high frequency spikes if the *Nyquist* sampling rate is not reached.

Statistical analysis is an important tool for understanding data. Its legacy, to a certain extent, is deeply embedded in the foundation of today's scientific data visualizations. Statistical measures such as mean, median, and standard deviation are considered to be healthy and consistent for scientific data analysis. By continuously applying a pairwise average function to one-dimensional data, we generate a multiresolution hierarchy with a *decomposition ratio* (that is, the size of output divided by the size of input) of 0.5 at each level. The same idea can be applied to higher-dimensional data. In each resolution, we take *averages* of each item's closest neighbors. In Figure 19.3, one-dimensional data

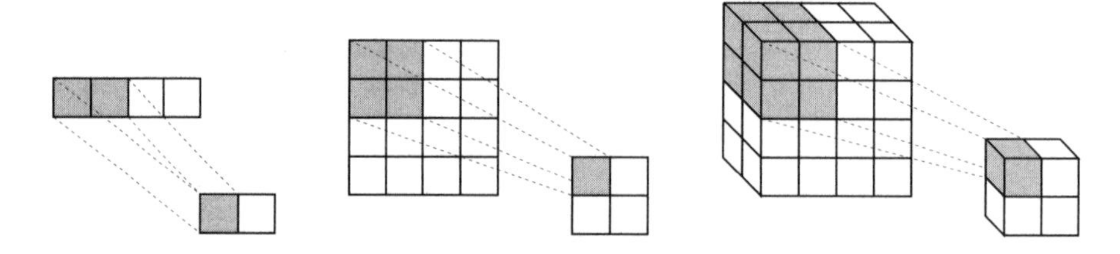

Figure 19.3: A pairwise averaging of one-, two-, and three-dimensional scientific data.

of four items is zoomed down to two by taking the average of every two items. In two-dimensional data, a total of four items is averaged in each resolution. Similarly, eight items are combined during each step in decomposing three-dimensional data. Note that with nD data, each average operation takes 2^n data items. This implies a decomposition ratio of $\frac{1}{2^n}$.

The pairwise average operation, however, can be harmful if the data oscillates like sinusoids. That is, a discrete negative signal can be weakened or even eliminated by a positive neighbor with a similar amplitude. This implies a violation of the critical requirement that a zooming function maintain the distinctiveness of the pattern in all resolutions.

In our multiresolution representation, more data is generated during the zooming process. Although this new lower-resolution representation may be useful for improving the performance of analysis, we are actually increasing total memory requirements by generating it. We improve memory utilization only if the new representation allows us to discard the higher-resolution representation. If we need to revisit the higher-resolution data later, we must be able to either recreate it or restore it (such as from secondary memory). This dilemma occurs at every level of resolution. Wavelets provide us with unique opportunities for building multiresolution hierarchies such that the intermediate levels of the hierarchy can be reconstructed from lower levels.

19.6 Wavelet Overview

In this section, we briefly describe the idea behind wavelet transforms and summarize the properties of an orthogonal wavelet. See [7, 18, 15, 16] for detailed discussions of this topic. More powerful wavelets, which work well in a wide variety of applications, are also discussed in [1, 19].

The wavelet transform is a mathematical tool that can be utilized to extract or encode information. It represents a data stream by cutting it down into different frequency components. Since their first appearance in 1980, wavelets have been applied to a variety of studies such as two-dimensional image processing [7], three-dimensional volume rendering [10, 11, 4], radiosity [3], speech analysis [19], data compression [1, 13], progressive analysis [17], and statistics [5].

19.6.1 Hierarchical Representation Using Wavelet

Given a 1D data stream with n items, an application of an orthogonal wavelet transform generates $\frac{n}{2}$ coefficients of low-frequency *approximations*, and $\frac{n}{2}$ coefficients of high-frequency *details*, as shown in Figure 19.4. The operation can be applied iteratively on the

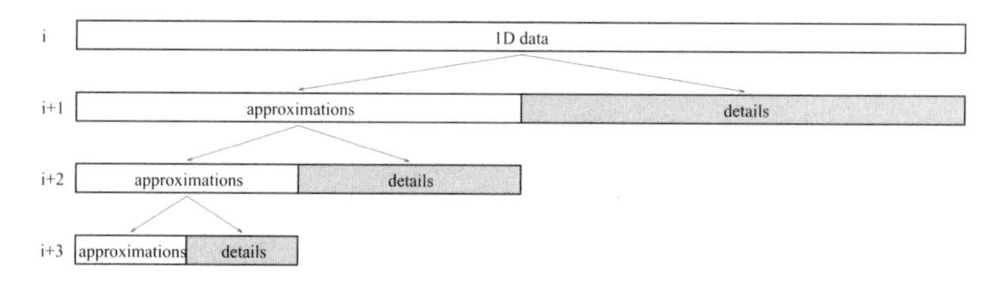

Figure 19.4: Wavelet transforms on one-dimensional data stream.

approximation part to get increasingly coarse zooming data, until the size of the approximation is less than or equal to the number of wavelet filter coefficients [1]. Figure 19.5 shows a hierarchy of multiresolution wavelet operations. The size of each level is reduced by exactly 50 percent.

Wavelet transforms are invertible. The magic lies in the low-frequency details of each resolution. If both the approximations and details of any one stage are available, it is possible to have a *lossless* reconstruction of the approximation part of the next higher resolution.

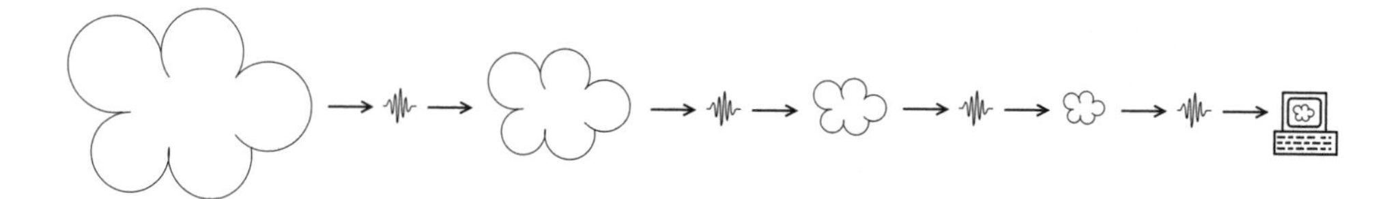

Figure 19.5: Multiresolution wavelet decompositions. This figure does not show the details, only the approximations.

For example, if both the approximation and detail of stage $i + 3$ are available, the approximation of stage $i + 2$ can be reconstructed without loss. In fact, the whole hierarchy can be rebuilt in this manner. Our experiments show that, even with the floating-point approximations of digital computers, the reconstruction and the original achieve a perfect *correlation* of 1 [18]. A correlation of 0.8 or higher is considered to be strong evidence of highly correlated data.

The pairwise average operation we discussed in the previous section is, in fact, the approximation part of a *Haar* wavelet [1, 18]. In wavelet terminology, this pairwise average operation is only part of the decomposition of a wavelet transform. The detail part of the Haar wavelet is computed from pairwise differences of the input signal. The simple pairwise average operation does not by itself allow reconstruction.

19.6.2 Memory Management for Wavelet Transforms

As we mentioned above, the decomposition actually generates more data to build the multiresolution hierarchy. For one-dimensional data with n data items, the first decomposition generates $\frac{n}{2}$ approximations and $\frac{n}{2}$ details. The rest of the hierarchy (including approximations and details of every stage) occupies

$$\frac{n}{2} + \frac{n}{4} + \frac{n}{8} + \frac{n}{16} + \cdots \leq n$$

units of memory. Thus an upper bound on memory is $\frac{n}{2} + \frac{n}{2} + n = 2n$ units. In this example, we also assume the input is destructable, otherwise another n memory units are required to keep the input data. This scenario supports the fastest data access, but also requires the largest amount of memory.

We can still achieve lossless reconstruction with only n memory units at the expense of additional computation. In this design, we only store the details of each stage except the lowest resolution, for which we store both the approximation and the detail. The maximum memory required is

$$\frac{n}{2} + \frac{n}{4} + \frac{n}{8} + \frac{n}{16} + \cdots \leq n$$

units. Since none of the approximations are kept, extra calculations are required to reconstruct them sequentially (from coarse to fine.)

An optimal design of a hierarchy representation is a combination of intermediate stage elimination and data truncation:

Intermediate stage eliminations – We only keep some intermediate stages of the hierarchy. Since a scientist may not need all intermediate stages for data analysis, some of them can be eliminated in order to save memory, as long as enough information is retained to rebuild the eliminated stages later. For example, in Figure 19.4, stage $i + 2$ can be eliminated as its approximation can always be reconstructed on the fly from stage $i + 3$. In case the detail of $i + 2$ is needed, it can be decomposed from stage $i + 1$.

Data truncations – One of the most important properties of a wavelet transform is its highly localized coefficients in the space domain. Many small coefficients can be discarded with only minimal effect, yielding *lossy* representations at significantly reduced storage costs. We demonstrate test results of an extreme case in [18]. A reconstruction of three-dimensional isosurface data from only three percent of the wavelet coefficients has a correlation $\simeq 0.74$ compared to the original data. When we are dealing with terabyte-sized data, data truncation may be the most important part of the hierarchy representation design.

19.7 Applications of Wavelet Transforms

In this section, we present an example of a multiresolution data hierarchy. The data we use contains concentrations of seven chemical elements provided by the Greenland Ice Sheet Project Two (GISP2) [8] from the Institute for the Study of Earth, Oceans, and Space of the University of New Hampshire. The GISP2 project recovered a two-mile-long ice core from the central plateau of the Greenland ice sheet. The resultant record covers a period of 200,000 years, and provides a multivariate time-series record documenting climatic and atmospheric change and forcing.

The ice core data used here covers the past 13,000 years. The data has seven variates. In the following discussion, we extract one of the them (8,192 records of calcium concentration) to illustrate a complete multiresolution hierarchy of one-dimensional one-variate data. The data is first decomposed to form a fine to coarse hierarchy. A partial reconstruction is then presented to simulate a multiresolution analysis process. Some of the characteristics of this process are highlighted.

19.7.1 Wavelet Decomposition

In Figure 19.6, the graph at the top is the original data, and the rest are approximations at various resolutions. Polylines are used to plot the approximation data. The first peaks moving from left to right of these graphs are identified by scientists as the Younger Dryas Events, a 1,300-year period of cold and glacial conditions that ends about 11,000 years ago. These five graphs illustrate a fine to coarse decomposition of 8,192 data items. The top graph was generated from the entire data set. The rest of the graphs have 4,096, 2,048, 1,024, and 512 data values correspondingly. While megabyte-sized data is not considered very large data, this simple example does show the zooming capability of an orthogonal wavelet. Bear in mind that some of the fine details are not visible in the higher-resolution graphs because the data has higher resolution than the laser printer output.

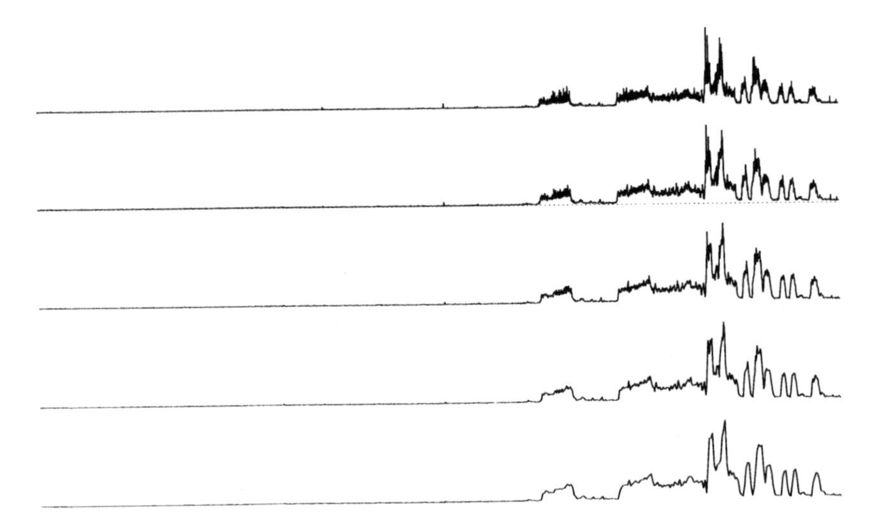

Figure 19.6: Wavelet decomposition of GISP2 calcium data. The top graph is the original data; the rest are approximations with 4,096, 2,048, 1,024, and 512 data values.

After five resolutions of decomposition with 256 approximations left, the graph at the top of Figure 19.7 still retains an excellent approximation of the original data. Even in the bottom graph of Figure 19.7, we can still see the peak of the Little Ice Age with only 16 data values left.

19.7.2 Wavelet Reconstructions

Figure 19.8 depicts a hierarchy of a partial reconstruction of wavelet coefficients from the previous decomposition. In this example, we start with a very coarse resolution with only 32 coefficients. After the interesting spot (the major peaks on the right half) is identified, we study the marked data at a resolution that is three levels up in the hierarchy using 256 coefficients ($2^3 \times 32 = 2^3 \times 2^5 = 2^8 = 256$). The zooming to finer resolutions (that is, more coefficients) continues until it reaches a resolution defined by 4,096 coefficients. At this point we are positioned to study the true original data of the first major peak of the Young Dryas Events. This subset of the original data is presented in the bottom graph.

One of the important points illustrated with this example is the power of multiresolution hierarchy generated by orthogonal wavelets. In this case, we identify the interesting spots from a very fast but somewhat less accurate display of 32 data items. Even though the original data is a relatively small data set, the important idea is the number of resolutions in the hierarchy. Theoretically, for data with 8,192 items, a total of 10 resolutions can be generated with a highly localized Daubechies wavelet. Terabyte-sized data, however, can be scaled down to gigabyte-sized data, also in the same number of resolutions with the same wavelet. Larger-sized data can be considered as a longer version of this example.

Another point we wish to illustrate is the ability to access any subset of the data in any resolution. Instead of saturating available resources by sending/receiving huge amounts of data, only a subset of data (small enough to be displayed on screen) is transported at a time.

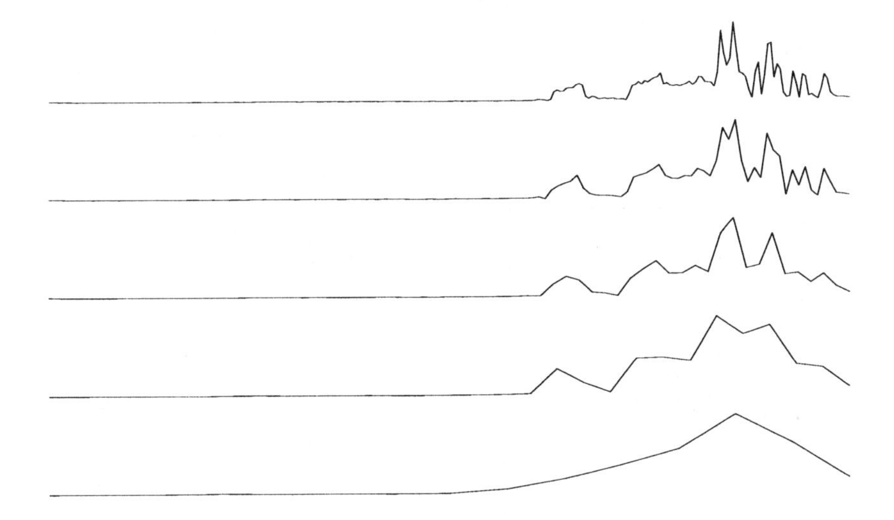

Figure 19.7: Wavelet decomposition of GISP2 calcium data with 256, 128, 64, 32, and 16 data values.

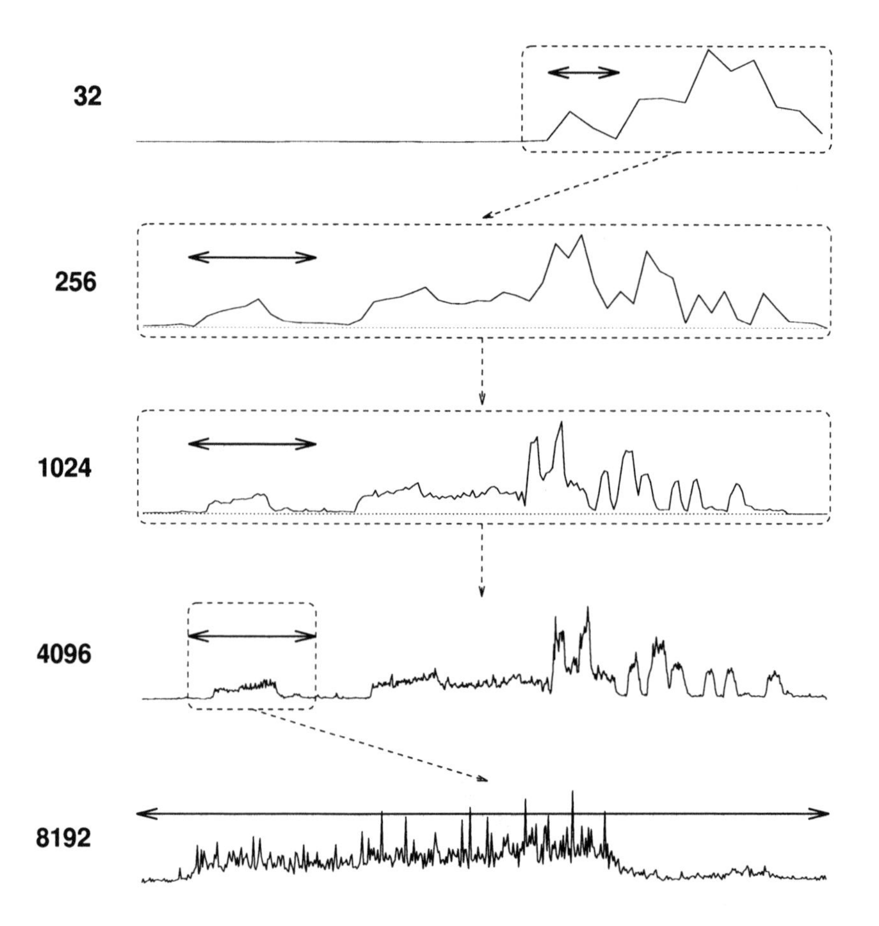

Figure 19.8: A partial reconstruction hierarchy. Dotted rectangles mark the zooming windows. Double-head arrows indicate the same time frame, which is the Younger Dryas Events, in every resolution.

19.8 Multiresolution Wavelet Analysis Tool

We developed a prototype system that supports multiresolution analysis of the data that is already available through the International Sun Earth Explorer-2 (ISEE-2) spacecraft. This data consists of low time-resolution plots on approximately one dozen rolls of microfilm that cover many years of data from two of the instruments on the ISEE-2 spacecraft.

The prototype uses orthogonal wavelet transforms to support hierarchical representation of very large data sets. A simplified front-end design layout is depicted in Figure 19.9. The top buttons support various user-interface functions such as data bank, colormap, sys-

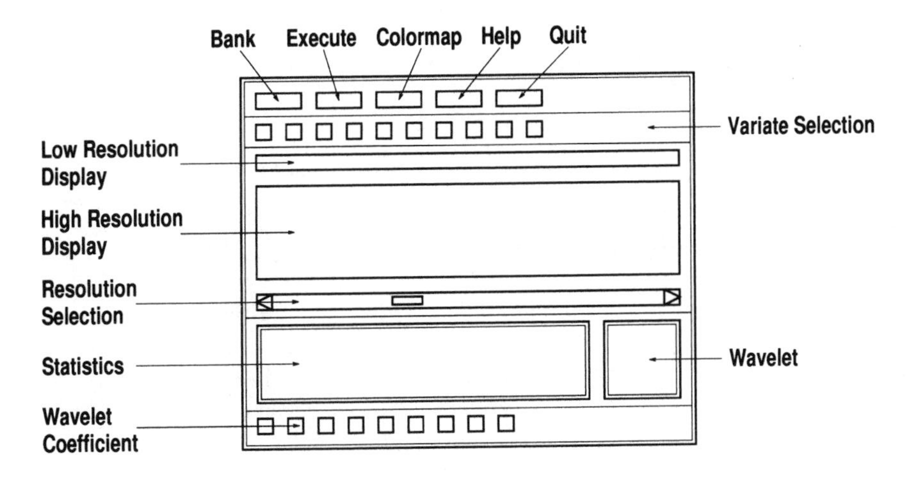

Figure 19.9: A simplified design layout of our system.

tem information, and execution. The widgets located right below are the variate selection buttons. There are two major display windows in this front end: a low-resolution window that displays all data, and a high-resolution window that displays data in the resolution of at least one pixel per data item. The resolution of the second display window is controlled by the slider below it. There is a message window to display information such as statistics, metadata, and error characteristics. The graph window located near the lower-right corner is the wavelet window. It displays the shape of the current wavelet, which is controlled by the buttons in the bottom widget. A display from the prototype front end is shown in Figure 19.10. This data, which is obtained from the Institute for the Study of Earth, Oceans, and Space at the University of New Hampshire, has a total of 31 variates. A polyline plot is not very effective with such a large number of variates. Instead, the value of each data item is represented by the color of a small rectangle. The low-resolution window displays 31,744 approximations which represent eight megabytes of floating-point numbers. The high-resolution window is set at the finest resolution. Each data item is represented, in this case, by 16 pixels. The number of pixels used for each data item can be configured through the system setup.

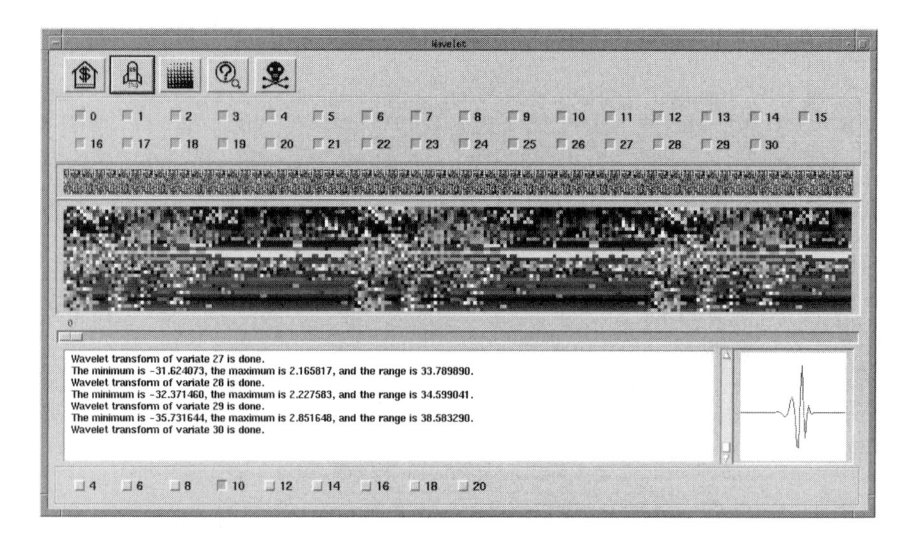

Figure 19.10: The front end of the multiresolution wavelet analysis prototype.

19.9 Conclusions and Future Directions

We propose a hierarchical multiresolution visualization model with the potential to visualize very large data sets using wavelets. Our model is described using examples of various scientific data sets. An experimental prototype implemented in C++ with a Motif front end is presented. This wavelet window module will be enhanced and eventually integrated into a scientific database visualization system [14, 6] currently under development at the University of New Hampshire.

Acknowledgments

The authors' work has been supported in part by the National Science Foundation under grant IRI-9117153. The GISP2 ice core data is courtesy of Dr. Paul Mayewski, and the GGS data is courtesy of Dr. Terry Onsager; both of the Institute for the Study of Earth, Oceans, and Space at the University of New Hampshire. This work was first presented at the Dagstuhl Workshop of Scientific Visualization in April, 1994. Further related research is described in technical reports and conference papers that are available publicly in ftp://ftp.cs.unh.edu/pub/vis.

Bibliography

[1] Ingrid Daubechies, *Ten Lectures on Wavelets*, SIAM, Philadelphia, Pa., 1992.

[2] Jose Encarnação, Jim Foley, Steve Bryson, Steven K. Feiner, and Nahum Gershon, "Research Issues in Perception and User Interfaces," *IEEE Computer Graphics and Applications*, Vol. 14, No. 2, Mar. 1994, pp. 67–69.

[3] Steven J. Gortler, Peter Schröder, Michael F. Cohen, and Pat Hanrahan, "Wavelet Radiosity," *SIGGRAPH 93 Conference Proceedings*, Aug. 1993, pp. 221–230.

[4] Taosong He, Sidney Wang, and Arie Kaufman, "Wavelet-Based Volume Morphing," *Proceedings of IEEE Visualization '94*, R. Daniel Bergeron and Arie E. Kaufman, editors, IEEE Computer Society Press, Los Alamitos, Calif., Oct. 1994, pp. 85–92.

[5] Christian Houdre, "Wavelets, Probability, and Statistics: Some Bridges," *Wavelets, Mathematics and Applications*, John J. Benedetto and Michael W. Frazier, editors, CRC Press, 1994, chapter 9, pp. 365–398.

[6] David T. Kao, R. Daniel Bergeron, and Ted M. Sparr, "An Extended Schema Model for Scientific Data," *Database Issues for Data Visualization*, John P. Lee and Georges G. Grinstein, editors, Springer-Verlag, IEEE Visualization '93 Workshop, Oct. 1994, pp. 69–82.

[7] Stephane G. Mallat, "A Theory for Multiresolution Signal Decomposition: The Wavelet Representation," *IEEE Transactions on Pattern Analysis and Machine Intelligence*, Vol. 11, No. 7, July 1989, pp. 674–693.

[8] P.A. Mayewski, L.D. Meeker, M.C. Morrison, S. Whitlow, K.K. Ferland, D.A. Meese, M.R. Legrand, and J.P. Steffensen, "Greenland Icecore Signal Characteristics Offer Expanded View of Climate Change," *Journal of Geophysics*, Vol. 98, 1993, pp. 12839–12847.

[9] Bruce H. McCormick, Thomas A. DeFanti, and Maxine D. Brown, "Visualization in Scientific Computing," *Computer Graphics*, Vol. 21, No. 6, Nov. 1987, pp. 1–14.

[10] Shigeru Muraki, "Application and Rendering of Volume Data Using Wavelet Transforms," *Proceedings IEEE Visualization '92*, Arie E. Kaufman and Gregory M. Nielson, editors, IEEE Computer Society Press, Los Alamitos, Calif., Oct. 1992. pp. 21–28.

[11] Shigeru Muraki, "Volume Data and Wavelet Transforms," *IEEE Computer Graphics and Applications*, Vol. 13, No. 4, July 1993, pp. 50–56.

[12] *CDF User's Guide, Version 2.4*, National Space Science Data Center, Greenbelt, Md., Feb. 1994.

[13] William H. Press, Saul A. Teukolsky, William T. Vetterling, and Brian P. Flannery, *Numerical Recipes in C, The Art of Scientific Computing*, Cambridge University Press, second edition, 1992.

[14] Ted M. Sparr, R. Daniel Bergeron, and Loren D. Meeker, "A Visualization-Based Model for a Scientific Database System," *Focus on Scientific Visualization*, Hans Hagen, Heinrich Muller, and Gregory M. Nielson, editors, Springer-Verlag, 1992, pp. 103–121.

[15] Eric J. Stollnitz, Anthony D. DeRose, and David H. Salesin, "Wavelets for Computer Graphics: A Primer, Part 1," *IEEE Computer Graphics and Applications*, Vol. 15, No. 3, May 1995, pp. 77–84.

[16] Eric J. Stollnitz, Anthony D. DeRose, and David H. Salesin, "Wavelets for Computer Graphics: A Primer, Part 2," *IEEE Computer Graphics and Applications*, Vol. 15, No. 4, July 1995, pp. 75–85.

[17] Hai Tao and Robert J. Moorhead, "Progressive Transmission of Scientific Data Using Biorthogonal Wavelet Transform," *Proceedings of IEEE Visualization '94*, R. Daniel Bergeron and Arie E. Kaufman, editors, Washington, D.C., Oct. 1994, pp. 93–99.

[18] Pak Chung Wong and R. Daniel Bergeron, *A Child's Garden of Wavelet Transforms*, Technical Report 94-24, Computer Science Department, University of New Hampshire, Durham, N.H., 1994.

[19] Randy K. Young, *Wavelet Theory and Its Applications*, Kluwer Academic Publishers, 1993.

Chapter 20

Tools for Triangulations and Tetrahedrizations and Constructing Functions Defined over Them

Gregory M. Nielson

20.1 Introduction

This chapter is about triangulations and tetrahedrizations and functions defined over them. The original and main goal was to provide some information about tetrahedra and tetrahedrizations and functions defined over them, but it was quickly realized that many of these topics are easier to describe and understand with some background on their two-dimensional analogs. Therefore, it was decided to also include material on triangulations. While some of the material exists elsewhere in the literature, much is new and appears here for the first time. The intended purpose for this chapter is to serve as a survey/tutorial in the area of data modeling and visualization. As data modeling and visualization becomes more sophisticated in its application domains and begins to deal with data sets that are more complex than Cartesian grids, there will be the need for tools to deal with these data sets. We feel that the tools and techniques covered here are very basic and will prove to be useful in a variety of contexts in data visualization.

Now we have some comments about the organization of this chapter. While tetrahedrizations are the goal, researchers have dealt with triangulations for a much longer period of time than tetrahedrizations and so triangulations and related matters are much better understood. The next section is a survey of triangulations and related matters of interest in modeling and visualization. The following section is on tetrahedrizations and we attempt to follow the same flow of information as in the section on triangulations as well as possible. We use the phrase "as well as possible" because some aspects of triangulations do not generalize to tetrahedrization and some facts known about triangulations and triangular domains are yet to be known about tetrahedrizations and tetrahedral domains. On the other hand, there are topics of interest to tetrahedrization which have no 2D counterpart of interest—for example, visibility sorting for tetrahedrizations. The outline of this chapter is very simple. In Section 20.2 we go through a list of topics on triangulations and triangular domains and then in Section 20.3 we repeat these topics with reference to tetrahedrizations and tetrahedral domains.

20.2 Triangulations

20.2.1 Basics

Definitions, Data Structures, and Formulas for Triangulations

In order to avoid any possible confusion and problems later, it is usually best to be a little precise and formal about the definition of a triangulation. We start with a collection of points in the plane, P = $\{p_i = (x_i, y_i), i = 1, \ldots, N\}$ and a domain of interest, D, which contains all of the points of P. We assume that the boundary of D is a simple (does not intersect itself), closed polygon. Often D is the convex hull of P, but in general, it need not be convex. In fact the boundary does not have to be a single polygon so that the domain is not even simply connected. (*Connected* means that there is a path joining any two points and *simply connected* means that the complement is connected.) Roughly speaking, a triangulation is a decomposition of D into a collection of triangles which are formed from vertices of P. Since we are eventually interested in defining functions over D in a piecewise manner over each triangle, we must require that the triangles do not overlap so as not to have any ambiguities. Thus we require the collection of triangles of the triangulation to be mutually exclusive and collectively exhaustive. In order to continue this formalism to a precise definition, we need some additional notation. A single triangle with vertices p_i, p_j, and p_k is denoted by T_{ijk} and the list of triples which represents the triangulation is denoted by I_t. A triangle T_{ijk} is a closed 2D point set that includes its three edges which comprise its boundary. The interior of T_{ijk}, denoted by Int(T_{ijk}) is open and does not include the boundary. The edge joining p_i and p_j is denoted by e_{ij} and $N_e = \{ij : ijk$ in I_t for some $k\}$ is used to refer to the collection of all edges. Formally, the definition of a triangulation requires the following:

i) No triangle $T_{ijk}, ijk \in I_t$ is degenerate. That is, if $ijk \in I_t$ then p_i, p_j and p_k are not collinear.

ii) The interior of any two triangles do not intersect. That is, if $ijk \in I_t$ and $\alpha\beta\gamma \in I_t$ then Int(T_{ijk})$\cap$ Int($T_{\alpha\beta\gamma}$) $= \emptyset$.

iii) The boundary of two triangles can only intersect at a common edge.

iv) The union of all the triangles is the domain D $= \cup_{ijk \in I_t} T_{ijk}$.

Examples of valid triangulations are shown in Figure 20.1 and Figure 20.2. Note that the example of Figure 20.1 is not convex and that of 20.2 is not simply connected. Even though the diagrams of Figure 20.3 and Figure 20.4 look all right, the actual triangulations

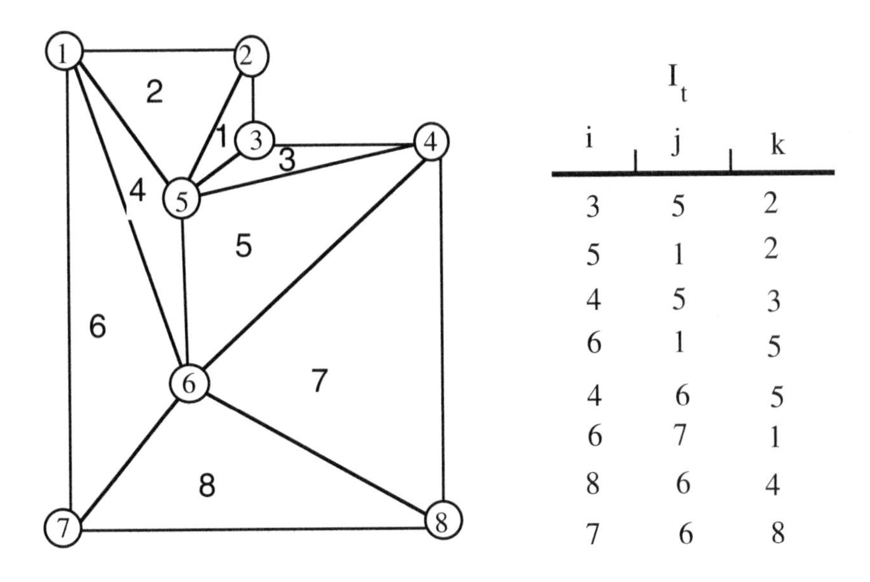

i	j	k
3	5	2
5	1	2
4	5	3
6	1	5
4	6	5
6	7	1
8	6	4
7	6	8

Figure 20.1: A triangulation of a nonconvex domain.

given by the corresponding I_t's do not represent valid triangulations. In the case of Figure 20.3 the triangle T_{465} is degenerate. Even if this triangle is eliminated, what remains is not a valid triangulation because condition iii) would then be violated since edge e_{46} contains p_5. This example would become a valid triangulation if the point p_5 were to be moved slightly to the right so as not to be on the edge e_{46}. The information of Figure 20.4 is not a valid triangulation because condition ii) is violated.

We now want to make some assertions about the possibility of triangulating a domain containing a collection of data points that is bounded by a simple, closed polygon. First we note that in the case that the domain contains no interior data points, it is always possible to form a triangulation. Just for the sake of interest, we mention two ways that this can be accomplished. The first way is based upon the fact that every simple closed polygon with more than three vertices can be split into two polygons. This leads to an algorithm that recursively splits each subpolygon until only triangles are left. The following argument which guarantees that each simple closed polygon has a diagonal has been discussed in [16]. A diagonal is an edge between two vertices that lies inside the polygon and does not intersect the polygon except at the endpoints.

Splitting a polygon: Let b be the vertex with minimum x-coordinate and ab and bc be its two incident edges as is shown in Figure 20.5. If ac is not cut by the polygon, then ac is a diagonal. Otherwise there must be at least one polygon vertex inside T_{abc}. Let d be the vertex inside abc furthest from the line through a and c. Now edge bd cannot be cut by the polygon, since any edge intersecting bd must have one endpoint further from line ac.

The second approach leads to an iterative algorithm. We first give a definition. A vertex, p_i, of a simple, closed polygon is called *protruding,* provided the following conditions hold:

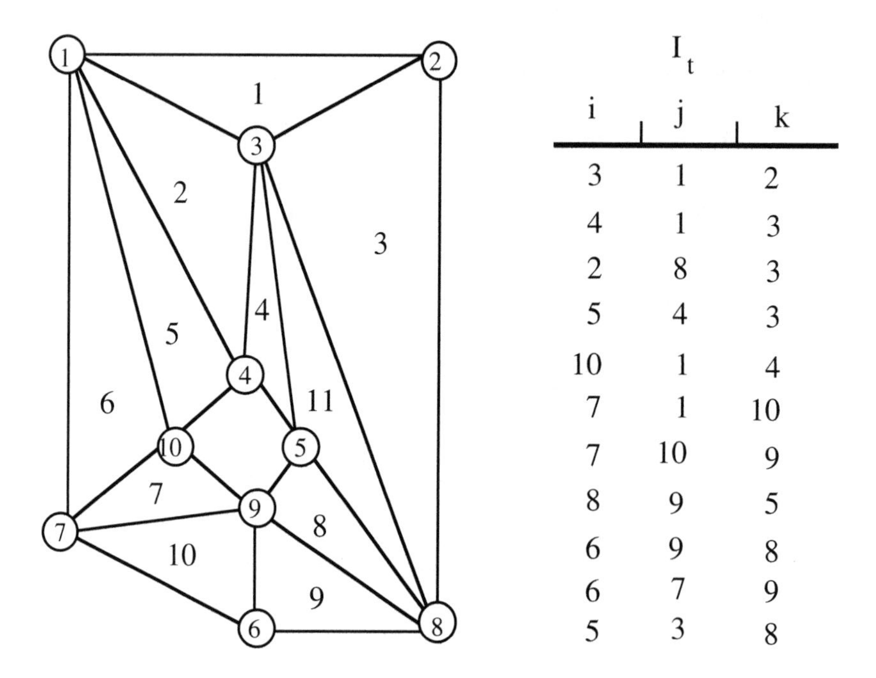

i	j	k
3	1	2
4	1	3
2	8	3
5	4	3
10	1	4
7	1	10
7	10	9
8	9	5
6	9	8
6	7	9
5	3	8

Figure 20.2: A triangulation of a domain which is not simply connected.

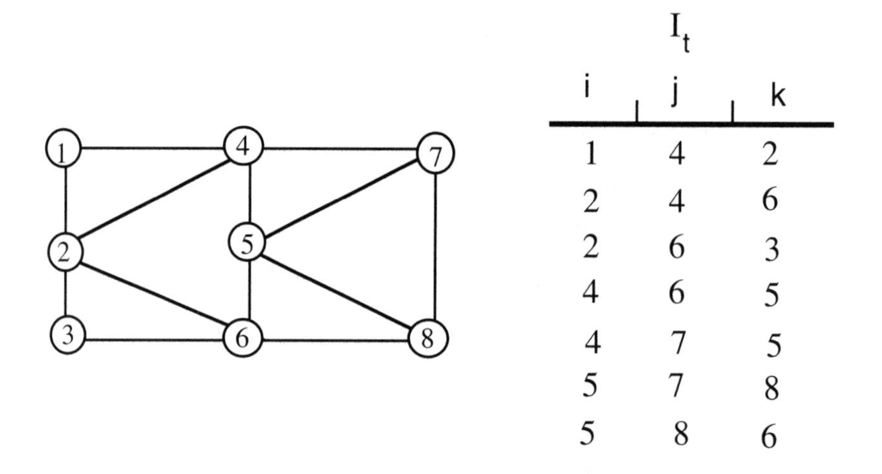

i	j	k
1	4	2
2	4	6
2	6	3
4	6	5
4	7	5
5	7	8
5	8	6

Figure 20.3: Not a valid triangulation.

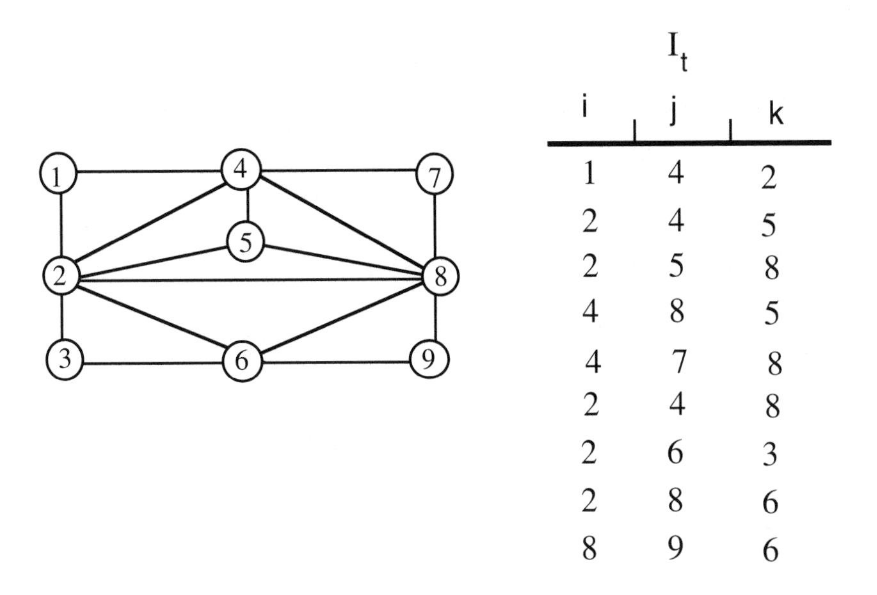

I_t

i	j	k
1	4	2
2	4	5
2	5	8
4	8	5
4	7	8
2	4	8
2	6	3
2	8	6
8	9	6

Figure 20.4: Not a valid triangulation.

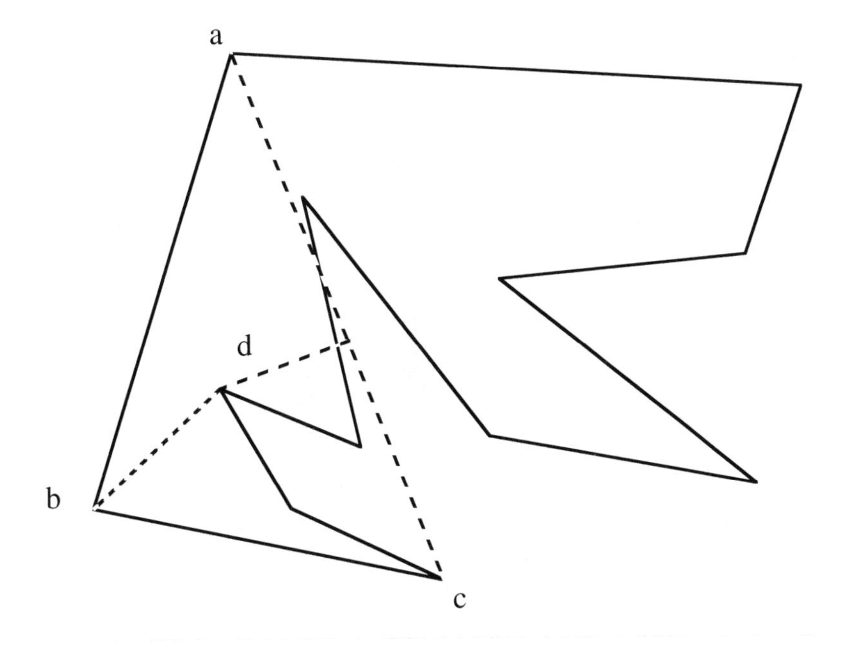

Figure 20.5: Any polygon with more than three vertices can be split.

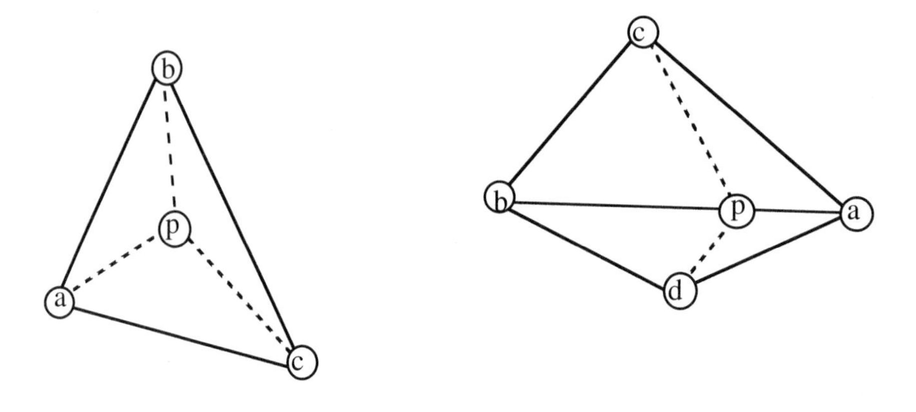

Figure 20.6: Insertion of an interior point.

i) The interior angle Θ_i, between the edges, $p_{i-1}p_i$ and p_ip_{i+1} is less than or equal to π. (Cyclic notation is used here so that $p_{N+1} = p_1$.)

ii) The triangle $T_{i-1,i,i+1}$ contains no other vertices of the polygon than p_{i-1}, p_i or p_{i+1}.

iii) The interior of $T_{i-1,i,i+1}$ is contained in the interior of D.

It is an easy matter to prove that every simple, closed polygon has at least one protruding vertex. (The proof is left to the reader. Some people call them ears and so there must be two of them!) We can triangulate the polygon-bounded domain by successively removing protruding vertices. This approach to triangulating the region bounded by a simple closed polygon is called the "boundary stripping algorithm." It is easy to implement, but in a theoretical sense, it is not competitive with other algorithms (see, for example, the papers of Narkhede and Manocha [175] and Fournier and Montuno [94], among others).

Once the boundary of D has been triangulated, it is a relatively simple matter to build a triangulation including the interior points. This can be done by simply inserting them sequentially in a manner which we now describe.

Insertion of an interior point: If the point to be inserted, p, lies in the interior of the triangle T_{abc}, we replace T_{abc} with the three triangles: $T_{abp}, T_{bcp}, T_{cap}$. If p lies on an edge shared by T_{abc} and T_{bad}, then we replace the two triangles T_{abc} and T_{bad} with the four triangles $T_{bcp}, T_{dbp}, T_{pca}, T_{pad}$. These two cases are illustrated in Figure 20.6.

It is also possible to generalize the insertion idea to include an edge. Once we are armed with this capability, we know that we can triangulate any polygon-bounded domain: simply connected or multiply connected (that is, with holes).

Insertion of an interior edge: Assume that the one endpoint, p, lies in the triangle T_{abc} and that the other endpoint, q, lies in the triangle T_{xyz}. See Figure 20.7. Collect all of the triangles from T_{abc} to T_{xyz} which are intersected by edge pq and form a region R

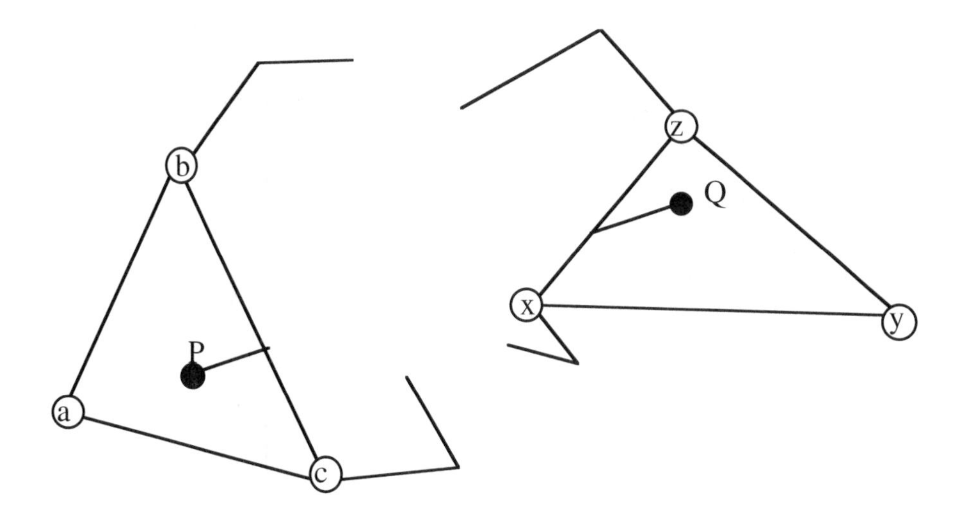

Figure 20.7: Insertion of an interior edge.

with polygon boundary D. We can split D with polygon $dpqw$, where d is the vertex of T_{abc} not on the edge common with the other triangles whose union is R, and w is the analogous vertex of T_{xyz}. Now we know that each of these two domains can be triangulated. The union of these two triangulations, which each contain the edge pq, can replace the previous triangulation of D.

In addition to I_t, which represents the triangulation, it is often worthwhile to generate and maintain some auxiliary information about the neighbors of each triangle. This information is useful for traversal algorithms and evaluation algorithms which have a searching component that determines the particular triangle containing a point where a function defined piecewise over the triangulation is to be evaluated. One very common and particularly useful data structure is that which is illustrated in Figure 20.8. The first three columns contain the data of I_t, with the additional constraint that in reading from left to right (cyclically), the vertices of each triangle are traversed in a clockwise order. The next three columns contain the indices of the triangles which are neighbors to this triangle. The character ϕ indicates that the triangle has an edge that is part of the boundary of D. The entries of these three columns are also in a special order. The fourth column contains the index of the triangle which shares the common edge with vertex indices specified in the second and third columns. Similar relationships hold for the fifth and sixth columns. The information represented by this data structure is called a "triangular grid." The neighborhood information contained in the last three columns does not contain any "new" information over that of I_t, but it is often the case (and this depends, of course, on the application) that it is useful data which is worth generating a priori.

Another data structure for representing a triangulation which is useful for some applications is illustrated by the example shown in Figure 20.9, which represents the same triangulation as that of Figure 20.8. Here, for each vertex, a list of all vertices which are joined by an edge of the triangulation is given. This list is given in counterclockwise order

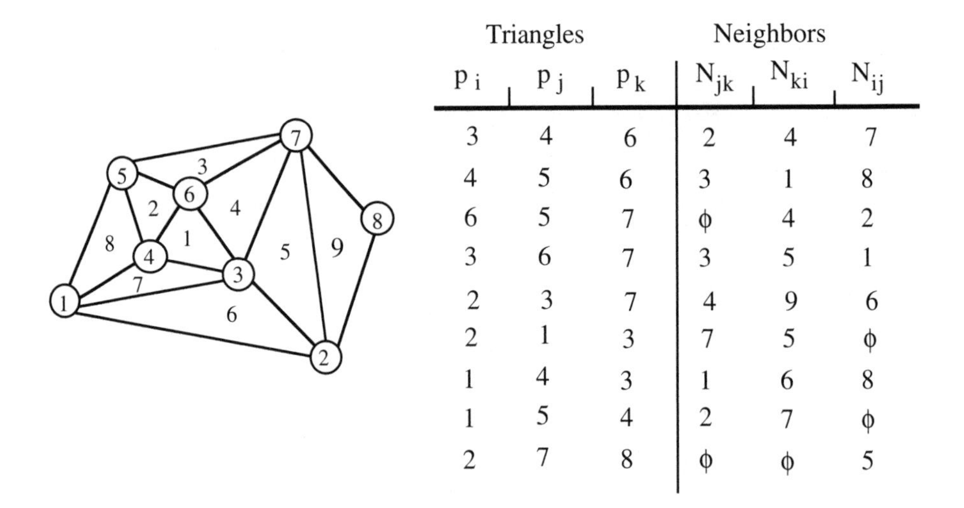

Triangles			Neighbors		
p_i	p_j	p_k	N_{jk}	N_{ki}	N_{ij}
3	4	6	2	4	7
4	5	6	3	1	8
6	5	7	ϕ	4	2
3	6	7	3	5	1
2	3	7	4	9	6
2	1	3	7	5	ϕ
1	4	3	1	6	8
1	5	4	2	7	ϕ
2	7	8	ϕ	ϕ	5

Figure 20.8: An example that defines a *triangular grid structure*.

Vertex	Joining Vertices
1	2, 3, 4, 5
2	8, 7, 3, 1
3	1, 2, 7, 6, 4
4	3, 6, 5, 1
5	1, 4, 6, 7
6	3, 7, 5, 4
7	6, 3, 2, 8, 5
8	2, 7

Figure 20.9: The data that defines the data point contiguity list.

around each vertex. This is called the *data point contiguity list.* We mention this particular data structure because of its convenience for dealing with the optimal Delaunay triangulation discussed in the next section. Also, it is very useful for computing the parameters of the Minimum Norm Network method [179], which is one of the most effective C^1 interpolation methods for scattered data.

Even though there are a number of possible triangulations for any given domain D, the number of triangles is fixed once the boundary has been specified. More precisely, if N_b represents the number of vertices on the boundary and N_i the number of interior vertices so that $N = N_b + N_i$, then the following formulas hold:

$$N_t = 2N_i + N_b - 2$$

and

$$N_e = 3N_i + 2N_b - 3,$$

where N_t is the total number of triangles and N_e is the total number of edges. The importance of these formulas (not so much what the values in the formulas are, but more the fact that some fixed formula holds) will show up in the next section. If we let M_i represent the number of points joining to p_i then it is easy to see that

$$\sum_{i=1}^{N} M_i = 2N_e$$

and so we have that the "average valence" of a point is given by

$$\bar{M} = \frac{\sum_{i=1}^{N} M_i}{N_i + N_b} = 6 - 2\frac{N_b + 3}{N}$$

which is approximately 6. For a sphere (or any domain homeomorphic to a sphere) we have no boundary points and so $N = N_i$ and the analogous formulas are

$$N_t = 2(N-1), \quad N_e = 3(N-1), \quad \bar{M} = \frac{6}{N}(N-1).$$

Some Special Triangulations

One of the simplest triangulations results from splitting the rectangles of a Cartesian grid. A Cartesian grid involves two monotonically increasing sequences, $x_i, i = 1, \ldots, n$ and $y_j, j = 1, \ldots, m$. The grid points have coordinates (x_i, y_j) and these points mark out a cellular decomposition of the domain consisting of rectangles. See Figure 20.10. Forming an edge with one of the diagonals of these rectangular cells leads to a triangulation of the domain. In Figure 20.11 is shown a triangulation where a consistent choice for the diagonal is made. In Figure 20.12 is shown a triangulation with mixed choices for the diagonals. In some applications where dependent ordinate values are known, it is possible to base the choice of the diagonal upon some criteria such as minimum jump in normal vector (see Section 20.2.4) or whether or not the diagonal vertices are separated or connected based upon the hyperbolic contours at the mean value (see the *asymptotic decider* criteria discussed in [186]). In general for this type of triangulation which results from a Cartesian grid, it is not necessary to maintain the triangular grid structure (see Figure 20.8) as this information can be directly inferred from the natural labeling of $p_{ij} = (x_i, y_j)$. Only the information which indicates which diagonal is selected needs to be made available.

We now want to discuss some special triangulations which result from curvilinear grids. A curvilinear grid is specified with two "geometry arrays" $(x_{ij}, y_{ij}), i = 1, \ldots, M; j = 1, \ldots, N$. A cell C_{ij} consists of the quadrilateral with the boundary delineated by (x_{ij}, y_{ij}) to (x_{i+1j}, y_{i+1j}) to (x_{ij+1}, y_{ij+1}) back to (x_{ij}, y_{ij}). It is assumed that these four points form a simple (nonintersecting) polygon so that the quadrilateral is actually well-defined. This condition obviously puts some geometric constraints on the values of the geometry arrays that specify a curvilinear grid.

An example of a curvilinear grid is shown in Figure 20.13. In this case the cell C_{73} degenerates to a triangle because (X_{83}, Y_{83}) and (X_{84}, Y_{84}) are the same point and the cell C_{83} degenerates to an edge because, in addition, (X_{93}, Y_{93}) and (X_{94}, Y_{94}) are the same point. The cells $C_{33}, C_{43}, C_{53}, C_{63}$, and C_{73} have been removed from the domain creating the hole in the interior.

The domain (the union of all of its cells) can be triangulated simply by triangulating each of the cells, by choosing a diagonal to an edge of the triangulation. An example related

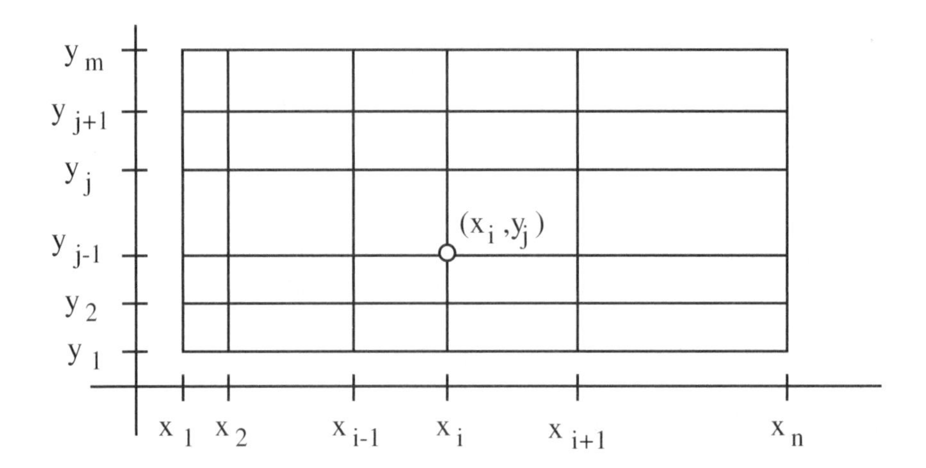

Figure 20.10: Cartesian grid.

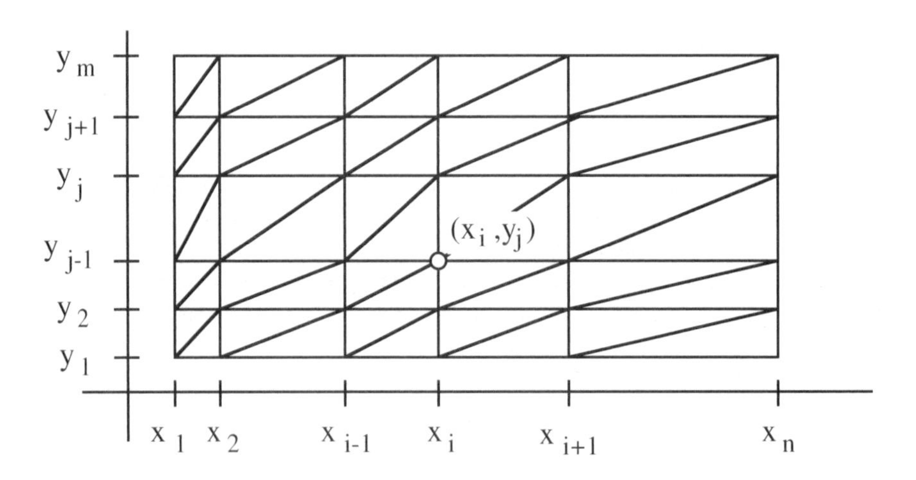

Figure 20.11: Triangulation from Cartesian grid with uniform diagonal choice.

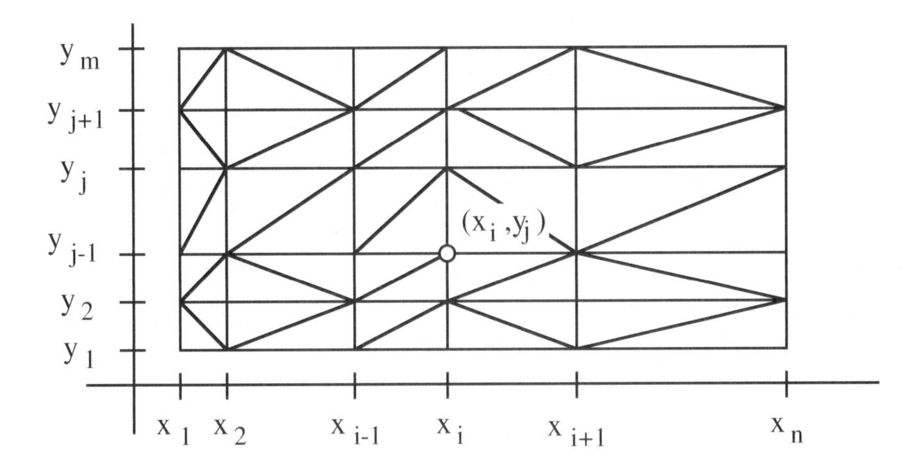

Figure 20.12: Triangulation from Cartesian grid with mixed diagonals.

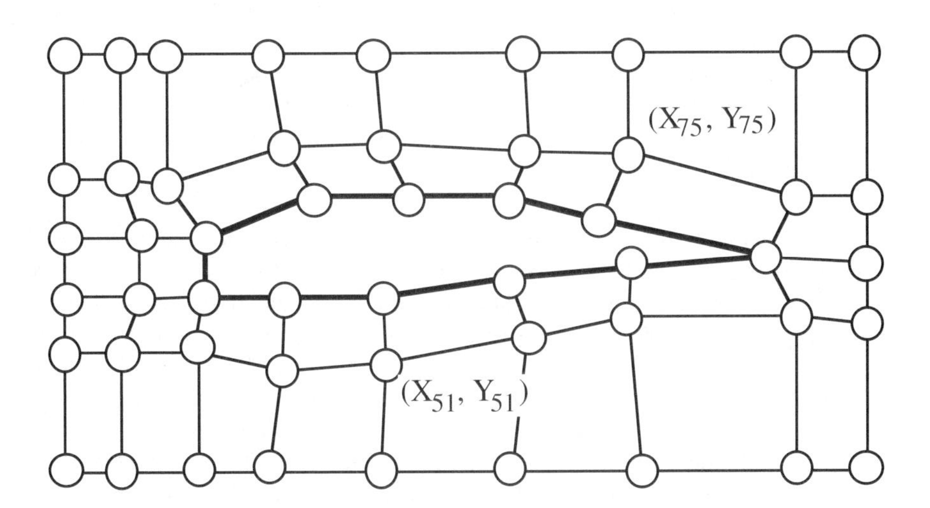

Figure 20.13: An example of a curvilinear grid.

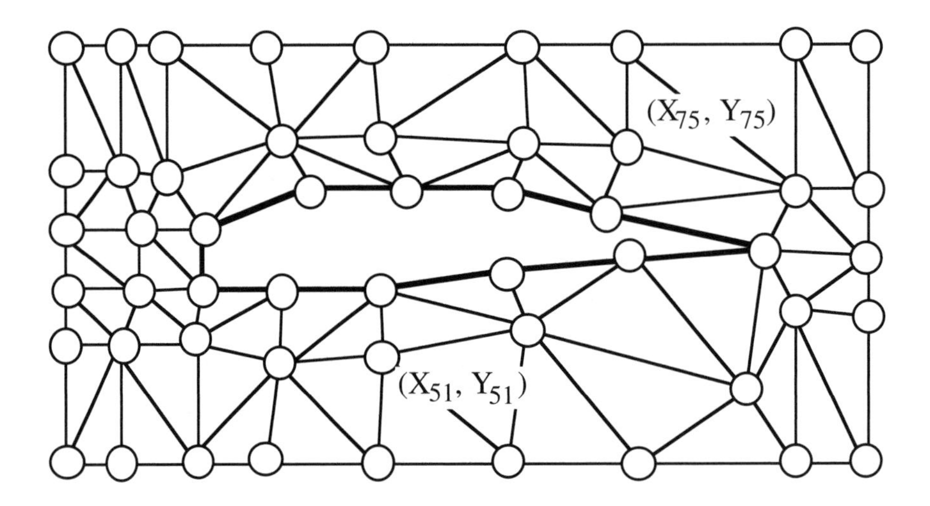

Figure 20.14: Triangulation resulting from curvilinear grid.

to the grid of Figure 20.13 is shown in Figure 20.14. Here we have modified the grid by moving the point (X_{72}, Y_{72}) a little. This serves to point out that if the cell is not convex, then there may be only one choice for the diagonal.

We now discuss some special triangulations obtained by subdividing an existing triangulation. We briefly mention a couple of possibilities. The first is based upon inserting an additional point into the interior of an existing triangle and thereby forming three new triangles. This is illustrated in Figure 20.15. This particular type of subdivision is sometimes referred to as the Clough-Tocher split because of its association with a very well known finite element shape function defined over a triangular domain.

Another way to subdivide an existing triangulation is to insert a new point on an existing edge and split the two triangles (unless the edge is on the boundary) which share this edge. If all edges are split simultaneously we obtain yet another triangulation where each previous triangle is replaced by four new ones. Two different ways for forming triangles from these points is shown in Figure 20.16 and Figure 20.17, respectively. These types of subdivision are particularly interesting due to the nested properties of function spaces which are defined in a piecewise manner over the embedded subdivisions. This can lead to wavelets and their related multiresolution analysis. For the efficient application of these triangulations, it is important to have a method of labeling the triangles which allows an efficient algorithm for finding the labels of all neighbors of a triangle. The labeling scheme illustrated in Figure 20.17 has these properties. We call it the *divide and flip* scheme and have found it to be very useful for implementations. It is related to the spherical quadtrees discussed by Fekete [85]. The first step of the subdivision applied to the triangulation of Figure 20.8 is shown in Figure 20.18.

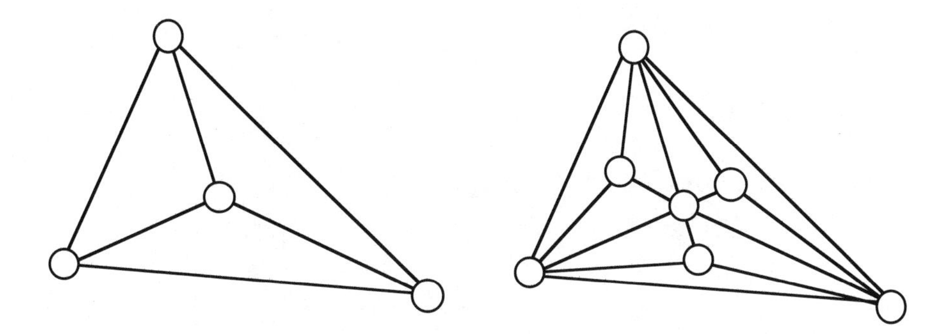

Figure 20.15: Subdivision by inserting a new point that is interior to an existing triangle.

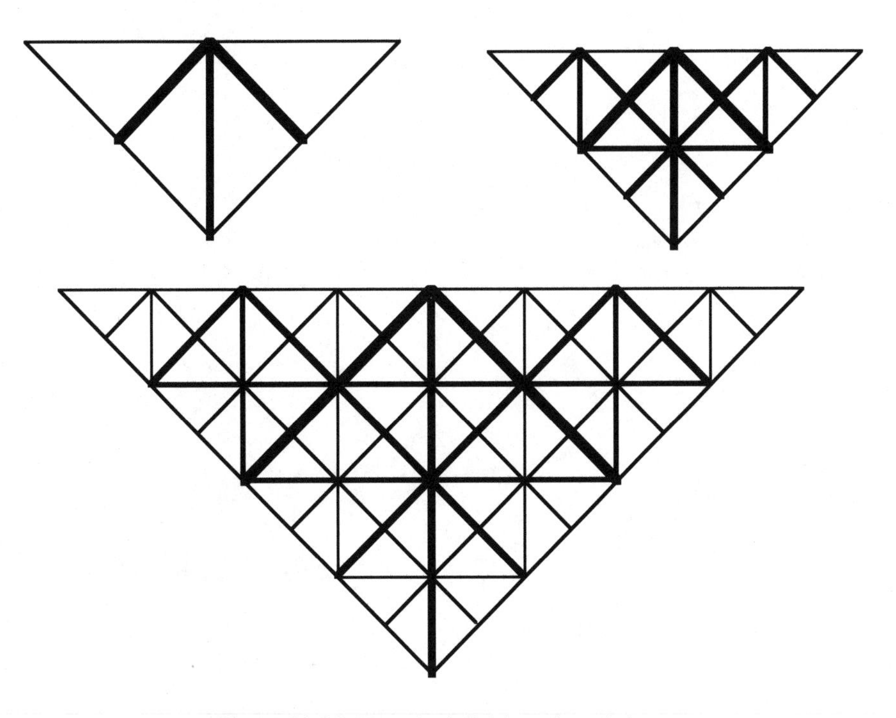

Figure 20.16: Nested subdivision triangulation.

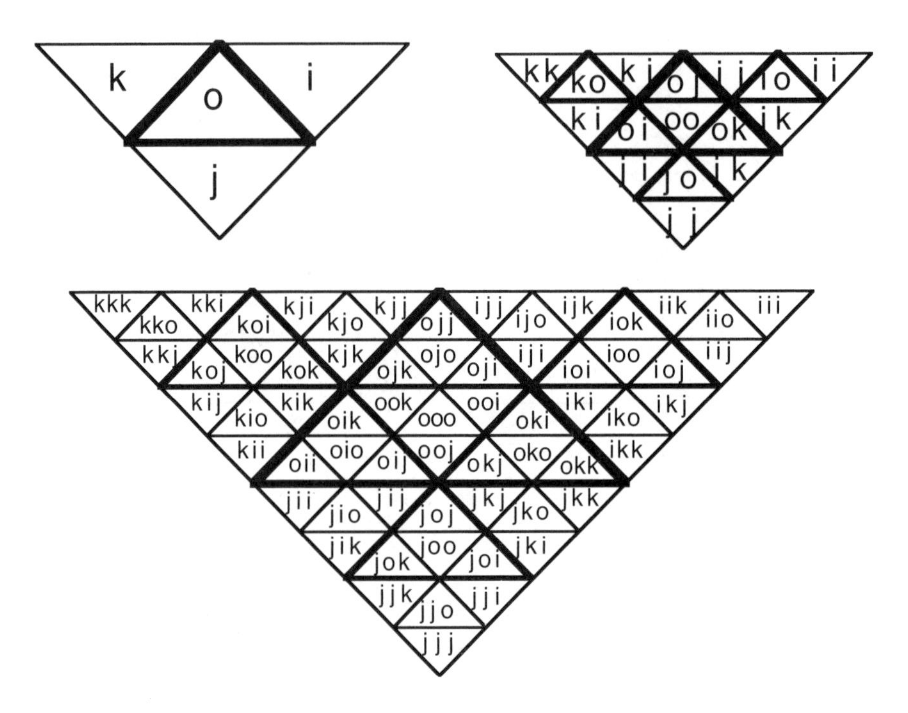

Figure 20.17: The *divide and flip* labeling scheme for a nested subdivision triangulation.

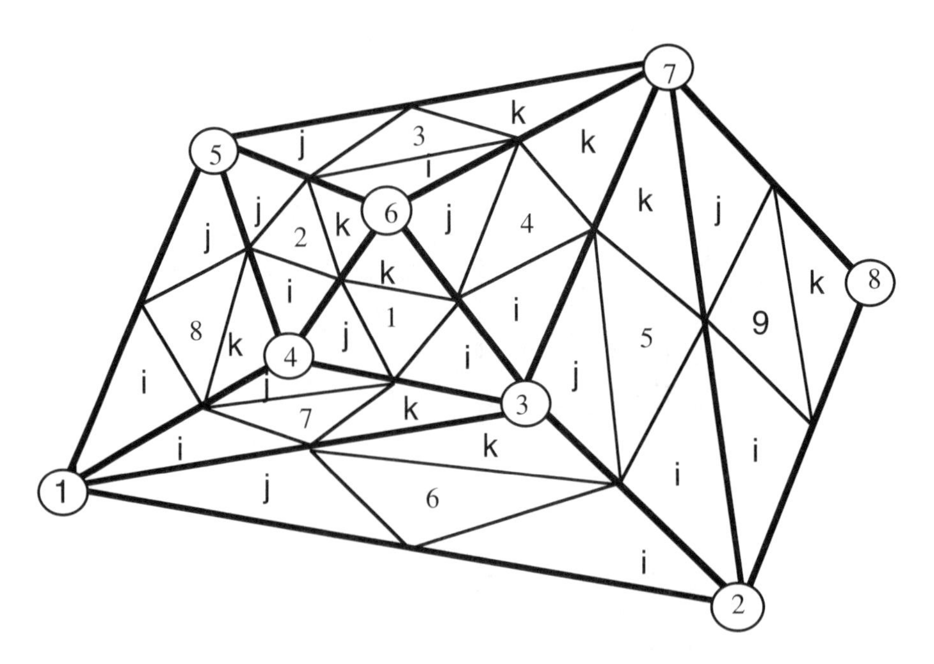

Figure 20.18: A triangulation obtained by splitting each edge of an existing triangulation and forming triangles as indicated in Figure 20.17.

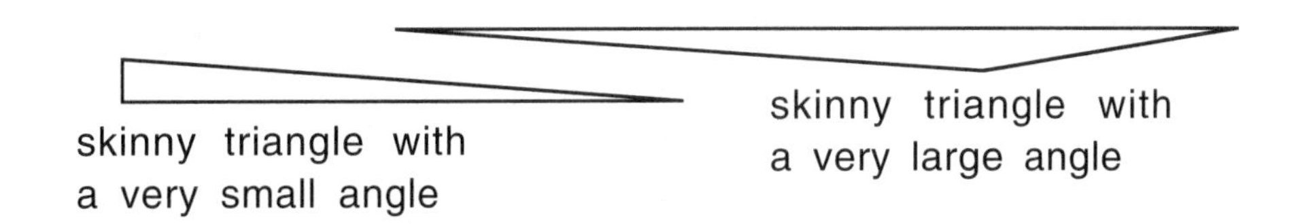

Figure 20.19: Examples of poorly shaped triangles.

20.2.2 Optimal Triangulations

Types and Characterizations

There are many possible triangulations of a given, polygon-bounded domain D. For some applications (but not all) it is desirable to avoid poorly shaped triangles. See Figure 20.19. These are triangles with very large angles or ones with very small angles. This gives rise to two types of optimal triangulations which have been discussed quite widely: the MaxMin and MinMax. Both of these optimal triangulations have a similar method of characterization. Associated with each triangulation there is a vector with N_t entries representing either the largest or smallest angle of each triangle. The entries of each vector are ordered and then a lexicographic ordering of the vectors is used to impose an ordering on the set of all triangulations. In the case of the MinMax criterion, A_i is the largest angle of a triangle and the entries of each vector, A_t, are ordered so that

$$A_t = (A_1, A_2, \ldots, A_{N_t}), A_i \geq A_j, i < j.$$

The smallest of these vectors, based on their lexicographic ordering, associates with the optimal triangulation. In the case of the MaxMin criteria, a_i is the smallest angle and the entries of each vector are ordered the other way so that

$$a_t = (a_1, a_2, \ldots, a_{n_t}), a_i \leq a_j, i < j.$$

The largest of these vectors represents the optimal triangulation in the MaxMin sense. In Figure 20.20, six data points are shown which have a total of ten possible triangulations, which are shown in Figure 20.21. The associated vectors for the MinMax criterion are

$$\begin{aligned}
A_{T_0} &= (2.84, 2.36, 1.99, 1.77, 1.57) \\
A_{T_1} &= (2.98, 2.84, 1.99, 1.91, 1.57) \\
A_{T_2} &= (2.98, 2.42, 1.91, 1.88, 1.57) \\
A_{T_3} &= (2.84, 2.36, 2.32, 1.99, 1.40) \\
A_{T_4} &= (2.42, 2.36, 1.88, 1.77, 1.57) \\
A_{T_5} &= (2.98, 2.42, 1.95, 1.91, 1.27) \\
A_{T_6} &= (2.42, 2.36, 2.32, 1.88, 1.40) \\
A_{T_7} &= (2.42, 2.36, 2.32, 1.50, 1.50) \\
A_{T_8} &= (2.42, 2.36, 1.95, 1.74, 1.50) \\
A_{T_9} &= (2.42, 2.36, 1.95, 1.77, 1.27)
\end{aligned}$$

which we rearrange into decreasing order to obtain

$$\begin{aligned}
A_{\tau_1} &= (2.98, 2.84, 1.99, 1.91, 1.57)\\
A_{\tau_5} &= (2.98, 2.42, 1.95, 1.91, 1.27)\\
A_{\tau_2} &= (2.98, 2.42, 1.91, 1.88, 1.57)\\
A_{\tau_3} &= (2.84, 2.36, 2.32, 1.99, 1.40)\\
A_{\tau_0} &= (2.84, 2.36, 1.99, 1.77, 1.57)\\
A_{\tau_6} &= (2.42, 2.36, 2.32, 1.88, 1.40)\\
A_{\tau_7} &= (2.42, 2.36, 2.32, 1.50, 1.50)\\
A_{\tau_9} &= (2.42, 2.36, 1.95, 1.77, 1.27)\\
A_{\tau_8} &= (2.42, 2.36, 1.95, 1.74, 1.50)\\
A_{\tau_4} &= (2.42, 2.36, 1.88, 1.77, 1.57)
\end{aligned}$$

which implies the following ordering

$$\tau_4 < \tau_8 < \tau_9 < \tau_7 < \tau_6 < \tau_0 < \tau_3 < \tau_2 < \tau_5 < \tau_1$$

and so τ_4 is the optimal triangulation in the MinMax sense. On the other hand, the associated vectors for MaxMin criteria sorted in increasing order are

$$\begin{aligned}
a_{\tau_1} &= (0.02, 0.04, 0.35, 0.46, 0.50)\\
a_{\tau_2} &= (0.02, 0.11, 0.42, 0.46, 0.50)\\
a_{\tau_5} &= (0.02, 0.11, 0.50, 0.58, 0.88)\\
a_{\tau_3} &= (0.04, 0.14, 0.35, 0.37, 0.66)\\
a_{\tau_0} &= (0.04, 0.14, 0.35, 0.46, 0.62)\\
a_{\tau_6} &= (0.11, 0.14, 0.37, 0.42, 0.66)\\
a_{\tau_7} &= (0.11, 0.14, 0.37, 0.46, 0.70)\\
a_{\tau_4} &= (0.11, 0.14, 0.42, 0.46, 0.62)\\
a_{\tau_8} &= (0.11, 0.14, 0.57, 0.58, 0.70)\\
a_{\tau_9} &= (0.11, 0.14, 0.58, 0.62, 0.88)
\end{aligned}$$

which results in the following ordering

$$\tau_1 < \tau_2 < \tau_5 < \tau_3 < \tau_0 < \tau_6 < \tau_7 < \tau_4 < \tau_8 < \tau_9$$

and so τ_9 is the optimal triangulation in the case of the MaxMin criterion.

In the case where D is the convex hull of the points of P, there is an important relationship between the MaxMin triangulation and the Dirichlet tessellation. The Dirichlet tessellation is a partition of the plane into regions $\mathrm{R}_i, i = 1, \ldots, N$ called Thiessen regions. The Thiessen region R_k consists of all points in the plane whose closest point among $p_i, i = 1, \ldots, n$ is p_k. A Dirichlet tessellation is usually illustrated by drawing the boundaries of the Thiessen regions. The collection of these edges is sometimes referred to as the Voronoi diagram (see [252]). An example is shown in the left image of Figure 20.24. In the right image of Figure 20.24 is shown the MaxMin triangulation, which is also called the Delaunay triangulation. It is dual to the Dirichlet tessellation in that the edges of this optimal triangulation join vertices which share a common Thiessen region boundary. We have included the great circles in the left image of this figure so as to point out another important property of the Dirichlet tessellation and its companion Delaunay triangulation.

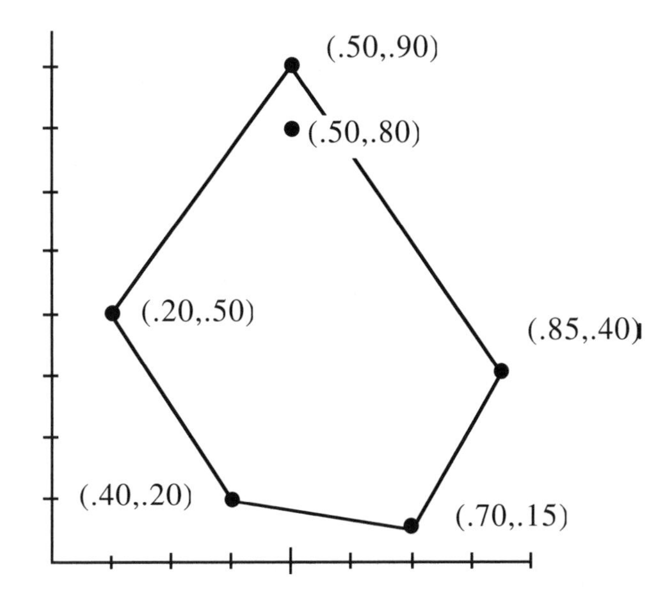

Figure 20.20: Six data points.

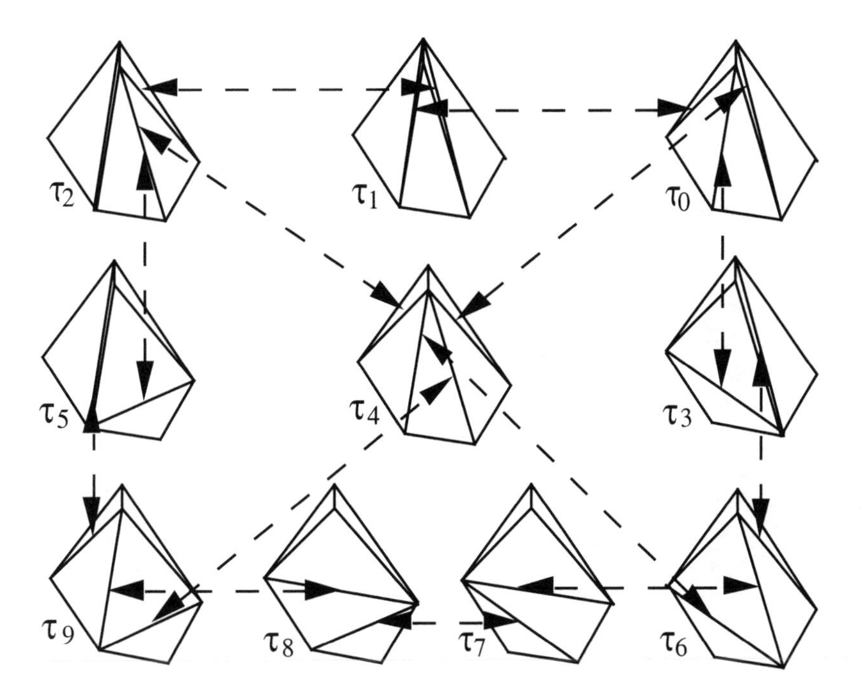

Figure 20.21: Ten triangulations of six data points.

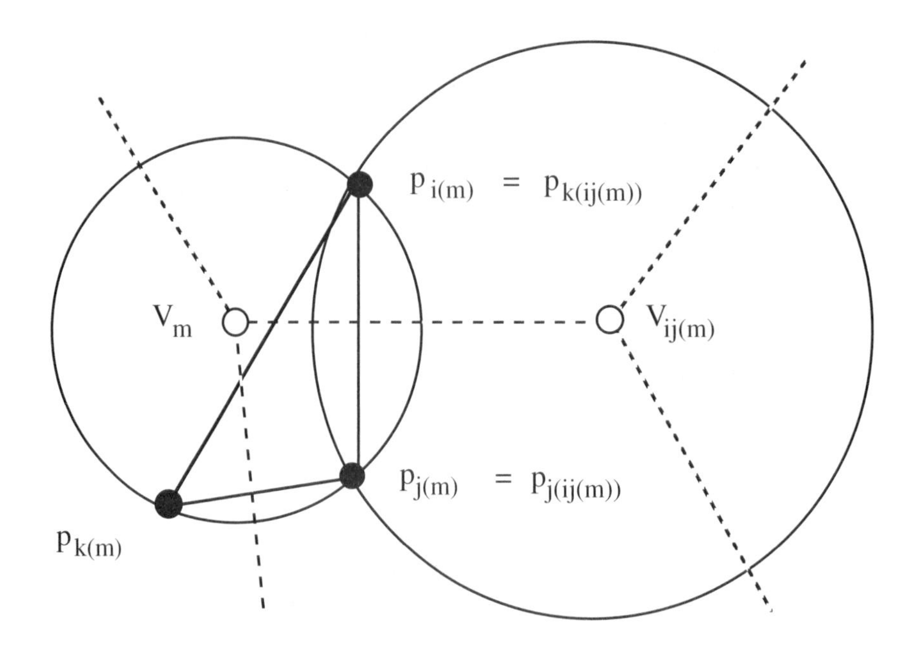

Figure 20.22: Drawing the Dirichlet tessellation from the triangular grid structure.

By definition, the edges of the Thiessen regions meet at triads (possibly more than three edges meet in some special, neutral/cyclic cases) which are equally distant to three points. These three points will form a triangle of the optimal triangulation and the great circle will not contain any other data points. See Figure 20.25.

We can be a little more formal about these properties if we introduce some notation. Recall that $I_t = \{(i(m), j(m), k(m)), m = 1, \ldots, N_t\}$ so that the three data points $p_{i(m)}, p_{j(m)}, p_{k(m)}$ will be the vertices of a triangle of the triangulation. We assume that the neighbor information of the triangular grid is given by three arrays $ij(m), jk(m)$, and $ki(m), m = 1, \ldots, N_t$. Let V_m be the point which is equidistant from $p_{i(m)}, p_{j(m)}$, and $p_{k(m)}$ and $C_m = \{p : ||p - V_m|| \leq ||V_m - p_{a(m)}||, a = i, j \text{ or } k\}$ be the circumcircle (disk) for this triangle which has V_m as its center. The Delaunay triangulation is characterized by the fact that C_m does not contain any other data points $p_i, i = 1, \ldots, N$ other than $p_{i(m)}, p_{j(m)}$, and $p_{k(m)}$. The points V_m are the vertices of the Voronoi diagram. In order to draw the Voronoi diagram we simply start with some V_m and draw the edges to the three points that are joined to it, namely, $V_{ij(m)}, V_{jk(m)}$, and $V_{ki(m)}$. If any one of $ij(m), jk(m)$, or $ki(m)$ is zero (say $ij(m)$, indicating the edge joining $p_{i(m)}$ and $p_{j(m)}$ is on the boundary of the convex hull), then we draw the ray emanating from V_m in the direction perpendicular to the appropriate edge (which is $p_{i(m)}p_{j(m)}$ if $ij(m) = 0, p_{j(m)}p_{k(m)}$ if $jk(m) = 0$, and $p_{k(m)}p_{i(m)}$ if $ki(m) = 0$). If we go through the list of triangles and draw three edges for each V_m we will actually be drawing each edge (not each ray) twice. We can avoid this duplication by testing (for example) whether or not $m > ij(m), m > jk(m), m > ki(m)$ before we draw the corresponding edge.

Because of this relationship between the Dirichlet tessellation and the optimal MaxMin

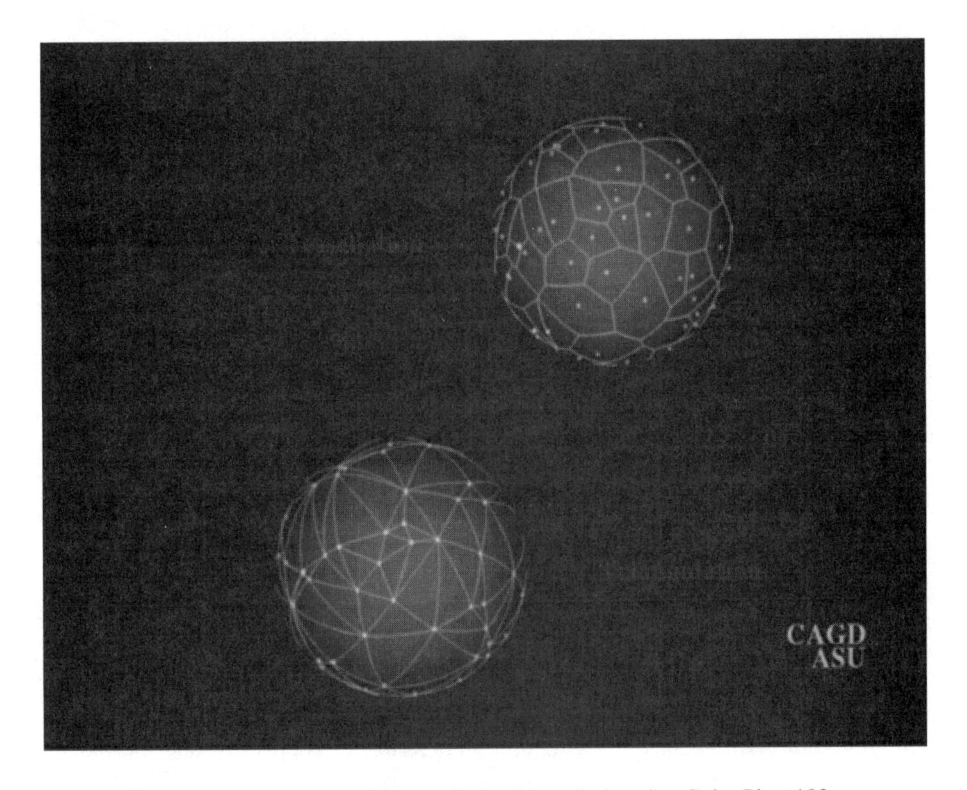

Figure 20.23: Spherical triangulation and tessellation. See Color Plate 103.

triangulation, we can extend the idea of MaxMin or Delaunay triangulation to any domain where we can compute the distance between two points. The sphere provides an interesting and useful example. Here the distance between two points p and q is easily computed as $\cos^{-1}(p \cdot q)$ so the Dirichlet tessellation is also easy to compute. An example is shown in the right image of Figure 20.23. The left image depicts the triangulation which is dual to this tessellation.

There have been many other criteria for characterizing optimal triangulations that have been studied and discussed in the literature. Some turn out to be equivalent to those we have mentioned here and some only appear to be similar, so one needs to be rather careful. Even though the terminology can be similar, the criterion of minimizing the maximum angle is not the same as the MinMax criterion we have described here. It is easily the case that the two quite different triangulations with different vectors A_τ (as defined above) could have the same maximum angle and could both be a triangulation which minimizes the maximum angle. The example of Figure 20.20 has this property. Each of the triangulations $\tau_6, \tau_7, \tau_9, \tau_8$, and τ_4 have a maximum angle of 2.42, which turns out to be a minimum, so any one of these triangulations would satisfy the criterion of minimizing the maximum angle, while only τ_4 satisfies the MinMax criterion described here. Overall, the topic of optimal triangulations can be rather technical, and one has to be careful when comparing results found in the literature.

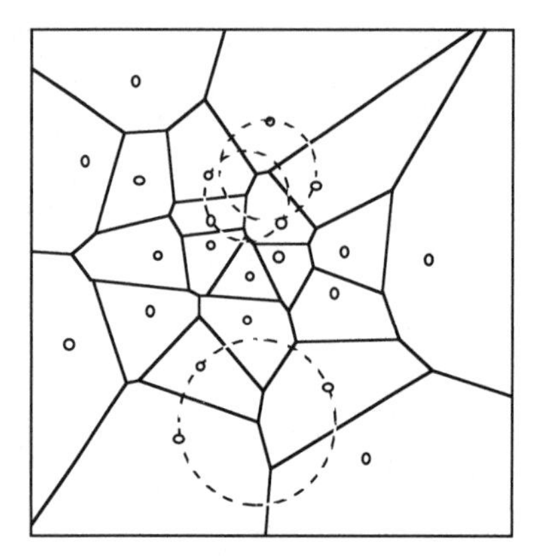

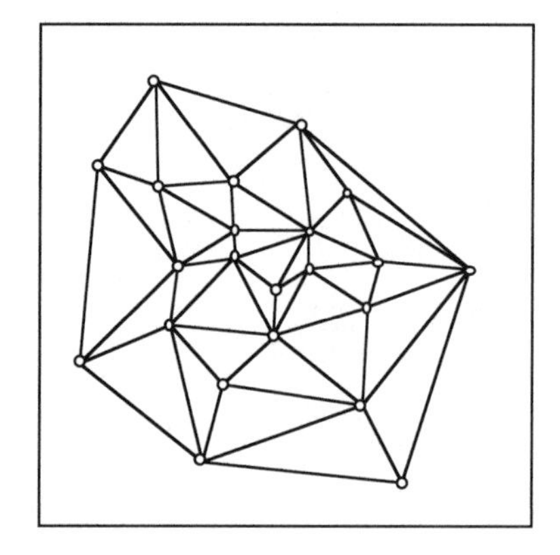

Figure 20.24: The Dirichlet tessellation and its dual triangulation.

Algorithms for Delaunay Triangulations

In this section we discuss some ideas and techniques leading to algorithms for computing the Delaunay triangulation of a set of points in the plane. In general, this is a very rich and full area of research and here we can only provide a glimpse. The literature is very abundant with both practical and theoretical papers on this subject. There is not a single "best" algorithm. The choice depends upon the particular application and the tools and resources available. It is a good strategy to be armed with a collection of ideas, tools, and techniques so that an effective algorithm can be custom-designed for the application at hand. Our approach for the material for this section is based upon a discussion of the ideas behind a few selected algorithms. Our selection is based upon potential usefulness of the ideas and on which would be representative. In addition, we are particularly interested in those ideas which extend most easily to three dimensions. But, just for the sake of interest, we have included the description of one 2D algorithm which does not extend at all to 3D!

The Swapping Algorithm of Lawson [139]: The basic operation of this algorithm consists of swapping the diagonal of a convex quadrilateral. Lawson [138] showed that any triangulation of the convex hull can be obtained from any other triangulation by a sequence of these operations. (Later this property was established for nonconvex domains by Dyn and Goren [66].) Furthermore, Lawson proved that if the choice of the diagonal is made on the basis of the MaxMin criterion for the quadrilateral only, eventually the global optimal triangulation will be obtained. In other words, for this criterion, a local optimum is a global optimum. A typical implementation of this type of algorithm would insert new points (say, in sorted x-order) in the interior of an existing triangulation or connect to all points on the boundary which are visible from the new point. This new triangulation is then optimized by testing and possibly swapping the diagonals of convex quadrilaterals. It is interesting to note that this type of algorithm will not necessarily produce the MinMax because for this criterion, a local extreme is not necessarily a global optimum. The example of Figure 20.20 illustrates this. Based upon the MinMax criterion, τ_4 is optimal and τ_8 is a local minimum. Locally optimal swaps of diagonals from τ_8 would never lead to τ_4. The algorithm could

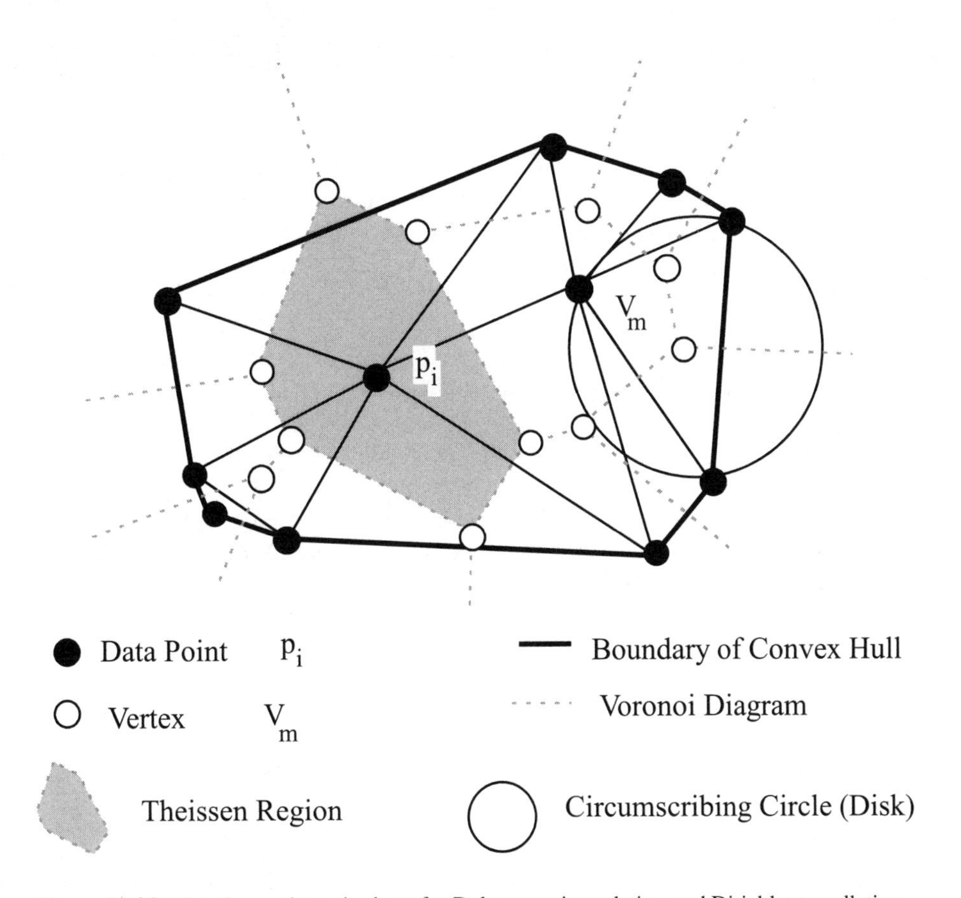

Figure 20.25: Notation and terminology for Delaunay triangulation and Dirichlet tessellation.

easily get trapped in a local extreme at τ_8. The ideas of simulated annealing can be used to develop algorithms which can escape from these local extrema. See Schumaker [225], for example.

The Algorithm of Green and Sibson [107]: This algorithm depends heavily upon a particular data structure used to store the Delaunay triangulation (or Dirichlet tessellation). For each object (a Dirichlet tile or window boundary constraint) is recorded in a "contiguity list" consisting of all objects with which it is contiguous. This data structure is very similar to the contiguity list structure we described in Figure 20.9 but it also includes some window boundary constraints. New points are inserted sequentially. We quote directly from [107] to how this is done.

> The contiguity list for the new point is then built up in reverse (that is, clockwise) order and subsequently standardised. We begin by finding where the perpendicular bisector of the line joining the new point to its nearest neighbour meets the edge of the nearest neighbour's tile, clockwise round the new point. Identifying the edge where this happens gives the next object contiguous with the new point and this is in fact the first to go onto its contiguity list. The new perpendicular bisector is then constructed and its incidence on the edge of this new tile is examined to obtain the subsequent contiguous object: successive objects are added to the contiguity list in this way until the list is completed by the addition of the nearest neighbour. Whilst this being done old contiguity lists are being modified: the new point is inserted in each and any contiguities strictly between the entry and exit points of the perpendicular bisector are deleted, the anticlockwise-cyclic arrangement of the lists making both this and the determination (sic) of the exit very easy.

This insertion algorithm requires the computation of the nearest existing data point to the data point that is to be inserted. The authors discuss an algorithm which takes advantage of the tessellation computed so far. In the authors' words: "Simply start at an arbitrary point and "walk" from neighbour to neighbour, always approaching the new point, until the point nearest to it is found."

The Algorithm of Bowyer [21]: Bowyer described an algorithm for inserting a new point (lying in the convex hull) into an existing Delaunay triangulation. An example given by Bowyer and which we include in Figure 20.27 serves to define this data structure. In the terminology of Bowyer, the forming points for a vertex are simply the vertices of the triangle which has this particular vertex as the center of its circumcircle. Since each triangle gives rise to a vertex, giving a list of indices of the forming points for each vertex (as Bowyer does) is equivalent to giving a list of indices of the data points which comprise each triangle of triangulation. Except for a change in ordering, the neighboring vertices are exactly the same as the indices of the triangle neighbors as given in the triangular grid data structure of Figure 20.26.

In order to insert a new point (Q in Figure 20.27) within the current convex hull of the data points, Bowyer [21] gives the following algorithm:

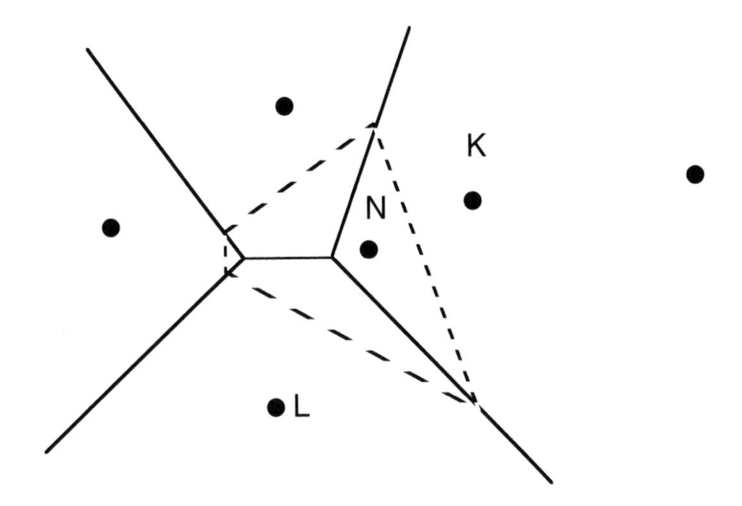

Figure 20.26: An aid to the Green and Sibson algorithm.

1. Identify a vertex currently in the structure that will be deleted by the new point (say V_4). Such a vertex is any that is *nearer to the new point than to its forming points.*

2. Perform a tree search through the vertex structure starting at the deleted vertex looking for others that will be deleted. In this case the list will be: $\{V_4, V_3, V_5\}$.

3. The points contiguous to Q are all the points forming the deleted vertices: $\{P_2, P_5, P_4, P_3, P_7\}$.

4. An old contiguity between a pair of those points will be removed ($P_2 - P_4$ say) if all of its vertices $\{V_4, V_3\}$ are in the list of deleted vertices.

5. In this case the new point has five new vertices associated with it: $\{W_1, W_2, W_3, W_4, W_5\}$. Compute their forming points and neighbouring vertices. The forming points for each will be the point Q and two of the points contiguous to Q. Each line in the tessellation has two points around it (the line $V_3 - V_2$, for example, is formed by P_3 and P_4). The forming points of the new vertices and their neighbouring vertices may be found by considering vertices pointed to by members of the deleted vertex list that are not themselves deleted, and finding the rings of points around them. Thus W_5 points outwards to V_2 from Q and is formed by $\{P_3, P_4, Q\}$.

6. The final step is to copy some of the new vertices, overwriting the entries of those deleted to save space.

The Algorithm of Watson [254]: This algorithm relies on the property of a Delaunay triangulation that a triple of data point indices (i, j, k) will be in I_t provided the circumcircle of p_i, p_j, and p_k contains no other data points. As with the other algorithms, this algorithm is based upon inserting a new point into an already existing Delaunay triangulation. The general philosophy of Watson's approach is described by the following two steps:

Vertex	Forming points			Neighboring vertices		
	1	2	3	1	2	3
V_1	P_6	P_4	P_5	V_4	ϕ	V_6
V_2	P_1	P_4	P_3	V_3	ϕ	V_7
V_3	P_2	P_3	P_4	V_2	V_4	V_5
V_4	P_2	P_5	P_4	V_1	V_3	ϕ
V_5	P_7	P_3	P_2	V_3	ϕ	ϕ
V_6	P_6	P_8	P_4	V_7	V_1	ϕ
V_7	P_1	P_8	P_4	V_6	V_2	ϕ

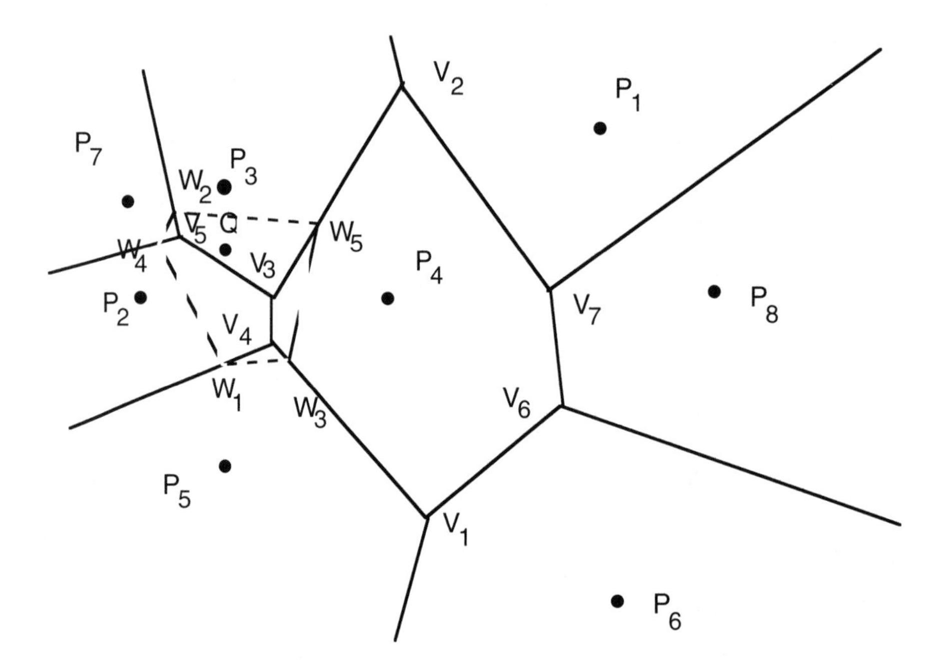

Figure 20.27: Illustrating the algorithm of Bowyer [21].

1. Find all triangles whose circumcircle contains the point to be inserted.
2. For each of these triangles, form three new triangles from the point to be inserted and the three edges of this triangle and test to see if any of these three new triangles contain any other data points. If not, then add this new triangle to the triangulation.

More details for this general approach are given in the flow diagram of Figure 20.28, which is based upon the flow diagram of [254].

Watson [254] describes a number of features and details to make the basic algorithm efficient and eventually discusses a particular implementation which he says has an expected running time which is observed to increase not more than $N^{3/2}$.

The Embedding/Lifting Approach: Algorithms of this type are based upon a very interesting relationship that exists between the three-dimensional convex hull of the lifted points $(x_i, y_i, x_i^2 + y_i^2)$ and the Delaunay triangulation. Faces on the convex hull are designated as being either in the upper or lower part. The lower part consists of faces which are supported by a plane that separates the point set from $(0, 0, -\infty)$. The Delaunay triangulation is obtained directly from the projection onto the xy-plane of the lower part of the convex hull. See [27] and [68]. An algorithm for computing the convex hull which is based on an initial sort followed by a recursive divide-and-conquer approach has been described by Preparata and Hong [202]. This algorithm is also covered in [68] and [203]. Theoretically the algorithm is optimal time $O(n \cdot \log(n))$, but Day [49] reports that empirical data implies a worst-case complexity of $O(N^2)$. Day covers many of the details and special-case issues of practical interest for implementation which are often brushed over in more theoretical papers.

Divide-and-Conquer Algorithms: The general structure of this type of algorithm is to divide the data set into subsets A and B, solve the problem for A and solve the problem for B and merge the results into a solution for A ∪ B. See Figure 20.29. Divide-and-conquer algorithms can lead to theoretically optimal algorithms, but often fail to be competitive in practical usage. The merging portion is often the most troublesome in trying to maintain bounds on the running times and complexity of the algorithm.

20.2.3 Visibility Sorting of Triangulations

This is an example of an area that is interesting in 3D but not in 2D. It is possible to make a definition of a visibility sort for a triangulation which is completely analogous to that of a tetrahedrization, but there does not appear to be any application or use for such a property. We defer further discussion on visibility sorting to Section 20.3.3.

20.2.4 Data-Dependent Triangulations

The topic of data-dependent triangulations arises within the context of determining a modeling function $F(x, y)$ for the data $(F_i; x_i, y_i), i = 1, \ldots, N$. A relatively simple approach to defining a modeling function is to first form a triangulation of the convex hull of the independent data $(x_i, y_i), i = 1, \ldots, N$ and then define F to be piecewise linear over this

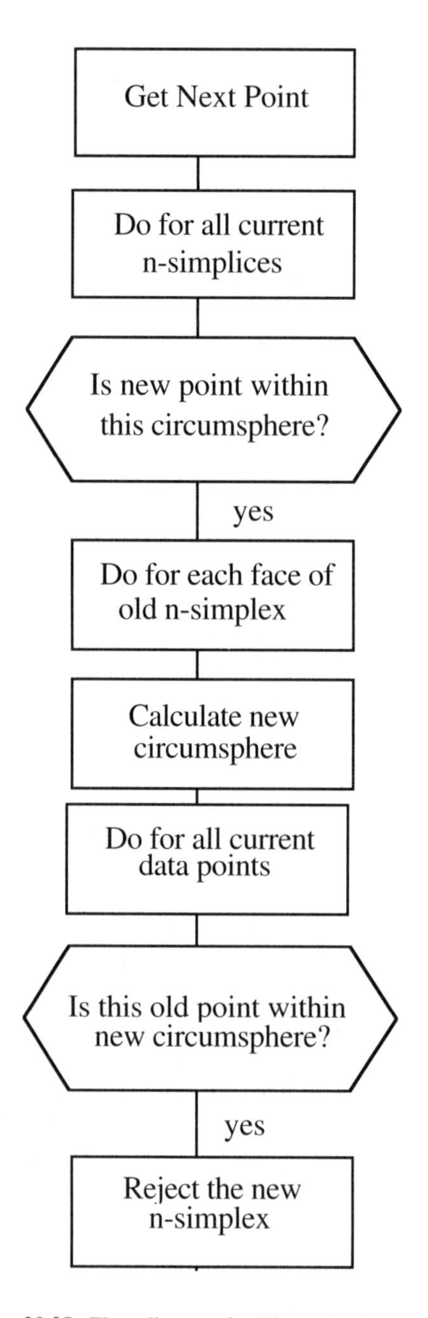

Figure 20.28: Flow diagram for Watson's algorithm.

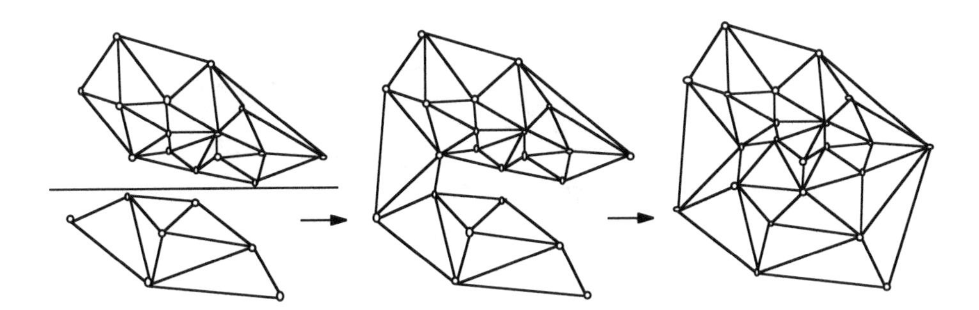

Figure 20.29: Divide-and-conquer algorithms.

triangulation. This will yield a C^0 (continuous) function which interpolates the data; that is, $F(x_i, y_i) = F_i, i = 1, \ldots, N$. We denote this function by $F_T(x, y)$. Any triangulation of the independent data $(x_i, y_i), i = 1, \ldots, N$ will suffice for this approach. While we are well aware of the many desirable properties of the Delaunay triangulation, it might very well be the case that some other triangulation whose choice would depend upon the values $F_i, i = 1, \ldots, N$ would lead to some desirable properties for the modeling function F. This is the basic idea of data-dependent triangulation. Of course, there are potentially many ways to accomplish this, but we choose for this discussion here to briefly describe the criteria called "nearly C^1" as proposed in [67]. An ordering is imposed on the collection of all possible triangulations of the convex hull in the following manner. First a local cost function (see Figure 20.30) for each edge $e_i = 1, \ldots, N_{ie} = N_e - N_b$ is defined and denoted by $S(F_T, e_i)$. (We will shortly describe the four examples of local cost functions covered in [67].) If T and T' are two triangulations, then

$$T \leq T'$$

provided the vector

$$(s(F_T, e_1), s(F_T, e_2), \ldots, s(F_T, e_{N_{ie}}))$$

is lexicographically less than or equal to

$$(s(F_{T'}, e_1), s(F_{T'}, e_2), \ldots, s(F_{T'}, e_{N_{ie}})).$$

It is assumed that the components of these vectors are arranged in nonincreasing order. The goal is then to find the optimal data-dependent triangulation which is defined by having the smallest associated vector under this lexicographical ordering. Since there are only a finite (albeit possibly very large) number of possible triangulations, we know that a global minimum exists even though it may not be unique and it may not be so easy to compute. The algorithm used in [67] is similar to the swapping algorithm of Lawson (which we have described above in Section 20.2.2) in that an initial triangulation is obtained and then an internal edge of a convex quadrilateral is considered. If $T' < T$, where T' is the same triangulation as T except the diagonal of the convex quadrilateral has been switched, then this switch is made and other edges are considered for potential swapping. Since each swap moves strictly lower in the lexicographic ordering, we are guaranteed that this algorithm

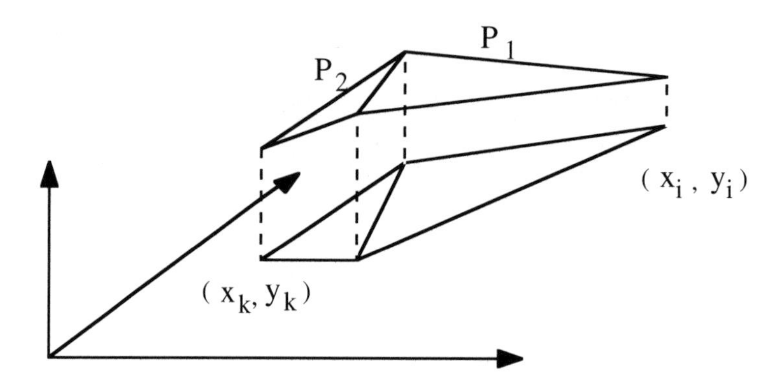

Figure 20.30: Notation for local cost function definitions.

will eventually converge after a finite number of steps. This means that swapping any edge would not move to a smaller triangulation. This limit triangulation may not be the global minimum; it is only guaranteed to be a local minimum and steps to find the global minimum must do more than swap diagonals which improve (with respect to the ordering) the triangulation.

We now describe the four local edge cost functions used in [67]. Let $P_1 = a_1x + b_1y + c_1$ and $P_2 = a_2x + b_2y + c_2$ be the two planes defined over the two triangles of a convex quadrilateral.

i) The angle between normals: The local cost function is taken as the acute angle between N_1 and N_2, which are the respective normals for P_1 and P_2.

$$s(F_T, e) = \cos^{-1}(A)$$

where

$$A = \frac{a_1a_2 + b_1b_2 + 1}{\sqrt{(a_1^2 + b_1^2 + 1)(a_2^2 + b_2^2 + 1)}}$$

ii) The jump in normal derivative: This cost function is the difference between the derivative of P_1 and P_2. This derivative is taken in the direction perpendicular to the edge dividing the two triangles.

$$s(F_T, e) = [n_x(a_1 - a_2) + n_y(b_1 - b_2)]$$

where (n_x, n_y) is a unit vector perpendicular to the edge e.

iii) The deviations from linear polynomials: The cost function measures the error between P_1 and P_2, evaluated at the other point of the quadrilateral.

$$s(F_T, e) = \sqrt{(P_1(x_i, y_i) - F_i)^2 + (P_2(x_k, y_k) - F_k)^2}$$

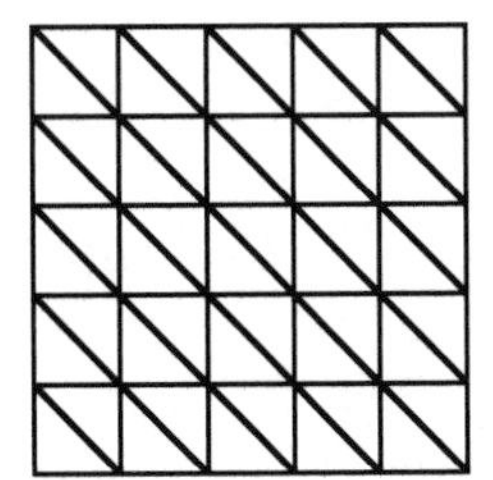
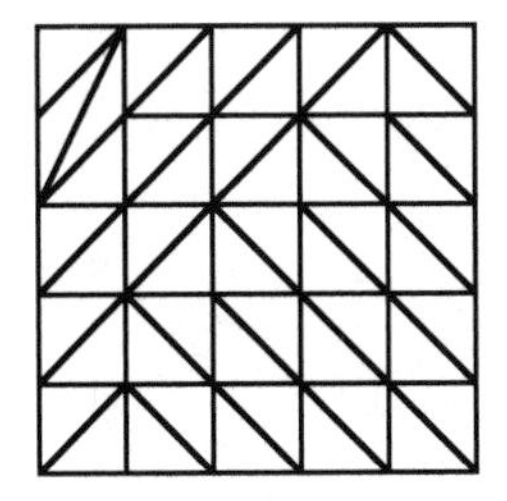
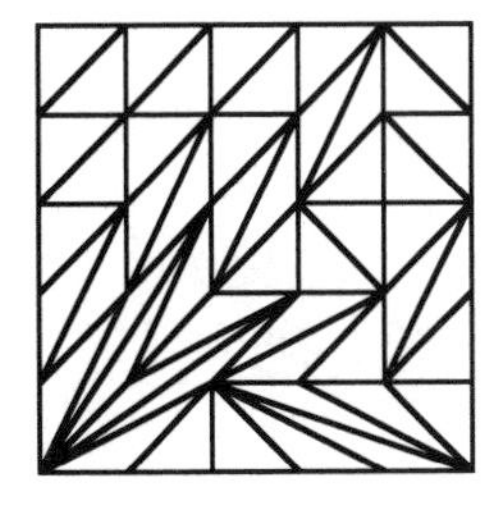

Figure 20.31: Examples of data-dependent triangulations.

iv) The distance from planes: This cost function measures the distance between the planes P_1 and P_2 and the corresponding vertex of the quadrilateral.

$$s(F_T, e) = \sqrt{\frac{(P_1(x_i, y_i) - F_i)^2}{a_1^2 + b_1^2 + 1} + \frac{(P_2(x_k, y_k) - F_k)^2}{a_2^2 + b_2^2 + 1}}$$

Some typical results are given in [67] which confirm the expectation that using the optimal data-dependent triangulation improves the overall fitting properties of F_T over that of the Delaunay triangulation, which, by the way, is used as the initial triangulation for the swapping algorithm. It is observed that long, thin triangles tend to appear where the data seems to indicate a function that is increasing (or decreasing) relatively rapidly in a certain direction. The use of the data-dependent triangulation generally gives an overall reduction in errors when certain test functions are used to generate the data.

As we have mentioned, the local swapping algorithm used in [67] can only find a local minimum. In order to move more closely to the globally optimal data-dependent triangulation, Schumaker [225] and Quak and Schumaker [204,205,206] have involved the tools of simulated annealing. More details on this are contained in Section 20.3.6 on data-dependent tetrahedrizations. We include here the results of one example described by Schumaker. The data consists of

$$(F_{ij}; x_i, y_j); \quad x_i, y_j = 0.0, 0.2, 0.4, 0.6, 0.8, 1.0;$$

where

$$F(x, y) = (y - x^2)_+.$$

Three triangulations are shown in Figure 20.31. The first is the Delaunay triangulation of the independent data. The next is the triangulation which results from the local swapping algorithm of [67] using the local cost function of "angle between normals." The last is the triangulation after simulated annealing has been applied. The associated vectors for each of these triangulations is given in Figure 20.33 and the piecewise linear approximations over these triangulations are shown in Figure 20.32.

20.2.5 Affine Invariant Triangulations

The desirable properties of the Delaunay triangulation have been previously discussed. Unfortunately, this optimal triangulation is not invariant under affine transformations, and this

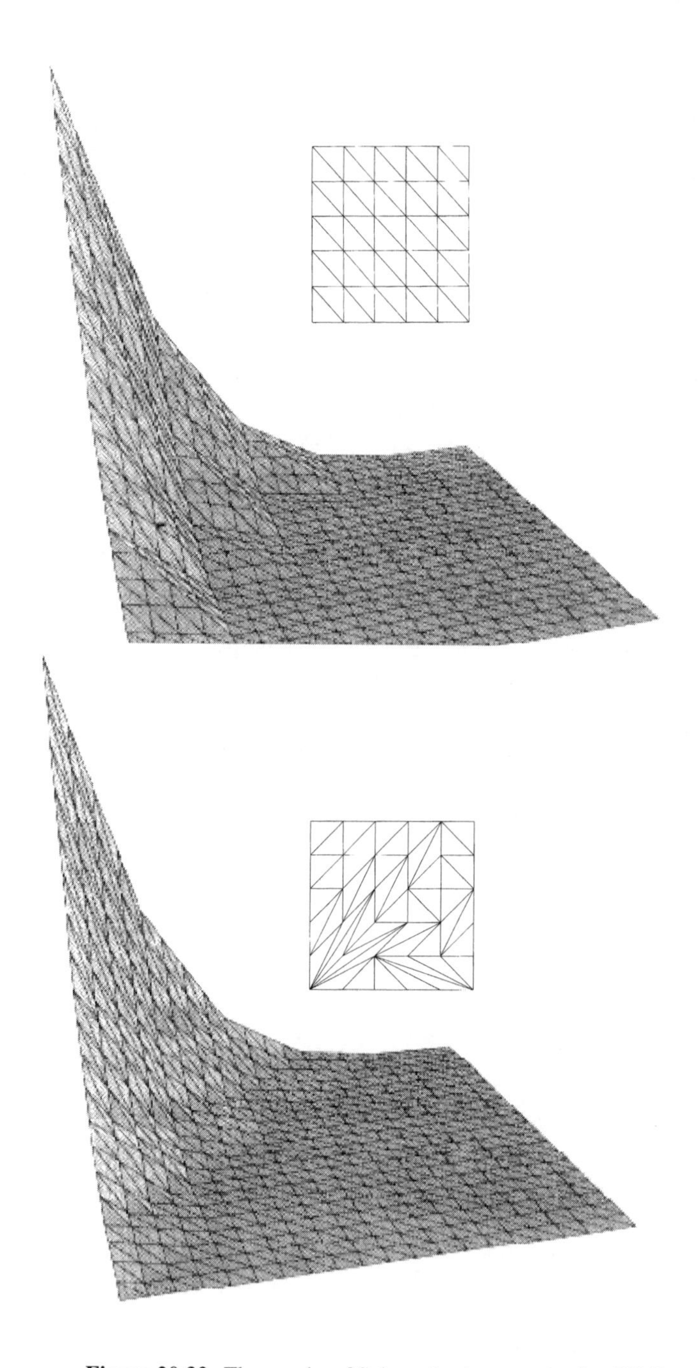

Figure 20.32: The graphs of Schumaker's example. See [225].

Angles between normals for Delaunay triangulation:

55.077 48.155 44.684 39.801 39.588 38.378 37.734 35.445 33.992
33.786 33.561 33.162 30.470 28.898 28.287 27.284 27.284 26.003
23.633 21.958 20.814 17.886 16.066 15.942 15.642 11.310 10.302
9.661 7.294 7.294 7.294 6.843 0.649 0.649 0.459 0.458
0.458 0.000 0.000 0.000 0.000 0.000 0.000 0.000 0.000
0.000 0.000 0.000 0.000 0.000 0.000 0.000 0.000 0.000
0.000 0.000 0.000 0.000 0.000 0.000 0.000 0.000 0.000
0.000 0.000

Angles between normals for locally optimal triangulation:

35.993 30.590 26.070 23.610 21.813 21.558 16.563 16.521 15.793
12.810 11.929 11.310 10.646 10.261 9.622 8.844 8.707 8.321
8.076 8.047 5.794 5.563 3.777 0.649 0.649 0.459 0.459
0.458 0.458 0.458 0.448 0.020 0.020 0.000 0.000 0.000
0.000 0.000 0.000 0.000 0.000 0.000 0.000 0.000 0.000
0.000 0.000 0.000 0.000 0.000 0.000 0.000 0.000 0.000
0.000 0.000 0.000 0.000 0.000 0.000 0.000 0.000 0.000
0.000 0.000

Angles between normals for annealed triangulation:

26.070 22.929 22.113 20.049 17.257 16.563 16.521 13.031 12.505
11.929 10.389 10.270 10.261 8.954 8.321 7.844 5.962 5.794
5.256 1.652 1.480 1.025 0.649 0.648 0.459 0.458 0.458
0.448 0.447 0.020 0.000 0.000 0.000 0.000 0.000 0.000
0.000 0.000 0.000 0.000 0.000 0.000 0.000 0.000 0.000
0.000 0.000 0.000 0.000 0.000 0.000 0.000 0.000 0.000
0.000 0.000 0.000 0.000 0.000 0.000 0.000 0.000 0.000
0.000 0.000

Figure 20.33: Angles for the data-dependent triangulation.

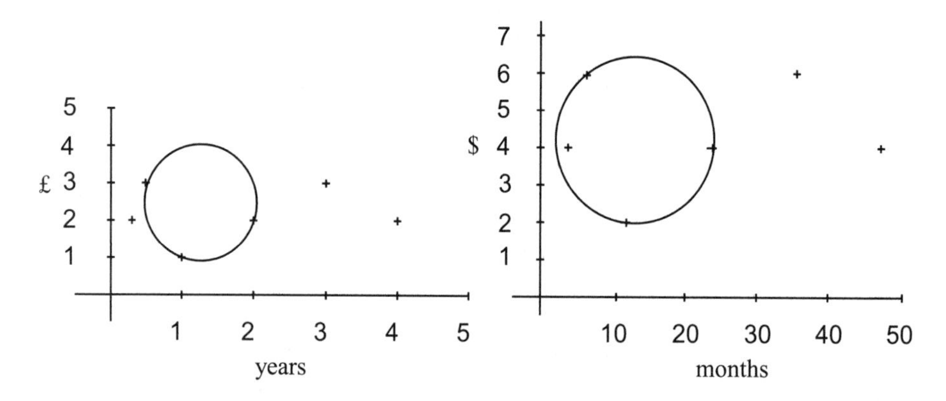

Figure 20.34: Two different units used to measure the same data lead to two different Delaunay triangulations.

means that methods for analyzing and visualizing data that use this particular triangulation can be affected by the choice of units used to measure the data. This could be considered an undesirable property. In this section we describe a relatively new method for characterizing and computing an optimal triangulation which is invariant under affine transformations. Before we proceed with the discussion of these techniques, we wish to motivate further the desirability of affine invariance.

As we have mentioned earlier, one of the main purposes for triangulations and tetrahedrizations is their use in defining functions in a piecewise manner over the domain of a data set. It would be undesirable if the happenstance of the choice of units used to measure the data were to affect the definition of a data modeling function. But this does happen with the Delaunay triangulation. The example of Figure 20.34 points this out. This data represents the independent data; the dependent data is not given as it is not important in this context. The data is the same in both the left and right graphs of Figure 20.34; the only difference is that in the left graph we have used years and £ (pounds, British monetary unit, approximately equal and assumed here to be exactly equal to two US dollars), and in the right graph we have used months and dollars. If we use the units of years and £ then we can see that the three vertices (1yr, 1£), (0.5yr, 3£), (2yr, 2£) will mark out a triangle to be included in the list of triangles for the Delaunay triangulation. But on the other hand, if we use months and \$ we can see that the circumcircle defined by these same three vertices (12mon, 2\$), (6mon, 6\$), (24mon, 4\$) contains the data point (4mon, 4\$). Therefore, these three vertices will not comprise a triangle of the Delaunay triangulation if these units are used. This simple example points out the possible effects of the choice of the units of measurement. The choice of the units of measurement is the same as a change in scale, $x \leftarrow ax$ and $y \leftarrow by$. Uniform scale changes of the type $x \leftarrow ax,\ y \leftarrow ay$ will not affect the Delaunay triangulation.

We now discuss how to avoid this problem. It would be possible to simply normalize all data ranges to one unit by scaling by the range. But this approach would mean that rotations of the data could have an effect on the Delaunay triangulation, meaning the final data model would be affected by rotations of the data. In other words, the placement and alignment of the axes for the measurement of the data would have an effect on the data modeling function and subsequently on our analysis of the data, and this we would like to have the opportunity to avoid. It would, in general, be useful to have a characterization (and subsequent algorithms) for an optimal triangulation which is not affected by affine transformation. An affine transformation is a map of the form

$$(x, y) = A(x, y) + c$$

where A is a 2×2 matrix and c is a two-dimensional point. Affine transformations include not only scale changes and rotations, but also translations, reflections, and shearing transformations. The approach to such an optimal triangulation covered here is through the duality that exists between the conventional Delaunay triangulation and the Dirichlet tessellation. As we described previously, the characterization of the Delaunay triangulation (as a MaxMin triangulation), it was heavily dependent upon angles, and angles are affected by scaling transformations; so it should be no surprise that the Delaunay triangulation is also affected by scaling transformations. But the definition of the Dirichlet tessellation uses only distance and we know that the Delaunday triangulation is dual to (a direct result of) the Dirichlet tessellation. The approach here is to use a method of measuring distance which is invariant under affine transformations. The Dirichlet tessellation based upon this new method of measuring distance will have a dual which will serve as our optimal triangulation. Rather than use the standard Euclidean norm $||(x, y)||^2 = \sqrt{x^2 + y^2}$ we propose the use of the following norm

$$||(x, y)||_V^2 = (x, y) \begin{pmatrix} \frac{\Sigma_y^2}{\Sigma_x^2 \Sigma_y^2 - (\Sigma_{xy})^2} & \frac{-\Sigma_{xy}}{\Sigma_x^2 \Sigma_y^2 - (\Sigma_{xy})^2} \\ \frac{-\Sigma_{xy}}{\Sigma_x^2 \Sigma_y^2 - (\Sigma_{xy})^2} & \frac{\Sigma_x^2}{\Sigma_x^2 \Sigma_y^2 - (\Sigma_{xy})^2} \end{pmatrix} \begin{pmatrix} x \\ y \end{pmatrix} \tag{20.1}$$

where

$$\Sigma_x^2 = \frac{\sum_{i=1}^N (x_i - \mu_x)^2}{N}, \qquad \mu_x = \frac{\sum_{i=1}^N x_i}{N}$$

$$\Sigma_y^2 = \frac{\sum_{i=1}^N (y_i - \mu_y)^2}{N}, \qquad \mu_y = \frac{\sum_{i=1}^N y_i}{N}$$

$$\Sigma_{xy} = \frac{\sum_{i=1}^N (x_i - \mu_x)(y_i - \mu_y)}{N}$$

and

$$V = \begin{pmatrix} x_1 - \mu_x & x_2 - \mu_x & \cdots & x_N - \mu_x \\ y_1 - \mu_y & y_2 - \mu_y & \cdots & y_N - \mu_y \end{pmatrix}.$$

We have used the subscript of V on the norm to explicitly indicate that this method of measuring distance is dependent upon the data set. Change the data set and you change how you measure distance, but the distance between any two data points will remain constant.

This norm and its use within the context of scattered data modeling was first described in [181]. This norm has the property that it is invariant under affine transformations. More precisely,

$$||P - Q||_V = ||T(P) - T(Q)||_{T(V)} \tag{20.2}$$

for any two points $P = (x, y)$ and $Q = (u, v)$ and any affine transformation

$$T(P) = \begin{pmatrix} a_{11} & a_{12} \\ a_{21} & a_{22} \end{pmatrix} \begin{pmatrix} x \\ y \end{pmatrix} + \begin{pmatrix} c_1 \\ c_2 \end{pmatrix}.$$

Here, $T(V)$ (used as a subscript in Equation (20.2)) is the transformed data

$$T(V) = \left(T\begin{pmatrix} x_1 - \mu_x \\ y_1 - \mu_y \end{pmatrix} \quad T\begin{pmatrix} x_2 - \mu_x \\ y_2 - \mu_y \end{pmatrix} \quad \cdots \quad T\begin{pmatrix} x_N - \mu_x \\ y_N - \mu_y \end{pmatrix} \right).$$

Figure 20.35 illustrates the properties of this new method of measuring distance. Each of the data sets shown in this figure are affine images of each other. Starting in the upper left and moving in a clockwise direction, the transformations are: counterclockwise rotation of 44 degrees; a scaling in x by a factor of 2; a scaling in y by a factor of 0.4. The four ellipses in each figure represent points which are $1/4$, $1/2$, $3/4$, and 1 unit(s) from their center point as measured with the affine invariant norm. In Figure 20.36 we show the Dirichlet tessellation of these four affinely related data sets, and in Figure 20.37 we show the corresponding dual triangulation. As one can see, the triangulation is unchanged by these transformations.

As a comparison, we have also included the Delaunay triangulation based upon the standard Euclidean norm in Figure 20.38. And as we indicated earlier, we can see that triangulation results are affected by the transformations. Not all triangles are changed, but some are.

And now we suggest some practical information on how to incorporate this feature into an algorithm for computing triangulations. If you already have a procedure for computing an optimal triangulation, then it is possible to modify it slightly to achieve the results we have described in this section. Say, for example, that the procedure is based upon Lawson's algorithm and that there is a subprocedure which decides whether or not to switch the diagonal of a quadrilateral formed from two triangles. It might be that this procedure is based solely on Euclidean distance. That is, the center and radius of the circumcircle of three points are determined and the distance to the center from the fourth point is computed so as to make this decision. In order to modify this subprocedure, we need only to replace the use of the Euclidean norm with the affine invariant norm described here. The equations for computing circumscribing circles (ellipses) for a quadratic norm in general are given in [182]. If, on the other hand, the procedure you are already using is known to be rotation invariant, then there is an even easier way to affect the results of the affine invariant triangulation. This is based upon the factorization of the matrix which defines the affine invariant norm. We denote this matrix by $A(V)$ so that we have

$$||(x, y)||_V^2 = (x, y)A(V) \begin{pmatrix} x \\ y \end{pmatrix}$$

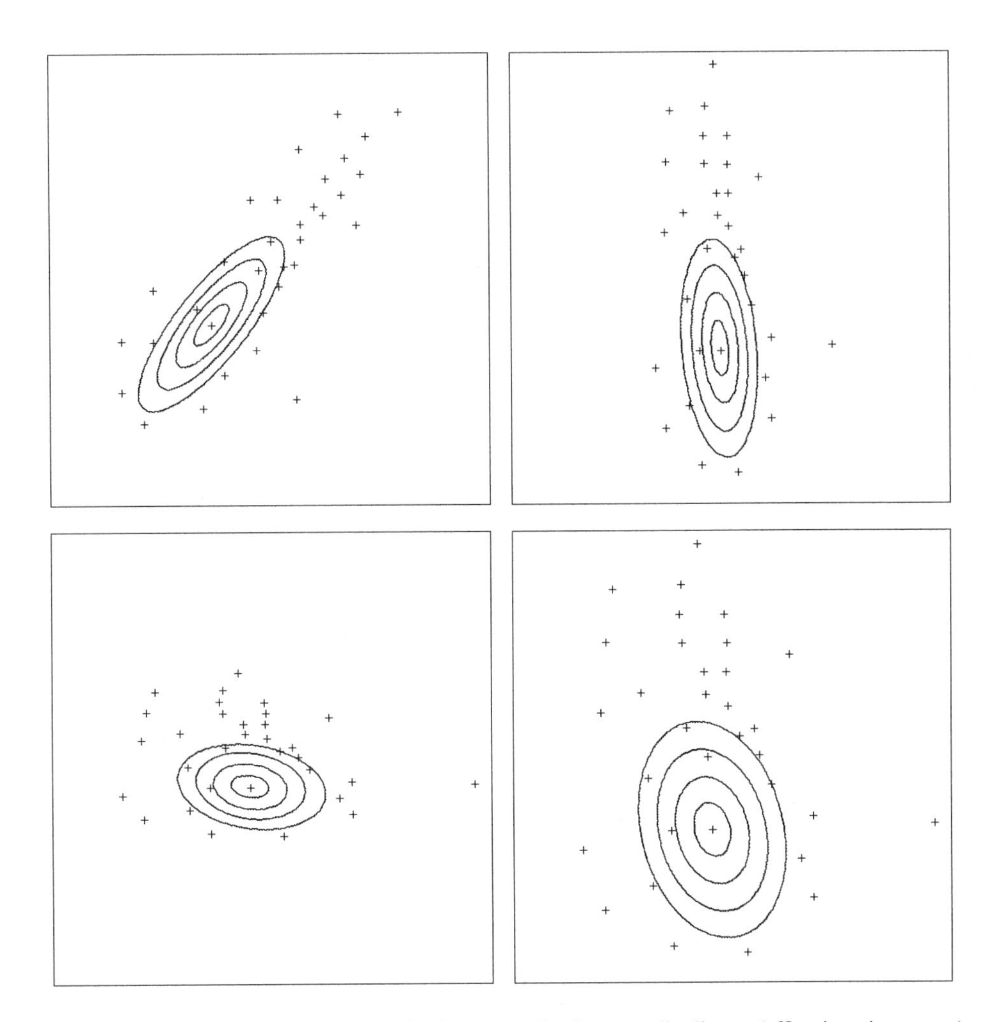

Figure 20.35: Affine transformations of a data set and points equally distant (affine invariant norm) from a point.

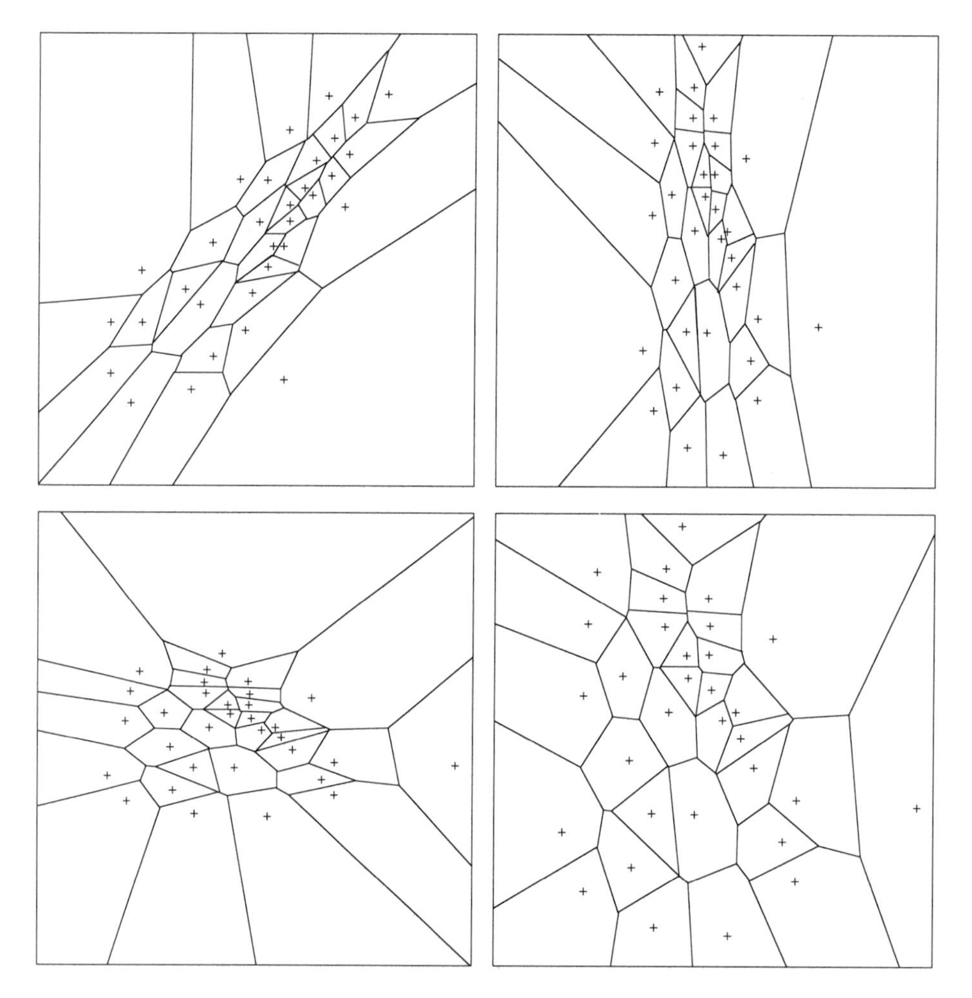

Figure 20.36: The Dirichlet tessellation (affine invariant norm) of affine transformations of a given data set.

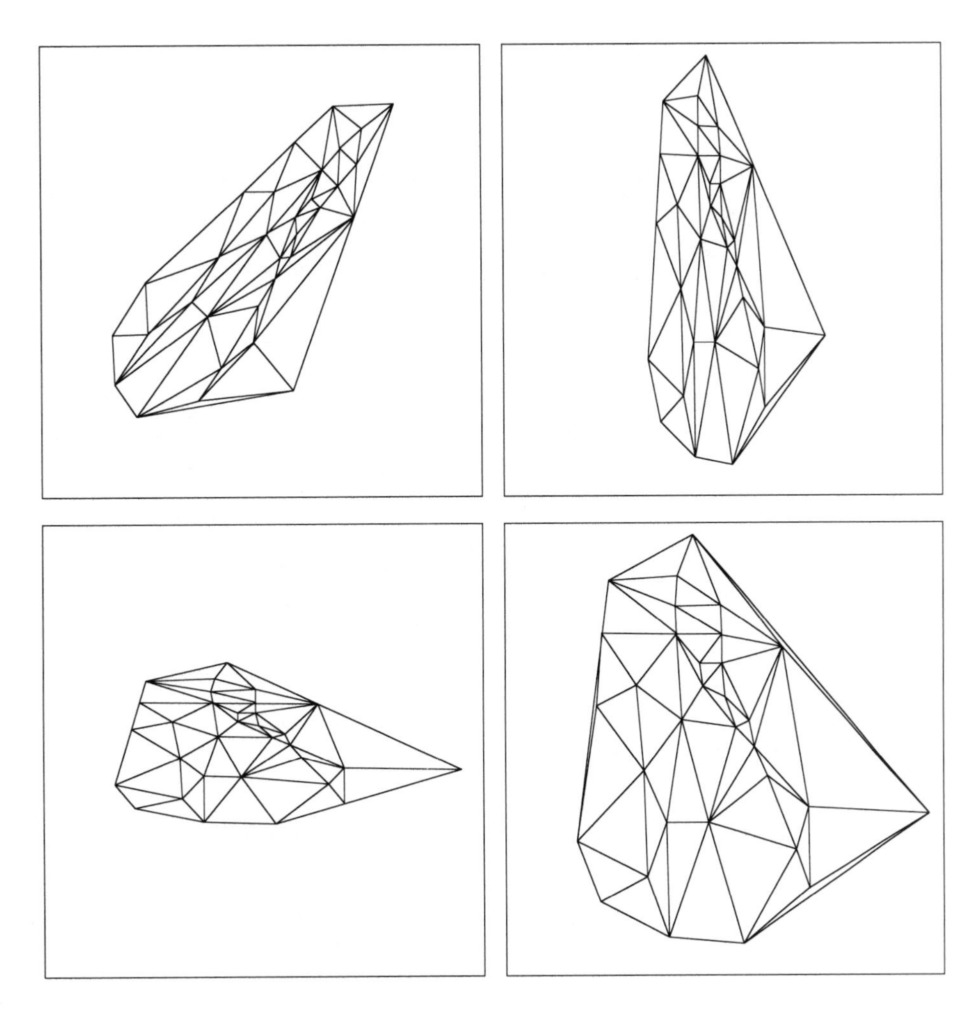

Figure 20.37: The triangulation dual to the Dirichlet tessellation (affine invariant norm) of a given data set and some affine transformations.

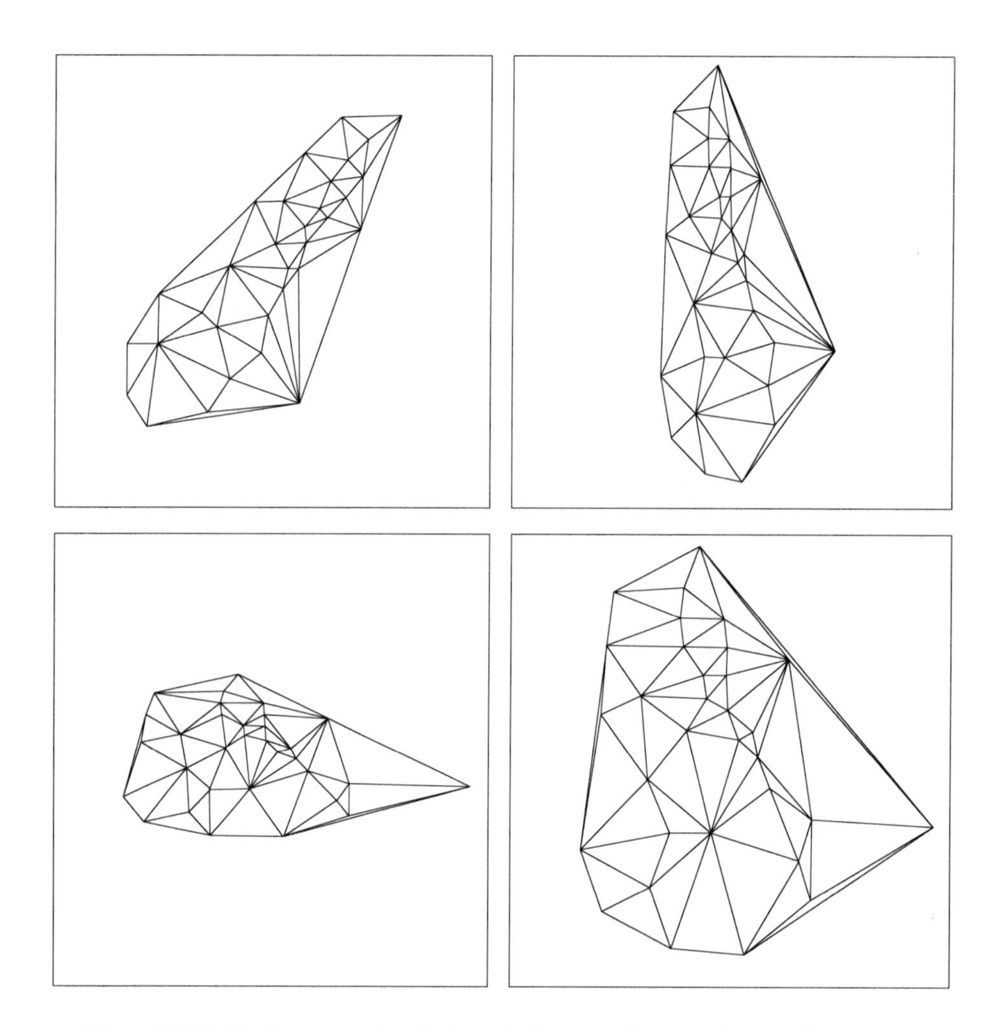

Figure 20.38: The Delaunay triangulation of a data set and some affine transformations.

The matrix $A(V)$ can be factored (Cholesky) into

$$A(V) = \begin{pmatrix} l_{11} & 0 \\ l_{21} & l_{22} \end{pmatrix} \begin{pmatrix} l_{11} & l_{21} \\ 0 & l_{22} \end{pmatrix} = L(V)L(V)^*.$$

Here the notation $L(V)^*$ denotes the transpose of $L(V)$.

Using this factorization, we have that

$$||(x,y)||_V^2 = (x,y)L(V)L(V)^* \begin{pmatrix} x \\ y \end{pmatrix} = ||(x,y)L(V)||^2$$

which means measuring distances with the affine invariant norm is the same as measuring distance in the standard Euclidean but with the points transformed by multiplying by $L(V)$. This means that we can achieve the result of the optimal affine invariant triangulation by computing the standard Delaunay triangulation on the transformed data

$$(X_i, Y_i) = (x_i, y_i)L(V)$$

In summary, we need only to compute

$$l_{11} = \sqrt{a_{11}}, \quad l_{21} = \frac{a_{21}}{\sqrt{a_{11}}}, \quad l_{22} = \sqrt{\frac{a_{11}a_{22} - a_{21}^2}{a_{11}}}$$

where $a_{11} = \dfrac{\Sigma_y^2}{\Sigma_x^2\Sigma_y^2 - (\Sigma_{xy})^2}$, $a_{21} = \dfrac{-\Sigma_{xy}}{\Sigma_x^2\Sigma_y^2 - (\Sigma_{xy})^2}$, and $a_{22} = \dfrac{\Sigma_x^2}{\Sigma_x^2\Sigma_y^2 - (\Sigma_{xy})^2}$ and to apply any rotation invariant triangulation algorithm to the transformed data

$$\begin{aligned} X_i &= l_{11}x_i + l_{21}y_i \\ Y_i &= l_{22}y_i, \qquad i = 1, \ldots, N. \end{aligned}$$

20.2.6 Interpolation in Triangles

We now take up the topic of interpolating into (or over) a single triangular domain. The interpolants we describe here form the basic building blocks for constructing the global interpolants which have piecewise definitions over the individual triangles of a triangulation. The domain here is a single triangle, $T = T_{ijk}$ with vertices V_i, V_j, and V_k, and the data consists of values given on the boundary of the triangular domain. We need to differentiate between two types of boundary data. If the data consists of function and certain derivative values specified only at the vertices (or possibly other points such as midpoints), then we call this *discrete data*. If, on the other hand, the data is provided on the entire boundary of the triangle, we refer to this type of data as *transfinite data*. The importance of an interpolant which will match transfinite data is that it serves as a prototype for developing a large variety of discrete interpolants. This is accomplished through the process of *discretization*, where the data required for a transfinite interpolant is provided by means of using some interpolation scheme only on the boundary, discrete data. For example, given only data values at the vertices, we can use linear interpolation along an edge to produce the transfinite data required by the transfinite interpolant, or if we also have data values at the midpoints, we could use quadratic interpolation.

There is a second concept which is rather important for interpolants defined over triangles and this has to do with the degree of continuity of the global interpolant. Often, we require that the global interpolant at least be continuous. We call such an interpolant a C^0 interpolant. If the global interpolant has continuous first-order derivatives, we say it is a C^1 interpolant. A C^0 interpolant for a single triangle is one which interpolates to boundary data consisting of only position values, either at the vertices only (and possibly points along the edges) or on the entire boundary. A C^1 interpolant for a single triangle is one which will interpolate to first-order derivative data specified on the boundary. But this must be done in a manner so as to guarantee C^1 continuity across the boundary edges. So, for example, if the cross-boundary derivative varies quadratically along an edge, then the data on this edge must be sufficient to uniquely determine this derivative, so that on an adjoining triangle we will have exactly the same cross-boundary derivative. For this reason, it is common for C^1 interpolants to have linearly varying cross-boundary derivatives which are determined by their values at the two endpoint vertices.

Combining the two concepts of discrete and transfinite data and C^0 and C^1 data leads to four types of triangular interpolants. This general area of interpolation in triangles is fairly rich and well developed, and we urge the really interested reader to follow the citations into the literature (for example [177], [178], and [189]) after taking a look at the sampling we have chosen to include here. We first cover C^0, discrete interpolants, then a sampling of three C^0, transfinite interpolants. This is followed by the description of a C^1, discrete interpolant. We have chosen to include a discretized version of the minimum norm triangular interpolant (see [178]). Another rather popular C^1, discrete interpolant, is the Clough/Tocher interpolant often mentioned in conjunction with the finite element method. Much has been written about this interpolant in the past and so we do not include it here. This section is concluded with a description of a C^1, transfinite interpolant, called the side-vertex interpolant [177]. It is one of the easiest to describe and the most versatile to use. It also generalizes rather nicely to a tetrahedral domain.

C^0, Discrete Interpolation in Triangles

The lowest-degree polynomial, C^0, discrete interpolant, is linear and unique. Given the data $F(V_i)$, $F(V_j)$, and $F(V_k)$, the coefficients of the linear function

$$F(x, y) = a + bx + cy$$

which interpolates this data can be found by solving the linear system of equations

$$\begin{aligned} a + bx_i + cy_i &= F(V_i) \\ a + bx_j + cy_j &= F(V_j) \\ a + bx_k + cy_k &= F(V_k). \end{aligned}$$

Another path to this basic linear interpolant is via barycentric coordinates. Given a point $V = (x, y)$, barycentric coordinates, b_i, b_j, and b_k of this point relative to the triangle T_{ijk}

are defined by the relationships

$$\begin{aligned} \begin{pmatrix} x \\ y \end{pmatrix} &= b_i V_i + b_j V_j + b_k V_k \\ 1 &= b_i + b_j + b_k. \end{aligned}$$

The linear interpolant now takes the form

$$F(x, y) = F(V) = b_i F(V_i) + b_j F(V_j) + b_k F(V_k).$$

There are several alternative ways of defining or determining the barycentric coordinates of a point. For example,

$$b_i = \frac{A_i}{A} \quad b_j = \frac{A_j}{A} \quad b_k = \frac{A_k}{A}$$

where A_i, A_j, and A_k represent the areas of the subtriangle shown in Figure 20.39 and A is the area of T_{ijk}. Also,

$$b_i = \frac{\begin{vmatrix} x - x_k & x_j - x_k \\ y - y_k & y_j - y_k \end{vmatrix}}{\begin{vmatrix} x_i - x_k & x_j - x_k \\ y_i - y_k & y_j - y_k \end{vmatrix}} \qquad b_j = \frac{\begin{vmatrix} x - x_i & x_i - x_k \\ y - y_i & y_i - y_k \end{vmatrix}}{\begin{vmatrix} x_j - x_i & x_i - x_k \\ y_j - y_i & y_i - y_k \end{vmatrix}}$$

$$b_k = \frac{\begin{vmatrix} x - x_j & x_i - x_j \\ y - y_j & y_i - y_j \end{vmatrix}}{\begin{vmatrix} x_k - x_j & x_i - x_j \\ y_k - y_j & y_i - y_j \end{vmatrix}}.$$

Given the values at the three vertices and the three midpoints of a triangle, there is a unique quadratic which interpolates this data,

$$\begin{aligned} Q(x, y) = \; & F(V_i) b_i (b_i - b_j - b_k) + F(M_{jk}) 4 b_j b_k \\ & + F(V_j) b_j (b_j - b_i - b_k) + F(M_{ik}) 4 b_i b_k \\ & + F(V_k) b_k (b_k - b_i - b_j) + F(M_{ij}) 4 b_i b_j \end{aligned}$$

where $M_{jk} = (V_j + V_k)/2$, $M_{ik} = (V_i + V_k)/2$ and $M_{ij} = (V_i + V_j)/2$.

A common way to specify a cubic along an edge is to use the Hermite form which involves the first-order directional derivatives along the edges

$$F'_{ki}(V_i) = (x_k - x_i) F_x(V_i) + (y_k - y_i) F_y(V_i)$$

which are further illustrated in Figure 20.40.

The six directional derivatives at the three vertices along with $F(V_i)$, $F(V_j)$ and $F(V_k)$ do not uniquely determine a cubic since the bivariate cubics are of dimension 10. The

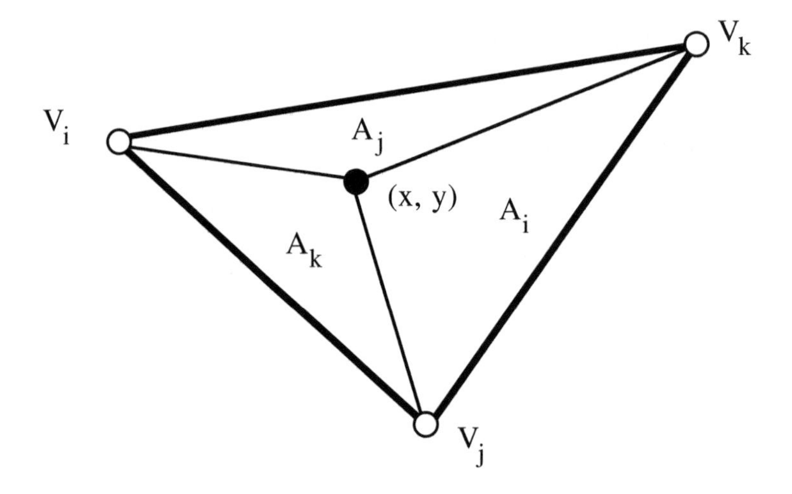

Figure 20.39: Areas leading to barycentric coordinates.

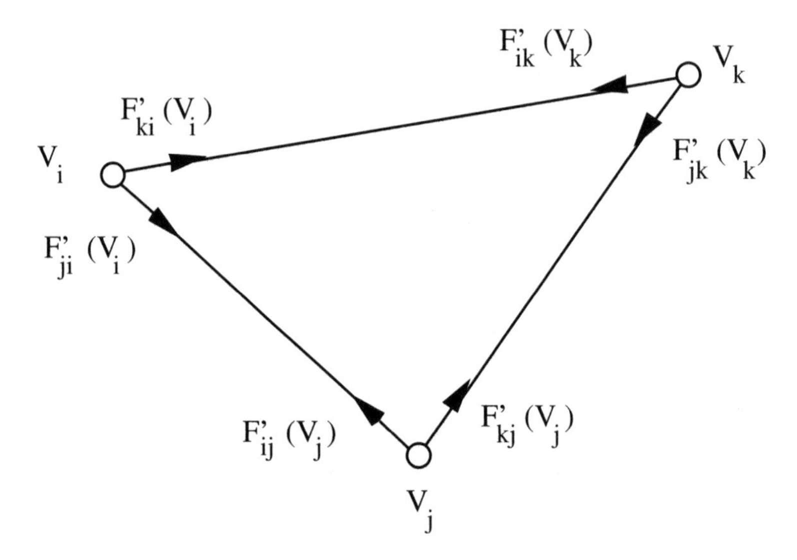

Figure 20.40: The notation for the six directional derivatives.

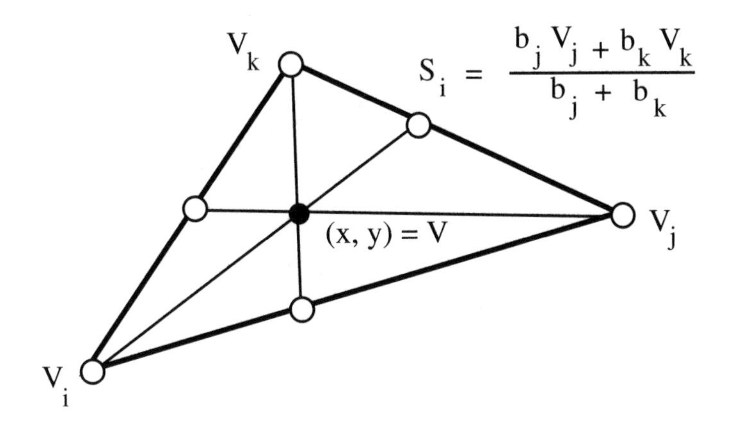

Figure 20.41: The side-vertex interpolant notation.

interpolant

$$\begin{aligned} C(x,y) &= F(V_i)b_i^2 + F'_{ki}(V_i)b_i^2 b_k + F'_{ji}(V_i)b_i^2 b_j \\ &\quad + F(V_j)b_j^2 + F'_{ij}(V_j)b_j^2 b_i + F'_{kj}(V_j)b_j^2 b_k \\ &\quad + F(V_k)b_k^2 + F'_{ik}(V_k)b_k^2 b_i + F'_{jk}(V_k)b_k^2 b_j \\ &\quad + w b_i b_j b_k \end{aligned}$$

will match this function and derivative data for any value of w. This remaining degree of freedom represented by w can be absorbed by a variety of conditions. For example, it can additionally be required that the interpolant match some predescribed value at the centroid. Another common choice is

$$\begin{aligned} w &= 2[F(V_i) + F(V_j) + F(V_k)] \\ &\quad + \frac{1}{2}[F'_{ki}(V_i) + F'_{ji}(V_i) + F'_{ij}(V_j) + F'_{kj}(V_j) + F'_{ik}(V_k) + F'_{jk}(V_k)] \end{aligned}$$

which guarantees quadratic precision and is a result of discretization of a number of transfinite interpolants (see [189]). Quadratic precision means that whenever the data comes from a bivariate quadratic function the interpolant will become this very same quadratic polynomial.

C^0, Transfinite Interpolation in Triangles

In this section, we give only a sampling of three interpolants which will interpolate to arbitrary function values on the boundary of a triangular domain, $T = T_{ijk}$. More information on this general topic can be found in [189].

The Side-Vertex Interpolant: The side-vertex interpolant is built from three basic interpolants which are defined by linear interpolation along line segments joining a vertex

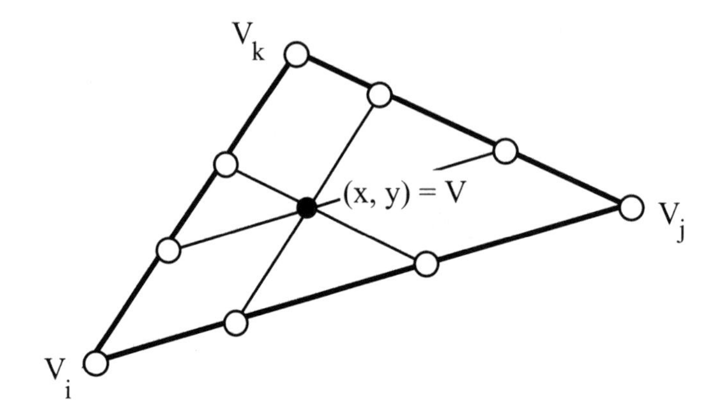

Figure 20.42: The evaluation points (stencil) for side-side interpolant.

and the opposing side. See Figure 20.41. In terms of barycentric coordinates, we have

$$\begin{aligned}
A_i[F] &= b_i F(V_i) + (1 - b_i) F(S_i), \\
A_j[F] &= b_j F(V_j) + (1 - b_j) F(S_j), \\
A_k[F] &= b_k F(V_k) + (1 - b_k) F(S_k)
\end{aligned}$$

where $S_i = \frac{b_j V_j + b_k V_k}{b_j + b_k}$, $S_j = \frac{b_i V_i + b_k V_k}{b_i + b_k}$, $S_k = \frac{b_i V_i + b_j V_j}{b_i + b_j}$. Each of these interpolants will interpolate to arbitrary function values on one edge of the triangular domain. In order to obtain an interpolant which matches arbitrary values on the entire boundary of T_{ijk}, we form the Boolean sum of these three interpolants:

$$\begin{aligned}
A[F] &= A_i \oplus A_j \oplus A_k[F] = A_i[F] + A_j[F] + A_k[F] \\
&\quad - A_i[A_j[F]] - A_j[A_k[F]] - A_k[A_j[F]] + A_i[A_j[A_k[F]]] \\
&= (1 - b_i) F(S_i) + (1 - b_j) F(S_j) + (1 - b_k) F(S_k) \\
&\quad - b_i F(V_i) - b_j F(V_j) - b_k F(V_k)
\end{aligned}$$

The Side-Side Interpolant: The side-side interpolant (Figure 20.42) is based upon the basic operation of linear interpolation along edges which are parallel to the edges of T_{ijk}. There are three of these interpolants:

$$\begin{aligned}
P_i[F] &= \frac{b_k F(b_i V_i + (1 - b_i) V_k + b_j F(b_i V_i + (1 - b_i) V_j}{b_k + b_j} \\
P_j[F] &= \frac{b_i F(b_j V_j + (1 - b_j) V_i + b_k F(b_j V_j + (1 - b_j) V_k}{b_i + b_k} \\
P_k[F] &= \frac{b_i F(b_k V_k + (1 - b_k) V_i + b_j F(b_k V_k + (1 - b_k) V_j}{b_i + b_j}
\end{aligned}$$

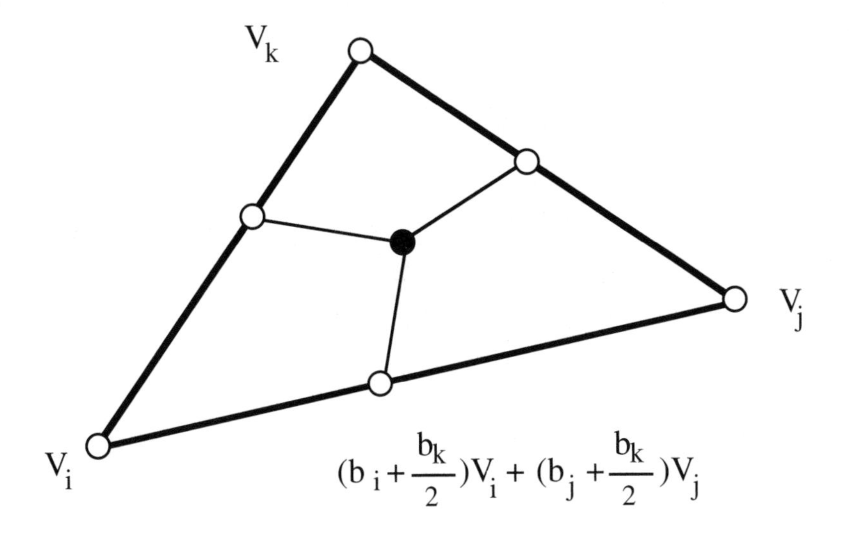

Figure 20.43: The stencil of the C* interpolant.

Unlike the basic interpolants of the side-vertex interpolant, these interpolants do not commute and so their triple Boolean sum is not well defined. However, it is possible to form the average of all double Boolean sums (each of which interpolates to the entire boundary) to arrive at the following affine invariant interpolant

$$
\begin{aligned}
Q^*[F] \quad = \quad & \frac{b_k F(b_i V_i + (1 - b_i)V_k) + b_j F(b_i V_i + (1 - b_i)V_j)}{b_k + b_j} \\
& + \frac{b_i F(b_j V_j + (1 - b_j)V_i) + b_k F(b_j V_j + (1 - b_j)V_k)}{b_i + b_k} \\
& + \frac{b_i F(b_k V_k + (1 - b_k)V_i) + b_j F(b_k V_k + (1 - b_k)V_j)}{b_i + b_j} \\
& - b_i F(V_i) - b_j F(V_j) - b_k F(V_k).
\end{aligned}
$$

The C* Interpolant: The third and final transfinite, C^0 interpolant we describe utilizes the stencil illustrated in Figure 20.43.

$$
\begin{aligned}
C^*[F](b_i, b_j, b_k) \quad = \quad & \frac{b_i b_j}{\left(b_i + \frac{b_k}{2}\right)\left(b_i + \frac{b_k}{2}\right)} F\left(\left(b_i + \frac{b_k}{2}\right) V_i + \left(b_j + \frac{b_k}{2}\right) V_j\right) \\
& + \frac{b_i b_k}{\left(b_i + \frac{b_j}{2}\right)\left(b_k + \frac{b_j}{2}\right)} F\left(\left(b_i + \frac{b_j}{2}\right) V_i + \left(b_k + \frac{b_j}{2}\right) V_k\right) \\
& + \frac{b_j b_k}{\left(b_j + \frac{b_i}{2}\right)\left(b_k + \frac{b_i}{2}\right)} F\left(\left(b_j + \frac{b_i}{2}\right) V_j + \left(b_k + \frac{b_i}{2}\right) V_k\right) \\
& - \frac{2 b_i b_j b_k}{(b_j + 2b_k)(b_k + 2b_j)} F(V_i) \\
& - \frac{2 b_i b_j b_k}{(b_i + 2b_k)(b_k + 2b_i)} F(V_j) \\
& - \frac{2 b_i b_j b_k}{(b_i + 2b_j)(b_j + 2b_i)} F(V_k)
\end{aligned}
$$

which can be written in the form

$$
\begin{aligned}
C^*[F](b_i, b_j, b_k) \quad = \quad & b_i F(V_i) + b_j F(V_j) + b_k F(V_k) \\
& + W_k \left\{ F(Q_k) - \left(b_i + \frac{b_k}{2}\right) F(V_i) - \left(b_j + \frac{b_k}{2}\right) F(V_j) \right\} \\
& + W_j \left\{ F(Q_j) - \left(b_i + \frac{b_j}{2}\right) F(V_i) - \left(b_k + \frac{b_j}{2}\right) F(V_k) \right\} \\
& + W_i \left\{ F(Q_i) - \left(b_j + \frac{b_i}{2}\right) F(V_j) - \left(b_k + \frac{b_i}{2}\right) F(V_k) \right\}
\end{aligned}
$$

where

$$
W_i = \frac{4 b_j b_k}{(2b_j + b_i)(2b_k + b_i)}, \quad W_j = \frac{4 b_i b_k}{(2b_i + b_j)(2b_k + b_j)}, \quad W_k = \frac{4 b_i b_j}{(2b_i + b_k)(2b_j + b_k)}
$$

$$
\begin{aligned}
Q_i &= \left(b_j + \frac{b_i}{2}\right) V_j + \left(b_k + \frac{b_i}{2}\right) V_k, \\
Q_j &= \left(b_i + \frac{b_j}{2}\right) V_i + \left(b_k + \frac{b_j}{2}\right) V_k, \\
Q_k &= \left(b_i + \frac{b_k}{2}\right) V_i + \left(b_j + \frac{b_k}{2}\right) V_j.
\end{aligned}
$$

In this form of C^* we can see that it consists of linear interpolation plus a correction term. It can easily be verified that C^* is precise for all quadratic functions. That is, if f is a quadratic, bivariate polynomial, then $C^*[f] = f$.

C¹, Discrete Interpolation in Triangles

A commonly used 9-parameter, C^1 interpolant is

$$\begin{aligned} C_\Delta[F](x,y) = \sum_{(i,j,k)\in I} & \{F(V_i)\left[b_i^2(3-2b_i)+6wb_i(b_k\alpha_{ij}+b_j\alpha_{ik}\right] \\ & +F'_{ki}(V_i)\left[b_i^2 b_k + wb_i(3b_k\alpha_{ij}+b_j-b_k)\right] \\ & +F'_{ji}(V_i)\left[b_i^2 b_j + wb_i(3b_j\alpha_{ik}+b_k-b_j)\right]\}, \end{aligned}$$

where

$$F'_{ki}(V_i) = (x_k - x_i)F_x(V_i) + (y_k - y_i)F_y(V_i),$$

$$F'_{ji}(V_i) = (x_j - x_i)F_x(V_i) + (y_j - y_i)F_y(V_i),$$

$$w = \frac{b_i b_j b_k}{b_i b_j + b_i b_k + b_j b_k}, \qquad I = \{(i,j,k),(j,k,i),(k,i,j)\},$$

and

$$\alpha_{ij} = \frac{||e_{jk}||^2 + ||e_{ik}||^2 - ||e_{ij}||^2}{2||e_{ik}||^2}$$

We use $||e_{ij}||$ to denote the length of edge e_{ij}. This 9-parameter, C^1 interpolant is a discretized version of a transfinite, C^1, triangular interpolant, which is described in [178]. The derivatives which are in a direction perpendicular to an edge vary linearly along an edge. This guarantees that when two of these interpolants share a common edge the two surface patches will join with continuous first-order derivatives. It is possible to discretize the same transfinite interpolant and use an additional three parameters consisting of cross-boundary derivatives at the midpoints of the three edges. This leads to an interpolant that has all first-order derivatives varying quadratically along the edges. For a comparison of the C_Δ interpolant to the Clough/Tocher interpolant within the context of triangle-based scattered-data models, see Franke and Nielson [97].

C¹, Transfinite Interpolation in Triangles

In this section, we extend the problem of interpolating to transfinite data on the boundary to include also the requirement that the interpolant match user-specified transfinite derivative data on the boundary. These types of interpolants can be used to construct surfaces over triangulated domains which are C^1—that is, functions which have continuous first-order partial derivatives. One of the most versatile and easily described C^1, transfinite interpolants is the C^1, side-vertex interpolant [177].

Earlier, we saw that the basic building blocks of the C^0, side-vertex interpolant consisted of linear interpolation along lines joining a vertex and its opposing side. In order to extend these ideas to C^1 data, we make use of the univariate cubic, Hermite interpolation applied along rays emanating from a vertex and joining to the opposing edge. See

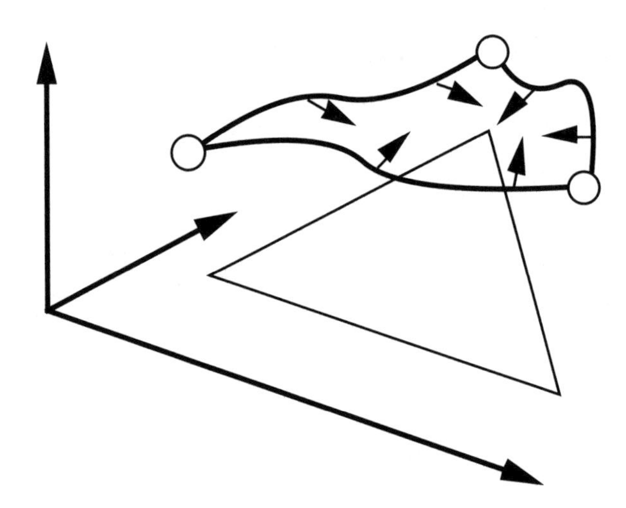

Figure 20.44: The data for C^1 interpolants position and derivative boundary values.

Figure 20.41. Cubic Hermite interpolation will match position and derivatives at the two ends of the interval. We assume that position and derivative information is available on the entire boundary of a triangle T_{ijk}.

$$\begin{aligned} S_i[F](p) &= b_i^2(3-2b_i)F(V_i) + b_i^2(b_i-1)F'(V_i) \\ &\quad +(1-b_i)^2(2b_i+1)F(S_i) + b_i(1-b_i)^2F'(S_i) \end{aligned}$$

where $F'(V_i) = \dfrac{(x-x_i)F_x(V_i) + (y-y_i)F_y(V_i)}{1-b_i}$ and

$F'(S_i) = \dfrac{(x-x_i)F_x(S_i) + (y-y_i)F_y(S_i)}{1-b_i}$. $S_i[F]$ has the property that it interpolates to the boundary data provided by F at V_i and on the entire opposing edge e_{kj}. It also matches first-order derivatives on this edge and at V_i. It does not necessarily interpolate F or its derivatives on the other two edges. In order to have an interpolant for the entire boundary of the triangular domain, we could try to construct one using the ideas of Boolean sums as was done earlier for the C^0, side-vertex interpolant. Even though the interpolants S_i, S_j, and S_k commute so that their Boolean sums are well defined, this approach does not work (see [177]) and so the use of convex combination techniques has been suggested. This leads to the interpolant

$$S[F] = \frac{b_j^2b_k^2S_i[F] + b_i^2b_k^2S_j[F] + b_j^2b_i^2S_k[F]}{b_i^2b_j^2 + b_j^2b_k^2 + b_i^2b_k^2}$$

which has the property that it matches F and its first order derivatives on the entire boundary of the triangular domain. In the case where the boundary information has been discretized with cubically varying (Hermite) position values and linearly varying cross-boundary derivatives, it is possible to obtain a final interpolant with simpler weights in the convex combination. Namely,

$$S[F] = \frac{b_jb_kS_i[F] + b_ib_kS_j[F] + b_jb_iS_k[F]}{b_ib_j + b_jb_k + b_ib_k}.$$

20.3 Tetrahedrizations

In this section we follow the outline of the previous section as well as possible. Since the dimension is one less and since bivariate problems have been considered for a much longer period of time, the development in the 3D domain is not as rich as it is in the 2D domain. So we cannot parallel the previous section exactly, but most everything generalizes or leads to something interesting and often useful.

20.3.1 Basics

Definitions, Data Structures, and Formulas for Tetrahedrizations

Our definition of a tetrahedrization follows very closely to that given for a triangulation at the beginning of Section 20.2.1. We start with a collection of points $p_i = (x_i, y_i, z_i), i = 1, \ldots, N$ which we assume are not collectively coplanar. We denote this collection of points by P. A tetrahedrization consists of a list of 4-tuples which we denote by I_t. Each 4-tuple, $ijkl \in I_t$ denotes a single tetrahedron with the four vertices p_i, p_j, p_k, and p_l. The following conditions must hold:

i) No tetrahedron $T_{ijkl}, ijkl \in I_t$ is degenerate. That is, if $ijkl \in I_t$ then p_i, p_j, p_k, and p_l are not coplanar.

ii) The interiors of any two triangles do not intersect. That is, if $ijkl \in I_t$ and $\alpha\beta\gamma\delta \in I_t$ then Int$(T_{ijkl})\cap$ Int$(T_{\alpha\beta\gamma\delta}) = \emptyset$.

iii) The boundary of two triangles can intersect only at a common triangular face.

iv) The union of all the triangles is the domain D $= \cup_{ijkl \in I_t} T_{ijkl}$.

We should point out that condition iii) must hold in the strictest sense, and so tetrahedra joining as shown on the right side of Figure 20.45 are not allowed. The reason for this condition (and all the others) is the same as the reason for the conditions of a triangulation; that is, we eventually wish to be able to define C^0 functions in a piecewise manner over the domain consisting of the union of all tetrahedra. The triangular grid data structure for representing triangulations (illustrated in Figure 20.8) generalizes very nicely to a structure for representing tetrahedrizations. For example, in Figure 20.46, we show a tetrahedrization of the cube into five tetrahedra. We saw earlier in the case of triangulations that once the boundary is specified, the number of triangles comprising the triangulation was fixed, and moreover, we had a simple approach for determining a formula for the number of triangles that existed in the triangulation. This property allowed for the definition of the vectors of angles which lead to the criterion for optimal triangulations and was therefore rather important. It would be nice if everything extended to 3D in a straightforward manner. That is, we would like to say that any polyhedron can be decomposed into tetrahedra and that there is a fixed formula of the following form $N_t = aN_b + bN_i + c$, whereas before, N_b and N_i are the number of vertices on the boundary and interior, respectively. Unfortunately, this is not the case and, in fact, the situation is much worse than that. We saw earlier that any polygon-bounded region can be triangulated using only the vertices of the polygon. This is one of the first areas where matters differ significantly when going from 2D to 3D. It turns

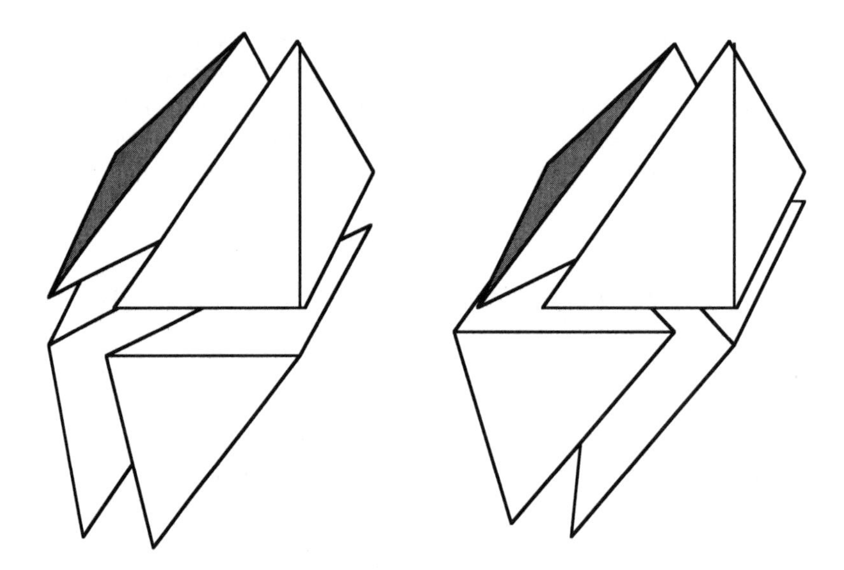

Figure 20.45: The configuration indicated by the diagram on the left is acceptable, while that on the right is not acceptable for a tetrahedrization. It is eliminated by condition iii) listed above.

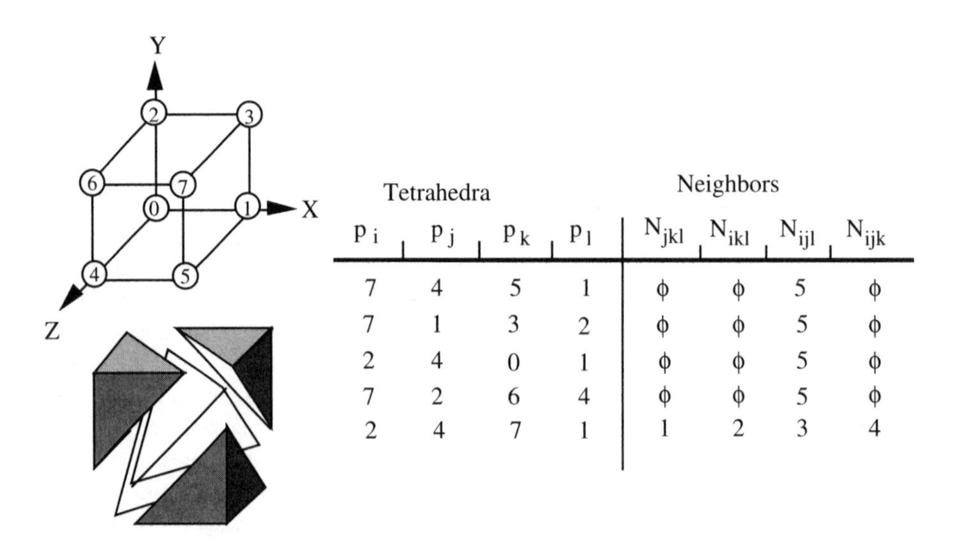

Tetrahedra				Neighbors			
p_i	p_j	p_k	p_l	N_{jkl}	N_{ikl}	N_{ijl}	N_{ijk}
7	4	5	1	ϕ	ϕ	5	ϕ
7	1	3	2	ϕ	ϕ	5	ϕ
2	4	0	1	ϕ	ϕ	5	ϕ
7	2	6	4	ϕ	ϕ	5	ϕ
2	4	7	1	1	2	3	4

Figure 20.46: An example which defines the tetrahedral grid data structure.

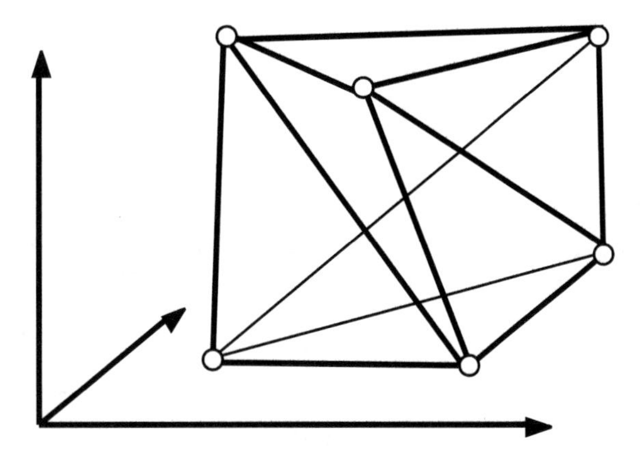

Figure 20.47: The twisted prism of Schoenhardt [221], which cannot be tetrahedrized.

out that not every polyhedron can be tetrahedrized. The example illustrated in Figure 20.47 is originally due to Schoenhardt [221]. It can be visualized as a prism which has been twisted until each face (a quadrilateral comprised of two triangles) has "buckled" inward. Any tetrahedron we form from these vertices must include an edge which lies outside the domain of the "twisted prism," so it is clear that the object cannot be tetrahedrized.

One very basic operation does carry over in a straightforward manner from 2D to 3D—the process of inserting an additional vertex into the interior of an existing tetrahedrization. See Figure 20.48. If the new vertex p lies interior to an existing tetrahedron, say T_{abcd}, then this tetrahedron is simply replaced with the four tetrahedron, $T_{abcp}, T_{abdp}, T_{bcdp}, T_{acdp}$, adding a net increase of three tetrahedra. If the new vertex p lies on the common triangular face of two tetrahedra, then these two tetrahedra are replaced with six new tetrahedra, $T_{abcp}, T_{bcdp}, T_{abdp}, T_{aecp}, T_{ecdp}, T_{aedp}$, resulting in a net increase of four new tetrahedra. This latter aspect of the number of tetrahedra increasing which is different here from the 2D case is that net increase in the number of tetrahedra depends on the actual location of the interior point to be inserted. This observation points out that not only can the number of ways that a data set is tetrahedrized vary, but also the number of tetrahedra can vary. We will illustrate this further with some examples that do not even have interior points.

We have already seen (Figure 20.46) the decomposition of a cube into five tetrahedra. It is also possible to tetrahedrize the cube into six tetrahedra. This is illustrated in Figure 20.49.

It is interesting to note that from the exterior, the tetrahedrization of Figure 20.49 looks exactly the same as that of Figure 20.46 because all external edges are the same. Another interesting connection between these two tetrahedrizations of the cube is that one can be obtained from the other by "swapping" operations, similar to those used in the Lawson algorithm for computing optimal triangulations. Previously, in the case of triangulations, there was the possibility of two triangulations of a convex quadrilateral. The analogous situation in 3D is the tetrahedrization of the region formed by five vertices when two tetrahedra meet at a common triangular face. If the line segment joining the two vertices not on

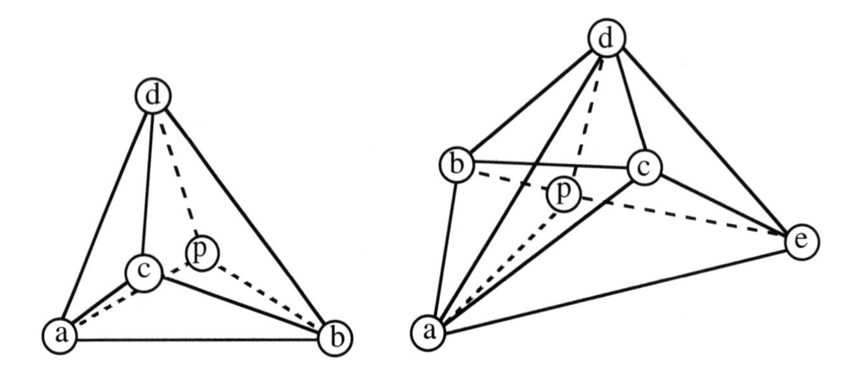

Figure 20.48: Inserting a point interior to an existing tetrahedrization. On the left, the new point is interior to a tetrahedron, and on the right it is on a common face of two tetrahedra.

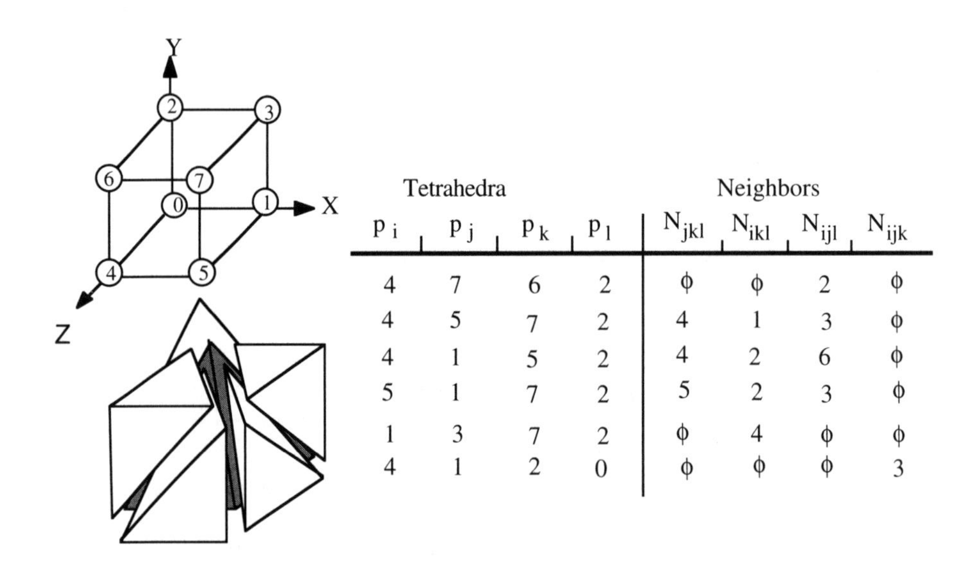

Tetrahedra				Neighbors			
p_i	p_j	p_k	p_l	N_{jkl}	N_{ikl}	N_{ijl}	N_{ijk}
4	7	6	2	ϕ	ϕ	2	ϕ
4	5	7	2	4	1	3	ϕ
4	1	5	2	4	2	6	ϕ
5	1	7	2	5	2	3	ϕ
1	3	7	2	ϕ	4	ϕ	ϕ
4	1	2	0	ϕ	ϕ	ϕ	3

Figure 20.49: A tetrahedrization of the cube into six tetrahedra.

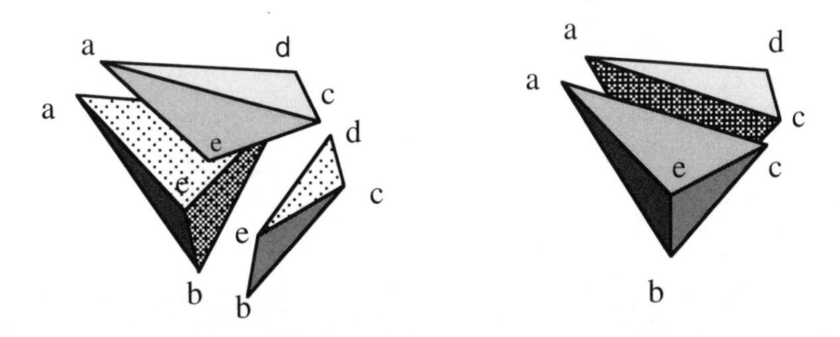

Figure 20.50: Two different tetrahedrizations of five points.

the common face intersects the interior of the common face, then, analogous to the convex quadrilateral case in 2D, there is the possibility of an alternate tetrahedrization. But what is really different from the 2D case is that the number of tetrahedra changes from two to three! This is illustrated in Figure 20.50. This basic operation was applied to the center and upper, back-right tetrahedra of Figure 20.46 to arrive at the tetrahedrization of Figure 20.49.

Another example worth noting in this context is the case where $p_i = (i, i^2, i^3), i = 1, \dots, N$. The (Delaunay) tetrahedrization of the convex hull of this set of points consists of the tetrahedra with vertices p_i, p_{i+1}, p_j, and p_{j+1}, of which there are a total of $((N-2)(N-1))/2$ tetrahedra. Bern and Eppstein [16] point out that this example provides an upper bound on the number of tetrahedra in a tetrahedrization of an N-vertex polyhedron, and that a lower bound is provided by the fact that any tetrahedrization of a simple polyhedron has at least $N - 3$ tetrahedra.

Some Special Tetrahedrizations

Following the pattern established in the earlier sections on triangulations, we first discuss tetrahedrizations related to Cartesian grids followed by tetrahedrizations associated with curvilinear grids. A 3D Cartesian grid (Figure 20.51) involves three monotonically increasing sequences, $x_i, i = 1, \dots, N_x$, $y_j, j = 1, \dots, N_y$ and $z_k, k = 1, \dots, N_z$. The grid points have coordinates (x_i, y_j, z_k) and these points mark out a cellular decomposition of the domain consisting of regular parallelepipeds. Each of these cells can be tetrahedrized in a manner similar to that given for the cube in the previous section. Probably the most popular is the tetrahedrization involving five tetrahedra shown in Figure 20.46. So as to not end up with a nontetrahedrization with problems similar to those shown on the right side of Figure 20.45, it is necessary to "alternate" the tetrahedrization from one cell to the next so that adjoining cells have the same diagonal on the common faces. This alternate tetrahedrization is not really different but is just a rotation of its companion. It is shown in Figure 20.52. Another popular choice is the tetrahedrization shown in the upper-left corner of Figure 20.53. It has the advantage that all of the tetrahedra are the same shape (up to mirror images). Actually, it turns out that there are six different tetrahedrizations of a cube (parallelepiped). See Nielson [183]. We have previously shown pictures of two of them in Figure 20.46 and Figure 20.49. The other four are shown in Figure 20.53.

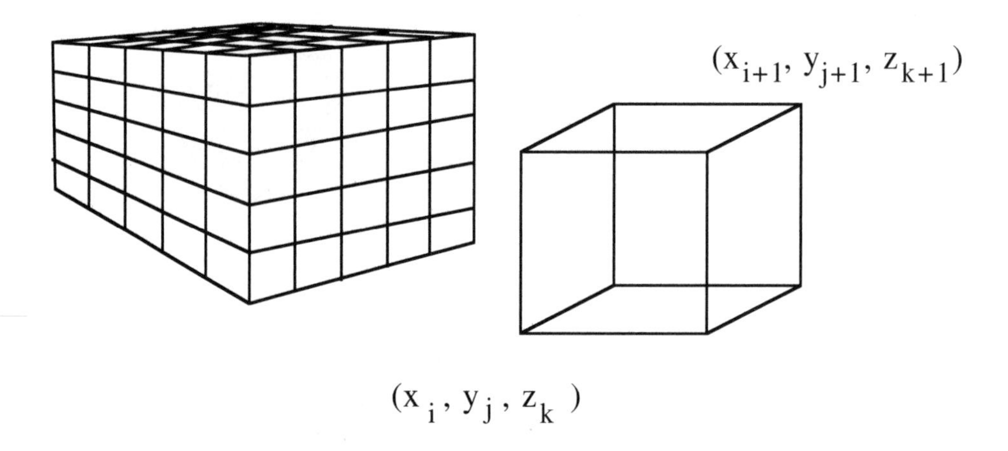

Figure 20.51: Three-dimensional Cartesian grid.

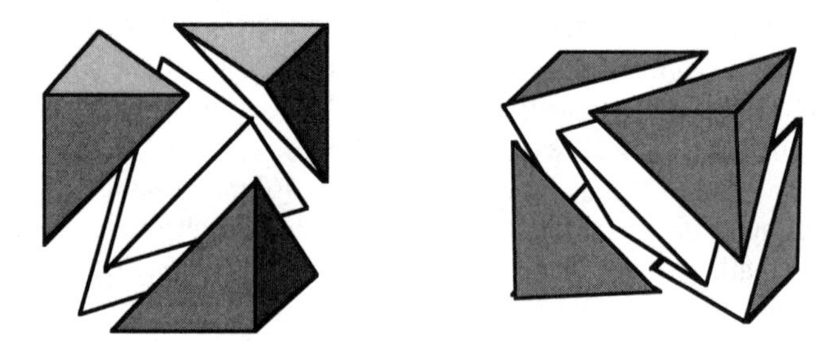

Figure 20.52: The two alternating tetrahedrizations with five tetrahedra of the cell of a 3D Cartesian grid. (One can be rotated to the other.)

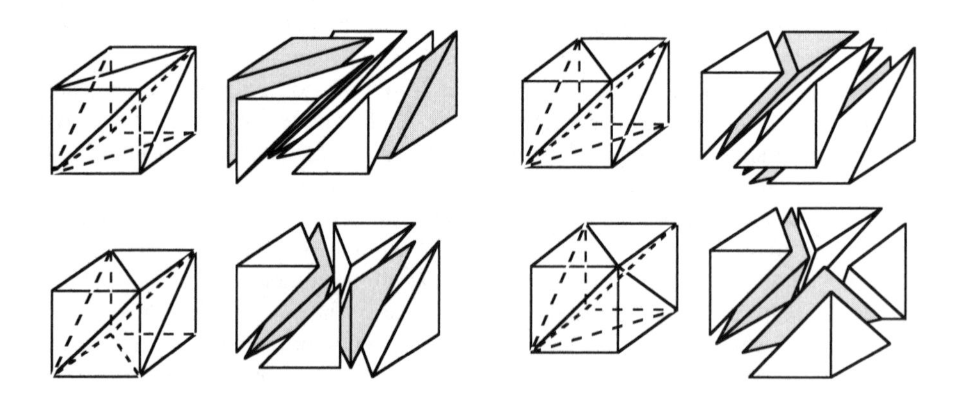

Figure 20.53: Four different tetrahedrizations of the cube, each with six tetrahedra.

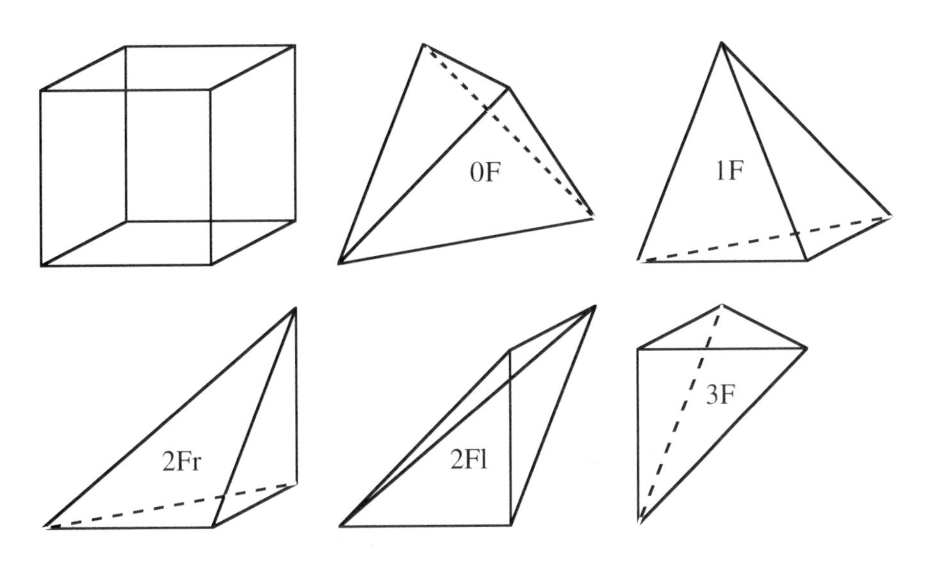

Figure 20.54: The five primitive tetrahedra comprising the tetrahedrizations of the cube.

All six tetrahedrizations of the cube are comprised of five primitive tetrahedra, which are shown in Figure 20.54. We use the names 0F, 1F, 2Fr, 2Fl and 3F for these tetrahedra to indicate the number of exterior faces for each tetrahedra. There are two different primitive tetrahedra with two exterior faces; one is a mirror-image version of the other and so it cannot be rotated to the other. The tetrahedron 0F has volume $1/3$ and all the others have volume $1/6$. During informal discussion we most often use the names 3F = "corner," 2Fr or 2Fl = "right wedge" or "left wedge," 1F = "kite" and 0F = "equi" or "fatboy."

In a joining similar to that shown in Figure 20.50, three 1F tetrahedra can come together to form the same exact shape formed by a 0F and a 3F together. Also a 2Fl and 2Fr together form the same shape as a 1F and a 3F, but two 2Fr's or two 2Fl's cannot share a common face and remain inside a unit cube. There are four tetrahedrizations (each composed of three primitive tetrahedra) of the prism making up half of the cube. They are 3F, 1F, 2Fl; 3F, 1F, 2Fr; 2Fr, 2Fl, 2Fr; and 2Fl, 2Fr, 2Fl. In Figure 20.55 we show the dual graphs of the six tetrahedrizations of the cube. A node is a primitive tetrahedron and an arc is a common triangular face. As expected, in each case the "names" add to twelve.

Each of these six tetrahedrizations has unique and interesting properties. The tetrahedrization of Figure 20.46 and Figure 20.49 both "swap" diagonals on all three pairs of opposing faces. The tetrahedrization shown in the lower right of Figure 20.53 swaps the diagonals of two pair of opposing faces and that of the upper right swaps one pair. The two tetrahedrizations on the left of Figure 20.53 do not swap any diagonals of any opposing faces. The tetrahedrization of the upper left of Figure 20.53 can be realized with three cuts of the entire cube, while the others cannot. This particular tetrahedrization also has the unique property of being composed of only 2F primitives whose faces are all right triangles, and all six of them share the diagonal of the cube as a common edge. This tetrahedrization has been discussed and used widely. It is called the CFK triangulation of the cube after Coxeter [47], Freudenthal [79], and Kuhn [137]. A replacement rule can be

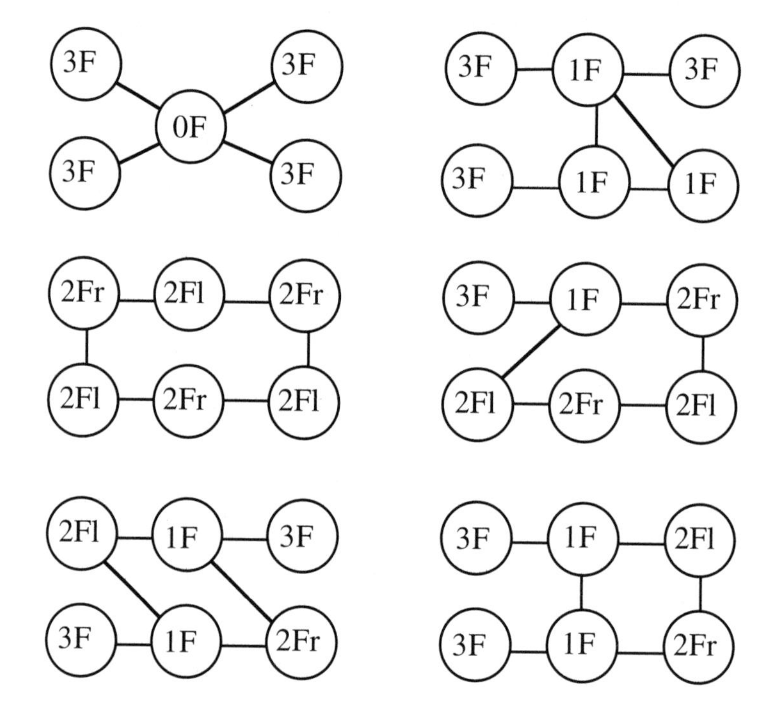

Figure 20.55: The six tetrahedrizations of the cube shown as dual graphs. (These are the only tetrahedrizations of the cube.)

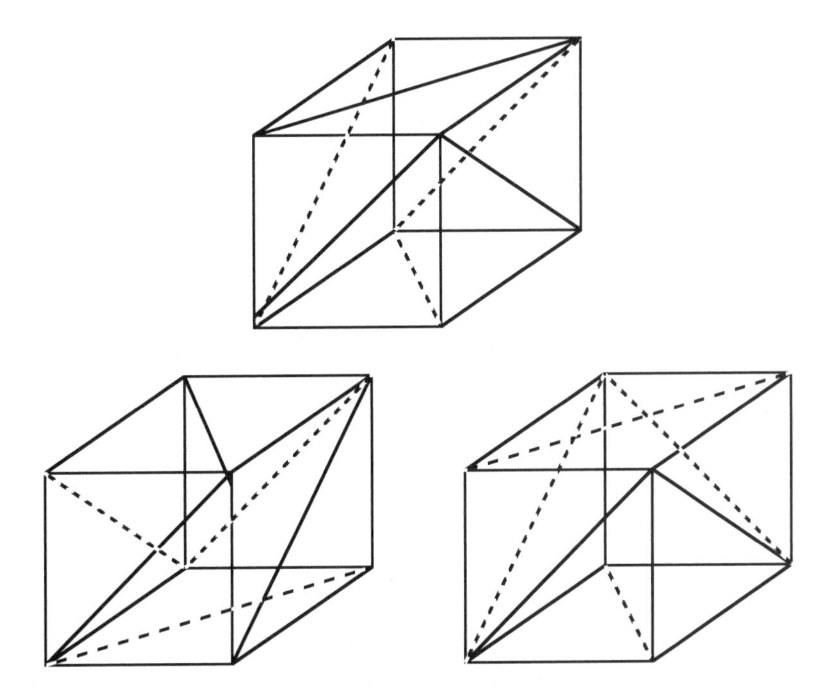

Figure 20.56: Face triangulations that are not consistent with any tetrahedrization of the cube.

used to generate this tetrahedrization. Using the labeling scheme of Figure 20.46, we start with the four vertices $P_{2^i-1}, i = 0, 1, 2, 3$ and replace each vertex V_j, other than V_0 and V_7, with $V_{j+1} + V_{j-1} - V_j$. Explicitly, this will successively generate the six tetrahedra: $p_0p_1p_3p_7$; $p_0p_2p_3p_7$; $p_0p_2p_6p_7$; $p_0p_4p_6p_7$; $p_0p_4p_5p_7$; $p_0p_1p_5p_7$. The CFK triangulation generalizes to n-dimensions as does the "replacement" algorithm for generating the simplicial decomposition.

It is interesting to note that not all possible face triangulations are realized by the six possible tetrahedrizations of the cube. In addition to the five different face triangulations which are realizable (note that two tetrahedrizations have the same face triangulations) there are three others which cannot be realized. They are shown in Figure 20.56. In order to determine these eight unique face triangulations, we start with the $64 = 2^6$ face triangulations and then group them into these eight equivalence classes by rotations.

Theorem: It is impossible to tetrahedrize a cube and yield face triangulations as shown in Figure 20.56.

Proof: We give only the proof for the case in the top center, as the others are similar. We use the same labeling as shown in Figure 20.49. We start with the face 457. Only vertex 0 can be attached to the face 457, which gives the tetrahedron 0457. The internal face 047 must be shared by some other tetrahedron. Any vertex, however, cannot be joined to the face of 457 without violating the conditions of the face triangulations, so this completes the argument.

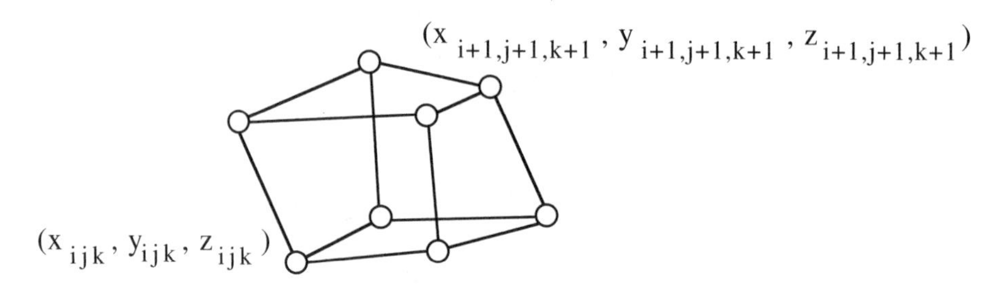

Figure 20.57: Single cell of a 3D curvilinear grid.

Earlier we discussed triangulations related to curvilinear grids. We now take up the topic of tetrahedrization of 3D curvilinear grids. Analogous to the 2D situation, a 3D curvilinear grid is specified by three geometry arrays $x_{ijk}, y_{ijk}, z_{ijk},\ i = 1, \ldots, N_x;\ j = 1, \ldots, N_y;\ k = 1, \ldots, N_z$. In the 2D case a cell C_{ij} consisted of the quadrilateral with vertices $(x_{ij}, y_{ij}), (x_{i+1,j}, y_{i+1,j}), (x_{i,j+1}, y_{i,j+1}), (x_{i+1,j+1}, y_{i+1,j+1})$, and the cells serve as a decomposition of the domain.

In the 3D case, matters are not as straightforward as we might expect, and there are some areas where we need to be concerned. These have mainly to do with just exactly what comprises a cell. In 3D the cell C_{ijk} has the eight vertices $(x_{abc}, y_{abc}, z_{abc}), a = i, i+1, b = j, j+1, c = k, k+1$, but there is not always a consistent definition for the cell boundaries. We mention briefly some possible choices. If the geometry arrays are constrained so that each collection of four vertices of the six "faces" of the cells are coplanar, then an obvious choice for the cell boundaries is this common planar quadrilateral. In this case the cells are hexahedron, and it is relatively easy to determine whether or not an arbitrary point (x, y, z) is in a particular cell or not. Often this planarity condition does not hold, and cell boundaries are taken to be the parametrically defined (hyperbolic) surface obtained by substituting 0 or 1 for any of the parameter values s, t, u in the following trilinear mapping:

$$\begin{aligned} C_{i,j,k}(s,t,u) &= (1-s)(1-t)(1-u)P_{i,j,k} + (1-s)(1-t)uP_{i,j,k+1} \\ &\quad +(1-s)t(1-u)P_{i,j+1,k} + (1-s)tuP_{i,j+1,k+1} \\ &\quad +s(1-t)(1-u)P_{i+1,j,k} + s(1-t)uP_{i+1,j,k+1} \\ &\quad +st(1-u)P_{i+1,j+1,k} + stuP_{i+1,j+1,k+1} \end{aligned}$$

where

$$P_{i,j,k} = (x_{i,j,k}, y_{i,j,k}, z_{i,j,k})$$

Given a point (x, y, z) in the cell C_{ijk}, the value (s, t, u) which associates with it via the trilinear mapping is called the corresponding *computational coordinate*. In fact, in order to determine whether or not an arbitrary point is in this type of cell or not requires that we solve the three nonlinear equations which represent this association. This can be a considerable problem from a computational point of view. Most methods use some heuristics to obtain an initial approximation for some type of Newton's method. Another choice for the cell boundaries, in the event the four vertices of a face are not coplanar,

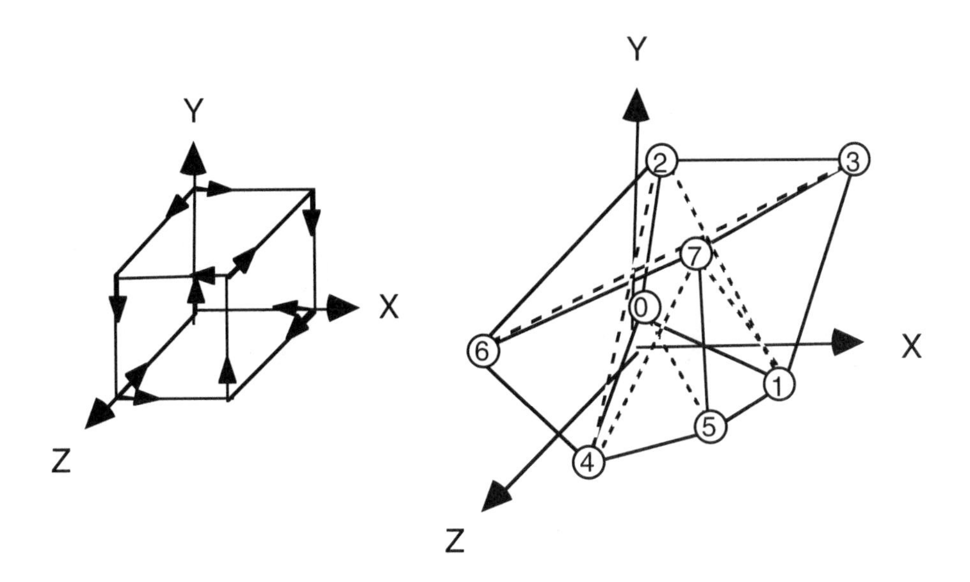

Figure 20.58: A curvilinear grid cell (polyhedron) that cannot be tetrahedrized.

is to choose them to be piecewise planar. That is, a diagonal edge is selected and the boundary between the two cells consists of the two triangles which result. Often the cell would be further decomposed into tetrahedra, thus leading to an overall tetrahedrization of the curvilinear grid. We should point out that not all choices for the diagonals can lead to a tetrahedrization of the cell. In order to be specific about this, consider the cell illustrated in Figure 20.58. This cell was created from a unit cube by cutting notches in the faces so as to force the diagonal edges $p_2p_7, p_4p_1, p_3p_5, p_3p_0, p_0p_6, p_6p_5$ to be exterior to the cell. If the depth of the notches is e, then this results in the points $p_0 = (0, e, 0)$, $p_1 = (1 - e, 0, e)$, $p_2 = (e, 1, e)$, $p_3 = (1, 1 - e, 0)$, $p_4 = (e, 0, 1 - e)$, $p_5 = (1, e, 1)$, $p_6 = (0, 1 - e, 1)$, $p_7 = (1 - e, 1, 1 - e)$. Note that p_6, p_3, p_4 and p_1 all lie in the plane $x + z - 1 = 0$ and p_2, p_7, p_0 and p_5 are in the plane $x - z = 0$.

Theorem: The polyhedron of Figure 20.58 cannot be tetrahedrized.

Proof: Consider the triangle face with vertices p_6, p_4, and p_7. In any tetrahedrization, this face must be joined to some vertex to form a tetrahedron. By considering the remaining five vertices p_5, p_0, p_2, p_1, and p_3 we find that only p_3 would not lead to a tetrahedron with an edge which is outside the cell. If the tetrahedron p_6, p_4, p_7, and p_3 is included in the list of tetrahedra, then the interior triangle face $p_3p_4p_7$ must connect to another vertex (besides p_6) to form a tetrahedron. But a consideration of each of the possible vertices p_5, p_1, p_2, and p_0 each lead to an edge which is exterior to the cell and this concludes the argument.

We conclude this discussion on the tetrahedrization of the cells of a curvilinear grid by pointing out that some hexahedra will decompose into seven tetrahedra. Consider the cell of Figure 20.57 and let the six faces be planar, but assume that the four diagonal points p_{ijk}, $p_{i+1,j+1,k+1}$, $p_{i,j,k+1}$ and $p_{i+1,j+1,k}$ are not coplanar, so that they will form

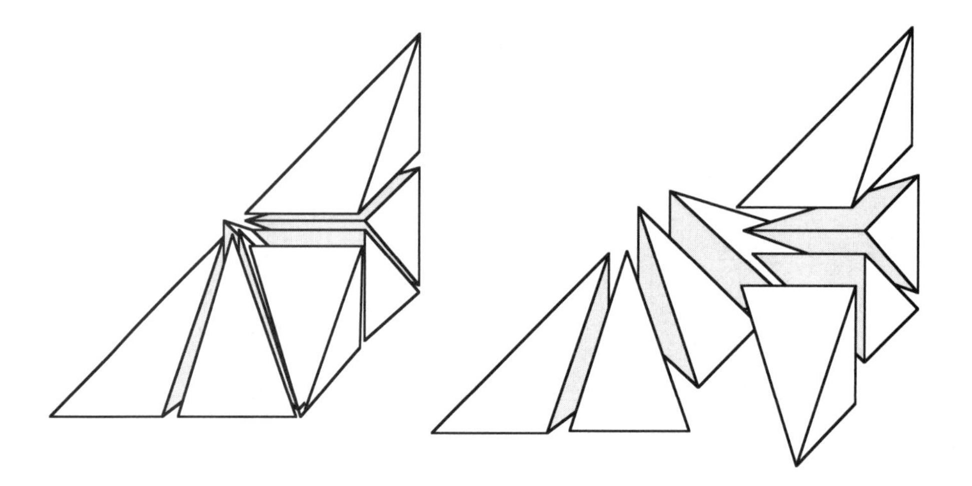

Figure 20.59: Nested tetrahedral subdivision analogous to that of Figure 20.16.

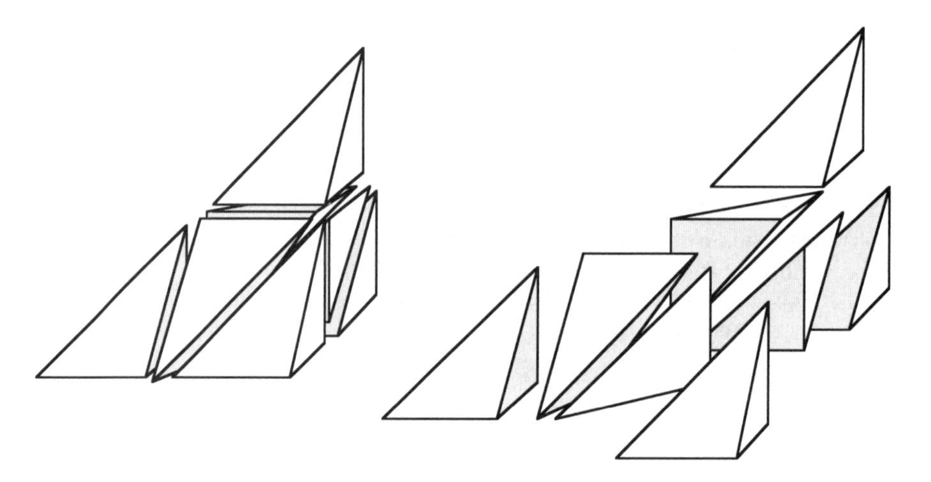

Figure 20.60: Symmetric nested tetrahedral subdivision.

a tetrahedron. Remove this tetrahedron, leaving two prisms with two planar quadrilateral faces which can each be decomposed into three tetrahedra. We should point out that we have observed cases where this decomposition was the Delaunay tetrahedrization.

In Section 20.2.1 we described two different approaches leading to nested subdivision triangulations and pointed out their potential value in multiresolution approximations. These both have analogs in 3D and are shown in Figures 20.59 and 20.60, respectively. The first one is based upon recursive subdivision and the second one is called "symmetric" subdivision and is related to the CFK tetrahedrization of the cube [170]. It is composed of six 2Fr's and two 2Fl's and is the same shape and twice the size of one 2Fr.

It should be noted that if primitive tetrahedra of the shape shown in Figure 20.61 are assembled as in Figure 20.60, then we obtain a composite tetrahedron which is twice the size

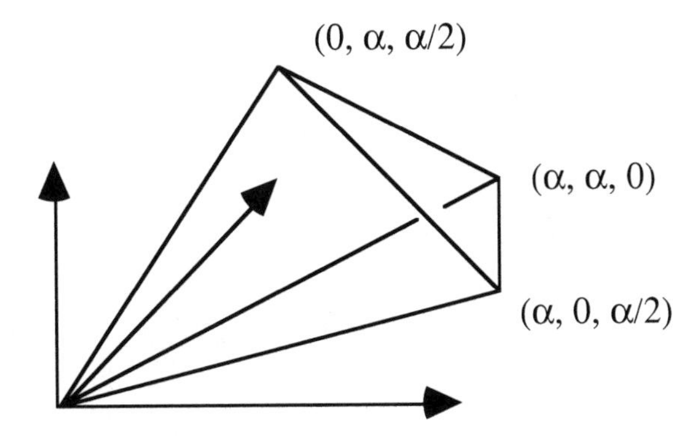

Figure 20.61: A tetrahedron that can be tetrahedrized into eight tetrahedra, each of which are the same shape as the original yet half the size.

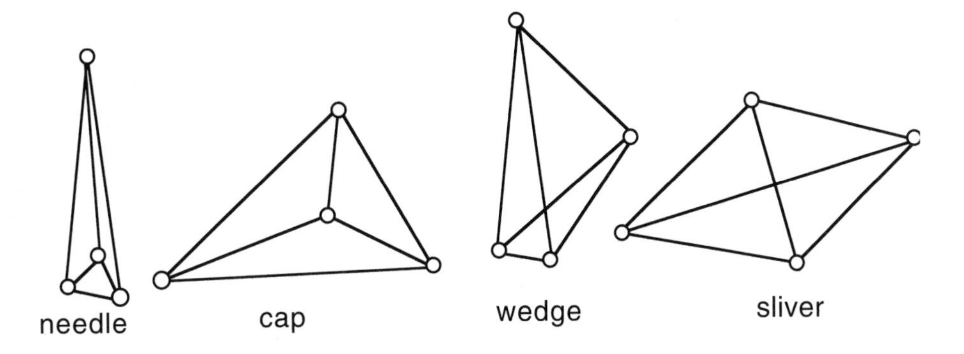

Figure 20.62: Examples of poorly shaped tetrahedra.

and exactly the same shape as the primitive tetrahedron. This particular tetrahedrization of tetrahedra is related to the Delaunay tetrahedrization of the BCC lattice, which is the union of the lattices $\{(i, j, k) : i, j,$ and k are integers$\}$ and $\{(i + \frac{1}{2}, j + \frac{1}{2}, k + \frac{1}{2}) : i, j,$ and k are integers$\}$. See also Senechal [229] for a discussion of tetrahedra that can be decomposed into similar tetrahedra.

20.3.2 Algorithms for Delaunay Tetrahedrizations

Analogous to the examples of Figure 20.19, examples of poorly shaped tetrahedra are shown in Figure 20.62. The sliver has small dihedral angles, but need not have any small planar angles. Several measures of the quality of tetrahedrizations have been proposed. See Baler [12] and Field [86]. Take as an example the ratio of the inradius (radius of inscribed sphere) and the circumradius. The problem here is that there is no apparent way to order the collection of all tetrahedrizations of a point set. The approach of lexicographically ordering the associated vectors of angles, as we described in Section 20.2.2, does not extend to 3D because the number of tetrahedra in a tetrahedrization is not necessarily fixed. Nev-

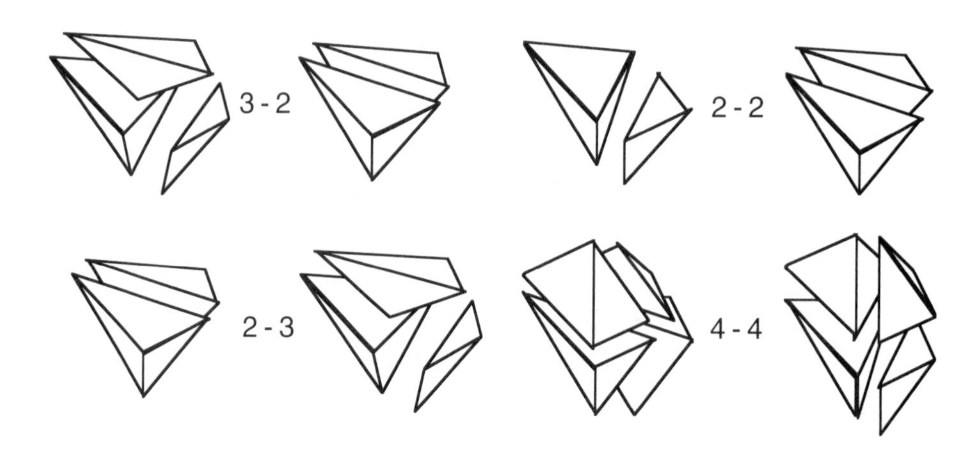

Figure 20.63: Different cases of swapping for 3D version of Lawson's algorithm.

ertheless, the Delaunay tetrahedrization of the convex hull which is dual to the Dirichlet tessellation is well defined (in the absence of neutral cases where points lie on a common sphere), so the remainder of this section is devoted to a discussion of the extension of the previously discussed 2D algorithms for computing the Delaunay triangulations to the case of 3D tetrahedrizations.

Extension of Lawson's Algorithm (Incremental Flipping): It is possible to extend this algorithm to 3D, but the extension is not as simple as one might expect. The first major difference that one encounters is the character of the basic swapping step. In 2D we take an edge and consider the quadrilateral formed by the two triangles which share this edge. If the quadrilateral is convex, we can swap the diagonal if this step moves us closer to the optimal solution, which can easily be determined by applying the circle inclusion test. Two triangles are replaced by two other triangles. But the analogous steps in 3D can lead to a situation where the two tetrahedra sharing a face can be replaced with three tetrahedra. See Figure 20.63 for an example.

Joe [122] showed that if the points are inserted in a particular manner, then incremental flipping will lead to the optimal Delaunay tetrahedrization. Edelsbrunner and Shah have generalized these results [72]. See also [82]. Software based upon these ideas is provided by the Software Development Group at the National Center for Supercomputing Applications and is available at the World Wide Web site:
http://www.ncsa.uiuc.edu/SDG/Brochure/Overview/ALVIS.overview.html.

Extensions of the Algorithm of Green and Sibson: There does not seem to be an apparent method of extending this type of algorithm to 3D. The algorithm is dependent upon the "contiguity list," and here lies the difficulty to extend to 3D. We included this algorithm in our selection of 2D algorithms so that this very point could be made. Some concepts extend easily to 3D and others do not.

Bowyer's Algorithm for 3D: It is a straightforward exercise to extend Bowyer's 2D algorithm to 3D. In fact, the original paper of Bowyer [21] describes the algorithm for arbitrary dimensions. Bowyer also mentions that with some care, the algorithm can be extended to other domains. In [164] there is a brief discussion of Bowyer's algorithm along with some code.

Watson's Algorithm for 3D: The original description of Watson's algorithm applies to arbitrary dimension. In Watson's paper [254] results for 2, 3, and 4 dimensions are reported. Information on implementing this algorithm in 3D is given by Field in [86] and [87]. It is also the basis for the 3D algorithms discussed in [29].

Embedding/Lifting Algorithms for 3D: Software for computing general dimension convex hulls and Delaunay tetrahedrizations, based on the relationship mentioned earlier in Section 20.2.2, are provided by the Geometry Center, University of Minnesota at the WWW site:
http://freeabel.geom.umn.edu/software/download/qhull.html.

20.3.3 Visibility Sorting of Tetrahedra

We first give a motivation for the definition of and the need for a visibility sort. We use the example of volume rendering, which is a means of graphing (visualizing) a density function (cloud) $\delta(x, y, z)$ defined over a 3D domain (which is often a cube). A viewpoint V is selected along with a projection plane. A rectangular portion of the projection plane is subdivided into a rectangular array of subrectangles which associate directly with the pixels of an image to be generated. The value for each pixel is defined by

$$F(i,j) = \int_0^D \delta(s)C(s)e^{-\int_s^D \delta(u)\,du}\,ds + F_0 e^{-\int_0^D \delta(u)\,du} \tag{20.3}$$

where the integral is taken along the ray emanating from the viewpoint and passing through the center of the subrectangle associated with the pixel at location (i, j), F_0 is the background intensity, and D is a distance along the ray sufficiently large so that the ray completely passes through the domain of interest. The function C, also defined over the same domain as δ, is called the color function and governs the color of light emanating (by reflection, say) from a point within the density cloud. In actual application the integrals are approximated by numerical schemes based upon sampled values of the integrand. The sample values are often obtained by some simple interpolation into the cells covering the domain. And these cells are often a result of the positions where δ has been measured. If we let $0 = x_0 < x_1 < x_2 < \cdots < x_{n-1} < x_n = D$ be the distances from the viewpoint to each sampled value along the ray, then the upper Riemann sum approximation to this integral is

$$F_n = \sum_{i=0}^{n} \Delta x_i \delta(x_i) C_i \prod_{j=i+1}^{n} t_j, \tag{20.4}$$

where $C_i = C(x_i), t_j = e^{-\Delta x_i \delta(x_i)}$ and $\Delta x_i = x_i - x_{i-1}$. This discrete approximation can be computed by the *compositing process*

$$F_i = t_i F_{i-1} + I_i, \tag{20.5}$$

where $I_i = \Delta x_i \delta(x_i) C_i$.

Another way to view this compositing process is as a simple model of transparency, where an object of thickness Δx_i attenuates the incoming light intensity F_{i-1} by the factor t_i, and this object emits light of intensity I_i. Algorithms which accumulate these values into a frame buffer (with each location holding the value for a pixel) can either be image-space-oriented or object-space-oriented. Image-space algorithms proceed along the lines of our development here and accumulate all contributions for a pixel along a particular ray. Object-space algorithms compute exactly the same values but the calculations are done in a different order. These algorithms sequentially process each cell by accumulating into the proper location of the frame buffer all contributions of a particular cell. Due to the nature of the compositing process, it is mandatory that these accumulations be done in the proper order. It is this latter approach which motivates the definition of and need for visibility sorting in this context.

Definition of Visibility Order: Let T and T' be tetrahedra of a tetrahedrization and let V be the center of perspective projection. If there is a ray emanating from V which intersects T' before T, then T is said to precede T' and we write $T < T'$.

The purpose of a visibility sort is to find a linear ordering of all of the tetrahedra of a tetrahedrization so that the ordering relation is never violated.

Definition of Visibility Ordering: A visibility ordering of a tetrahedrization is a sequence, $n_1, n_2, \ldots, n_T$ which has the property that whenever $T_{n_i} < T_{n_j}$ then $i < j$. The implication of the definition of visibility ordering for splatting or object-space traversal algorithms for volume rendering is that a tetrahedron T must be processed (sampled and composited into the frame buffer) before T' whenever $T < T'$.

A couple of items should be noted at this point. The relation of visibility order is not, in the strict mathematical sense, a partial ordering. A partial ordering is required to be i) transitive: $x < y,\ y < z$ implies $x < z$; ii) antisymmetric: $x < y$ and $y < x$ implies $x = y$; and iii) reflexive: $x < x$. It is entirely possible that a visibility order could not exist at all due to the presence of cycles as shown in Figure 20.64.

Knuth [136] has discussed in some detail (including MIX programs) the topological sort algorithm as a means of "embedding a partial order in a linear order." A linear ordering is a partial ordering where either $x < y$ or $y < x$ for all x, y. Even though this does not strictly apply in the context of a general tetrahedrization, the basic ideas (mainly due to the manner in which it is described) are very useful for developing visibility-sorting algorithms for specific applications, so we include a description of the topological sort algorithm here.

Topological Sort Algorithm: The topological sort algorithm as described by Knuth [136] starts with a directed, acyclic graph (DAG). The DAG can be represented with a diagram using nodes and arrows. See Figure 20.65. The nodes represent the elements of the set to be ordered, and an arrow from node x to node y represents the relation of the partial ordering, $x < y$. The algorithm is simple. Any node that has no incoming arrow is

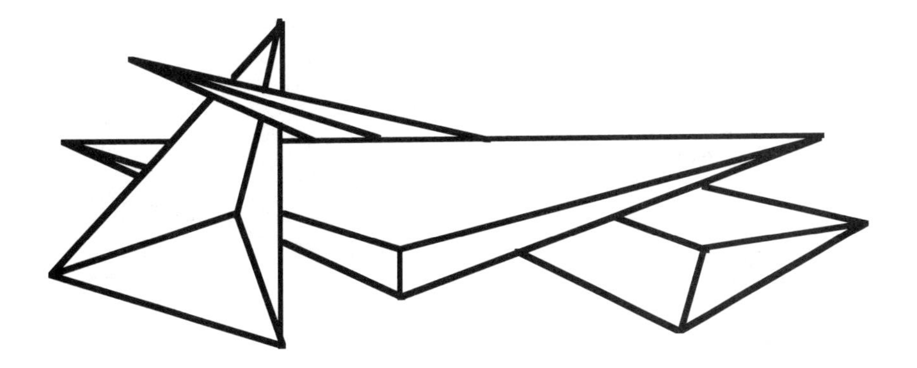

Figure 20.64: An example of three tetrahedra that cannot be visibility ordered.

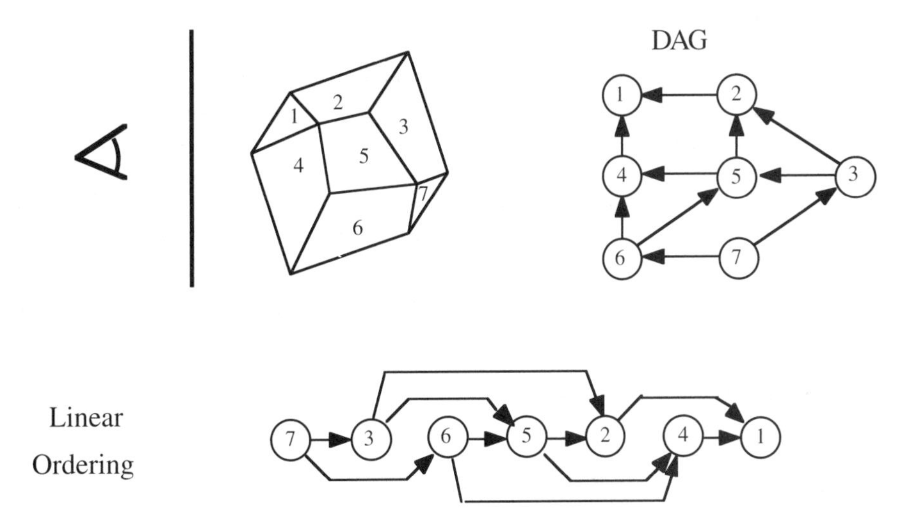

Figure 20.65: An example of the topological sort algorithm.

removed from the DAG (with all of its attached arrows) and placed in the linear ordering. This process is repeated until the DAG is empty. It is easy to prove (left to the reader) that if the DAG represents a partial ordering, a linear ordering will always be produced by this algorithm.

Max [166] has discussed the application of the ideas of the topological sort algorithm to the problem of producing a visibility sort for a cellular decomposition of a domain. Max defines the order relation in the following way. The DAG contains an arrow for each face common to two cells x and y. The arrow is directed from x to y if the viewpoint is on the same side of the face as x, meaning that y must be processed before x. Max mentions that the topological sorting algorithm will be successful "if every ray through the data volume intersects it in a single sequence of adjacent cells." Of course, if the cell complex contains cycles (see Figure 20.64), then a visibility sort is not possible. Williams [257] discusses

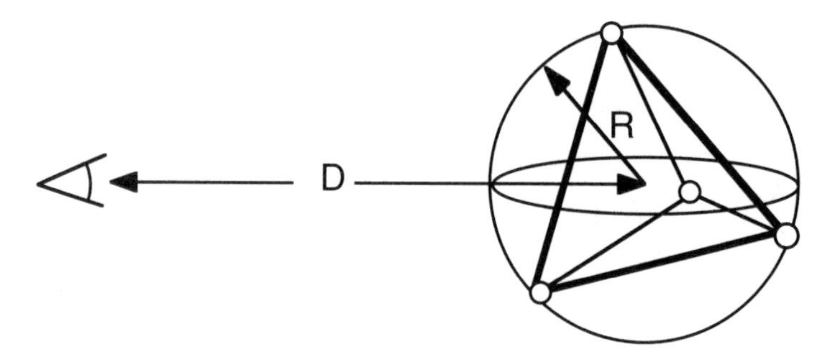

Figure 20.66: Elements of the definition of the power of a tetrahedron.

similar algorithms applied to a very general cellular decomposition which may contain empty cavities.

We conclude this section with some rather interesting properties about the special case of the Delaunay tetrahedrization of the convex hull of a collection of 3D points. The *power* of a tetrahedra is defined as $D^2 - R^2$, where D is the distance to the viewpoint from the center of the circumsphere of the tetrahedron and R is the radius of the circumscribing sphere. See Figure 20.66. A visibility sort can be accomplished by a simple sort based upon the power. This property is covered in [69] and used by Max, Hanrahan, and Crawfis [167]. We caution the reader that this approach breaks down in the presence of neutral cases where possibly several tetrahedra have the same power (as in the case of decomposing the cube). One additional interesting observation in this context is that a sort based upon the power of the tetrahedra does not require the neighborhood information that is required for the algorithms using the ideas of topological sorting. Another method which does not use adjacency information is described by Stein, Becker, and Max [240].

20.3.4 Data-Dependent Tetrahedrizations

Lee [148] has investigated the topic of data-dependent tetrahedrizations. This work generalizes from 2D to 3D the ideas and techniques of [67] and [225]. Similar to the algorithms of [225], simulated annealing is used. The initial tetrahedrization is the Delaunay tetrahedrization of the convex hull of the independent data site locations. Local swapping of tetrahedra is performed based upon random values compared to an annealing schedule and a cost function. This "randomness" of the simulated annealing approach allows the algorithm to escape local extrema of the cost function. Local swapping for 2D simply involves the choice of one or the other of the diagonals of a quadrilateral. In 3D the situation is more complex. There are four cases shown in Figure 20.63 which are the same as those used in the 3D version of Lawson's algorithm. In the first case, three triangles are swapped for two. The second case is the reverse of the first—two tetrahedra are replaced by three. The third case is where two triangles are on the boundary of the convex hull and the two tetrahedra can be swapped for two other tetrahedra. In the last case four tetrahedra are swapped for four other tetrahedra.

In Section 20.2.4, we described the cost function used by Dyn, Levin, and Rippa [67]. Analogous to these cost functions for 2D, Lee [148] uses the following criterion for 3D:

Method	RMS Error
Delaunay	.007475
Difference in Gradient	.005445
Jump in Normal Derivative	.004361

Figure 20.67: Errors for the piecewise linear interpolant using different tetrahedrization.

Gradient Difference: Let T_1 and T_2 be two tetrahedra with a common triangular face. Let G_1 be the gradient of the linear function which interpolates the data at the four vertices of T_1, and let G_2 be the similar gradient for the linear interpolant of T_2. The gradient difference is defined as $||G_1 - G_2||$.

Jump in Normal Direction Derivatives: Let $L_1(x, y, z) = a_1x + b_1y + c_1z + d_1$ be the linear function which interpolates to the data at the four vertices of T_1 and let $L_2(x, y, z) = a_2x + b_2y + c_2z + d_2$ be the similar function for T_2. Let $N = (n_x, n_y, n_z)$ be the normal (normalized) of the common triangular face of T_1 and T_2. $D_1 = a_1n_x + b_1n_y + c_1n_z$ is the directional derivative of L_1 in the direction of N. $D_2 = a_2n_x + b_2n_y + c_2n_z$ is the analogous value for T_2. The jump in normal direction criterion is $|D_1 - D_2| = |(a_1 - a_2)n_x + (b_1 - b_2)n_y + (c_1 - c_2)n_z)|$.

Some example results reported by Lee [148] are repeated here in Figure 20.67. This example involves a test function, $F(x, y, z) = (\tanh(9y - 9x - 9z) + 1)/9$, which provides the dependent data. The piecewise linear interpolant over the tetrahedrization is compared to the test function. The RMS errors are based upon evaluations of the functions and this approximation over a $20 \times 20 \times 20$ Cartesian grid. The dependent data site locations are taken to be 1000 random points in the unit cube.

In Figure 20.68 are some graphs which can be considered as 3D analogs of the graphs shown in Figure 20.31 of Section 20.2.4. Similar to the 2D case, the data-dependent tetrahedrization involves some badly shaped tetrahedra. This is the cost of having an optimal (or nearly optimal) piecewise linear approximation.

20.3.5 Affine Invariant Tetrahedrizations

In this section we extend the results of Section 20.2.5 on affine invariant triangulations to that of affine invariant tetrahedrizations. Prior to discussing the characterization and computation of this type of tetrahedrization, we make some comments about the need for such a tetrahedrization over and above those reasons for the 2D case. It appears that as the dimension of the independent data increases, our need to be concerned about lack of affine invariance also increases.

One source of 3D independent data is the case of time-varying 2D data. In some cases the data measurement locations might stay fixed over time and in some cases they may vary over time. Let us say, for example, that we have a vector field which is known (by means of a numerical simulation) at the locations of a 2D curvilinear grid $(x_{ij}, y_{ij}), i =$

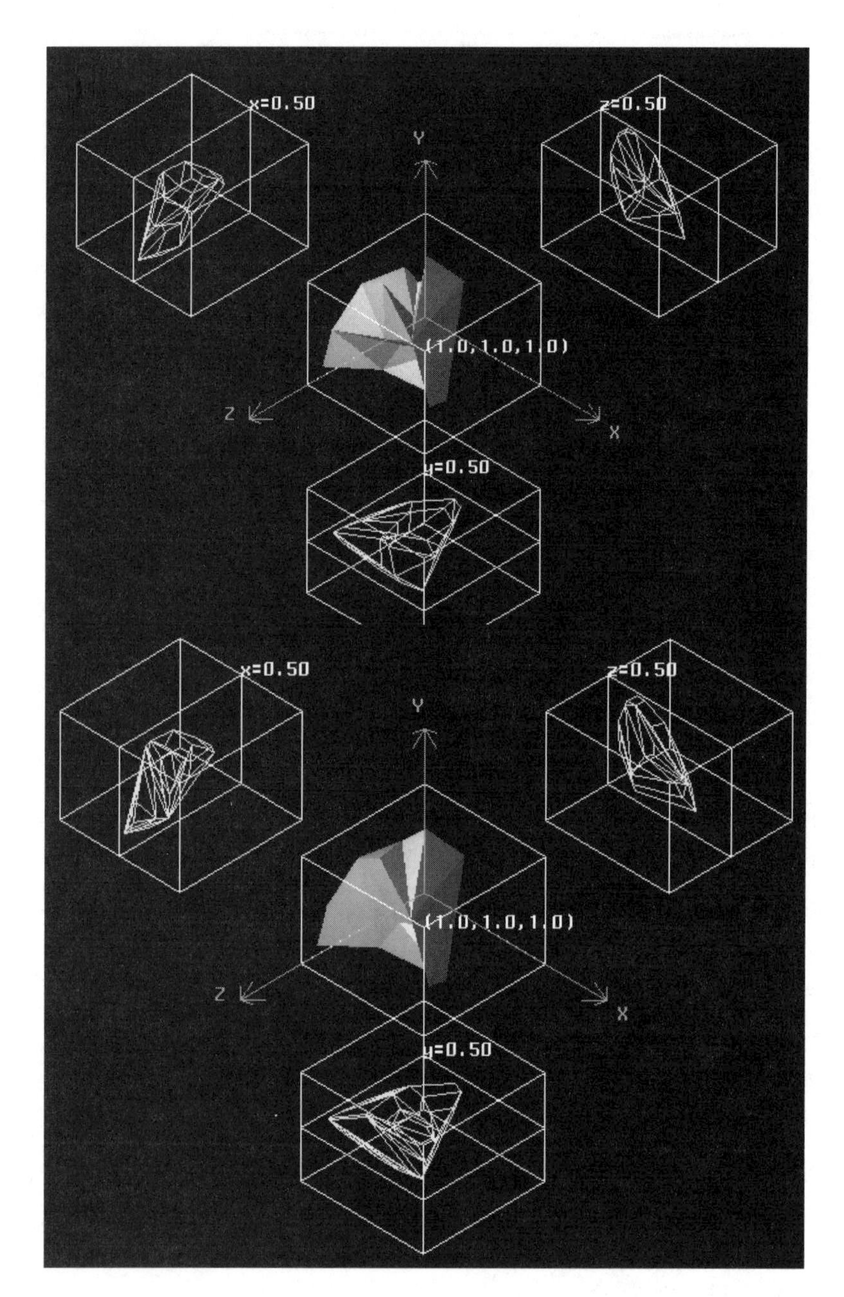

Figure 20.68: Data-dependent tetrahedrization compared to the Delaunay tetrahedrization. See Color Plate 104.

$1, \ldots, N_x;\ j = 1, \ldots, N_y$. As time proceeds, the vector field varies, but the dependent data site locations stay fixed. So in this case, we have data which can be represented as $(V_{ijk}; x_{ij}, y_{ij}, t_k),\ i = 1, \ldots, N_x,\ j = 1, \ldots, N_y,\ k = 1, \ldots, N_t$. If the definition of a modeling function $F(x, y, t)$, designed to interpolate the data, $F(x_{ij}, y_{ij}, t_k) = F_{ijk}$, is based upon a tetrahedrization of the 3D independent data (x_{ij}, y_{ij}, t_k), then this model will not necessarily be affine invariant, and the units used to measure and represent the physical coordinates and time could have an effect on the modeling function $F(x, y, t)$ and subsequently an effect on the visualization and analysis. The same problem could also occur for a time-varying vector field over a curvilinear grid which also varies over time—that is, data of the type $(F_{ijk}; x_{ijk}, y_{ijk}, t_k),\ i = 1, \ldots, N_x,\ j = 1, \ldots, N_y,\ k = 1, \ldots, N_t$. In general, any tetrahedrization of the independent data of (F_{ijk}, x_i, y_j, z_k), where the choice of the units of measurement used for the independent data could lead to a nonuniform scaling, could have the problem of being dependent on the choice of the units used. If each of the variables use the same units there will be no problems of this type, because a scale transformation of the form $x \leftarrow ax,\ y \leftarrow ay,\ z \leftarrow az$, where the scale change is uniform for each variable, will not affect the tetrahedrization. It is only the nonuniform scaling $x \leftarrow ax,\ y \leftarrow by,\ z \leftarrow cz$ which creates the problem. An example of a scale change affecting the tetrahedrization is shown in Figure 20.69. Here there are 10 data points. In the right image, the data has been scaled in the y variable by a factor of 2. Not only does the tetrahedrization change, but even the number of tetrahedra changes. The Delaunay tetrahedrization of the original 10 data points has 18 tetrahedra and the scaled data has 13 tetrahedra.

We now describe the 3D version of the affine invariant norm, which leads (by way of the Dirichlet tessellation) to an affine invariant tetrahedrization. Actually, we can define it so that it is clear what the generalization is for any dimension. Let

$$\|(x, y, z)\|_V^2 = (x, y, z)(VV^*)^{-1} \begin{pmatrix} x \\ y \\ z \end{pmatrix}$$

where V is the $3 \times N$ matrix of translated data values

$$V = \begin{pmatrix} x_1 - \mu_x & x_2 - \mu_x & \ldots & x_N - \mu_x \\ y_1 - \mu_y & y_2 - \mu_y & \ldots & y_N - \mu_y \\ z_1 - \mu_z & z_2 - \mu_z & \ldots & z_N - \mu_z \end{pmatrix}.$$

As with the 2D case, there are some different approaches to modifying an existing tetrahedrization procedure. Probably the simplest is to preprocess the data with the transformation given by the lower triangular matrix, $L(V)$ which results from the Cholesky decomposition of $(VV^*)^{-1}$

$$L(V)L(V)^* = (VV^*)^{-1}.$$

Explicitly in the 3D case, we use the transformed data

$$\begin{aligned} X_i &= l_{11}x_i + l_{21}y_i + l_{31}z_i \\ Y_i &= l_{22}y_i + l_{32}z_i \\ Z_i &= l_{33}z_i \end{aligned}$$

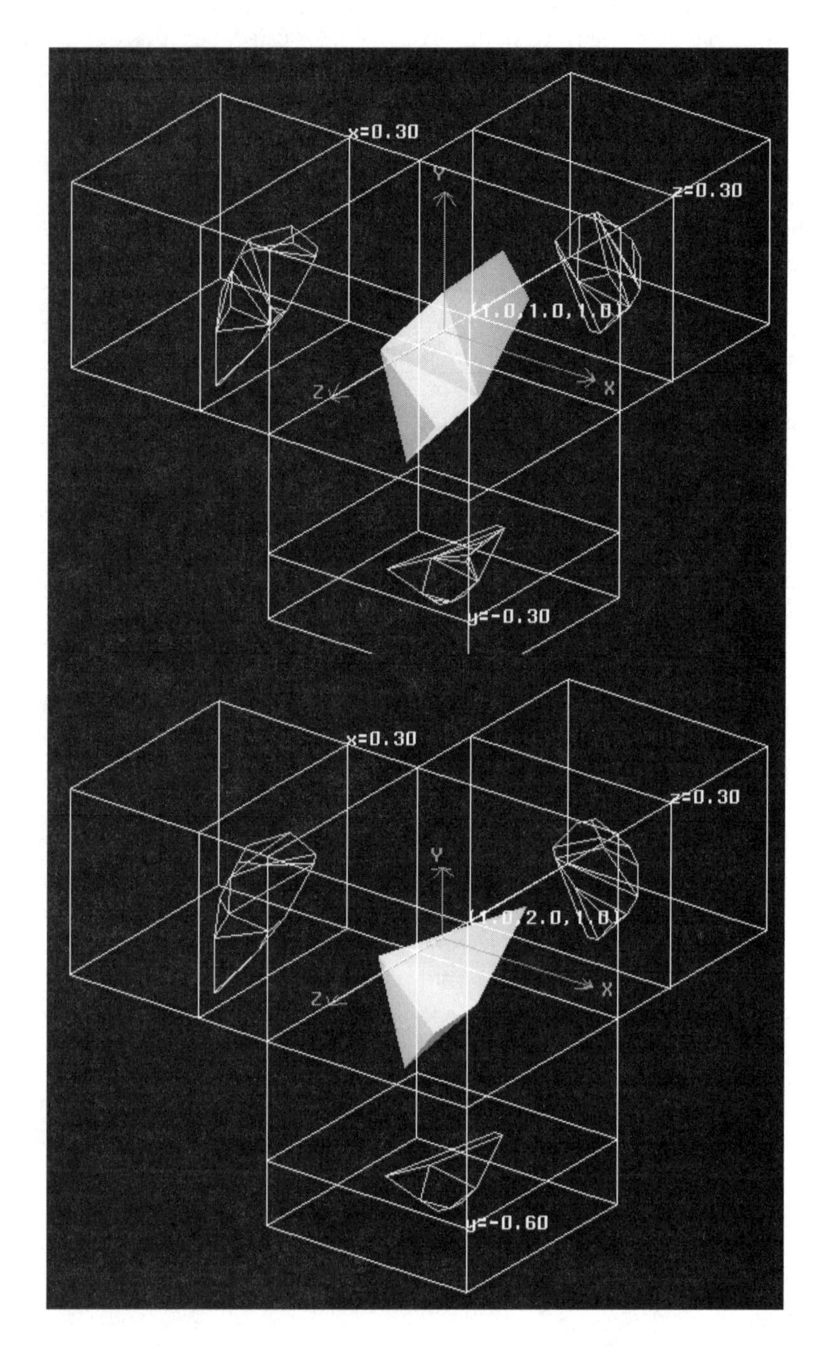

Figure 20.69: Delaunay tetrahedrization of 10 data points and a scaled version of the same data points. See Color Plate 105.

where

$$l_{11} = \sqrt{a_{11}}, \quad l_{21} = \frac{a_{11}}{l_{11}}, \quad l_{31} = \frac{a_{13}}{l_{11}},$$

$$l_{22} = \sqrt{a_{22} - (l_{21})^2}, \quad l_{32} = \frac{a_{33} - l_{21}l_{31}}{l_{22}},$$

$$l_{33} = \sqrt{a_{33} - (l_{31})^2 - (l_{32})^2}$$

$$\begin{aligned} A &= (a_{ij}) = (VV^*)^{-1} = L(V)L(V)^* \\ &= \frac{1}{\det}\begin{pmatrix} \Sigma_y^2\Sigma_z^2 - \Sigma_{yz}^2 & -(\Sigma_{xy}\Sigma_z^2 - \Sigma_{xz}\Sigma_{yz}) & \Sigma_{xy}\Sigma_{yz} - \Sigma_{xz}\Sigma_y^2 \\ -(\Sigma_{xy}\Sigma_z^2 - \Sigma_{xz}\Sigma_{yz}) & \Sigma_x^2\Sigma_z^2 - \Sigma_{xz}^2 & -(\Sigma_{yz}\Sigma_x^2 - \Sigma_{xz}\Sigma_{xy}) \\ \Sigma_{xy}\Sigma_{yz} - \Sigma_{xz}\Sigma_y^2 & -(\Sigma_{yz}\Sigma_x^2 - \Sigma_{xz}\Sigma_{xy}) & \Sigma_x^2\Sigma_y^2 - \Sigma_{xy}^2 \end{pmatrix} \end{aligned}$$

where

$$\begin{aligned} \Sigma_x^2 &= \frac{\sum_{i=1}^N (x_i - \mu_x)^2}{N}, & \mu_x = \frac{\sum_{i=1}^N x_i}{N} \\ \Sigma_y^2 &= \frac{\sum_{i=1}^N (y_i - \mu_y)^2}{N}, & \mu_y = \frac{\sum_{i=1}^N y_i}{N} \\ \Sigma_z^2 &= \frac{\sum_{i=1}^N (z_i - \mu_z)^2}{N}, & \mu_z = \frac{\sum_{i=1}^N z_i}{N} \\ \Sigma_{xy} &= \frac{\sum_{i=1}^N (x_i - \mu_x)(y_i - \mu_y)}{N}, \\ \Sigma_{yz} &= \frac{\sum_{i=1}^N (y_i - \mu_y)(z_i - \mu_z)}{N}, \\ \Sigma_{xz} &= \frac{\sum_{i=1}^N (x_i - \mu_x)(z_i - \mu_z)}{N}, \end{aligned}$$

and

$$\det = \Sigma_x^2\Sigma_y^2\Sigma_z^2 + 2\Sigma_{xy}\Sigma_{yz}\Sigma_{xz} - \Sigma_x^2(\Sigma_{yz})^2 - \Sigma_y^2(\Sigma_{xz})^2\Sigma_z^2(\Sigma_{xy})^2.$$

We conclude this section with some examples illustrating this affine invariant norm and its use in characterizing affine invariant tetrahedrizations. In Figure 20.70 there are four graphs of 13 data points. The transparent ellipsoids represent all the points that are 0.50 and 1.0 units from the center point using the affine invariant norm. The different graphs show the data after it has undergone an affine transformation. The original data is displayed in the upper left. The upper right shows the data after it has been rotated by 44 degrees about the z axis. The lower right is after it has subsequently been scaled in the x variable by a factor of 1.5. The lower left is after it has been scaled in y by a factor of 0.6. A close examination of these graphs will show that the relative distances (as measured by the affine invariant norm) between points is unchanged by these transformations. Figure 20.71 shows an affine invariant tetrahedrization. In comparison, the conventional Delaunay tetrahedrization is shown in Figure 20.72.

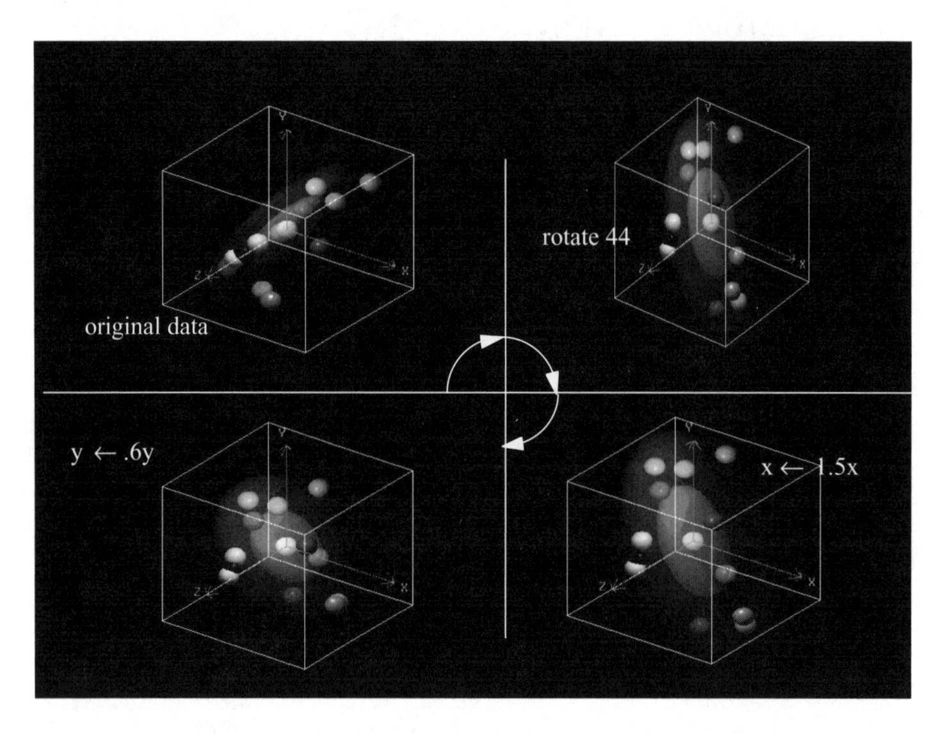

Figure 20.70: Examples illustrating the affine invariant norm. The ellipsoids are 0.50 and 1.0 units from the center point. See Color Plate 106.

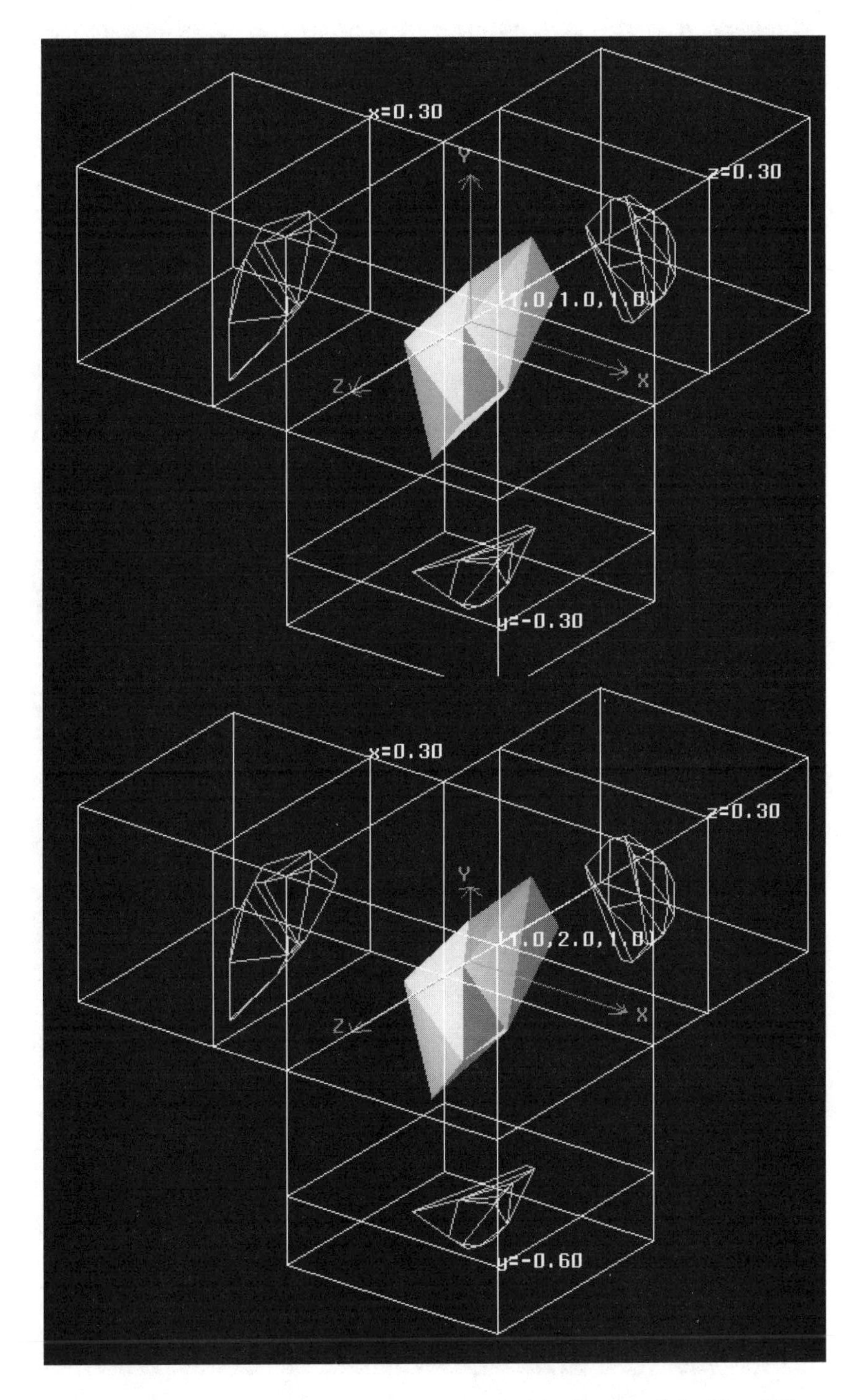

Figure 20.71: Examples of affine invariant tetrahedrization. See Color Plate 107.

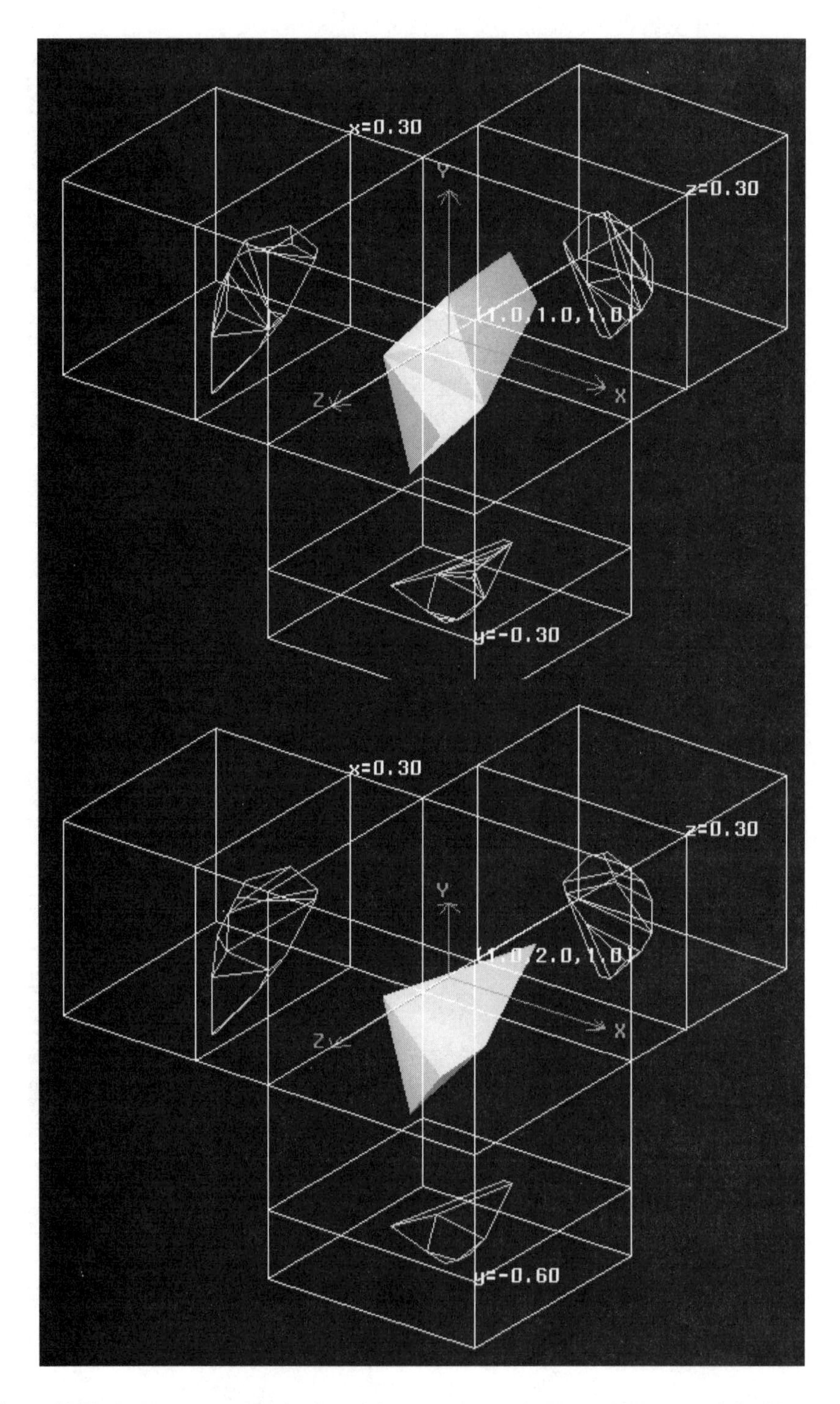

Figure 20.72: Delaunay tetrahedrization of the same data as in Figure 20.71. See Color Plate 108.

20.3.6 Interpolation in Tetrahedra

As with the bivariate case covered in Section 20.2.6, there are two concepts of interest for interpolation in tetrahedra. The first is concerned with the amount of boundary data that is provided or assumed to be available. This can be discrete data provided at a finite number of locations (usually the vertices or midpoints) or transfinite data where boundary data values are assumed to be available at all locations on the boundary. The second concept relates to the degree of continuity of a piecewise defined interpolant using the local interpolants described here. C^0 interpolants use only boundary position data and lead to overall interpolants which are continuous. C^1 interpolants utilize first-order derivative information and lead to global interpolants which have all first-order derivative continuous. These two concepts lead to four possibilities which we discuss below.

C^0, Discrete Interpolation in Tetrahedra

Analogous to the bivariate linear interpolant which will match predescribed values at the three vertices of a triangle, there is a unique trivariate linear interpolant which will match data at the four vertices of a tetrahedra, T_{ijkl}. Given $F(V_i), F(V_j), F(V_k)$ and $F(V_l)$, the coefficients of this linear function which interpolates this data

$$F(x, y, z) = a + bx + cy + dz$$

can be found by solving the linear system of equations

$$\begin{aligned} a + bx_i + cy_i + dz_i &= F(V_i) \\ a + bx_j + cy_j + dz_j &= F(V_j) \\ a + bx_k + cy_k + dz_k &= F(V_k) \\ a + bx_l + cy_l + dz_l &= F(V_l) \end{aligned}$$

As before, it is also possible to use barycentric coordinates. The barycentric coordinates of a point $V = (x, y, z)$ are defined by the relationships

$$\begin{aligned} V &= b_i V_i + b_j V_j + b_k V_k + b_l V_l \\ 1 &= b_i + b_j + b_k + b_l \end{aligned}$$

and the linear interpolant has the form

$$F(x, y, z) = F(V) = b_i F(V_i) + b_j F(V_j) + b_k F(V_k) + b_l F(V_l). \tag{20.6}$$

As before, there are several ways of defining or computing barycentric coordinates. The analog of the ratios of areas we saw before is the ratio of volumes of subtetrahedra,

$$b_i = \frac{Vol(T_{Vjkl})}{Vol(T_{ijkl})}, \quad b_j = \frac{Vol(T_{iVkl})}{Vol(T_{ijkl})}, \quad b_k = \frac{Vol(T_{ijVl})}{Vol(T_{ijkl})}, \quad b_l = \frac{Vol(T_{ijkV})}{Vol(T_{ijkl})}$$

where T_{Vjkl} is the tetrahedron with vertices V, V_j, V_k, and V_l and similar definitions for the other subtetrahedra. The volume of a tetrahedron, T_{abcd}, with vertices a, b, c, and d is

$$Vol(T_{abcd}) = \frac{1}{6}[(d - a) \cdot ((b - a) \times (c - a))]$$

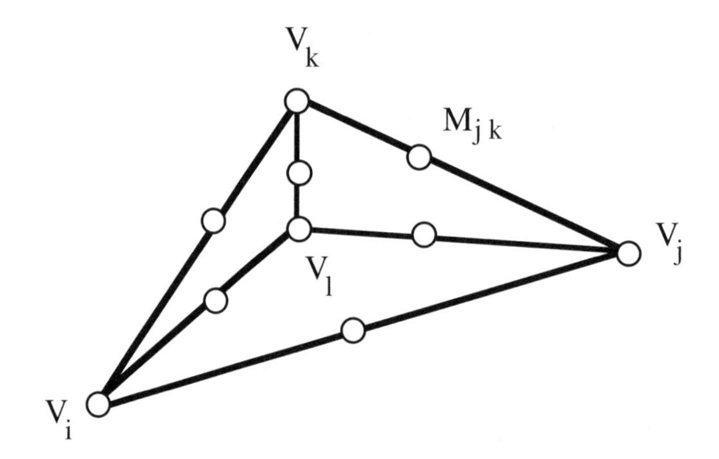

Figure 20.73: Data site locations for trivariate quadratic interpolation.

Also determinants can be used,

$$b_i = \frac{\begin{vmatrix} x - x_j & x - x_k & x - x_l \\ y - y_j & y - y_k & y - y_l \\ z - z_j & z - z_k & z - z_l \end{vmatrix}}{\begin{vmatrix} x_i - x_j & x_i - x_k & x_i - x_l \\ y_i - y_j & y_i - y_k & y_i - y_l \\ z_i - z_j & z_i - z_k & z_i - z_l \end{vmatrix}}, \quad b_j = \frac{\begin{vmatrix} x - x_i & x - x_k & x - x_l \\ y - y_i & y - y_k & y - y_l \\ z - z_i & z - z_k & z - z_l \end{vmatrix}}{\begin{vmatrix} x_j - x_i & x_j - x_k & x_j - x_l \\ y_j - y_i & y_j - y_k & y_j - y_l \\ z_j - z_i & z_j - z_k & z_j - z_l \end{vmatrix}},$$

$$b_k = \frac{\begin{vmatrix} x - x_i & x - x_j & x - x_l \\ y - y_i & y - y_j & y - y_l \\ z - z_i & z - z_j & z - z_l \end{vmatrix}}{\begin{vmatrix} x_k - x_i & x_k - x_j & x_k - x_l \\ y_k - y_i & y_k - y_j & y_k - y_l \\ z_k - z_i & z_k - z_j & z_k - z_l \end{vmatrix}}, \quad b_l = \frac{\begin{vmatrix} x - x_i & x - x_j & x - x_k \\ y - y_i & y - y_j & y - y_k \\ z - z_i & z - z_j & z - z_k \end{vmatrix}}{\begin{vmatrix} x_l - x_i & x_l - x_j & x_l - x_k \\ y_l - y_i & y_l - y_j & y_l - y_k \\ z_l - z_i & z_l - z_j & z_l - z_k \end{vmatrix}}$$

Given the values at the four vertices and the six midpoints of a tetrahedron, there is a unique trivariate quadratic which interpolates this data,

$$\begin{aligned} Q(x,y,z) \quad = \quad & F(V_i)b_i(b_i - b_j - b_k - b_l) + F(V_j)b_j(b_j - b_i - b_k - b_l) \\ & + F(V_k)b_k(b_k - b_i - b_j - b_l) + F(V_l)b_l(b_l - b_i - b_j - b_k) \\ & + F(M_{ik})4b_ib_k + F(M_{jl})4b_jb_l + F(M_{ij})4b_ib_j \\ & + F(M_{jk})4b_jb_k + F(M_{il})4b_ib_l + F(M_{kl})4b_kb_l \end{aligned} \tag{20.7}$$

where $M_{ij} = (V_i + V_j)/2$ and the other midpoints are defined similarly. See Figure 20.73.

C^0, Transfinite Interpolation in Tetrahedra

As before in Section 20.2.6, we give a sampling of interpolants. One is a generalization of the side-vertex interpolant and the other is a generalization of the C* interpolant. Both of

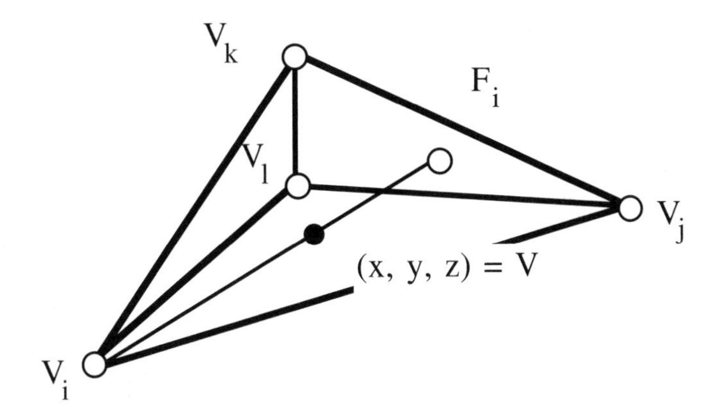

Figure 20.74: Notation for the face-vertex interpolant.

these bivariate interpolants were discussed previously in Section 20.2.6.

The C^0, Face-Vertex Interpolant: Analogous to the basic interpolants used to construct the side-vertex interpolant, we have the interpolants which consist of linear interpolation along edges joining a vertex and the opposing face

$$\begin{aligned}
A_i[F] &= b_iF(V_i) + (1-b_i)F(F_i) \\
A_j[F] &= b_jF(V_j) + (1-b_j)F(F_j) \\
A_k[F] &= b_kF(V_k) + (1-b_k)F(F_k) \\
A_l[F] &= b_lF(V_l) + (1-b_l)F(F_l)
\end{aligned} \tag{20.8}$$

where $F_i = \dfrac{b_jV_j + b_kV_k + b_lV_l}{b_j + b_k + b_l}$, $F_j = \dfrac{b_iV_i + b_kV_k + b_lV_l}{b_i + b_k + b_l}$, $F_k = \dfrac{b_iV_i + b_jV_j + b_lV_l}{b_i + b_j + b_l}$
and $F_l = \dfrac{b_iV_i + b_jV_j + b_kV_k}{b_i + b_j + b_k}$. See Figure 20.74. Computing the Boolean sum of these four interpolants leads to

$$\begin{aligned}
A[F] = \; & (1-b_i)F(F_i) + (1-b_j)F(F_j) + (1-b_k)F(F_k) + (1-b_l)F(F_l) \\
& -(b_k + b_l)F(S_{kl}) - (b_i + b_l)F(S_{il}) - (b_j + b_l)F(S_{jl}) \\
& -(b_j + b_k)F(S_{jk}) - (b_i + b_k)F(S_{ik}) - (b_i + b_j)F(S_{ij}) \\
& +b_iF(V_i) + b_jF(V_j) + b_kF(V_k) + b_lF(V_l)
\end{aligned} \tag{20.9}$$

where $S_{mn} = \dfrac{b_mV_m + b_nV_n}{b_m + b_n}, mn = kl, il, jl, jk, ik, ij.$

The C* Interpolant (for a tetrahedron): The analog of the bivariate C* interpolant described in Section 20.2.6 is

$$\begin{aligned} C^*[F] &= b_i F(V_i) + b_j F(V_j) + b_k F(V_k) + b_l F(V_l) \qquad (20.10) \\ &+W_l \left\{ F(Q_l) - \left(b_i + \frac{b_l}{3}\right) F(V_i) - \left(b_j + \frac{b_l}{3}\right) F(V_j) - \left(b_k + \frac{b_l}{3}\right) F(V_k) \right\} \\ &+W_k \left\{ F(Q_k) - \left(b_i + \frac{b_k}{3}\right) F(V_i) - \left(b_j + \frac{b_k}{3}\right) F(V_j) - \left(b_l + \frac{b_k}{3}\right) F(V_l) \right\} \\ &+W_j \left\{ F(Q_j) - \left(b_i + \frac{b_j}{3}\right) F(V_i) - \left(b_k + \frac{b_j}{3}\right) F(V_k) - \left(b_l + \frac{b_j}{3}\right) F(V_l) \right\} \\ &+W_i \left\{ F(Q_i) - \left(b_j + \frac{b_i}{3}\right) F(V_j) - \left(b_k + \frac{b_i}{3}\right) F(V_k) - \left(b_l + \frac{b_i}{3}\right) F(V_l) \right\} \end{aligned}$$

where $Q_l = \left(b_i + \frac{b_l}{3}\right) V_i + \left(b_j + \frac{b_l}{3}\right) V_j + \left(b_k + \frac{b_l}{3}\right) V_k$,

$W_l = \dfrac{27 b_i b_j b_k}{(3b_i + b_l)(3b_j + b_l)(3b_k + b_l)}$ and the other Q's and W's are defined in a similar manner.

C¹, Transfinite Interpolation in Tetrahedra

The C¹, Face-Vertex Interpolant: It is a straightforward process to extend the C^1, transfinite side-vertex interpolant to a tetrahedral domain, T_{ijkl}. It is called the C^1, face-vertex interpolant and we assume that position and derivative information is available at all locations on the four faces which make up the boundary of the tetrahedron T_{ijkl}. The basic face-vertex operator is defined as

$$\begin{aligned} S_i[F](p) &= b_i^2(3 - 2b_i)F(V_i) + b_i^2(b_i - 1)F'(V_i) \\ &+(1 - b_i)^2(2b_i + 1)F(F_i) + b_i(1 - b_i)^2 F'(F_i) \qquad (20.11) \end{aligned}$$

where $F'(V_i) = \dfrac{(x - x_i)F_x(V_i) + (y - y_i)F_y(V_i) + (z - z_i)F_z(V_i)}{1 - b_i}$ and

$F'(F_i) = \dfrac{(x - x_i)F_x(S_i) + (y - y_i)F_y(S_i) + (z - z_i)F_z(S_i)}{1 - b_i}$. The point F_i is the intersection point of the ray from V_i through V and the face opposite V_i, and the derivatives are taken in the direction of this same ray. See Figure 20.75. If we form the convex combination

$$S[F] = \frac{b_j^2 b_k^2 b_l^2 S_i[F] + b_i^2 b_k^2 b_l^2 S_j[F] + b_j^2 b_l^2 b_i^2 S_k[F] + b_j^2 b_k^2 b_i^2 S_l[F]}{b_j^2 b_k^2 b_l^2 + b_i^2 b_k^2 b_l^2 + b_j^2 b_l^2 b_i^2 + b_j^2 b_k^2 b_i^2}$$

then $S[F]$ will match position and derivative values on the entire boundary of T_{ijkl}.

C¹, Discrete Interpolation in Tetrahedra

For a C^1, discrete interpolant, we assume that position and first-order derivative information is given at all four vertices of the tetrahedron T_{ijkl}. Since there are three (linearly independent) directional derivatives at each vertex, this amounts to a total of 16 data values. The

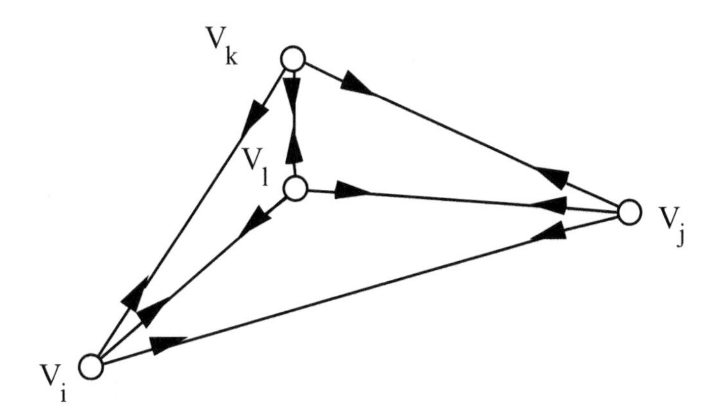

Figure 20.75: The data for a 16 parameter, C^1 interpolant over a tetrahedron.

method for describing an interpolant that will match these 16 pieces of data—and which also has the property that all first-order derivatives across a face with common data will be continuous—is somewhat different from the previous interpolants we have described so far. Our description (and subsequent implementation) is based upon a two-step procedural discretization process. We use the transfinite interpolant of the previous section. In order to apply this transfinite interpolant, we need to define position and derivative values on the entire boundary of T_{ijkl}. First we assume that information is known on all the edges of the tetrahedra, and we describe how to extend it to the entire boundary. Second, we describe how to provide this transfinite edge data from only the discrete data at the vertices. If we know both position and derivative information on the edges, then we can use any C^1 transfinite planar triangular interpolant to define position values on the interior points of the face triangles. For example, the side-vertex method itself could be used. Specifying position information on a face also implies some information about the derivatives on the interior of a triangle. Namely, all directional derivatives in a direction parallel to the face triangle are determined; so, in order to completely specify all derivatives, we need only provide a definition for the derivative perpendicular to the face. For this we use the C^0 version of the side-vertex interpolant which interpolates position data only and not derivatives, but we apply it to the edge data consisting of derivatives normal to a face. We now describe the second step of the discretization, which is how to compute edge information when only the point and derivative values are known at the four vertices. For position only on an edge, we simply use univariate cubic Hermite interpolation. This will also specify one directional derivative on the edge—namely $\frac{\partial F}{\partial e_{ij}}$, which will vary as a quadratic polynomial. In order to get a C^1 join from one tetrahedron to the next, the other two directional derivatives must vary linearly along this edge. This is accomplished by specifying the gradient, ∇F, by the relationship

$$\begin{aligned}\nabla F_{ij}(p) \quad &= \quad (1-t)\nabla F_i + t\nabla F_j \\ &+ \left[\frac{\partial F}{\partial e_{ij}}(p)((1-t)\nabla F_i + t\nabla F_i, e_{ij})\right] e_{ij} \qquad (20.12)\end{aligned}$$

where $\nabla F_i = (F_x(p_i), F_y(p_i), F_z(p_i))$ and $t = \frac{||p-p_i||}{||p_j-p_i||}$. This interpolation of the gradient is consistent with the value $\frac{\partial F}{\partial e_{ij}}$ already specified because $(\nabla F_{ij}(p), e_{ij}) = \frac{\partial F}{\partial e_{ij}}$ and it also has the property that for $(n, e_{ij}) = 0$,

$$(\nabla F_{ij}(p), n) = (1-t)(\nabla F_i, n) + t(\nabla F_j, n),$$

and so we have linear interpolation for any derivative in a direction perpendicular to e_{ij}. This completes the definition of the 16-parameter, C^1, tetrahedral interpolant which is based upon the face-vertex interpolant. Examples and more discussion on this interpolant can be found in [187]. The Clough-Tocher interpolant has been generalized to n-dimensional by Worsey and Farin [261]. Other C^1, discrete interpolants for a tetrahedral domain are discussed in [2], [3], and [260], but each have some problem or drawback. The method of [2] is based upon the side-side, transfinite method of interpolation and apparently it has a problem with the linear independence of the discretized data. The method of [3] requires C^2 data for a C^1 interpolant and the method of [260] has a problem similar to its bivariate precursor [199] and [198]. This problem lies in the constraint that the center of the circumcircle of each triangle must lie interior to the triangular domain.

Acknowledgments

We wish to acknowledge the support of the National Aeronautical and Space Administration under NASA-Ames Grant, NAG 2-990. Also, partial support was provided by the North Atlantic Treaty Organization under grant RG 0097/88. We wish to thank Herbert Edelsbrunner for the idea of the dual graphs of Section 20.3.1 and other insightful discussions about tetrahedrizations. We wish to thank Kun Lee for his help in generating the images of Sections 20.3.4 and 20.3.5.

Bibliography

[1] A. Aggarwal, L.J. Guibas, J. Saxe, and P.W. Shor, "A Linear Time Algorithm for Computing the Voronoi Diagram of a Convex Polygon," *Disc. and Comp. Geometry 4*, 1989, pp. 591–604.

[2] P. Alfeld, "A Discrete C^1 Interpolant for Tetrahedral Data," *The Rocky Mountain Journal of Mathematics*, Vol. 14, No. 1, Winter 1984, pp. 5–16.

[3] P. Alfeld, "A Trivariate Clough-Tocher Scheme for Tetrahedral Data," *Computer Aided Geometric Design*, Vol. 1, No. 2, 1984, pp. 169–181.

[4] T. Asano and R. Pinter, "Polygon Triangulation: Efficiency and Minimality," *J. Algorithms*, Vol. 7, 1986, pp. 221–231.

[5] D. Avis, and B.K. Bhattacharya, "Algorithms for Computing d-Dimensional Voronoi Diagrams and their Duals," *Advance in Computing Research*, Vol. 1, pp. 159–180.

[6] D. Avis and K. Fukuda, "A Pivoting Algorithm for Convex Hulls and Vertex Enumeration of Arrangements and Polyhedra," *Proceedings 7th Annual ACM Symposium Computational Geometry*, 1991, pp. 98–104.

[7] D. Avis and G.T. Toussaint, "An Efficient Algorithm for Decomposing a Polygon into Star-Shaped Polygons," *Pattern Recogn.*, Vol. 13, No. 6, 1981, pp. 395–398.

[8] F. Aurenhammer, "Voronoi Diagrams—A Survey of a Fundamental Geometric Data Structure," *ACM Computing Surveys*, Vol. 23, 1991, pp. 345–405.

[9] I. Babuska and A. Aziz, "On the Angle Condition in the Finite Element Method," *SIAM J. Numer. Analysis*, Vol. 13, 1976, pp. 214–227.

[10] P.I. Baegmann, M.S. Shepard, and J.E. Flaherty, "A Posterori Error Estimation for Triangular and Tetrahedral Quadratic Elements Using Interior Residuals," *Internat. J. Numer. Meth. Eng.*, Vol. 34, 1992, pp. 979–996.

[11] F. Bagemihl, "On Indecomposable Polyhedra," *American Mathematical Monthly*, Sep. 1948, pp. 411–413.

[12] T.J. Baker, "Automatic Mesh Generation for Complex Three-Dimensional Regions Using a Constrained Delaunay Triangulation," *Eng. with Computers*, Vol. 5, 1989, pp. 161–175.

[13] B.S. Baker, E. Grosse, and C.S. Rafferty, "Nonobtuse Triangulation of Polygons," *Disc. and Comp. Geom.*, Vol. 3, 1988, pp. 147–168.

[14] G. Baszenski and L.L. Schumaker, "Use of Simulated Annealing to Construct Triangular Facet Surfaces," *Curves and Surfaces*, P.-J. Laurent, A. Le Mehaute and L.L. Schumaker, editors, Academic Press, Boston, Mass., 1991, pp. 27–32.

[15] M. Bern and D. Eppstein, "Polynomial-Size Nonobtuse Triangulation of Polygons," *Proc. 7th ACM Symp. Comp. Geometry*, 1991, pp. 342–350.

[16] M. Bern, and D. Eppstein, "Mesh Generation and Optimal Triangulation," *Computing in Euclidean Geometry*, F. K. Hwang and D.-Z. Du, editors, World Scientific, Singapore, 1992, pp. 23–90.

[17] M. Bern, D. Dobkin, and D. Eppstein, "Triangulating Polygons with Large Angles," *Proc. 8th ACM Sym. Comp. Geometry*, 1992.

[18] M. Bern, D. Eppstein, and F. Yao, "The Expected Extremes in a Delaunay Triangulation," *Int. J. Comp. Geometry and Applications*, Vol. 1, 1991, pp. 79–92.

[19] J. Bloomenthal, "Polygonization of Implicit Surfaces," *CAGD*, Vol. 5, 1988, pp. 341–355.

[20] C. Borgers, "Generalized Delaunay Triangulations of Nonconvex Domains," *Computers & Mathematics with Applications*, Vol. 20, No. 7, 1990, pp. 45–49.

[21] A. Bowyer, "Computing Dirichlet Tessellations," *Computer J.*, Vol. 24, 1981, pp. 162–166.

[22] C. Bradford, D. Barber, D.P. Dobkin, and H. Huhdanpaa, *The Quickhull Algorithm for Convex Hull*, Technical Report GCG53-93, Geometry Center, University of Minnesota, July 1993.

[23] J. Bramble and M. Zlamal, "Triangular Elements in the Finite Element Method," *Math. Comp.*, Vol. 24, 1970, pp. 809–820.

[24] K.E. Brassel and D. Reif, "A Procedure to Generate Thiessen Polygons," *Geograph. Anal.*, Vol. 11, 1979, pp. 289–303.

[25] W. Brostow, J.P. Dussault, and B.L. Fox, "Construction of Voronoi Polyhedra," *J. Comp. Physics*, Vol. 29, 1978, pp. 81–92.

[26] J.L. Brown, "Vertex Based Data Dependent Triangulations," *Computer Aided Geometric Design*, Vol. 8, 1991, pp. 239–251.

[27] K.Q. Brown, "Voronoi Diagrams from Convex Hulls," *Inform. Process. Lett.*, Vol. 9, 1979, pp. 223–228.

[28] J.C. Cavendish, "Automatic Triangulation of Arbitrary Planar Domains for the Finite Element Method," *Int. J. for Numer. Methods in Engr.*, Vol. 8, 1974, pp. 679–696.

[29] J.C. Cavendish, D.A. Field, and W.H. Frey, "An Approach to Automatic Three-Dimensional Finite Element Mesh Generation," *Int. J. Numer. Meth. Eng.*, Vol. 21, 1985, pp. 329–347.

[30] M.S. Chang, N.-F. Huang, and C.Y. Tang, "Optimal Algorithm for Constructing Oriented Voronoi Diagrams and Geographic Neighborhood Graphs," *Information Processing Letters*, Vol. 35, No. 5, Aug. 1990, pp. 255–260.

[31] R.C. Chang and R.C.T. Lee, "On the Average Length of Delaunay Triangulations," *BIT*, Vol. 24, 1984, pp. 269–273.

[32] S. Chattopodhyay and P.P. Das, "Counting Thin and Bushy Triangulations," *Pattern Recognition Letters*, Vol. 12, No. 3, 1991, pp. 139–144.

[33] B. Chazelle, "Convex Partitions of Polyhedra: A Lower Bound and Worst-Case Optimal Algorithm," *SIAM J. Comput.*, Vol. 13, 1984, pp. 488–507.

[34] B. Chazelle, "Triangulating a Simple Polygon in Linear Time," *Disc. and Comp. Geometry*, Vol. 6, 1991, pp. 485–524.

[35] B. Chazelle and D. Dobkin, "Decomposing a Polygon into its Convex Parts," *ACM Proceedings of the 11th Symposium on Theory of Computing*, 1979, pp. 38–48.

[36] B. Chazelle, H. Edelsbrunner, L.J. Guibas, J.E. Hershberger, R. Reidel, and M. Sharir, "Selecting Multiply Covered Points and Reducing the Size of Delaunay Triangulations," *Proc. 6th ACM Symp. Comp. Geometry*, 1990, pp. 116–127.

[37] B. Chazell and J. Incerpi, "Triangulating a Polygon by Divide and Conquer," *Proceedings of the 21st Allerton Conference on Communications, Control and Computing*, 1983, pp. 447–456.

[38] B. Chazelle and J. Incerpi, "Triangulation and Shape Complexity," *ACM Trans on Graphics*, Vol. 3, 1984, pp. 135–152.

[39] B. Chazelle and L. Palios, "Triangulating a Nonconvex Polytope," *Disc. and Comp. Geometry*, Vol. 5, 1990, pp. 505–526.

[40] L.P. Chew, "Constrained Delaunay Triangulations," *Algorithmica*, Vol. 4, 1989, pp. 97–108.

[41] B.K. Choi, H.Y. Shin, Y.I. Yoon, and J.W. Lee, "Triangulation of Scattered Data in 3D Space," *CAD*, Vol. 20, 1988, pp. 239–248.

[42] K.L. Clarkson, R.E. Tarjan, and C.J. Van Wyk, "A Fast Las Vegas Algorithm for Triangulating a Simple Polygon," *Discrete and Computational Geometry*, Vol. 4, 1989, pp. 423–432.

[43] A.K. Cline and R.J. Renka, "A Constrained Two-Dimensional Triangulation and the Solution of Closest Node Problems in the Presence of Barriers," *SIAM Journal on Numerical Analysis*, Vol. 27, No. 5, 1990, pp. 1305–1321.

[44] A.K. Cline and R.L. Renka, "A Storage-Efficient Method for Construction of a Thiessen Triangulation," *Rocky Mountain Journal of Mathematics*, Vol. 14, No. 1, Winter 1984, pp. 119–140.

[45] H.E. Cline, W.E. Lorensen, S. Ludke, C.R. Crawford, and B.C. Teeter, "Two Algorithms for the Reconstruction of Surfaces from Tomograms," *Medical Physics*, June 1988.

[46] Y. Correc and E. Chapuis, "Fast Computation of Delaunay Triangulations," *Advances in Engineering Software*, Vol. 9, No. 2, 1987, pp. 77–83.

[47] H.S.M. Coxeter, "Discrete Groups Generated by Reflections," *Ann. Math.*, Vol. 35, 1934, pp. 588–621.

[48] J.R. Davy and P.M. Dew, "A Note on Improving the Performance of Delaunay Triangulation," *New Advances in Computer Graphics: Proceedings of Computer Graphics International 89*, R.A. Earnshaw and B. Wyvill, editors, Springer, Tokyo, 1989, pp. 209–226.

[49] A.M. Day, "The Implementation of an Algorithm to Find the Convex Hull of a Set of Three-Dimensional Points," *ACM Transactions on Graphics*, Vol. 9, No. 1, Jan. 1990, pp. 105–132.

[50] L. De Floriani, "A Pyramidal Data Structure for Triangle-Based Surface Representation," *IEEE Computer Graphics and Applications*, Vol. 9, Mar. 1989, pp. 67–78.

[51] L. De Floriani, B. Falcidieno, and C. Pienovi, "Delaunay-Based Representation of Surfaces Defined over Arbitrarily Shaped Domains," *Computer Vision, Graphics, and Image Processing*, Vol. 32, 1985, pp. 127–140.

[52] L. De Floriani and E. Puppo, "An On-Line Algorithm for Constrained Delaunay Triangulation," *CVGIP: Graphical Models and Image Processing*, Vol. 54, No. 3, 1992, pp. 290–300.

[53] L. De Floriani, B. Falcidieno, G. Nagy, and C. Pienovi, "On Sorting Triangles in a Delaunay Tessellation," *Algorithmica*, Vol. 6, 1991, pp. 522–532.

[54] B. Delaunay, "Sur la sphere vide," *Izvestia Akademii Nauk SSSR, Otdelenie Matematicheskii i Estestvennyka Nauk 7, [Bull. Acad. Sci. U.S.S.R.(VII), Classe Sci. Mat. Nat]*, 1934, pp. 793–800.

[55] P.A. Devijver and M. Dekesel, "Insert and Delete Algorithms for Maintaining Dynamic Delaunay Triangulations," *Pattern Recogn. Lett.*, Vol. 1, 1982, pp. 73–77.

[56] T. Dey, "Triangulation and CSG Representation of Polyhedra with Arbitrary Genus," *Proc. 7th ACM Symp. Comp. Geometry*, 1991, pp. 793–800.

[57] T. Dey, K. Sugihara and C.L. Bajaj, "Delaunay Triangulations in Three Dimensions with Finite Precision Arithmetic," *Computer Aided Geometric Design*, Vol. 9, No. 6, 1992, pp. 457–470.

[58] M.B. Dillencourt, "Realizability of Delaunay Triangulations," *Information Processing Letters*, Vol. 33, No. 6, 1990, pp. 283–287.

[59] G.L. Dirichlet, "Ueber die Reduktion der positiven quadratischen Formen mit drei unbestimmten ganzen Zahlen," *J. Reine u. Angew. Math.*, Vol. 40, 1850, pp. 209–227.

[60] H. Djidjev and A. Lingas, "On Computing the Voronoi Diagram for Restricted Planar Figures," *Proc. 2nd Workshop Algorithms Data Struct. Volume 519 of Lecture Notes in Computer Science*, Springer, 1991, pp. 54–64.

[61] D. P. Dobkin, "Computational Geometry and Computer Graphics," *Proc. IEEE*, Vol. 80, No. 9, Sep. 1992, pp. 1400–1411.

[62] D.P. Dobkin and M.J. Laszlo, "Primitives for the Manipulation of Three-Dimensional Subdivisions," *Algorithmica*, Vol. 4, 1989, pp. 3–32.

[63] D. Dobkin, S. Friedman, and K. Supowit, "Delaunay Graphs Are Almost as Good as Complete Graphs," *Disc. and Comp. Geometry*, Vol. 5, 1990, pp. 389–423.

[64] D. Dobkin, S. Levy, W. Thurston, and A. Wilks, "Contour Tracing by Piecewise Linear Approximations," *ACM Trans. on Graphics*, Vol. 9, 1990, 389–423.

[65] R.A. Dwyer, "A Faster Divide and Conquer Algorithm for Constructing Delaunay Triangulation," *Algorithmica*, Vol. 2, 1987, pp. 137–151.

[66] N. Dyn and I. Goren, "Transforming Triangulations in Polygon Domains," *Computer Aided Geometric Design*, Vol. 10, No. 6, Dec. 1993, pp. 531–536.

[67] N. Dyn, D. Levin, and S. Rippa, "Data Dependent Triangulations for Piecewise Linear Interpolation," *IMA Journal of Numerical Analysis*, Vol. 10, 1990, pp. 137–154.

[68] H. Edelsbrunner, *Algorithms in Combinatorial Geometry*, Springer-Verlag, 1987.

[69] H. Edelsbrunner, "An Acyclicity Theorem for Cell Complexes in d Dimensions," *Combinotorica*, Vol. 18, 1990, pp. 251–260. Also: *Proceedings of the 5th Annual ACM Symposium on Computation Geometry*, 1989, pp. 145–151.

[70] H. Edelsbrunner and E.P. Muecke, "Simulation of Simplicity, A Technique to Cope with the Degenerate Cases in Geometric Computations," *ACM Trans. Graphics*, Vol. 9, 1990, pp. 66–104.

[71] H. Edelsbrunner and E.P. Muecke, "Three Dimensional Alpha Shapes," *ACM Transactions on Graphics*, Vol. 13, 1994, pp. 43–72.

[72] H. Edelsbrunner and N.R. Shah, "Incremental Topological Flipping Works for Regular Triangulations," *Proceedings of the 8th Annual ACM Symposium on Computational Geometry*, June 1992, pp. 43–52.

[73] H. Edelsbrunner and T.S. Tan, "A Quadratic Time Algorithm for the Minmax Length Triangulation," *Proc. 32nd IEEE Symp. foundations of Comp. Science*, 1991, pp. 414–423.

[74] H. Edelsbrunner and T.S. Tan, "An Upper Bound for Conforming Delaunay Triangulations," *Proc. 8th Symp Comp. Geometry*, 1992, pp. 53–62.

[75] H. Edelsbrunner, T.S. Tan, and R. Waupotitsch, "A Polynomial Time Algorithm for the Minmax Angle Triangulation," *Proc. 5th Symp Comp. Geometry*, 1990.

[76] H. Edelsbrunner, T.S. Tan, and R. Waupotitsch, "O($N^2 \log N$) Time Algorithm for the Minmax Angle Triangulation," *SIAM Journal on Scientific and Statistical Computing*, Vol. 13, No. 4, July 1992, pp. 994–1008.

[77] H. Edelsbrunner, F.P. Preparata, and D.B. West, "Tetrahedrizing Point Sets in Three Dimensions," *J. Symbolic Comp.*, Vol. 10, 1990, pp. 335–347.

[78] M. Elbaz and J.-C. Spehner, "Construction of Voronoi Diagrams in the Plane by Using Maps," *Theoretical Computer Science*, Vol. 77, No. 3, 1990, pp. 331–343.

[79] H. ElGindy and G.T. Toussaint, "On Geodesic Properties of Polygons Relevant to Linear Time Triangulation," *Visual Computer*, Vol. 5, No. 1, 1989, pp. 68–74.

[80] D. Eppstein, "The Farthest Point Delaunay Triangulation Minimizes Angles," *Comput. Geom. Theory Appl.*, Vol. 1, 1992, pp. 143–148.

[81] G. Erlebacher and P.R. Eiseman, "Adaptive Triangular Mesh Generation," *AIAA Journal*, Vol. 25, 1987, pp. 1356–1364.

[82] M.A. Facello, "Implementation of a Randomized Algorithm for Delaunay and Regular Triangulations in Three Dimensions," *Computer Aided Geometric Design*, Vol. 12, pp. 351–370, 1995.

[83] T.P. Fang and L.A. Piegl, "Delaunay Triangulation Using a Uniform Grid," *IEEE Computer Graphics and Application*, Vol. 13, No. 3, pp. 36–47, May 1993.

[84] G. Farin, "A Modified Clough-Tocher Interpolant," *Computer Aided Geometric Design*, Vol. 2, Nos. 1–3, pp. 19–27.

[85] G. Fekete, "Rendering and Managing Spherical Data with Sphere Quadtrees," *Proceedings of Visualization '90*, IEEE Computer Society Press, 1990, pp. 176–186.

[86] D. Field, "Implementing Watson's Algorithm in Three Dimension," *Pro. 2nd ACM Symp. Comp. Geometry*, 1986, pp. 246–259.

[87] D. Field, "A Generic Delaunay Triangulation Algorithm for Finite Element Meshing," *Adv. Eng. Software*, Vol. 13, 1991, pp. 263–272.

[88] D. Field, "Laplacian Smoothing and Delaunay Triangulations," *Comm. in Applied Numer. Analysis*, Vol. 4, 1988, pp. 709–712.

[89] D. Field and W.D. Smith, "Graded Tetrahedral Finite Element Meshes," *Int. J. Numer. Meth. Eng.*, Vol. 31, 1991, pp. 413–425.

[90] R. Forrest, "Computational Geometry," *Proc. Royal Society London*, Volume 321, Series 4, 1971, pp. 187–195.

[91] S. Fortune, "Numerical Stability of Algorithms for 2-d Delaunay Triangulations and Voronoi Diagrams," *Proc. 8th Annual. ACM Symposium. Comput. Geom.*, 1992, pp. 83–92.

[92] S. Fortune, "Voronoi Diagrams and Delaunay Triangulations," *Computing in Euclidean Geometry*, F.K. Hwang and D.-Z. Du, editors, World Scientific, Singapore, 1992, pp. 193–233.

[93] S. Fortune, "Sweepline Algorithm for Voronoi Diagrams," *Algorithmica*, Vol. 2, No. 2, 1987, pp. 153–174.

[94] A. Fournier and D.Y. Montuno, "Triangulating Simple Polygons and Equivalent Problems," *ACM Transaction on Graphics*, Vol. 3, No. 2, Apr. 1984, pp. 153–174.

[95] R.J. Fowler and J.J. Little, "Automatic Extraction of Irregular Network Digital Terrain Models," *Computer Graphics*, Vol. 13, No. 2, Aug. 1979, pp. 199–207.

[96] R. Franke, "Scattered Data Interpolation: Tests of Some Methods," *Math. Comp.*, Vol. 38, 1982, pp. 181–200.

[97] R. Franke and G. Nielson, "Surface Construction Based upon Triangulations," *Surfaces in Computer Aided Geometric Design*, Springer, 1983, pp. 163–179.

[98] R. Franke and G. Nielson, "Scattered Data Interpolation and Applications: A Tutorial and Survey," *Geometric Modelling: Methods and their Application*, H. Hagen and D. Roller, editors, Springer, 1990.

[99] H. Freudenthal, "Simplizialzerlegungen von beschraenkter Flachheit," *Ann. Math.*, Vol. 43, 1942, pp. 580–582.

[100] W.H. Frey and D.A. Field, "Mesh Relaxation: A New Technique for Improving Meshes," *Int. J. Numer. Neth. Eng.*, Vol. 31, 1991, pp. 1121–1133.

[101] M.R. Garey, D.S. Johnson, F.P. Preparata, and R.E. Tarjan, "Triangulating a Simple Polygon," *Inform. Process. Lett.*, Vol. 7, 1978, pp. 175–179.

[102] P.L. George and F. Hermeline, "Delaunay's Mesh of a Convex Polyhedron in Dimension d. Application to Arbitrary Polyhedra," *International Journal for Numerical Methods in Engineering*, Vol. 33, No. 5, Apr. 1992, pp. 975–995.

[103] J. Gleue, *Triangulierung und Interpolation von im R^2 unregelmaessig verteilten Daten*, HMI B 357, 1981.

[104] M.T. Goodrich, "Efficient Piecewise-Linear Function Approximation Using the Uniform Metric," *Proc. 10th Annu. ACM Sympos. Comput. Geom.*, 1994, pp. 322–331.

[105] S. Goldman, "A Space Efficient Greedy Triangulation Algorithm," *Information Processing Letters*, Vol. 31, No. 4, 1989, pp. 191–196.

[106] T. Gonzalez and M. Razzazi, "Properties and Algorithms for Constrained Delaunay Triangulations," *Proc. 3rd Canad. Conf. Comput. Geom.*, 1991, pp. 114–117.

[107] P.J. Green and R. Sibson, "Computing Dirichlet Tessellations in the Plane," *The Computer Journal*, Vol. 21, 1978, pp. 168–173.

[108] P.J. Green and B.W. Silverman, "Constructing the Convex Hull of a Set of Points in the Plane," *The Computer Journal*, Vol. 22, No. 3, 1979, pp. 262.

[109] J.A. Gregory, "Error Bounds for Linear Interpolation on Triangles," *The Mathematics of Finite Elements and Application II*, J.R. Whiteman, editor, Academic Press, London, 1975, pp. 163–170.

[110] J.A. Gregory, "A Blending Function Interpolant for Triangles," *Multivariate Approximation*, D.G. Handscomb. editor, Academic Press, London.

[111] J.A. Gregory, "Interpolation to Boundary Data on the Simplex," *Computer Aided Geometric Design*, Vol. 2, Nos. 1–3, pp. 43–52.

[112] J.A. Gregory, "Error Bounds for Linear Interpolation on Triangles," *The Mathematics of Finite Elements and Applications II*, J. Whiteman, editor, Academic Press, London, 1975, pp. 163–170.

[113] L. Guibas and J. Stolfi, "Primitives for the Manipulation of General Subdivisions and the Computation of Voronoi Diagrams," *ACM Trans. Graphics*, Vol. 4, 1985, pp. 74–123.

[114] L.J. Guibas, D.E. Knuth, and M. Sharir, "Randomized Incremental Construction of Delaunay and Voronoi Diagrams," *Automata, Languages and Programming*, LNCS N.443, Springer-Verlag, 1990, pp. 414–431.

[115] A.J. Hansen and P.L. Levin, "On Conforming Delaunay Mesh Generation," *Adv. Engineering Software*, Vol. 14, No. 2, 1992, pp. 129–135.

[116] D. Hansford, "The Neutral Case for the Min-Max Triangulation," *CAGD*, Vol. 7, 1990, pp. 431–438.

[117] F. Hermeline, "Triangulation automatique d'un polyedre in dimension n," *RAIRO Anal. Numer.*, Vol. 76, 1982, pp. 211–242.

[118] C. Hazelwood, "Approximating Constrained Tetrahedrizations," *Computer Aided Geometric Design*, Vol. 10, No. 1, pp. 67–87.

[119] S. Hertel and K. Mehlhorn, "Fast Triangulation of Simple Polygons," *4th Conf. Foundations of Computation Theory*, Springer LNCS 158, 1983, pp. 207–218.

[120] H. Jin and R.I. Tannel, "Generation of Unstructured Tetrahedral Meshes by Advancing Front Technique," *Internat. J. Numer. Meth. Eng.*, Vol. 36, 1993, pp. 1805–1823.

[121] B. Joe, "ree-Dimensional Triangulations from Local Transformations," *SIAM Journal Sci. Stat. Comput.*, Vol. 10, pp. 718–741, 1989.

[122] B. Joe, "Construction of Three Dimensional Delaunay Triangulations Using Local Transformations," *Computer Aided Geometric Design*, Vol. 8, No. 2, pp. 123–142, 1991.

[123] B. Joe and C.A. Wang, "Duality of Constrained Voronoi Diagrams and Delaunay Triangulations," *Algorithmica*, Vol. 9, No. 2, 1993, pp. 149–155.

[124] D.-M. Jung, "An Optimal Algorithm for Constrained Delaunay Triangulation," *Proceedings Twenty-Sixth Annual Allerton Conference on Communication, Control and Computing*, Urbana, Ill., 1988, pp. 85–86.

[125] Y.H. Jung and K. Lee, "Tetrahedron-Based Octree Encoding for Automatic Mesh Generation," *Computer Aided Design*, Vol. 25, 1993, pp. 141–153.

[126] T.C. Kao and D.M. Mount, "An Algorithm for Computing Compacted Voronoi Diagrams Defined by Convex Distance Functions," *Proc. 3rd Canad. Conf. Comput. Geom.*, 1991, pp. 104–109.

[127] T.C. Kao and D.M. Mount, "Incremental Construction and Dynamic Maintenance of Constrained Delaunay Triangulations," *Proc. 4th Canad. Conf. Comput. Geom.*, 1992, pp. 170–175.

[128] M.D. Karasick, D. Lieber, and L.R. Nackman, "Efficient Delaunay Triangulation Using Rational Arithmetic," *ACM Transactions on Graphics*, Vol. 10, No. 1, Jan. 1991, pp. 71–91.

[129] J. Katajainen and M. Koppinen, "Constructing Delaunay Triangulations by Merging Buckets in Quadtree Order," *Annales Societatis Mathematicae Polonae, Series IV, Fundamenta Informaticae*, Vol. 11, No. 3, 1988, pp. 275–288.

[130] D.G. Kirkpatrick, "A Note on Delaunay and Optimal Triangulations," *Inform. Process. Lett.*, Vol. 10, 1990, pp. 127–128.

[131] D.G. Kirkpatrick, M.M. Klawe, and R.E. Tarjan, "Polygon Triangulation in O(n log log n) Time with Simple Data Structures," *Proc. 6th Annual ACM Symposium. Comput. Geom.*, 1990, pp. 34–43.

[132] V. Klee, "On the Complexity of d-Dimensional Voronoi Diagrams," *Arch. Math.*, Vol. 34, 1980, pp. 75–80.

[133] R. Klein, "Concrete and Abstract Voronoi Diagrams," *Volume 400 of Lecture Notes in Computer Science*, Springer, 1989.

[134] R. Klein and A. Lingas, "A Note on Generalizations of Chew's Algorithm for the Voronoi Diagram of a Simple Polygon," *Proc. 9th Annu. ACM Sympos. Comput. Geom.*, 1993, pp. 124–132.

[135] G.T. Klincsek, "Minimal Triangulations of Polygonal Domains," *Ann. Disc. Math.*, Vol. 9, 1980, pp. 121–123.

[136] D. Knuth, *The Art of Computer Programming, Volume 1; Fundamental Algorithms*, Addison Wesley, Reading, Mass., 1973.

[137] H.W. Kuhn, "Simplicial Approximation of Fixed Points," *Proc. Nat. Acad. Sci. USA*, Vol. 61, 1968, pp. 1238–1242.

[138] C. Lawson, "Transforming Triangulations," *Discrete Mathematics*, Vol. 3, 1972, pp. 365–372.

[139] C. Lawson, "Software for C^1 Surface Interpolation," *Mathematical Software III*, J.R. Rice, editor, Academic Press, New York, 1977, pp. 161–194.

[140] C. Lawson, "Properties of n-Dimensional Triangulations," *Computer Aided Geometric Design*, Vol. 3, No. 4, pp. 231–246.

[141] C. Lawson, "C^1 Surface Interpolation for Scattered Data on a Sphere," *Rocky Mountain Journal of Mathematics*, Vol. 14, No. 1, Winter 1984, pp. 177–202.

[142] C. Lee, "Regular Triangulations of Convex Polytopes," *Applied Geometry and Discrete Mathematics: The Victor Klee Festschrift*, Gritzmann and B. Strumfels, editors, Amer. Math. Soc., Providence, RI, 1991, pp. 443–456.

[143] D.T. Lee, "Two Dimensional Voronoi Diagram in the L_p-Metric," *J. ACM*, Vol. 27, 1980, pp. 604–618.

[144] D.T. Lee and A. Lin, "Generalized Delaunay Triangulation for Planar Graphs," *Disc. and Comp. Geometry*, Vol. 1, 1986, pp. 201–217.

[145] D.T. Lee and C.K. Wong, "Voronoi Diagrams in L_1 (L_∞) Metrics with 2-Dimensional Storage Applications," *SIAM J. Comput.*, Vol. 9, 1980, pp. 200–211.

[146] D.T. Lee, and B.J. Schacter, "Two Algorithms for Constructing a Delaunay Triangulation," *Int. J. of Computer and Information Science*, Vol. 9, No. 3, 1980, pp. 219–242.

[147] J. Lee, "Comparison for Existing Methods for Building Triangular Irregular Network Models of Terrain from Grid Digital Elevation Models," *Int. J. of Geographical Information Systems*, Vol. 5, No. 2, July–Sep. 1991, pp. 267–285.

[148] K. Lee, *Data Dependent Tetrahedrizations*, Ph.D. thesis, Arizona State University, 1995.

[149] N.J. Lennes, "Theorems on the Simple Finite Polygon and Polyhedron," *American Journal of Mathematics*, Vol. 33, 1911, pp. 37–62.

[150] C. Levcopoulos and A. Lingas, "On Approximation Behavior of the Greedy Triangulation for Convex Polygons," *Algorithmica*, Vol. 2, 1987, pp. 175–193.

[151] B.A. Lewis and J.S. Robinson, "Triangulation of Planar Regions with Applications," *Computer J.*, Vol. 21, 1978, pp. 324–332.

[152] A. Lingas, *Advances in Minimum Weight Triangulation*, Ph.D. thesis, Linkoeping Univ., 1983.

[153] A. Lingas, "Voronoi Diagrams with Barriers and the Shortest Diagonal Problem," *Inform. Process. Lett.*, Vol. 32, 1989, pp. 191–198.

[154] D. Lischinski, "Incremental Delaunay Triangulation,"*Graphic Gems IV*, Paul S. Heckbert, editor, Academic Press, 1994, pp. 47–59.

[155] E. L. Lloyd, On Triangulations of a Set of Points in the Plane," *Proc. 18th IEEE Symp. Found. Comp. Sci.*, 1977, pp. 228–240.

[156] S. Lo, "Delaunay Triangulations of Nonconvex Planar Domains," *Int. J. Numer. Meth. Eng.*, Vol. 28, 1989, pp. 2695–2707.

[157] S. Lo, "Volume Discretizations Into Tetrahedra. I. Verification and Orientation of Boundary Surfaces," *Computers and Structures*, Vol. 39, 1991, pp. 493–500.

[158] R. Loehner and P. Parikh, Generation of Three-Dimensional Unstructured Grids by the Advancing Front Method, Internat. J. Numer. Meht. Fluids, Vol. 8, 1988, pp. 1135–1149.

[159] M.K. Loze and R. Saunders, "Two Simple Algorithms for Constructing a Two-Dimensional Constrained Delaunay Triangulation," *Applied Numerical Mathematics*, Vol. 11, 1993, pp. 403–418.

[160] W. Lorensen and H.E. Cline, "Marching Cubes: A High-Resolution 3D Surface Construction Algorithm," *SIGGRAPH 87 Conference Proceedings, Computer Graphics*, Vol. 21, No. 4, July 1987, pp. 163–169.

[161] G. Macedonio and M.T. Pareschi, "An Algorithm for the Triangulation of Arbitrarily Distributed Points: Applications to Volume Estimate and Terrain Fitting," *Computers & Geosciences*, Vol. 17, No. 7, 1991, pp. 859–874.

[162] G.K. Manacher, and A.L. Zobrist, "Neither the Greedy nor the Delaunay Triangulation Approximates the Optimum," *Inform. Process. Lett.*, Vol. 9, 1979, pp. 31–34.

[163] L. Mansfield, "Interpolation to Boundary Data in Tetrahedra with Applications to Compatible Finite Elements," *J. Mat. Anal. Appl.*, Vol. 56, pp. 137–164.

[164] G. Marton, "Acceleration of Ray Tracing Via Voronoi Diagrams," *Graphic Gems V*, Alan Paeth, editor, Academic Press, 1995, pp. 268–284.

[165] A. Maus, "Delaunay Triangulation and the Convex Hull of n Points in Expected Linear Time," *BIT*, Vol. 24, 1984, pp. 151–163.

[166] N. Max, "Sorting for Polyhedron Compositing," *Focus on Scientific Visualization*, H. Hagen, H. Mueller, G.M. Nielson, editors, Springer, 1993, pp. 259–268.

[167] N. Max, P. Hanrahan, and R. Crawfis, "Area and Volume Coherence for Efficient Visualization of 3D Scalar Functions," *Computer Graphics*, Vol. 24, Nov. 1990, pp. 27–33.

[168] A. Mirante and N. Weingarten, "The Radial Sweep Algorithm for Constructing Triangulated Irregular Networks," *IEEE Computer Graphics and Applications*, May 1982, pp. 11–21.

[169] G.H. Meisters, "Polygons Have Ears," *Amer. Math. Monthly*, Vol. 82, 1975, pp. 648–651.

[170] D. Moore, "Subdividing Simplices," *Graphics Gems III*, D. Kirk, editor, Academic Press, 1992, pp. 244–249.

[171] D. Moore, "Understanding Simploids," *Graphics Gems III*, D. Kirk, editor, Academic Press, 1992, pp. 250–255.

[172] J.-M. Moreau and P. Volino, "Constrained Delaunay Triangulation Revisited," *Proc. 5th Canad. Conf. Comput. Geom.*, 1993, pp. 340–345.

[173] D.E. Muller and F.P. Preparata, "Finding the Intersection of Two Convex Polyhedra," *Theoretical Computer Science*, Vol. 7, 1978, pp. 217–236.

[174] E.J. Nadler, *Piecewise Linear Approximation on Triangulations of a Planar Region*, Ph.D. thesis, Brown University, Division of Applied Mathematics, 1985.

[175] A. Narkhede and D. Manocha, "Fast Polygon Triangulation Based on Seidel's Algorithm," *Graphic Gems V*, Academic Press, 1995, pp. 394–397.

[176] J.M. Nelson, "A Triangulation Algorithm for Arbitrary Planar Domains," *Appl. Math. Modelling*, Vol. 2, 1978, pp. 151–159.

[177] G.M. Nielson, "The Side-Vertex Method for Interpolation in Triangles," *Journal of Approx. Theory*, Vol. 25, 1979, pp. 318–336.

[178] G.M. Nielson, "Minimum Norm Interpolation in Triangles," *SIAM Journal Numer. Analysis*, Vol. 17, 1980, pp. 46–62.

[179] G.M. Nielson, "A Method for Interpolating Scattered Data Based upon a Minimum Norm Network," *Mathematics of Computation*, Vol. 40, 1983, pp. 253–271.

[180] G M. Nielson, *An Example with a Local Minimum for the Minmax Ordering of Triangulations*, Arizona State University Computer Science Technical Report TR-87-014, 1987.

[181] G.M. Nielson, "Coordinate Free Scattered Data Interpolation," *Topics in Multivariate Approximation*, C. Chui, F. Utreras, L. Schumaker, editors, Academic Press, New York, 1987, pp. 175–184.

[182] G.M. Nielson, "A Characterization of an Affine Invariant Triangulation," *Geometric Modelling*, Computing Supplementum 8, G. Farin, H. Hagen, H. Noltemeier, W. Knoedel, editors, Springer, 1993, pp. 191–210.

[183] G.M. Nielson, *How Many Ways Can a Cube Be Subdivided Into Tetrahedra?*, Arizona State University Computer Science Department Technical Report TR-95-13, 1995.

[184] G.M. Nielson and T. Foley, "A Survey of Applications of an Affine Invariant Metric," *Mathematical Methods in Computer Aided Geometric Design*, T. Lyche and L.L. Schumaker, editors, Academic Press, New York, 1989, pp. 445–467.

[185] G.M. Nielson and R. Ramaraj, "Interpolation over a Sphere," *Computer Aided Geometric Design*, Vol. 4, 1987, pp. 41–57.

[186] G.M. Nielson and B. Hamann, "The Asymptotic Decider: Resolving the Ambiguity in Marching Cubes," *Proceedings of Visualization '91*, IEEE Computer Society Press, Los Alamitos, Calif., 1990, pp. 83–91.

[187] G.M. Nielson and K. Opitz, "The Face-Vertex Method for Interpolating in Tetrahedra," *Workshop on Computational Geometry*, A. Conte, V. Demichelis, F. Fontanella and I. Galligani, editors, World Scientific, 1993, pp. 231–244.

[188] G.M. Nielson and J. Tvedt, "Comparing Methods of Interpolation for Scattered Volumetric Data," *State of the Art in Computer Graphics—Aspects of Visualization*, D. Rogers and R.A. Earnshaw editors, Springer-Verlag, 1994, pp. 67–86.

[189] G.M. Nielson, D.H. Thomas, and J.A. Wixom, "Interpolation in Triangles," *Bull. Austral. Math. Soc.*, Vol. 20, 1979, pp. 115–130.

[190] T. Ohya, M. Iri, and K. Murota, "Improvements of the Incremental Method for the Voronoi Diagram with Computational Comparison of Various Algorithms," *Journal of the Operations Research Society of Japan*, Vol. 27, No. 4, 1984, pp. 306–336.

[191] A. Okabe, B. Boots, and K. Sugihara, *Spatial Tessellations: Concepts and Applications of Voronoi Diagrams*, Wiley & Sons, 1992.

[192] A.A. Oloufa, "Triangulation Applications in Volume Calculation," *Journal of Computing in Civil Engineering*, Vol. 5, No. 1, Jan. 1991, pp. 103–121.

[193] T.K. Peucker, R.J. Fowler, and J.J. Little, "The Triangulated Irregular Network," *Proceedings ASP-ACSM Symposium on Digital Terrain Models*, 1978.

[194] C.S. Peterson, "Adaptive Contouring of Three-Dimensional Surfaces," *CAGD*, Vol. 1, 1984, pp. 61–74.

[195] L.A. Piegl and A.M. Richard, "Algorithm and Data Structure for Triangulating Multiply Connected Polygonal Domains," *Computers & Graphics*, Vol. 17, No. 5, 1993, pp. 563–574.

[196] D.A. Plaisted and J. Hong, "A Heuristic Triangulation Algorithm," *J. Algorithms*, Vol. 8, 1987, pp. 405–437,

[197] M. Pourazady and M. Radhakrishnan, "Optimization of a Triangular Mesh," *Computers and Structures*, Vol. 40, No. 3, 1991, pp. 795–804.

[198] M.J.D. Powell, "Piecewise Quadratic Approximation on Triangles," *Software for Numerical Mathematics*, D.J. Evans, editor, Academic Press, New York, 1974.

[199] M.J.D. Powell and M.A. Sabin, "Piecewise Quadratic Approximation on Triangles," *ACM Trans. on Mathematical Software*, Vol. 3, 1977, pp. 316–325.

[200] P.L. Power, "Minimal Roughness Property of the Delaunay Triangulation: A Shorter Approach," *Computer Aided Geometric Design*, Vol. 9, 1992, pp. 491–494.

[201] P.L. Power, "The Neutral Case for the Min-Max Angle Criterion: A Generalized Approach," *Computer Aided Geometric Design*, Vol. 9, 1992, pp. 413–418.

[202] F.P. Preparata and S.J. Hong, "Convex Hull of a Finite Set of Points in Two and Three Dimension," *Commun. ACM*, Vol. 20, No. 2, Feb. 1977, pp. 87–93.

[203] F.P. Preparata and M.I. Shamos, *Computational Geometry: An Introduction*, Springer-Verlag, New York, 1985.

[204] E. Quak and L.L. Schumaker, "Cubic Spline Fitting Using Data Dependent Triangulations," *Computer Aided Geometric Design*, Vol. 7, Nos. 1–4, 1990, pp. 293–301.

[205] E. Quak and L.L. Schumaker, "C^1 Surface Fitting Using Data Dependent Triangulations," *Approximation Theory VI*, C. Chui, L.L. Schumaker, and J. Ward, editors, Academic Press, 1989, pp. 545–548.

[206] E. Quak and L.L. Schumaker, "Least Squares Fitting by Linear Splines on Data Dependent Triangulations," *Curves and Surfaces*, P.-J. Laurent, A. Le Mehaute and L.L. Schumaker, editors, Academic Press, 1991, pp. 387–390.

[207] R.J. Renka, "Algorithm 624: Triangulation and Interpolation of Arbitrarily Distributed Points in the Plane," *ACM TOMS*, Vol. 10, 1984, pp. 440–442.

[208] V.T. Rajan, "Optimality of the Delaunay Triangulation in R^d," *Proc. 7th ACM Symp. Comp. Geometry*, 1991, pp. 357–363.

[209] P.N. Rathie, "A Census of Simple Planar Triangulations," *J. Comb. Theory B*, Vol. 16, 1974, pp. 134–138.

[210] D. Rhynsburger, "Analytic Delineation of Thiessen Polygons," *Geograph. Anal.*, Vol. 5, 1973, pp. 133–144.

[211] S. Rippa, "Minimal Roughness Property of the Delaunay Triangulation," *Computer Aided Geometric Design*, Vol. 7, 1990, pp. 489–497.

[212] S. Rippa, "Long and Thin Triangles Can Be Good for Linear Interpolation," *SIAM Journal on Numerical Analysis*, Vol. 29, No. 1, Feb. 1992, pp. 257–270.

[213] S. Rippa, *Piecewise Linear Interpolation and Approximation Schemes over Data Dependent Triangulations*, Ph.D. thesis, Tel Aviv, 1989.

[214] S. Rippa and B. Schiff, "Minimum Energy Triangulations for Elliptic Problems," *Comp. Meth. in Applied Mech. and Eng.*, Vol. 84, 1990, pp. 257–274.

[215] C.A. Rogers, *Packing and Covering*, Cambridge University Press, 1964.

[216] J. Ruppert and R. Seidel, "On the Difficulty of Tetrahedralizing 3-Dimensional Nonconvex Polyhedra," *Proc. 5th ACM Symp. Comp. Geometry*, 1989, pp. 380–393.

[217] N. Sapidis and R. Perucchio, "Delaunay Triangulation of Arbitrarily Shaped Planar Domains," *Computer Aided Geometric Design*, Vol. 8, 1991, pp. 421–438.

[218] V. Sarin and S. Kapoor, "Algorithms for Relative Neighbourhood Graphs and Voronoi Diagrams in Simple Polygons," *Proc. 4th Canad. Conf. Comput. Geom.*, 1992, pp. 292–298.

[219] L. Scarlatos and T. Pavlidis, "Optimizing Triangulation by Curvature Equalization," *Proceedings of Visualization '92*, IEEE Computer Society Press, Oct. 1992, pp. 333–339.

[220] B. Schachter, "Decomposition of Polygons Into Convex Sets," *IEEE Transactions on Computing C-27*, Vol. 11, Nov. 1978, pp. 1078–1082.

[221] E. Schoenhardt, "Ueber die Zerlegung von Dreieckspolyedern in Tetraeder," *Math. Annalen*, Vol. 98, 1928, 309–312.

[222] W.J. Schroeder and M.S. Shephard, "Geometry-Based Fully Automatic Mesh Generation and the Delaunay Triangulation," *Int. J. Numer. Meth. Eng.*, Vol. 26, 1988, pp. 2503–2515.

[223] W.J. Schroeder, J.A. Zarge, and W.E. Lorensen, "Decimation of Triangle Meshes," *SIGGRAPH '92*, Vol. 26, July 1992, pp. 65–70.

[224] L.L. Schumaker, "Fitting Surfaces to Scattered Data," *Approximation Theory II*, G.G. Lorentz, C.K. Chui, and L.L. Schumaker, editors, Academic Press, 1976, pp. 203–268.

[225] L.L. Schumaker, "Computing Optimal Triangulations Using Simulated Annealing," *Computer Aided Geometric Design*, Vol. 10, Nos. 3–4, pp. 329–345.

[226] L.L. Schumaker, "Triangulation Methods," *Topics in Multivariate Approximation*, L.L. Schumaker, C. Chui and F. Utreras, editors, Academic Press, New York, 1987, pp. 219–232.

[227] L.L. Schumaker, "Triangulations Methods in CAGD," *IEEE Computer Graphics and Applications*, Vol. 13, Jan. 1993, pp. 47–52.

[228] A. Seidel, "Constrained Delaunay Triangulations and Voronoi Diagrams with Obstacles," *1978–1988 Ten Years IIG*, H.S. Poingratz and W. Schinnerl, editors, 1988, pp. 178–191.

[229] M. Senechal, "Which Tetrahedra Fill Space?," *Math. Magazine*, Vol. 54, 1981, pp. 227–243.

[230] M.I. Shamos, *Computational Geometry*, Ph.D. dissertation, Yale University, 1978.

[231] M. Shapiro, "A Note on Lee and Schachter's Algorithm for Delaunay Triangulation," *International Journal of Computer and Information Sciences*, Vol. 10, No. 6, 1981, pp. 413–418.

[232] D N. Shenton and Z.J. Cendes, "Three-Dimensional Finite Element Mesh Generation Using Delaunay Tessellation," *IEEE Trans. on Magnetics, MAG-21*, 1985, pp. 2535–2538.

[233] D. Shirley and A. Tuchman, "A Polygonal Approximation to Direct Scalar Volume Rendering," *Computer Graphics*, Vol. 24, Nov. 1990, pp. 63–70.

[234] G.M. Shute, L.L. Deneen, and C.D. Thomborson, "An O(N log N) Plane-Sweep Algorithm for L_1 and L_∞ Delaunay triangulations," *Algorithmica*, Vol. 6, 1978, pp. 207–221

[235] R. Sibson, "Locally Equiangular Triangulations," *Computer J.*, Vol. 21, 1978, pp. 243–245.

[236] R. Sibson, "A Brief Description of Natural Neighbour Interpolation," *Chapter 2 of Interpreting Multivariate Data*, Wiley, New York, 1981.

[237] C.T. Silva, J.S.B. Mitchell, and A.E. Kaufman, "Automatic Generation of Triangular Irregular Networks Using Greedy Cuts," *Proceedings of Visualization '95*, IEEE Computer Society Press, Oct. 1995, pp. 201–208.

[238] S.W. Sloan, "A Fast Algorithm for Constructing Delaunay Triangulations in the Plane," *Advances in Engineering Software*, Vol. 9, Jan. 1987, pp. 34–55.

[239] S.W. Sloan and G.T. Houlsby, "An Implementation of Watson's Algorithm for Computing 2-Dimensional Delaunay Triangulations," *Advances in Engineering Software*, Vol. 6, 1984, pp. 192–197.

[240] C. Stein, B. Becker, and N. Max, "Sorting and Hardware Assisted Rendering for Volume Visualization," *1994 Symposium on Volume Visualization*, Washington, D.C., Oct. 1994, pp. 83–89.

[241] K. Sugihara and M. Iri, "Construction of the Voronoi Diagram for "One Million" Generators in Single-Precision Arithmetic," *Proceedings of the IEEE*, Vol. 80, 1992, pp. 1471–1484.

[242] M. Tanemura, T. Ogawa, and W. Ogita, "A New Algorithm for Three-Dimensional Voronoi Tessellation," *Journal of Computational Physics*, Vol. 51, 1983, pp. 191–207.

[243] R.E. Tarjan and C.J. Van Wyk, "An O(n log log n)-Time Algorithm for Triangulating a Simple Polygon," *SIAM J. Comput.*, Vol. 17, 1988, pp. 143–178.

[244] A.H. Thiessen, "Precipitation Averages for Large Areas," *Monthly Weather Review*, Vol. 39, 1911, pp. 1032–1034.

[245] J.F. Thompson, *Numerical Grid Generation*, North-Holland, 1982.

[246] J.C. Tipper, "Straightforward Iterative Algorithm for the Planar Voronoi Diagram," *Information Processing Letters*, Vol. 34, No. 3, Apr. 1990, pp. 155–160.

[247] J.C. Tipper, "FORTRAN Programs to Construct the Planar Voronoi Diagram," *Computers & Geosciences*, Vol. 17, 1991, pp. 597–632.

[248] G. Toussaint, "Efficient Triangulation of Simple Polygons," *Visual Comput.*, Vol. 7, 1991, pp. 280–295.

[249] G.T. Toussaint, C. Verbrugge, C. Wang, and B. Zhu, "Tetrahedrization of Simple and Not Simple Polyhedra," *CCCG Proc. of the Fifth Canadian Conference on Computational Geometry*, 1994.

[250] V.J.D. Tsai, "Delaunay Triangulation in TIN Creation: An Overview and a Linear-Time Algorithm," *Int. J. Geographical Information Systems*, Vol. 7, 1993, pp. 501–524.

[251] W.T. Tutte, "A Census of Planar Triangulations," *Canadian J. Math.*, Vol. 14, 1962, pp. 21–38.

[252] G. Voronoi, "Nouvelles applications des paramatres continusala theorie des formes quadratiques, Deuxieme Memoire, Recherches sur les parallelloedres primitifs," *J. reine angew. Mathe.*, Vol. 134, 1908, pp. 198–287.

[253] C. Wang and L. Schubert, "An Optimal Algorithm for Constructing the Delaunay Triangulation of a Set of Line Segments," *Proc. 3rd ACM Symp. Comp. Geometry*, 1987, pp. 223–232.

[254] D.F. Watson, "Computing the n-Dimensional Delaunay Tessellation with Application to Voronoi Polytopes," *Comp. J.*, Vol. 24, 1981, pp. 167–172.

[255] D.F. Watson and G.M. Philip, "Systematic Triangulations," *Computer Vision, Graphics, and Image Processing*, Vol. 26, 1984, pp. 217–223.

[256] N.D. Weatherhill and O. Hassan, "Efficient Three-Dimensional Delaunay Triangulation with Automatic Point Creation and Imposed Boundary Constraints," *Internat. J. Numer. Meth. Eng.*, Vol. 37, 1994, pp. 2005–3039.

[257] P. Williams, "Visibility Ordering Meshed Polyhedra," *ACM Transactions on Graphics*, Vol. 11, No. 2, 1992, pp. 103–126.

[258] B. Woerdenweber, *Automatic Mesh Generation of 2- and 3-Dimensional Curvilinear Manifolds*, Ph.D. dissertation, University of Cambridge, 1981.

[259] B. Woerdenweber, "Finite-Element Analysis for the Naive User," *Solid Modeling by Computers from Theory to Applications*, M.S. Pickett and J. Boyse, editors, Plenum, 1984, pp. 81–100.

[260] A.J. Worsey and B. Piper, "A Trivariate Powell-Sabin Interpolant," *Computer Aided Geometric Design*, Vol. 5, No. 3, 1988 pp. 177–186.

[261] A.J. Worsey and G. Farin, "An n-Dimensional Clough-Tocher Interpolant," *Constructive Approximation*, Vol. 3, 1987, pp. 99–110.

[262] F.F. Yao, "Computational Geometry," *Handbook of Theoretical Computer Science*, Vol. A, Chapter 7, J. van Leeuwen, editor, Elsevier and MIT Press, 1990, pp. 343–389.

[263] A. Zenisek, "Polynomial Approximation on Tetrahedrons in the Finite Element Method," *J. Approximation Theory*, Vol. 7, 1973, pp. 334–351.

Chapter 21

Tools for Computing Tangent Curves and Topological Graphs for Visualizing Piecewise Linearly Varying Vector Fields over Triangulated Domains

Gregory M. Nielson, Il-Hong Jung, Nat Srinivasan, Junwon Sung, and Jong-Beum Yoon

Abstract. *We describe some methods for computing tangent curves and topological graphs for a vector field defined over a two-dimensional domain. We assume that the vector field is piecewise linearly defined over a triangulation of the domain. Piecewise explicit representations in terms of elementary transcendental functions form the basis of algorithms for determining and displaying tangent curves. Topological methods which link critical values with separating tangent curves are developed and discussed.*

21.1 Introduction

Streamlines are a well-established visualization technique for investigating a vector field. In this chapter we describe some new techniques for computing these invariant tangent curves. Conventional methods for computing tangent curves consist of using numerical methods for solving vector-valued initial valued problems. Euler's method is the simplest and Runge-Kutta-type methods are often used in practice (see [3,24,39,41,49]). The issues involved in selecting a particular algorithm usually focus on the two, often opposing, requirements of accuracy and speed. Accuracy is especially important in the computation of tangent curves for topological methods. Erroneous results can easily occur unless special provisions are taken to control errors. Speed is particularly important for interactive methods, but accuracy should not be discounted too much in this context since the goal of the visualization is to impart meaningful and valid information. The methods described in this chapter allow for a reasonable balance to be achieved between accuracy and speed.

A streamline (or tangent) curve is a parametric curve

$$P(t) = \begin{pmatrix} x(t) \\ y(t) \end{pmatrix}$$

that is everywhere tangent to the vector field. If

$$V(x, y) = \begin{pmatrix} u(x, y) \\ v(x, y) \end{pmatrix}$$

represents a static vector field, then P is characterized by the equations

$$P'(t) = V(P(t)) = \begin{pmatrix} \dot{x}(t) \\ \dot{y}(t) \end{pmatrix} = \begin{pmatrix} u(x(t), y(t)) \\ v(x(t), y(t)) \end{pmatrix}. \tag{21.1}$$

Typically there is an entire family of solutions for Equation (21.1) and a particular solution is selected with the initial condition

$$P(0) = \begin{pmatrix} x(0) \\ y(0) \end{pmatrix} = P_0 = \begin{pmatrix} x_0 \\ y_0 \end{pmatrix}.$$

In this chapter, we discuss methods for the special case where $V(x, y)$ is a piecewise linear function. The domain is decomposed into a collection of triangles and V has the form

$$V(x, y) = \begin{pmatrix} u(x, y) \\ v(x, y) \end{pmatrix} = \begin{pmatrix} a_{11}x + a_{12}y + b_1 \\ a_{21}x + a_{22}y + b_2 \end{pmatrix} = A \begin{pmatrix} x \\ y \end{pmatrix} + B \tag{21.2}$$

over each triangle. With this assumption, the tangent curve becomes a piecewise concatenation of curve segments with each segment associated with a particular triangle. The entry point on a particular triangle provides the initial conditions for the constant coefficient ODE which characterizes the curve segment for each triangle. The exit point serves as the entry point for the next adjoining triangle domain. This basic idea is further illustrated in Figure 21.1. In this way, it is possible to completely characterize and know a tangent curve by a sequence of entry (exit) points.

There are a variety of sources for the type of data covered here. If a 2D flow (or a 3D flow assumed to be constant in one direction) is measured at a collection of scattered, planar points, then the domain points can serve as the vertices for a triangulation of the domain. The Delaunay triangulation of the convex hull would be a possibility. See Nielson [33]. We will show examples later of a flow over a spherical domain which represents wind speed and direction over the Earth. If the data is measured at scattered locations, these points can serve as the vertices of a triangulation of the sphere and the methods of this chapter can be applied. If the data is modeled, then the model can be evaluated on a triangular grid and again the methods of this chapter will apply. Curvilinear grids which normally associate

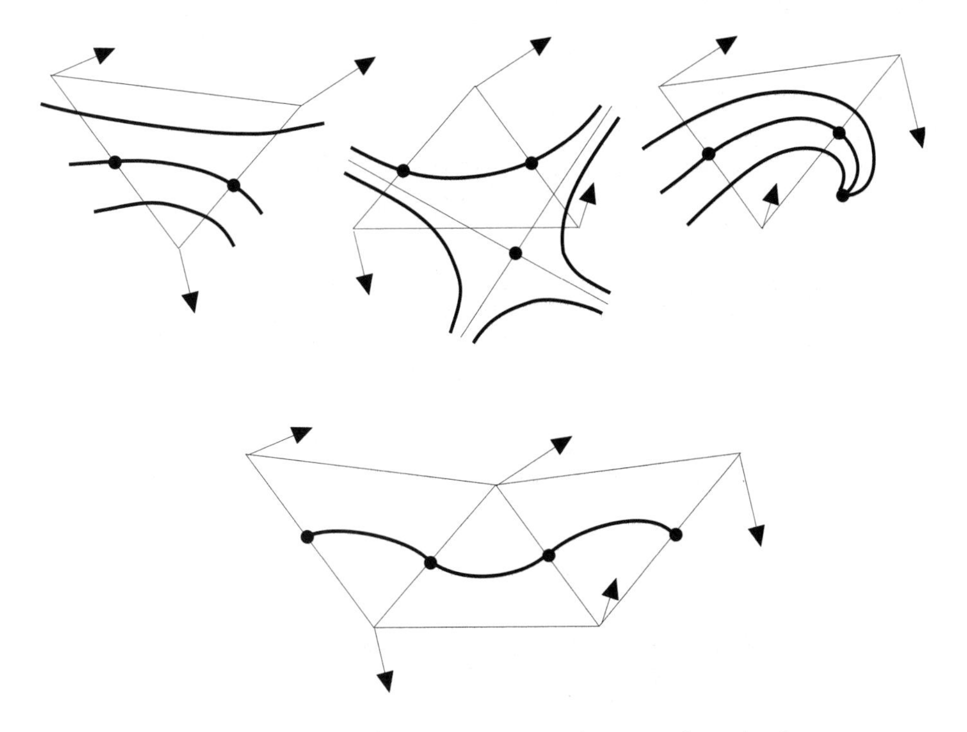

Figure 21.1: A composite tangent curve running across three triangles.

with simulated data can also be triangulated by simply inserting one diagonal or the other into the quadrilateral cells. See Figure 21.2.

The values A and B of Equation (21.2) for a particular triangle are determined from the values of the vector field at the vertices of the triangle. If the vertices of the triangle are labeled as in Figure 21.3, then the equations

$$\begin{aligned} a_{11}x_i + a_{12}y_i + b_1 &= u(x_i, y_i) \\ a_{11}x_j + a_{12}y_j + b_1 &= u(x_j, y_j) \\ a_{11}x_k + a_{12}y_k + b_1 &= u(x_k, y_k) \\ \\ a_{21}x_i + a_{22}y_i + b_2 &= v(x_i, y_i) \\ a_{21}x_j + a_{22}y_j + b_2 &= v(x_j, y_j) \\ a_{21}x_k + a_{22}y_k + b_2 &= v(x_k, y_k) \end{aligned} \tag{21.3}$$

will yield the values of A and B. It is also possible to use barycentric coordinates to find these values.

The remainder of the chapter is organized as follows. In Section 21.2, we discuss explicit methods which are based upon the fact that the differential equation which characterizes a tangent curve is first order with constant coefficients and so an explicit solution in terms of common transcendental functions is possible. Section 21.3 is concerned with incremental methods which produce a sequence of points on the curve (or approximations)

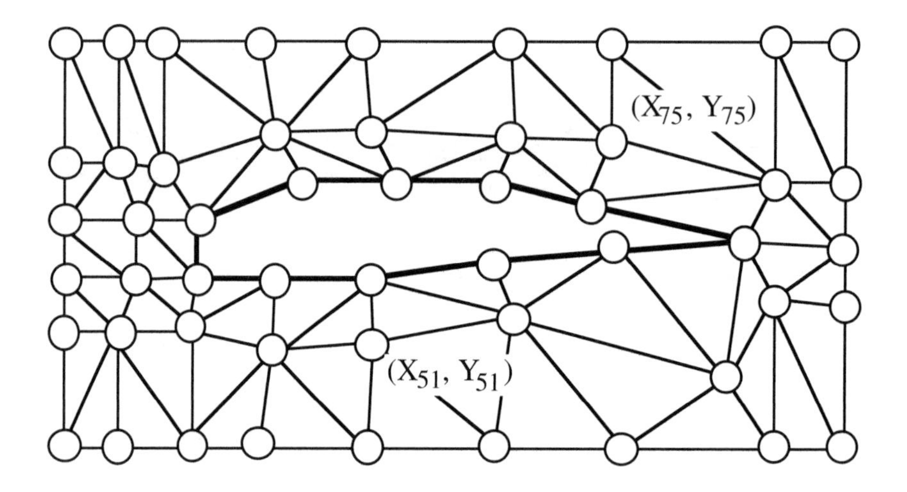

Figure 21.2: Triangulated curvilinear grid.

with the computation of each subsequent point based upon the previous point. Topological methods are covered in Section 21.4.

21.2 Explicit Methods

In this section we describe what we call "explicit" methods. Because of the special nature of the ODE which characterizes a tangent curve, it is possible to obtain explicitly defined solutions and therefore, there is no need to use numerical methods to compute the values of these curves. Library routines for computing common transcendental functions can be used instead. Computing the exit point of a particular curve segment, which will serve as the initial point for the next curve segment, amounts to computing the intersection of a curve and a line segment. This is equivalent to a univariate root computation problem. We discuss two general approaches to this intersection point calculation problem. Each approach is covered in a subsequent subsection. In Section 21.2.1 we discuss the approach which is based upon a parametric representation of the tangent curve and an implicit representation of a line containing one of the edges of the triangle. The parametric curve is substituted into the implicit line equation yielding a single equation in the parameter of the tangent curve. The root of this equation yields the parameter value of the intersection point and this point is subsequently tested to determine if it is actually on the edge (a subset of the line). In Section 21.2.2, the second general approach is covered. It is based upon an implicit representation of the tangent curve and a parametric representation of the edge of the triangle. In this approach, the parametric representation of the edge is substituted into the implicit representation of the tangent curve, yielding again a single, univariate equation. The root is computed and used to evaluate the parametric representation of the edge so as to obtain the intersection point.

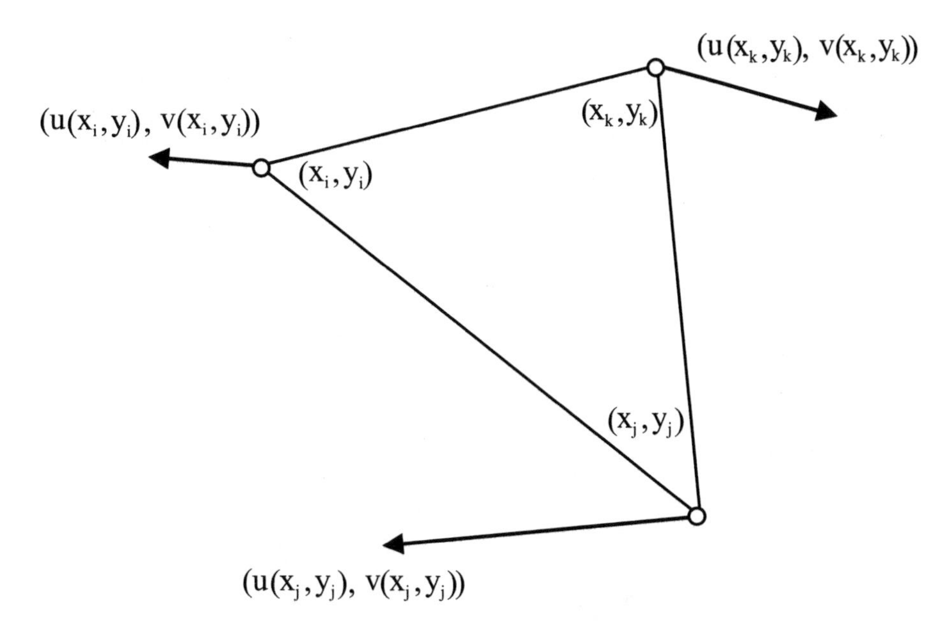

Figure 21.3: Notation and conventions used for data.

21.2.1 Parametric Tangent Curve, Implicit Edge

In this section, we discuss the general properties of a tangent curve for a linearly varying vector field over a single triangular domain. In particular we give the details of a parametric representation,

$$P(t) = \begin{pmatrix} x(t) \\ y(t) \end{pmatrix}.$$

(Implicit representations are covered in the next section, Section 21.2.2.) The curve enters the triangle at a point $P(0) = P_0$ and then either attaches to a critical point in the triangle or exits from the triangle at a point $P(t_e) = P_e$. We assume that the entry point is given and so we need to compute, t_e, P_e, and the various coefficients of a parametric representation of the curve from P_0 to P_e. The value t_e along with the parametric representation are used for displaying the curve and the value P_e serves as the entry point for the next triangle. If P_e lies on one of the edges, then t_e must satisfy one of the equations

$$f_a(x(t_e), y(t_e)) = 0, \qquad a = i, j, k \tag{21.4}$$

where

$$\begin{aligned} f_i(x,y) &= (x - x_j)(y_k - y_j) - (y - y_j)(x_k - x_j) \\ f_j(x,y) &= (x - x_k)(y_i - y_k) - (y - y_k)(x_i - x_k) \\ f_k(x,y) &= (x - x_i)(y_j - y_i) - (y - y_i)(x_j - x_i). \end{aligned}$$

But not all of the roots of $f_a(x(t), y(t))$, $a = i, j, k$ are candidates to be t_e. It must be the case that t_e is the smallest of these roots and that P_e is actually on the appropriate edge and not simply on the line containing this edge. The point P_e could potentially be on any of the edges, including the edge containing P_0, but some simple tests and modifications can possibly eliminate some edges and portions of edges where P_e might be found. Along an edge, the flow is either always in, always out, or there is a special point where the flow is parallel to the edge and the direction of flow relative to the triangle changes at this point. This is further illustrated in Figure 21.4. The change of direction point P_Δ is easy to compute. For example, if the direction changes on edge e_k, then

$$P_\Delta = tP_i + (1-t)P_j \tag{21.5}$$

where

$$tV_i + (1-t)V_j = c(P_j - P_i)$$

for some constant, c, and $0 < t < 1$.

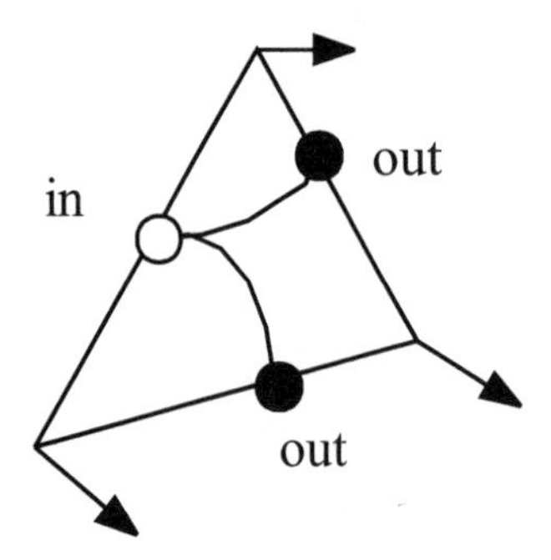

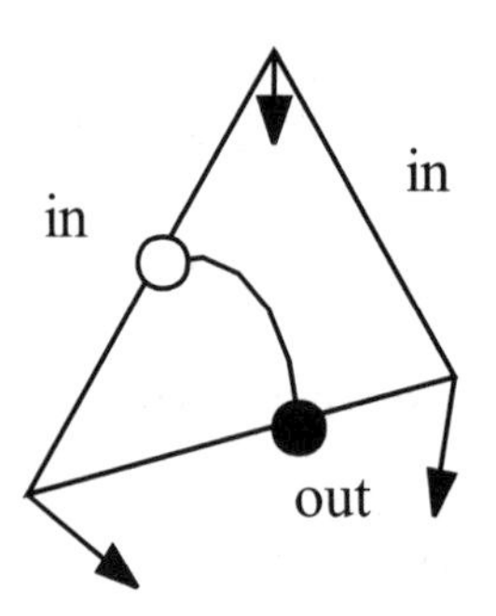

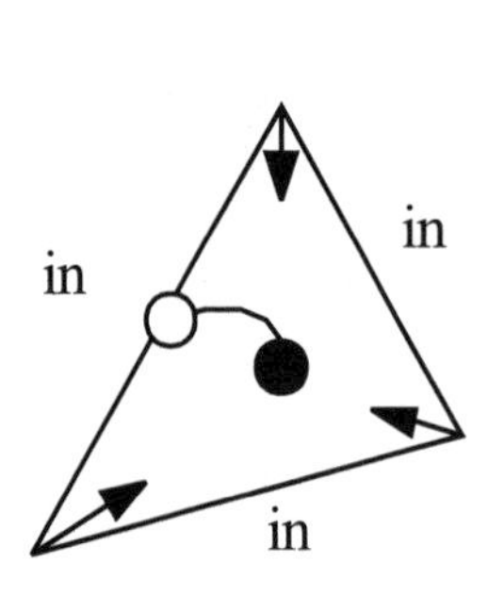

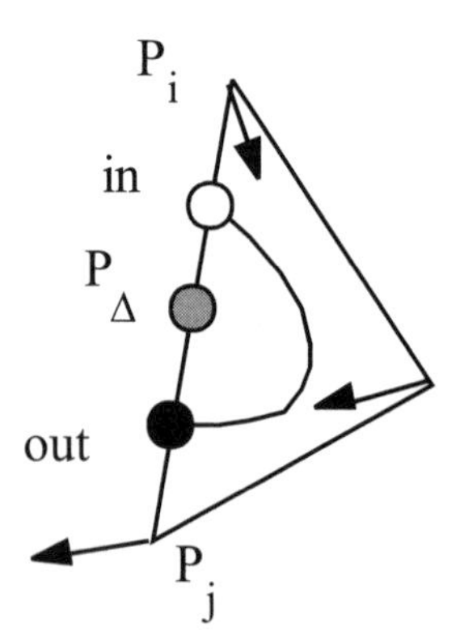

Figure 21.4: Exit point search strategy regions.

We now turn our attention to the details of the parametric representation of the tangent curve. The tangent curve

$$P(t) = \begin{pmatrix} x(t) \\ y(t) \end{pmatrix}$$

over a particular triangular cell is defined as

$$\begin{pmatrix} \dot{x}(t) \\ \dot{y}(t) \end{pmatrix} = A \begin{pmatrix} x(t) \\ y(t) \end{pmatrix} + B \tag{21.6}$$

with the initial conditions

$$\begin{pmatrix} x(0) \\ y(0) \end{pmatrix} = \begin{pmatrix} x_0 \\ y_0 \end{pmatrix}.$$

The 2×2 matrix A and the 2×1 vector B are determined as in Equation (21.3) or equivalent other means. The general solution of this initial value problem is of the form

$$\begin{pmatrix} x(t) \\ y(t) \end{pmatrix} = P(t) = \Phi_1(t)E_1 + \Phi_2(t)E_2 + C \tag{21.7}$$

where the particular functions Φ_1, Φ_2, and the coefficients E_1, E_2, and C depend on the eigenvalues of A. There are five separate cases:

Case 1. A has two real, nonzero eigenvalues, $0 \neq r_1 \neq r_2 \neq 0$.

$$P(t) = e^{tr_1}E_1 + e^{tr_2}E_2 + P_c, \tag{21.8}$$

$$E_1 = \left(\frac{A - r_2 I}{r_1 - r_2}\right)(P_0 - P_c), \qquad E_2 = \left(\frac{A - r_1 I}{r_2 - r_1}\right)(P_0 - P_c), \qquad AP_c + B = 0.$$

Case 2. A has one zero and one nonzero eigenvalue, $0 = r_1, r_2 \neq 0$.

$$P(t) = tE_1 + \left(e^{tr_2} - 1\right)E_2 + P_0, \tag{21.9}$$

$$E_1 = \left(I - \frac{A}{r_2}\right)B, \qquad E_2 = \frac{A}{r_2}\left(P_0 + \frac{B}{r_2}\right).$$

Case 3. A has only a zero eigenvalue, $r_1 = 0 = r_2$.

$$P(t) = tE_1 + t^2 E_2 + P_0, \tag{21.10}$$

$$E_1 = AP_0 + B, \qquad E_2 = \frac{AB}{2}.$$

Case 4. A has a single real, nonzero eigenvalue, $r_1 = r_2 \neq 0$.

$$P(t) = e^{tr_1} E_1 + te^{tr_1} E_2 + P_c, \tag{21.11}$$

$$E_1 = P_0 - P_c, \qquad E_2 = (A - r_1 I)(P_0 - P_c), \qquad AP_c + B = 0.$$

Case 5. A has complex eigenvalues, $\mu + \lambda i, \mu - \lambda i, \lambda \neq 0$.

$$P(t) = cos(\lambda t) e^{\mu t} E_1 + sin(\lambda t) e^{\mu t} E_2 + P_c \tag{21.12}$$

$$E_1 = P_0 - P_c, \qquad E_2 = \left(\frac{A - \mu I}{\lambda}\right)(P_0 - P_c), \qquad AP_c + B = 0.$$

In Cases $1, 4$, and 5, A is nonsingular and it is guaranteed that there is a unique critical value satisfying $AP_c + B = 0$. In Cases 2 and 3, it is possible that no critical values exist or even that an entire line of critical values exist. Not all cases occur with equal frequency. Cases 1 and 5 are the predominate ones, Cases 2 and 4 are much less likely to occur, and Case 3 is extremely rare. This is explained by taking a look at Figure 21.5. In this example, we determined the case for a triangle where we varied the value of the vector field at one vertex of the triangle. The flow at two of the vertices is fixed and illustrated by the arrows drawn at these vertices. The flow at the vertex marked with the white circle is taken be a vector which is based at the vertex and emanating out to an arbitrary point in the plane. We classify this point on the basis of the case classification associated with this flow. The interior to the parabolic bounded region is Case 5. The boundary is Case 4 except for the degenerate subcase of Case 3 indicated by the black box on the boundary. Case 2 consists of a line tangent at this point. Of course the tolerances used for determining these regions affect their relative occurrence in actual practice. It is interesting and instructive to look at some examples of tangent curves for the various cases. We have included samples for each of the cases in Figure 21.6. The data for these examples is the same as in Figure 21.5. The boxes of Figure 21.5 indicate the tip of the one vector of the triangle for the particular example of Figure 21.6. In each of these images, the triangular domain is shown along with arrows at each vertex indicating the flow at these points. Particular tangent curves are determined by using initial values along a particular edge. We use 10 equally spaced points along these edges. The flow is shown at each of these initial points and the tangent curve is traversed out in both negative and positive parameter domain for a fixed amount. Some additional examples are shown in Figure 21.7. For these examples, the flow data at each vertex is indicated by the line segment and again we use initial values at 10 equally spaced values along an edge and the tangent curves are traversed out in both positive and negative parameter directions.

Displaying the tangent curve and computing the roots for the determination of t_e depends upon the evaluation of $P(t)$ given in its various forms by the five cases above. Depending upon the computing resources available, it is possible to structure efficient methods for performing these computations. In some cases, it is desirable to compute $P(i\Delta t)$, $i = 0, 1, 2, \ldots$, where Δt is some fixed increment of the parameter t. This may be the case in a situation where these values are used to display the curve and the curve is determined

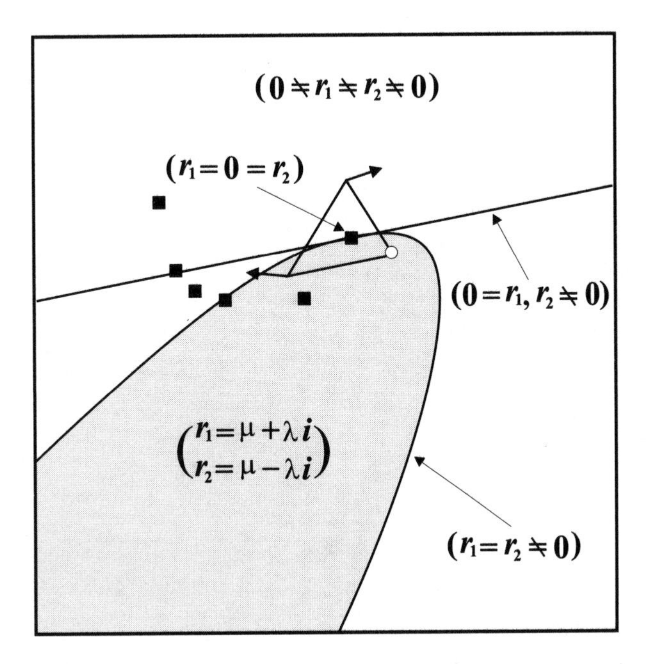

Figure 21.5: The five case regions.

to exit the triangle once one of the functions f_i, f_j, or f_k (of Equation (21.4)) is found to change sign from $a\Delta t$ to $(a+1)\Delta t$. The exponentials of Cases 1, 2, 4, and 5 can be computed by the computation of a single exp(Δt) and subsequent multiplications. The sin and cos functions of Case 5 can be computed with the formulas

$$sin(t+\Delta t) = cos(t)sin(\Delta t) + sin(t)cos(\Delta t)$$

$$cos(t+\Delta t) = cos(t)cos(\Delta t) - sin(t)sin(\Delta t).$$

We should also point out that the formulas which form the basis of the incremental methods of Section 21.3.2 can also be used to evaluate the curve at equally spaced parameter values for display purposes.

21.2.2 Parametric Edge, Implicit Tangent Curve

As we have previously mentioned, there are two general approaches to computing the exit point $P(t_e)$. A parametric representation for the tangent curve can be substituted into an implicit representation for the line segment as in Section 21.2.1. Or a parametric (or explicit) representation of the line segment can be substituted into an implicit representation of the curve. In this section, we discuss the latter approach. In this approach, either an explicit, $y = ax + b$ or $x = cy + d$, or a parametric $(x(s), y(s)) = sP_i + (1-s)P_j$, is used to represent the line segment and an implicit representation

$$F(x(t), y(t)) = 0$$

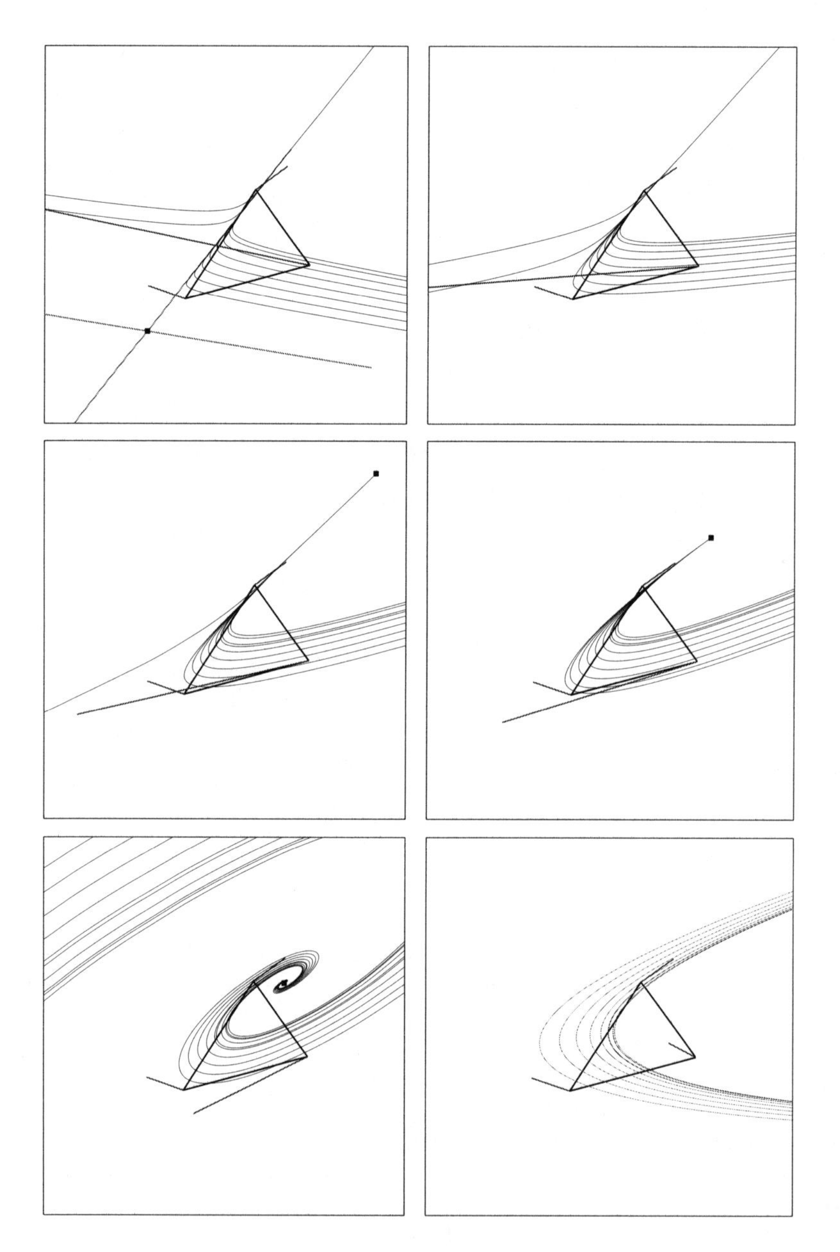

Figure 21.6: Examples of tangent curves for all five cases.

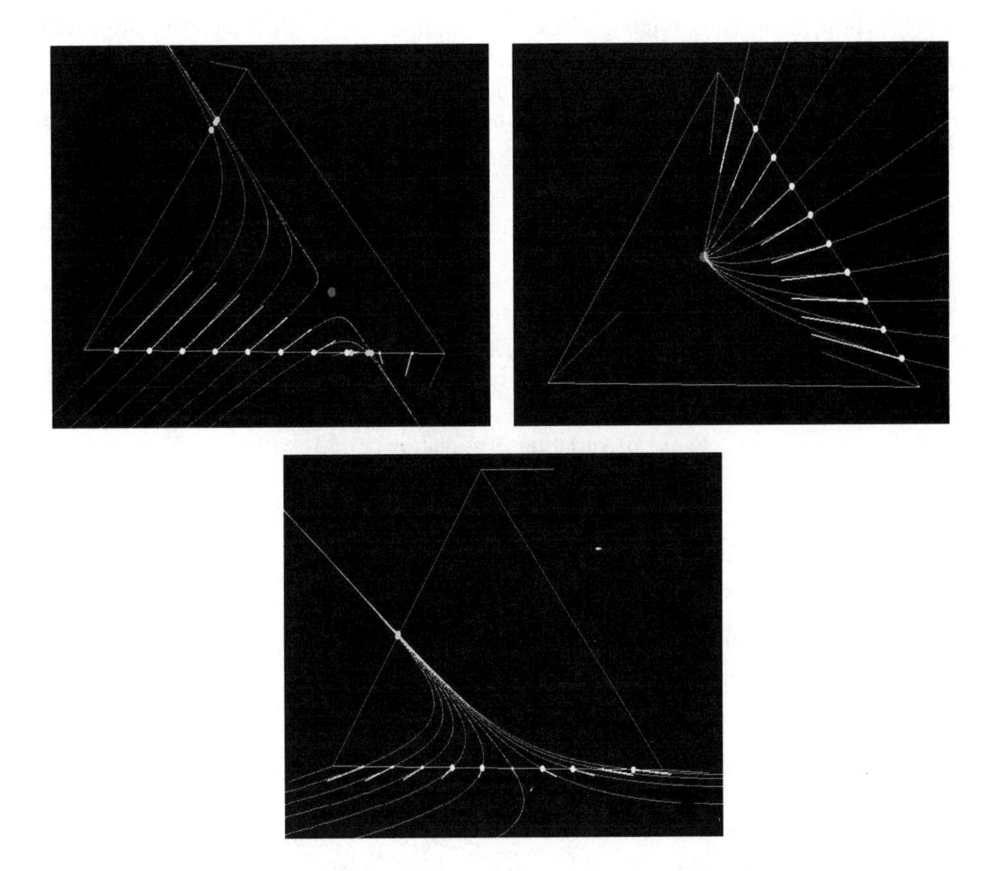

Figure 21.7: Additional examples of tangent curves. *(Upper-left)* Case 1 with $r_1 \cdot r_2 < 0$. *(Upper-right)* Case 1 with $r_1 \cdot r_2 > 0$. *(Lower)* Case 2. See Color Plate 109.

is required for the curve. For the edge e_k, this leads to one of the following root finding problems:

i) $F(x, ax + b)$, $\quad min\{x_i, x_j\} \leq x \leq max\{x_i, x_j\}$

ii) $F(cy + d, y)$, $\quad min\{y_i, y_j\} \leq y \leq max\{y_i, y_j\}$

iii) $F(x(s), y(s))$, $\quad 0 \leq s \leq 1$.

Up to now, the missing component for this approach is an implicit representation of the streamline curve. We now take up this topic. Again, there are five separate cases. In each of the five cases covered below, there are specified two vectors, E_1 and E_2. We let $E = (E_1, E_2)$. If $|E| \neq 0$ then the two vectors E_1 and E_2 are linearly independent, and we can use the change of variables

$$\begin{pmatrix} x \\ y \end{pmatrix} = E \begin{pmatrix} X \\ Y \end{pmatrix} + C. \tag{21.13}$$

The point C is either P_0 for Cases 2 and 3 or P_c for Cases 1, 4, and 5. We now give the details for each separate case.

Case 1. $(0 \neq r_1 \neq r_2 \neq 0)$

If $|E| \neq 0$ then the change of variables of Equation (21.13) with $C = P_c$ changes the parametric curve to

$$X(t) = e^{r_1 t}$$
$$Y(t) = e^{r_2 t}$$

which leads to the implicit equation

$$F(X,Y) = X^{r_2} - Y^{r_1} = 0 \tag{21.14}$$

with $X, Y \geq 0$ and $t = \frac{ln(X)}{r_1} = \frac{ln(Y)}{r_2}$.

If $|E| = 0$ then P_0 is on one of the lines passing through P_c in the direction of the eigenvectors of A, and either $E_1 = 0$ or $E_2 = 0$ (or $E_1 = E_2 = 0$ if $P_0 = P_c$). The tangent curve will be a straight line. If $E_2 = 0$ then an implicit equation for the tangent curve is

$$F(x, y) = (x - x_c)e_{21} - (y - y_c)e_{11} = 0$$

and in the case that $E_1 = 0$, we have

$$F(x, y) = (x - x_c)e_{22} - (y - y_c)e_{12} = 0$$

Note that these last two implicit equations are in terms of the original variables, (x, y). See Figure 21.8.

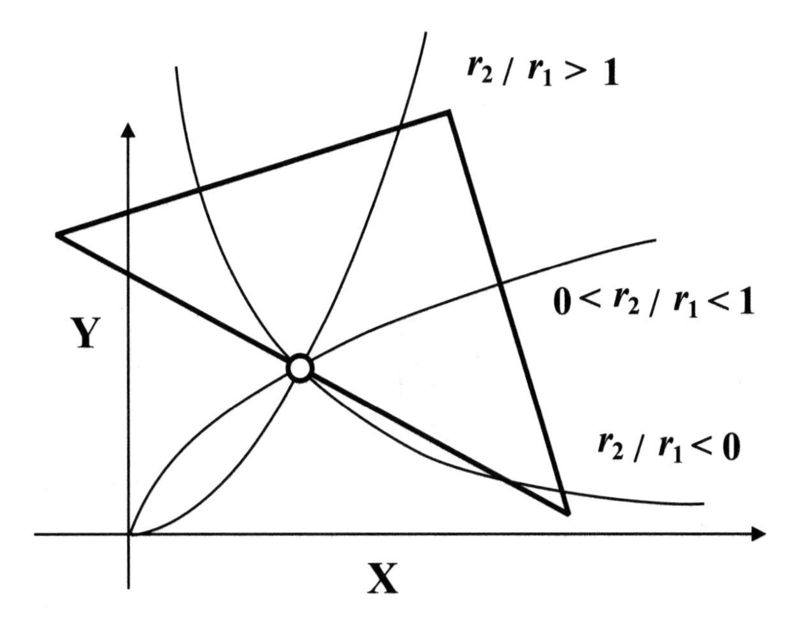

Figure 21.8: Typical curves in the transformed variables for Case 1.

Case 2. $(0 = r_1, r_2 \neq 0)$

If $|E| \neq 0$, then the change of variables given by Equation (21.13) with $C = P_0$ will yield

$$X(t) = t$$

$$Y(t) = e^{tr_2} - 1$$

and

$$F(X, Y) = Y - e^{r_2 X} + 1 = 0 \tag{21.15}$$

with $X \geq 0, \quad Y \geq -1, \quad t = X.$

If $|E| = 0$, then either $E_2 = 0$ and (in the original variables) we have the implicit equation

$$(x - x_0)e_{21} - (y - y_0)e_{11} = 0,$$

or $E_1 = 0$ and we have the equation

$$(x - x_0)e_{22} - (y - y_0)e_{12} = 0.$$

See Figure 21.9.

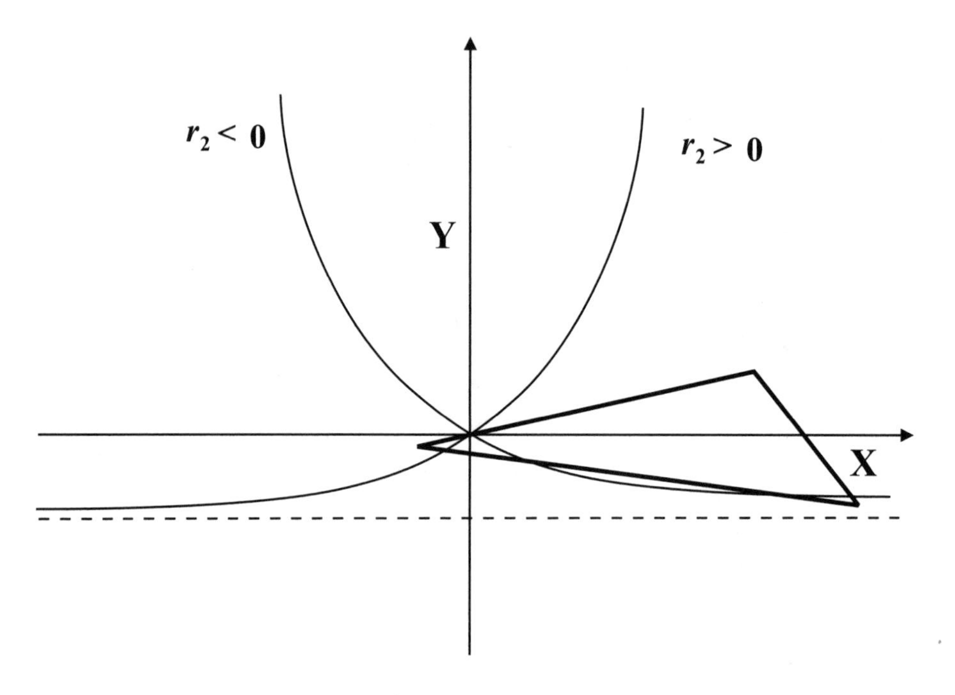

Figure 21.9: Typical curve in the transformed variable for Case 2.

Case 3. $(r_1 = 0 = r_2)$
If $|E| \neq 0$, then Equation (21.13) with $C = P_0$, can be used to obtain

$$Y - X^2 = 0, \qquad X \geq 0, \qquad Y \geq 0; \qquad t = X = \sqrt{Y}. \tag{21.16}$$

If $|E| = 0$, then $E_2 = 0$ and (in the original variables) we have the implicit equation

$$(x - x_0)e_{21} - (y - y_0)e_{11} = 0.$$

Case 4. $(r_1 = r_2 \neq 0)$
If $|E| \neq 0$, then the change of variables of Equation (21.13) leads to

$$Xln(X) - rY = 0, \qquad X, Y \geq 0, \qquad t = \frac{Y}{X}. \tag{21.17}$$

If $|E| = 0$, then the implicit equation (in the original coordinates) is

$$(x - x_c)(y_0 - y_c) - (y - y_c)(x_0 - x_c) = 0.$$

Case 5. $(\mu + \lambda i, \mu - \lambda i, \lambda \neq 0)$
The change of variable given by Equation (21.13) leads to

$$\frac{Y}{X} = tan\left(\frac{\lambda}{2\mu} ln(X^2 + Y^2)\right), \qquad t = \frac{ln(X^2 + Y^2)}{\mu}. \tag{21.18}$$

21.3 Incremental Methods

The basic idea behind incremental methods consists of starting at the entry point $P(0) = P_0$ and then successively computing $P(t_i)$, $i = 1, 2, 3, \ldots$ for increasing values of the parameter t and stopping when $P(t_i)$ leaves the triangle. These successive values that are computed along the way to finding the exit point can also be used as the basis for displaying the curve. Often it is efficient and convenient to use equally spaced values of the parameter; that is, $t_i = i\Delta t, i = 1, 2, 3, \ldots$. At least this can be the default strategy and the value of Δt can be adjusted every so often, depending on accuracy estimates or the spacing requirements between $P(t)$ and $P(t + \Delta t)$.

In this section, we cover two types of incremental methods. In Section 21.3.1 we discuss briefly the application of conventional methods such as Euler's and Runge-Kutta to the special case at hand of linearly varying vector fields. In Section 21.3.2 we discuss some incremental methods for computing values on the curve which are based upon the exact solutions which were explicitly given in Section 21.2. Before we proceed, we wish to briefly discuss two topics that come up in the implementation of incremental methods. Namely, how to choose Δt and how to test when the curve leaves the triangle. We take up the latter topic first.

The functions $f_i(x, y)$, $f_j(x, y)$, and $f_k(x, y)$ of Equation (21.4) were carefully defined so that $f_a(x, y) > 0$ for points (x, y) on the same side of e_a as P_a, $a = i, j,$ or k. This assumes that the points P_i, P_j, and P_k are listed in counterclockwise order. So by testing these functions, we can determine if the next point leaves the triangle and also which edge it leaves from and this can then be used to determine the data for the next triangle it enters. A particularly efficient way to compute all three values is based upon the identity

$$f(x, y) = x(f(1, 0) - f(0, 0)) + y(f(0, 1) - f(0, 0)) + f(0, 0) \tag{21.19}$$

where

$$f(x, y) = \begin{pmatrix} f_i(x, y) \\ f_j(x, y) \\ f_k(x, y) \end{pmatrix}.$$

As we previously mentioned, it is sometimes useful to be able to approximately control the distance between $P(t)$ and $P(t+\Delta t)$ by choosing an appropriate Δt. Say, for example, it is desired that

$$||P(t + \Delta t) - P(t)|| \approx \delta.$$

The mean value theorem implies

$$P(t + \Delta t) - P(t) = \Delta t P'(\tau), \qquad t \leq \tau \leq t + \Delta t.$$

Equation (21.6) then yields

$$P(t + \Delta t) - P(t) = \Delta t(A(P(\tau) + B), \qquad t \leq \tau \leq t + \Delta t.$$

Now a local estimate as in

$$\Delta t = \frac{\delta}{||AP(t) + B)||}$$

can be used or an overall estimate as in

$$\Delta t = \frac{\delta}{max\{||V_i||, ||V_j||, ||V_k||\}}.$$

This last estimate is based upon the property that $||AP + B||$ is bounded by (max$\{||V_i||, ||V_j||, ||V_k||\}$) for P in the triangle of interest.

21.3.1 Conventional Methods Applied to Linearly Varying Vector Fields

It is possible to treat Equation (21.1) as an ordinary differential equation and to use standard and conventional numerical methods to compute approximations to $P(t)$. The most popular numerical methods in this context are incremental methods. The particular computational formulas for a sampling of these methods are:

Euler's:

$$P(t + \Delta t) \cong P(t) + \Delta t \cdot V(p(t))$$

Runge-Kutta 2nd Order:

$$P(t + \Delta t) \cong P(t) + \tfrac{1}{2}(V_1 + V_2)$$

$$V_1 = \Delta t \cdot V(P(t))$$

$$V_2 = \Delta t \cdot V(P(t) + V_1)$$

$$P(t + \Delta t) = P(t) + \tfrac{1}{2}(V_1 + \Delta t \cdot V(P(t) + V_1))$$

Runge-Kutta 4th Order:

$$P(t + \Delta t) \cong P(t) + \tfrac{1}{6}(V_1 + 2V_2 + 2V_3 + V_4)$$

$$V_1 = \Delta t \cdot V(p(t))$$

$$V_2 = \Delta t \cdot V(p(t) + \tfrac{1}{2}V_1)$$

$$V_3 = \Delta t \cdot V(p(t) + \tfrac{1}{2}V_2)$$

$$V_4 = \Delta t \cdot V(p(t) + \tfrac{1}{2}V_3)$$

Adaptive Runge-Kutta-Fehlberg:

$$P(t + \Delta t) \cong P(t) + \tfrac{16}{135}V_1 + 0V_2 + \tfrac{6656}{12825}V_3 + \tfrac{28561}{56430}V_4 - \tfrac{9}{50}V_5 + \tfrac{2}{55}V_6$$

$$\text{error estimate } = \tfrac{1}{360}V_1 + 0V_2 - \tfrac{128}{4275}V_3 - \tfrac{2197}{75240}V_4 + \tfrac{1}{50}V_5 + \tfrac{2}{55}V_6$$

$$V_1 = \Delta t \cdot V(p(t))$$

$$V_2 = \Delta t \cdot V(p(t) + \tfrac{1}{4}V_1)$$

$$V_3 = \Delta t \cdot V(p(t) + \tfrac{3}{32}V_1 + \tfrac{9}{32}V_2)$$
$$V_4 = \Delta t \cdot V(p(t) + \tfrac{1932}{2197}V_1 - \tfrac{7200}{2197}V_2 + \tfrac{7296}{2197}V_3)$$
$$V_5 = \Delta t \cdot V(p(t) + \tfrac{439}{216}V_1 - 8V_2 + \tfrac{3680}{513}V_3 - \tfrac{845}{4104}V_4)$$
$$V_6 = \Delta t \cdot V(p(t) - \tfrac{8}{27}V_1 + 2V_2 - \tfrac{3544}{2565}V_3 + \tfrac{1859}{4104}V_4 - \tfrac{11}{40}V_5)$$

The above formulas can be simplified somewhat when the special property that $V(P) = AP + B$ is utilized. The formulas in this case become:

Euler's:

$$P(t + \Delta t) \cong (I + \Delta t A)P(t) + \Delta t B$$

Runge-Kutta 2nd Order:

$$P(t + \Delta t) \cong (I + \Delta t A + \tfrac{(\Delta t A)^2}{2})P(t) + \Delta t(I + \tfrac{\Delta t A}{2})B$$

Runge-Kutta 4th Order:

$$P(t + \Delta t) \cong \left(\sum_{i=0}^{4} \frac{(\Delta t A)^i}{i!}\right) P(t) + \Delta t \left(\sum_{i=0}^{3} \frac{(\Delta t A)^i}{(i+1)!}\right) P(t)$$

Adaptive Runge-Kutta-Fehlberg:

$$P(t + \Delta t) \cong \sum_{i=0}^{5} \frac{(\Delta t A)^i}{i!} + \frac{(\Delta t A)^6}{2080} P(t) + \Delta t \left(\sum_{i=0}^{4} \frac{(\Delta t A)^i}{(i+1)!} + \frac{(\Delta t A)^5}{2080}\right) B$$

$$\text{Error Estimate} \cong (\Delta t A)^5 (\tfrac{\Delta t A}{2080} - \tfrac{1}{780})$$

21.3.2 Incremental Methods for the Exact Solution

Here we are looking for a formula of the form

$$P(t + \Delta t) = G(\Delta t)P(t) + H(\Delta t)$$

which will allow computation of points at equal parameter values on the curve to be computed with only the application of an affine transformation. Repeated application of the basic differential equation which characterizes $P(t)$ leads to

$$P'(t) = AP(t) + B$$
$$P''(t) = AP'(t) = A^2P(t) + AB$$
$$P'''(t) = AP''(t) = A^3P(t) + A^2B$$

which leads to the expansion

$$\begin{aligned} P(t + \Delta t) &= (I + \Delta t A + \frac{(\Delta t A)^2}{2!} + \cdots)P(t) \\ &\quad + (\Delta t + (\Delta t)^2\frac{A}{2!} + (\Delta t)^3\frac{A^2}{3!} + \cdots)B \\ &= E(\Delta t)P(t) + F(\Delta t)B \end{aligned}$$

where $E(\Delta t) = I + F(\Delta t)A$. Both of these series for $E(\Delta t)$ and $F(\Delta t)$ have limits that can be approximately calculated in a variety of ways. One, of course, is to simply truncate the series. If we utilize the expressions for the solution given in Section 21.2.1, we arrive at the following particularly efficient methods which rely only upon the need to compute the functions e^t, $sin(t)$, and $cos(t)$.

The matrix E referred to below is (E_1, E_2) where these two vectors are defined in the different cases covered in Sections 21.2.1 and 21.2.2.

Case 1. $(0 \neq r_1 \neq r_2 \neq 0)$
If $|E| \neq 0$ then

$$P(t+\Delta t) - P_c = E \begin{pmatrix} e^{\Delta t r_1} & 0 \\ 0 & e^{\Delta t r_2} \end{pmatrix} E^{-1}(P(t) - P_c).$$

If $|E| = 0$ then P_0 is on one of the lines passing through P_c in the direction of the eigenvectors of A and either $E_1 = 0$ or $E_2 = 0$ (or $E_1 = E_2 = 0$ if $P_0 = P_c$). The tangent curve will be a straight line and so

$$P(t+\Delta t) = P(t) \pm \Delta t(P_c - P_0).$$

Case 2. $(0 = r_1, r_2 \neq 0)$

$$P(t+\Delta t) = [I + \frac{(e^{\Delta t r_2} - 1)}{r_2} A]P(t) + \Delta t[I + \frac{(e^{\Delta t r_2} - \Delta t r_2 - 1)}{\Delta t r_2^2} A]B.$$

Case 3. $(r_1 = 0 = r_2)$

$$P(t+\Delta t) = (I + \Delta t A)P(t) + (\Delta t + \frac{(\Delta t)^2 A}{2})B.$$

Case 4. $(r_1 = r_2 \neq 0)$
If $|E| \neq 0$, then

$$P(t+\Delta t) - P_c = E e^{\Delta t r_1} \begin{pmatrix} 1 & 0 \\ \Delta t & 1 \end{pmatrix} E^{-1}(P(t) - P_c).$$

If $|E| = 0$, then $P(t)$ is a straight line and so

$$P(t+\Delta t) = P(t) \pm \Delta t(P_c - P_0).$$

Case 5. $(\mu + \lambda i, \mu - \lambda i, \lambda \neq 0)$

$$P(t+\Delta t) - P_c = E e^{\mu \Delta t} \begin{pmatrix} cos(\lambda \Delta t) & -sin(\lambda \Delta t) \\ sin(\lambda \Delta t) & cos(\lambda \Delta t) \end{pmatrix} E^{-1}(P(t) - P_c).$$

21.4 Topological Methods

Topological methods provide a clear and uncluttered means of visualizing a two-dimensional vector field. They give a good overview of the flow with relatively little information, but they can require some effort to compute. They were first introduced into the visualization literature by Helman and Hesselink [17]. They consist of a collection of tangent curves which separate the flow into regions. The tangent curve boundaries of these regions connect together certain critical points. Critical points are locations where the flow is zero.

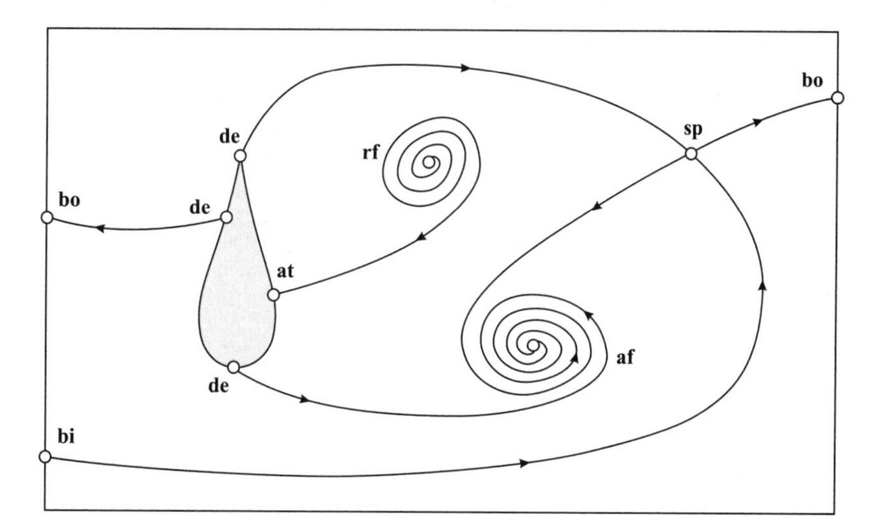

Figure 21.10: A topological graph of a two dimensional vector field.

In a nutshell, there are two main steps to computing a topological graph. First, all critical points are computed and classified on the basis of the nature of the flow near the critical point. In the second step, certain tangent curves are started at critical points and traversed out until they either link up with other critical points or leave the domain. In this section, we will discuss what is necessary to compute a topological graph within the special context of this chapter which is piecewise linear vector fields over triangulated domains. An example of a topological graph is shown in Figure 21.10.

We now discuss some details of the critical point computation and classification step of the algorithm. In a general context with the vector field

$$V(x,y) = \begin{pmatrix} u(x,y) \\ v(x,y) \end{pmatrix},$$

a critical point, P_c, is simply a position where

$$V(P_c) = \begin{pmatrix} u(x_c, y_c) \\ v(x_c, y_c) \end{pmatrix} = \begin{pmatrix} 0 \\ 0 \end{pmatrix}.$$

The local behavior close to a critical point is determined by the Jacobian

$$J(x_c, y_c) = \begin{pmatrix} u_x(x_c, y_c) & u_y(x_c, y_c) \\ v_x(x_c, y_c) & v_y(x_c, y_c) \end{pmatrix}.$$

This is due to the fact that the flow field is approximated by the linear terms of its expansion near the critical point. That is,

$$V(P) \cong V(P_c) + (P - P_c)J(P_c) = (P - P_c)J(P_c).$$

Using techniques similar to those used in Section 21.2.1, it is determined that there are six different types of critical points (see Figure 21.11):

i)	Saddle Point	$J(P_c)$ has two real, nonzero eigenvalues which differ in sign
ii)	Attracting Node	$J(P_c)$ has two negative eigenvalues
iii)	Repelling Node	$J(P_c)$ has two positive eigenvalues
iv)	Attracting Focus	$J(P_c)$ has complex eigenvalues $\mu + \lambda i$, $\mu - \lambda i$, $\lambda \neq 0$, $\mu < 0$
v)	Repelling Focus	$J(P_c)$ has complex eigenvalues $\mu + \lambda i$, $\mu - \lambda i$, $\lambda \neq 0$, $\mu > 0$
vi)	Center	$J(P_c)$ has purely imaginary eigenvalues $+\lambda i$ and $-\lambda i$, $\lambda \neq 0$

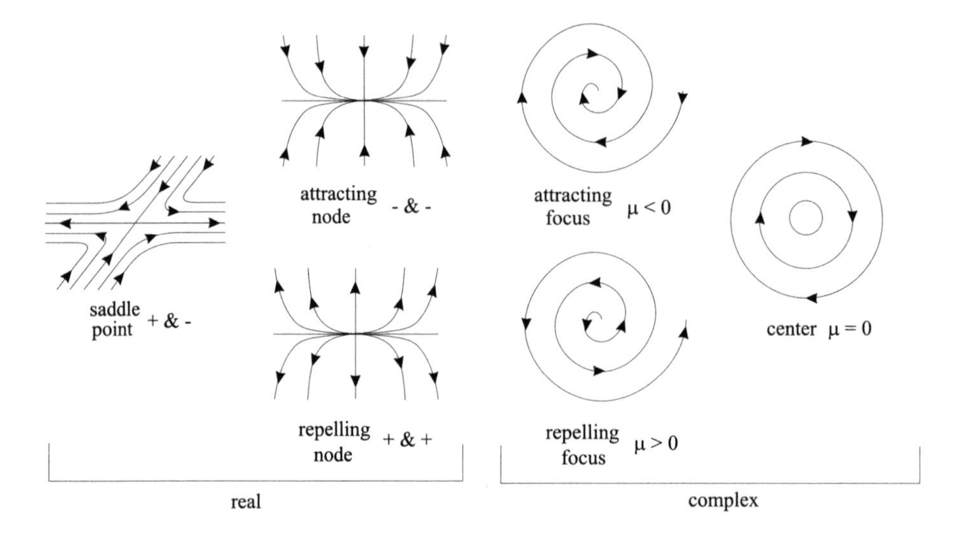

Figure 21.11: Characterization of critical values.

In general, the computation of critical points can be a rather formidable problem. For example, in the case of curvilinear grids, cells must be searched for possible candidate cells and then a nonlinear system (two equations in two unknowns) of equations must be solved. This normally requires some iterative method which could possibly fail to converge or converge to a point which is actually not a critical point. In the case of a linearly varying vector field, the situation is much simpler. Only the linear system

$$A\begin{pmatrix} x_c \\ y_c \end{pmatrix} + B = 0$$

must be solved. Note that A and B are defined the same as earlier in Equation (21.2). We next determine if this is a "real" critical point by deciding whether or not P_c is actually in the triangle. This determination can be made by computing the barycentric coordinates of P_c (or possibly a scaling of them) and checking to see if all three of them are positive.

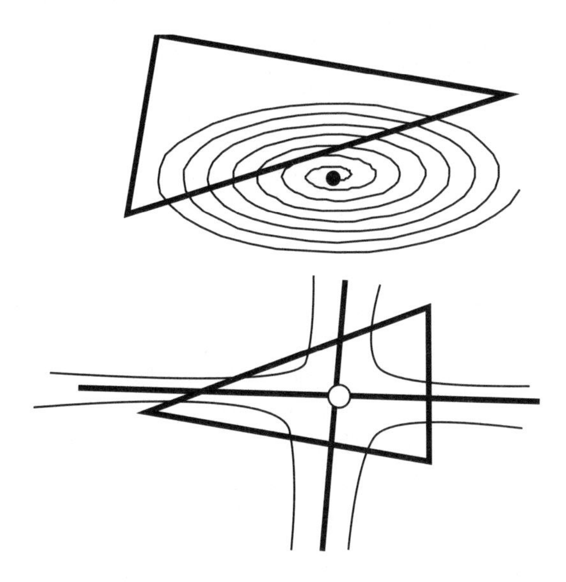

Figure 21.12: Bottom critical point is "real," but not the top one.

Often, a flow field will have an interior boundary which represents an object about which the flow is being analyzed. All the points on this interior boundary are critical points as the flow is forced to be zero here. Some of these points are of special interest. These are the points of attachment (at) and detachment (de) which belong to tangent curves which separate the flow along and near the inner boundary. In some previous discussion in the literature, the characterization of these points has not been so rigorous, but here in the context of piecewise linear flow fields, we can be very precise. Consider a triangle with only one vertex on the inner boundary. This vertex will necessarily be a critical point. If the eigenvalues of the Jacobian for this triangle indicate that this critical point is classified as a saddle point, then it is a candidate for a point of attachment or detachment. It will be a point of attachment if the eigenvector (or its negative) associated with the negative eigenvalue lies between the two edges of the triangle sharing this critical point. If an eigenvector associated with the positive root lies in the triangle, then this critical point will be a point of detachment. See Figure 21.13.

It should be noted that it is possible for a single point to be both a point of attachment and a point of detachment. This is illustrated in Figure 21.14. We should also point out that it is possible for a single point on the inner boundary to belong to two different triangles which have only this point on the inner boundary and because of this, this single point could be classified as a certain type of critical point for one triangle and another (or the same) type of critical point as a vertex for a different triangle. An example of this is shown

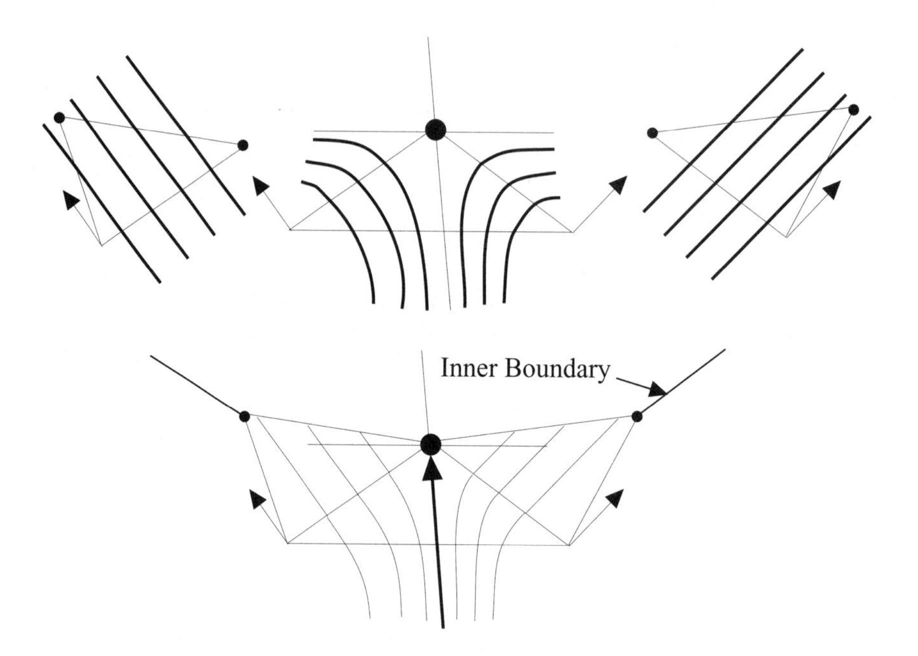

Figure 21.13: Diagram to support definition of point of attachment.

in Figure 21.15 where the critical point on the upper-left portion of the inner boundary is a point of attachment for one triangle and a point of detachment for an adjacent triangle.

The second major step in computing a topological graph consists of the linking algorithm. From each saddle point which is interior to a triangle, four tangent curves will emanate—two associated with each eigenvector of the Jacobian. One curve emanates in the direction of the eigenvector and one in the negative direction of the eigenvector. The curves emanating in the directions of the eigenvector associated with the positive eigenvalue will be traversed in positive parameter direction and these curves will move along with the flow. They have the chance to link up with attracting nodes or foci or possibly other saddle points along the eigenvectors of inward flow to the saddle point. The two curves emanating in the direction of the eigenvectors associated with the negative eigenvalue will be traversed in a negative parameter direction and will move along in a direction opposite (or negative) to the direction of the flow field. They have the chance to link up with repelling nodes or foci or other saddle points. From each point of detachment, one curve will emanate and be traversed in positive parameter direction. From each point of attachment, one curve will emanate and be traversed in a negative or opposite direction to the flow. The algorithm is complete when each of these curves has linked to another critical point or leaves the domain either by encountering the outer boundary or the inner boundary. An example of the results of this algorithm are shown in Figure 21.15. We should point out that this linking algorithm does not produce all separating tangent curves for it is possible that in addition to the center points, some repelling or attracting nodes or foci could be left unconnected to any tangent curve. However, this does not occur when the domain is a triangulated approximation to the sphere.

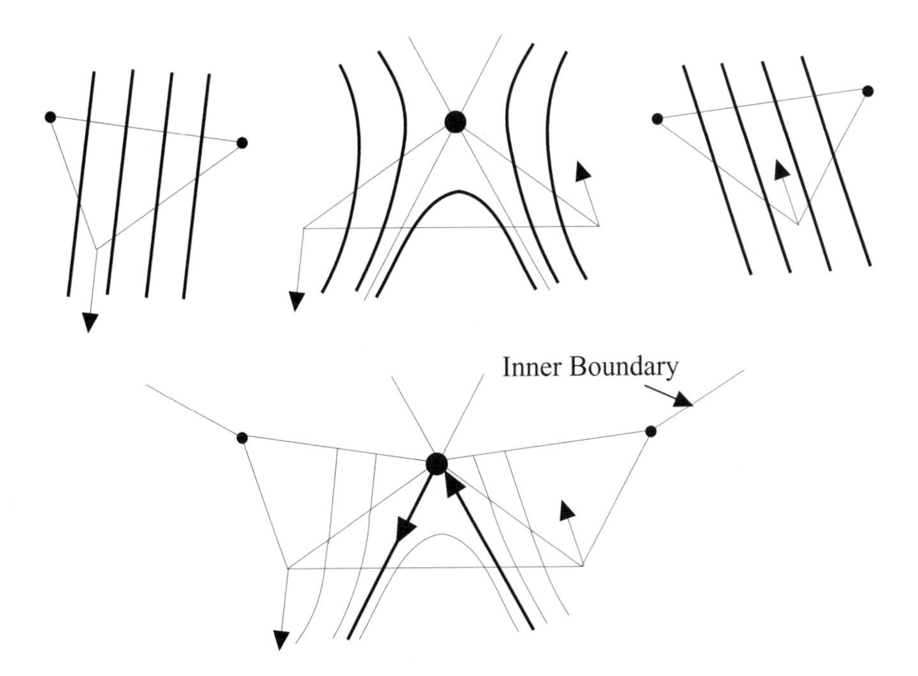

Figure 21.14: A point on the inner boundary which is both a point of attachment and a point of detachment.

21.5 Examples

The first example that we include uses a simple quadratic equation to specify the vector field. The equations are given in Equation (21.20). One reason for including this type of example is to allow other implementors to easily verify and compare their software results.

$$
\begin{aligned}
U(x, y) &= -0.103209 + 0.051511x - 0.302699y \\
&\quad +0.037546xy - 0.232875x^2 + 0.611528y^2 \\
V(x, y) &= 0.143656 + 0.687847x - 0.144779y \\
&\quad -0.213010xy - 1.029676x^2 + 0.246278y^2
\end{aligned}
\tag{21.20}
$$

As we have mentioned earlier, the methods described here can be applied to any triangulated domain. In Figure 21.17 we show the topological graph and some additional tangent curves (in cyan) for a vector field defined over a triangulation of a spherical domain. Similar to Figure 21.16, two different resolutions of the triangulation are shown. In the right column, each triangle of the left column has been replaced by four subtriangles. We have intentionally used flat shading rather than Gourard shading for the rendering of the sphere so that the triangulation is apparent.

In the next example, we illustrate the use of the methods developed here to visualize a multiresolution model of a vector field over a curvilinear grid. The topological graph is

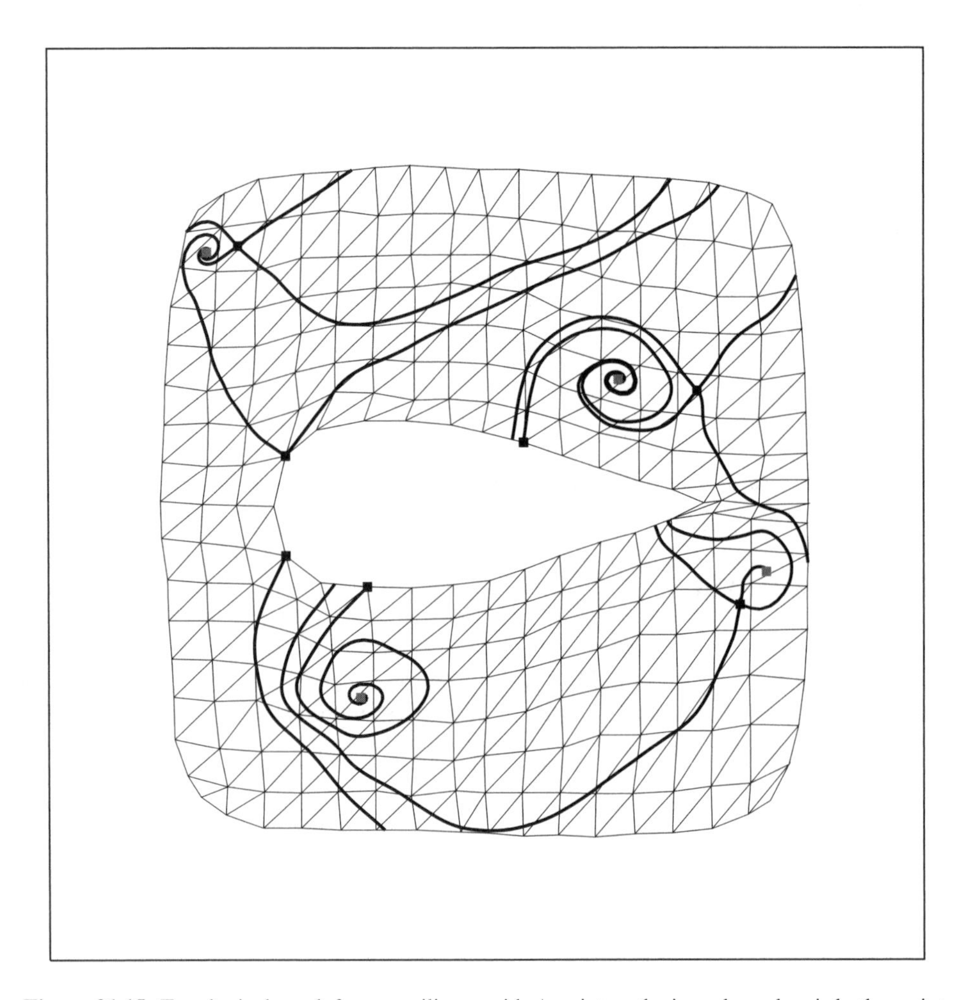

Figure 21.15: Topological graph for a curvilinear grid. A point on the inner boundary is both a point of attachment and a point of detachment.

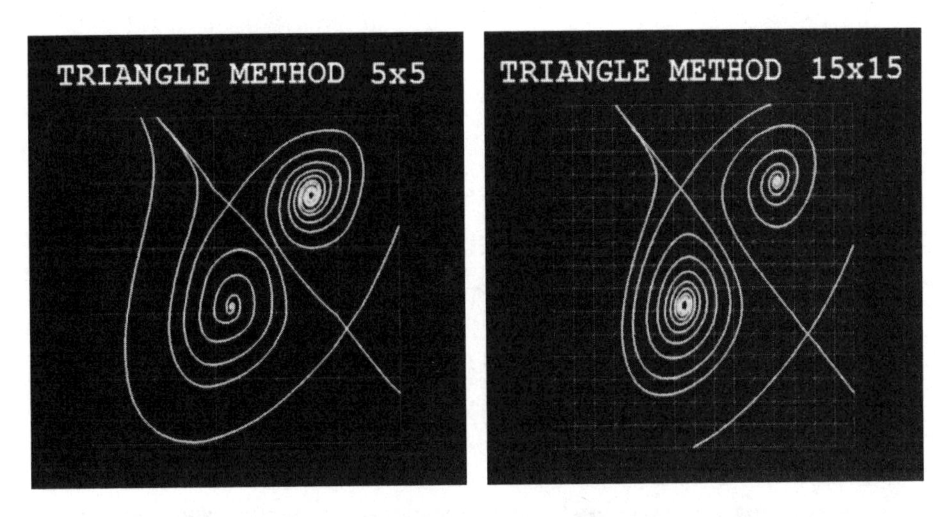

Figure 21.16: Explicit method used to compute topological graph for vector field given by Equation (21.20). Two different resolutions for the triangulation are shown. See Color Plate 110.

shown at four different levels of approximation in Figure 21.18. One unique feature allowed by these methods is that no matter how coarse the resolution becomes, the boundaries have not changed from the original data. This is a particularly important aspect for the inner boundary which is often the focus of attention for a flow analysis. In Figure 21.19, we show the topological graph for some partial reconstructions of the flow field. These methods allow the user to zoom in and out and only reconstruct the portion of interest.

In Figure 21.20 we show results similar to those of Figures 21.18 and 21.19 except that now the domain is a triangulated sphere and the reconstruction is not done on the basis of regions, but on the basis of the magnitude of the coefficients of the wavelet basis functions. This data represents "real" data provided to us by Roger Crawfis and Nelson Max of Lawrence Livermore National Laboratory. It is one time step and one tier of simulated wind velocity data.

The data used for the examples shown in the images of Figure 21.21 and Figure 21.22 was provided to us by Yasuo Nakajima, Nissan Motor Company, Japan. Actually, the domain of this data is three-dimensional and not two-dimensional as required for the algorithms covered here. Figure 21.22 shows one slice through this 3D data. The 3D velocity vectors were projected into the plane of the slice. Many of the results of this chapter have been extended to 3D domains where the vector field is assumed to vary linearly over a tetrahedrization of the domain. See [33] for more discussion on tetrahedrizing a 3D curvilinear grid. The tangent curves and critical points shown in Figure 21.21 were computed using extensions of the algorithms discussed here. See [22] for more details.

21.6 Conclusions and Remarks

1. One of the advantages of the explicit methods we have developed here is that accuracy can be monitored and controlled. Runge-Kutta and other incremental methods

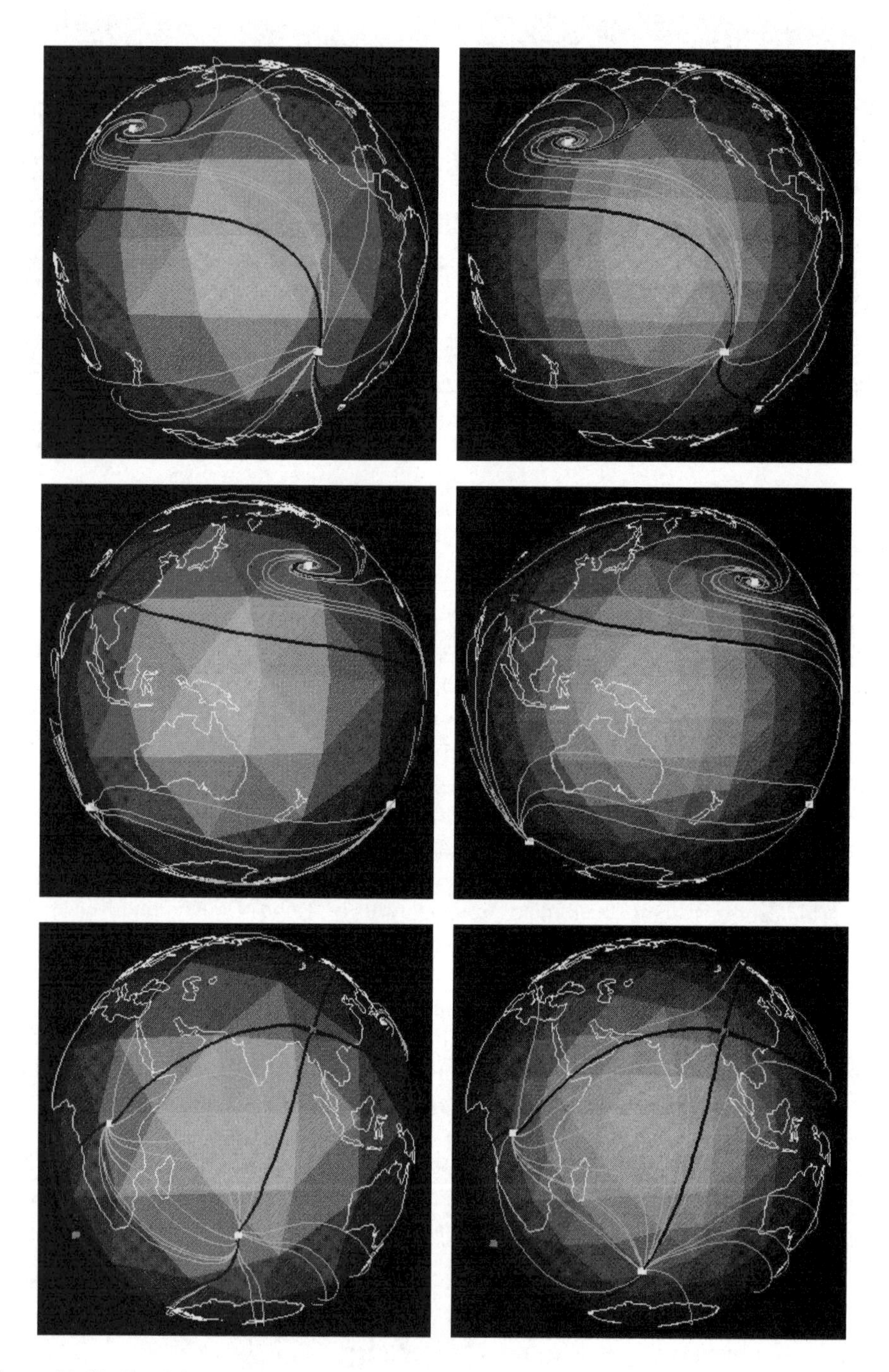

Figure 21.17: Explicit method used to compute critical values and tangent curves for a vector field defined over a spherical domain. See Color Plate 111.

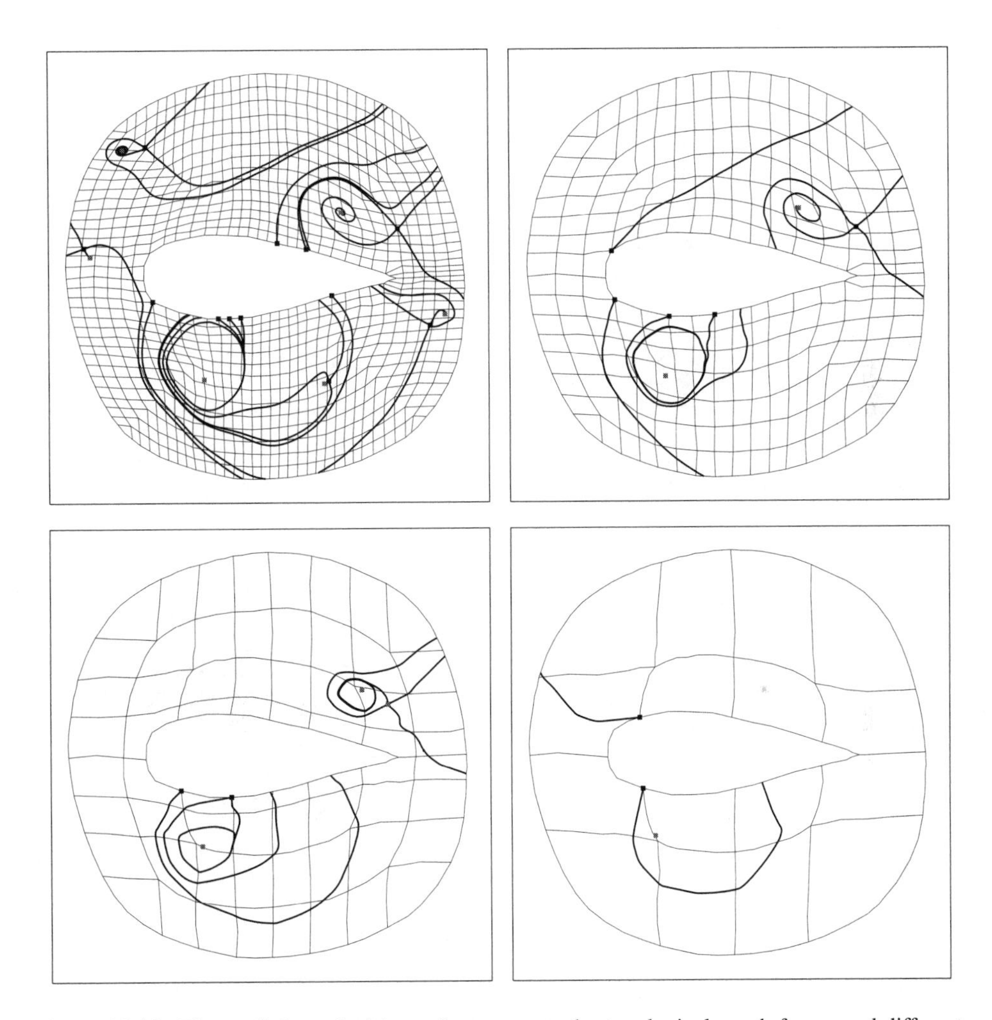

Figure 21.18: The explicit method is used to compute the topological graph for several different resolutions of a curvilinear grid.

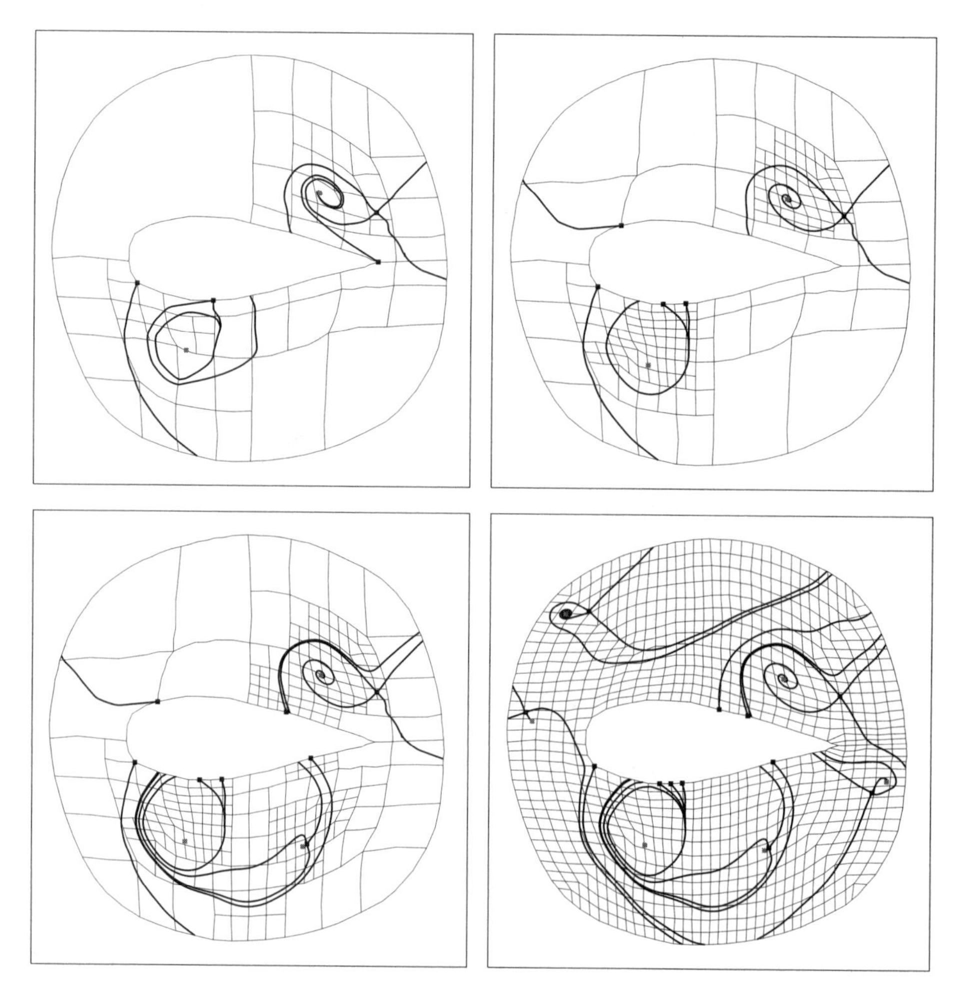

Figure 21.19: Partial wavelet reconstruction of the flow over a curvilinear grid indicating the efficiencies of zooming in or out.

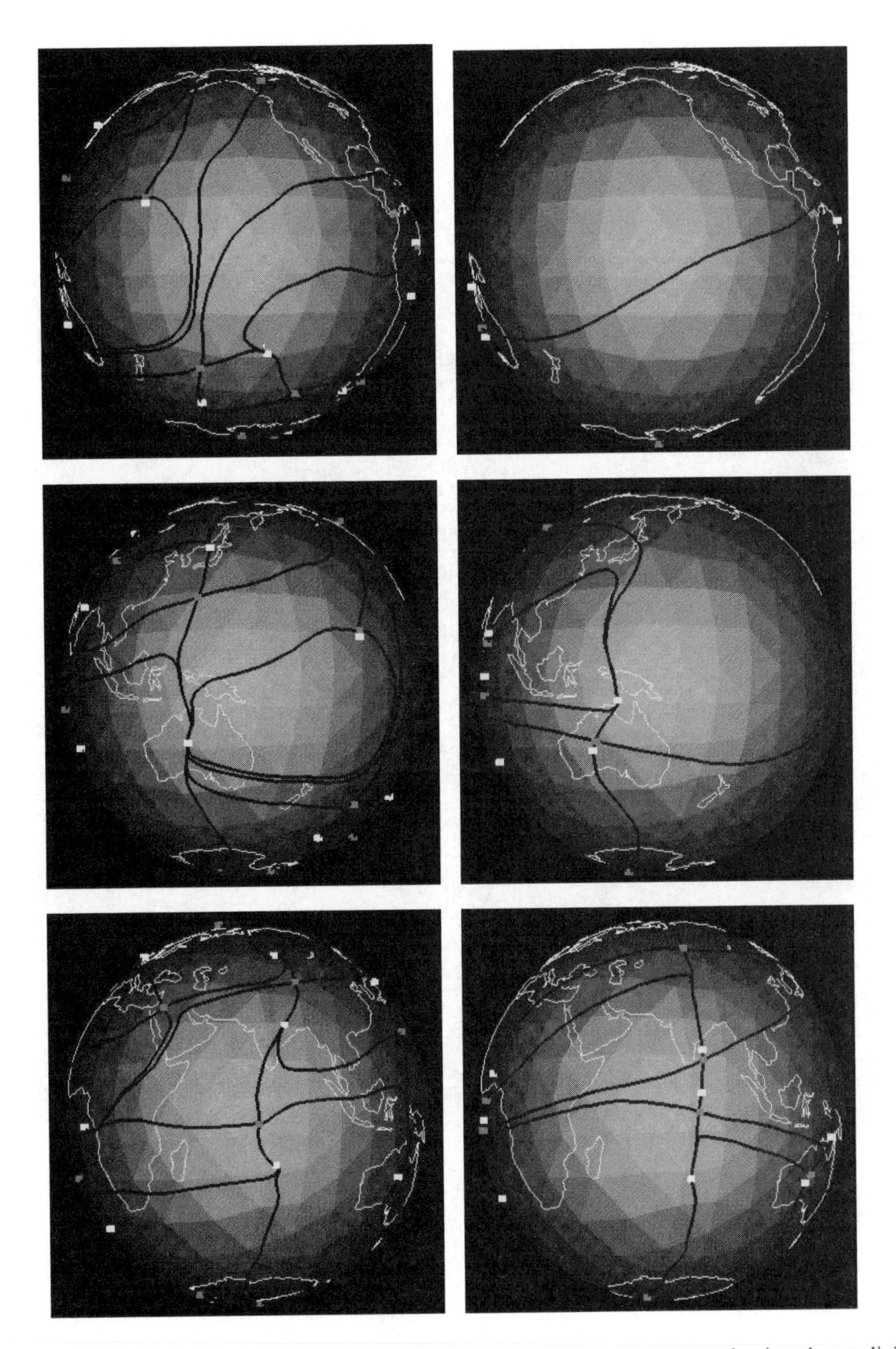

Figure 21.20: Topological graphs for wind data over the earth are computed using the explicit method. The right column is wavelet reconstruction based on the largest 3% of the wavelet coefficients. See Color Plate 112.

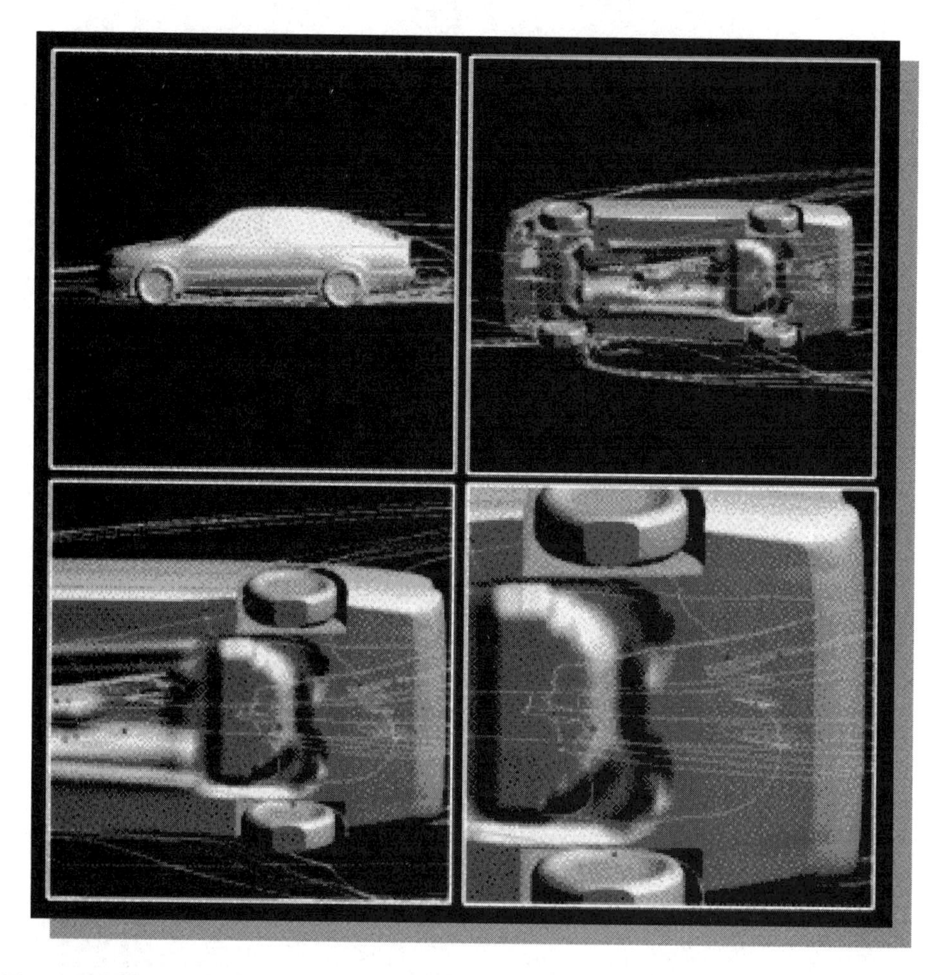

Figure 21.21: The explicit method is used to compute tangent curves linking critical values for a 3D vector field. See Color Plate 113.

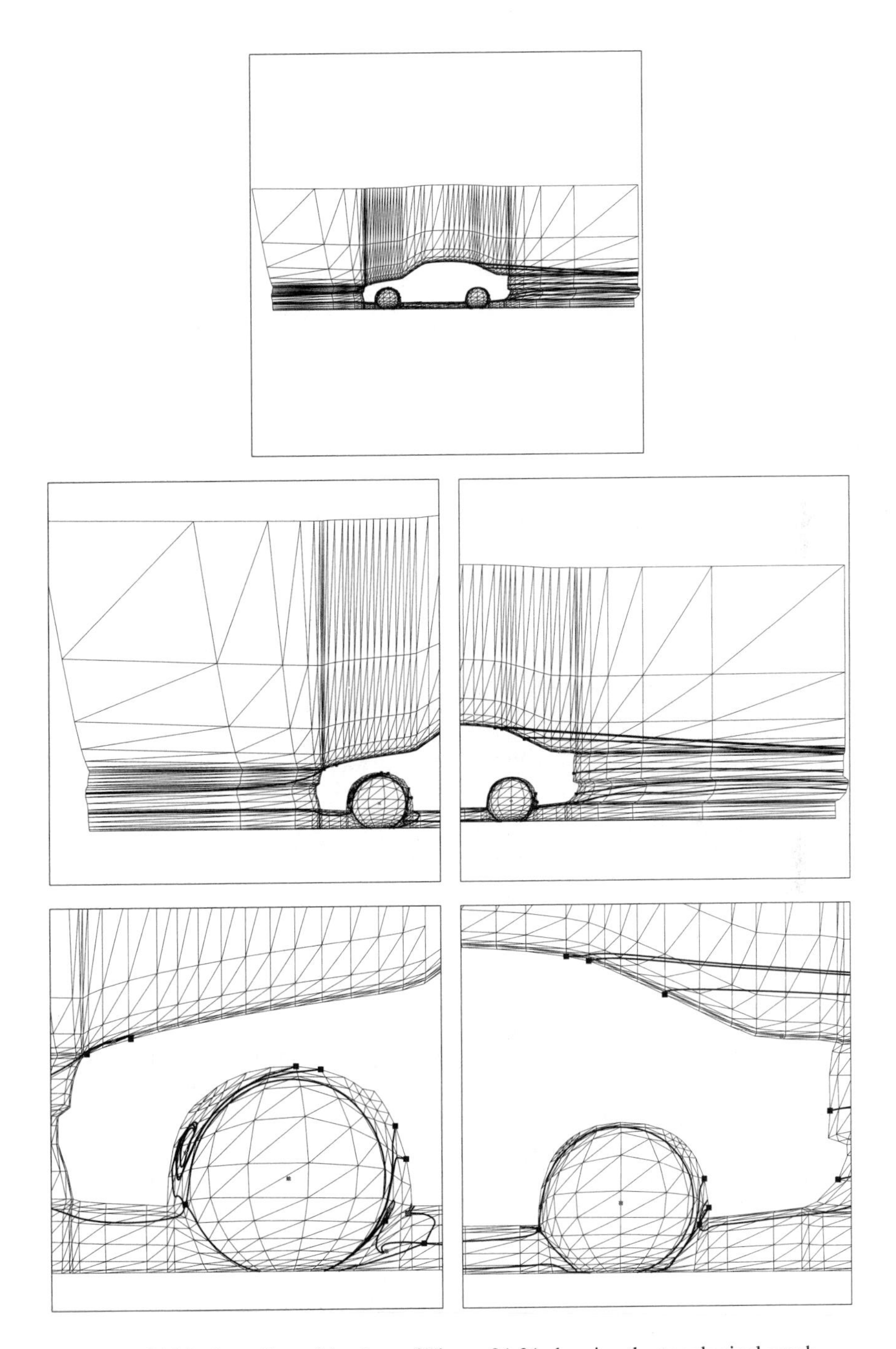

Figure 21.22: One slice of the data of Figure 21.21 showing the topological graph.

for solving ODEs are notorious for "wandering off" the true solution and once an error is made, there is no way to recover because from the erroneous point forward, the method is attempting to solve a different (wrong) problem than the one it set out to solve in the beginning. Variable step size methods which estimate the error and attempt to control it by adaptively changing the step size are helpful in this regard, but the problem is that the error is only estimated and only guesses about the proximity of the computed solution to the desired solution can be made. This is the case for the R-F-K method covered in Section 21.3.1. In the proposed method, the accumulated error is a result of how accurately the intersection of a cell boundary and a particular explicitly defined tangent curve are computed. This is formulated as a root finding problem and so this computation can be done as accurately as deemed necessary.

2. Another advantage of the present method is speed and efficiency. Since the tangent curve is known explicitly for each triangle, parameters pertaining to this definition can be precomputed and stored and then used when a particular tangent curve penetrates this triangle. The global shape of the tangent curve is known by its sequence of entry and exit points for each triangle it intersects and so the overall appearance of the curve is not seriously degraded if a local approximation for each cell is used. We have used a parametric cubic Hermite curve on each triangle with good success.

3. Because of the general nature of the approach of the methods covered here, they can be extended to any domain which consists of a collection of triangles. This includes manifolds of arbitrary topological type if a triangulated approximation of the domain is acceptable.

4. The computation of critical points is greatly simplified. Critical points are crucial to topological methods. If the vector field which has been defined by a particular cell and extended to the entire domain of E^2 has a critical value, then it can be computed by solving a linear system of equations. A solution to the linear system of equations is a "real" critical value for the piecewise defined vector field if it lies within the cell, otherwise it is not. Normally, the computation of critical values requires a rather tedious computation involving testing whether or not a cell might have a critical value, followed by the solution to a nonlinear system of equations by Newton's methods or some other iterative scheme. See [17], for example.

5. The last advantage we mentioned is that the present method does all of its computations in physical coordinates as opposed to computational coordinates. To some this might initially appear to be a disadvantage since computational coordinates are introduced for the specific purpose of their namesake. But if the domain is triangulated and linear variation is assumed over each triangle, we not only gain a method that is affine invariant, but there is no need to map the data to computational coordinates, solve the problem and map the solution back; we simply compute the solution directly in physical space.

6. We first mentioned the basic ideas of these methods and reported on some preliminary results at a presentation given in 1991 at the first Dagstuhl Seminar on Scientific Visualization. At this time, Nelson Max pointed out that he had mentioned the idea in an earlier paper [29]. Sawada [42] has also mentioned the idea of using explicit

representations. We presented some preliminary results on the 3D extension of the methods covered here at a technology assessment workshop held the Summer of 1993 in Darmstadt, Germany. See [40] for the proceedings of this workshop. Figure 21.21 has previously appeared in [34].

Acknowledgments

This research was supported by the National Aeronautical and Space Administration under NASA-Ames Grant, NAG 2-990. Partial support was also provided by the North Atlantic Treaty Organization under grant RG0097/88. The data for the example of Figures 21.21 and 21.22 was provided by Yasuo Nakajima of Nissan, Japan. The data for the example of Figure 21.20 was provided by Roger Crawtis and Nelson Max of Lawrence Livermore National Laboratory.

Bibliography

[1] F.H. Bertrand and P.A. Tanguy, "Graphical Representation of Two-Dimensional Fluid Flow by Stream Vectors," *Communications in Applied Numerical Methods*, Vol. 4, 1988, pp. 213–217.

[2] J. Buckmaster, "Perturbation Technique for the Study of Three-Dimensional Separation," *Phys. Fluids*, Vol. 15, 1972, pp. 2106–2113.

[3] P.G. Buning, "Numerical Algorithms in CFD Post-Processing," *von Karman Institute for Fluid Dynamics*, Lecture Series 1989-07.

[4] B.J. Cantwell, "Similarity Transformations of the Two-Dimensional Unsteady Stream Function Equations," *J. Fluid Mech.*, Vol. 85, 1978, pp. 257–271.

[5] M.S. Chong, A.E. Perry and B.J. Cantwell, "A General Classification of Three-Dimensional Flow Fields," *Physics of Fluids A*, Vol. 2, No. 5, 1990, pp. 765–777.

[6] D. Darmofal and R. Haimes, "An Analysis of 3D Particle Path Integration Algorithms," *Proc. 1995 AIAA CFD Meeting*, San Diego, Calif., 1995.

[7] A. Davey, "Boundary Layer Flow at a Point of Attachment," *J. Fluid Mech.*, Vol. 10, 1961, pp. 593–610.

[8] R. Denzer, "Application of Visualization in Environmental Protection," *Focus on Scientific Visualization*, H. Hagen, M. Muller and G.M. Nielson, editors, Springer, 1993, pp. 73–82.

[9] R.R. Dickinsion, "Interactive Analysis of the Topology of 4D Vector Fields," *IBM Journal of Research and Development*, Vol. 35, No. 1/2, 1991, pp. 59–66.

[10] D.S. Ebert and R.E. Parent, "Rendering and Animation of Gaseous Phenomena by Combining Fast Volume and Scanline A-Buffer Techniques," *Computer Graphics*, Vol. 24, No. 4, 1990, pp. 357–366.

[11] M. Fruehauf, "Combining Volume Rendering with Line and Surface Rendering," *Eurographics '91*, F.H. Post and W. Barth, editors, North Holland, 1991, pp. 21–32.

[12] R.S. Gallagher, "Span Filtering: An Optimization Scheme for Volume Visualization of Large Finite Element Models," *Proceedings of Visualization '91*, G.M. Nielson and L. Rosenblum, editors, IEEE Computer Society Press, 1991, pp. 68–75.

[13] R.S. Gallagher and J.C. Nagtegaal, "An Efficient 3-D Visualization Technique for Finite Element Models and Other Coarse Volumes," *Computer Graphics*, Vol. 23, No. 3, 1989, pp. 185–194.

[14] M. Geiben and M. Rumpf, "Visualization of Finite Elements and Tools for Numerical Analysis," *Advances in Scientific Visualization*, F.H. Post and A.J.S. Hin, editors, Springer, 1992.

[15] R.B. Haber and D.A. McNabb, "Visualization Idioms: A Conceptual Model for Scientific Visualization Systems," *Visualization in Scientific Computing*, G.M. Nielson, L. Rosenblum and B. Shriver, editors, IEEE Computer Society Press, 1990, pp. 74–92.

[16] B. Hamann, D. Wu, and R. Moorhead, "Flow Visualization with Surface Particles," *IEEE Transactions on Visualization and Computer Graphics*, Vol. 1, No. 3, Sep. 1995, pp. 210–217.

[17] J. Helman, and L. Hesselink, "Representation and Display of Vector Field Topology in Fluid Flow Data Sets," *IEEE Computer*, Vol. 22, No. 8, 1989, pp. 27–36.

[18] J. Helman and L. Hesselink, "Surface Representation of Two- and Three-Dimensional Fluid Flow Topology," *Proceedings Visualization '90*, A. Kaufman, editor, IEEE Computer Society Press, 1990, pp. 6–13.

[19] J. Helman and L. Hesselink, "Visualizing Vector Field Topology in Fluid Flows," *IEEE Computer Graphics and Applications*, Vol. 11, No. 3, pp. 36–46.

[20] W. Hibbard and D. Santek, "Visualizing Large Data Sets in the Earth Sciences," *IEEE Computer*, Vol. 22, No. 8, pp. 53–57.

[21] S. Hultquist, "Interactive Numeric Flow Visualization Using Stream Surfaces," *Computer Systems in Engineering*, Vol. 1, Nos. 2–4, 1990, pp. 349–353.

[22] I.-H. Jung, *Topological Visualizing Method of Vector Fields in Fluid Data Sets*, M.S. thesis, Arizona State University, Department of Computer Science and Engineering, June 1993.

[23] J.C.R. Hunt, C.J. Abell, J.A. Peterka and H. Woo, "Kinematical Studies of the Flows Around Free or Surface-Mounted Obstacles; Applying Topology to Flow Visualization," *J. Fluid Mech.*, Vol. 86, 1978, pp. 179–200.

[24] D.N. Kenwright, *Dual Stream Function Methods for Generating Three-Dimensional Stream Lines*, Ph.D. thesis, University of Auckland, Department of Mechanical Engineering, Aug. 1993.

[25] D.N. Kenwright and D.A. Lane, "Interactive Time-Dependent Particle Tracing Using Tetrahedral Decompostion," *IEEE Transaction on Visualization and Computer Graphics*, Vol. 2, No. 2, June 1996, pp. 120–129.

[26] M.J. Lighthill, "Attachment and Separation in Three-Dimensional Flow," *Laminar Boundary Layers*, L. Rosenhead, editor, Oxford University Press, 1963, pp. 72–82.

[27] R. Lohner and J. Ambrosiano, "A Vectorized Particle Tracer for Unstructured Grids," *J. Computational Physics*, Vol. 91, 1990, pp. 22–31.

[28] K.-L. Ma and P.J. Smith, "Cloud Tracing in Convection-Diffusion Systems," *Proceedings of Visualization '93*, G. Nielson and L. Rosenblum, editors, IEEE Computer Society Press, 1993, pp. 253–260.

[29] N. Max, *Private Communication*, 1991.

[30] W. Merzkirch, *Flow Visualization*, Academic Press, 1987.

[31] C. Moler and C. Van Loan, "Nineteen Dubious Ways to Compute the Exponential of a Matrix," *SIAM Review*, Vol 20, No. 4, Oct. 1978, pp. 801–836.

[32] G.M. Nielson, "Modeling and Visualizing Volumetric and Surface-on-Surface Data," *Focus on Scientific Visualization*, H. Hagen, M. Muller and G.M. Nielson, editors, Springer, 1993, pp. 191–240.

[33] G.M. Nielson, "Tools for Triangulation and Tetrahedrization," Chapter 20 in this volume.

[34] G.M. Nielson and A. Kaufman, "Visualization Graduates," *Computer Graphics and Applications*, Vol. 14, No. 5, Sep. 1994, pp. 17–18.

[35] T.V. Pathomas, J.A. Schiavone and B. Julesz, "Applications of Computer Graphics to The Visualization of Meteorological Data," *Computer Graphics*, Vol. 22, No. 4, pp. 327–335.

[36] A.E. Perry and B.D. Fairlie, "Critical Points in Flow Patterns," *Adv. Geophys.*, Vol. 18B, 1974, pp. 299–315.

[37] A.E. Perry and D.K.M. Tan, "Simple Three-Dimensional Motions in Coflowing Jets and Wakes," *J. Fluid Mech.*, Vol. 141, 1984, pp. 197–231.

[38] A.E. Perry and M.S. Chang, "A Description of Eddying Motions and Flow Patterns Using Critical-Point Concepts," *Ann. Rev. Fluid Mech.*, Vol. 19, 1987, pp. 125–155.

[39] F.H. Post and T. van Walsum, "Fluid Flow Visualization," *Focus on Scientific Visualization*, H. Hagen, M. Muller and G.M. Nielson, editors, Springer, 1993, pp. 1–40.

[40] L. Rosenblum et al., editors, *Scientific Visualization: Advances and Challenges*, Academic Press and IEEE Computer Society Press, 1994.

[41] A. Sadarjoen, T. van Walsum, A.J.S. Hin, and F.H. Post, "Particle Tracing Algorithms for Curvilinear Grids," *Proceedings Fifth Eurographics Workshop on Visualization in Scientific Computing*, Rostock, Germany, May 1994.

[42] K. Sawada, "Visualization of Unsteady Vortex Motion in the Flow over a Delta Wing," *Proc. of the 5th Int. Symp. on Computational Fluid Dynamics*, Sendai, Vol. III, 1993, ISCFD.

[43] W.J. Schroeder, C.R. Volpe and W.E. Lorensen, "The Stream Polygon: A Technique for 3D Vector Field Visualization," *Proceedings Visualization '91*, L. Rosenblum and G.M. Nielson, editors, IEEE Computer Society Press, 1991, pp. 126–132.

[44] N. Srinivasan, *An Efficient Method for the Computation of Tangent Curves*, M.S thesis, Arizona State University, Department of Computer Science and Engineering, May 1995.

[45] J. Stolk and J.J. van Wijk, "Surface-Particles for 3D Flow Visualization," *Advances in Scientific Visualization*, Springer, 1992.

[46] T. Strid, A. Rizzi and J. Oppelstrup, "Development and Use of some Flow Visualization Algorithms," *von Karman Institute for Fluid Dynamics*, Lecture Series 1989-07.

[47] M. Tobak and D.J. Peake, "Topology of Two-Dimensional and Three-Dimensional Separated Flows," *AIAA 12th Fluid and Plasma Dynamics Conference*, Williamsburg, Va., July 23–25, 1979.

[48] M. Tobak and D.J. Peake, "Topology of Three-Dimensional Separated Flows," *Ann. Rev. Fluid Mech.*, Vol. 14, 1982, pp. 61–85.

[49] S.K. Ueng, K. Sikorski, and K.-L. Ma, "Efficient Streamline, Streamribbon, and Streamtube Constructions on Unstructured Grids," *IEEE Transactions on Visualization and Computer Graphics*, Vol. 2, No. 2, 1996, pp. 100–110.

[50] J.J. van Wijk, "A Raster Graphics Approach to Flow Visualization," *Eurographics '90*, D.A. Duce and C.E. Vandoni, editors, North Holland, pp. 251–259.

[51] J.J. van Wijk, "Spot Noise—Texture Synthesis for Data Visualization," *Computer Graphics*, Vol. 25, No. 4, 1991, pp. 309–318.

[52] W.J. Yang, editor, *Handbook of Flow Visualizaton*, Hemisphere Publishing, 1989.

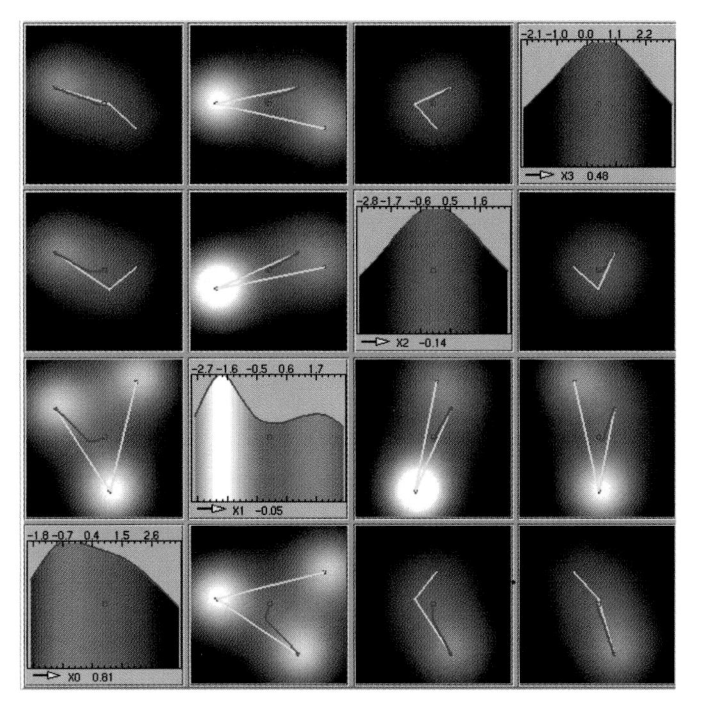

Color Plate 1: Figure 1.5, page 12.

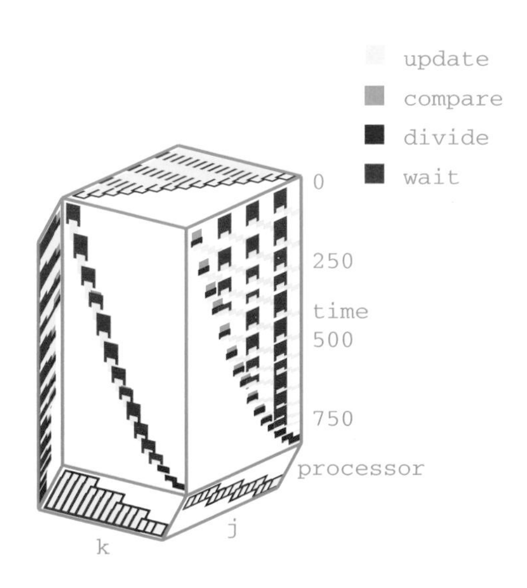

Color Plate 2: Figure 1.9, page 15.

Color Plate 3: Figure 1.11, page 16.

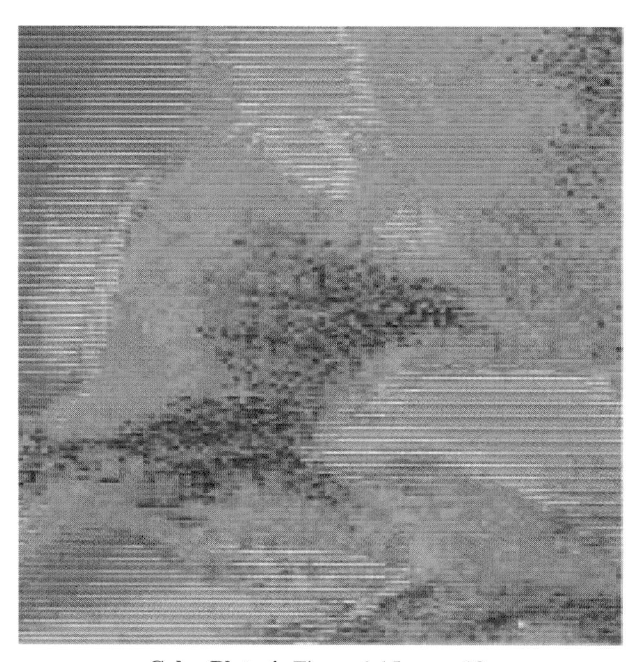

Color Plate 4: Figure 1.15, page 18.

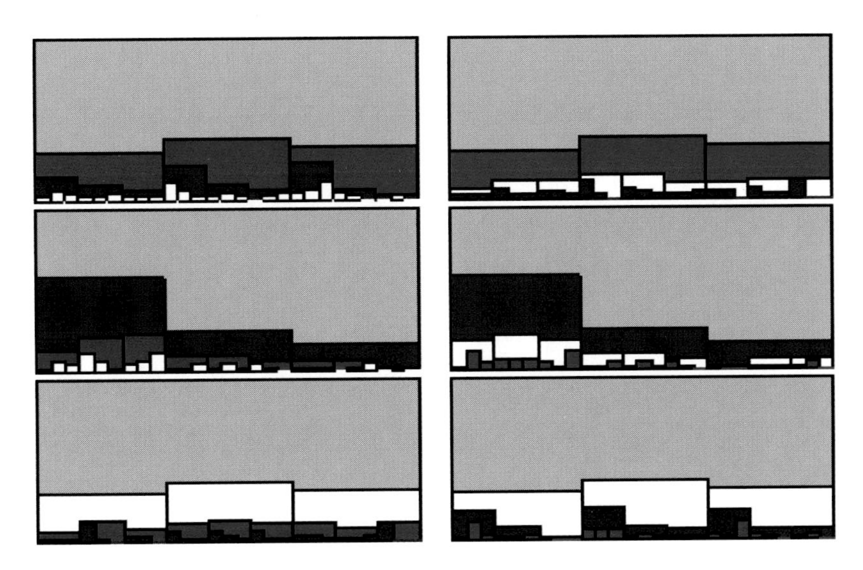

Color Plate 5: Figure 1.18, page 20.

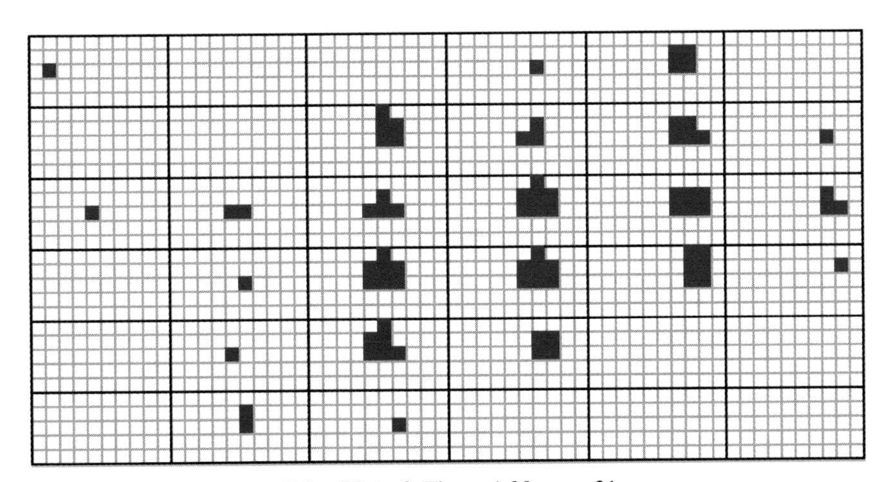

Color Plate 6: Figure 1.20, page 21.

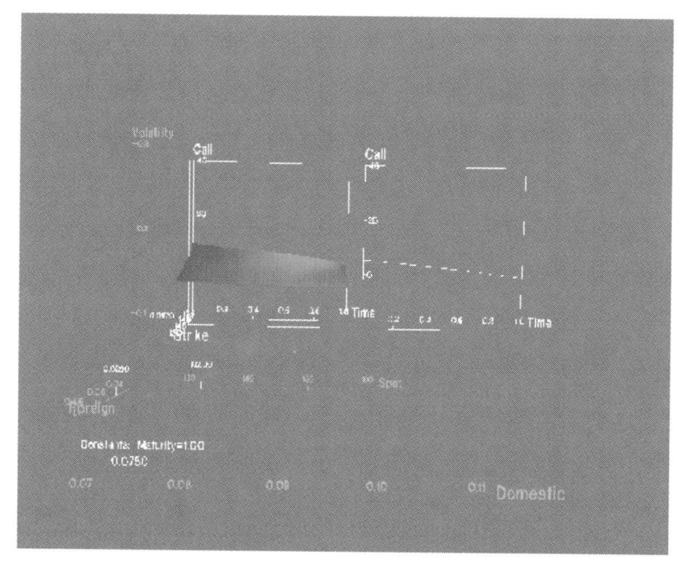

Color Plate 7: Figure 1.22, page 22.

Color Plate 8: Figure 1.28, page 25.

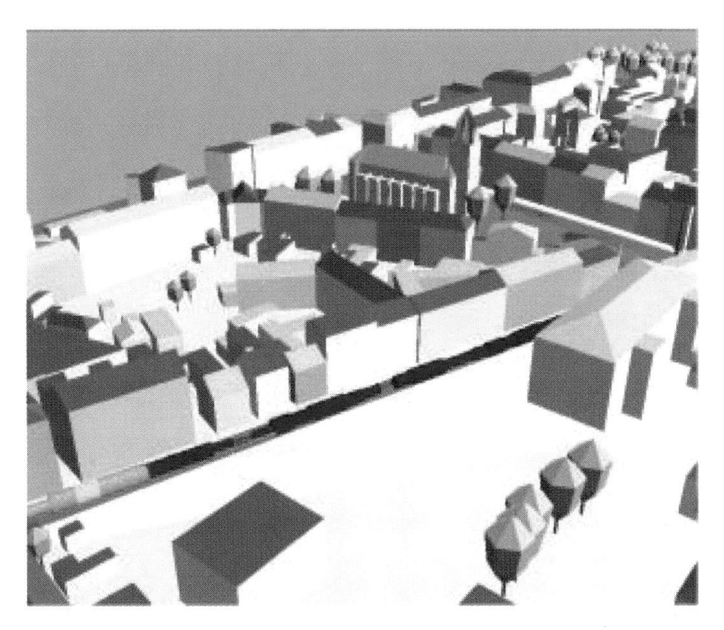

Color Plate 9: View of measured values of a small-scale cellular radio network in Darmstadt. Measured data are represented by cylinders colored according to received power. The diameter of the cylinders is proportional to delay spread. Data courtesy of German Telekom, Forschungs und Technologiezentrum. (Figure 2.4, page 49.)

Color Plate 10: Particle tracing in virtual environment: A finite element data set containing the air flow around a vehicle (100,000 elements) is presented in VE. A particle source can be positioned interactively with the data glove. This position is transmitted to a visualization system which simulates the movement of the particle and transmits the resulting positions back to the VE system for visualization. (Figure 2.5, page 50.)

Color Plate 11: Detail view of a Trypsin molecule in ball-and-stick representation. The "sticks" consist of only five surfaces; smooth appearance is achieved by Gouraud shading. Spheres have varying resolution depending on their distance from the viewer (level of detail). Data courtesy of Wolfgang Heiden, Department of Chemistry, Technical University Darmstadt. (Figure 2.6, page 52.)

Color Plate 12: Overall view of an enzyme and a substratum (Trypsin and Arginin). Ball-and-stick representation and surface representation combined. Size and shape of the molecules are also shown by the shadows which are implemented by means of precomputed textures. Data courtesy of Wolfgang Heiden, Department of Chemistry, Technical University Darmstadt. (Figure 2.7, page 53.)

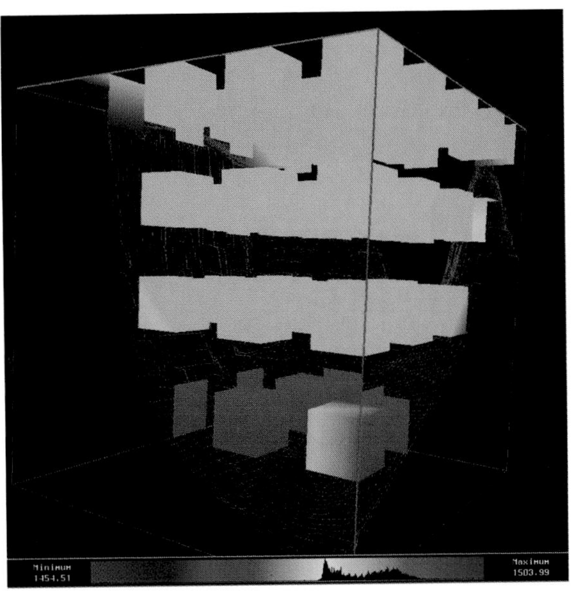

Color Plate 13: Minicubes method used for rendering sound speed with OVIRT system (uniform spacing and uniform cubes). (Figure 3.16, page 80.)

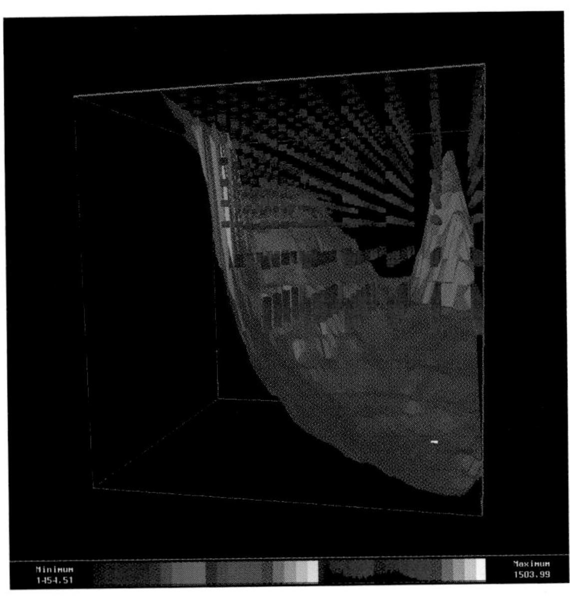

Color Plate 14: Minicubes method used for rendering sound speed with OVIRT system (nonuniform spacing and nonuniform cubes). (Figure 3.16, page 80.)

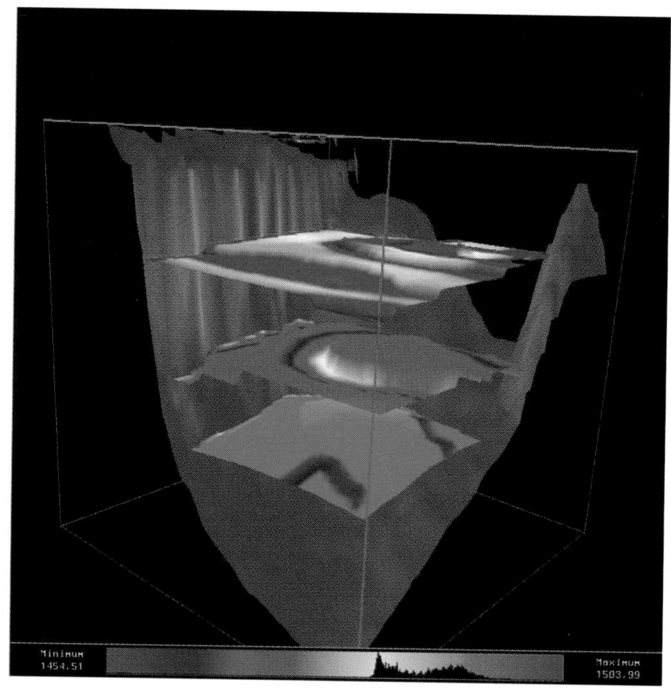

Color Plate 15: Sonic layer depth, deep sound channel axis, and critical depth generated and rendered with OVIRT system. (Figure 3.19, page 82.)

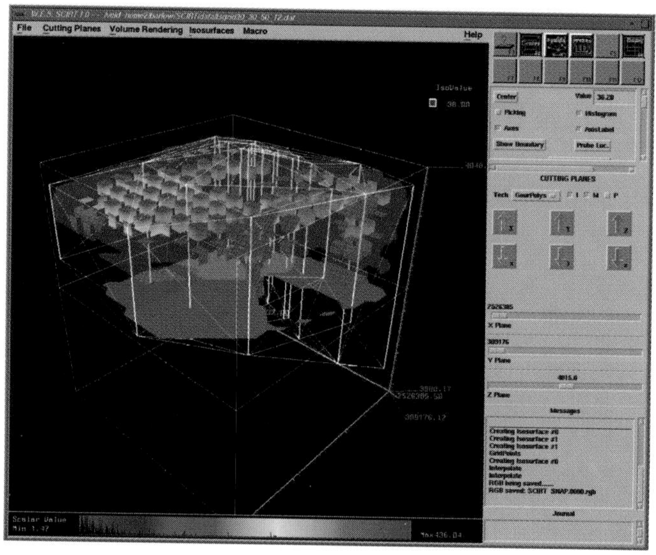

Color Plate 16: Transparent isosurfaces, minicubes, and cutting planes (clipped against convex hull) in a single image generated with SCIRT. (Figure 3.20, page 84.)

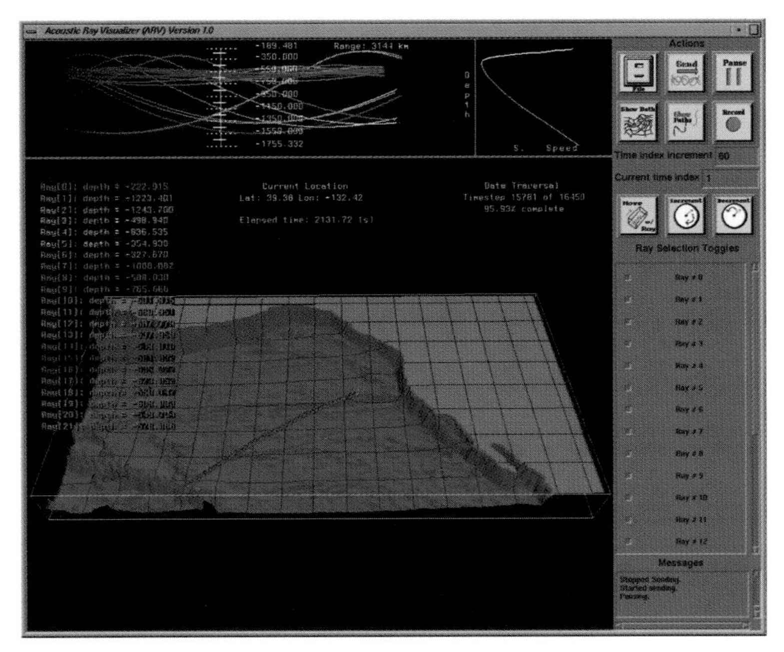

Color Plate 17: Snapshot of Acoustic Ray Visualizer (ARV). (Figure 3.21, page 86.)

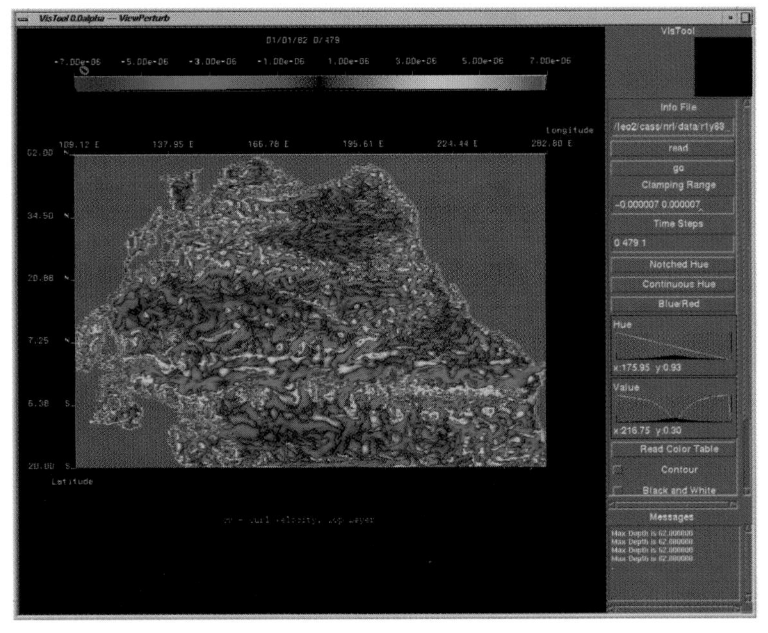

Color Plate 18: Relative vorticity for Pacific Ocean on January 1, 1982. (Figure 3.22, page 86.)

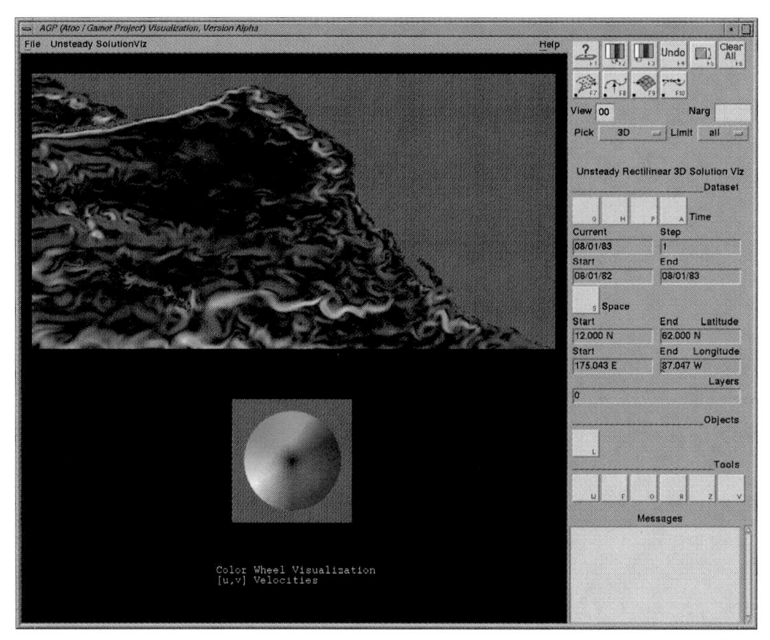

Color Plate 19: Flow in NE Pacific on August 1, 1983; direction of flow is mapped to a hue and magnitude to saturation and value; "rotated" color wheels close to coastline representing eddies. (Figure 3.23, page 87.)

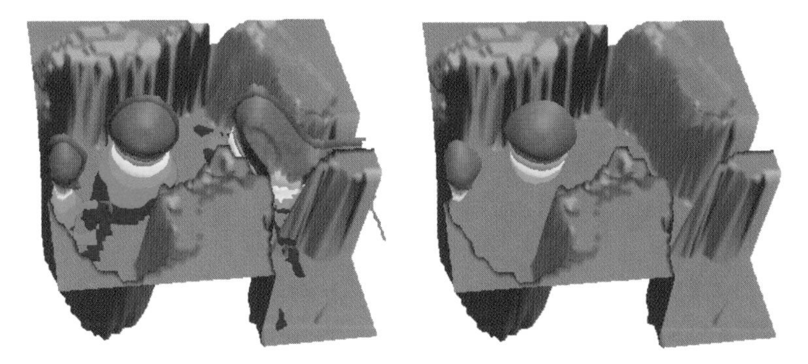

Color Plate 20: Difference in straightforward isosurfacing and enhanced edge-detection scheme. (Figure 3.24, page 88.)

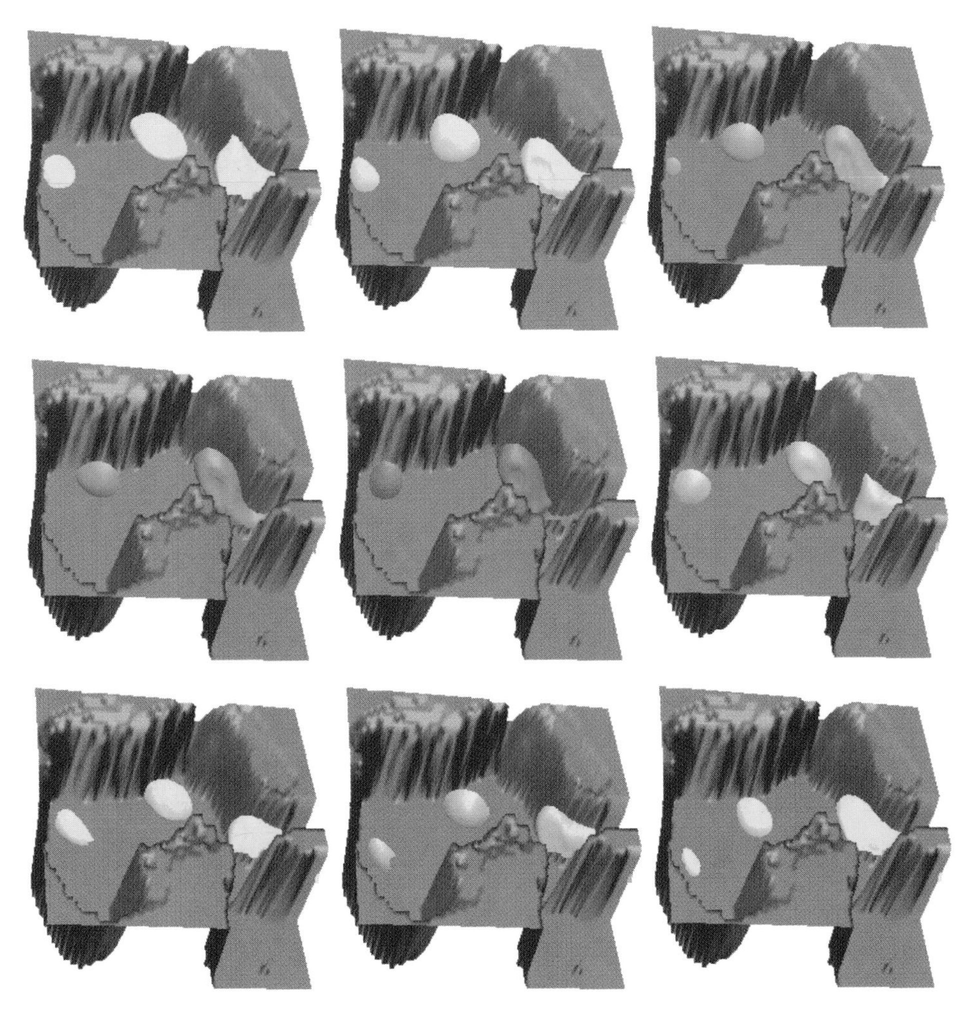

Color Plate 21: Example of eddies propagating through the Gulf of Mexico; sequence progresses from left to right and from top to bottom. (Figure 3.25, page 89.)

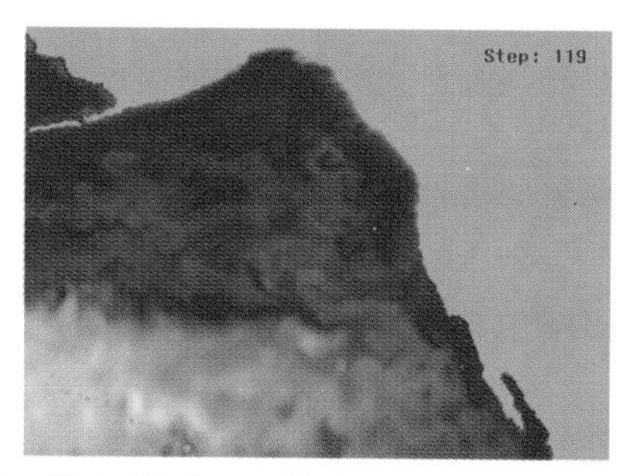

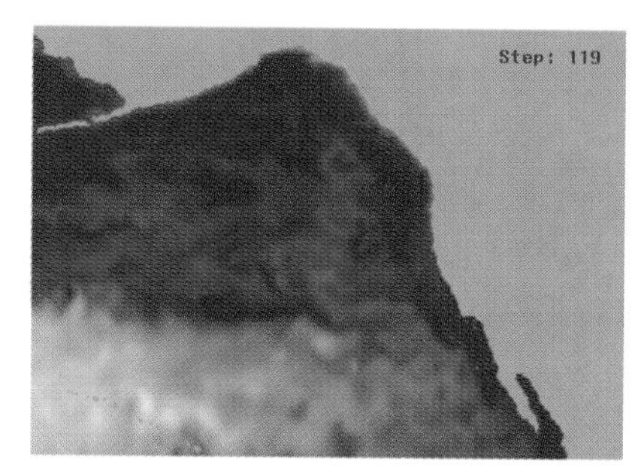

Color Plate 22: Layer thickness for NE Pacific Ocean. *(left)* original data; *(right)* reconstructed data (compression ratio 50:1). (Figure 3.28, page 93.)

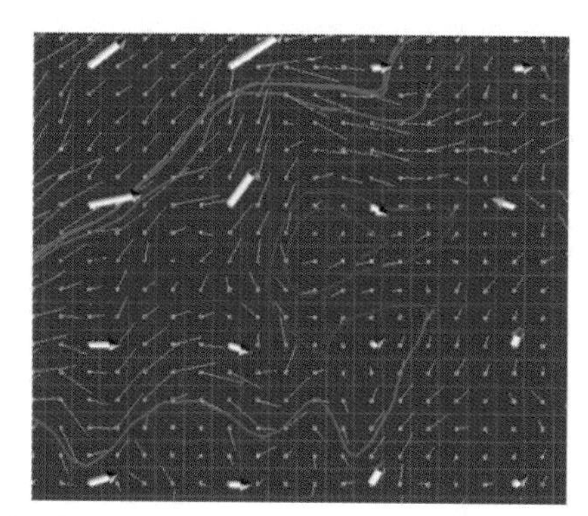

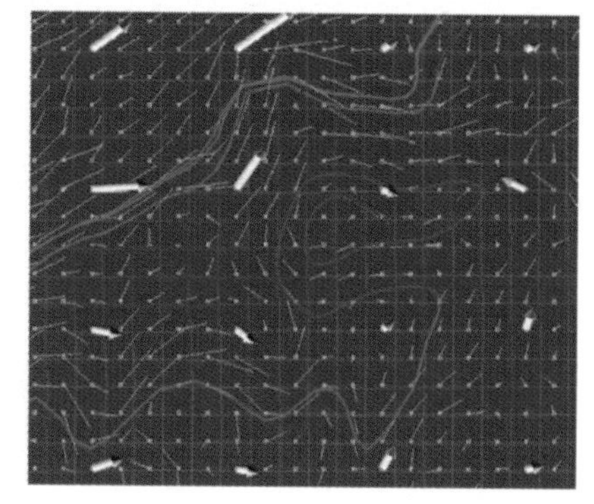

Color Plate 23: Velocity data for NE Pacific Ocean. *(left)* original data; *(right)* reconstructed data (compression ratio 50:1). (Figure 3.29, page 94.)

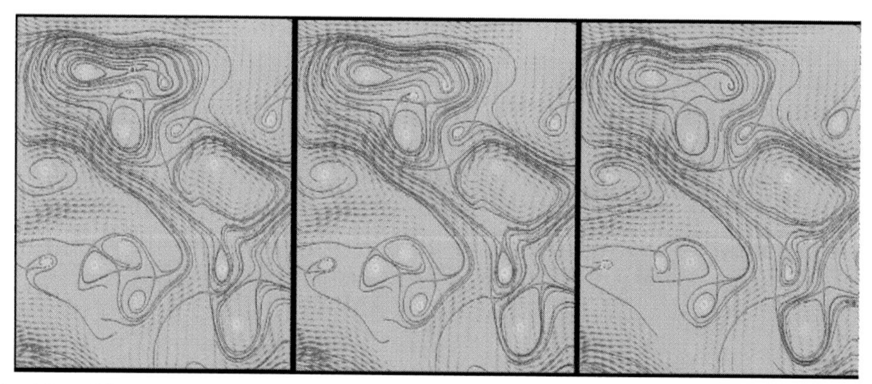

Color Plate 24: Vector field topology extraction on reconstructed data. *(left)* original data; *(middle)* reconstructed data using 1/4 of the WT coefficients; *(right)* reconstructed data using 1/16 of the WT coefficients. (Figure 3.30, page 94.)

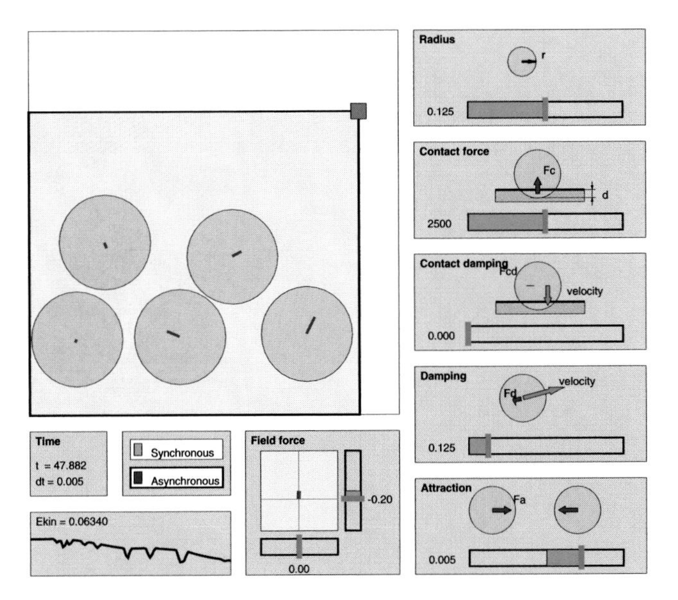

Color Plate 25: Simulation of bouncing balls. (Figure 4.2, page 105.)

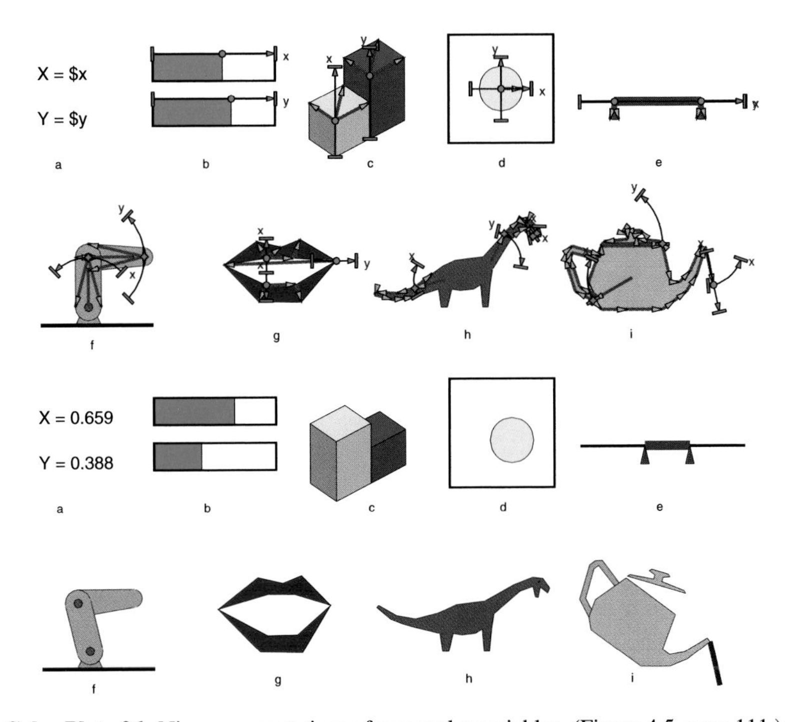

Color Plate 26: Nine representations of two scalar variables. (Figure 4.5, page 111.)

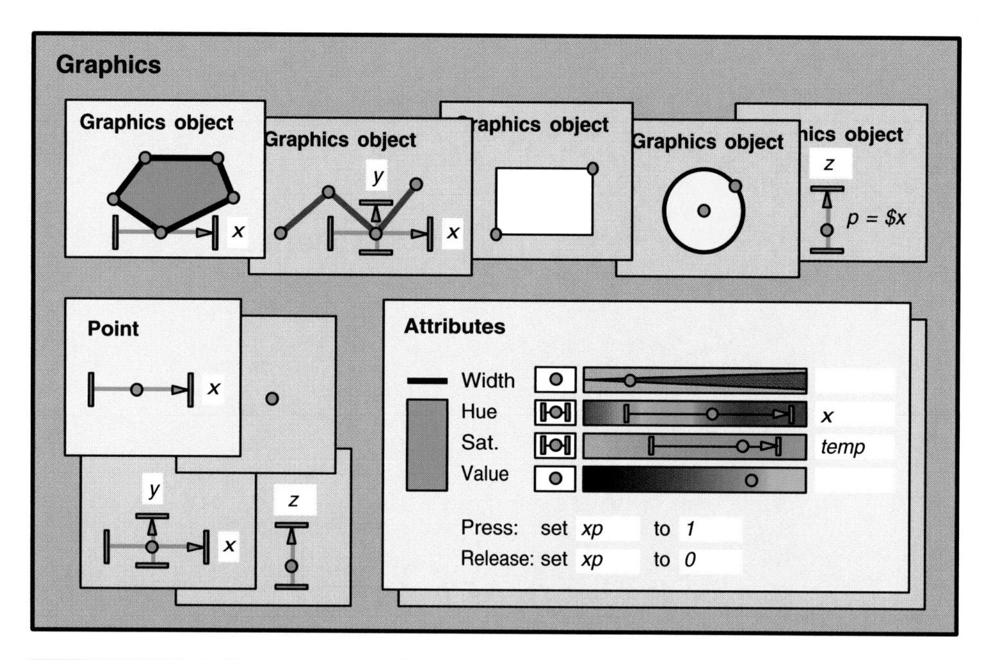

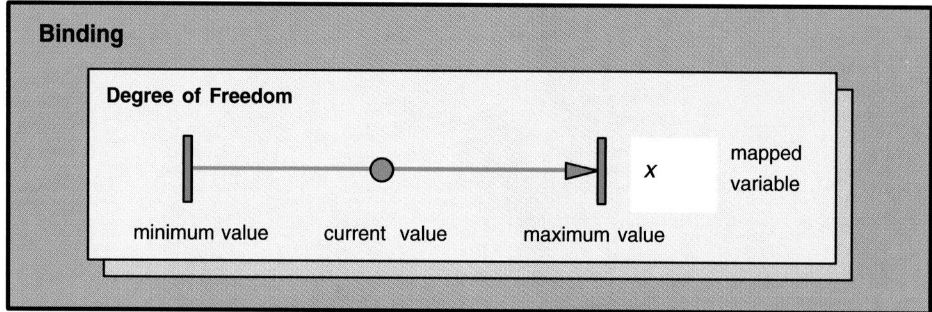

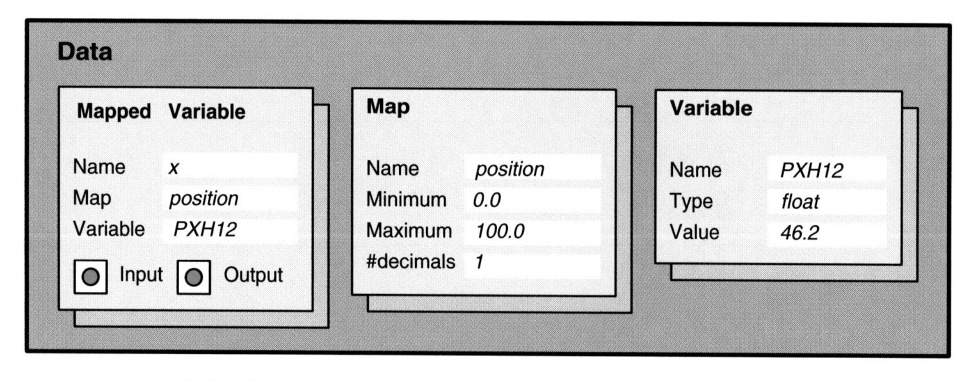

Color Plate 27: Objects in the PGO editor. (Figure 4.6, page 113.)

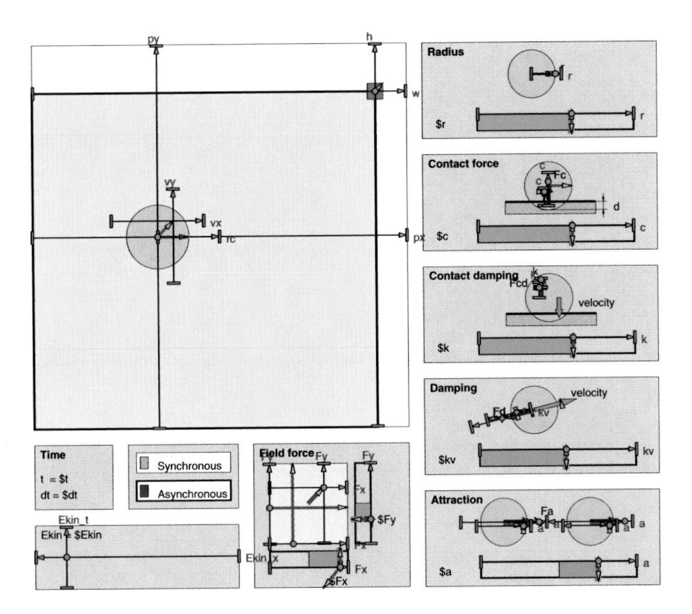

Color Plate 28: Simulation of bouncing balls, edit-mode. (Figure 4.10, page 118.)

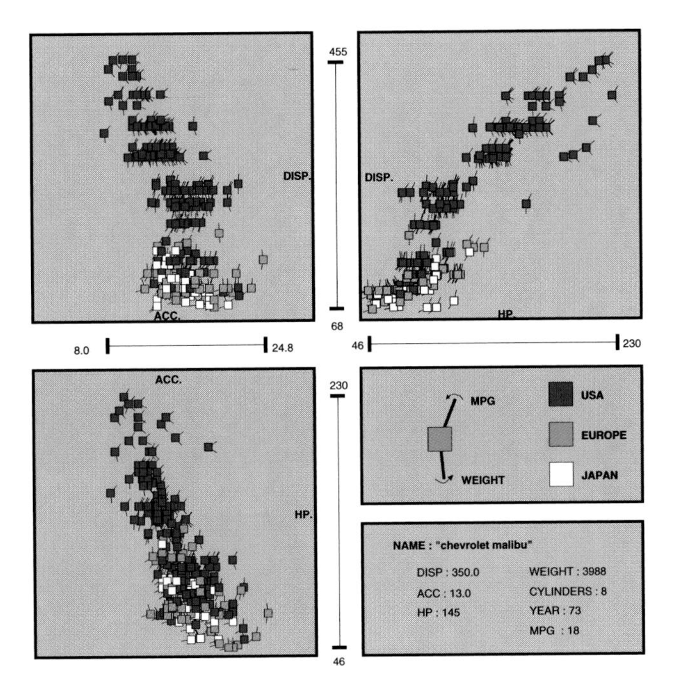

Color Plate 29: Visual front end to a multidimensional database. (Figure 4.11, page 119.)

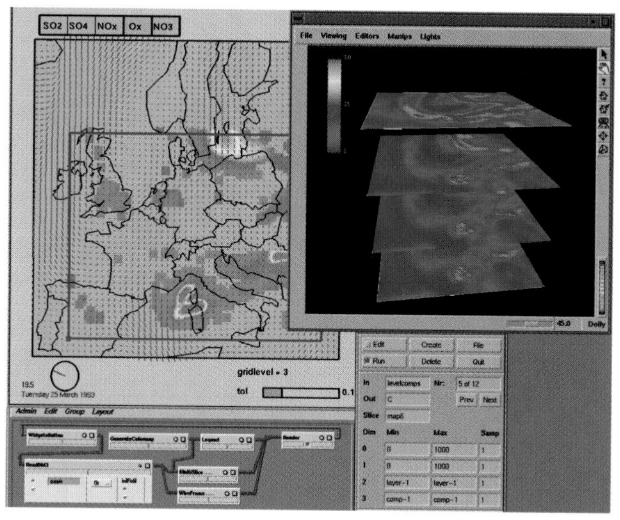

Color Plate 30: Visualization of a smog prediction model with the PGO editor and IRIS Explorer. (Figure 4.12, page 120.)

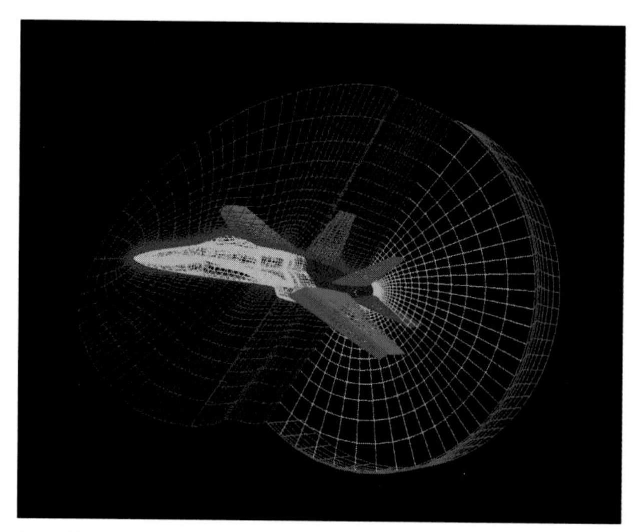

Color Plate 31: A multizoned curvilinear grid surrounding a F18 aircraft, generated by Yehia Rizk and Ken Gee, NASA Ames Research Center. Each sub-block is distinguished by color. (Figure 5.1, page 126.)

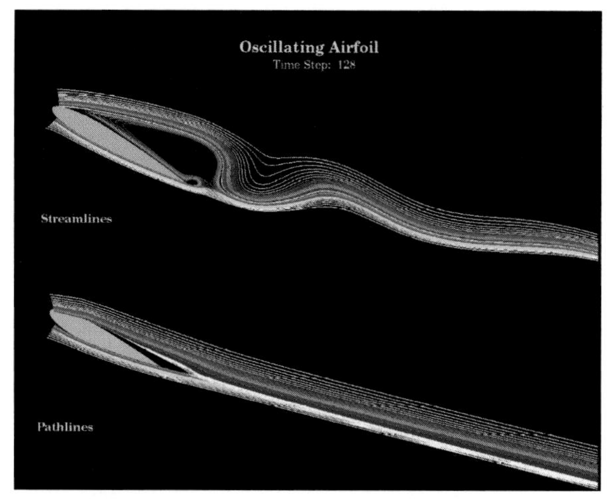

Color Plate 32: Streamlines and pathlines pass through an oscillating airfoil. Particle traces are colored by release location. (Figure 5.2, page 131.)

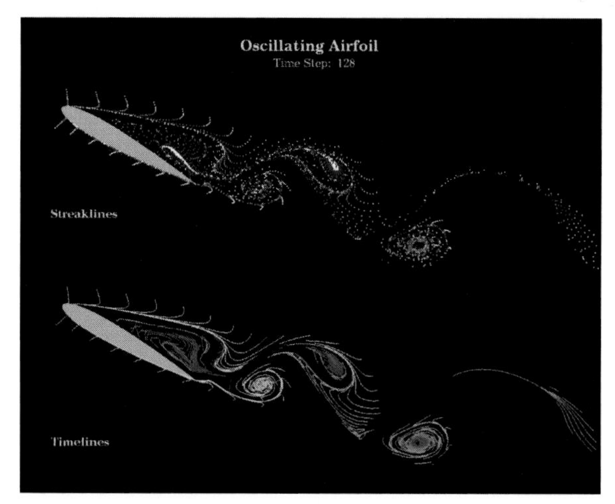

Color Plate 33: Streaklines and timelines pass through an oscillating airfoil. Streaklines are colored by release location and timelines are colored by release time. (Figure 5.3, page 131.)

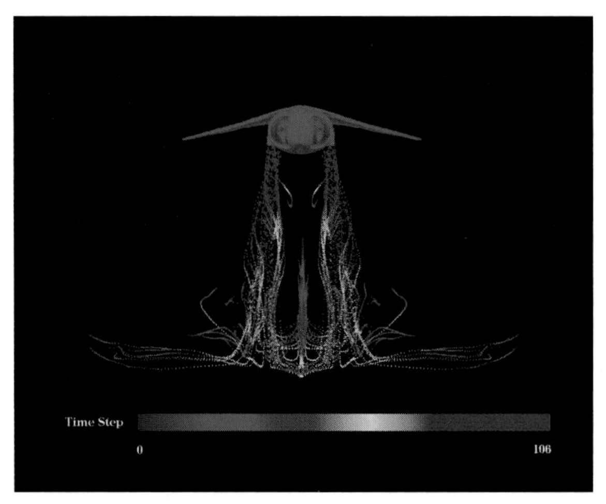

Color Plate 34: Streaklines released from two jet exits of the Harrier jet. Particles are colored by time at release. (Figure 5.4, page 140.)

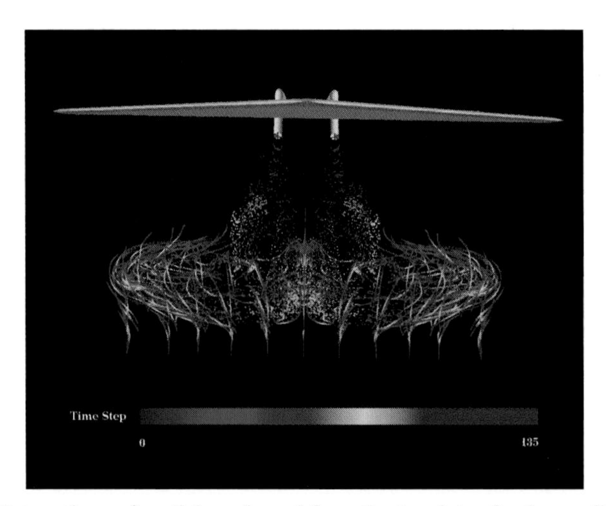

Color Plate 35: Interactions of particles released from the two jets of a descending delta wing and those released near the surface of the ground are shown. Particles are colored by time at release. (Figure 5.5, page 141.)

Color Plate 36: Streaklines surrounding the V-22 tilt rotor aircraft after three blade revolutions. The V-22 is colored by pressure and the particles are colored by time at release. (Figure 5.6, page 142.)

Color Plate 37: Particles released near the noise of a delta wing with wing rock motion. Spiral flow is evident above the left wing and vortex breakdown is visible above the right wing. (Figure 5.7, page 142.)

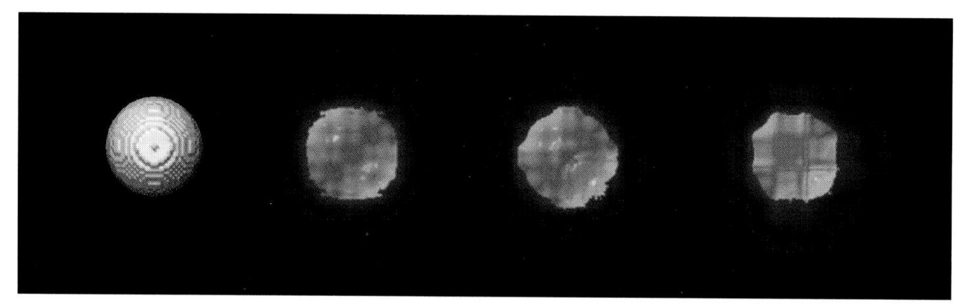

Color Plate 38: Images obtained from a Gaussian density distribution of 32^3 voxels: *(a)* Isosurface at $\tau = 0.5$ with a marching cubes algorithm. *(b)* Isosurfaces and translucent hull obtained from a Battle-Lemarie wavelet with 961 coefficients. *(c)* Isosurfaces and translucent hull obtained from a Coiflet wavelet with 1006 coefficients. *(d)* Isosurfaces and translucent hull obtained from a Daubechies wavelet with 1154 coefficients. (Figure 6.15, page 171.)

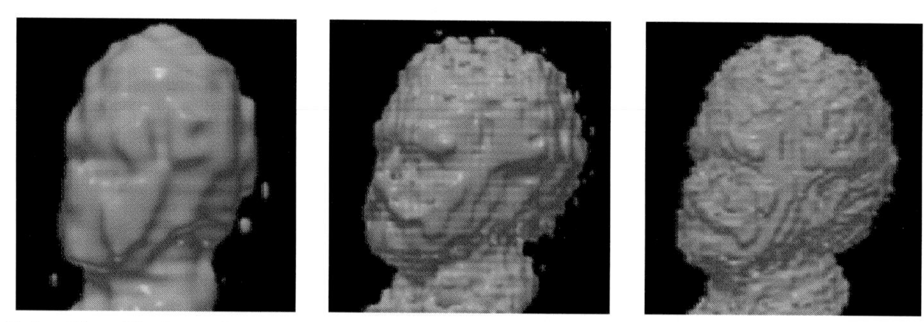

Color Plate 39: Isosurface reconstruction using Kalra's method and 3D wavelets with $\tau = 0.5$: *(a)* 1180 coefficients *(b)* 2832 coefficients *(c)* 9601 coefficients. (Figure 6.17, page 173.)

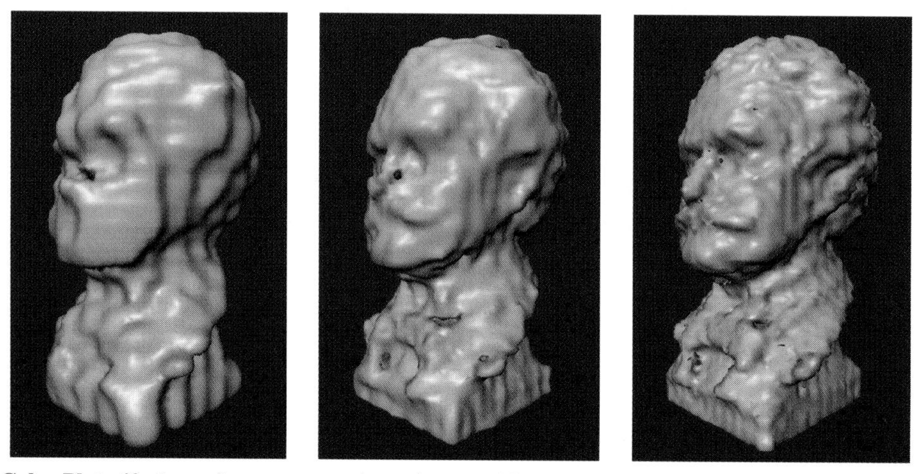

Color Plate 40: Isosurface reconstruction using marching cubes methods with $\tau = 0.5$ and Battle-Lemarie wavelets: *(a)* 1180 coefficients *(b)* 2832 coefficients *(c)* 9601 coefficients. (Figure 6.18, page 173.)

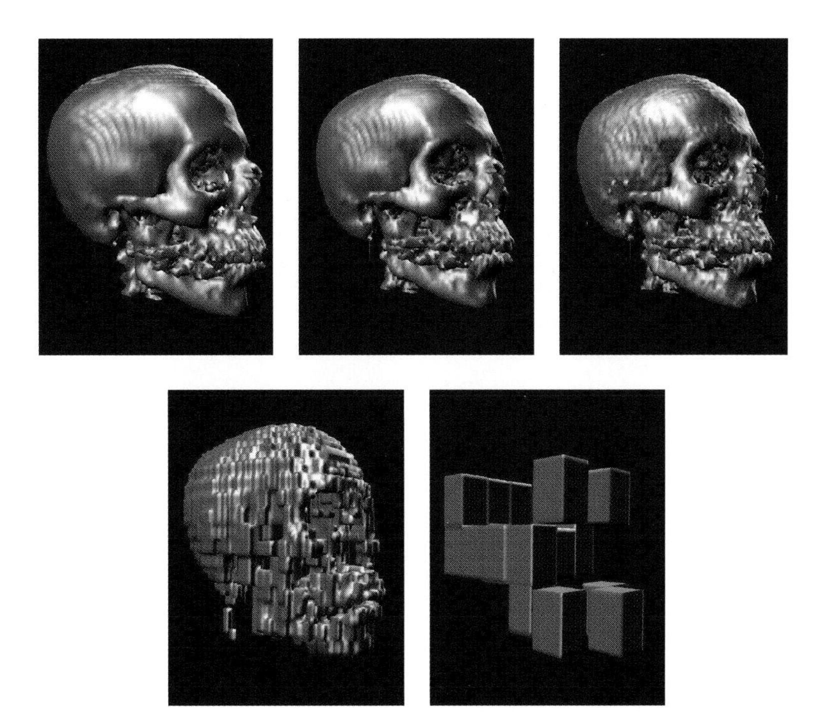

Color Plate 41: Isosurfaces from a human scull obtained by Haar decompositions of the data with different levels of approximation: *(a)* 100% coefficients *(b)* 15.5% coefficients *(c)* 6.8% coefficients *(d)* 0.3% coefficients *(e)* 0.02% coefficients. (Figure 6.19, page 174.)

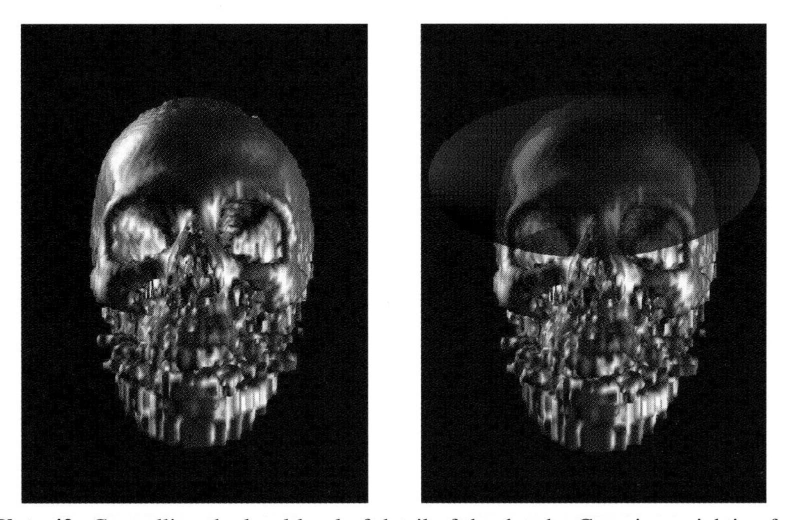

Color Plate 42: Controlling the local level-of-detail of the data by Gaussian weighting functions: *(a)* Result obtained with Haar wavelets. *(b)* Illustration of the weighting function (red ellipsoid). (Figure 6.20, page 174.)

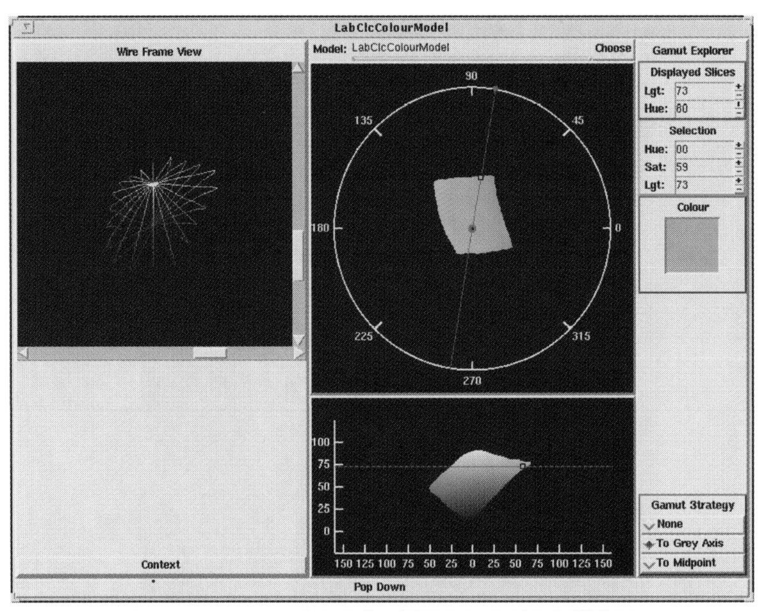

Color Plate 43: Gamut explorer. (Figure 7.1, page 182.)

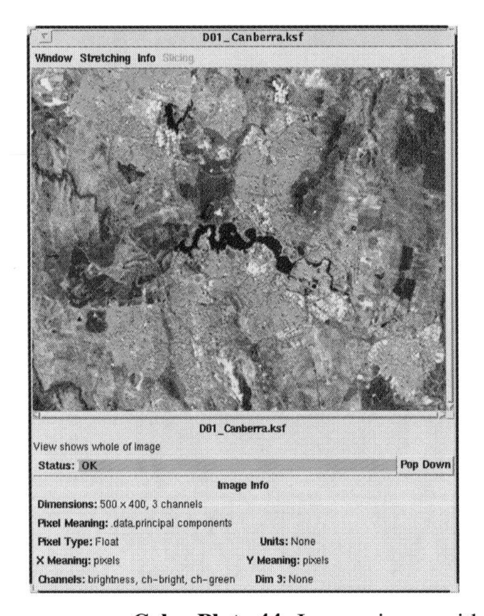

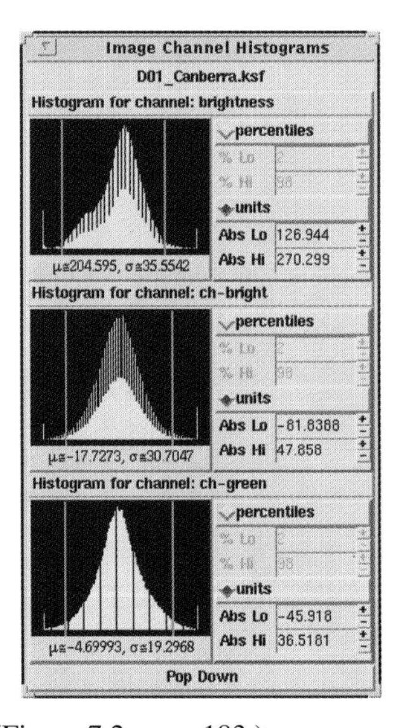

Color Plate 44: Image viewer with channel histograms. (Figure 7.2, page 183.)

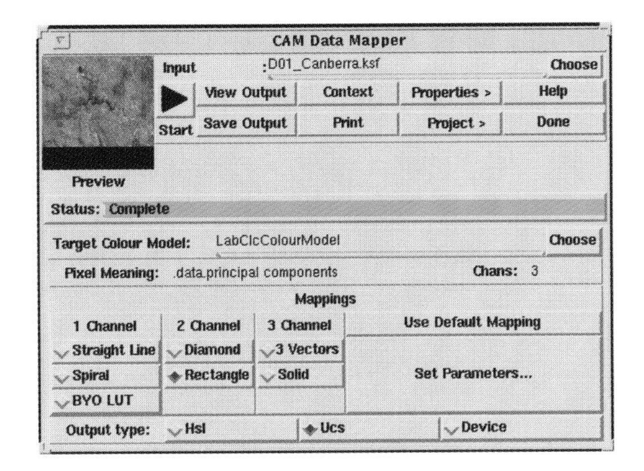

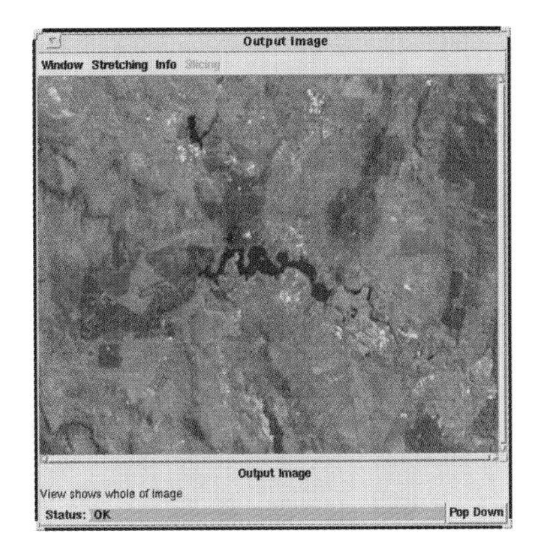

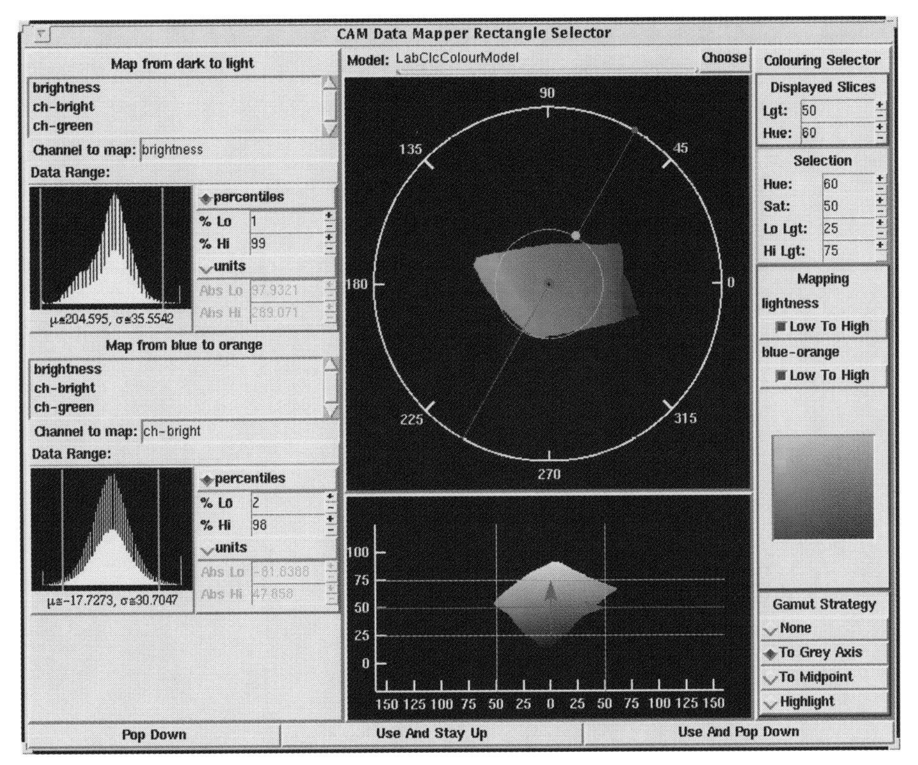

Color Plate 45: Data mapper, output image view, and coloring selector. (Figure 7.3, page 184.)

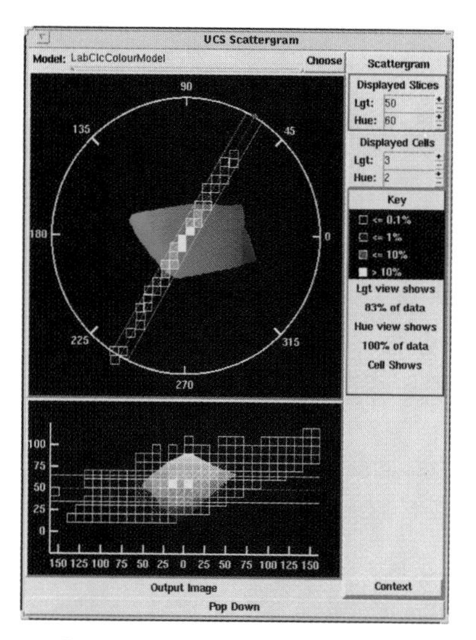

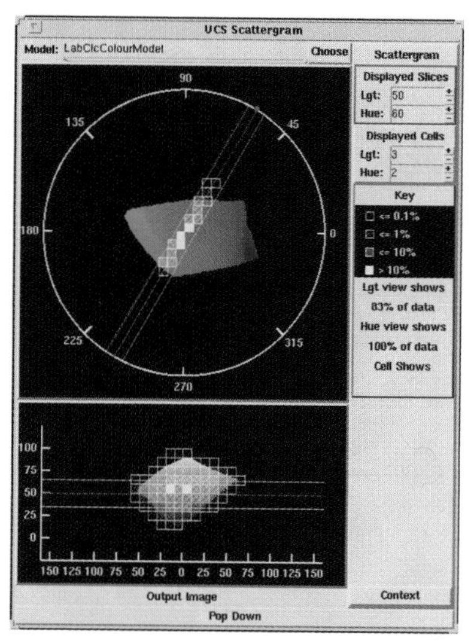

Color Plate 46: Color scattergrams, before and after return-to-gamut. (Figure 7.4, page 185.)

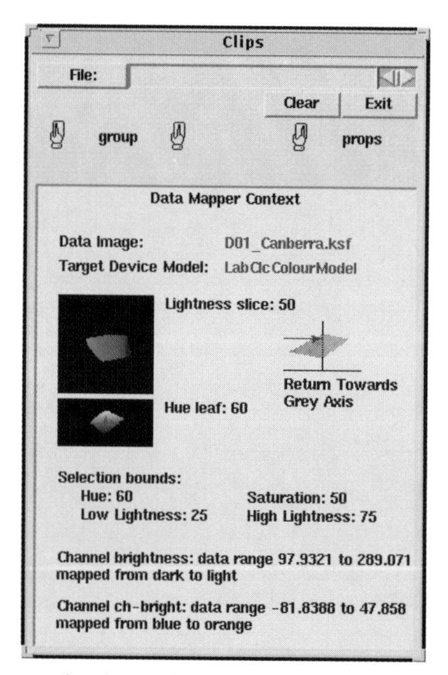

Color Plate 47: Encapsulated metavisualizations as context. (Figure 7.5, page 186.)

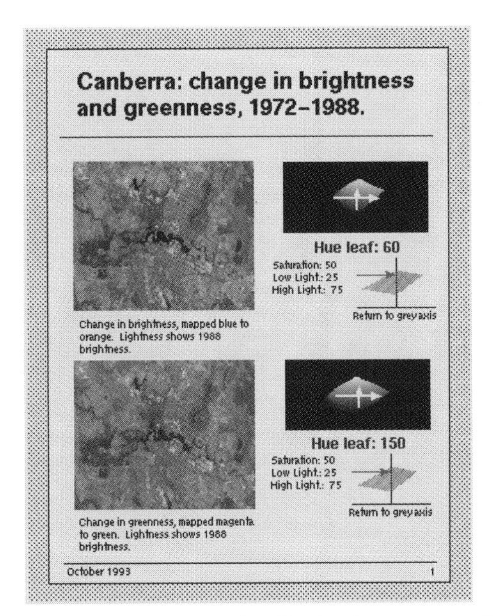

Color Plate 48: Integration of data and metavisualization into a report. (Figure 7.6, page 187.)

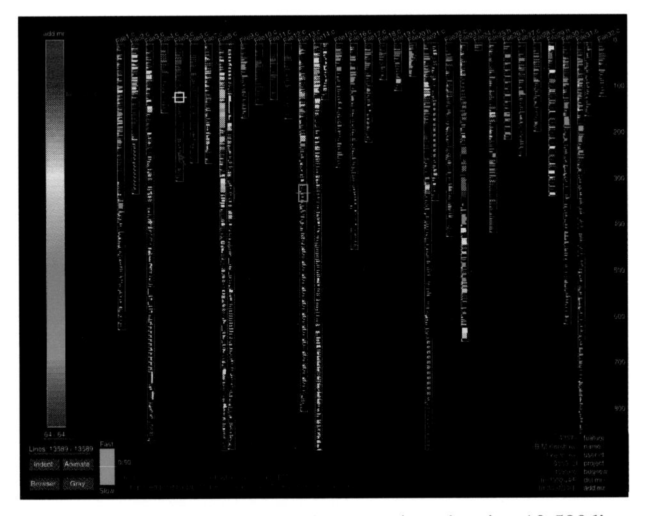

Color Plate 49: SeeSoft reduced representation overview showing 13,589 lines of code, color coded by age. The newest lines are in red and the oldest in blue with a color spectrum in between. Lines in the same hue were written at approximately the same time and are likely related. (Figure 8.1, page 194.)

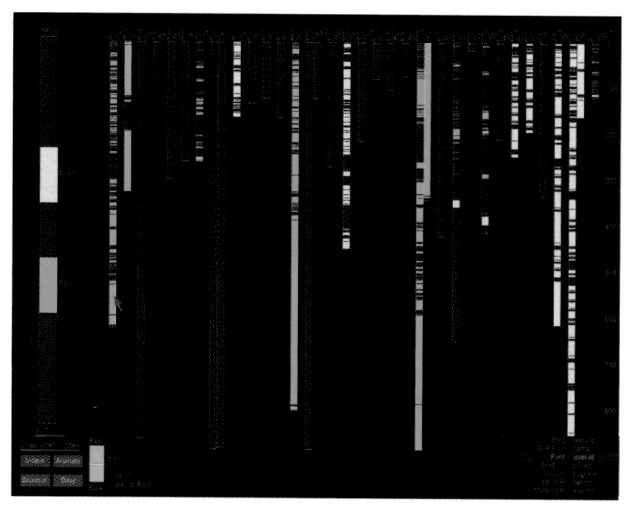

Color Plate 50: SeeSoft line indentation has been turned off and here the lines are color coded to show the `user_id` of the author. `User_id` *Ford* wrote the green lines and *Bush* wrote the yellow lines. (Figure 8.2, page 195.)

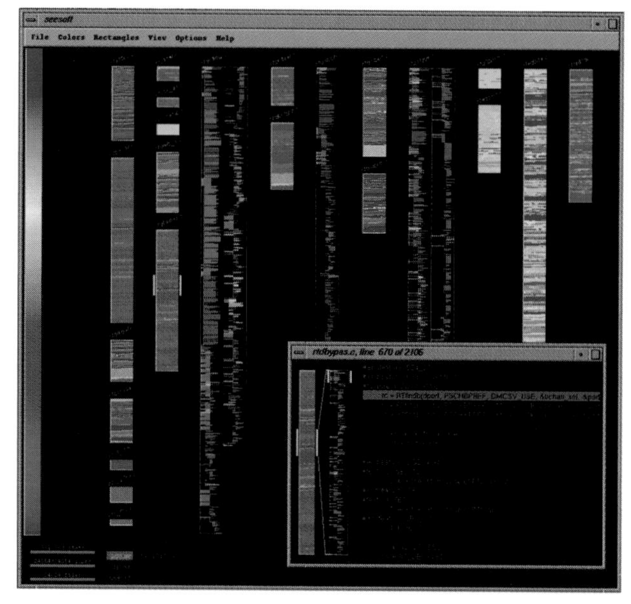

Color Plate 51: An experimental version of SeeSoft showing three views of code. The filled columns represent each line of code using a few pixels positioned horizontally along each row, the indented columns use the traditional SeeSoft representation with one line per row, and the browser window combines both views. By representing more than one line in a row, more code can be shown on a single view. (Figure 8.3, page 196.)

Color Plate 52: SeeNet showing AT&T calling volume by switch Christmas morning, 1994. The size of each rectangle shows the volume with the x-extent coding the inbound volume, the y-extent coding the outbound volume, and color showing direction. Red rectangles indicate sinks and green rectangles indicate source. In this time period there is heavy calling volume both into and out of Florida and also heavy volume into the five red gateway (Figure 8.4, page 197.)

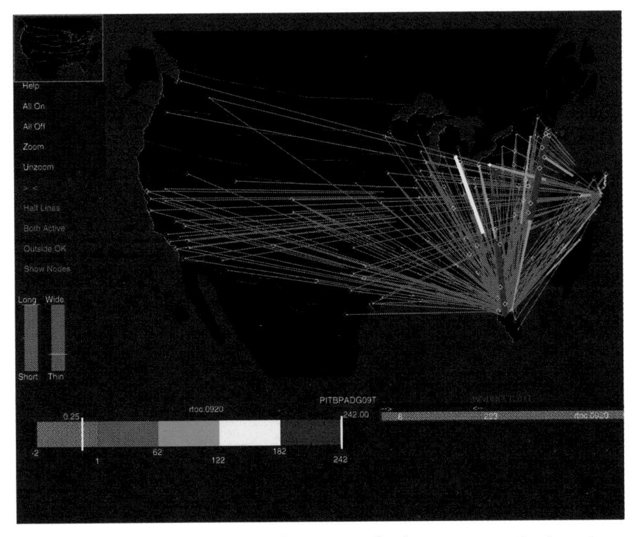

Color Plate 53: SeeNet display of the long distance telephone network showing overload between the nodes. (Figure 8.5, page 198.)

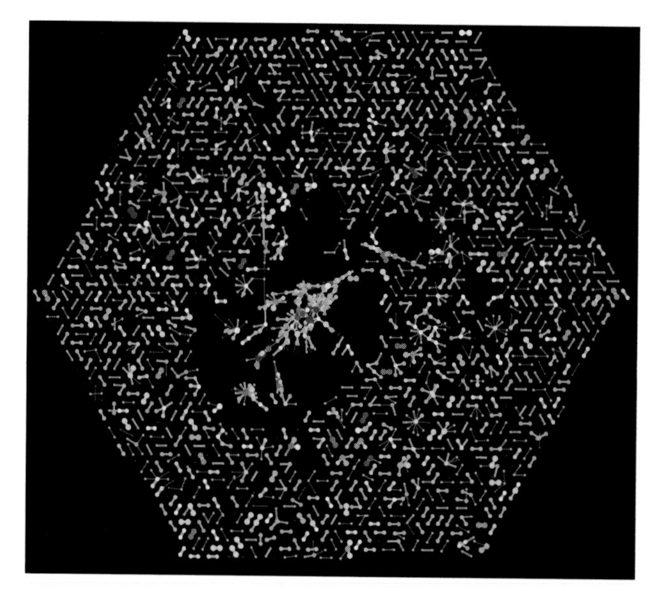

Color Plate 54: NicheWorks visualization of local exchange calling patterns. Each subscriber making a call is represented by a node with lines drawn between the nodes showing individual conversations. The nodes are positioned to show the calling patterns. The cluster of nodes in the center of the diagram represents callers who all talk to each other. The time spent talking is coded by the node colors and size. The link colors code the length of individual conversations. (Figure 8.6, page 199.)

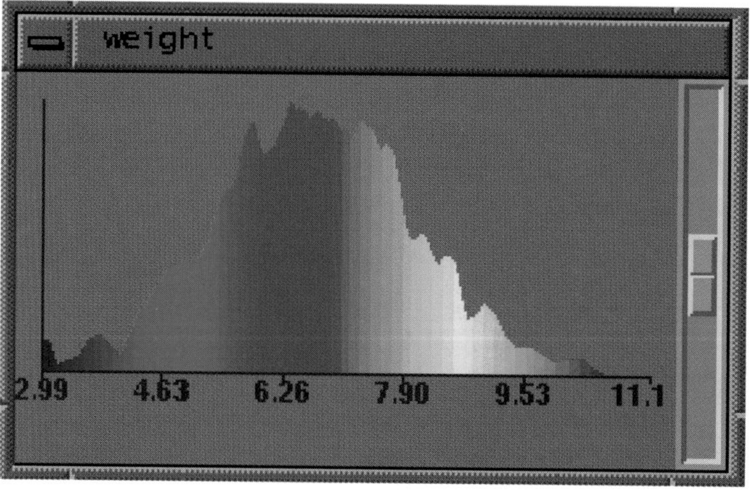

Color Plate 55: Histograms show the distribution of the statistics; users may interactively filter out nodes or links that fall in uninteresting portions of the distribution. The histogram also maps colors from the main display to values. (Figure 8.7, page 200.)

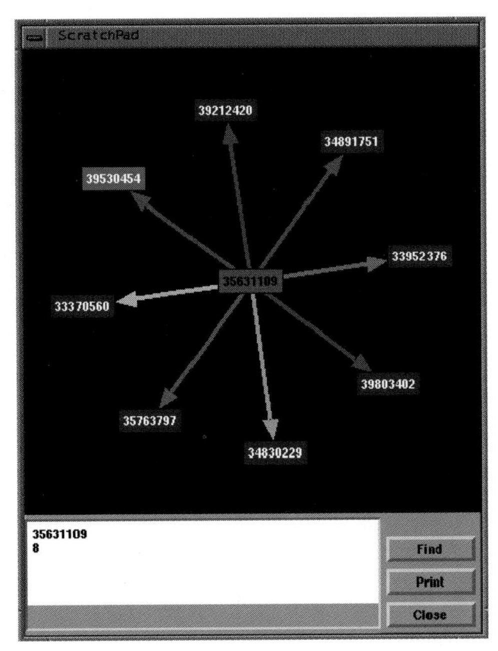

Color Plate 56: A `scratch pad` shows the details for any node. (Figure 8.8, page 201.)

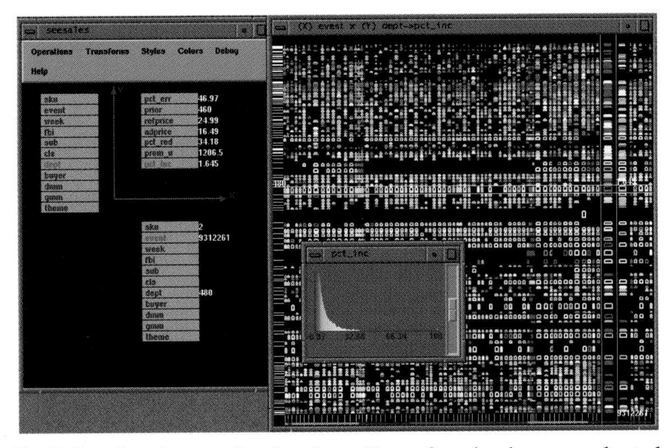

Color Plate 57: SeeSales showing retail sales data. Items for viewing are selected using the control panel (left) and represented as a color-coded rectangle showing on an item-by-week grid. "Hotter" colors (pinks and whites) represent large increases and "colder" colors (blues, greens) represent smaller increases. The two widest columns on the right are the Thanksgiving weekend and day after Christmas sales. The histogram shows the distribution of the percentage increases in sales. (Figure 8.9, page 202.)

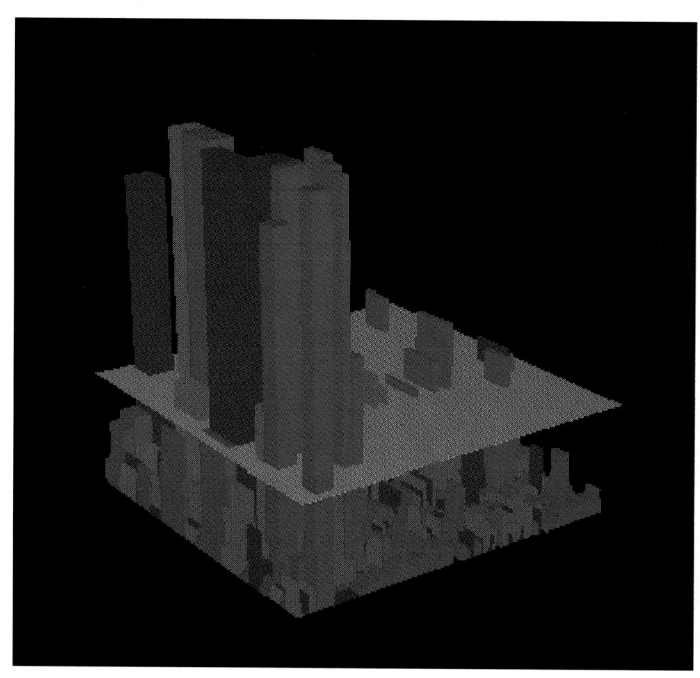

Color Plate 58: An experimental 3D representation with a cutting plane showing all sales increases above a threshold. (Figure 8.10, page 203.)

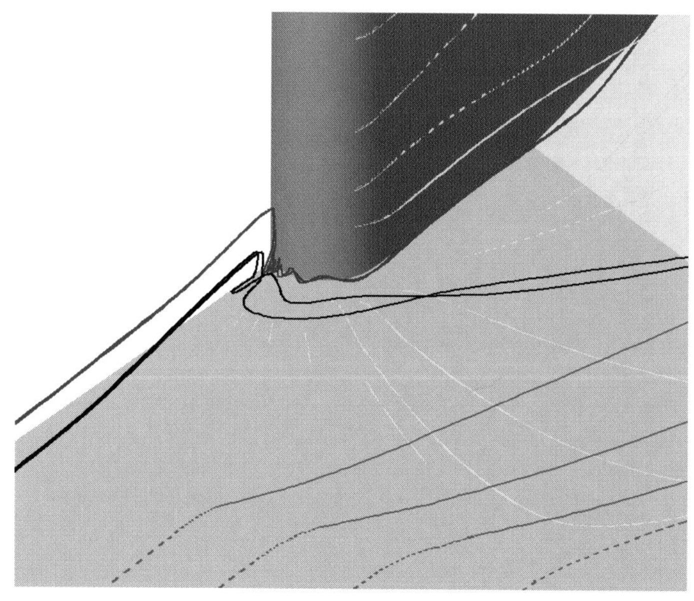

Color Plate 59: Poor representation of a vector field by streamlines. (Figure 9.2, page 214.)

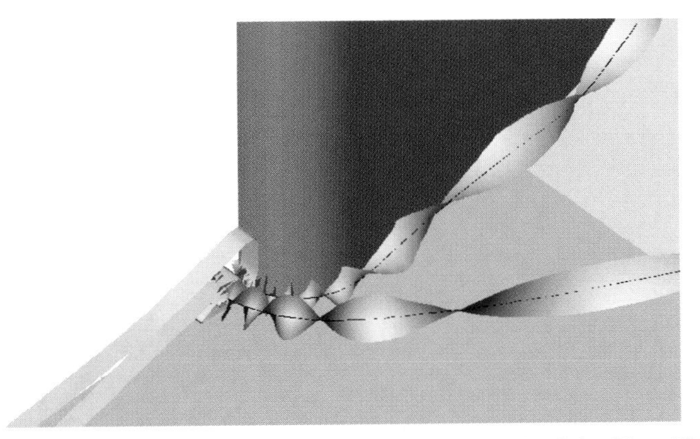

Color Plate 60: Two TSR streamribbons visualize the same data as in Color Plate 59 and clearly depict the axis of two vortices as well as the swirl in the flow. The scale is identical to Color Plate 59. (Figure 9.3, page 214.)

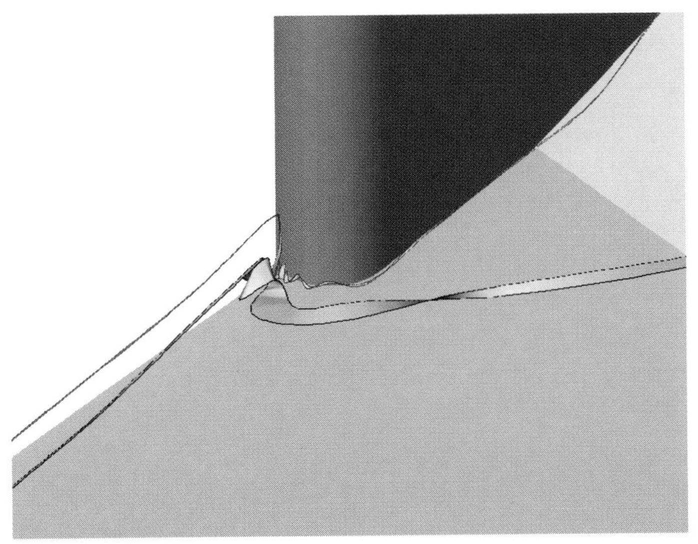

Color Plate 61: The situation of Color Plate 60 visualized by constructing a ribbon from adjacent streamlines. (Figure 9.4, page 215.)

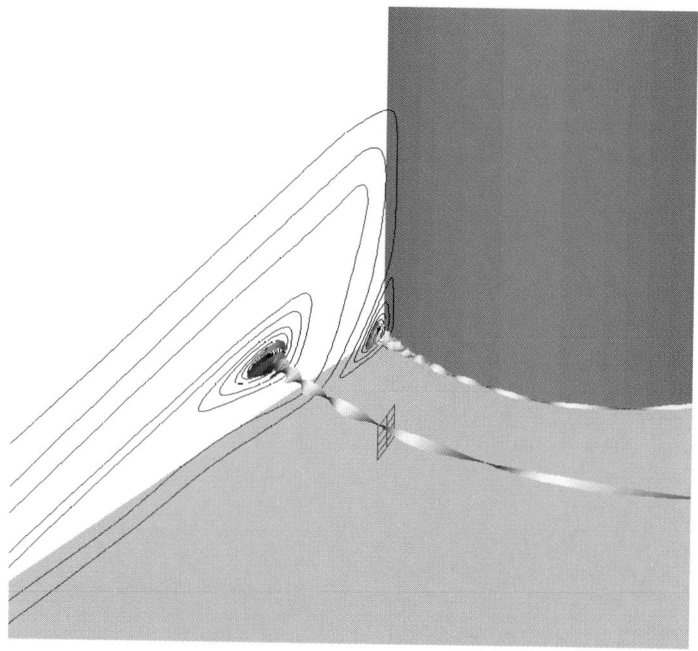

Color Plate 62: ASL streamlines provide useful visualization if they are used within the proximity of the vortex core where the flow is clearly dominated by vortical effects. Some cells of the numerical grid are shown in the image to emphasize that the relevant region is restricted to a very small scale of a few cells in these data. (Figure 9.5, page 216.)

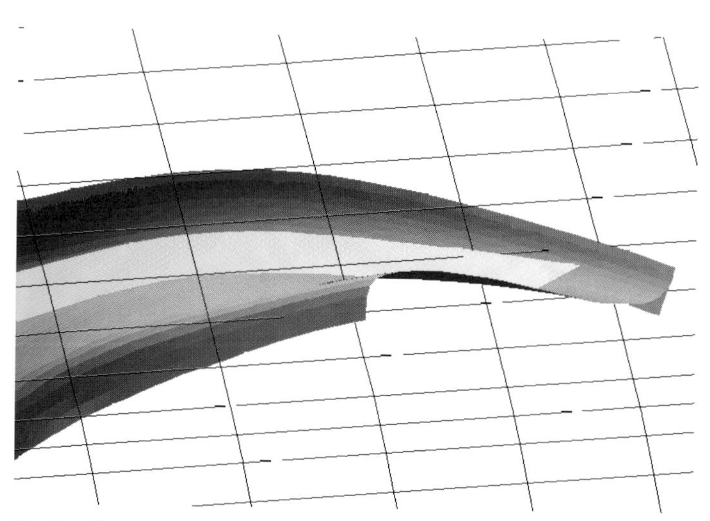

Color Plate 63: The blue stream surface is constructed from 30 adjacent streamlines. While the red ribbon picks up the vortical motion of the flow correctly, the green ribbon fails because of the limited numerical resolution. (Figure 9.6, page 217.)

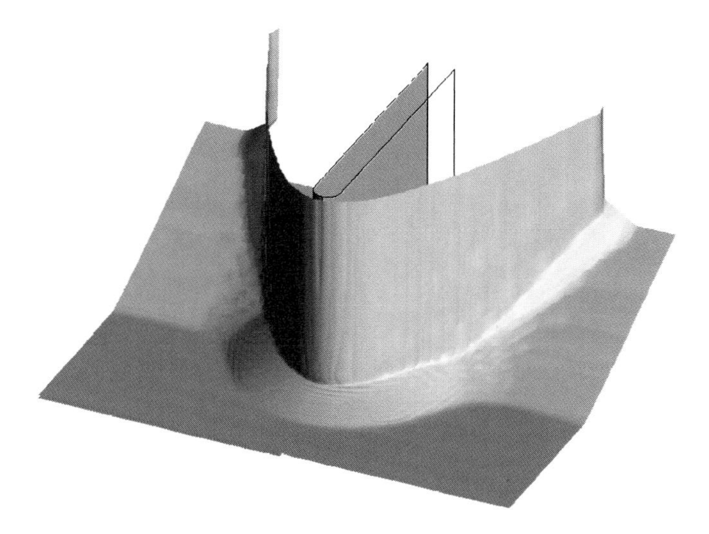

Color Plate 64: Front shock wave visualized using an isosurface close to the free-stream Mach-number. The flow field is the same as in Color Plates 59–63. (Figure 9.8, page 218.)

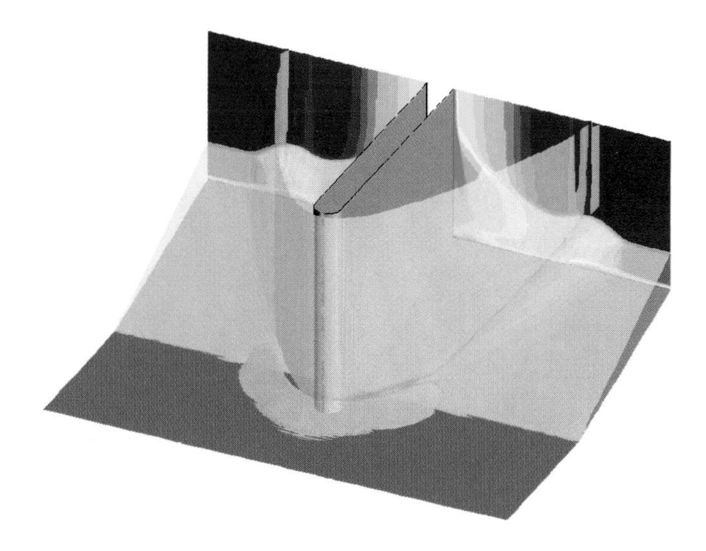

Color Plate 65: Shock wave computed from the location of the maximum local gradient. This method allows us to visualize a secondary shock wave which is visible behind the transparent front shock. (Figure 9.9, page 218.)

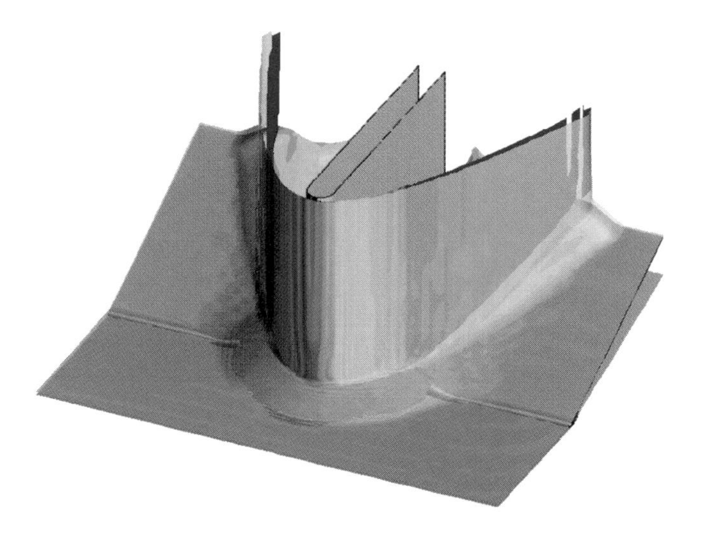

Color Plate 66: Direct comparison of the shock visualization resulting from two alternative methods shows significant spatial displacement of the extracted features. The blue transparent surface is equivalent to the one in Color Plate 64, the surface behind it matches the golden transparent surface shown in Color Plate 64, and is pseudocolored with Mach-number. (Figure 9.10, page 219.)

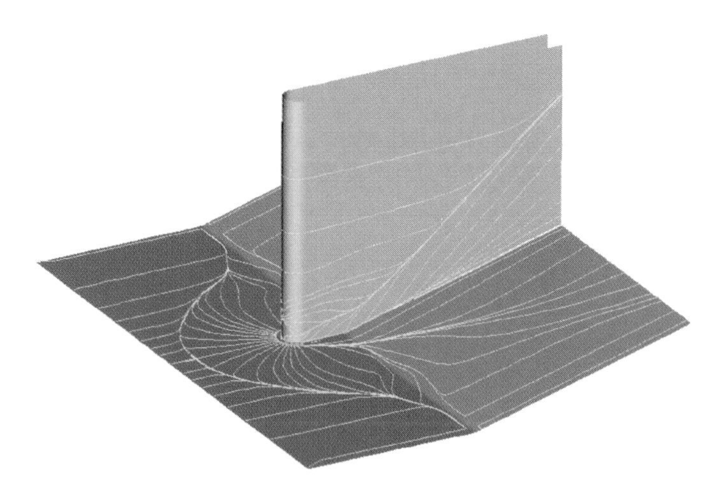

Color Plate 67: Skin-friction lines on walls of the blunt fin and wedge. (Figure 9.12, page 222.)

Color Plate 68: Skin-friction lines on walls of the blunt fin and wedge. (Figure 9.13, page 222.)

Color Plate 69: The oil-flow pattern shows a second weak separation trace s_2, which is not visible in the numerical data. (Figure 9.14, page 223.)

Color Plate 70: Ribbons represent vortices in the flow field. The image allows the examination of three-dimensional phenomena with respect to their traces on the solid walls. (Figure 9.15, page 224.)

Color Plate 71: Three-dimensional vortex cores from the numerical simulation visualized in combination with oil-flow traces from wind tunnel experiments. (Figure 9.16, page 224.)

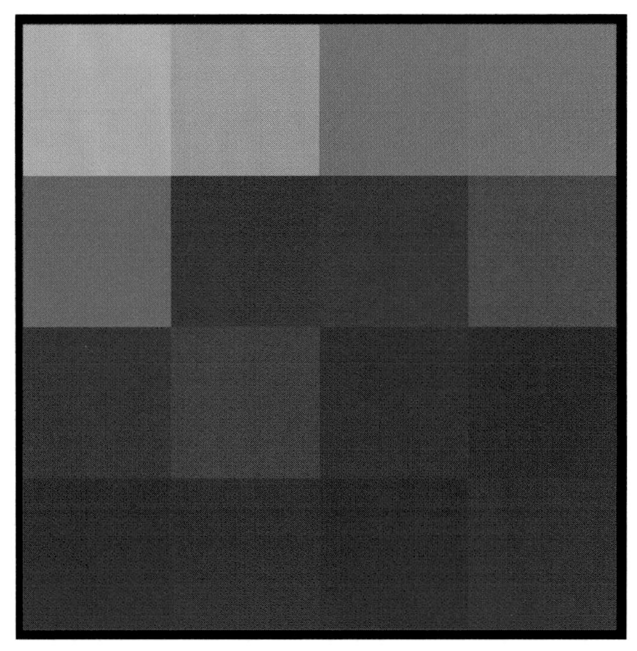

Color Plate 72: Least precise image in sequence of four. (Figure 10.1, page 231.)

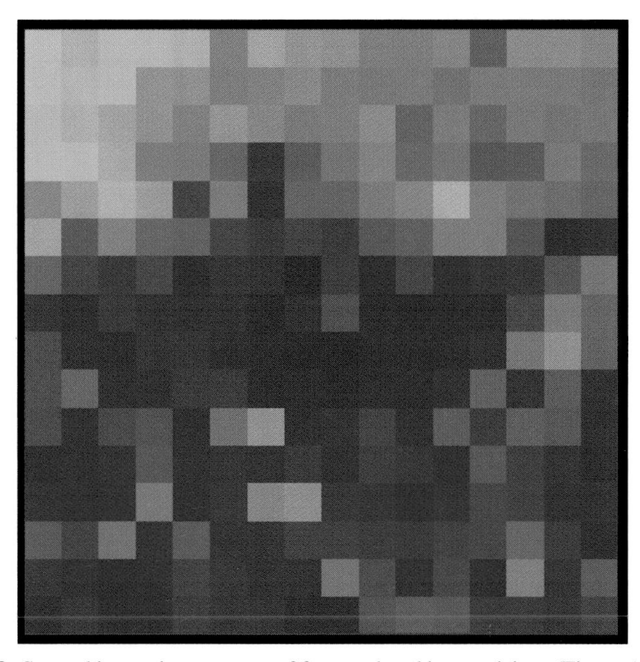

Color Plate 73: Second image in sequence of four, ordered by precision. (Figure 10.2, page 232.)

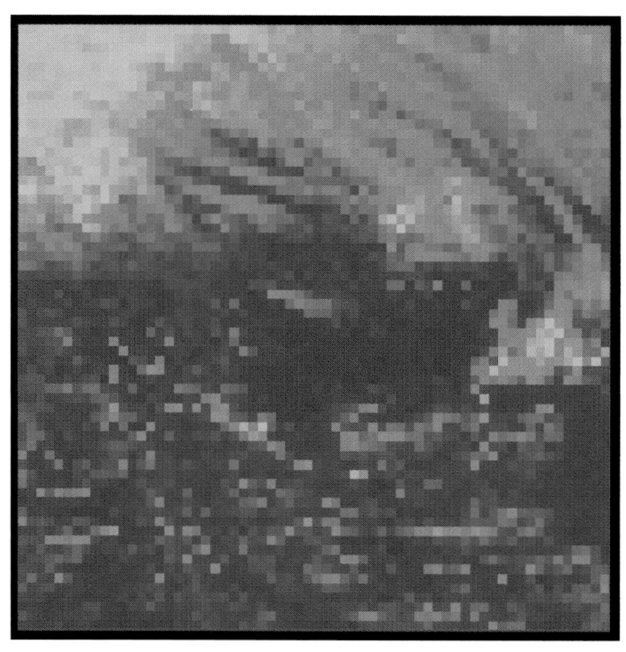

Color Plate 74: Third image in sequence of four, ordered by precision. (Figure 10.3, page 231.)

Color Plate 75: Most precise image in sequence of four. (Figure 10.4, page 233.)

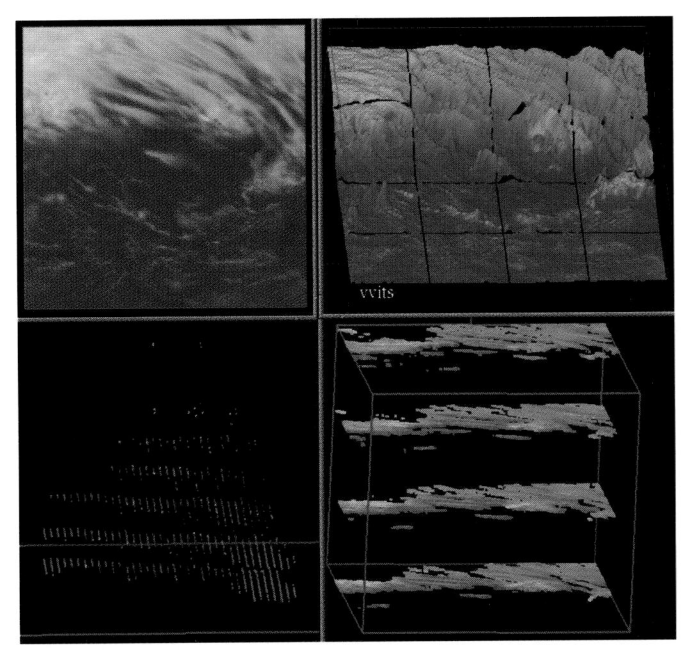

Color Plate 76: A time sequence of multivariate images displayed according to four different sets of mappings. The top-right window uses the mappings shown in Figure 10.6 (page 245), the top-left maps *ir* (red) and *vis* (blue-green) to *color*, the bottom-right maps *ir* to *selector* (only pixels with selected *ir* radiances are displayed) and *time* to the *y* axis, and the bottom-left maps *ir*, *vis*, and *variance* to the *x*, *y*, and *z* axes, maps *textures* to *color* (*variance* and *texture* are pixel fields derived from *ir* radiance). (Figure 10.7, page 246.)

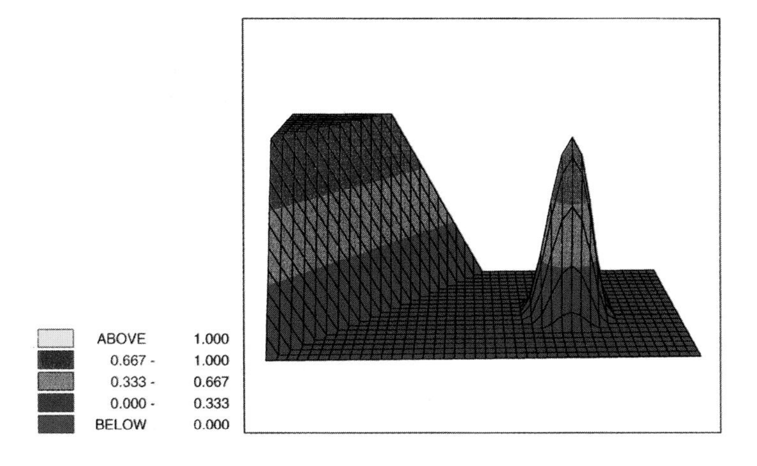

Color Plate 77: The test function. (Figure 11.5, page 262.)

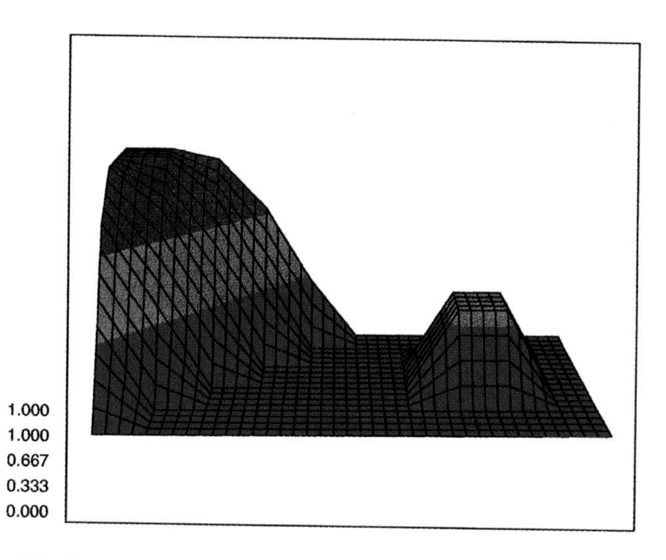

Color Plate 78: Piecewise bilinear interpolant. (Figure 11.6, page 263.)

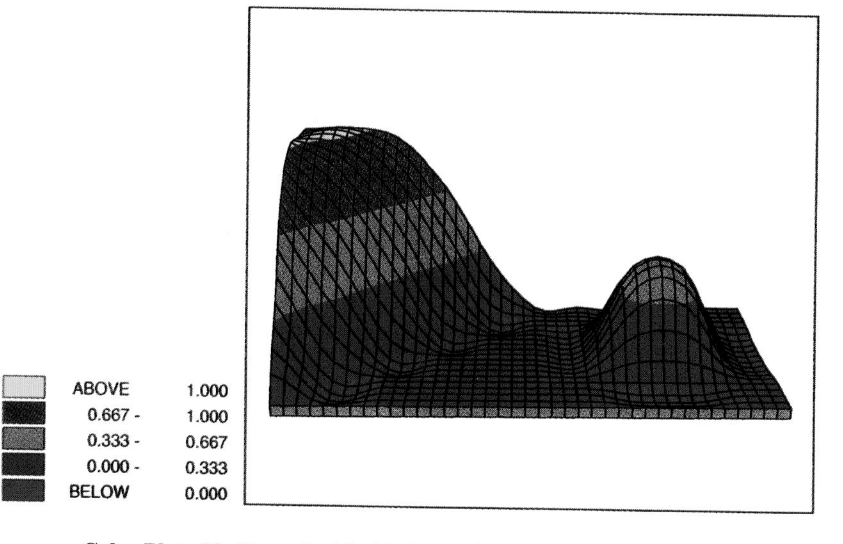

Color Plate 79: Piecewise bicubic interpolant. (Figure 11.7, page 264.)

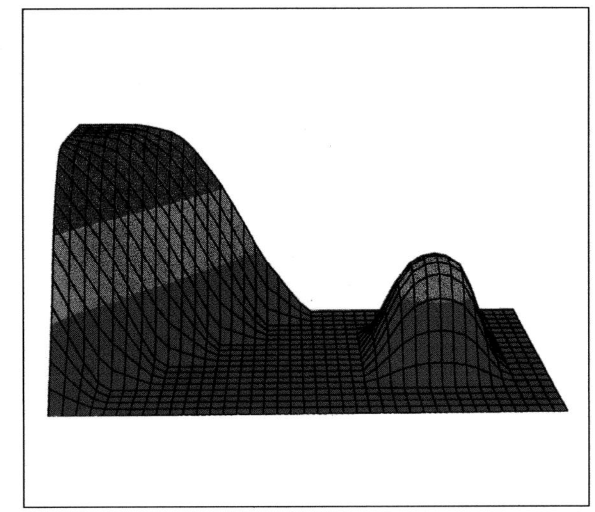

Color Plate 80: Bounded piecewise bicubic interpolant. (Figure 11.8, page 264.)

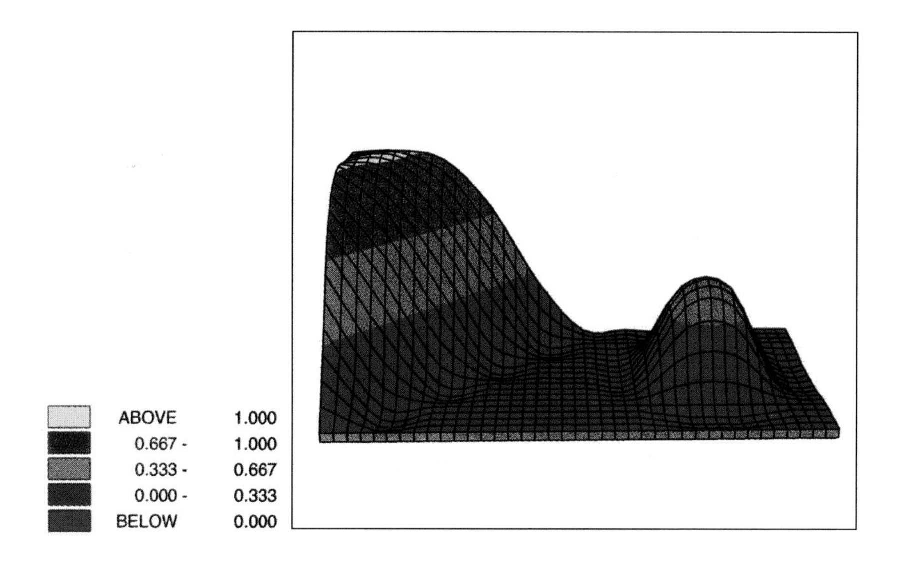

Color Plate 81: Blending function interpolant. (Figure 11.9, page 269.)

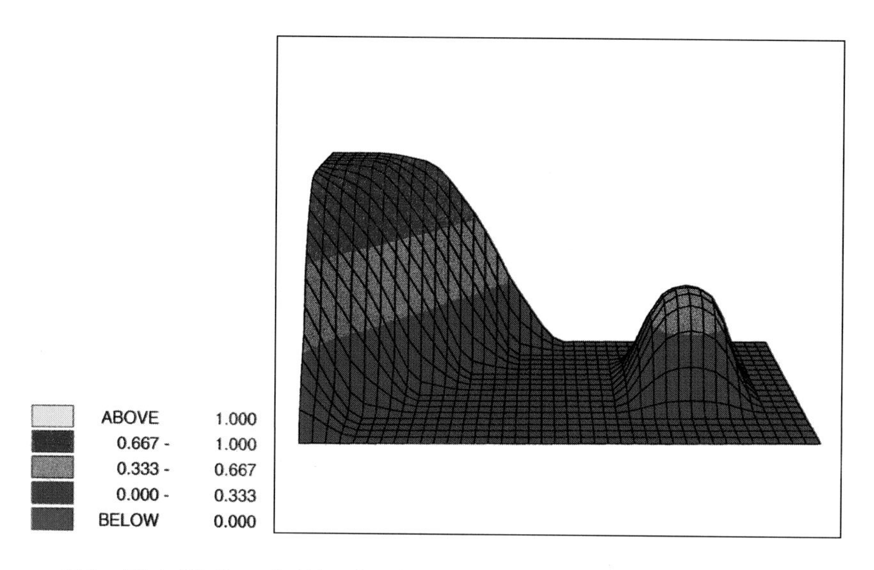

Color Plate 82: Bounded blending function interpolant. (Figure 11.10, page 269.)

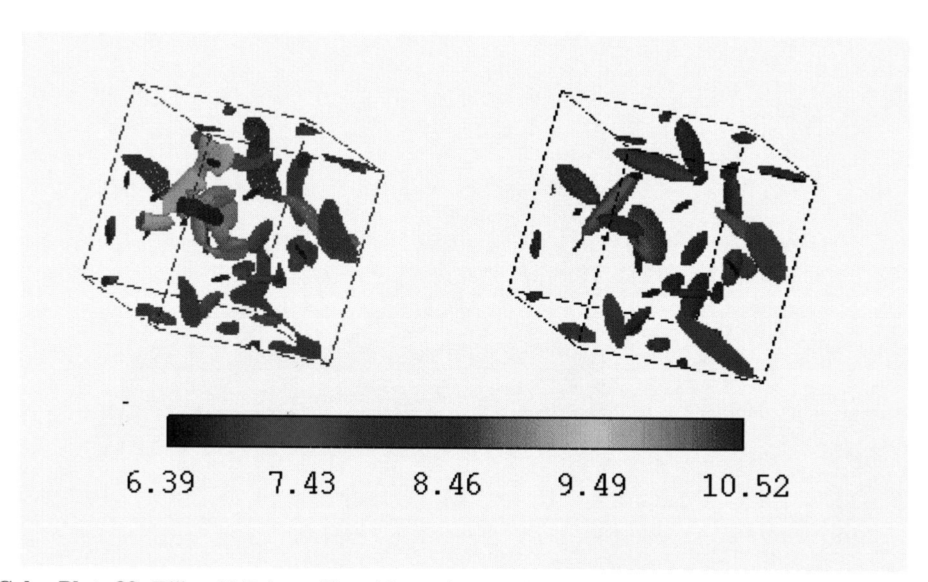

Color Plate 83: Ellipsoid fitting: ellipsoids are fit to regions segmented from a vorticity magnitude data set. Regions are colored by their interior local maxima. (Figure 12.4, page 286.)

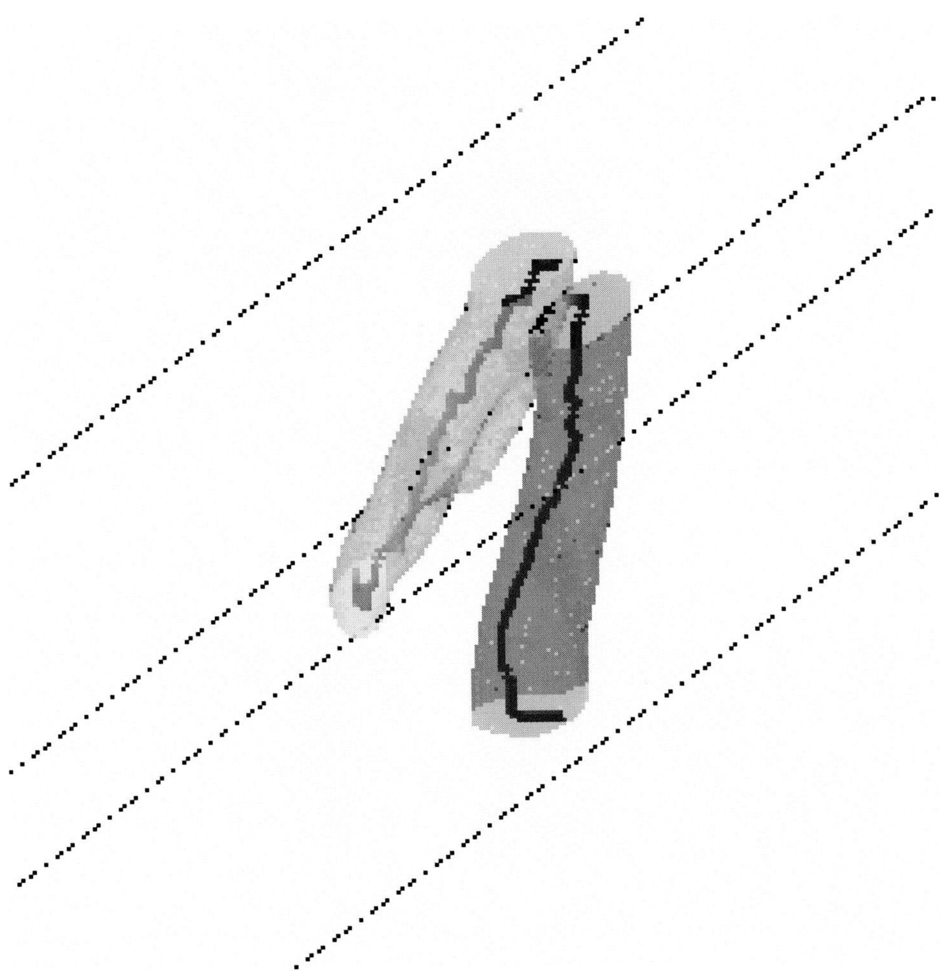

Color Plate 84: Skeleton using local maxima lines and vortex direction. (Figure 12.5, page 287.)

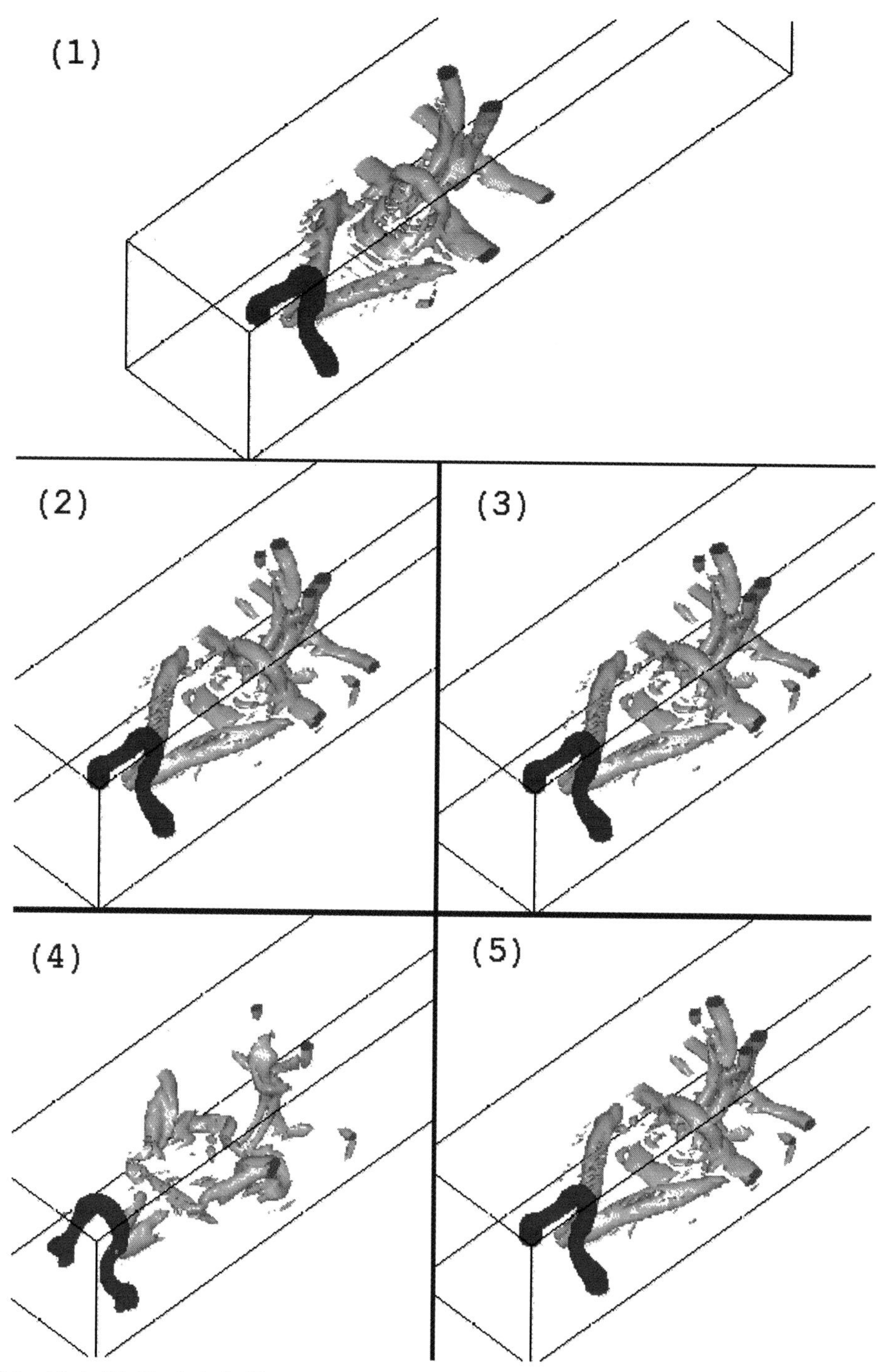

Color Plate 85: The hairpin-like vortex has been isolated, tracked, and shaded to highlight its topology. (Figure 12.8, page 290.)

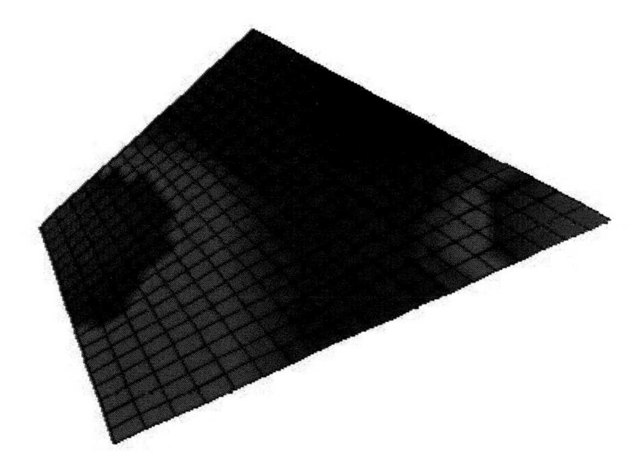

Color Plate 86: A seed plane with vector magnitudes mapped to colors (red represents large magnitudes, blue represents small magnitudes, and green represents those in the middle). (Figure 13.2, page 302.)

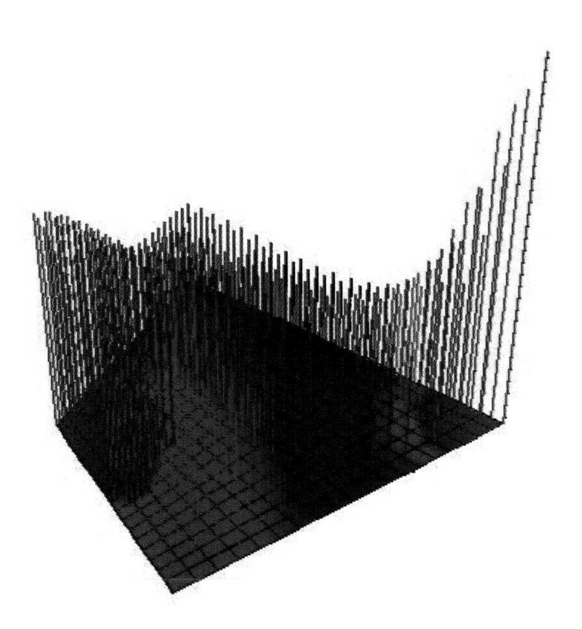

Color Plate 87: Hedgehogs have been added to Color Plate 86. (Figure 13.3, page 303.)

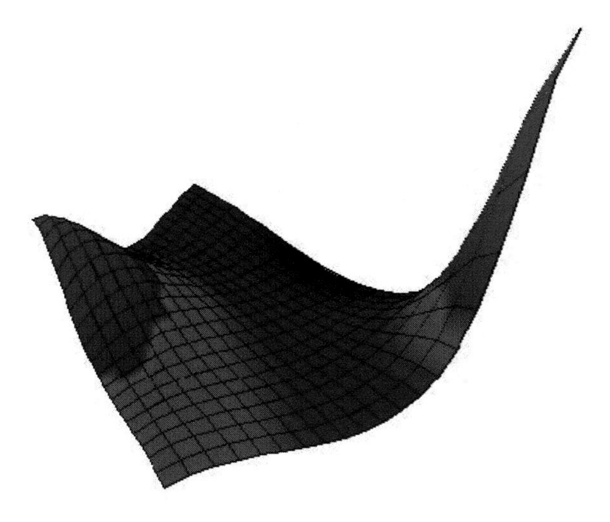

Color Plate 88: An animation frame of a flow surface probe evolving from the planar seed surface shown in Color Plate 86. (Figure 13.4, page 303.)

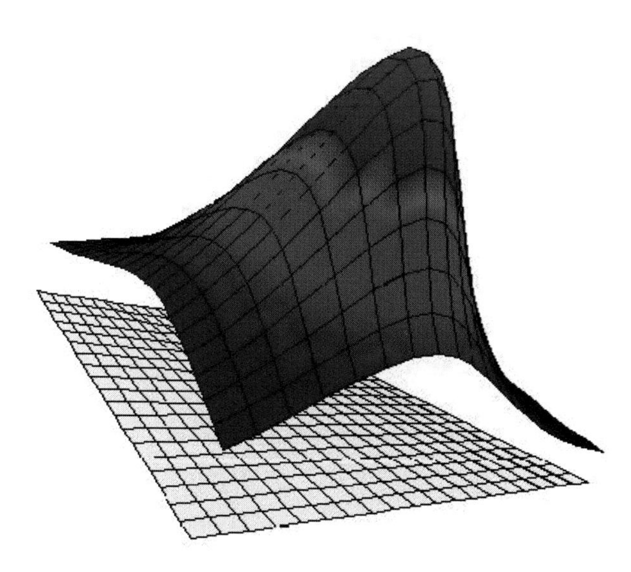

Color Plate 89: An animation frame of a flow surface in another vector field (the seed plane is shown at the bottom in white). (Figure 13.5, page 304.)

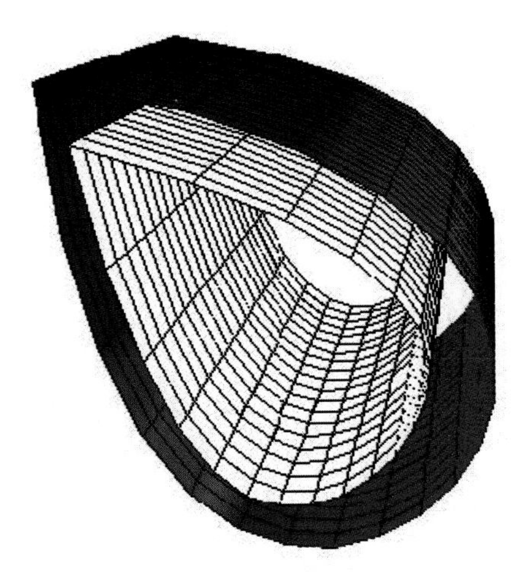

Color Plate 90: An animation frame of a flow surface probe with a seed surface (not shown) as a Bézier surface wrapped around the object shown in white. (Figure 13.6, page 305.)

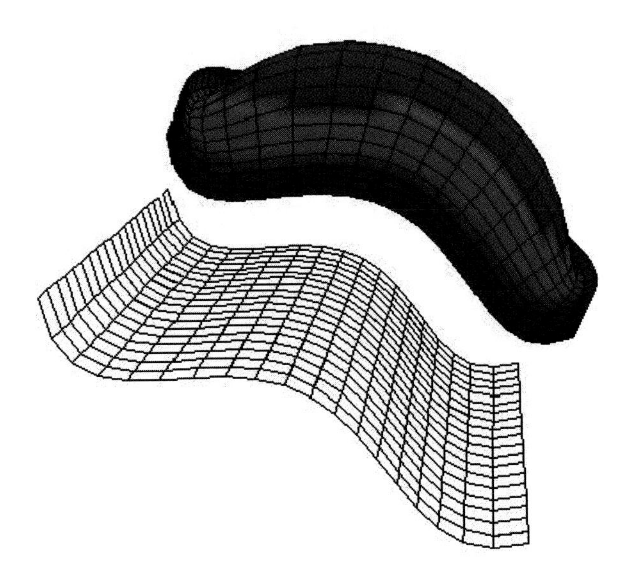

Color Plate 91: An animation frame of a flow surface with the seed Bézier surface at the bottom in white. (Figure 13.7, page 305.)

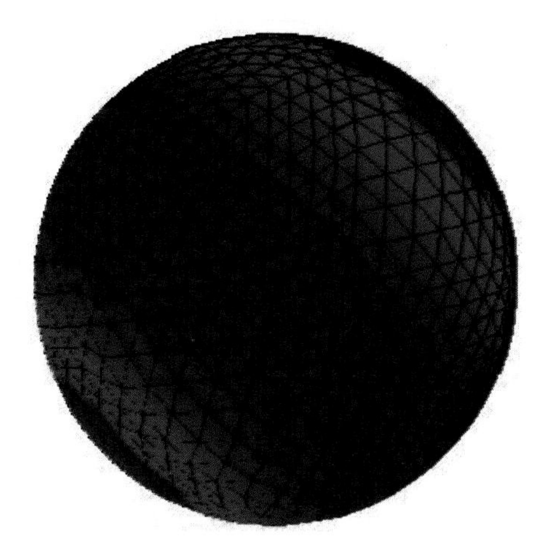

Color Plate 92: A seed spherical surface. (Figure 13.8, page 306.)

Color Plate 93 An animation frame of a flow surface with the spherical seed surface shown in Color Plate 92. (Figure 13.9, page 306.)

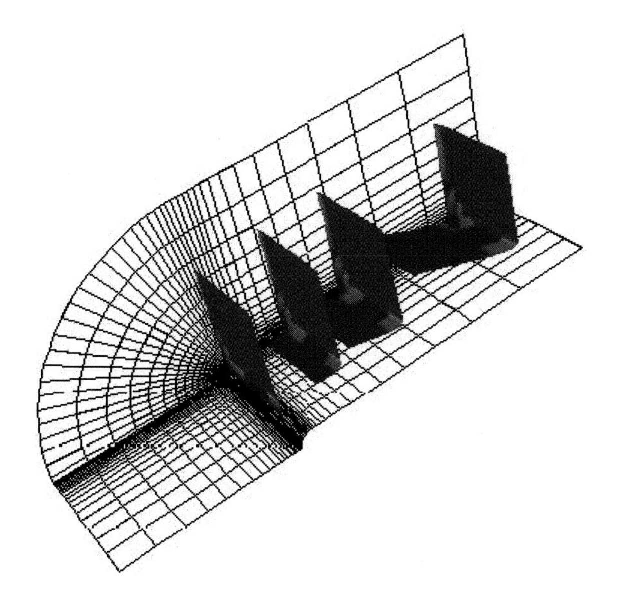

Color Plate 94: A sequence of flow surfaces at animation times 0, 10, 20, and 40, respectively, with the grid mesh in white. (Figure 13.10, page 307.)

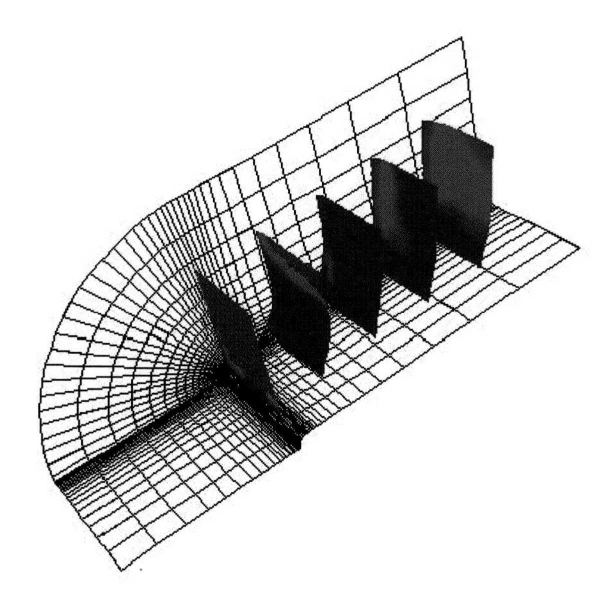

Color Plate 95: A sequence of time surfaces at animation times 0, 10, 20, 30, and 40, respectively, with the grid mesh in white. (Figure 13.11, page 308.)

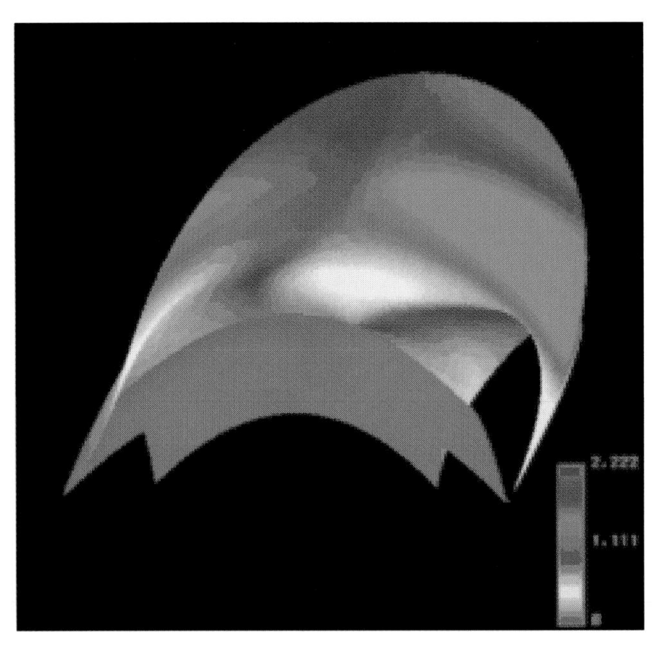

Color Plate 96: Focal surface for maximum eigenvalue. (Figure 16.1, page 362.)

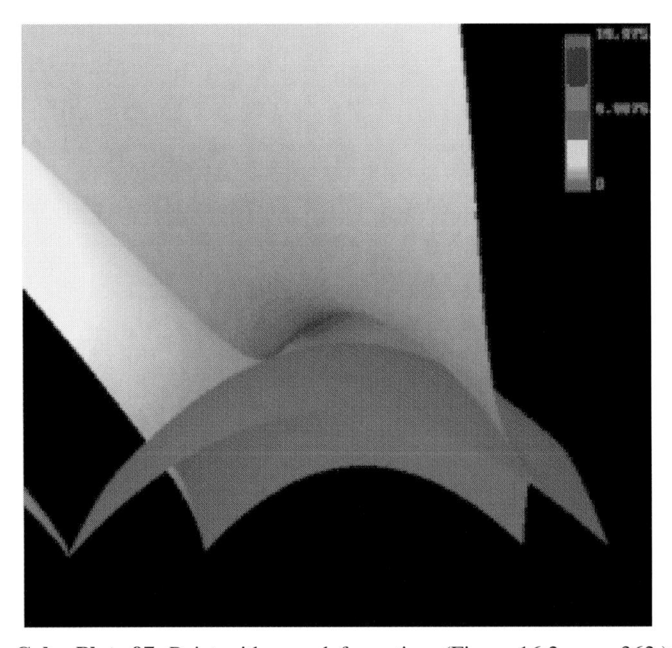

Color Plate 97: Point with zero deformation. (Figure 16.2, page 362.)

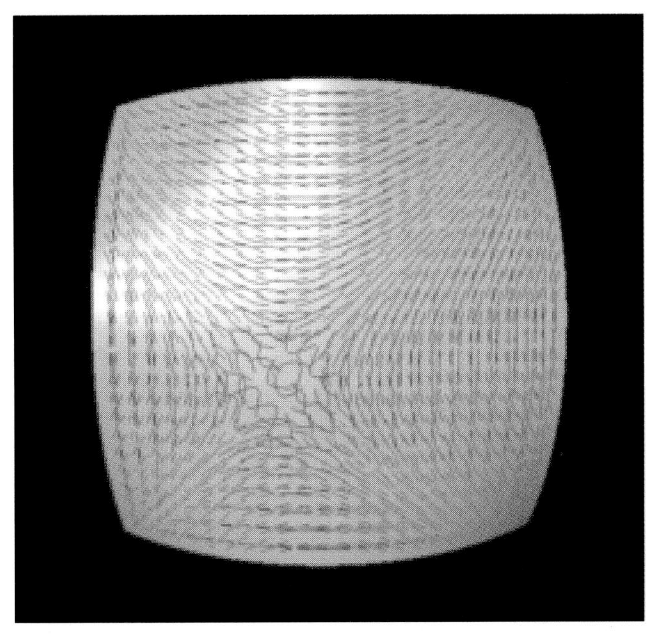

Color Plate 98: Characteristic curves for visualization of a deformation tensor field. (Figure 16.3, page 365.)

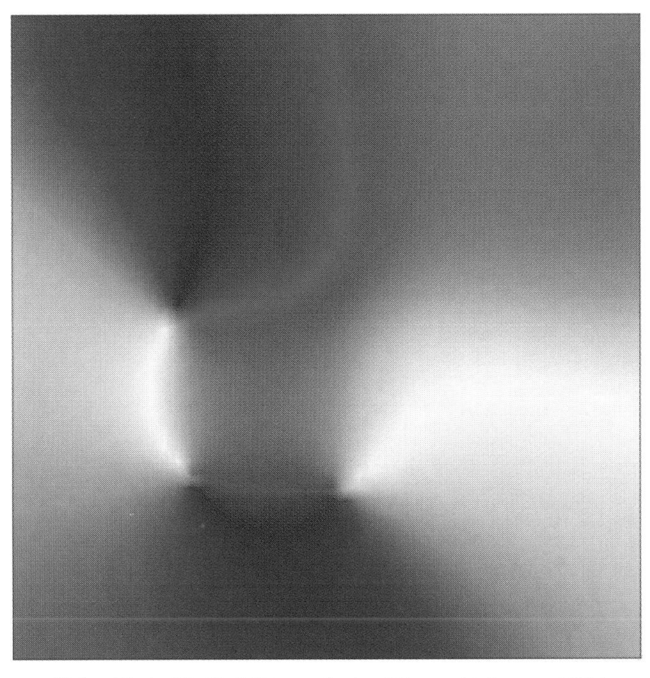

Color Plate 99: Stability analysis. (Figure 16.8, page 370.)

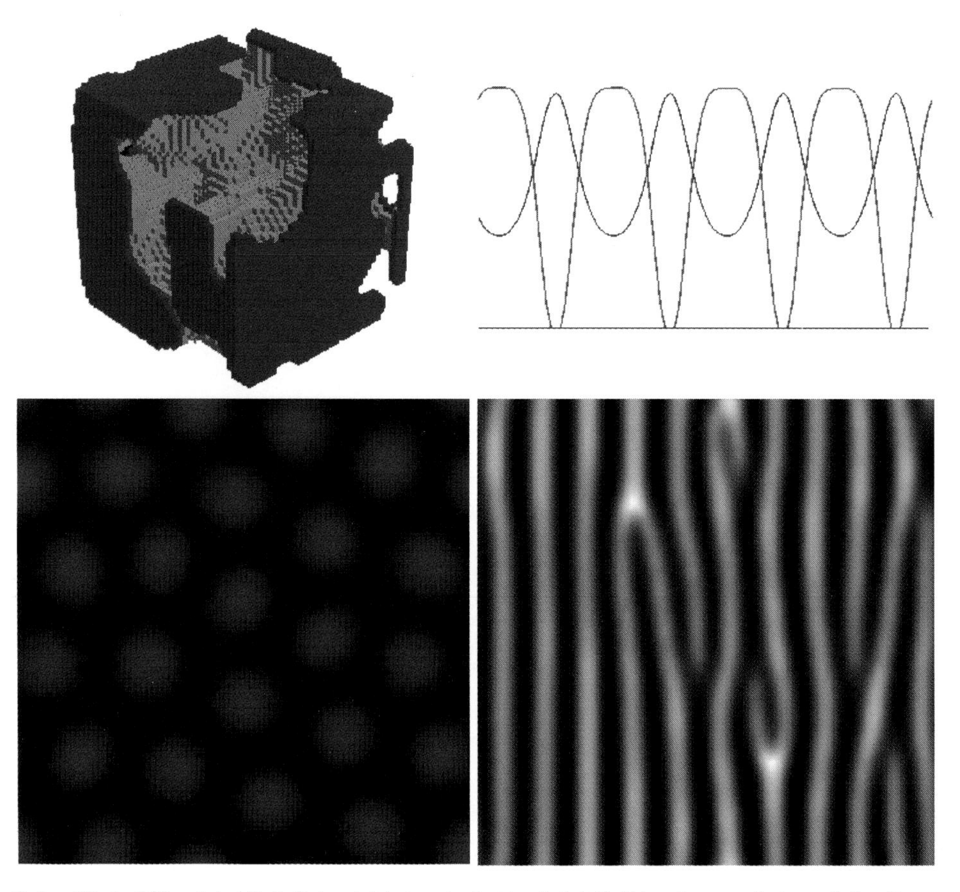

Color Plate 100: *(a)–(d)* (left to right, top to bottom) *(a)* Solid cubes rendering of the three-dimensional cellular automaton data. *(b)* x-y plot of the one-dimensional reaction-diffusion data set. *(c)* Intensity map of the two-dimensional reaction-diffusion data. *(d)* Intensity map of the anisotropic two-dimensional reaction-diffusion data. (Figure 17.1, page 376.)

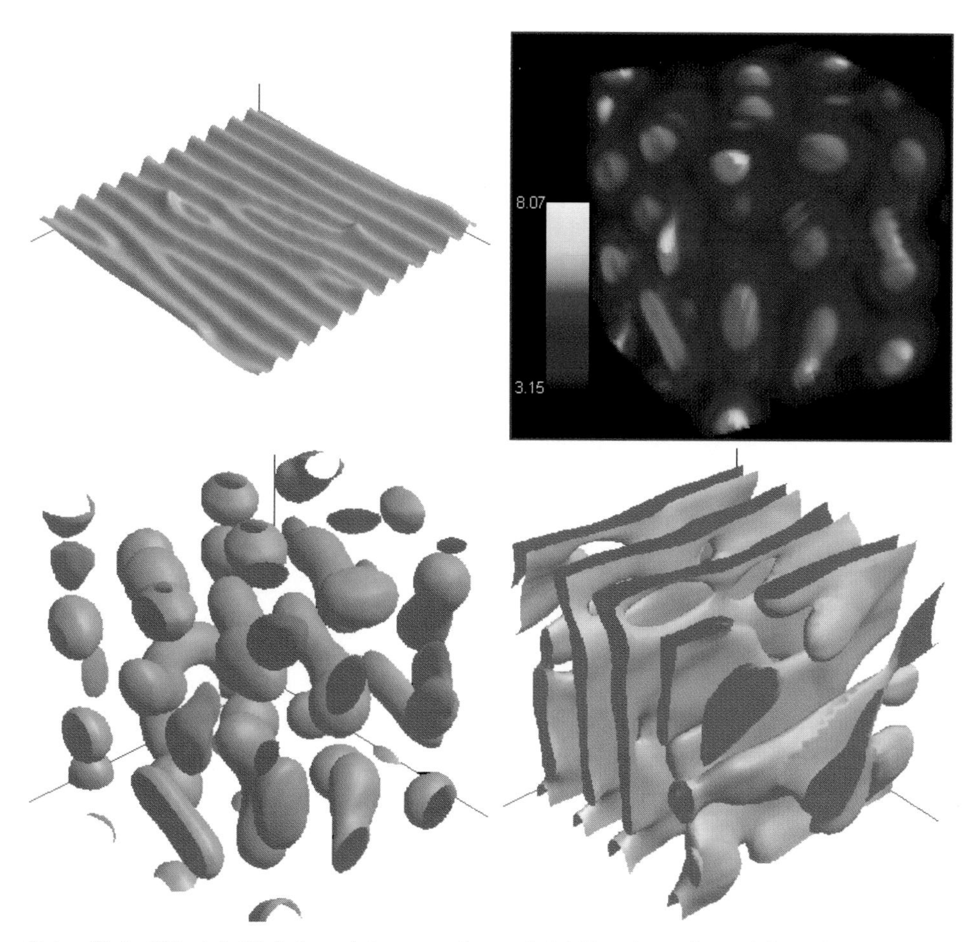

Color Plate 101: *(a)–(d)* (left to right, top to bottom) *(a)* Hermite surface of the anisotropic two-dimensional reaction-diffusion data set. *(b)* Volume rendering of the three-dimensional reaction-diffusion data set. *(c)* Isosurface rendering of the three-dimensional reaction-diffusion data set. *(d)* Isosurface rendering of the anisotropic three-dimensional reaction-diffusion data set. (Figure 17.2, page 381.)

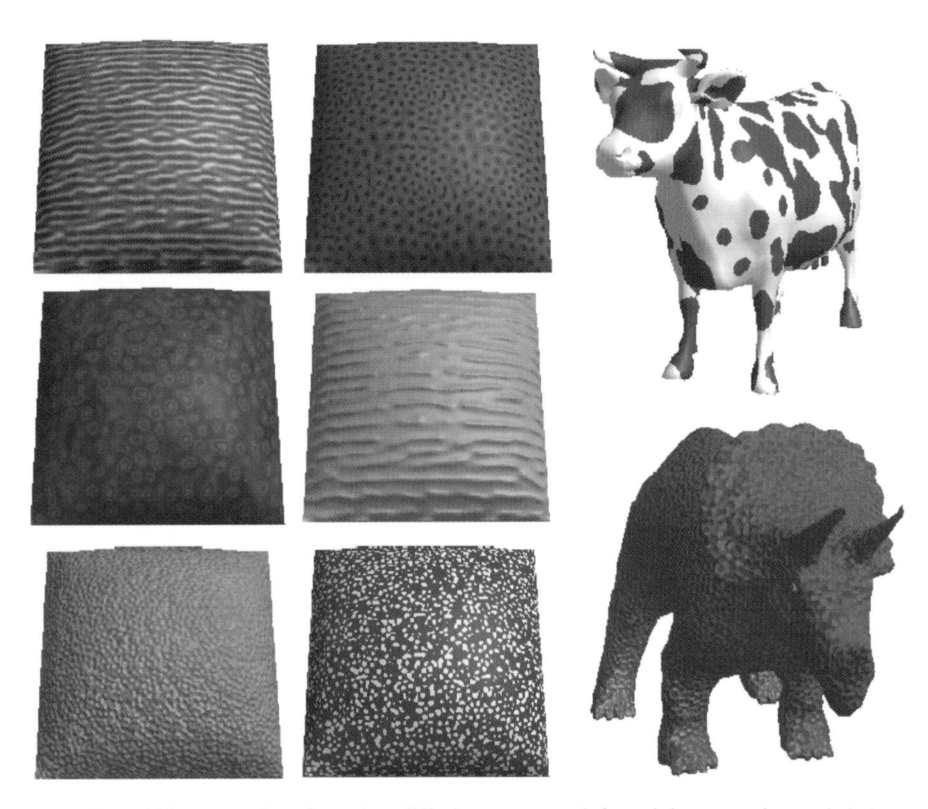

Color Plate 102: Examples of reaction-diffusion textures, (left to right, top to bottom) *Anisotropic color map:* varying morphogen concentrations map to different colors. *Spatial variation of reaction rate:* the reaction rate parameter s is varied, resulting in features of different sizes. *Cow:* Three-dimensional discrete color map. *Interpolated-pair color map:* With a linearly-interpolated color set, smooth transitions occur. *Sand dune pillow:* by applying bump-mapping to anisotropic reaction-diffusion data, the characteristic ridge formation emerges. *Triceratops:* Bump-mapped triceratops, three-dimensional texture. *Jade pillow:* "Dent-mapping," the inverse of bump-mapping, with normalized perturbation vectors. *Discrete color map:* by partitioning morphogen concentrations into two ranges, a pattern defined by the morphogen's isogram appears. (Figure 17.3, page 388.)

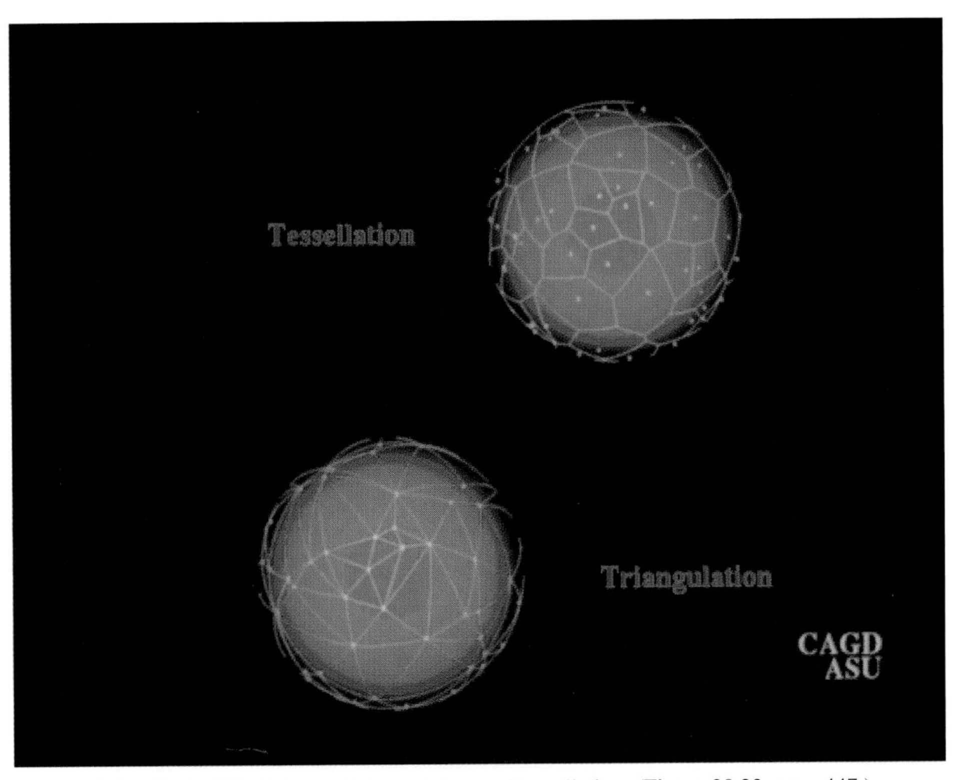

Color Plate 103: Spherical triangulation and tessellation. (Figure 20.23, page 447.)

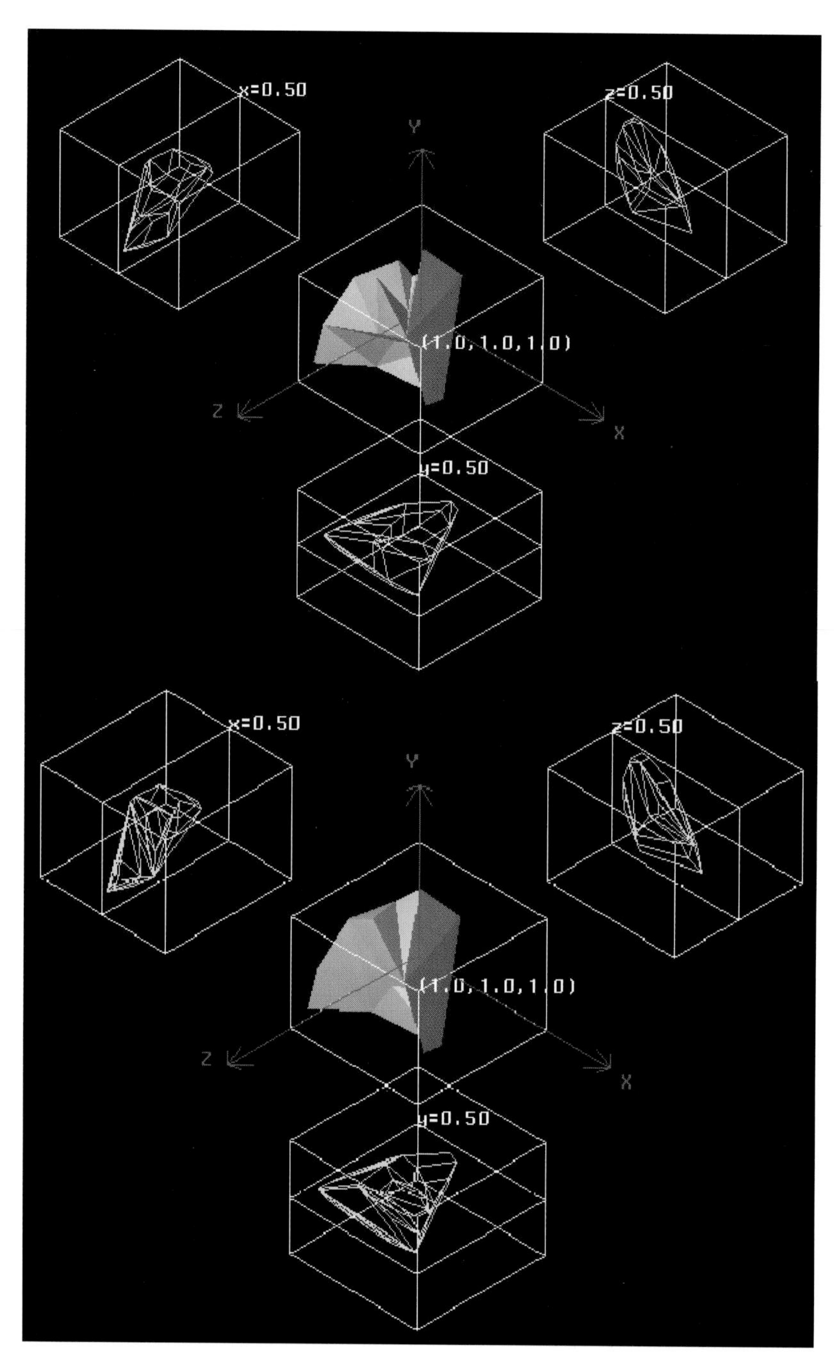

Color Plate 104: Data-dependent tetrahedrization compared to the Delaunay tetrahedrization. (Figure 20.68, page 496.)

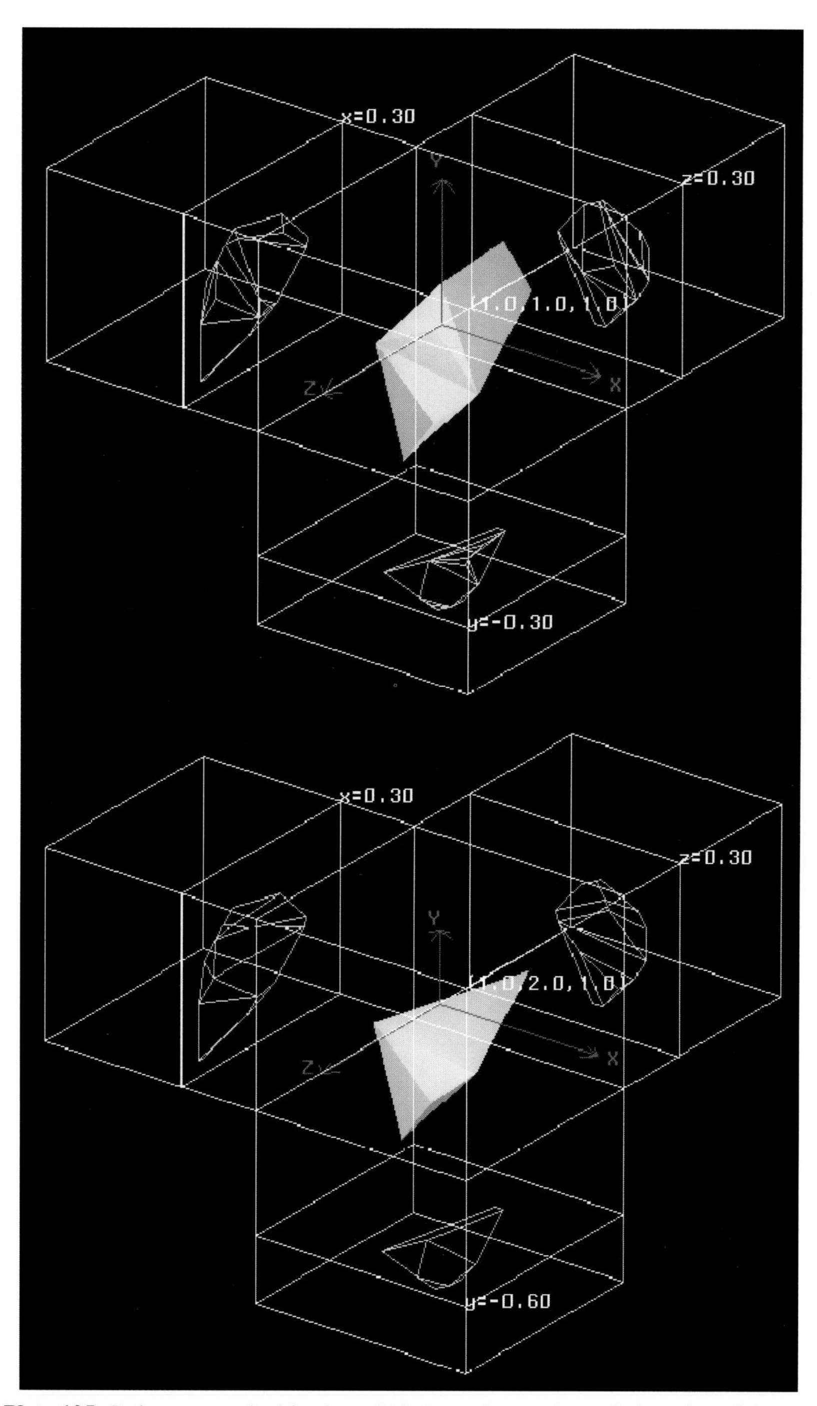

Color Plate 105: Delaunay tetrahedrization of 10 data points and a scaled version of the same data points. (Figure 20.69, page 498.)

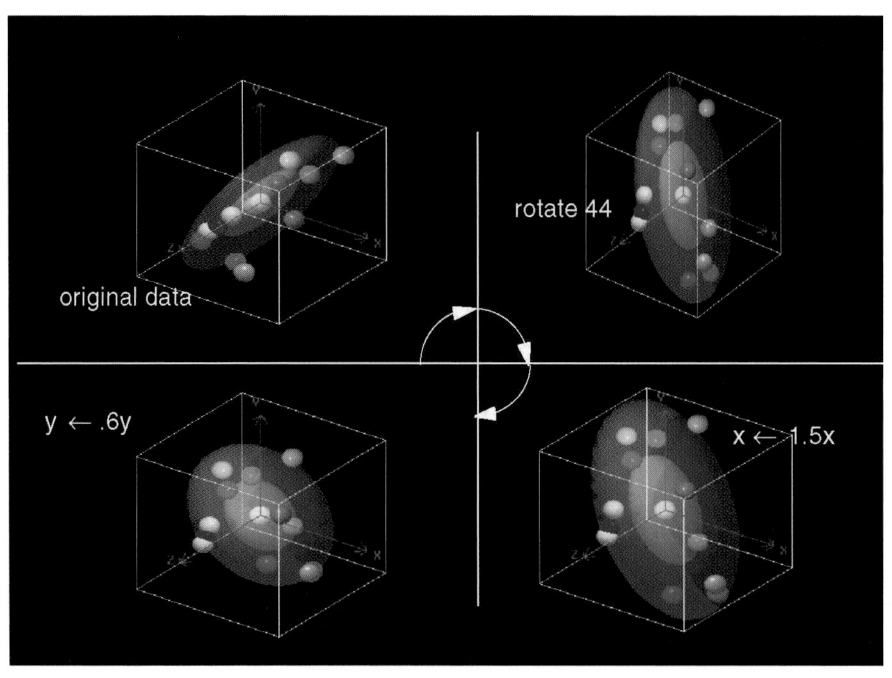

Color Plate 106: Examples illustrating the affine invariant norm. The ellipsoids are 0.50 and 1.0 units from the center point. (Figure 20.70, page 500.)

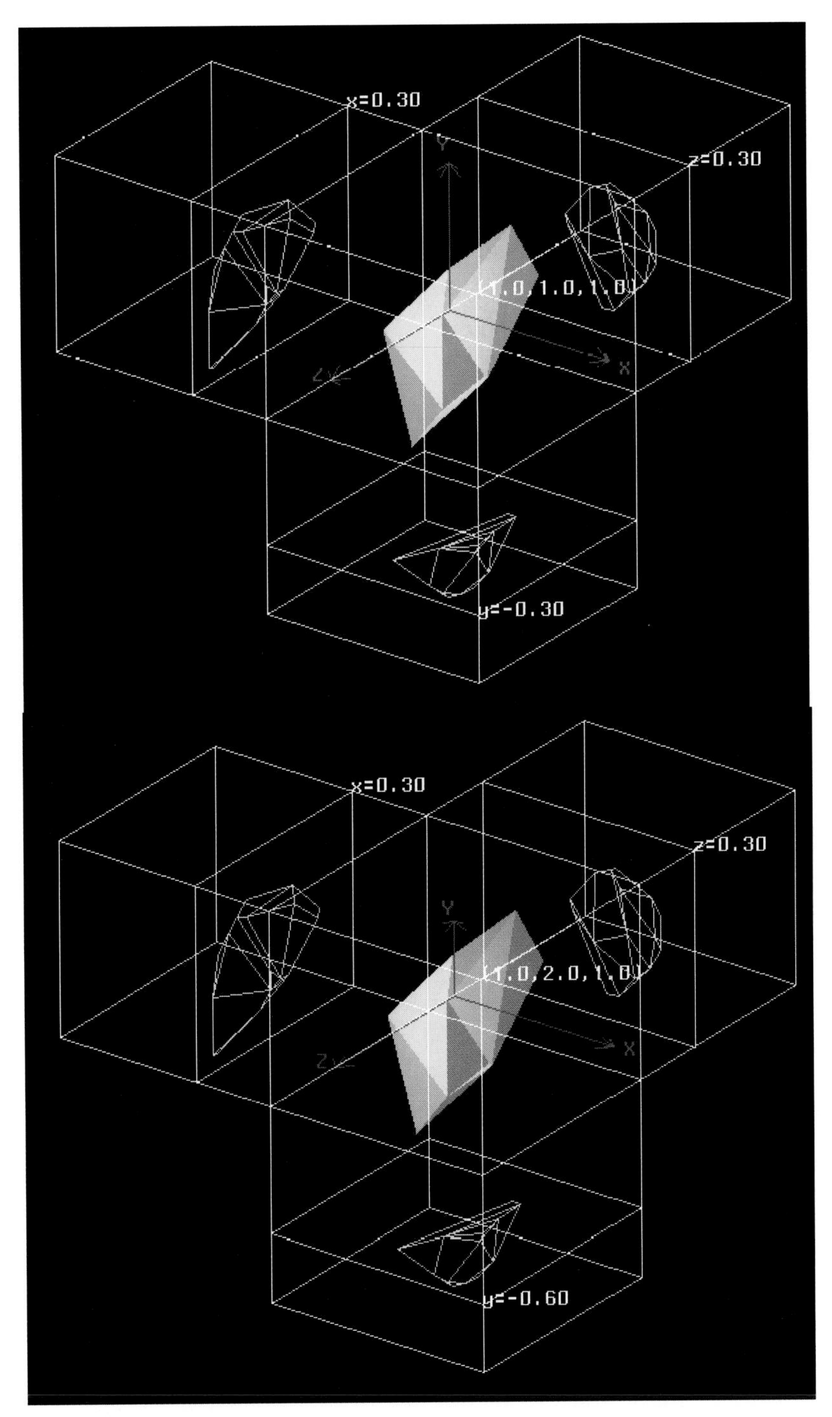

Color Plate 107: Examples of affine invariant tetrahedrization. (Figure 20.71, page 501.)

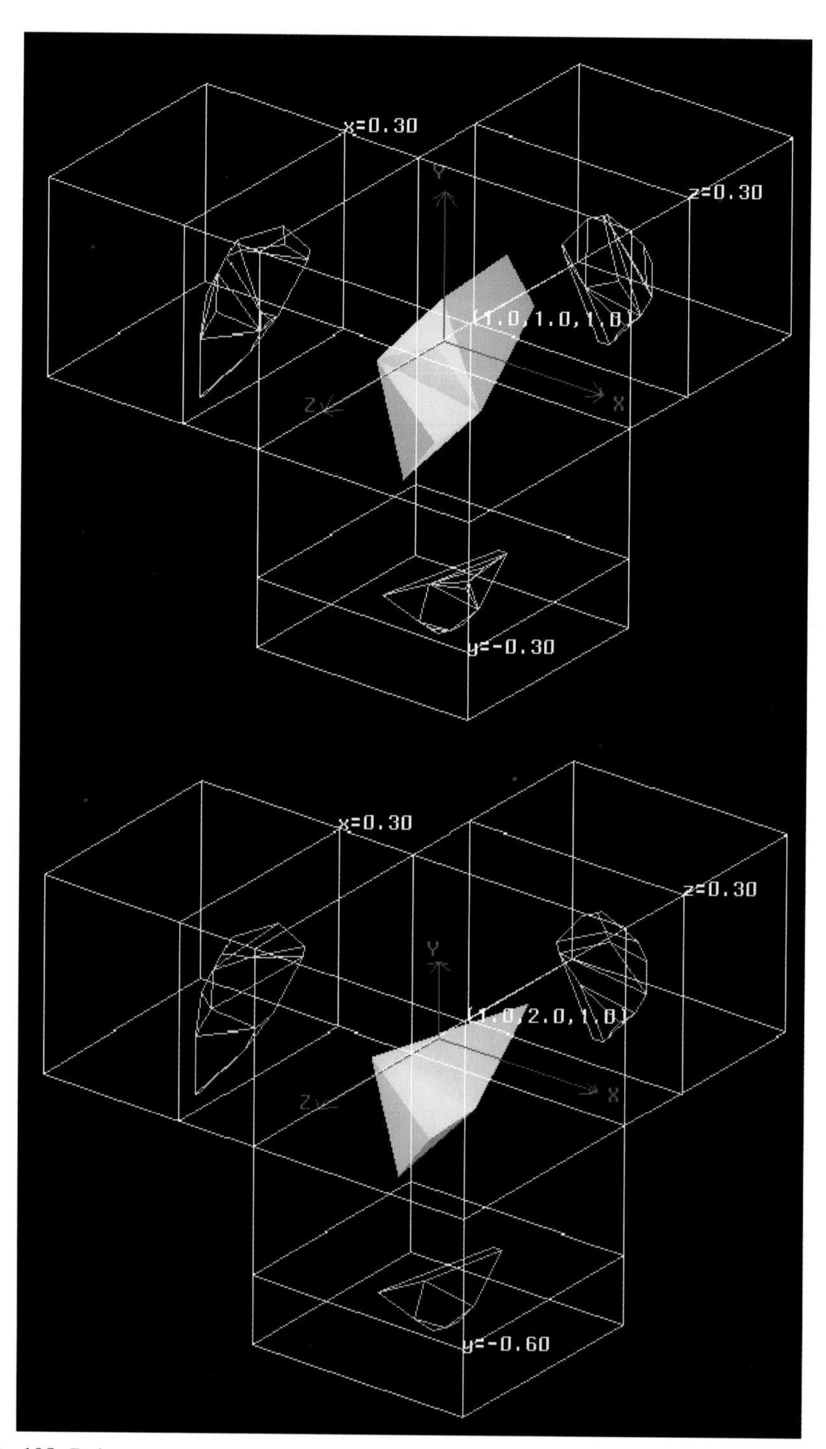

Color Plate 108: Delaunay tetrahedrization of the same data as in Color Plate 107. (Figure 20.72, page 502.)

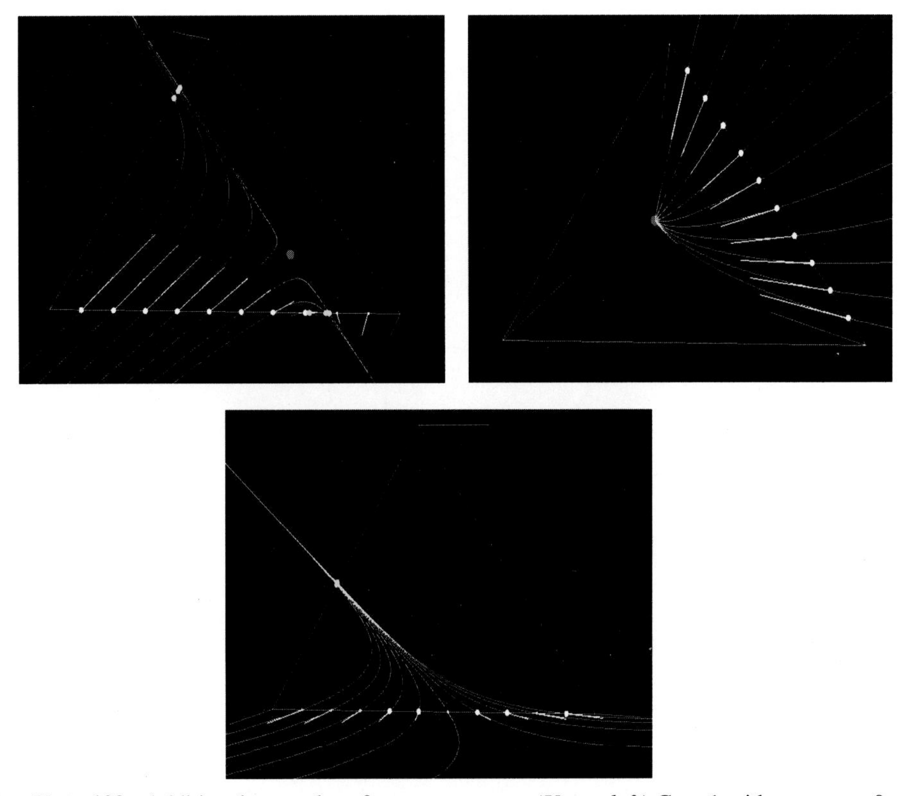

Color Plate 109: Additional examples of tangent curves. *(Upper-left)* Case 1 with $r_1 \cdot r_2 < 0$. *(Upper-right)* Case 1 with $r_1 \cdot r_2 > 0$. *(Lower)* Case 2. (Figure 21.7, page 537.)

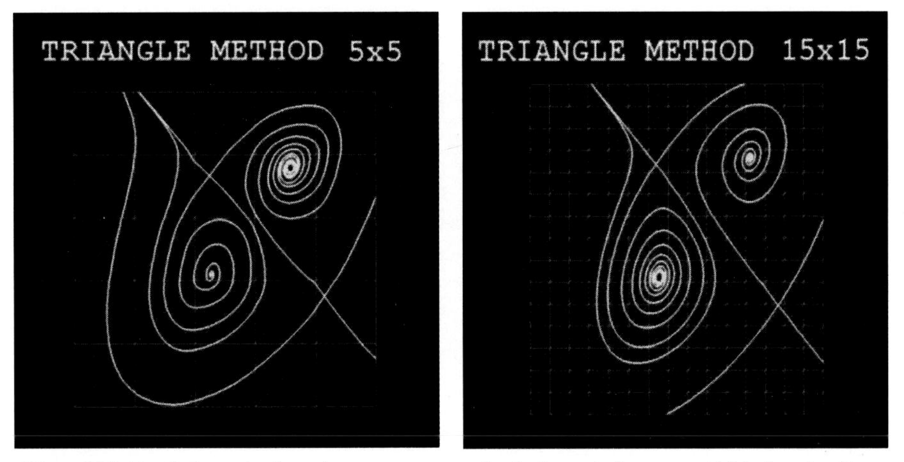

Color Plate 110: Explicit method used to compute topological graph for vector field given by Equation (21.20). Two different resolutions for the triangulation are shown. (Figure 21.16, page 551.)

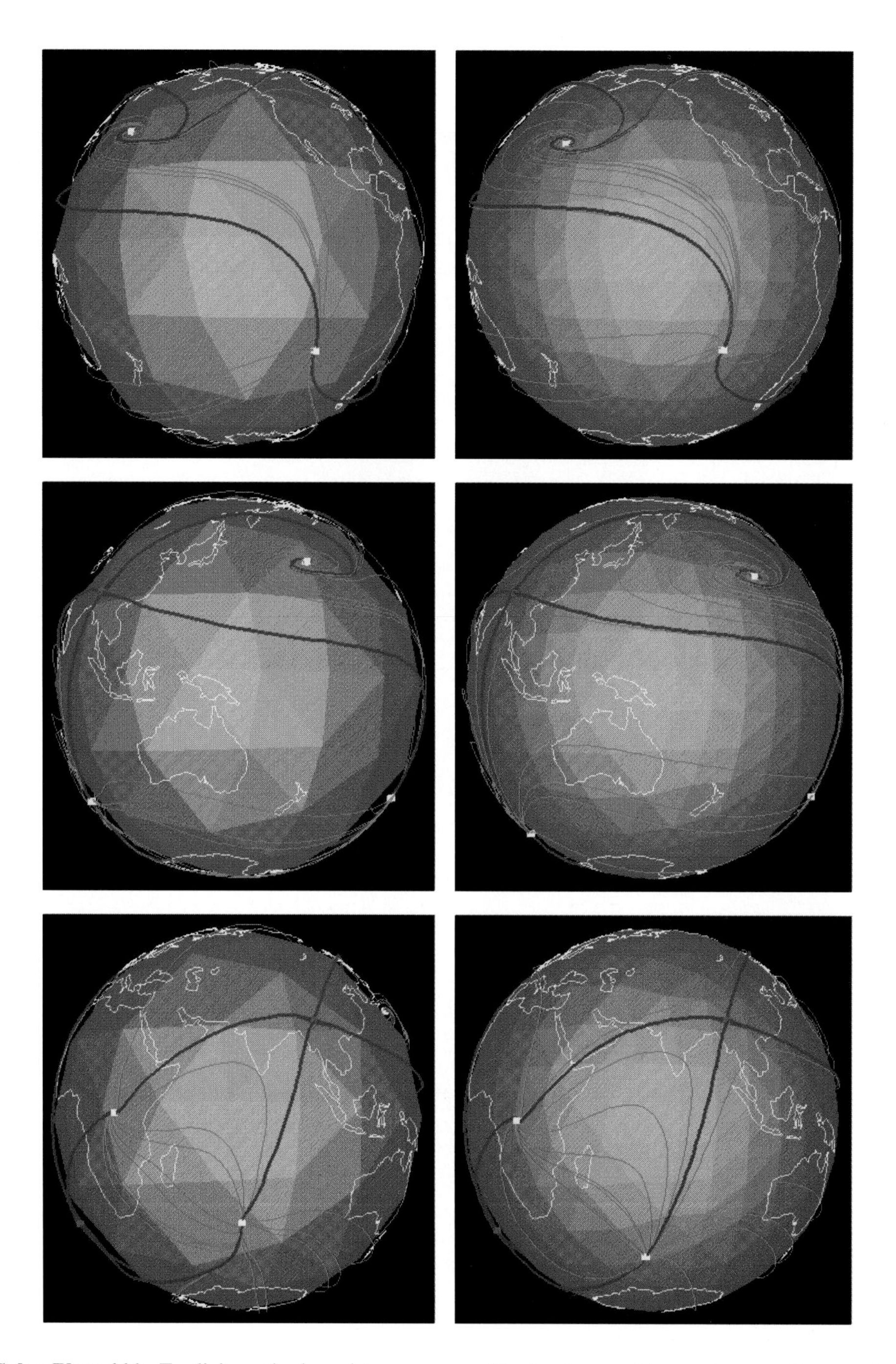

Color Plate 111: Explicit method used to compute critical values and tangent curves for a vector field defined over a spherical domain. (Figure 21.17, page 552.)

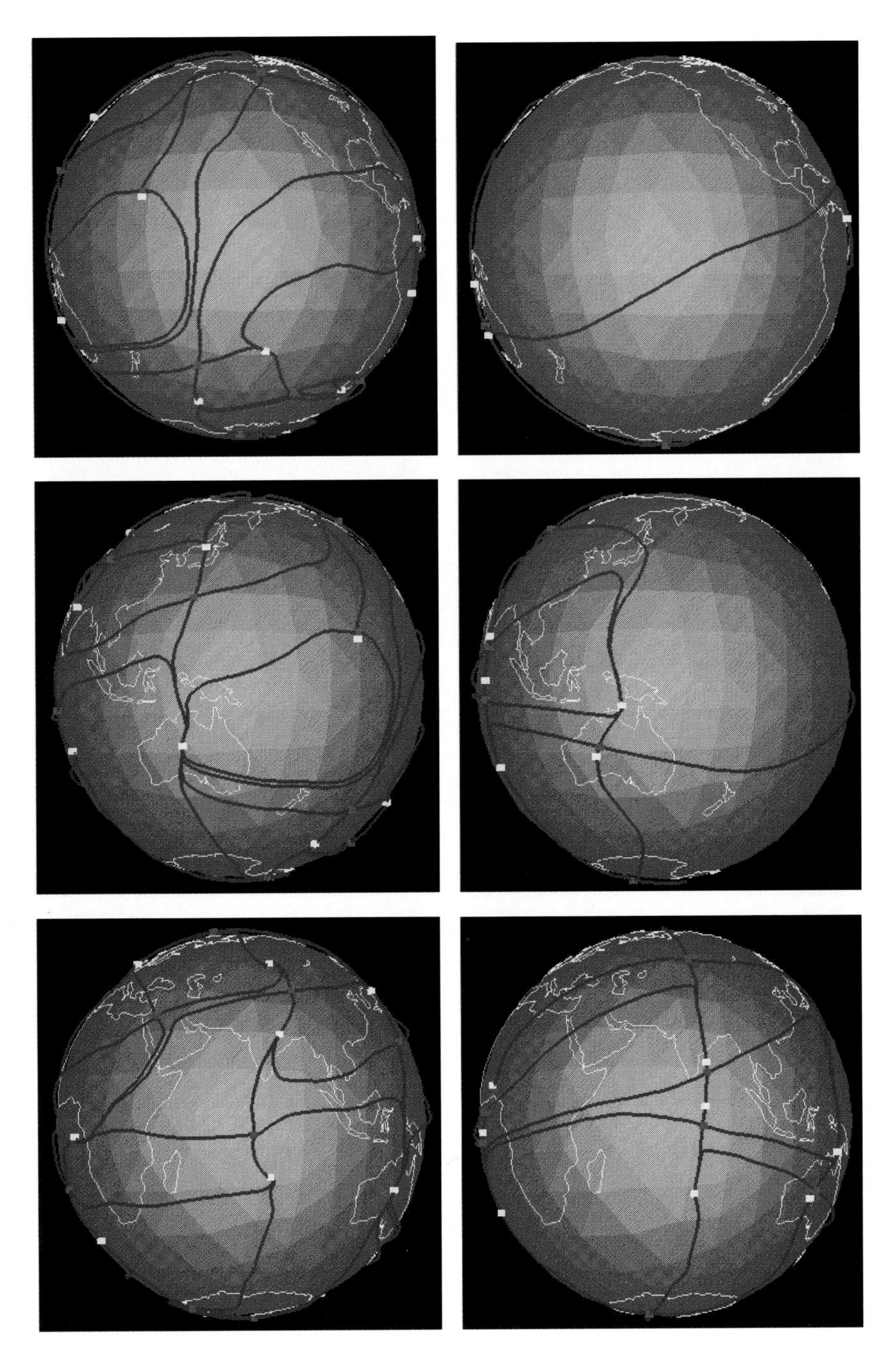

Color Plate 112: Topological graphs for wind data over the earth are computed using the explicit method. The right column is wavelet reconstruction based on the largest 3% of the wavelet coefficients. (Figure 21.20, page 555.)

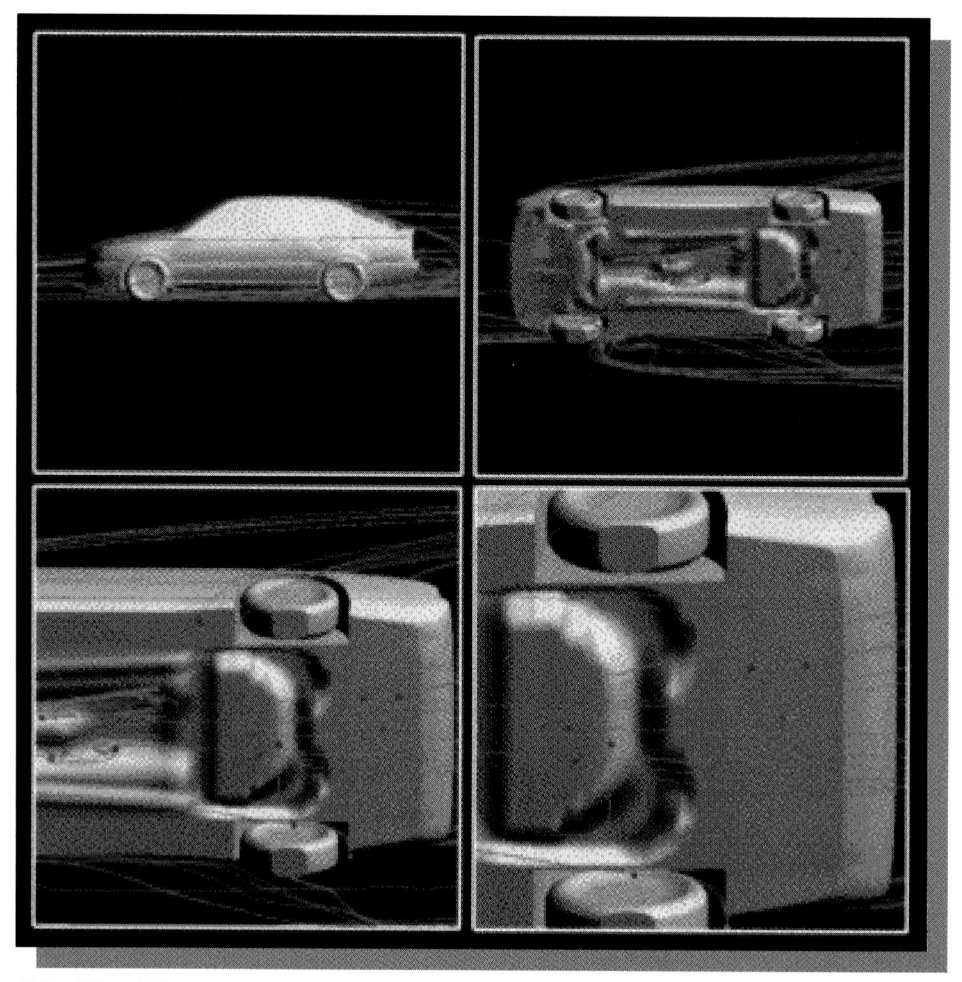

Color Plate 113: The explicit method is used to compute tangent curves linking critical values for a 3D vector field. (Figure 21.21, page 556.)

Index

Q

R

S

T

U

Contributors

R. Daniel Bergeron Department of Computer Science, University of New Hampshire, Durham, New Hampshire 03824, USA, http://www.cs.unh.edu/~rdb

Ken Brodlie School of Computer Studies, University of Leeds, Leeds, LS2 9JT, UK

Peter Chambers VLSI Technology, Inc., 8375 River Parkway, Tempe, AZ 85284, USA

Fan Dai Fraunhofer Institute for Computer Graphics (Fraunhofer IGD), Wilhelminenstr. 7, D-64283 Darmstadt, Germany

V.V. Dyachin Institute for High Energy Physics, Protvino, Russia

Charles R. Dyer Department of Computer Sciences, University of Wisconsin-Madison, Madison, WI, USA

Stephen G. Eick AT&T Bell Laboratories, Rm 1G-351, 1000 East Warrenville Road, Naperville, IL 60566, USA, email: eick@research.att.com

Martin Göbel Fraunhofer Institute for Computer Graphics (Fraunhofer IGD), Wilhelminenstr. 7, D-64283 Darmstadt, Germany

M.H. Groß Computer Science Department, Swiss Federal Institute of Technology, (ETH-Zürich)

Helmut Haase Fraunhofer Institute for Computer Graphics (Fraunhofer IGD), Wilhelminenstr. 7, D-64283 Darmstadt, Germany, email: haase@igd.fhg.de, http://www.igd.fhg.de/~haase

H. Hagen University of Kaiserslautern, Computer Science Department, D-67653 Kaiserslautern, Germany

S. Hahmann Laboratoire de Modélisation et Calcul, University INP de Grenoble, 38041 Grenoble, France

Bernd Hamann Acting Associate Professor and Co-Director of the Center for Image Processing and Integrated Computing (CIPIC), Department of Computer Science, University of California, Davis, CA 95616-8562, USA, hamann@cs.ucdavis.edu

William L. Hibbard Space Science and Engineering Center and Department of Computer Sciences, University of Wisconsin-Madison, Madison, WI, USA, email: whibbard@macc.wisc.edu

Andrea J.S. Hin Delft University of Technology, Faculty of Technical Mathematics and Informatics, Julianalaan 132, 2628 BL Delft, The Netherlands

Lichan Hong Department of Computer Science, State University of New York at Stony Brook, Stony Brook, NY 11704-4400, USA

Matthew A. Hutchins CSIRO Division of Information Technology, GPO Box 664, Canberra ACT 2601, Australia

Il-Hong Jung Computer Science and Engineering, Arizona State University, Tempe, AZ 85287-5406, USA

Arie Kaufman Department of Computer Science, State University of New York at Stony Brook, Stony Brook, NY 11704-4400, USA

S.V. Klimenko Institute for High Energy Physics, Protvino, Russia

David A. Lane MRJ, Inc., NASA Ames Research Center, M/S T27A-2, Moffett Field, CA 94035-1000, USA

Petros Mashwama School of Computer Studies, University of Leeds, Leeds, LS2 9JT, UK

Robert J. Moorhead II NSF Engineering Research Center for Computational Field Simulation, Mississippi State University, P.O. Box 6176, Mississippi State, MS 39762, USA and Department of Electrical and Computer Engineering, Mississippi State University, P.O. Drawer EE, Mississippi State, MS 39762, USA

Heinrich Müller Universität Dortmund, Informatik VII, 44221 Dortmund, Germany, email: mueller@ls7.informatik.uni-dortmund.de

Gregory M. Nielson Computer Science and Engineering, Arizona State University, Tempe, AZ 85287-5406, USA, email: nielson@asu.edu

I.N. Nikitin Institute for High Energy Physics, Protvino, Russia

Hans-Georg Pagendarm Deutsche Forschungsanstalt für Luft- und Raumfahrt, DLR, Bunsenstrasse 10, D37073 Göttingen, Germany

Brian E. Paul Space Science and Engineering Center, University of Wisconsin-Madison, Madison, WI, USA

Frits H. Post Delft University of Technology, Faculty of Technical Mathematics and Informatics, Zuidplantsoen 4, 2628 BZ Delft, The Netherlands

Philip K. Robertson CSIRO Division of Information Technology, GPO Box 664, Canberra ACT 2601, Australia, email: philip.robertson@cbr.dit.csiro.au

Alyn Rockwood Department of Computer Science, Arizona State University, Tempe, AZ 85287, USA

I. Ari Sadarjoen Delft University of Technology, Faculty of Technical Mathematics and Informatics, Julianalaan 132, 2628 BL Delft, The Netherlands

Cláudio Silva Department of Computer Science, State University of New York at Stony Brook, Stony Brook, NY 11704-4400, USA

D. Silver Department of Electrical and Computer Engineering, P.O. Box 909, Rutgers University, Piscataway, NJ 08855, USA, email: silver@vizlab.rutgers.edu, http://www.caip.rutgers.edu/vizlab.html

Nat Srinivasan Computer Science and Engineering, Arizona State University, Tempe, AZ 85287-5406, USA

Michael Stark Universität Dortmund, Informatik VII, 44221 Dortmund, Germany, email: stark@ls7.informatik.uni-dortmund.de

Johannes Strassner Fraunhofer Institute for Computer Graphics (Fraunhofer IGD), Wilhelminenstr. 7, D-64283 Darmstadt, Germany

Junwon Sung Computer Science and Engineering, Arizona State University, Tempe, AZ 85287-5406, USA

Robert van Liere Centrum voor Wiskunde en Informatica, P.O. Box 4097, 1009 AB Amsterdam, The Netherlands

Theo van Walsum Delft University of Technology, Faculty of Technical Mathematics and Informatics, Julianalaan 132, 2628 BL Delft, The Netherlands

Jarke J. van Wijk Netherlands Energy Research Foundation ECN, P.O. Box 1, 1755 ZG Petten, The Netherlands and Centrum voor Wiskunde en Informatica, P.O. Box 4097, 1009 AB Amsterdam, The Netherlands

H. Weimer University of Kaiserslautern, Computer Science Department, D-67653 Kaiserslautern, Germany

Ulrike Welsch Universität Dortmund, Informatik VII, 44221 Dortmund, Germany

Pak Chung Wong Department of Computer Science, University of New Hampshire, Durham, New Hampshire 03824, USA, http://www.cs.unh.edu/~pcw

Jong-Beum Yoon Computer Science and Engineering, Arizona State University, Tempe, AZ 85287-5406, USA

IEEE Computer Society Publications

The world-renowned Computer Society publishes, promotes, and distributes a wide variety of authoritative computer science and engineering texts. These books are available in two formats: 100 percent original material by authors preeminent in their field who focus on relevant topics and cutting-edge research, and reprint collections consisting of carefully selected groups of previously published papers with accompanying original introductory and explanatory text.

Submission of proposals: For guidelines and information on Computer Society books, send e-mail to cs.books@computer.org or write to the Acquisitions Editor, IEEE Computer Society, P.O. Box 3014, 10662 Los Vaqueros Circle, Los Alamitos, CA 90720-1314. Telephone +1 714-821-8380. FAX +1 714-761-1784.

IEEE Computer Society Proceedings

The Computer Society also produces and actively promotes the proceedings of more than 130 acclaimed international conferences each year in multimedia formats that include hard and softcover books, CD-ROMs, videos, and on-line publications.

For information on Computer Society proceedings, send e-mail to cs.books@computer.org or write to Proceedings, IEEE Computer Society, P.O. Box 3014, 10662 Los Vaqueros Circle, Los Alamitos, CA 90720-1314. Telephone +1 714-821-8380. FAX +1 714-761-1784.

Additional information regarding the Computer Society, conferences and proceedings, CD-ROMs, videos, and books can also be accessed from our web site at http://computer.org/cspress

4/15/97